Lecture Notes in Networks and Systems 1897

The series "Lecture Notes in Networks and Systems" publishes the latest developments in Networks and Systems—quickly, informally and with high quality. Original research reported in proceedings and post-proceedings represents the core of LNNS.

Volumes published in LNNS embrace all aspects and subfields of, as well as new challenges in, Networks and Systems.

The series contains proceedings and edited volumes in systems and networks, spanning the areas of Cyber-Physical Systems, Autonomous Systems, Sensor Networks, Control Systems, Energy Systems, Automotive Systems, Biological Systems, Vehicular Networking and Connected Vehicles, Aerospace Systems, Automation, Manufacturing, Smart Grids, Nonlinear Systems, Power Systems, Robotics, Social Systems, Economic Systems and other. Of particular value to both the contributors and the readership are the short publication timeframe and the world-wide distribution and exposure which enable both a wide and rapid dissemination of research output.

The series covers the theory, applications, and perspectives on the state of the art and future developments relevant to systems and networks, decision making, control, complex processes and related areas, as embedded in the fields of interdisciplinary and applied sciences, engineering, computer science, physics, economics, social, and life sciences, as well as the paradigms and methodologies behind them.

Indexed by SCOPUS, EI Compendex, INSPEC, WTI Frankfurt eG, zbMATH, SCImago.

All books published in the series are submitted for consideration in Web of Science.

For proposals from Asia please contact Aninda Bose (aninda.bose@springer.com).

Mohd. Aminudin bin Jamlos ·
Syed Zulkarnain Syed Idrus · Soh Ping Jack ·
Wan Azani Mustafa · Rashidah Binti Che Yob ·
Liyana binti Zahid
Editors

Computational Intelligence in Modern Engineering Systems

Proceedings of the 5th International Conference on Science, Engineering, and Technology ICSET 2025

Springer

Editors
Mohd. Aminudin bin Jamlos
Faculty of Electronic Engineering
and Technology
Universiti Malaysia Perlis
Arau, Malaysia

Soh Ping Jack
Faculty of Information Technology
and Electrical Engineering
University of Oulu
Oulu, Finland

Rashidah Binti Che Yob
Faculty of Electronic Engineering
and Technology
Universiti Malaysia Perlis
Arau, Malaysia

Syed Zulkarnain Syed Idrus
Faculty of Business & Communication (FPK)
Universiti Malaysia Perlis
Arau, Malaysia

Wan Azani Mustafa
Faculty of Electrical Engineering
and Technology
Universiti Malaysia Perlis
Arau, Malaysia

Liyana binti Zahid
Faculty of Electronic Engineering
and Technology
Universiti Malaysia Perlis
Arau, Malaysia

ISSN 2367-3370 ISSN 2367-3389 (electronic)
Lecture Notes in Networks and Systems
ISBN 978-3-032-20721-0 ISBN 978-3-032-20722-7 (eBook)
https://doi.org/10.1007/978-3-032-20722-7

This Springer imprint is published by the registered company Springer Nature Switzerland AG
The registered company address is: Gewerbestrasse 11, 6330 Cham, Switzerland

Contents

Performance Comparison of Machine Learning Models for Brain Tumor Classification Using Feature-Extracted MRI and CT Images

Aceng Sambas[1,2,3](✉), Volodymyr Rusyn[4], Sundarapandian Vaidyanathan[5], Hani Rubiani[6], Saqib Murtaza[1], and Rama Krishna Mani Kanta Yalla[1]

[1] Faculty of Informatics and Computing, Universiti Sultan Zainal Abidin, Campus Besut, 22200 Besut, Terengganu, Malaysia
acengsambas@unisza.edu.my, lncs@springer.com

[2] Artificial Intelligence Research Centre for Islam Sustainability (AIRIS), Universiti Sultan Zainal Abidin, 21300 Gongbadak, Kuala Nerus, Terengganu, Malaysia

[3] Department of Mechanical Engineering, Universitas Muhammadiyah Tasikmalaya, Tamansari Gobras, 46196 Tasikmalaya, Indonesia

[4] Department of Radio Engineering and Information Security, Yuriy Fedkovych Chernivtsi National University, 58012 Chernivtsi, Ukraine

[5] Centre for Control Systems, Vel Tech University, Avadi, 600062, Tamil Nadu Chennai, India

[6] Department of Electrical Engineering, Universitas Muhammadiyah Tasikmalaya, Tamansari Gobras, 46196 Tasikmalaya, Indonesia

Abstract. This study conducts a comprehensive comparison of six machine learning models—Support Vector Machine (SVM), Random Forest (RF), K-Nearest Neighbor (KNN), Logistic Regression (LR), Naive Bayes (NB), and Extreme Gradient Boosting (XGBoost)—for the classification of brain tumors using multimodal medical imaging. The analysis utilizes 8,029 images obtained from the Kaggle Brain Tumor Multimodal Image Dataset, comprising 3,009 MRI tumor, 2,000 MRI healthy, 2,318 CT tumor, and 702 CT healthy images. All images were preprocessed and represented through 20 handcrafted features derived from statistical and texture-based descriptors. To mitigate the imbalance in class distribution, particularly within the CT modality, the Synthetic Minority Over-sampling Technique (SMOTE) was applied to ensure a more balanced dataset and improve model learning. The experimental findings reveal that ensemble-based models deliver the strongest performance. Both Random Forest and XGBoost achieved the highest accuracy of 96.01%, with precision, recall, and F1-scores of 0.97 for tumor classification. KNN followed with an accuracy of 94.27%, demonstrating stable performance across both modalities. Meanwhile, SVM produced moderate results with 87.55% accuracy. Logistic Regression and Naive Bayes yielded lower accuracy levels of 75.59% and 70.24%, respectively, showing noticeable difficulty in correctly identifying healthy cases.

Keywords: Brain tumor classification · Machine learning · MRI · CT scan · feature extraction · SMOTE · Random Forest · XGBoost

M. A. b. Jamlos et al. (Eds.): ICSET 2025, LNNS 1897, pp. 1–12, 2026.
https://doi.org/10.1007/978-3-032-20722-7_1

1 Introduction

Brain tumors are life-threatening conditions caused by the abnormal growth of cells within the brain or surrounding tissues [1]. Depending on their type and location, these tumors can lead to severe neurological complications such as headaches, seizures, vision impairment, and motor dysfunction [2]. Early and accurate diagnosis is essential for determining appropriate treatment strategies and improving patient outcomes [3]. Currently, Magnetic Resonance Imaging (MRI) and Computed Tomography (CT) are the primary imaging techniques used in clinical settings to detect and monitor brain tumors due to their high-resolution visualization of internal brain structures [4].

Despite advancements in imaging technology, the manual interpretation of MRI and CT scans remains prone to human error and subjectivity [5]. Moreover, the complexity of tumor characteristics often requires significant expertise to analyze, which may not be readily available in all healthcare facilities. As a result, the integration of artificial intelligence—particularly machine learning—into medical imaging analysis has attracted growing attention [6, 7]. Machine learning algorithms can automatically extract patterns and classify images with high precision, thereby enhancing diagnostic accuracy and assisting radiologists [8–10].

Numerous studies have applied machine learning models to brain tumor detection, demonstrating promising results. Anantharajan et al. [11] improve brain tumor detection from MRI images using a deep learning approach. It combines preprocessing techniques, fuzzy c-means segmentation, and GLCM-based feature extraction. Mathivanan et al. [12] studied brain tumor diagnosis using deep transfer learning by evaluating four pre-trained models—ResNet152, VGG19, DenseNet169, and MobileNetv3—on MRI data. Agarwal et al. [13] developed a robust, automated, and hybrid system for the early diagnosis and classification of brain tumors using MRI images. They designed to outperform existing models in terms of accuracy and robustness, achieving 98.89% accuracy on a public MRI dataset, demonstrating its effectiveness across varying image conditions. Islam et al. [14] developed a highly accurate brain tumor detection and classification system using deep learning models, including 2D CNNs and CNN-LSTM. Appiah et al. [15] evaluated a low-order machine learning approach for brain tumor detection by integrating POD with CNN on 2D MRI images and compared with state-of-the-art pre-trained models like MobileNetV2, Inception-v3, ResNet101, and VGG-19. However, many of these studies either focus on a limited range of algorithms or lack comprehensive system implementations that can be tested in practical scenarios. Furthermore, issues such as class imbalance and limited feature interpretability often affect model performance.

This study addresses these gaps by developing and comparing six machine learning algorithms—Support Vector Machine, Random Forest, K-Nearest Neighbor, Logistic Regression, Naive Bayes, and XGBoost—for classifying brain tumors using MRI and CT images. A feature extraction approach is employed using handcrafted statistical and texture features, and the Synthetic Minority Over-sampling Technique (SMOTE) is applied to manage data imbalance. Additionally, a web-based interactive dashboard is introduced to visualize prediction results and support future clinical deployment.

This paper is organized as follows. Section 1 introduces the importance of automated brain tumor detection and the role of machine learning in enhancing MRI and CT image analysis. Section 2 describes the dataset, preprocessing steps, and handcrafted feature extraction, including the use of SMOTE for class balancing. Section 3 presents the implementation of six machine learning models and compares their performance using standard evaluation metrics. Section 4 discusses confusion matrix results and error patterns across the models. Section 5 introduces the interactive dashboard developed for visualization and real-time prediction. Section 6 concludes the paper by summarizing the key findings and highlighting future research opportunities.

2 Methodology

2.1 Image Dataset and Classification Workflow

This study utilizes the Brain Tumor Multimodal Image Dataset from Kaggle, consisting of MRI and CT images labeled as either *tumor* or *healthy*. The dataset contains 8,029 high-resolution images, including 3,009 MRI tumor, 2,000 MRI healthy, 2,318 CT tumor, and 702 CT healthy images. To ensure consistent processing across modalities, all images were treated within a unified binary classification framework: *tumor* versus *healthy*.

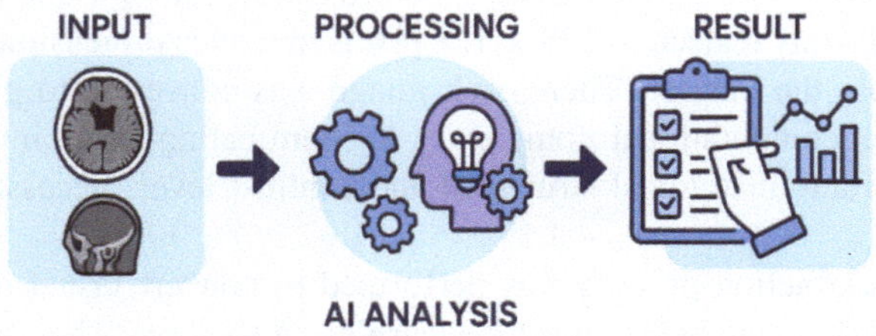

Fig. 1. Workflow of MRI-Based Brain Tumor Analysis

The overall workflow is illustrated in Fig. 1. The process begins with dataset loading, followed by several preprocessing steps designed to enhance image quality and standardize input dimensions. These prepared images then undergo feature extraction to obtain representative statistical and texture descriptors, which serve as inputs for machine learning models. The final stage involves model training, performance evaluation, and visualization of prediction results using the developed dashboard interface.

Figure 2 illustrates the distribution of MRI and CT images across the tumor and healthy classes in the dataset. As shown, the MRI modality contains a larger number of samples, with 3,009 tumor images and 2,000 healthy images, while the CT dataset is more imbalanced, consisting of 2,318 tumor and only 702 healthy images. This uneven distribution highlights the inherent class imbalance present in the raw dataset, especially within the CT modality. Understanding this imbalance is important because it can negatively affect model training by biasing predictions toward the majority class. Therefore, a balancing technique such as SMOTE is required to ensure that both tumor and healthy categories are adequately represented during the learning process.

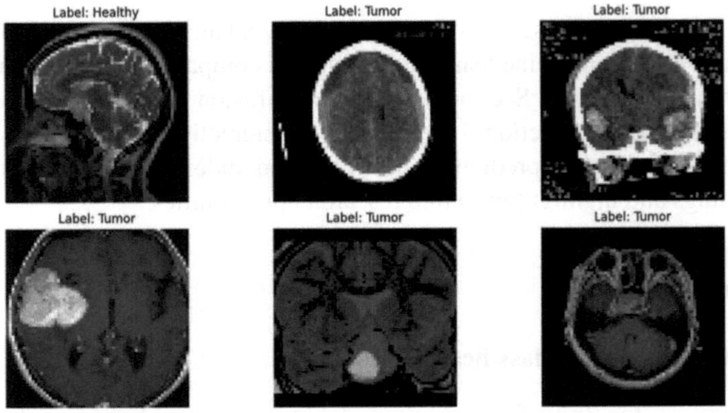

Fig. 2. Class-wise Distribution of CT and MRI Images

2.2 Image Preprocessing and Feature Extraction

The preprocessing phase of this study involved several crucial steps to standardize and enhance the quality of brain tumor images before classification. First, all CT and MRI images were uniformly resized to 128 × 128 pixels to reduce dimensionality and ensure consistency across the dataset. Then, each image was converted to grayscale, a process that reduces computational complexity by eliminating color information while preserving important anatomical structures and contrast levels necessary for accurate analysis.

The feature extraction process was performed to convert visual information into quantitative attributes suitable for machine learning. A total of 20 hand-crafted features were extracted and grouped into four main categories: intensity statistics, distribution statistics, morphological features, and texture features, as detailed in Table 1. These features collectively represent intensity, shape, pixel distribution, and texture patterns within the images, providing a robust foundation for model training and classification.

Table 1. Categories and Types of Extracted Image Features

Feature categories	Feature name	Amount
Intensity Statistics	Contrast Entropy Mean Intensity	3
Distribution Statistics	Skewness Kurtosis	2
Morphology	Perimeter Area Ratio	2
Texture (Haralick Features)	13 texture features of mahotas	13
Total		20

Figure 3 shows the distribution of brain tumor and healthy image data before and after applying the SMOTE. Initially, the dataset exhibited a class imbalance, where one class had significantly more samples than the other, which could lead to biased model predictions. To address this, SMOTE was applied to generate synthetic samples for the minority class by interpolating between existing data points, rather than duplicating them. As illustrated in the figure, the post-SMOTE distribution is more balanced, ensuring that both classes are equally represented during model training. This step enhances the model's ability to generalize and improves its performance in correctly classifying both tumor and healthy images.

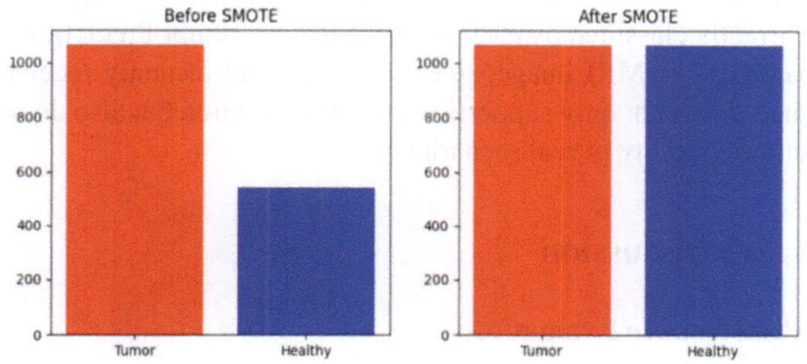

Fig. 3. Distribution of class data before and after applying the SMOTE

2.3 Model Training and Development

The dataset was split into 80% training and 20% testing using stratified sampling to maintain balanced class distribution. Six machine learning algorithms were implemented for classification: SVM, RF, KNN, LR, NB, and XGBoost. These models were selected due to their popularity and strong performance in medical image classification tasks.

Each model applies a distinct classification strategy: SVM finds the optimal separating hyperplane; RF uses ensemble decision trees with majority voting; KNN classifies based on the closest training samples; LR estimates class probabilities using a sigmoid function; NB relies on probabilistic assumptions of feature independence; and XGBoost employs a boosting approach to optimize model accuracy. These algorithms were trained using extracted features to evaluate their effectiveness in distinguishing between tumor and healthy brain images.

2.4 Model Evaluation and Visualization

This section outlines how the performance of the trained machine learning models was assessed and visualized. The evaluation was conducted using a confusion matrix, which records the number of correct and incorrect predictions for both tumor and healthy classes. From this matrix, key metrics such as accuracy, precision, recall, specificity, and F1-score were calculated to provide a comprehensive view of model effectiveness. These metrics help measure how well the models identify true positives (e.g., correctly predicted tumors) and avoid false predictions. Additionally, an interactive dashboard was developed to visualize the results. The dashboard includes two key features: (1) "Correct vs Wrong Predictions" — showing side-by-side comparisons of images that were correctly and incorrectly classified by each model, and (2) "Tumor Prediction" — allowing users to upload a CT or MRI image, select a model, and instantly receive the prediction. This visualization not only supports technical evaluation but also demonstrates the system's practical usability in real scenarios.

3 Results and Discussion

3.1 Model Performance Evaluation

The performance of the six machine learning models—SVM, Random Forest, Logistic Regression, KNN, XGBoost, and Naive Bayes—was evaluated using accuracy, precision, recall, and F1-score. Table 2 summarizes the overall results. Among all models, Random Forest and XGBoost achieved the highest performance, each reaching an accuracy of 96.01% with balanced precision, recall, and F1-scores of 0.97 for the tumor class and above 0.94 for the healthy class.

The KNN model also performed competitively with 94.27% accuracy and strong classification balance across both categories. In contrast, SVM yielded moderate performance at 87.55% accuracy, showing weaker sensitivity in detecting healthy images. Logistic Regression and Naive Bayes produced the lowest results—75.59% and 70.24% accuracy, respectively—indicating difficulties in learning the feature distributions and a tendency to misclassify healthy samples. These findings confirm that ensemble-based models are more robust when handling handcrafted features extracted from multimodal medical images.

Table 2. Machine Learning Model Performance Evaluation Results

Model	Accuracy	Precision (Tumor)	Recall (Tumor)	F1-Score (Tumor)	Precision (Healthy)	Recall (Healthy)	F1-Score (Healthy)
SVM-RBF	0.8755	0.94	0.87	0.90	0.78	0.89	0.83
RF	0.9601	0.97	0.97	0.97	0.93	0.95	0.94

(continued)

Table 2. *(continued)*

Model	Accuracy	Precision (Tumor)	Recall (Tumor)	F1-Score (Tumor)	Precision (Healthy)	Recall (Healthy)	F1-Score (Healthy)
LR	0.7559	0.85	0.77	0.81	0.61	0.74	0.67
KNN	0.9427	0.97	0.94	0.96	0.89	0.95	0.92
XGBoost	0.9601	0.97	0.97	0.97	0.93	0.95	0.94
Naive Bayes	0.7024	0.77	0.79	0.78	0.56	0.52	0.54

3.2 Model Performance Evaluation

The confusion matrix provides a detailed breakdown of each model's classification behavior through four key components: True Positive (TP), True Negative (TN), False Positive (FP), and False Negative (FN). In this study, TP represents tumor cases correctly identified as tumor, TN indicates healthy cases correctly recognized as healthy, FP refers to healthy images incorrectly predicted as tumor, and FN corresponds to tumor cases that the model failed to detect.

Overall, the confusion matrices (Fig. 4) reveal clear performance differences across the six models. Random Forest and XGBoost show the strongest classification ability, achieving high TP and TN values while maintaining very low FP and FN counts. These results indicate excellent reliability in detecting both tumor and healthy cases. KNN also performs well, although with slightly more FP and FN than the ensemble models, reflecting modest reductions in sensitivity and specificity.

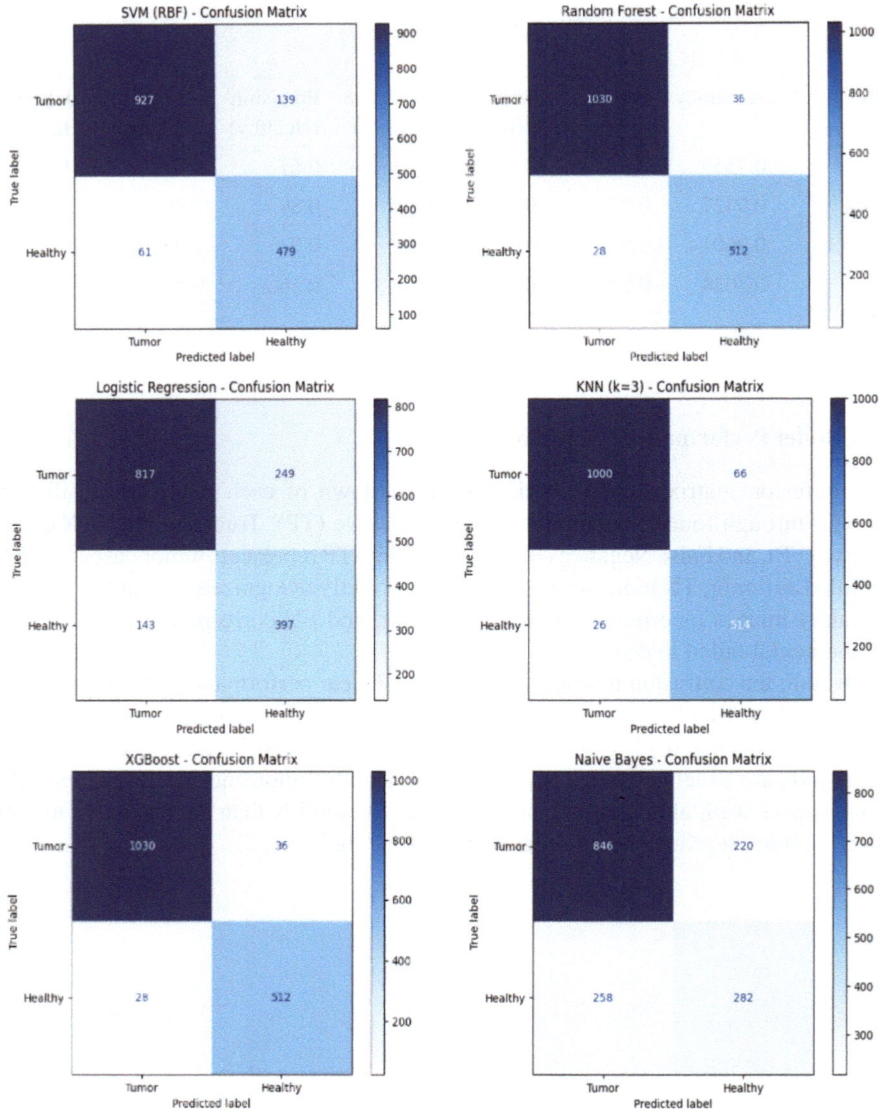

Fig. 4. Confusion Matrix of Each Model

SVM demonstrates moderate performance but tends to generate higher FN values, indicating lower sensitivity in detecting tumor cases. This means SVM is more likely to miss existing tumors, which is a critical limitation for medical diagnosis. Naive Bayes records the weakest results, characterized by the highest FP and FN values among all models, suggesting substantial difficulty in distinguishing between tumor and healthy features. These observations reinforce the conclusion that ensemble-based approaches outperform simpler probabilistic or linear models in brain tumor classification tasks.

3.3 Correct and Incorrect Prediction Comparison

Figure 5 presents a comparative analysis of the total number of correct and incorrect predictions made by each machine learning model during the brain tumor classification task. This visualization clearly illustrates the reliability and stability of the models based on the 1,606 test samples (calculated from correct + incorrect predictions for each model).

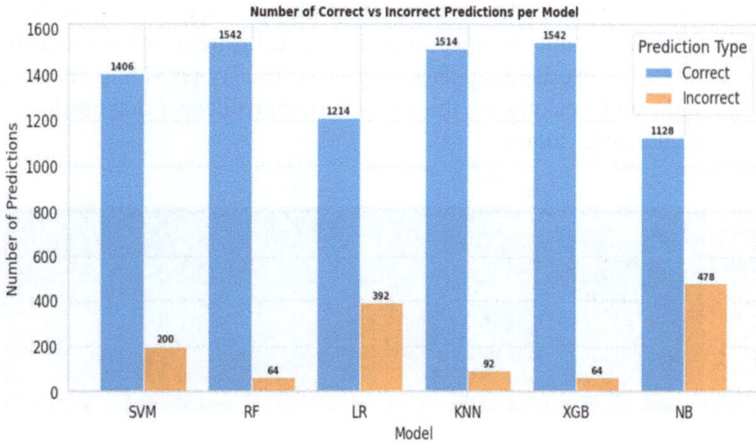

Fig. 5. Number of Correct and Incorrect Predictions from Each Model

The updated results show that the ensemble models—Random Forest (RF) and XGBoost—achieved the best performance, each producing 1,542 correct predictions and only 64 incorrect predictions, corresponding to an error rate of approximately ±4%. This strong performance highlights the robustness of ensemble learning methods, especially when dealing with complex and heterogeneous feature sets extracted from medical images. Both models demonstrate consistent predictive accuracy across tumor and healthy classes.

The KNN model also performed competitively, with 1,514 correct and 92 incorrect predictions, making it slightly less accurate than the ensemble models but still a strong performer within the overall comparison. In contrast, the SVM model recorded 1,406 correct and 200 incorrect predictions, indicating a higher error rate and lower sensitivity, particularly when distinguishing between subtle differences in feature distributions.

The weakest performance was observed in Logistic Regression (LR) and Naive Bayes (NB). LR achieved 1,214 correct and 392 incorrect predictions, while NB had only 1,128 correct and 478 incorrect predictions, with error rates approaching 30%. The poorer results of these two models stem from their linear assumptions and the independence assumptions in NB, which are not well-suited for the complex statistical and texture-based features used in this study.

3.4 Visualization and Model Testing Interface (Revised Version)

Figure 6 illustrates the "Correct vs. Wrong Predictions" visualization module integrated into the classification dashboard. This interface enables users to select any of the trained machine learning models and view a side-by-side comparison of images that were correctly classified and those that were misclassified. Each displayed image is annotated with its true label (True) and predicted label (Pred), allowing users to easily identify the types of errors produced by each model. This visualization is particularly useful for analyzing the behavior of lower-performing models such as Logistic Regression and Naive Bayes—where misclassifications often reveal patterns related to feature variability, image noise, or overlapping class characteristics. By providing visual evidence of model strengths and weaknesses, this module supports deeper diagnostic evaluation beyond numerical metrics alone.

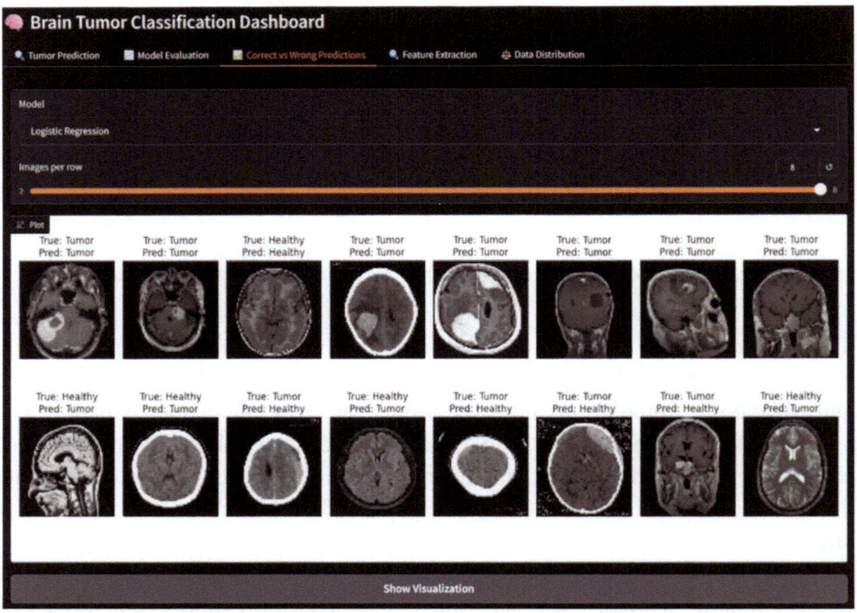

Fig. 6. Visualization of correct and incorrect prediction results

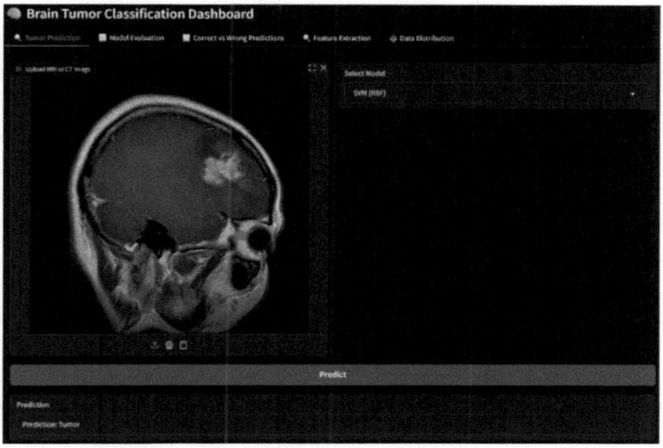

Fig. 7. Prediction test for each model

Figure 7 presents the "Tumor Prediction" module of the dashboard, which allows users to upload a CT or MRI image and select a model to perform real-time inference. After the "Predict" button is pressed, the system immediately returns the classification result, identifying the image as either *Tumor* or *Healthy*. This interactive feature demonstrates the practical usability of the developed system, enabling clinicians, researchers, or students to test models directly on new and unseen data. The module thus serves as a prototype interface for potential clinical decision-support applications, offering an intuitive and efficient way to validate model predictions in real-world scenarios.

4 Conclusion

This study successfully developed and evaluated a brain tumor classification system using feature-extracted MRI and CT images with six machine learning algorithms. The results revealed that ensemble-based models, particularly Random Forest and XGBoost, achieved the highest accuracy of 96.01%, demonstrating strong and balanced performance in distinguishing tumor and healthy cases. K-Nearest Neighbor also performed competitively with 94.27% accuracy, while simpler models like Logistic Regression and Naive Bayes showed lower reliability. The application of handcrafted features and the SMOTE technique effectively improved model generalization and addressed class imbalance issues. Furthermore, the developed web-based dashboard enables real-time prediction and visualization, supporting potential integration into clinical diagnostic workflows

Acknowledgments. This research was funded by Universiti Sultan Zainal Abidin, Malaysia through **Pelan Strategik Universiti (PSU)**.

Disclosure of Interests The authors have no competing interests to declare that are relevant to the content of this article.

References

1. Seetha, J., Raja, S.S.: Brain tumor classification using convolutional neural networks. Biomed. Pharmacol. J. **11**(3), 1457 (2018)
2. Wang, J., Lu, S.Y., Wang, S.H., Zhang, Y.D.: RanMerFormer: Randomized vision transformer with token merging for brain tumor classification. Neurocomputing. **573**, 127216 (2024)
3. Islam, M.M., Talukder, M.A., Uddin, M.A., Akhter, A., Khalid, M.: Brainnet: Precision brain tumor classification with optimized EfficientNet architecture. Int. J. Intell. Syst. **2024**(1), 3583612 (2024)
4. Pacal, I., Celik, O., Bayram, B., Cunha, A.: Enhancing EfficientNetv2 with global and efficient channel attention mechanisms for accurate MRI-based brain tumor classification. Clust. Comput. **27**(8), 11187–11212 (2024)
5. Rabby, S.F., Arafat, M.A., Hasan, T.: BT-Net: An end-to-end multi-task architecture for brain tumor classification, segmentation, and localization from MRI images. Array. **22**, 100346 (2024)
6. Nhlapho, W., Atemkeng, M., Brima, Y., Ndogmo, J.C.: Bridging the gap: Exploring interpretability in deep learning models for brain tumor detection and diagnosis from MRI images. Information. **15**(4), 182 (2024)
7. Amin, J., Sharif, M., Haldorai, A., Yasmin, M., Nayak, R.S.: Brain tumor detection and classification using machine learning: A comprehensive survey. Complex Intell. Syst. **8**(4), 3161–3183 (2021)
8. Sambas, A., et al.: ANN-based chaotic PRNG in the novel jerk chaotic system and its application for the image encryption via 2-D Hilbert curve. Sci. Rep. **14**(1), 29602 (2024)
9. Missaoui, R., Hechkel, W., Saadaoui, W., Helali, A., Leo, M.: Advanced deep learning and machine learning techniques for MRI brain tumor analysis: A review. Sensors. **25**(9), 2746 (2025)
10. Celik, M., Inik, O.: Development of hybrid models based on deep learning and optimized machine learning algorithms for brain tumor multi-classification. Expert Syst. Appl. **238**, 122159 (2024)
11. Anantharajan, S., Gunasekaran, S., Subramanian, T.: MRI brain tumor detection using deep learning and machine learning approaches. Measur. Sens. **31**, 101026 (2024)
12. Mathivanan, S.K., Sonaimuthu, S., Murugesan, S., Rajadurai, H., Shivahare, B.D., Shah, M.A.: Employing deep learning and transfer learning for accurate brain tumor detection. Sci. Rep. **14**(1), 7232 (2024)
13. Agarwal, M., Rani, G., Kumar, A., Kumar, P., Manikandan, R., Gandomi, A.H.: Deep learning for enhanced brain tumor detection and classification. Results Eng. **22**, 102117 (2024)
14. Islam, M.N., Azam, M.S., Islam, M.S., Kanchan, M.H., Parvez, A.S., Islam, M.M.: An improved deep learning-based hybrid model with ensemble techniques for brain tumor detection from MRI image. Inform. Med. Unlocked. **47**, 101483 (2024)
15. Appiah, R., et al.: Brain tumor detection using proper orthogonal decomposition integrated with deep learning networks. Comput. Methods Prog. Biomed. **250**, 108167 (2024)

Coastal Urbanization Around a Mega Liquid Bulk Port in Thailand: Is the Growth Perpetual?

Htet Khaing Kyi Lin[1], Naruphun Chotechuang[1], Worawut Poma[1], Cherdvong Saengsupavanich[1](✉), Effi Helmy Ariffin[2], Abdullah Açık[3], Ikha Magdalena[4,5], and Resista Vikaliana[6]

[1] Faculty of International Maritime Studies, Kasetsart University, Sri Racha Campus, Sri Racha, Thailand
cherdvong.sa@ku.th, cherdvong.saengsupavanich@hotmail.com
[2] Institute of Oceanography and Environment, Universiti Malaysia Terengganu, Kuala Nerus, Malaysia
[3] Maritime Faculty, Dokuz Eylül University, İzmir, Turkey
[4] Faculty of Mathematics and Natural Sciences, Bandung Institute of Technology, Bandung, Indonesia
[5] Center for Coastal and Marine Development, Institut Teknologi Bandung, Bandung, Indonesia
[6] Logistics Department, Faculty of Industrial Engineering, Universitas Pertamina, Jakarta Selatan, Indonesia

Abstract. A mega port is thought to quickly and continually transform its surrounding areas. However, there is limited research confirming that these ports indeed drive long-term land-use development. Therefore, it is essential to explore land-use changes before and after a port's establishment to assess the connection and impact between the ports and their nearby regions. This study focuses on analyzing historical land-use changes from 1975 to 2021 around Map Ta Phut port, the largest liquid bulk port in Thailand. Satellite images from Google Earth and the U.S. Geological Survey were utilized for this analysis, employing standard spatiotemporal analysis techniques. The key findings reveal that urbanization surged during the port's planning and construction stages but then slowed down significantly afterward. Thus, it can be deduced that mega ports act mainly as a temporary catalyst for urban growth, rather than fostering permanent urbanization. This research is significant as it enhances our understanding of the link between urban development and ports, assisting in crafting strategies that balance coastal urban growth with environmental conservation.

Keywords: Coastal management · Land use change · Sustainable development

1 Introduction

Seaports are primarily human-made structures that extend land functions into the ocean, serving as vital links in global logistics and supply chains [1–3]. Over time, these ports have evolved not only to accommodate ships but also to provide value-added services,

M. A. b. Jamlos et al. (Eds.): ICSET 2025, LNNS 1897, pp. 13–24, 2026.
https://doi.org/10.1007/978-3-032-20722-7_2

such as packaging and labeling, while creating employment opportunities in their surrounding areas [4]. Port locations are strategically selected to optimize land access in navigable waters, offering secure sites for commercial trade where vessels can be safely loaded, unloaded, refueled, or repaired. Seaports play a crucial role in maritime transportation and trade facilitation, significantly impacting economic, social, and environmental development at local, regional, and national levels. Today, ports contribute to the expansion of the global economy and act as engines of economic growth for nations by connecting them to the expanding landscape of international trade [5]. Historically, the influence of ports on the economy has been examined in various studies. Chang et al. [6] and Mlambo [7] concluded that ports stimulate economic growth at both national and regional levels. Furthermore, port development significantly increases the land value of adjacent areas, as businesses and industries thrive and capitalize on the benefits provided [8, 9].

Ports, shipping activities, environmental concerns, and urban growth are interconnected [10–13]. The operations of ports influence sustainability factors such as human well-being, green areas, ecosystem health, biodiversity conservation, and pollution control. The well-being of coastal communities is a critical element of sustainable development [14–16]. As a result, numerous researchers have examined how port development impacts urban growth and changes in land use [17–19]. Significant changes have been observed in these port areas, leading to numerous negative consequences that have been investigated more thoroughly by various researchers, including deterioration of water quality, degradation of coastal resources, and air quality problems [20–22]. Urbanization driven by ports can adversely impact the distribution of coastal resources. Economically, Deng et al. [23] and Jung [24] determined that these ports do not play a substantial role in fostering economic growth. Consequently, the concentration of industries around major ports has a pronounced effect on the lives of workers and their families, sometimes resulting in unprecedented land-use conflicts [25, 26], primarily between residential and recreational services competing for limited coastal space. Our conventional view that port-driven urbanization is invariably beneficial for coastal residents may need re-evaluation in light of these findings.

Urban growth necessitates sufficient infrastructure development [27–29], which includes coastal facilities, transport systems, water supply, and electricity, and these infrastructures carry costs and environmental consequences [30, 31]. Therefore, it is vital to accurately understand the urban growth patterns surrounding mega ports. Misjudging whether a mega port leads to a temporary or a permanent influence on urban expansion can lead to inefficient investments. This brings us to our research question: Does a port consistently boost urbanization in its vicinity, or does it only produce a short-lived effect on coastal urban development? This study aims to address this question by examining land-use transformations around Map Ta Phut (MTP) Port, one of the mega ports in Thailand. The significance of this research lies in its potential to assist the government, urban planners, and coastal managers in formulating a strategic urban development plan for the regions adjacent to the ports.

2 Study Site and Methodology

Thailand has a coastline of 3,148 km with over 4,000 km of waterways for both domestic and coastal transportation. For this study, we aim to focus on the MTP port, Rayong province.

Fig. 1. The MTP port and the 5-km radius

2.1 Map Ta Phut Port Description

Map Ta Phut (MTP) port is the largest high-capacity industrial port of Thailand, which is fully equipped with a wide range of services and facilities in the Rayong Province. In this study, we focus on land use change within a radius of 5 km from the port (Fig. 1). The MTP port was developed with the intention of promoting heavy chemical and petrochemical industries, and is operated by the Industrial Estate Authority of Thailand. The port is approximately 3.5 km wide, and extends roughly 4.0 km seaward. The construction of the port began in 1989, and its operation started in 1992 [32]. The

MTP port area and its industrial estate contain about 50 M-sized industrial factories and accommodate over 5,000 ships annually. After the port construction, the coastal zone around the MTP port has changed drastically, becoming more urbanized than inland areas [33]. This is primarily because of an increase in industrial land use. At the time of construction, agricultural and forest areas surrounded most industrial areas, forming a buffer zone between the industrial facilities and residential areas. As of 2025, the MTP port is expanding its capacity by reclaiming more ocean. Like other ports in the world that have applied modern technology to increase their performance [34] and the fact that coastal communities are the key to sustainable coastal development [35], the MTP port has undertaken measures to improve its performance as well as the surrounding communities' well-being.

2.2 Methodology

Remote sensing (RS) and geographical information systems (GIS) have become essential tools for acquiring timely and precise spatial data regarding land cover and for assessing the changes in marine and coastal resources over time [36, 37]. Remotely sensed imagery effectively tracks land-use conditions, making it a valuable resource for analyzing coastal landforms and changes in land use. GIS offers a versatile platform for collecting, storing, visualizing, and analyzing digital data vital for detecting changes. Its application is prevalent in port-related studies, as illustrated in the research by Gao et al., who utilized GIS to evaluate the accessibility of ports in China from a land-sea coordination viewpoint [38]. Additionally, Ayotamuno et al. [39] explored shifts in land use and economic activities from 1986 to 2005 in a residential complex in Port Harcourt, Nigeria—initially intended for senior civil servants and military personnel—highlighting a swift increase in commercial activities within the area. Similarly, Sonde et al. [40] utilized RS satellite data to study the rapid, uncontrolled growth driven by domestic needs and industrialization around Jawaharlal Nehru Port Trust in India. Therefore, employing RS and GIS to investigate historical urbanization trends at the MTP port in Thailand is both viable and suitable.

High-resolution satellite images of the MTP port were readily accessible from two main sources: the U.S. Geological Survey (USGS) and Google Earth. These images date back to 1975 and were selected for the study based on the absence of cloud cover over the study area, ensuring clear visibility of land use characteristics. Information regarding the port's construction year was also collected. Although LANDSAT imagery provided by USGS had a relatively low resolution, it has been widely used in examining coastal urban land cover transformations. For more recent images, Google Earth was used due to its superior resolution. To identify significant changes over time, satellite images from either USGS or Google Earth were gathered at 5-year intervals (ranging from 4 to 6 years) from 1975 to 2021, as shown in Table 1, based on image availability and cloud cover. These images were subsequently processed to determine land-use cover percentages before and after the port's development. The study's methodology framework is illustrated in Fig. 2.

Georeferencing a satellite image is a systematic process that allows the image to be accurately positioned within a geographical context, ensuring that its features correspond to actual coordinates on earth. The procedure begins with preparing relevant data and selecting appropriate geospatial software. After loading the satellite image, identifiable control points were selected and inputted, linking locations on the image to their corresponding geographic coordinates. Following this, a transformation model was applied to align the image with the reference data. Once satisfied with the results, the georeferenced images were saved, and a final review was conducted to check for visual alignment and accuracy. The images in Table 1 were carefully rectified and geo-referenced, followed by field surveys and ground truthing. The authors then employed visual interpretation to differentiate between residential, sea, and agricultural/forest areas. Finally, land-use maps covering the period from 1975 to 2021 were developed, allowing for an analysis of land use changes surrounding the MTP port.

Table 1. Data used in this study

Focus Area	Image sources	Satellite	Image Date
MTP Port	Google Earth		23/1/2021
			8/2/2016
			2/1/2011
			29/1/2004
			31/12/2001
			31/12/1996
			31/12/1991
			31/12/1986
	USGS	Landsat 2	8/10/1979
		Landsat 2	13/12/1975

3 Results

An increasing percentage of land development was observed in neighboring municipalities, with urban expansion in areas within a radius of 5 km from the MTP port (Table 2). The residential and building areas were expanded from 0.9 sq. km to 43.0 sq. km from 1975 to 2021. In contrast, a decrease in other land categories (e.g., agricultural and forest) was observed, dropping from 80.7% in 1975 to 26.3% in 2021. Effectively, 43.1 sq. km of agricultural and empty lands were urbanized in these 46 years. However, the transformation rate was not linear. The high rate of urbanization is seen to be clustered around the year of port development (5 years before and 5 years after the port construction). The time period over which high urbanization is observed is considered consistent with any infrastructure project wherein the initial few years see increased growth of urbanization for implementation of the project. Subsequently, once the project is completed, there is again a major influx and increased rate of urbanization due to high movement of project users and their support systems. Once this movement and these requirements are met, the rate of urbanization slows down, as observed in this study.

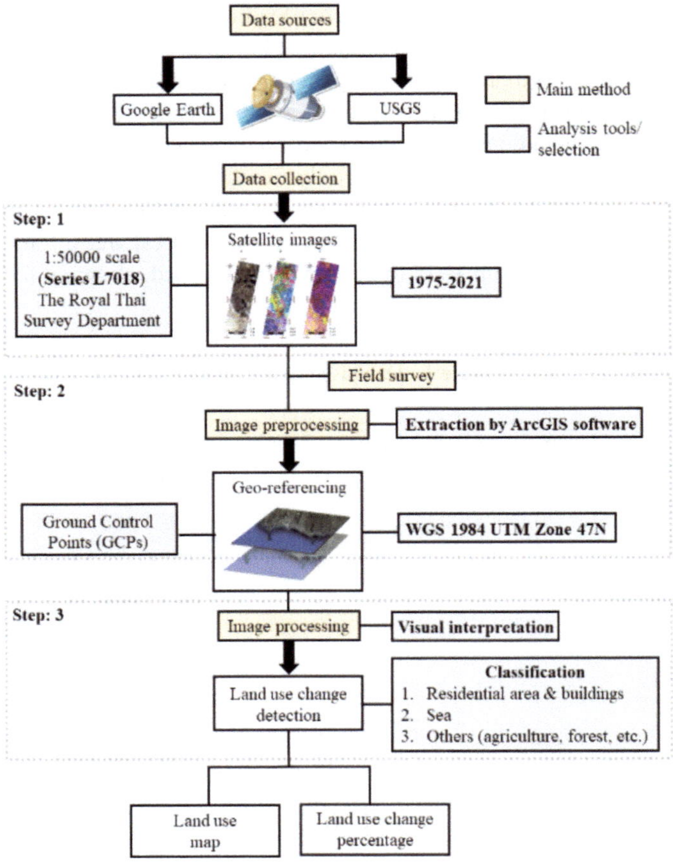

Fig. 2. A Flowchart illustrating steps to monitor land-use changes at the MTP port

When the historical trend of urbanization in MTP is studied from 1975 to 2021, it is observed that the percentage of residential and building areas constantly increased from 1.1% in 1975, to 6.3% in 1986 (approximately 0.47% per year) prior to the port construction. There was a sharp rise to 24.2% from 1986 to 1996 (approximately 2.42% per year), when the MTP port development plan began, and after it started to operate (Fig. 3). This spike in urbanization is considered to be what one would expect for an infrastructure project. If the MTP port had not been constructed in 1989, the percentage of urbanization around it would have increased steadily, instead of the observed sharp increase. After the MTP port operated for a while, urbanization returned to increase steadily again from 1996 to 2016 (at around 1.07% per year). Later, after the MTP port announced its additional reclamation expansion of 1,000,000 sq. m. and started to undertake an environmental impact assessment study, the urbanization rate jumped as much as 1.74% per year (from 2016 to 2021). A summary of urbanization transformation is shown in Fig. 4, displaying strong evidence that mega port development had a non-linear impact on its surrounding urbanization.

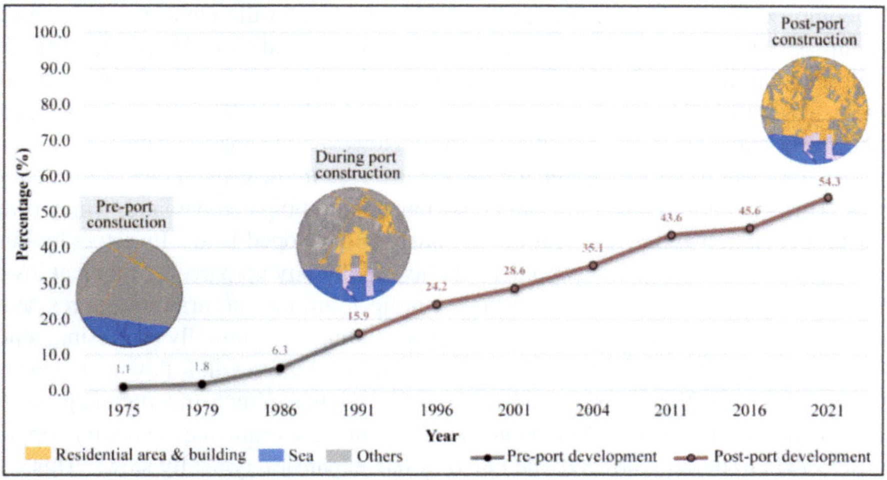

Fig. 3. Percentage of urbanization in the radius of 5 km around the MTP port

Table 2. Urbanization rate (1975 to 2021) in a radius of 5 km around the MTP port

Type of land use									
Year	Port		Residential area and buildings		Sea		Others		Total area (km^2)
	Area (km^2)	%	Area (km^2)	%	Area (km^2)	%	Area (km^2)	%	
2021	3.8	4.8	43.0	54.3	11.6	14.7	20.8	26.3	79.1
2016	3.8	4.8	36.1	45.6	11.6	14.7	27.6	34.9	
2011	3.8	4.8	34.5	43.6	11.6	14.7	29.2	37.0	
2004	3.8	4.8	27.8	35.1	11.6	14.7	36.0	45.4	
2001	3.8	4.8	22.6	28.6	11.6	14.7	41.2	52.0	
1996	3.4	4.3	19.1	24.2	11.9	15.1	44.6	56.3	
1991	3.4	4.3	12.6	15.9	11.9	15.1	51.2	64.7	
1986	–	–	5.0	6.3	14.3	18.2	59.8	75.6	
1975	–	–	1.4	1.8	14.3	18.2	63.3	80.0	

4 Discussion

Land-use change, a dynamic process influenced by both natural phenomena and human activities, significantly impacts human settlements and coastal environments [41]. Among the key players in this transformation are ports, which serve as vital connections in a complex network of interactions between society and the environment, particularly in coastal regions. This phenomenon is particularly evident in Thailand, where the development of a mega liquid bulk port has marked significant changes in urbanization patterns. Prior to the planning and construction of the MTP port, the urban expansion rate in the surrounding areas was relatively modest. However, the announcement of a major port development project triggered a dramatic increase in the growth rate of residential zones. This spike in urbanization can be attributed to anticipated demand, which often attracts speculation and investment in previously undeveloped lands. Investors looking to capitalize on the expected economic benefits rapidly buy up parcels of land at lower prices, anticipating substantial returns as the project unfolds and draws more residents and businesses to the area. This pattern of urbanization, while initially promising, tends to encounter limitations over time. After the initial flurry of growth following the establishment of ports, which mirrors trends seen globally with similar infrastructure projects, the momentum often wanes. As demand stabilizes and the immediate economic opportunities begin to be fully utilized, the rate of urban expansion typically slows. This was clearly illustrated in this case study in Thailand.

As ports create environmental impacts during their construction and operations [42–44], incorrectly forecasting the rate of port expansion can have profound implications for effective environmental management. An underestimation of this rate may result in insufficient planning and resource allocation for critical infrastructure, ultimately leading to overstressed systems unable to accommodate increased activity, thus exacerbating environmental degradation. Moreover, excessive development derived from miscalculation can disrupt sensitive coastal ecosystems, precipitating a loss of biodiversity and altering established habitats. Conversely, overestimating the expansion rate may cultivate complacency regarding the necessity of pollution controls, permitting unchecked increases in air and water pollution that adversely affect both local communities and ecological health. This mismanagement arising from inaccurate forecasts can further engender conflicts with stakeholders committed to environmental preservation, complicating future expansion endeavors. Additionally, erroneous projections heighten vulnerability to the impacts of climate change, as inadequate resilience planning may amplify risks to port infrastructure and surrounding areas. Ultimately, these considerations underscore the paramount importance of accurate forecasting in facilitating a balanced approach to economic development while ensuring the protection of environmental integrity and the well-being of coastal communities [45].

Technological advancements in port management are pivotal for enhancing environmental protection as they facilitate improved resource management, pollution reduction, and the integration of sustainable infrastructure [46]. Contemporary innovations, including the Internet of Things (IoT) and artificial intelligence (AI), enable real-time monitoring of resource consumption and emissions, thereby optimizing energy utilization and minimizing waste generation. Advanced environmental monitoring systems allow for continuous tracking of air and water quality [47], aiding in the identification of pollution sources and the implementation of effective mitigation strategies. Together with correctly projecting the growth rate surrounding a mega port, technology can bring a brighter future to coastal communities.

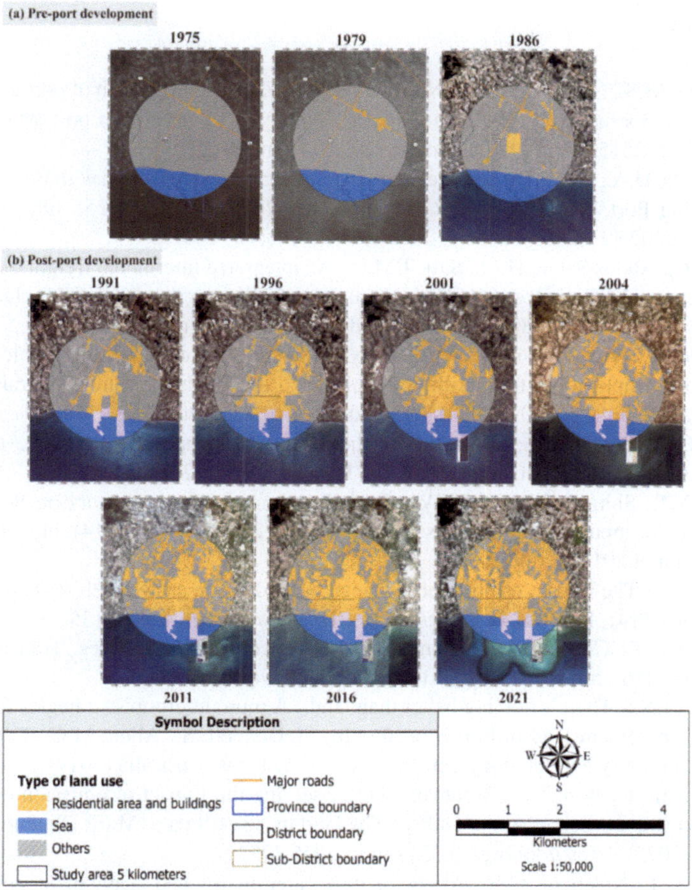

Fig. 4. Land cover map around the MTP port in a 5-km radius

5 Conclusions

Urban growth accelerated in the period leading up to the port's construction, this momentum quickly diminished after operations began, only to temporarily spike again later. This pattern aligns with trends observed in other infrastructure projects, where a rush in urbanization occurs as opportunists set up businesses in anticipation of the project, followed by a period where development caters to the needs of the workforce and their families. Once these needs are sufficiently addressed, any further growth in urbanization is typically linked to the expansion of the project itself. Regardless of the reasons behind this trend, it is crucial to ensure that developments are thought out and implemented in a way that is sustainable, environmentally friendly, and does not adversely affect the coastal ecosystem, air quality, or local wildlife. Failure to accurately assess urbanization impacts can result in poor investment decisions and various negative social and environmental outcomes.

References

1. Abubakar, M.S., Nwoloziri, C.N., Newman, E., Hlali, A.: A correlation of port users' patronage and tariff structure of Nigerian seaports: A study of Lagos Apapa port. Marit. Technol. Res. **7**(1), 272515 (2025). https://doi.org/10.33175/mtr.2025.271519
2. Syahrianda, D.A., Suadi, D.: Service excellence at sea: User satisfaction with Belawan Samudera Fishing Port, Indonesia. Marit. Technol. Res. **7**(1), 270067 (2025). https://doi.org/10.33175/mtr.2025.270067
3. Saha, R.C., Abdus Sabur, H.M., Saif, T.M.R.: An integrated intermodal freight transportation system to avoid container supply chain disruptions in Chattogram Port of Bangladesh. Marit. Technol. Res. **6**(4), 269380 (2024). https://doi.org/10.33175/mtr.2024.269380
4. Fernando, H., Wickramasinghe, V.: Employability skills of maintenance technicians in container ports: Implications for maritime technical and vocational education and training. Marit. Technol. Res. **6**(4), 269909 (2024). https://doi.org/10.33175/mtr.2024.269909
5. Vinod, G.K., Prakash, G.: Evaluation of the potential for maritime cluster development in South Asia. Aust. J. Marit. and Ocean Aff. **16**(4), 431–455 (2024)
6. Chang, Y.T., Shin, S.H., Lee, P.T.W.: Economic impact of port sectors on South African economy: An input–output analysis. Transp. Policy. **35**, 333–340 (2014). https://doi.org/10.1016/j.tranpol.2014.04.006
7. Mlambo, C.: The impact of port performance on trade: The case of selected African states. Economies. **9**(4), 135 (2021). https://doi.org/10.3390/economies9040135
8. Notteboom, T.: Concession agreements as port governance tools. Res. Transp. Econ. **17**, 437–455 (2016). https://doi.org/10.1016/S0739-8859(06)17019-5
9. Rooney, M.N.: 'There's nothing better than land': A migrant group's strategies for accessing informal settlement land in Port Moresby. In: McDonnell, S., Allen, M.G., Filer, C. (eds.) Kastom, Property and Ideology, pp. 111–144. ANU Press, Australia (2017)
10. Premsamarn, P., Win, T.K., Watanabe, D.: Assessing the impact of emission control areas policy on ship emissions in the Gulf of Thailand using AIS data. Marit. Technol. Res. **7**(3), 275311 (2025). https://doi.org/10.33175/mtr.2025.275311
11. Onyena, A.P., Nwaogbe, O.R.: Assessment of water quality and heavy metal contamination in ballast water: Implications for marine ecosystems and human health. Marit. Technol. Res. **6**(4), 270227 (2024). https://doi.org/10.33175/mtr.2024.270227
12. Hoang, A.T., et al.: Technological solutions for boosting hydrogen role in decarbonization strategies and net-zero goals of world shipping: Challenges and perspectives. Renew. Sust. Energ. Rev. **188**, 113790 (2023)
13. Hoang, A.T., et al.: Energy-related approach for reduction of CO2 emissions: A critical strategy on the port-to-ship pathway. J. Clean. Prod. **355**, 131772 (2022). https://doi.org/10.1016/j.jclepro.2022.131772
14. Saengsupavanich, C.: Which eroding site is more urgent for the government?: A reflection from coastal communities. J. Coast. Conserv. **24**(1), 9 (2020). https://doi.org/10.1007/s11852-020-00729-9
15. Saengsupavanich, C., Gallardo, W.G., Sajor, E., Murray, W.W.: Constraints influencing stakeholder participation in collective environmental management. Environ. Earth Sci. **66**(7), 1817–1829 (2012). https://doi.org/10.1007/s12665-011-1407-z
16. Bulengela, G.: "I am a Fisher": Identity and livelihood diversification in Lake Tanganyika Fisheries, Tanzania. Marit. Technol. Res. **6**(3), 268718 (2024). https://doi.org/10.33175/mtr.2024.268718
17. Nebot, N., Rosa-Jiménez, C., Ninot, R.P., Perea-Medina, B.: Challenges for the future of ports. What can be learnt from the Spanish Mediterranean ports? Ocean Coast. Manag. **137**, 165–174 (2017). https://doi.org/10.1016/j.ocecoaman.2016.12.016

18. Hawash, E., El-Hassanin, A., Amer, W., El-Nahry, A., Effat, H.: Change detection and urban expansion of Port Sudan, Red Sea, using remote sensing and GIS. Environ. Monit. Assess. **193**(11), 1–22 (2021). https://doi.org/10.1007/s10661-021-09486-0
19. Shu, B., Zhang, H., Li, Y., Qu, Y., Chen, L.: Spatiotemporal variation analysis of driving forces of urban land spatial expansion using logistic regression: A case study of port towns in Taicang City, China. Habitat Int. **43**, 181–190 (2014). https://doi.org/10.1016/j.habitatint.2014.02.004
20. Agarwala, N.: Project Green Ports: Are Indian ports on the right track? Marit. Aff.: J. National Marit. Found. India. **18**(2), 15–36 (2022). https://doi.org/10.1080/09733159.2022.2143134
21. Agarwala, N., Saengsupavanich, C.: Oceanic environmental impact in seaports. Oceans. **4**(4), 360–380 (2023). https://doi.org/10.3390/oceans4040025
22. Mathew, V., Agarwala, N.: Impact of climate change on coastal communities and security. Marit. Technol. Res. **7**(4), 277917 (2025). https://doi.org/10.33175/mtr.2025.277917
23. Deng, P., Lu, S., Xiao, H.: Evaluation of the relevance measure between ports and regional economy using structural equation modeling. Transp. Policy. **27**, 123–133 (2013). https://doi.org/10.1016/j.tranpol.2013.01.008
24. Jung, B.M.: Economic contribution of ports to the local economies in Korea. The Asian J. Ship. Logist. **27**(1), 1–30 (2011). https://doi.org/10.1016/S2092-5212(11)80001-5
25. Dwarakish, G.S., Salim, A.M.: Review on the role of ports in the development of a Nation. Aquat. Procedia. **4**, 295–301 (2015). https://doi.org/10.1016/j.aqpro.2015.02.040
26. Haralambides, H.: Globalization, public sector reform, and the role of ports in international supply chains. Marit. Econ. Logist. **19**(1), 1–51 (2017). https://doi.org/10.1057/s41278-017-0068-6
27. Saengsupavanich, C.: Coastal revetment design process in Thailand. WIT Trans. Built Environ. **170**, 33–44 (2017). https://doi.org/10.2495/CC170041
28. Saengsupavanich, C.: Successful coastal protection by step concrete revetments in Thailand. IOP Conf. Ser.: Earth Environ. Sci. **1072**(1), 012002 (2022). https://doi.org/10.1088/1755-1315/1072/1/012002
29. Saengsupavanich, C., Pranzini, E.: The 2021-procedure for coastal protection by revetments in Thailand. J. Appl. Water Eng. Res. **11**(2), 303–316 (2023). https://doi.org/10.1080/23249676.2022.2110529
30. Saengsupavanich, C.: Unwelcome environmental impact assessment for coastal protection along a 7-km shoreline in Southern Thailand. Ocean Coast. Manag. **61**, 20–29 (2012). https://doi.org/10.1016/j.ocecoaman.2012.02.008
31. Saengsupavanich, C.: A current Environmental Impact Assessment of a port in Thailand: Marine physical aspects. Ocean Coast. Manag. **54**(2), 101–109 (2011). https://doi.org/10.1016/j.ocecoaman.2010.11.001
32. Industrial Estate Authority of Thailand. Available online: https://ieat.go.th/en/about-the-industrial-port (2025). Accessed 21 April 2025
33. Tontisirin, N., Anantsuksomsri, S.: Economic development policies and land use changes in Thailand: From the eastern seaboard to the eastern economic corridor. Sustainability. **13**(11), 6153 (2021). https://doi.org/10.3390/su13116153
34. Tan, A.Y.N., Loh, H.S., Hsieh, C.-H., Lopez, M.C.R.: Adoption of digital technologies in the maritime industry: Insights from Singapore. Marit. Technol. Res. **7**(3), 275821 (2025). https://doi.org/10.33175/mtr.2025.275821
35. Bulengela, G.: Perception of marine fisheries resources in Tanzania from past to present: Evidence through local knowledge. Marit. Technol. Res. **6**(2), 267008 (2024). https://doi.org/10.33175/mtr.2024.267008
36. Abualhin, K., Abushaban, S.: Predictive models of non-optically active coastal water quality parameters by remote sensing imagery. Marit. Technol. Res. **7**(3), 274944 (2025). https://doi.org/10.33175/mtr.2025.274944

37. Saengsupavanich, C.: Morphological evolution of sand spits in Thailand. Mar. Geod. **44**(5), 432–453 (2021). https://doi.org/10.1080/01490419.2021.1893873
38. Gao, J., Wang, Z., Yu, X.: Accessibility measurement of China's coastal ports from a land-sea coordination perspective - An empirical study. J. Transp. Geogr. **105**, 103479 (2022). https://doi.org/10.1016/j.jtrangeo.2022.103479
39. Ayotamuno, A., Gobo, A.E., Owei, O.B.: The impact of land use conversion on a residential district in Port Harcourt, Nigeria. Environ. Urbanization. **22**(1), 259–265 (2010). https://doi.org/10.1177/956247810363529
40. Sonde, P., Balamwar, S., Ochawar, R.S.: Urban sprawl detection and analysis using unsupervised classification of high resolution image data of Jawaharlal Nehru Port Trust area in India. Remote Sens. Appl.: Soc. Environ. **17**, 100282 (2020). https://doi.org/10.1016/j.rsase.2019.100282
41. Dibba, B., Yaffa, S., Sawaneh, M., Adzawla, W.: Land cover transformation and population growth: Impacts on coastal environment of The Gambia (1990–2020). Sustainability. **17**(5), 1853 (2025). https://doi.org/10.3390/su17051853
42. Saengsupavanich, C., et al.: Land-based physical and biological environmental mitigation measures of a mega port construction in Thailand. E3S Web of Conferences. **422**, 03002 (2023). https://doi.org/10.1051/e3sconf/202342203002
43. Saengsupavanich, C., et al.: Mitigation measures to protect the quality of life in an expansion of Thailand's mega port. In: Lecture Notes in Civil Engineering 539 LNCE, pp. 231–243 (2025). https://doi.org/10.1007/978-981-97-5910-1_18
44. Abessa, D.M.S., et al.: Not all that glitters is gold: Glitter causes acute toxicity to nauplii of Artemia sp. Marit. Technol. Res. **6**(4), 270722 (2024). https://doi.org/10.33175/mtr.2024.270722
45. Saengsupavanich, C., Agarwala, N., Magdalena, I., Ratnayake, A.S., Ferren, V.: Impacts of a growing population on the coastal environment of the Bay of Bengal. Anthr. Coasts. **7**, 24 (2024). https://doi.org/10.1007/s44218-024-00055-9
46. Alamoush, A.S., Ölçer, A.I.: Harnessing cutting-edge technologies for sustainable future shipping: An overview of innovations, drivers, barriers, and opportunities. Marit. Technol. Res. **7**(4), 277313 (2025). https://doi.org/10.33175/mtr.2025.277313
47. Rahaman., T.: Smart environmental monitoring system for air and water quality management. American J. Adv. Technol. Eng. Solutions. **1**(01), 1–19 (2025). https://doi.org/10.63125/a08zay47

Rapid Visual Screening (RVS) of Buildings for Potential Seismic Hazards: Case Study of High-Rise Buildings in Bangkok, Thailand

Afif Ramadhan(✉), Teuku Dimaz Rahmatna, Anggit Hatta Roshindra, Jafar, and Pradipta Nandi Wardhana

Department of Civil Engineering, Universitas Islam Indonesia, Yogyakarta, Indonesia
afif.ramadhan@students.uii.ac.id

Abstract. This research presents the results of Rapid Visual Screening (RVS) of the seismic vulnerability of high-rise buildings in Bangkok, Thailand. Although it is not located in the Ring of Fire, Bangkok experienced significant earthquake impacts due to the Mw 7.7 earthquake that occurred in Myanmar on March 28, 2025. The RVS method was applied to four buildings using FEMA P-154 guidelines. Structural data from each building were collected and assessed based on their seismic resistance characteristics. Results showed that the buildings were generally in the low-risk category, but some showed structural vulnerabilities, especially those built before 2010. This study confirms the importance of regular seismic evaluation even in moderate risk zones.

Keywords: Seismic Risk · RVS · FEMA P-154 · High-Rise Building · Bangkok

1 Introduction

On March 28, 2025, a moment magnitude (Mw) 7.7 earthquake struck Myanmar at 12:50:52 MMT (06:20:52 UTC), with its epicenter located near Mandalay at a shallow focal depth of 10 km. The earthquake caused extensive damage not only within Myanmar but also in neighboring regions, including Thailand, and as far as Bangkok [1]. Although Thailand is not geographically located in a major earthquake zone, the vibrations significantly affected many high-rise buildings in Bangkok. Several structures were reported to have experienced violent shaking. This natural event serves as a strong reminder of the importance of considering earthquake risks, particularly for high-rise buildings that dominate the skyline of the Thai capital.

FEMA developed the Rapid Visual Screening (RVS) method in the United States to address the need for quick post-earthquake building evaluations. RVS is a field observation technique that assesses the potential for seismic damage by visually inspecting buildings and completing standardized forms. Documented in FEMA P-154, this method offers a rapid and cost-effective way to screen large numbers of buildings with limited access, aiming to identify structures that may require further, detailed seismic evaluation

M. A. b. Jamlos et al. (Eds.): ICSET 2025, LNNS 1897, pp. 25–32, 2026.
https://doi.org/10.1007/978-3-032-20722-7_3

[2]. Once a building is flagged as potentially vulnerable, it should undergo a more comprehensive assessment by a structural engineer with expertise in earthquake-resistant design. The RVS procedure typically involves walking around the building and filling out a Data Collection Form based on visual observations of both the exterior and, when possible, the interior.

In recent years, various studies have utilized the Rapid Visual Screening (RVS) method to assess the seismic vulnerability of buildings in different high-risk regions. Shahzada et al. applied a modified RVS approach in Pakistan to evaluate 100 buildings in Seismic Zone III, highlighting the susceptibility of non-engineered structures to severe damage, particularly those located on slopes or composed of stone masonry [1]. Similarly, Ahmed et al. conducted a large-scale RVS-based vulnerability assessment of over 2,900 RC buildings in Northern Algeria, emphasizing the practicality of RVS in urban-scale disaster risk mapping [2]. In Peru, Cardenas et al. assessed the seismic risk of 205 public school buildings using FEMA P-154 and found that a significant proportion were unfit for post-earthquake emergency use [3]. Dogan et al. compared various rapid seismic assessment techniques, including FEMA P-154 and regional alternatives. They concluded that although RVS is less conservative than capacity-based methods, it remains highly effective for preliminary screening in large inventories [4]. Finally, Bektas and Brassai emphasized the global application of traditional RVS methods and reviewed enhancements, such as machine learning integration, to improve reliability in varying building typologies [5]. These studies demonstrate the versatility and relevance of RVS in identifying seismic vulnerabilities, especially in dense urban settings with limited structural documentation.

Thus, this research aims to apply the RVS method to four high-rise buildings in Bangkok to determine their vulnerability to earthquake hazards. The assessment is based on field observations and the completion of standardized forms that have been adapted to the characteristics of the region and building typology. RVS (Rapid Visual Screening) assessment can be used in this research to assess the potential seismic risk of high-rise buildings with such irregularities. This method is known for its effectiveness, practicality, and affordability to carry out, especially in conditions where a lot of buildings need to be evaluated within a short period [6]. In this study, the term Rapid Visual Screening (RVS) refers to the guidelines outlined in FEMA P-154, titled *Rapid Visual Screening of Buildings for Potential Seismic Hazards: A Handbook*. The approach used follows the procedures described in that document.

2 Research Method

In this research, the type of building as an object that will be observed using the RVS method is High-Rise Buildings in Bangkok, Thailand. Pre-field planning needs to be conducted, which involves preparing a survey schedule, logistics, and supporting tools. After knowing the seismic risk of the location, the team must collect basic information about the building. The information collected includes the building's purpose, year of construction, number of stories, and structural system. This stage aims to ensure that the surveyed buildings are relevant to the criteria and can be assessed using RVS forms that need to be selected. This research used a field observation approach with the Rapid

Visual Screening (RVS) method based on FEMA P-154 guidelines. The team conducted the activities on four high-rise buildings located in Bangkok, Thailand, which are the Summit Tower Project (JRK Tower), Baan Klang Krung Siam-Pathumwan, The Line Ratchathewi, and Ideo Q Ratchathewi. The selection of objects is based on the variety of building functions (commercial and residential), as well as locations that represent the central city.

The RVS method includes five distinct Data Collection Forms, each tailored to one of the five seismicity zones: Low, Moderate, Moderately High, High, and Very High. Every form contains a Level 1 page and an optional Level 2 page. Appendix B provides full-size versions of each form. The structural evaluation system utilizes a matrix made up of Basic Scores assigned to each FEMA Building Type and its corresponding seismic force-resisting system and Score Modifiers, which adjust the score based on observed structural features that influence seismic behavior. The primary differences among the five forms lie in the assigned values of these Basic Scores, the Score Modifiers, and the Level 2 pounding assessment criteria. These values are determined by (1) regional seismic design and construction standards over time, (2) structural characteristics known to influence seismic capacity, and (3) the highest expected ground motion levels in the relevant seismic zone. The Basic Score, adjusted by the Modifiers to yield a Final Score, reflects the likelihood of structural collapse under maximum anticipated ground shaking. Final Scores generally fall between 0 and 7, with higher values indicating superior seismic performance.

The final stage of the RVS method is the calculation of the final score and interpretation of the results to determine the vulnerability level of the building. The score is calculated by summing the basic score of the structural type with all modifiers obtained during the observation. This sum is compared with the minimum threshold value (SMIN), which for the moderate seismic zone is set at 0.3. If the final score is greater than the SMIN, the building is declared relatively safe and does not require urgent further evaluation. Conversely, if the score is close to or below the threshold, then the building is recommended for a more detailed examination by a professional structural engineer.

3 Results and Discussion

3.1 Determination of RVS Form

The seismicity level of the study area was determined based on two main parameters in FEMA P-154, which are short-period (Ss) and 1-s period (S1) earthquake accelerations. The Ss and S1 values were estimated using satellite images and overlay data from Google Earth Pro, with coordinate points in Ratchathewi District, Bangkok [7, 8].

From the map in Fig. 1, the Ss value of 0.15 g and S1 of 0.20 g were obtained, which were then classified based on the FEMA seismicity category table. Based on the classification limits, the Ss value ($0.15\ g < 0.25\ g$) is in the Low Seismicity category, while the S1 value ($0.20\ g \geq 0.1\ g$ and $< 0.2\ g$) indicates Moderate Seismicity. Since one of the parameters reaches the Moderate category, the Moderate Seismicity form type is selected, according to the principle of determination with the most dominant value.

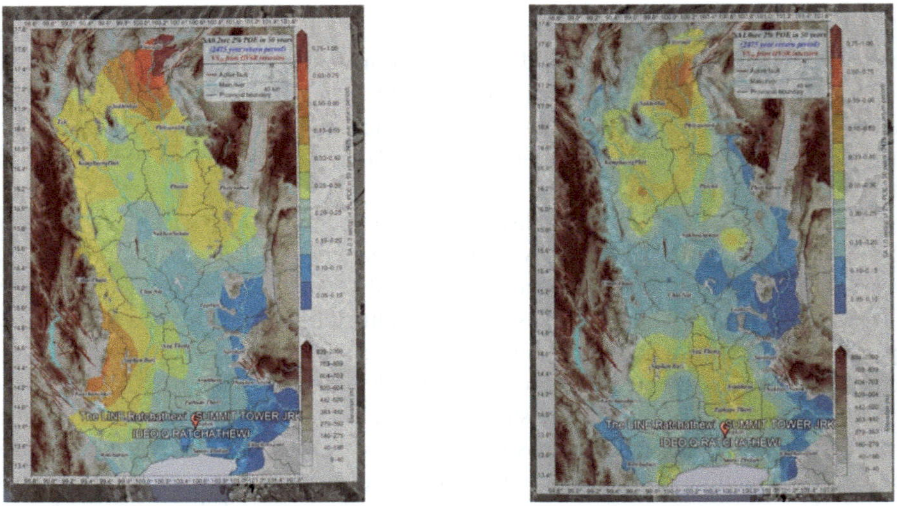

(a) (b)

Fig. 1. Buildings Location in (a) Ss Distribution Area and (b) S1 Distribution Area

3.2 Completion of RVS Assessment Form

As an example, one of the selected buildings, Ideo Q Ratchathewi, a 37-story condominium located on Phetchaburi Road, Bangkok, was assessed using the FEMA P-154 Level 1 form under moderate seismicity conditions. In the initial section of the form, the address of the building is filled in completely to ensure precise identification of the site location. The latitude and longitude fields are used to pinpoint the building's exact coordinates, which is useful for seismic zoning and mapping purposes. The Ss and S1 values are entered based on regional seismic hazard data, serving as the foundation for evaluating the seismic risk level.

In the Fema Building Type, the C2 building type was selected because the structure visibly utilizes a reinforced concrete moment-resisting frame system combined with shear walls, which is typical for modern high-rise residential buildings. This classification aligns with FEMA P-154 definition for high-rise concrete structures that rely on both moment frames and shear elements for lateral resistance. Choosing the correct building type is essential because it determines the initial (basic) score used in the risk evaluation. For a C2 structure under moderate seismicity, the FEMA P-154 guideline assigns a basic score of 2.5. This score represents the starting point before applying any score modifiers based on observed vulnerabilities.

Vertical irregularity was identified based on significant changes in floor geometry and setbacks observed between the podium and the tower portion of the building. The transition from a wider base to a slimmer upper structure creates a clear vertical discontinuity in stiffness and mass distribution. Plan irregularity was noted due to the asymmetric floor layout, especially the protruding balconies and varying facade depths along the building's elevation. These irregularities can amplify torsional response and concentrate seismic demand in specific areas of the structure. For these reasons, the form correctly selects moderate vertical irregularity and plan irregularity as score modifiers, reflecting observable architectural discontinuities that may affect seismic performance (Fig. 2).

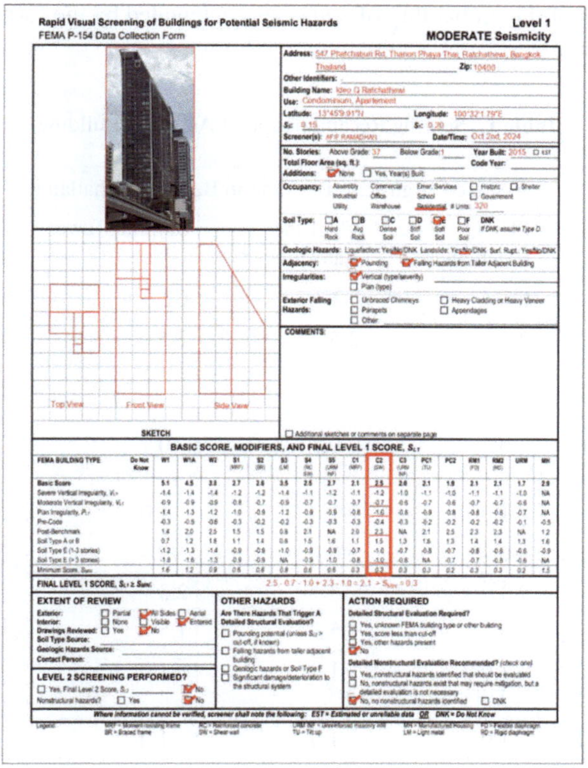

Fig. 2. Ideo Q Ratchathewi RVS Form

The post-benchmark year box is checked because the building was constructed in 2015, well after Thailand's most recent major seismic code updates, indicating improved design standards. For the Soil Type, soft soil (Type E) is selected based on regional geotechnical data indicating deep, unconsolidated layers common in central Bangkok. Soft soils can amplify ground motion, which justifies a negative score modifier in the FEMA assessment. After all applicable modifiers are added to the basic score of 2.5 for the C2 building type, including deductions for vertical and plan irregularities, pounding potential, and soft soil, the Final Level 1 Score is calculated to be 2.1. This score is below the FEMA safety threshold of 2.0, indicating that the building may require more detailed evaluation if designated for critical use. Table 1 below are Scoring Recapitulation of each building.

The results of the final calculation show that the highest score belongs to Summit Tower (3.1) because it is a new building (in 2021), without major irregularities, and post-benchmark code. Baan Klang Krung Siam scored 2.2 because it was built in 2007 and showed a combination of vertical and horizontal irregularities. The Line Ratchathewi and Ideo Q Ratchathewi scored 2.1 each, mainly due to the older age of the building than the benchmark and the irregular shape of the facade. All of these scores were compared against a SMIN (Minimum Score) of 0.3, and all buildings were declared Safe from potential earthquake-induced structural collapse according to FEMA P-154. In addition, the interpretation of the probability of collapse is calculated by the formula: 10^{-S} where S is the final score.

Table 1. Scoring Recapitulation of Assessed Building

Parameters	Assessed High-Rise Building in Bangkok, Thailand			
	Ideo Q Ratchathewi	The Line Ratchathewi	Summit Tower	Baan Klang Krung Siam
Building type	C2	C2	C2	C1
Basic score	2.5	2.5	2.5	2.1
Severe vertical irregularity	−1.2	−1.2	−1.2	−1.1
Moderate vertical irregularity	−0.7	−0.7	−0.7	−0.7
Plan irregularity	−1.0	−1.0	−1.0	−0.8
Pre-code	−0.4	−0.4	−0.4	−0.3
Post-benchmark	2.3	2.3	2.3	2.0
Hard rock soil	1.5	1.5	1.5	1.1
Soft soil (1–3 stories)	−1.0	−1.0	−1.0	−0.7
Soft soil (>3 stories)	−1.0	−1.0	−1.0	−0.8
Minimum score	0.3	0.3	0.3	0.3
Final score	2.1	2.1	3.1	2.2
Collapse Probability	0.79%	0.79%	0.079%	0.63%
C1 concrete moment-resisting frame				
C2 concrete shear wall				

All buildings achieved final scores resulting in collapse probabilities below FEMA's 1% threshold, indicating that they meet the recommended safety standards for public occupancy. However, this finding aligns with previous research that utilized RVS for assessing seismic vulnerability in various urban contexts. For instance, similar to the moderate to severe vulnerability observed in Baan Klang Krung Siam, Shahzada et al. reported that 61% of buildings in Dir City, Pakistan, were within the damage grade D2–D5 range, requiring immediate attention [1]. In Northern Algeria, Ahmed et al. found that soft-story irregularities and lack of seismic code compliance were critical contributors to low RVS scores in RC buildings, which echoes the irregularities identified in Bangkok mid-rise structures [2]. Moreover, the weighted average scoring approach used by Cardenas et al. in Lima reflects the necessity to prioritize schools and critical infrastructure based on occupancy and exposure levels, a strategy that could be adapted to prioritize residential and commercial towers in Bangkok [3]. Compared to these cases, buildings in the study area showed relatively safer RVS scores, likely due to newer construction years and concrete frame types. However, Dogan et al. cautioned that even seemingly robust RC buildings might still present underestimated risk when assessed only by first-level methods. This underlines the need for further investigation or detailed analysis for buildings near the safety threshold [4]. Overall, the consistency of the results with global RVS-based research confirms the validity of this approach for initial vulnerability mapping in Bangkok.

4 Conclusion

Based on the results of the form filling and RVS calculations using FEMA P-154 guidelines, the four buildings under review are structurally safe. Summit Tower obtained the highest score of 3.1, indicating that the building has a very low vulnerability to earthquakes. Baan Klang Krung with a score of 2.2, The Line and Ideo Q, which scored 2.1 each, are considered safe. In general, all buildings meet the minimum threshold score (SMIN = 0.3) and do not require urgent further technical evaluation. However, the Myanmar earthquake on March 28, 2025 with a magnitude of Mw 7.7 proved that the Bangkok area cannot be completely ignored from seismic risks. The vibrations of the earthquake were felt in the city center and caused noticeable impacts such as the spilling of water from rooftop swimming pools and panic among residents of high-rise buildings. This emphasizes the importance of implementing RVS assessments even if the area is not classified as a high seismic zone. Periodic structural evaluations should be considered as part of disaster mitigation strategies in urban areas. Therefore, the use of the RVS method can be a practical solution for early detection and reduction of building seismic risk.

Acknowledgement. The authors gratefully acknowledge the International Undergraduate Program in Civil Engineering, Universitas Islam Indonesia, for the support and opportunity provided through the International Mobility Program.

References

1. Shahzada, K., Haq, I.U., Noor, U.A., Xu, Z.-D.: Seismic vulnerability assessment of buildings in high-risk areas: A modified rapid visual screening approach in Pakistan. J. Dyn. Disasters. **1**(3), 100018 (2025). https://doi.org/10.1016/j.jdd.2025.100018
2. Ahmed, S., Abarca, A., Perrone, D., Monteiro, R.: Large-scale seismic assessment of RC buildings through rapid visual screening. Int. J. Disaster Risk Reduction. **80** (2022). https://doi.org/10.1016/j.ijdrr.2022.103219
3. Cardenas, O., Farfan, A., Huaco, G.: Seismic Risk Assessment of Peruvian Public School Buildings Using FEMA P-154 Rapid Visual Screening. In: 2020 Congreso Internacional de Innovacion y Tendencias en Ingenieria, CONIITI 2020- Conference Proceedings. Institute of Electrical and Electronics Engineers Inc. (2020). https://doi.org/10.1109/CONIITI51147.2020.9240369

4. Doğan, T.P., Kızılkula, T., Mohammadi, M., Erkan, H., Tekeli Kabaş, H., Arslan, M.H.: A comparative study on the rapid seismic evaluation methods of reinforced concrete buildings. Int. J. Disaster Risk Reduct. **56** (2021). https://doi.org/10.1016/j.ijdrr.2021.102143
5. Bektaş, N., Kegyes-Brassai, O.: Conventional RVS Methods for Seismic Risk Assessment for Estimating the Current Situation of Existing Buildings: A State-of-the-Art Review. MDPI (2022). https://doi.org/10.3390/su14052583
6. M. Mahoney et al., Rapid Visual Screening of Buildings for Potential Seismic Hazards: A Handbook Prepared for Federal Emergency Management Agency, Third Edition FEMA P-154, 2015. [Online]. www.ATCouncil.org
7. P. Warnitchai, C. Sangarayakul, and S. A. Ashford, Seismic hazard in Bangkok due to long-distance earthquakes
8. Pornsopin, P., et al.: Seismic hazard microzonation map for the central plain of Thailand. Prog. Appl. Sci. Technol. **14**(1), 42–53 (2024). https://doi.org/10.60101/PAST.2024.251551

Physics-Informed Hybrid Learning Model for Predictive Maintenance of Food Processing Equipment

Hua Yi Chai[1] and Chye Cheah Tan[2](✉)

[1] School of Computer Science, University of Nottingham, Jubilee Campus, Nottingham NG8 1BB, UK

[2] School of Computer Science, University of Nottingham Malaysia, Jalan Broga, 43500 Semenyih, Selangor Darul Ehsan, Malaysia

ChyeCheah.Tan@nottingham.edu.my

Abstract. Predictive Maintenance (PdM) advances traditional maintenance strategies by forecasting potential equipment failures and estimating the Remaining Useful Life (RUL) through modeling techniques. Despite its strong theoretical potential, PdM adoption in industrial applications remains challenging. Existing approaches are generally classified into two categories: model-based and data-driven methods. Model-based methods use physics-based equations and system knowledge to predict failures without requiring large datasets, offering strong interpretability. However, developing accurate physical models is often complex, costly, and dependent on expert knowledge, especially when degradation mechanisms are not fully understood. In contrast, data-driven methods leverage machine learning to identify failure patterns from sensor data, but they often ignore underlying physical principles, resulting in "black-box" models. These limitations are particularly evident in food processing equipment such as cookie depositors, where variability in dough properties, recipes, and process conditions complicates modeling. Additionally, limited industrial data due to confidentiality further hinders PdM model development. To address these challenges, this study proposes a hybrid machine learning framework that integrates physical knowledge into data-driven models. Using two public datasets—AI4I 2020 for classification and NASA C-MAPSS for RUL prediction—as proxies for a cookie depositor, we developed hybrid models incorporating physics-informed features such as mechanical power and temperature difference. Comparative analysis shows that these hybrid models outperform baseline data-driven models, achieving higher F1-scores for fault diagnosis and lower RMSE for RUL prediction. The findings confirm that embedding simple physical insights enhances accuracy and reliability, providing a practical pathway for PdM implementation in data-limited industrial environments.

Keywords: Predictive Maintenance · Remaining Useful Life · Physics-Informed Machine Learning · Fault Diagnosis

M. A. b. Jamlos et al. (Eds.): ICSET 2025, LNNS 1897, pp. 33–39, 2026.
https://doi.org/10.1007/978-3-032-20722-7_4

1 Introduction

With the increasing complexity of modern industrial systems and the continuous improvement of the degree of automation, the manufacturing industry's dependence on equipment uptime and ultimate quality standards has become more critical than ever. According to the data from National Institute of Standards and Technology (NIST), losses attributed to preventable maintenance issues reached as high as $119.1 billion in 2016 [1]. This context has given rise to intelligent maintenance strategies based on real-time data and condition monitoring, among which Predictive Maintenance (PdM) is the most prominent. By analyzing historical and real-time sensor data, PdM performs tasks such as anomaly detection, fault classification, and RUL prediction.

Although Predictive Maintenance (PdM) demonstrates immense theoretical potential, its application in industrial practice still faces numerous challenges. Recent literature reviews generally categorize predictive maintenance methods into three main types: model-based methods, data-driven methods, and hybrid methods that aim to combine the strengths of both [2].

Model-based methods rely on physics-based equations and system knowledge to predict failures without the need for large datasets, offering good interpretability. However, this approach has significant limitations. Building precise physical models is often highly complex and costly, as their development relies on deep expert experience. For complex systems, it can be difficult to build an accurate degradation model due to an incomplete understanding of the underlying degradation mechanisms [3]. Furthermore, this reliance on prior knowledge can limit the model's generalization, as it might not account for unforeseen operating conditions or evolving degradation behaviors [4].

In contrast, data-driven methods use machine learning to learn failure patterns from collected sensor data instead of relying on explicit physical principles [5], but this approach is highly dependent on data to learn failure patterns. If key components lack sensors or if data is sparse due to the rarity of failures, the credibility of the predictions is reduced [6].

These shortcomings are particularly prominent in some specific food processing equipment, such as the cookie depositor. On one hand, establishing an accurate physical model is difficult due to the high variability of the dough's rheological properties, recipes, and process conditions. On the other hand, the food industry maintains a high degree of confidentiality regarding process parameters and sensor data [7], which leads to an extreme scarcity of open, high-quality datasets. This raises the question of whether a methodological approach can be developed that mitigates excessive reliance on both precise physical modeling and extensive, high-quality labeled datasets.

Therefore, this research aims to explore a hybrid physics-guided, data-driven modeling approach. Hybrid paradigm combines physics-based interpretability with data-driven adaptability to capture complex behaviors [8]. This paper has two primary contributions. First, while existing PIML applications are concentrated in fields like aerospace and manufacturing, which benefit from well-defined physical models and rich datasets, this study applies the hybrid modeling paradigm to the food processing industry. We focus on the cookie depositor, a typical yet previously less-studied piece of complex equipment, thereby addressing a significant gap in the literature. Second, to address the poor interpretability and lack of trust in many data-driven 'black-box' models, this study designs and validates a set of interpretable features based on physical processes. Given the data scarcity for the cookie depositor, this feature engineering method is first tested and validated on two public industrial datasets.

2 Methodology

To establish our methodology, we first analyzed the cookie depositor. The analysis began with its system structure, which consists of components like a feed hopper, a screw, and a drive system [9], and extended to its potential anomalies, which in the field of predictive maintenance are typically divided into two basic modes: catastrophic failures and progressive degradation [10].

Considering the high costs and long experimental periods required to obtain real fault data for a cookie depositor, and the strong physical similarity of its failure modes to phenomena captured in public industrial datasets, this project selects two industry-recognized public datasets as proxies for methodology development and validation. For the classification task, this study uses the AI4I 2020 Predictive Maintenance Dataset; for analyzing progressive degradation, we have selected NASA's C-MAPSS dataset.

To ensure the validity and interpretability of models trained on general datasets when applied to food processing equipment, this study establishes a physics-based analogy mapping from the proxy datasets to the target system: the Cookie Depositor.

By comparing the structural and functional similarities between systems, we map the rotational mechanical system in AI4I, characterized by Torque and Rotational Speed, to the dough extrusion drive system of the Cookie Depositor, which includes the motor, gearbox, and screw. The thermal system in AI4I corresponds to the equipment's heat dissipation module, and the Tool Wear feature is mapped to the nozzle wear process. Meanwhile, the degradation process of the engine and its key rotating parts in the C-MAPSS dataset is mapped to the extrusion screw and drive bearing system of the Cookie Depositor, which are the components most susceptible to wear.

At the failure level, we associate Overstrain Failure (OSF), Heat Dissipation Failure (HDF), and Tool Wear Failure (TWF) in AI4I with blockage, ventilation obstruction, and nozzle wear in the Cookie Depositor, respectively. Furthermore, the degradation mode observed in the C-MAPSS dataset, characterized by gradual multi-sensor drift in variables such as temperature, pressure, and speed, is analogized to the fatigue and wear-induced degradation of the screw and drive bearings in the Cookie Depositor.

Although the physical carriers differ, both systems exhibit similar data-level patterns characterized by multi-sensor synergistic drift and progressive performance degradation. This ensures that the model transfer from generic industrial equipment to the food processing system is physically meaningful and that its predictions stay interpretable.

2.1 Classification: Physics-Informed Feature Engineering

Prior to model construction, the AI4I dataset was cleaned and inspected. The original dataset contains 14 fields, including identification information (UDI, Product ID, and Type), operational variables (air temperature, process temperature, rotational speed, torque, and tool wear), and failure labels (Machine failure and five specific fault types:

HDF, OSF, PWF, TWF, and RNF). No missing values were found. However, the binary label Machine failure was highly imbalanced, with 9,661 normal samples and 339 failures. To mitigate the bias toward the majority class, class-balancing strategies such as class_weight = 'balanced' (for Random Forest) and scale_pos_weight (for XGBoost) were applied during training.

To validate the value of physical knowledge, this study constructed four comparative models based on two algorithms: Random Forest and XGBoost, which were divided into two groups.

1) **Baseline Models (RF-Baseline, XGB-Baseline):** Trained using only the five original sensor features (Air temperature, Process temperature, Rotational speed, Torque, Tool wear).
2) **Hybrid Models (RF-Hybrid, XGB-Hybrid):** Trained using an extended feature set. Based on the original variables, two derived features with physical meaning were constructed:
   a) **Temperature difference:** Temp_diff = Process temperature – Air temperature, representing heat dissipation efficiency.
   b) **Mechanical power:** Power_W = Torque × Angular_speed, serving as a proxy for machine load and energy consumption.

2.2 Regression: RUL Prediction with Physical Constraints

For the regression task, the target variable is the Remaining Useful Life (RUL).

To focus the model's learning on the clear degradation phase, the RUL is capped at 125 cycles. To extract degradation information from the raw time-series data, this study implemented a multi-dimensional feature engineering strategy:

1) **Data Cleaning:** Removed sensor signals with constant values.
2) **Normalized Lifecycle (cycle_norm):** Normalized each engine's operational cycles to [0, 1] to provide a consistent degradation reference.
3) **Trend and Dynamic Features:** The first and second derivatives of each sensor signal were calculated to represent the velocity and acceleration of change.
4) **Sliding Windows:** Multiple sliding window sizes were used to compute the moving average and moving standard deviation of each sensor, thereby quantifying both long-term trends and short-term fluctuations.
5) **Interaction Features:** Pairwise product terms among the operational settings (op12, op13, and op23) were generated to explore potential non-linear effects.

Traditional regression models cannot guarantee that their predictions will follow the fundamental physical law that remaining useful life monotonically decreases over time. To solve this problem, we utilized the monotone_constraints parameter built into XGBoost, applying a negative constraint to the cycle_norm (normalized lifecycle) feature. This constraint forces the model to ensure its output RUL prediction has a non-increasing relationship with the lifecycle feature.

2.3 Evaluation Metrics

The models' performance was evaluated using standard metrics. For classification, we used Precision, Recall, and the F1-Score. For regression, we used the Root Mean Square

Error (RMSE), Mean Absolute Error (MAE), Mean Variance Ratio (MVR) and the Coefficient of Determination (R^2).

3 Results and Discussion

3.1 Classification: Fault Diagnosis

Table 1. Classification Evaluation Results

	Accuracy	Precision	Recall	F1-score
RF_Baseline	0.985	0.913	0.617	0.736
RF_Hybrid	0.989	0.896	0.764	0.825
XGB_Baseline	0.983	0.727	0.823	0.772
XGB_Hybrid	0.988	0.797	0.867	0.831

Table 1 clearly demonstrates that the Hybrid methods outperform the Baseline methods across the board. For both Random Forest and XGBoost, the introduction of physical features like Temp_diff and Power_W led to significant improvements in the F1-Score for the critical failure class. This directly proves that feature engineering incorporating physical knowledge was the core driver of performance enhancement in this study.

It can be observed that the Recall for the failure class in RF_Hybrid increased substantially from 0.617 to 0.764, meaning the model's ability to correctly identify true failures improved by approximately 23.8%. The XGB_Hybrid model achieved the highest Recall and F1-score among the four models, at 0.867 and 0.831 respectively, making it the optimal model in this study. This also indicates that the Gradient Boosting algorithm can more effectively utilize the new physical features to construct strong classification boundaries.

In summary, the evaluation results are perfectly consistent with our findings from the EDA phase and confirm our core hypothesis: by introducing physics-based domain knowledge, we can build more powerful and reliable fault diagnosis models.

3.2 Regression: RUL Prediction

In terms of the core accuracy metrics, the XGB_Hybrid model achieved the optimal results (RMSE = 10.254, $R^2 = 0.940$), slightly outperforming its baseline version, XGB_Baseline (Table 2).

In the RUL prediction for C-MAPSS engines, a purely data-driven model can easily get lost in noisy, multi-dimensional sensor signals. However, our method successfully extracted the degradation trajectory that is strongly correlated with the equipment's health status by imposing a physical constraint to ensure the RUL monotonically decreases. This led to a significant reduction in the RUL prediction error (RMSE), proving the indispensable role of physical knowledge in handling complex, long-term, and slowly changing time-series data.

Table 2. Regression Evaluation Results

	RMSE	MAE		MVR
SVR_Baseline	10.736	8.001	0.934	0.400
RF_Baseline	10.611	6.888	0.935	0.324
XGB_Baseline	10.298	6.787	0.939	0.378
XGB_Hybrid	10.254	6.972	0.940	0.366

4 Conclusion

This study systematically explored how to enhance the performance of Predictive Maintenance models in data-scarce industrial scenarios by incorporating physical knowledge. Facing the dual challenges where physics-based methods are difficult to model and data-driven methods depend on massive datasets, this research proposed and validated a hybrid physics-guided, data-driven modeling framework.

In the fault diagnosis task, experimental results demonstrated that hybrid models incorporating physics-informed features (e.g., mechanical power and temperature difference) significantly outperformed baseline models that used only raw sensor data. For the RUL prediction task, we successfully addressed the issue of physically implausible outcomes by imposing a monotonic physical constraint on the XGBoost model, achieving the best performance in terms of prediction error (RMSE) and model fit (R^2).

In summary, this research systematically demonstrates that incorporating physical knowledge is a key strategy for building more powerful and reliable fault diagnosis and RUL prediction models. By showing that physical insight can effectively compensate for data scarcity, this work provides a robust and interpretable framework that lowers the barrier for adopting advanced PdM The primary task for future work is to transfer this validated framework to real data collected from a Cookie Depositor for localized model calibration and final performance verification.

References

1. Thomas, D., Weiss, B.: Economics of manufacturing machinery maintenance. In: NIST AMS 100–34. NIST, Gaithersburg (2020)
2. Fassi, A., Du, X., Ghimire, P., Blaabjerg, F.: Physics-informed ML-based predictive maintenance for power converters. IEEE Trans. Power Electron. (2024)
3. Luo, W., Hu, T., Ye, Y., Zhang, C., Wei, Y.: Hybrid predictive maintenance for CNC machine tool using Digital Twin. Robotics Comput.-Integr. Manuf. **65**, 101974 (2020)
4. Achouch, M., Dimitrova, M., Ziane, K., et al.: Predictive maintenance in Industry 4.0: Overview and challenges. Appl. Sci. **12**(16), 8081 (2022)
5. García-Méndez, S., de Arriba-Pérez, F., Leal, F., et al.: Explainable ML for railway predictive maintenance. Sci. Rep. **15**, 27495 (2025)
6. Olesen, J.F., Shaker, H.R.: Predictive maintenance for pump systems and thermal power plants. Sensors. **20**(8), 2425 (2020)
7. Qian, C., Liu, Y., Barnett-Neefs, C., et al.: A perspective on data sharing in digital food safety systems. Crit. Rev. Food Sci. Nutr. **63**(33), 12513–12529 (2023)

8. Sepe, M., Graziano, A., Badora, M., et al.: Physics-informed ML framework for predictive maintenance. J. Glob. Power Propuls. Soc. **5**, 1–15 (2021)
9. Adekola, K.A.: Food extrusion technology and applications. J. Food Sci. Eng. **6**(3), 149–167 (2016)
10. Cha, J.H., Guida, M., Pulcini, G.: Competing risks model with degradation and catastrophic failures. Int. J. Performability Eng. **10**(1), 63–74 (2014)

A Decision-Oriented Analysis of Time-Cost Trade-Off in Accelerated Structural Construction

Fadil Adomas Fauzananda and Tri Nugroho Sulistyantoro(✉)

Universitas Islam Indonesia, Yogyakarta, Indonesia
tri.nugroho.sulistyantoro@uii.ac.id

Abstract. A construction project is a complex and temporary activity. In its implementation, various conditions often arise that do not align with the initial plan. One of the most common issues is increased project duration, leading to delays in completion. This study aims to determine the optimal time and cost for the construction of Tower C at Dharmais Cancer Hospital after acceleration, and to analyse the extent to which the project duration can be shortened from the planned schedule at minimal additional cost. The research employs a Time–Cost Trade-Off (TCTO) analysis with two acceleration alternatives: (1) additional working hours (overtime) and (2) additional labor deployment. This method is applied to shorten the completion time by reducing the duration of activities on the critical path, thereby determining the optimal combination of project duration and total cost. The study results show that the most efficient alternative is the addition of labor, yielding a total duration of 399 working days with a total project cost of Rp 91,384,674,621. This represents a time efficiency of 68 days (14.66%) from the original duration of 468 days and a cost efficiency of Rp 1,393,018,083 (1.50%) from the total normal cost of Rp 92,777,692,704.

Keywords: Construction project · Acceleration · Time-Cost Trade-Off · Labor productivity

1 Introduction

The rapid growth of Indonesia's construction industry is evident from the increasing number and complexity of infrastructure and building projects. This development presents both opportunities and challenges for construction service providers. Project delays remain a common problem, often resulting in cost overruns and extended completion times. When delays occur, they can significantly disrupt project performance, leading to financial losses, additional resource costs, and potential penalties due to missed deadlines. One effective approach is project acceleration, which can be implemented using techniques such as fast tracking, crashing, and the Time–Cost Trade-Off (TCTO) method [1].

In 2022, Dharmais Cancer Hospital initiated the construction of Tower C as part of its Women and Children Cancer Care Building project, organized by the Indonesian

M. A. b. Jamlos et al. (Eds.): ICSET 2025, LNNS 1897, pp. 40–46, 2026.
https://doi.org/10.1007/978-3-032-20722-7_5

Ministry of Health. The 76-m-high building will comprise 18 floors and 3 basements with a total area of 35,680 m^2, scheduled from 2022 to 2024 [2]. This study aims to determine the optimal time and cost for the construction of Tower C at Dharmais Cancer Hospital after acceleration, and to analyse the extent to which the project duration can be shortened from the planned schedule at minimal additional cost.

The Critical Path Method (CPM) was developed in the 1950s by Morgan R. Walker (DuPont) and James E. Kelly (Remington Rand) [3], as a scheduling tool to manage interrelated project activities [4]. The Time–Cost Trade-Off is a deliberate and systematic process that analyzes project activities on the critical path to minimize duration at the lowest possible additional cost [5]. Through TCTO analysis, critical activities may be accelerated by adding resources or working hours to achieve earlier completion without excessive cost increases. Project trade-off decisions typically involve more than schedule duration and expenditure. In response, the conventional time–cost trade-off problem has been broadened to incorporate quality considerations, resulting in multi-objective time–cost–quality optimization methods [6–9].

Based on the Ministry of Public Works and Housing Regulation No. 28/PRT/M/2016, direct costs include labor, materials, and equipment, while indirect costs comprise general expenses and profit margins [10]. The relationship between time and cost is interdependent: as project duration decreases, direct costs typically increase, whereas indirect costs decrease, forming the basis of TCTO optimization [11].

2 Research Methodology

The TCTO approach is commonly used to accelerate construction schedules while evaluating the corresponding increase in project cost [11]. The objective of this research is to identify the most efficient project acceleration alternative between overtime working hours and additional labor, in terms of both time reduction and total project cost.

The data collected includes the project time schedule, Bill of Quantities, unit price list for materials and labor, and work unit price analysis. Only activities on the critical path are considered for crashing, in line with standard TCTO principles [12, 13].

a. Alternative 1 – Overtime Work:

Overtime refers to working hours beyond 8 h per day or 40 h per week, as regulated by the Ministerial Decree No. KEP.102/MEN/VI/2004 [14]. Provided productivity coefficients indicating that longer overtime hours correlate with reduced work performance [11], can be observed in Table 1 below.

Table 1. Productivity Reduction Coefficient [11]

Overtime Hours	Productivity Index Reduction	Work Performance
1 h	0,1	90
2 h	0,2	80
3 h	0,3	70

$$\text{Daily productivity after crashing} = (c) + (a \times b \times d) \quad (1)$$

Where: a = additional overtime hours; b = productivity per hour; c = daily productivity; and d = productivity reduction coefficient after the addition of overtime hours.

$$\text{Crash duration} = \text{Volume}/(\text{Daily productivity after crashing}) \quad (2)$$

b. Alternative 2 – Additional Labor:

Adding labor can improve project efficiency when supported by realistic planning and adequate supervision [15]. Crash cost is the additional cost required to accelerate the completion of a project activity compared to its normal duration. This cost is typically higher than the normal cost because it includes expenses related to extended working hours (overtime), additional labor, or the use of extra resources.

$$\text{Crash labor cost per day} = (\text{Working hours per day} \times \text{Normal labor cost}) + (\text{Number of overtime hours} \times \text{Overtime cost per hour}) \quad (3)$$

$$\text{Cost Slope} = (\text{Crash Cost} - \text{Normal Cost})/(\text{Normal Duration} - \text{Crash Duration}) \quad (4)$$

3 Results and Discussion

3.1 Development of the Network Diagram

The concept of project acceleration using the Time-Cost Trade-Off (TCTO) method can only be applied to activities located on the critical path.

3.2 Application of the Time-Cost Trade-Off Method

Table 2 summarizes the cost slope values for activities on the critical path:

Table 2. Cost Slope Values for Each Activity on the Critical Path

No.	Work Description	Cost Slope (Rp/day)
1	Termite prevention for B3 floor slab	237,270
15	Column reinforcement B3	4,857,083
35	Floor beam casting L1	11,163,437
	...and other critical activities	

These values indicate that activities with the lowest cost slope should be prioritized for crashing, as they yield the greatest time reduction per additional cost incurred.

3.3 Calculation of Direct, Indirect, and Total Costs

After identifying the critical path and calculating the cost slopes, project duration was compressed step-by-step starting with the activity with the lowest slope. The following Table 3 shows the results of the time compression analysis for the overtime alternative and labor addition alternative.

Table 3. Summary of Direct, Indirect, and Total Costs (Overtime Alternative)

Duration (days)	Direct Cost (Rp)	Indirect Cost (Rp)	Total Cost (Rp)
468 (normal)	83,499,923,434	9,277,769,270	92,777,692,704
445	83,569,174,273	8,820,240,283	92,389,414,556
417	83,723,744,640	8,276,151,758	91,999,896,398
399	83,907,208,425	7,917,547,957	91,824,756,382

The table shows that the minimum total project cost under the overtime acceleration scenario is Rp 91,824,756,382 at a duration of 399 days. The trend of total project cost resulting from the overtime-based acceleration scenario is illustrated in Fig. 1.

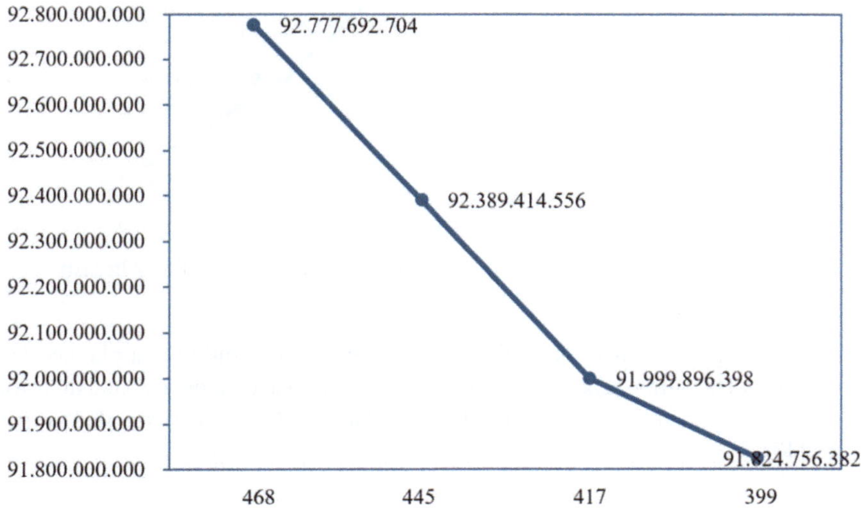

Fig. 1. Total Project Cost Trend under the Overtime Acceleration Alternative.

Literature indicates that extended overtime often induces productivity declines and safety/quality risks that increase effective direct costs, thereby reducing the practical advantage of overtime beyond short durations [16].

The summary of direct, indirect, and total project costs under the labor addition acceleration scenario is presented in Table 4.

Table 4. Summary of Direct, Indirect, and Total Costs (Labor Addition Alternative)

Duration (days)	Direct Cost (Rp)	Indirect Cost (Rp)	Total Cost (Rp)
468 (normal)	83,499,923,434	9,277,769,270	92,777,692,704
445	83,493,701,382	8,820,240,283	92,313,941,665
417	83,480,309,230	8,276,151,758	91,756,460,987
399	83,467,126,664	7,917,547,957	91,384,674,621

For the labor addition scenario, the optimal total project cost is Rp 91,384,674,621 with the same accelerated duration of 399 days. The trend of total project cost under the labor-addition acceleration scenario is illustrated in Fig. 2.

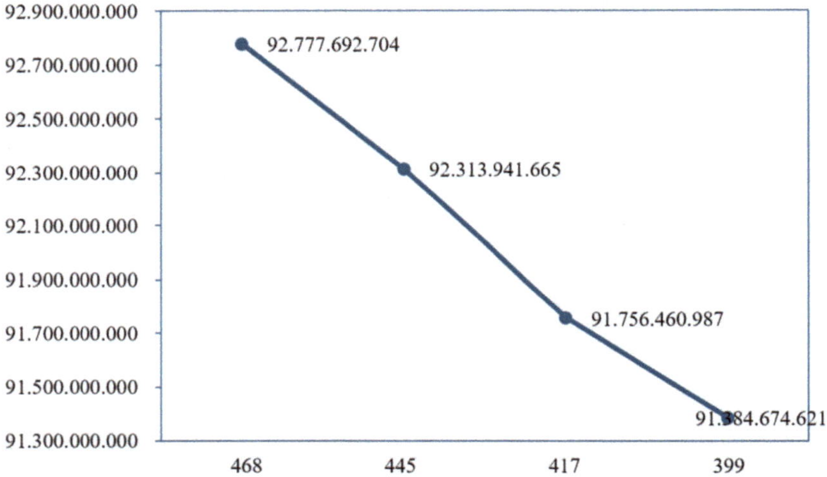

Fig. 2. Total Project Cost Trend under the Labor-Addition Acceleration Alternative.

These findings are consistent with recent studies demonstrating that planned labor reallocation and takt-time scheduling reduce both duration and labor cost, and that workforce augmentation tends to preserve per-hour productivity better than extended overtime regimes [17].

3.4 Discussion

From the results of the project acceleration calculations using the alternatives of added working hours (overtime) and additional labor, a comparison of total cost and time & cost efficiency of the project after acceleration was obtained, as shown in Table 5.

The difference in outcomes between adding overtime hours and augmenting the workforce can be explained by factors related to cost efficiency and labor productivity. In terms of cost efficiency, adding workforce is more effective and efficient compared

Table 5. Comparison Between Both Acceleration Alternatives

Acceleration Alternative	Duration (days)	Total Cost (Rp)	Time Efficiency	Cost Efficiency
Normal schedule	468	92,777,692,704	–	–
Overtime work	399	91,824,756,382	14.66%	1.03%
Labor addition	399	91,384,674,621	14.66%	1.50%

to overtime. Increasing the number of workers directly improves productivity: task allocation becomes more optimal while workers still operate under normal working hours, which helps reduce fatigue. Labor addition and overtime were applied to increase work output and maintain schedule targets, while bulk procurement of materials was used to reduce unit costs and prevent budget escalation [18]. Moreover, recent studies emphasize that effective cost planning supported by BIM-based cash flow management can further reduce the likelihood of cost overruns, thereby ensuring that schedule acceleration does not compromise financial performance [19].

Studies have shown that prolonged overtime significantly reduces labor productivity and increases fatigue-related risks, whereas controlled workforce augmentation and improved scheduling maintain better work efficiency and safety [20]. Furthermore, fatigue modeling studies indicate that redistributing workload across more workers rather than extending hours results in higher productivity and fewer safety incidents [21]. Similarly, systematic reviews on construction productivity highlight that schedule compression is best achieved through strategic labor allocation instead of excessive overtime [22]. These international findings are consistent with the results of this study, confirming that labor addition provides the most effective and sustainable acceleration alternative in construction project management.

4 Conclusion

Among the two alternatives analyzed, the addition of labor proved to be the most optimal acceleration strategy, yielding a time efficiency of 68 days (14.66%) and a cost difference of Rp 440,081,762 compared to the overtime alternative. Therefore, increasing the number of workers has a more positive impact on both project acceleration and work productivity, whereas overtime poses a higher risk of worker fatigue and productivity loss. For future research, it is recommended to explore other acceleration alternatives, such as equipment optimization, or lean construction-based methods to enable further comparison and identification of the optimal project duration and cost efficiency.

References

1. Caesaron, D., Thio, A.: Analisa Penjadwalan Waktu dengan Metode Jalur Kritis dan PERT pada Pro yek Pembangunan Ruko (Jl. Pasar Lama No. 20 Glodok). (JIEMS) J. Ind. Eng. Manage. Syst. **8**(2), 59–82 (2015)
2. Kemenkes RSK Dharmais. Groundbreaking Tower C. Retrieved from dharmais.co.id: https://dharmais.co.id/news/872/Ground-Breaking-Tower-C. 2022

3. Kelley, J.E., Walker, M.R.: Critical-path planning and scheduling. In: Papers presented at the December 1–3, 1959, eastern joint IRE-AIEE-ACM computer conference on - IRE-AIEE-ACM '59 (Eastern), pp. 160–173. ACM Press, New York, New York, USA (1959)
4. Rosanti, N., Setiawan, E., Ayuningtyas, A., et al.: Penggunaan Metode Jalur Kritis Pada Manajemen, vol. 8, p. 1 (2016)
5. Moch, C.: Studi Optimasi Waktu dan Biaya dengan Metode Time Cost Trade Off pada Proyek Pembangunan Gebung Tipe B SMPN Baru Siwalankerto. Rekayasa Teknik Sipil. **2**(2), 9–15 (2015)
6. Monghasemi, S., Nikoo, M.R., Khaksar Fasaee, M.A., Adamowski, J.: A novel multi criteria decision making model for optimizing time–cost–quality trade-off problems in construction projects. Expert Syst. Appl. **42**(6), 3089–3104 (2015 Apr)
7. Tavana, M., Abtahi, A.R., Khalili-Damghani, K.: A new multi-objective multi-mode model for solving preemptive time–cost–quality trade-off project scheduling problems. Expert Syst. Appl. **41**(4), 1830–1846 (2014 Mar)
8. Mungle, S., Benyoucef, L., Son, Y.J., Tiwari, M.K.: A fuzzy clustering-based genetic algorithm approach for time–cost–quality trade-off problems: A case study of highway construction project. Eng. Appl. Artif. Intell. **26**(8), 1953–1966 (2013 Sep)
9. Zhang, H., Xing, F.: Fuzzy-multi-objective particle swarm optimization for time–cost–quality tradeoff in construction. Autom. Constr. **19**(8), 1067–1075 (2010 Dec)
10. Ministry of public works and housing. In: Regulation No. 28/PRT/M/2016: Guidelines for Construction Implementation, Jakarta (2016)
11. Soeharto, I.: Manajemen Proyek (Dari Konseptual sampai Operasional). Erlangga, Jakarta (1999)
12. El-Rayes, K., Kandil, A.: Time-cost-quality trade-off analysis for highway construction. J. Constr. Eng. Manag. **131**(4), 477–486 (2005 Apr)
13. Hegazy, T.: Optimization of construction time-cost trade-off analysis using genetic algorithms. Can. J. Civ. Eng. **26**(6), 685–697 (1999)
14. Tenaga M, Dan K, Republik T. bphn.go.id. 2014
15. Virgarif F, Ariesta S, Arena A, Haris DMN. Tinjauan Waktu dan Biaya Menggunakan Metode Percepatan dengan Penambahan Jumlah Tenaga Kerja Pada Proyek Pembangunan Kantor. 2022;3(1):1–15
16. Hanna, A.S., Taylor, C.S., Sullivan, K.T.: Impact of extended overtime on construction labor productivity. J. Constr. Eng. Manag. **131**(6), 734–739 (2005 Jun)
17. Ding, F., Liu, M., Hsiang, S.M., Hu, P., Zhang, Y., Jiang, K.: Duration and labor resource optimization for construction projects—a conditional-value-at-risk-based analysis. Buildings. **14**(2), 553 (2024)
18. Jaenudin, D., Sulistyantoro, T.N.: Pengendalian waktu dan biaya menggunakan metode earned value concept pada proyek pembangunan sarana penunjang perkantoran Lengkong Wetan. In: Abma, V. (ed.) Civil Engineering Research Forum, pp. 402–413. Universitas Islam Indonesia, Yogyakarta (2025)
19. Abma, V., Farhana, A., Rachmawati, S.: Cash flow simulation planning based on building information modeling for construction projects. AIP Conf. Proc., 030010 (2024)
20. Qi, K., Owusu, E.K., Francis Siu, M.F., Albert Chan, P.C.: A systematic review of construction labor productivity studies: Clustering and analysis through hierarchical latent dirichlet allocation. Ain Shams Eng. J. **15**(9), 102896 (2024 Sep)
21. Abuwarda, Z., Mostafa, K., Morita, P., Hegazy, T.: Optimizing construction work–rest schedules and worker reassignment utilizing wristband physiological data. J. Constr. Eng. Manag. **150**(11) (2024)
22. Mirnayani, P.H.: The impact of overtime duration variations on labor productivity in apartment construction projects: A study on engineering and disaster management in the construction sector. Calamity: J. Disaster Technol. Eng. **2**(2) (2025)

Machine Learning-Based Enhancement of CMORPH Satellite Rainfall Estimates: A Case Study in Kuala Lumpur, Selangor and Putrajaya

Ren Jie Chin[1(✉)], Eugene Zhen Xiang Soo[2], Lloyd Ling[1], Naga Malleswari T Y J[3], Ushasukhanya S[3], and Kai Li Chai[1]

[1] Department of Civil Engineering, Lee Kong Chian Faculty of Engineering and Science, Universiti Tunku Abdul Rahman, 43000 Kajang, Malaysia
chinrj@utar.edu.my

[2] Civil Engineering Programme, Faculty of Engineering, Universiti Malaysia Sabah, 88400 Kota Kinabalu, Sabah, Malaysia

[3] Department of Networking And Communications, Faculty of Engineering and Technology, SRM Institute of Science and Technology, 603203 Kattankulathur, Chennai, India

Abstract. Accurate precipitation data is crucial for hydrological applications, yet the reliability of satellite precipitation products (SPPs) remains a challenge due to regional hydro-climatic variations and inherent biases. This study aims to enhance the accuracy of CMORPH satellite precipitation estimates using machine learning (ML) techniques in Kuala Lumpur, Selangor, and Putrajaya, Malaysia. The study employs three ML models: Artificial Neural Network (ANN), Random Forest (RF), and Support Vector Machine (SVM), integrating meteorological parameters such as temperature and wind speed to refine precipitation estimates. Daily precipitation data from 97 rain gauge stations (2011–2020) were obtained from the Department of Irrigation and Drainage Malaysia, while CMORPH data were retrieved from the Climate Engine platform. The ML models were trained and evaluated ussing quantitative (Root Mean Square Error [RMSE], Mean Absolute Error [MAE], Percentage Bias [PBIAS]) and categorical verification (Accuracy [ACC], Probability of Detection [POD], False Alarm Ratio [FAR], Critical Success Index [CSI]) statistical measures. Results indicate that ANN consistently outperforms RF and SVM across all monsoon seasons, achieving the lowest RMSE (6.71–15.23 mm/day) and MAE (4.05–8.47 mm/day), while maintaining minimal bias. RF demonstrated moderate accuracy, whereas SVM exhibited higher underestimation errors. Categorical verification analysis revealed ANN's superior classification accuracy and event detection performance. These findings suggest that ANN is the most reliable ML approach for enhancing CMORPH precipitation estimates, offering potential improvements for flood forecasting and hydrological modeling in Malaysia's diverse climatic conditions.

Keywords: CMORPH · Kuala Lumpur · Machine Learning · Putrajaya · Rainfall Estimates · Selangor

M. A. b. Jamlos et al. (Eds.): ICSET 2025, LNNS 1897, pp. 47–54, 2026.
https://doi.org/10.1007/978-3-032-20722-7_6

1 Introduction

Rain gauge stations provide direct point-based rainfall measurements, while satellite precipitation products (SPPs) offer wider spatial coverage using remote sensing. Both are essential, but SPP accuracy varies due to local climate, terrain, and changing weather patterns [1–4]. In Malaysia, with complex landscapes and strong monsoon effects, SPP performance is often inconsistent [5–7].

Climate Prediction Center Morphing Technique (CMORPH) a widely used SPP with high spatial (8 km) and temporal (30-min) resolution, still struggles with light and heavy rainfall detection, leading to false positives and false negatives [8]. These uncertainties reduce its effectiveness for applications like flood forecasting and early warnings.

To improve reliability, post-processing techniques such as machine learning are increasingly used to correct biases and capture local rainfall behavior [9]. By integrating meteorological variables such as wind speed, temperature, and humidity, machine learning models can correct systematic errors and capture local rainfall characteristics more effectively [10–12].

Therefore, this study aims to improve CMORPH precipitation estimates in Malaysia using machine learning and key meteorological variables, supporting more reliable hydrological modelling, flood forecasting, and climate risk assessments.

2 Methodology

2.1 Study Area

The study area covers Kuala Lumpur, Selangor, and Putrajaya, as shown in Fig. 1. Kuala Lumpur spans 243 km^2, Selangor is the largest with 8,104 km^2, and Putrajaya is the smallest at 49 km^2. This region is a major part of Malaysia's Central Region, where rapid urbanization and land use changes have the potential to influence rainfall patterns and contribute to climate change. The timing of each monsoon season is summarised in Table 1.

Table 1. Different seasonal periods over Malaysia.

Monsoon Seasons	Period
NEM	November to March
SWM	May to September
IM1	April
IM2	October

2.2 Data Acquisition

Daily rain gauge rainfall data for 2011–2020 were obtained from the Department of Irrigation and Drainage (DID) Malaysia. Out of 510 rain gauges in West Malaysia,

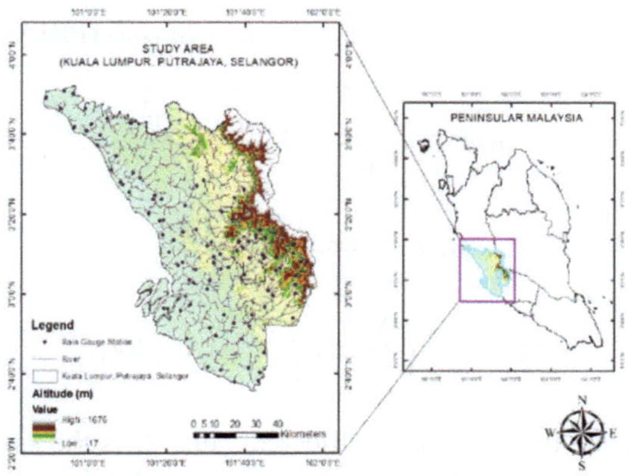

Fig. 1. Topography map of Kuala Lumpur, Selangor and Putrajaya.

this study used 97 stations: 84 in Selangor and Putrajaya and 13 in Kuala Lumpur. This study uses the CMORPH dataset, which provides global precipitation at 0.25° spatial resolution and daily time steps for regions between 60°N–60°S. The data were downloaded from the Climate Engine platform. (https://app.climateengine.org/climateEngine). Minimum and maximum air temperature were selected as meteorological inputs. These data were obtained from Global Land Data Assimilation System (GLDAS) for the period 2011–2020.

2.3 Model Development

This study uses three machine learning models, Artificial Neural Network (ANN), Random Forest (RF), and Support Vector Machine (SVM), to predict rainfall. The rainfall dataset was split into 70% training and 30% testing. CMORPH, minimum temperature, and maximum temperature were used as input variables, while raingauge data served as the output.

The ANN model has three layers (input, hidden, output) and was trained using backpropagation. It uses 20 nodes per layer, the ReLU activation function, and the Adam optimizer to balance accuracy and training efficiency [13]. The RF model improves predictions by combining many decision trees [14]. Using scikit-learn, the model was built with 10 estimators, max depth = 30, and min samples per leaf = 2. These settings were optimized through repeated testing. The SVM model (regression) uses the RBF kernel to find the best-fitting hyperplane. Three main parameters were tuned: Gamma = auto, C = 1, and epsilon = 5, which together control the model's complexity, margin, and error tolerance. Overall, the three models were trained and evaluated seasonally to determine their predictive performance.

2.4 Model Evaluation

This study used two data analysis methods: Quantitative and Categorical Verification Statistics. Three quantitative statistical measures were used: Root Mean Square Error

(RMSE), Mean Absolute Error (MAE), and Percentage Bias (PBIAS).

$$RMSE = \sqrt{\frac{\sum_{i=1}^{n} (S_i - G_i)^2}{n}} \tag{1}$$

$$MAE = \frac{\sum_{i=1}^{n} |S_i - G_i|}{n} \tag{2}$$

$$PBIAS = \left(\frac{\sum_{i=1}^{n} S_i}{\sum_{i=1}^{n} G_i} - 1 \right) \times 100\% \tag{3}$$

where G represents the rain gauge observed value, S is the satellite precipitation estimation prediction value, while *i* indicates the index of data and n is the number of measurements.

In this study, four categorical verification statistical measures were used: Accuracy (ACC), Probability of Detection (POD), False Alarm Ratio (FAR), and Critical Success Index (CSI).

$$Accuracy, ACC = \frac{A + D}{n} \tag{4}$$

$$Hit\ Rate/Probability\ of\ detection, POD = \frac{A}{A + C} \tag{5}$$

$$False\ Alarm\ Ratio, FAR = \frac{B}{A + B} \tag{6}$$

$$Critical\ Success\ Index, CSI = \frac{A}{A + B + C} \tag{7}$$

The quantities A, B, C, and D are obtained from the contingency table (Table 2). Here, A represents hits, meaning events that were predicted to occur and actually occurred. B refers to false alarms, where an event was forecasted but did not happen. C represents misses, which occur when an event was predicted not to happen but did occur. D denotes correct negatives, meaning events that were predicted not to happen and indeed did not occur. The total number of cases is given by n = A + B + C + D.

Table 2. Contingency table for comparing rain gauge and satellite precipitation estimate.

	Gauge ≥ Threshold	Gauge < Threshold
Satellite ≥ threshold	*A*	*B*
Satellite < threshold	*C*	*D*

3 Results and Discussion

3.1 Quantitative Analysis

Table 3 compares ANN, RF, and SVM for predicting seasonal rainfall in Kuala Lumpur, Selangor, and Putrajaya. Overall, ANN consistently shows the lowest errors, making it the most accurate model.

For NEM, ANN records the lowest RMSE 10.76 mm/day and MAE 6.01 mm/day. RF performs moderately (RMSE 11.96, MAE 6.33), while SVM has the highest errors (RMSE 12.15, MAE 7.09). RF shows the smallest bias (PBIAS 0.53%).

For SWM, ANN again performs best (RMSE 15.23, MAE 8.47). RF shows slightly higher errors (RMSE 16.06), and SVM performs the worst (RMSE 19.09) with a large underestimation (PBIAS −33.06%).

For IM1, ANN remains the most accurate (RMSE 6.71, MAE 4.05). RF (RMSE 15.87) and SVM (RMSE 13.97) show larger errors. ANN slightly overestimates (PBIAS 19.28%) while RF (−1.68%) and SVM (−15.34%) underestimate.

For IM2, ANN again performs best (RMSE 10.56, MAE 7.31). RF (RMSE 12.06) and SVM (RMSE 12.10) show higher errors. RF has strong overestimation (PBIAS 54.55%), while SVM underestimates (−29.75%).

In summary, ANN is the most reliable model across all monsoon seasons, while RF shows mixed performance and SVM consistently struggles, especially with bias.

Table 3. Quantitative analysis of different ML models on seasonal scale.

Monsoon Seasons	Performance Indicators	ML Models		
		ANN	RF	SVM
NEM	RMSE (mm/day)	10.76	11.96	12.15
	MAE (mm/day)	6.01	6.33	7.09
	PBIAS (%)	1.99	0.53	3.82
SWM	RMSE (mm/day)	15.23	16.06	19.09
	MAE (mm/day)	8.47	9.00	9.36
	PBIAS (%)	4.73	1.93	-33.06
IM1	RMSE (mm/day)	6.71	15.87	13.97
	MAE (mm/day)	4.05	8.21	8.06
	PBIAS (%)	19.28	−1.68	−15.34
IM2	RMSE (mm/day)	10.56	12.06	12.10
	MAE (mm/day)	7.31	8.71	8.13
	PBIAS (%)	38.16	54.55	−29.75

3.2 Categorical Verification Analysis

Table 4 shows the categorical verification results for detecting rainfall events across the monsoon seasons.

All models achieve a perfect POD of 1.00 in every season, meaning they detect all actual rainfall events. However, overall accuracy (ACC) is moderate, ranging from 0.42–0.60, due to false positives. The highest ACC occurs in IM2 (ANN: 0.60, RF: 0.58, SVM: 0.61).

False alarm rates (FAR) are relatively high, between 0.39–0.58. The best (lowest) FAR appears for SVM in IM2 (0.39), while the highest FAR is in IM1 for ANN (0.58).

For CSI, which reflects overall skill, scores range from 0.42–0.61. The best CSI values occur in IM2 (ANN: 0.60, RF: 0.58, SVM: 0.61). The weakest CSI is in IM1, especially for ANN (0.42).

In summary, all models detect rainfall well (POD = 1.00), but accuracy is moderate and false alarms are common. IM2 shows the best overall performance, especially for SVM, which achieves the highest ACC and lowest FAR. IM1 performs the worst, particularly for ANN. Overall, SVM appears slightly more balanced across seasons..

Table 4. Categorical verification analysis of different ML models on seasonal scale.

Monsoon Seasons	Performance Indicators	ML Models		
		ANN	RF	SVM
NEM	POD	1.00	0.97	1.00
	ACC	0.49	0.54	0.51
	FAR	0.51	0.48	0.49
	CSI	0.49	0.51	0.51
SWM	POD	1.00	1.00	1.00
	ACC	0.55	0.54	0.58
	FAR	0.45	0.46	0.42
	CSI	0.55	0.54	0.58
IM1	POD	1.00	1.00	1.00
	ACC	0.42	0.50	0.46
	FAR	0.58	0.51	0.54
	CSI	0.42	0.49	0.46
IM2	POD	1.00	1.00	1.00
	ACC	0.60	0.58	0.61
	FAR	0.40	0.42	0.39
	CSI	0.60	0.58	0.61

4 Conclusion

This study aimed to enhance the accuracy and reliability of CMORPH satellite precipitation estimates using machine learning techniques, specifically ANN, RF, and SVM. The results demonstrate that ANN consistently outperforms RF and SVM across all monsoon seasons, delivering the lowest RMSE and MAE values, particularly during the NEM and SWM. RF shows moderate performance but exhibits inconsistencies in bias, while SVM tends to struggle with significant underestimation in several monsoon periods. Furthermore, categorical verification analysis reveals that all models exhibit a perfect Probability of Detection (POD) but suffer from moderate accuracy and high false alarm rates, particularly during the inter-monsoon seasons. Overall, the findings highlight the potential of ANN as the most reliable model for seasonal precipitation forecasting in Kuala Lumpur, Selangor, and Putrajaya.

Acknowledgments. This research was supported by the Ministry of Higher Education (MoHE) Malaysia through the Fundamental Research Grant Scheme project (FRGS/1/2023/WAB02/UTAR/02/1). The authors are grateful to the Department of Irrigation and Drainage (DID) Malaysia for providing the rain gauge data used in this study.

Disclosure of Interests The authors have no competing interests to declare that are relevant to the content of this article.

References

1. Wang, P., Bai, X., Wu, X., Lai, C., Zhang, Z.: Spatially continuous assessment of satellite-based precipitation products using triple collocation approach and discrete gauge observations via geographically weighted regression. J. Hydrol. **608**, 127640 (2022)
2. Silwal, P., Upadhyay, S., Duwal, S., Lakhe, H., Prajapati, R., Talchabhadel, R.: Evaluation and comparison of satellite based-rainfall product and ground rain gauge observations over five basins of Nepal. AGU Fall Meeting Abstracts. **2021**, H45M–H1326M (2021)
3. Nguyen, G.V., Le, X.H., Van, L.N., Jung, S., Lee, G.: Machine learning approaches for reconstructing gridded precipitation based on multiple source products. J. Hydrol.: Reg. Stud. **48**, 101475 (2023)
4. Tan, M.L., Ibrahim, A.L., Duan, Z., Craknell, A.P., Chaplot, V.: Evaluation of six high-resolution satellite and ground-based precipitation products over Malaysia. Remote Sens. **7**(2), 1504–1528 (2015)
5. Chin, R.J., Lai, S.H., Chang, K.B., Wan Zurina, W.J., Faridah, O.: Relationship between minimum inter-event time and the number of rainfall events in Peninsular Malaysia. Weather. **71**(9), 213–218 (2016)
6. Chin, R.J., Lai, S.H., Chang, K.B., Faridah, O., Wan Zurina, W.J.: Analysis of rainfall events over Peninsular Malaysia. Weather. **71**(5), 118–123 (2016)
7. Chai, V.H., Chin, R.J., Ling, L., Huang, Y.F., Soo, E.Z.X.: Regional and seasonal assessment of biases on high-resolution satellite precipitation estimations in peninsular Malaysia: 2011–2020. J. Hydroinf. **26**(7), 1715 (2024)
8. Furl, C., Ghebreyesus, D., Sharif, H.O.: Assessment of the performance of satellite-based precipitation products for flood events across diverse spatial scales using GSSHA modeling system. Geosciences. **8**(6), 191 (2018)
9. Faramarzzadeh, M., et al.: Application of machine learning and remote sensing for gap-filling daily precipitation data of a sparsely gauged basin in East Africa. Environ. Process. **10**(1), 8 (2023)
10. Nguyen, G.V., Le, X.H., Van, L.N., Jung, S., Yeon, M., Lee, G.: Application of random forest algorithm for merging multiple satellite precipitation products across South Korea. Remote Sens. **13**(20), 4033 (2021)

11. Zhang, L., et al.: Merging multiple satellite-based precipitation products and gauge observations using a novel double machine learning approach. J. Hydrol. **594**, 125969 (2021)
12. Lu, X., Li, J., Liu, Y., Li, Y., Huo, H.: Quantitative precipitation estimation in the Tianshan Mountains based on machine learning. Remote Sens. **15**(16), 3962 (2023)
13. Zhang, Z., Wang, D., Qiu, J., Zhu, J., Wang, T.: Machine learning approaches for improving near-real-time IMERG rainfall estimates by integrating cloud properties from NOAA CDR PATMOS-x. J. Hydrometeorol. **22**(10), 2767–2781 (2021)
14. Wang, L.: Support Vector Machines: Theory and Applications. Springer, Berlin (2005)

Challenges of Implementing Green Features for Buildings in Malaysian Construction Industry

Yoke-Lian Lew(✉), Jenn-Xean Wong, Jeffrey-Boon-Hui Yap, Wah-Peng Lee, and Li-Ping Yow

Universiti Tunku Abdul Rahman, Selangor 43000 Kajang, Malaysia
lewyl@utar.edu.my

Abstract. Malaysia has been actively promoting the adoption of green building features to reduce the negative environmental impacts of construction. However, knowledge on the implementation of green features in developing countries remains limited. This research aims to enhance the integration of green features in Malaysian buildings. The objectives are to examine the willingness of construction practitioners to implement green features, identify the key factors that promote their adoption, and explore the challenges hindering implementation. A quantitative research approach was adopted through a questionnaire survey conducted among two respondent groups: construction stakeholders and building end users. A total of 100 valid responses were collected. The findings revealed that most Malaysians expressed strong willingness to adopt green features in buildings. The five most significant factors identified were: improved energy efficiency; enhanced indoor environmental quality and comfort; better water efficiency; life-cycle cost savings; and improved productivity, safety, and health. Conversely, the top five critical challenges were: lack of government incentives; funding limitations; unfamiliarity with green technology and low availability of green products; hesitation or lack of awareness of the green concept; and insufficient knowledge, skills, and information. The Mann-Whitney U Test further indicated that both respondent groups shared a homogeneous perception of green feature implementation. These findings can assist policymakers in designing effective awareness programs, incentives, and regulatory frameworks to encourage green building adoption. For construction stakeholders, the results highlight a growing market demand and willingness among Malaysians to invest in sustainable buildings, offering opportunities to integrate green practices as part of their competitive strategies.

Keywords: Green Building · Malaysia · Construction

1 Introduction

In today's rapidly developing world, global warming and climate change have become critical environmental issues. The construction and building sectors contribute significantly to these challenges, with building construction and operation accounting for

M. A. b. Jamlos et al. (Eds.): ICSET 2025, LNNS 1897, pp. 55–62, 2026.
https://doi.org/10.1007/978-3-032-20722-7_7

approximately 37% of global carbon dioxide emissions and 36% of global energy consumption [1]. These impacts highlight the urgent need for sustainable practices within the construction industry.

Green buildings are recognized as a key solution, as they can reduce environmental costs while improving the quality of the living environment [2]. Beyond environmental protection, green buildings also promote economic and social sustainability [3]. By incorporating green features, buildings can provide healthy and comfortable living spaces with optimal energy use and minimal environmental impact throughout their life cycle. Recent research in this field has grown rapidly, focusing mainly on technology applications, digital systems, design, cost–benefit analyses, and sustainable material selection [4]. However, studies on human perspectives—particularly the willingness and concerns of construction professionals—remain limited.

Most previous studies have examined challenges and barriers to green building adoption in developed countries such as Singapore, the United States, and England [5]. This leaves a knowledge gap regarding the implementation of green features in developing countries. To address this gap, the present research investigates the willingness of construction players and end users in Malaysia to implement green features in buildings, identifies the factors that promote adoption, and examines the challenges hindering implementation. Malaysia records more than 1,000 residential construction projects annually over the past five years. However, the adoption of green buildings remains limited, with only 214 residential buildings certified under the Green Building Index (GBI) system [6].

Fortunately, this situation may soon improve. The Coronavirus Disease 2019 (COVID-19) pandemic, which confined people to their homes due to remote working arrangements and reduced social contact, has reshaped housing preferences. As a result, 73% of Malaysians indicated a desire to improve their home conditions, and nearly 90% expressed a preference for sustainable living, particularly in green buildings designed for energy and electricity efficiency [7].

Sustainability has become a key priority, yet green building implementation in Malaysia remains limited despite increasing public awareness of its economic, environmental, societal, and external benefits. Although green buildings are crucial for sustainable development, their adoption is hindered by challenges such as economic, societal, technical, and exogenous, stemming from the greater complexity, cost, and risk compared to conventional construction.

2 Methodology

Questionnaire survey is one of the most conventional data collection instruments that commonly used in quantitative research [8]. Questionnaire survey is not only an instrument used to assemble information from widely scattered sources, but also a device managed to secure the reliable answer to the questions set and ensure impartiality of results and findings. Databases and information from CIDB, Real Estate & Housing Developers' Association (REDHA) Malaysia, and the Board of Quantity Surveyors Malaysia (BQSM) were referred. All contacts information was compiled in an Excel file, and random numbers were generated in order to pick the respondents randomly from the list using random sampling method. However, the targeted population is still too large to be included. Thus in order to find out the bare minimum sample size for this research, Cochran's formula of calculating sample size was implemented, as it is suitable to be applied when the population is too large. Hence, the minimum sample size was 384, when critical value of the normal distribution at 95% confidence level is 1.96. For building end users, convenient sampling was applied where the residents/home owners from area surrounding the campus of UTAR Sungai Long were selected.

3 Results and Discussions

A total of 400 sets of questionnaires had been distributed to targeted respondents, which are 200 numbers of construction players and 200 numbers of building end users in Malaysia through online Google Form and by hands. Only 200 sets of questionnaires are collected, including 100 sets from the construction players and 100 sets from the building end users respectively. Therefore, 50.0% of valid response rate has been signified. Table 1 disclosed the frequency and distribution percentage of respondents in terms of profession as a construction player and role as a building end user simultaneously.

Table 1. Profession and Role of Respondents.

Attributes	Description	Frequency (N)	Percentage (%)
Profession as Construction Player	Developer	14	7.0
	Contractor and Subcontractor	84	42.0
	Consultant	1	0.5
	Others	1	0.5
Role as Building End User	Home Buyer	48	24.0
	Household and Tenant	48	24.0
	Property Investor	4	2.0

3.1 Willingness of Green Features Implementation

In this research, the first objective targets to investigate the construction players' and building end users' willingness of implementing green features in buildings. The first objective is accomplished with the results displayed in Table 2, which exhibits the willingness of green features implementation of respondents.

Based on the result obtained, there was 98.5% of respondents (N = 197) expressed that they are willing to implement green features in buildings. Whereas, 1.5%, of the respondents (N = 3) showed unwillingness or reluctance on green features implementation in buildings. This denotes that plenty of Malaysia citizens are willing to implement green features in buildings and their houses. After experiencing COVID-19 pandemic and reducing time to go out from home, most Malaysians consider and deliberate to

Table 2. Willingness of Green Features Implementation of Respondents.

Attributes	Description	Frequency (N)	Percentage (%)
The Willingness of Implementing Green Features in Buildings	Yes	197	98.5
	No	3	1.5
The Willingness to Pay More for Implementing Green Features in Buildings	Yes	192	96.0
	No	8	4.0

convert their original home situation to sustainable home environment. Approximately 90% of Malaysians prefer to live sustainably, and a building designed with green features to achieve energy efficiency is the main preference and takes priority over any other choices [7]. However, a small amount of citizens is reluctant to implement green features in buildings and their houses.

The result obtained shows that there was 96% of respondents (N = 192) are willing to pay more for implementing green features in buildings. However, the remaining 4% of respondents (N = 8) are not willing and refuse to pay more for implementing green features in buildings. This reveals that some people repudiate extra charge or additional money to be paid for green features implementation, although they are willing to implement green features in buildings. In addition, lack of awareness about green concept and knowledge about green buildings significantly lower down the willingness to pay more to reside in green buildings.

3.2 Factors Promoting Green Features Implementation in Buildings

The second objective aspires to identify the factors that are promoting the implementation of green features in buildings. The second objective is achieved successfully by evaluating the ranking of the ten factors that are promoting the implementation of green features in buildings via mean analysis.

According to the analysed results of the research, the Cronbach's Alpha value of 0.918 indicated an excellent internal consistency and ensured the reliability of data pertaining to the factors that are promoting the implementation of green features in buildings. Table 3 showed the findings for ten factors that are promoting green features implementation in buildings.

Factor F1 = '*Improvement of energy efficiency*' is indisputably an important factor to promote green feature implementation in buildings. According to Vyas and Jha [9], although the initial incremental cost of green buildings is higher than other conventional buildings, a short payback period could be enjoyed and experienced via reimbursement of energy and electrical bills, as the energy efficiency performance of green features in buildings is outstanding. Besides, the factor F2 = '*Better indoor environmental quality and comfort level*' are the important pursuance of buildings in this modern society. Green features implemented in buildings could create a better indoor environment by providing acoustic comfort and visual comfort to occupants, while reduce energy consumption concurrently [10, 11]. Moreover, green building which sustainably designed with green features could protect the well-being of occupants, as thermal comfort has been taken

Table 3. Factors Promoting Green Features Implementation in Buildings

Ref	Factors	Mean	SD	Ranking	Mann-Whitney U	Asymp. Sig. (2-tailed)
F1	Improvement on energy efficiency	4.05	0.556	*1*	4691.000	.292
F2	Better indoor environmental quality and comfort level	4.05	0.528	*1*	4706.000	.312
F3	Improvement on water efficiency	4.04	0.534	*2*	4770.000	.395
F4	Life cycle cost saving	3.98	0.609	*3*	4965.500	−.905
F5	Improvement of productivity, safety and health	3.98	0.622	*3*	4859.000	.642
F6	Reduction of pollution and waste	3.97	0.633	4	4668.500	.310
F7	Reduction of resource depletion	3.92	0.588	5	4997.000	.992
F8	Increasing on market demand or building value	3.92	0.596	5	4922.000	.805
F9	Reduction of greenhouse gases emissions	3.92	0.632	5	4968.000	.921
F10	Government policy or incentive	3.90	0.634	*6*	4914.000	.783

into consideration. According to Zhang, Wu and Liu [12], occupants are more willing to pay additional expenditure for buildings with green features, since a better indoor environment is promoted with increased comfort level. After experiencing COVID-19 pandemic, Malaysians nowadays have more concern about sustainable lifestyle and comfortable interior spaces in buildings.

In contrast, the factor F10 = ‘*Government policy or incentive*’ is the lowest ranked of the factors that are promoting green features implementation in buildings. There are less government policy and incentives relevant to green practices in Malaysian construction and building sectors, hence more incentives should be provided to promote green features implementation in the industry.

3.3 Challenges of Promoting Green Features Implementation in Buildings

In this research, the third objective intends to determine the challenges of green features implementation in buildings. The third objective is attained by evaluating the ranking of the six challenges of green features implementation in buildings through mean analysis.

As shown by the Cronbach's Alpha value of 0.893 indicated a good internal consistency and secured the reliability of data regarding to the challenges of green features implementation in buildings. Table 4 showed the findings for six challenges of green features implementation in buildings.

Table 4. Challenges of Green Features Implementation in Buildings

Ref	Factors	Mean	SD	Ranking	Mann-Whitney U	Asymp. Sig. (2-tailed)
C1	Lack of government incentive	4.04	0.600	*1*	4991.000	.977
C2	Issues on funding	4.01	0.622	*2*	4912.500	.781
C3	Unfamiliarity with green technology and low availability of green products	3.98	0.580	*3*	4752.500	.418
C4	Hesitation or lack of awareness on green concept	3.98	0.609	*3*	4871.000	.675
C5	Lack of relevant knowledge, skills and information	3.98	0.645	*3*	4899.000	.757
C6	Lack of management competency	3.90	0.585	*4*	4434.500	.064

The challenge C1 = '*Lack of government incentive*' is perceived as the most critical challenge of green features implementation in buildings. Inactive government involvement had caused citizens to have reduced participation in green practices and less awareness on public policy and incentives pertaining to green concept, ultimately the green features implementation in buildings had been hindered and become more challenging [13]. Above and beyond, Nguyen, et al. [14] urged that the incomplete development of regulatory environment triggered the unwise enforcement of the enacted rules and regulations regarding green buildings. In Malaysian construction industry, the legislative system about green features implementation and green building projects is underdeveloped and not well known by publics.

Moreover, the challenge C2 = '*Issues on funding*' could also hinder the green features implementation in buildings seriously. The high initial funding is perceived as the major challenge to green buildings development [13]. In fact, both construction players and building end users need to contribute additional costs initially to invest either green design or green technologies for green features implementation in buildings, but these additional costs needed to be borne and the return was unlikely to be retrieved within a very short period [14]. Undeniably, in Malaysia, the stakeholders might be unwilling to implement green features in buildings although considerable cost saving could be realised throughout the buildings' life cycle, as such features normally incurred high initial cost and financial burden needed to be borne.

A serious challenge of green features implementation in buildings is the challenge C3 = '*Unfamiliarity with green technology and low availability of green products*'. In actual fact, outcome and performance of green features in building project could be adversely disturbed by the unfamiliarity with green technologies [5]. More risks and uncertainties of the green features implementation in buildings would be arisen throughout its life cycle. In Malaysia, the green features implementation in buildings is highly dependent to the supply chain in Malaysian construction industry, with its function of supplying green products, green techniques and services in market.

4 Conclusion

The findings of the research indicated that most of the Malaysian are willing to implement green features of buildings, even though they need to pay more for the implementation of green features in buildings. The society and construction stakeholders should take more attention on the practices of green concept in construction and building sectors, in terms of green features implementation and green buildings in future. The findings can be potentially used by policymakers to design effective awareness programs, incentives, and regulations that encourage the adoption of green building features. For construction stakeholders, the results highlight a growing market demand and willingness among Malaysians to invest in sustainable buildings, providing opportunities to integrate green practices as part of their competitive strategy.

References

1. United Nations Environment Programme: Global Status Report for Buildings and Construction: Towards a Zero-emission. Efficient and Resilient Buildings and Construction Sector, Nairobi, Kenya (2021)
2. Li, Q.W., Long, R.Y., Chen, H., Chen, F.Y., Wang, J.Q.: Visualized analysis of global green buildings: Development, barriers and future directions. J. Clean. Prod. **245**, 118775 (2020)
3. Khoshbakht, M., Gou, Z., Lu, Y., Xie, X., Zhang, J.: Are green buildings more satisfactory? A review of global evidence. Habitat Int. **74**, 57–65 (2018)
4. Wu, Z., He, Q., Chen, Q., Xue, H., Li, S.: A topical network based analysis and visualization of global research trends on green building from 1990 to 2020. J. Clean. Prod. **320**, 128818 (2021)
5. Darko, A., Chan, A.P.C., Yang, Y., Shan, M., He, B.J., Gou, Z.: Influences of barriers, drivers, and promotion strategies on green building technologies adoption in developing countries: The Ghanaian case. J. Clean. Prod. **200**, 687–703 (2018)
6. Green Building Index, Executive Summary as of 30 December 2021., https://www.greenbuildingindex.org/how-gbi-works/gbi-executive-summary/ (2022). Accessed 13 Dec 2022

7. PropertyGuru Group, Malaysia Consumer Sentiment Study H2 2021., https://www.propertyguru.com.my/property-guides/malaysia-consumer-sentiment-study-h2-2021-53736 (2022). Accessed 13 Dec 2022
8. Kuada, J.: Research methodology: A project guide for university students. Samfundslitteratur. (2012)
9. Vyas, G.S., Jha, K.N.: What does it cost to convert a non-rated building into a green building? Sustain. Cities Soc. **36**, 107–115 (2018)
10. Abdulaali, H.S., Usman, I., Hanafiah, M., Abdulhasan, M., Hamzah, M., Nazal, A.: Impact of poor indoor environmental quality (IEQ) to inhabitants' health, wellbeing and satisfaction. Int. J. Adv. Sci. Technol. **29**(3), 1–14 (2020)
11. Ganesh, G.A., Sinha, S.L., Verma, T.N., Dewangan, S.K.: Investigation of indoor environment quality and factors affecting human comfort: A critical review. Build. Environ. **204**, 108146 (2021)
12. Zhao, X.B., Hwang, B.G., Gao, Y.: A fuzzy synthetic evaluation approach for risk assessment: A case of Singapore's green projects. J. Clean. Prod. **115**, 203–213 (2016)
13. Mulligan, T.D., Korkmaz, S.M., Coyner, R., Goldsberry, A.D.: Public policy and impacts on adoption of sustainable built environments: Learning from the constuction industry playmakers. J. Green Build. **9**(2), 182–202 (2014)
14. Nguyen, H., Skitmore, T., Gray, M., Zhang, X., Olanipekun, A.O.: Will green building development take off? An exploratory study of barriers to green building in Vietnam. Resour. Conserv. Recycl. **127**, 8–20 (2017)

Biomass Production of *Meyerozyma guilliermondii* Strain SO in Optimal Medium With Optimized Impeller Tip Speed in 30-L Stirred-Tank Bioreactor

Diveyan Mohan[1,2], Che Haznie Ayu Che Hussian[1,2], Si Jie Lim[2,3], Mohammad Rizal Kapri[2], Siti Nur Hazwani Oslan[4], Adam Thean Chor Leow[2,5], Abu Bakar Salleh[2], and Siti Nurbaya Oslan[1,2,3](✉)

[1] Department of Biochemistry, Faculty of Biotechnology and Biomolecular Sciences, Universiti Putra Malaysia, 43400 UPM Serdang, Selangor, Malaysia
snurbayaoslan@upm.edu.my

[2] Enzyme and Microbial Technology (EMTech) Research Centre, Faculty of Biotechnology and Biomolecular Sciences, Universiti Putra Malaysia, 43400 UPM Serdang, Selangor, Malaysia

[3] Enzyme Technology and X-Ray Crystallography Laboratory, VacBio 5, Laboratory of Vaccine and Biomolecules, Institute of Bioscience, Universiti Putra Malaysia, 43400 UPM Serdang, Selangor, Malaysia

[4] Faculty of Food Science and Nutrition, Universiti Malaysia Sabah, Jalan UMS, Kota Kinabalu 88400, Sabah, Malaysia

[5] Department of Cell and Molecular Biology, Faculty of Biotechnology and Biomolecular Sciences, Universiti Putra Malaysia, 43400 UPM Serdang, Selangor, Malaysia

Abstract. Yeasts have emerged as valuable recombinant protein expression hosts due to their rapid growth, genetic manipulability, and eukaryotic post-translational modifications. However, commercial yeast expression kits are often expensive and rely on hazardous compounds for induction, prompting the exploration of alternative yeast expression hosts. The non-conventional yeast *Meyerozyma guilliermondii* strain SO, isolated from spoiled orange, demonstrated its potential to efficiently express industrially important enzymes with more cost-effective and time-efficient enzyme production process than the commonly used commercial *Komagataella phaffii*. Despite these advantages, its relatively low biomass yield remains a major bottleneck for its commercial application as an alternative yeast expression system. Therefore, this study aims to improve the biomass production of *Meyerozyma guilliermondii* strain SO through sequential optimization of growth medium composition, and evaluation of impeller tip speed as a scale-up parameter. In shake-flask experiments, yeast extract-peptone-dextrose (YPD) medium at 30 °C and 250 rpm was identified as the most favorable growth medium condition. Further optimization of dextrose and peptone concentrations within the YPD medium using a one-factor-at-a-time (OFAT) approach revealed 2% (w/v) and 3% (w/v) as the optimum concentrations, respectively. These optimized conditions were subsequently applied in a 30-L bioreactor, where evaluation of various impeller tip speeds identified 0.84 m/s impeller tip speed (200 rpm) as the optimal condition, with 6 g/L biomass production. In conclusion, this study successfully optimized the medium composition and constant impeller tip speed as scale up

M. A. b. Jamlos et al. (Eds.): ICSET 2025, LNNS 1897, pp. 63–79, 2026.
https://doi.org/10.1007/978-3-032-20722-7_8

parameter for the cultivation of *Meyerozyma guilliermondii* strain SO, offering enhanced biomass production potential.

Keywords: *Meyerozyma guilliermondii* · Biomass production · Bioreactor · Impeller tip speed · Pilot scale production

1 Introduction

With the rapid development of industrial and biotechnological processes, yeast species have become indispensable platforms for a wide range of applications. They serve as versatile production and expression systems capable of synthesizing various compounds and proteins, including industrial enzymes and pharmaceutical proteins (Liu et al., 2012). The increased demand for robust expression hosts has prompted a resurgence in the utilization of yeast and fungi due to their unique combination of unicellular efficiency and eukaryotic capabilities. This integration of strength and functionality allows for scalability, high-density production, and the assurance of pathogen-free production (Ergün et al., 2019; Karbalaei et al., 2020).

From this perspective, *Meyerozyma guilliermondii* strain SO, a local isolate of non-conventional flavinogenic yeast, has emerged as a promising candidate (Oslan et al., 2012). Isolated from spoiled orange (SO), this yeast strain efficiently expresses industrially important enzymes 6.5 times faster and with 5.3 times lower methanol consumption compared to the commercially available yeast expression host, *Komagataella phaffii* (formerly known as *Pichia pastoris*) (Oslan et al., 2014; Oslan et al., 2015; Abu et al., 2017). In addition to efficiency, *Meyerozyma guilliermondii* strain SO is able to express recombinant proteins without the usage of methanol for induction; thus, rendering it a sustainable alternative (Nasir et al., 2020). Notably, it has excelled in the laboratory, and its biomass production has been maximized in both shake flask and 7-L bioreactor (Abu et al., 2020). Moreover, Abu et al. (2020) reported that the optimized parameters in the 7-L bioreactor are capable of scaling up the biomass production in pilot scales in industries.

Scaling up from small-scale to industrial-scale fermentation, however, presents formidable challenges. Heterogenous conditions in industrial fermenters can disrupt microbial growth, leading to issues such as overflow fermentation, mixed acid fermentation, and decreased biomass production (Xia et al., 2021; Mediboyina et al., 2022). In addition, several factors may influence the cell biomass production, including agitation rate, pH, temperature, dissolved oxygen tension (DOT), aeration rate, and feed nutrients. Considering industrial-scale bioreactor performance is significantly influenced by mixing, improvements in terms of the agitation speed is vital to ensure uniform distribution of oxygen, heat and nutrients (Stephenie et al., 2007; Oslan et al., 2022).

In stirred-tank bioreactors, impellers play a pivotal role in ensuring the uniform dispersion of microbial cells, maintaining homogeneous mixing, and enhancing aeration, heat, and mass transfer within the vessel (Oslan et al., 2019). By promoting efficient oxygen and nutrient transfer from the medium to the cells, impellers also prevent the formation of cell clumps that can hinder growth and productivity (Thamer & Issa Al haboubi, 2020). Among the critical parameters, impeller tip speed governs the balance between effective mixing and shear stress, as excessive shear can damage cells and negatively affect biomass production. Therefore, optimizing impeller tip speed at pilot-scale is essential to achieve uniform mixing, minimize shear stress, and maximize growth and productivity (Stephenie et al., 2007; Xia et al., 2021).

In addition to impeller tip speed optimization, it is essential to optimize the experimental culture conditions to improve system performance prior to scaling up to pilot-scale cell production. One of the most used optimization techniques is the One-Factor-at-a-Time (OFAT) approach, which has been successfully applied in various studies for effective growth optimization (Pallujam et al., 2020; Kumar & Aggarwal, 2024).

In this context, the present study seeks to address the dearth of information on scaling up biomass production of *Meyerozyma guilliermondii* strain SO in pilot-scale bioreactors. The study integrates sequential medium optimization using the OFAT approach, with the evaluation of impeller tip speed in a 30-L stirred-tank bioreactor to enhance cell growth and biomass yield. These efforts highlight the feasibility of *Meyerozyma guilliermondii* strain SO as a cost-effective and sustainable alternative yeast expression host for industrial applications.

2 Materials and Methods

2.1 Strain, Maintenance and Inoculum Preparation

Meyerozyma guilliermondii strain SO was obtained from a previous study (Oslan et al., 2012), and maintained on yeast extract-peptone-dextrose (YPD) agar plates, consisting of 1% (w/v) yeast extract, 2% (w/v) bacteriological peptone, 2% (w/v) dextrose and 2% (w/v) agar. For inoculum preparation, a single yeast colony was inoculated into 10 mL YPD broth and incubated at 30 °C and 250 rpm for 24 h. The starter culture was used for subsequent cultivation experiments.

2.2 Cultivation, Medium Selection and Optimization of Dextrose and Peptone

The growth of *Meyerozyma guilliermondii* strain SO was evaluated in different medium formulations (Table 1) to determine the optimum growth medium for biomass production. A 1% (v/v) inoculum of the starter culture, prepared as described in Sect. 2.1, was harvested by centrifugation (1,500 × g, 30 s), washed twice with sterile 1× phosphate-buffered saline (PBS) buffer solution (137 mM NaCl, 2.7 mM KCl, 10 mM Na_2HPO_4, 1.8 mM KH_2PO_4), pH 7.4, and inoculated into 100 mL of medium in 500 mL Erlen- meyer flasks. Cultures were incubated at 30 °C in an orbital incubator shaker (Hotech, Taiwan) at 250 rpm for 72 h, with three biological replicates for each formulation (Abu et al., 2017; Nasir et al., 2020). Samples were collected every 6 h to monitor cell growth and biomass production. Growth was measured spectrophotometrically at 600 nm (OD_{600}), and biomass was determined as dry cell weight (DCW). The medium that supported the highest biomass production within the shortest cultivation time was selected for further optimization of carbon and nitrogen sources.

Optimization of dextrose (carbon) and peptone (nitrogen) sources was performed using the one-factor-at-a-time (OFAT) approach. The pre-determined YPD medium was modified with different dextrose concentrations (0, 0.5, 1, 2, 4, 6, 8, and 10% w/v) and peptone concentrations (0, 0.5, 1, 2, 3, and 4% w/v). Each 100 mL modified YPD medium was inoculated with 1% (v/v) of inoculum and incubated at 30 °C and 250 rpm for 48 h. Time course study, as describe previously, was conducted at 12 h and 6 h interval for dextrose and peptone optimization, respectively.

Table 1. Composition of different culture media used (Abu et al., 2020).

Growth media (mL/100 mL)			
Component:	YPD	YPTG	BMGY
Buffer			10
10% Yeast Extract	10	10	10
20% Peptone	10	20	20
20% Dextrose	10		
1.34% YNB			10
0.002% Biotin		2	2
2% TSB		2	
10% Glycerol		10	10
Distilled Water	70	56	38

YPD: Yeast Extract-Peptone-Dextrose; YPTG: Yeast Extract-Peptone-Tryptic-Glycerol; BMGY: Buffered Minimal Glycerol Medium; YNB: Yeast nitrogen base with ammonium sulphate and without amino acids; TSB: Tryptic soy broth; Buffer: 1 M potassium phosphate buffer, pH 6.0. Dextrose, YNB, and biotin were filter sterilized using a 0.22 μm sterile syringe filter (Sigma Aldrich, US), whereas yeast extract, peptone, TSB, and glycerol were sterilized by autoclaving at 121 °C for 20 min.

2.3 Preliminary Batch Cultivation in a 7.5-L Stirred-Tank Bioreactor

Preliminary batch cultivation of *Meyerozyma guilliermondii* strain SO was performed using 7.5-L bioreactor (Infors-HT, Switzerland) with a 3 L working volume. The 10% (v/v) inoculum was cultivated as a seed culture in optimal medium composition of YPD for 24 h under sterile conditions. The fermentation was controlled at 30 °C at a fixed aeration rate of 1.5 L min-1 (vvm). Throughout the cultivation time, the fermentation process was carried out at 300 rpm. The culture pH was controlled at 7 using 3 N HCl and 3 N NaOH. Sterile diluted antifoam (10% v/v) of Struktol SB2121 (Schill and Schelinger, Germany) was used to control the foam production. The optical density and the biomass of the culture were measured *via* periodic sampling.

2.4 Biomass Production in a 30-L Stirred-Tank Bioreactor

Biomass production of *Meyerozyma guilliermondii* strain SO was evaluated in a 30-L stirred-tank bioreactor (Infors-HT, Switzerland; working volume 20 L) at three impeller tip speeds (0.56, 0.84, and 1.13 m/s) with 0.84 m/s serving as the scale-up reference from the 7.5- L bioreactor. Fermentation was conducted for 48 h at 30 °C and pH 6, with an aeration rate of 0.5 vvm. The pre-determined YPD medium was inoculated with 5% (v/v) seed culture. Silicon-based antifoam (Sigma), 1 M NaOH, and 1 M HCl were automatically added to control foaming and maintain pH. Biomass production was monitored throughout the process by measuring cell growth (OD_{600}) and dry cell weight (DCW) at 4-h intervals. The impeller tip speed was determined based on the following equation (Stephenie et al., 2007):

$$Impeller\ tip\ speed\ (m/s) = \frac{\pi ND}{60} \tag{1}$$

where,

π -	3.142
N -	Agitation speed (rev/min)
D -	Impeller diameter (m)

2.5 Analytical Methods

To obtain the dry cell weight (DCW), the cell biomass was harvested at 8,000 rpm for 1 min, and the pellet was washed twice with 1× PBS buffer solution, pH 7.4, prior to oven-drying at 80 °C until a constant weight was achieved. DCW was calculated based on the formula below:

$$DCW(g/L) = \frac{(M_1 - M)}{V} \tag{2}$$

where,

M -	Mass of empty tube (g)
M_1 -	Mass of empty tube + oven dried pellet (g)
V -	Volume pelleted (L)

Quantitative analysis *via* calculation of kinetic parameters such as the doubling time, D_t and specific growth rate, μ_{max} was done using the exponential growth equation (Olivares-Marin et al., 2018):

$$X = X_0 e^{\mu_{max} \triangleright} (from\ exponential\ graph) \tag{3}$$

$$Doubling\ time, D_t = \frac{In2}{\mu_{max}} \tag{4}$$

2.6 Statistical Analysis

The mean values (OD_{600} and DCW) and standard deviation for the suitable medium selection and optimization study were calculated from three biological replicates and technical replicates, respectively. The significance of the difference between the mean values was determined using the one-way analysis of variance (ANOVA) followed by Fisher's LSD and Duncan's MR test, with the IBM SPSS Statistics software [Version 26 (2019)]. The confidence interval for a difference in the means was set at 95% (p- value <0.05) for all comparisons.

3 Results and Discussion

3.1 Selection of Culture Medium for Growth and Biomass Production of *Meyerozyma guilliermondii* Strain SO

The composition of a culture medium plays a crucial role in cell growth, metabolite production, and biomass production. The purpose of selecting or designing a culture medium is to ensure the microorganisms are supplied with proper ratio and accessible forms of required sources for uptake, for optimum growth and production (Allikian et al., 2019). Therefore, the cell growth (OD_{600}) and biomass production (DCW) of *Meyerozyma guilliermondii* strain SO at 6 h interval for 72 h, were investigated across three different complex media: YPD, YPTG, and BMGY Fig. 1.

These media are suitable for this study because each one of them provides various nutritional environments due to the unique combination of carbon and nitrogen sources that highly impacts cell growth and metabolism. YPD is a prominent media for yeast cultivation as it is rich in both carbon (dextrose) and nitrogen (yeast extract and peptone), promoting rapid growth and high biomass production. In contrast, YPTG consists of glycerol as the sole carbon source, and tryptic soy broth as an additional nutrient source, on top of yeast extract and peptone; thus, offering a balanced supply of amino acids, peptides, vitamins and other growth factors (Gandhi et al., 2015). This medium functions to examine the effects of different carbon source (glycerol) towards yeast cell growth. Furthermore, BMGY with glycerol as the sole carbon source, and yeast nitrogen base as an addition of nitrogen source, serves similar functions as YPTG. However, the unique presence of buffer in this medium renders it a potential candidate as it is able to maintain a stable pH condition, ensuring efficient absorption and utilization of nutrients (Ling et al., 2010).

A steady increase in cell growth and biomass production was observed in all three media until the stationary phase was reached (Fig. 1). YPD medium was observed to be the best medium as compared to YPTG and BMGY at 24- h cultivation time. Within 24 h, YPD medium reached its peak of log phase with a 9.82 g/L of biomass and $OD_{600} = 15.92$. On the other hand, YPTG and BMGY had a lower OD_{600} value of 11.51 and 13.26, respectively, and a lower biomass production of 6.12 g/L and 6.56 g/L, respectively, at 24 h.

Analysis of growth curve provides only qualitative information; thus, it is imperative to obtain quantitative information, which can be achieved through the calculation of doubling time, D_t and specific growth rate, μ_{max}. The exponential growth rate kinetics

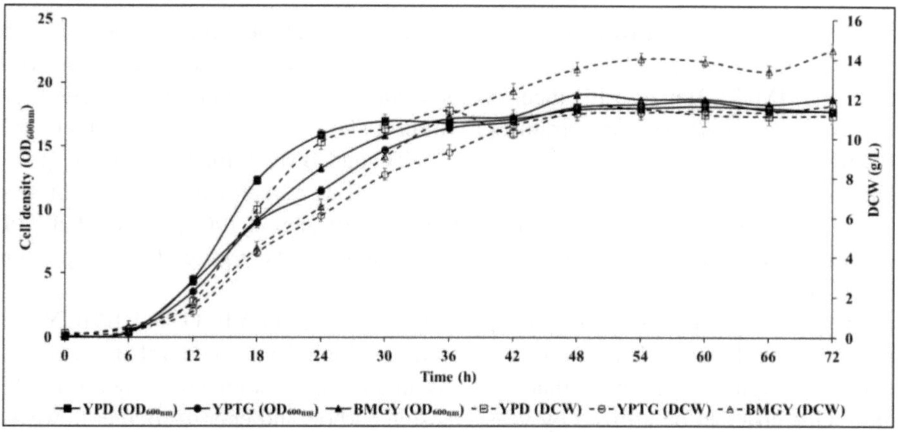

Fig. 1. Time course study for selection of culture medium for growth and biomass production of *Meyerozyma guilliermondii* strain SO in 500 mL Erlenmeyer flasks. Data were presented as mean ± SD of three technical replicates (n = 3). Different letters indicate significant differences ($p<0.05$) by using Fisher's LSD and Duncan's MR tests.

(Table 2) showed that YPD medium has the highest specific growth rate (0.1942 1/h) and lowest doubling time (3.57 h), making it the most efficient medium for the cultivation of *Meyerozyma guilliermondii* strain SO. The specific growth rate and doubling time are inversely proportional to each other. Therefore, the longer the doubling time, the slower the growth rate, as the doubling time indicates the time required for the cells to proliferate.

Table 2. Exponential growth rate kinetics of *Meyerozyma guilliermondii* strain SO in the tested growth media.

Media	Kinetic parameters	
	Specific growth rate, μmax (1/h)	Doubling time, Dt (h)
YPD	0.1942	3.57
YPTG	0.1858	3.73
BMGY	0.1891	3.67

During prolonged cultivation, it was observed that BMGY and YPTG media exhibited an increase in growth compared to YPD medium. Within 48 h of cultivation time, BMGY had the highest growth value of $OD_{600} = 19.07$, with a biomass production of 13.51 g/L. This may be due to the unique composition of BMGY medium that contains buffer to aid in maintaining the stable pH of the medium throughout the cultivation process; thus, improving absorption and uptake of nutrients, and permitting higher growth (Ling et al., 2010). However, the growth does not differ much between media during prolonged cultivation as at 48 h of cultivation time, the growth value in YPD medium is the lowest, where for $OD_{600} = 17.92$, and DCW = 11.59 g/L. This minimal difference of ±2 between YPD and BMGY media showed that YPD medium, despite being the simplest medium, is capable of cultivating growth at an approximately equal rate with YPTG and BMGY. Hence, YPD medium contributed to a high biomass production rate of *Meyerozyma guilliermondii* strain SO within a short period of time. Furthermore, YPD medium is cost-effective compared to YPTG and BMGY media, which required additional vitamins and buffer (for BMGY), as well as glycerol in exchange of dextrose in YPD as carbon and energy source. Besides, Struyf et al. (2017) stated that YPD medium is a rich medium that contains all sufficient nutrients required for the optimum cultivation of yeasts.

The growth of *Meyerozyma guilliermondii* strain SO in YPTG and BMGY was observed to be delayed. This is due to the composition of both media that consists of glycerol, an organic compound, that serves as an alternative source of carbon for cultivation. Glycerol can be utilized by yeasts as sole carbon and energy source; however, its degradation in yeasts is a two-step process: phosphorylation and oxidation, with L-glycerol 3-phosphate (G3P) as an intermediate (Klein et al., 2017). The first step of the pathway is glycerol phosphorylation, which involves the conversion of glycerol to G3P *via* glycerol kinase enzyme in the cytosol. G3P then enters the mitochondrion and is subsequently converted to dihydroxyacetone phosphate (DHAP) in a process known as oxidation (second step), catalysed by FAD-dependent glycerol 3-phosphate dehydrogenase enzyme. Finally, DHAP is channeled into the glycolysis or gluconeogenesis pathway, in the cytosol, as an intermediate metabolite (Klein et al., 2017; Xiberras et al., 2019).

The enzymes, glycerol kinase and glycerol-3-phosphate dehydrogenase, in the glycerol catabolism pathway are encoded by GUT1 and GUT2 genes, respectively. These genes are upregulated in the presence of non-fermentable carbon sources like glycerol and repressed in the presence of fermentable carbon sources like glucose (Xiberras et al., 2019). As a result, glycerol requires a longer time to metabolize than glucose in yeasts; thus, leading to a delayed growth. On the contrary, D-glucose or dextrose, a simple sugar classified as monosaccharide in YPD medium, is easily utilized by yeasts as it has a better binding specificity to the transporter involved in the glycolysis pathway of yeasts (Heitmann et al., 2018). Therefore, yeast cells readily consume glucose and produces energy during glycolysis.

3.2 Optimization of Dextrose Concentration in YPD Medium

An optimal balance of both carbon and nitrogen sources is crucial for achieving desired outcomes in bioprocess as these components function synergistically to provide source of nutrient and energy for microbial growth and metabolism; therefore, influencing protein synthesis and cellular functions. Optimization of these components can reduce production costs as cheaper carbon and nitrogen sources may be utilized while ensuring productivity and improving scalability (Pallujam et al., 2020). In this study, the traditional one-factor-at-a-time (OFAT) approach was employed by optimizing the media with 9 different concentrations of dextrose (0, 0.5, 1, 1.5, 2, 4, 6, 8, and 10% w/v) for high cell growth and biomass production towards a potential scale-up fermentation process. The findings offered fresh understandings of wild-type *Meyerozyma guilliermondii* strain SO physiology in different nutritional conditions.

The effect of various concentrations of dextrose on cultivation of *Meyerozyma guilliermondii* strain SO in YPD medium were evaluated, under conditions of 30 °C and 250 rpm for 48 h with 12 h intervals for cell harvesting and analysis. Based on the growth curve obtained (Fig. S1), it can be observed that at dextrose concentrations of 1.5, 2, 4, 6, 8, and 10% (w/v), the cell growth was approximately the same. *Meyerozyma guilliermondii* is an aerobic yeast and do not grow in a strict anaerobic environment (Fedorovych et al., 2015). This renders it a Crabtree negative species that do not exhibit Crabtree effect, whereby respiration is repressed, and aerobic fermentation occurs in high concentrations of substrate (Heitmann et al., 2018; Fedorovych et al., 2015).

Since it does not exhibit Crabtree effect; therefore, the phenomenon that is being observed could be substrate inhibition. Substrate inhibition occurred at the maximum reaction rate, where the excess substrate begins to impede the enzyme-catalyzed reaction, rather than achieving the steady-state equilibrium (Kokkonen et al., 2021). Besides, a study carried out by Bryan et al. (2018), proved that at glucose percentage of 40% and 80%, the rate of fermentation and production of carbon dioxide by *Saccharomyces cerevisiae* yeast was higher compared to 100% glucose percentage. It was also stated that the reduced carbon dioxide generation was due to the consequences of the cell achieving the saturation gradient, which occurred when all of the solute-binding sites were occupied, and the transport availability was at its highest.

Similarly, Zhang et al. (2015) and Canché-Collí et al. (2021) reported that excessive levels of sugar (substrate) concentration cause declines in yeast cell growth and size, as cells encounter a gradient from low osmotic pressure to high osmotic pressure (hypertonic), causing water loss through osmosis. As a result, the yeast cell volume decreases, and fermentation ability of the cell is halted; thus, the production does not increase, causing a waste of resources and energy.

Moreover, Zhang et al. (2015) reported that for the first 24 h, the cells rapidly divide and increases its concentrations considerably. However, when the glucose concentrations were increased, causing a hypertonic environment, the production of yeast (*S. cerevisiae*) eventually stopped rising and even slightly declined as the membrane fluidity and enzyme activity were damaged during prolonged cultivation. This justified the current findings, whereby the biomass production did not differ greatly between dextrose concentrations, ranging from 1.5, 2, 4, 6, 8, and 10% (w/v).

In this study, the biomass production was more pronounced compared to the cell growth. Based on Fig. S1, 24 h was the optimum cultivation period because during prolonged cultivation, the biomass production was unfavorable. Based on the one-way analysis of variance (ANOVA) (Table S1), 36 h and 48 h were significantly higher than 24 h for 1, 1.5, 2, 4, 6, 8 and 10% dextrose. However, increasing the incubation time by 1.5 times (24 h to 36 h) and 2 times (24 h to 48 h), did not warrant the increase in cell biomass at similar multiplication factors. Thus, prolonging the cultivation above 24 h is time consuming and unvaluable. Besides, when compared to 12 h (2 times), the biomass production was at least 2.25× higher (0.5% w/v dextrose), and exhibited the highest increment of approximately 5.8× (8% w/v dextrose). At all concentrations, the biomass production was more than 2 times at 24 h; thus, was concluded as the optimum cultivation duration.

At 24 h, the DCW at 2% dextrose was significantly higher than all other concentrations, except for 1.5, 4, 6 and 8% (Fig. 2). Despite doubling the concentration (4%), there was no significant difference in biomass production between both concentrations of 2 and 4%. Therefore, maintaining the concentration at 2% is more cost-efficient and valuable. Moreover, Gomar-Alba et al. (2015) reported that with glucose concentrations of more than 2%, carbon catabolite repression is prompted to occur, which halts the growth and biomass production. In addition, at 24 h and 2% glucose concentration, the biomass production reached 9.74 g/L; therefore, if the cultivation was run twice, the maximum biomass production can be increased to 19.48 g/L.

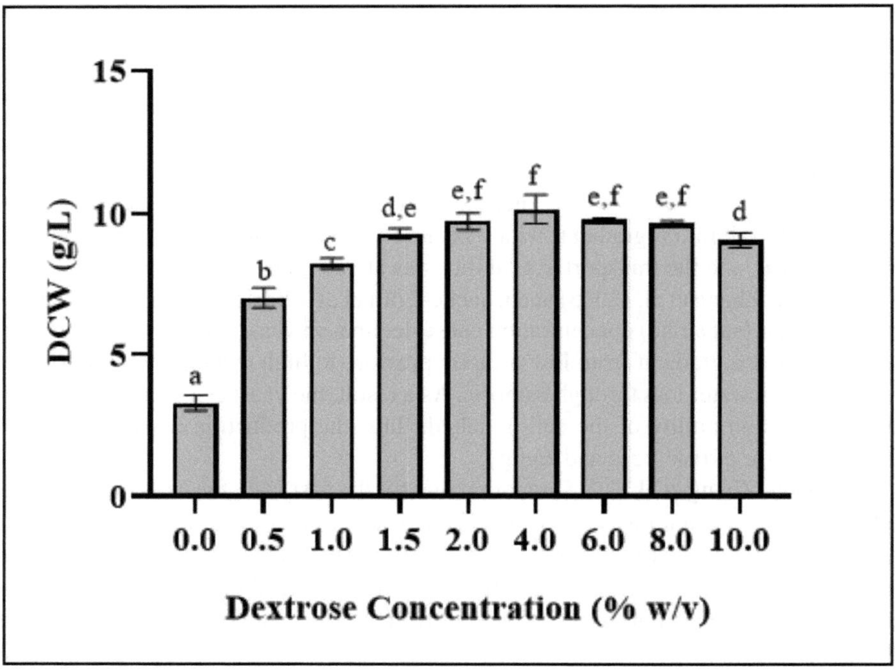

Fig. 2. Effect of different dextrose concentrations (0–10% w/v) on *Meyerozyma guilliermondii* strain SO biomass production at 24-h cultivation time in 500 mL Erlenmeyer flasks. Data were presented as mean ± SD of three technical replicates (n = 3). Different letters indicate significant differences ($p<0.05$) by using Fisher's LSD and Duncan's MR tests.

3.3 Optimization of Peptone Concentration in YPD Medium

In the same view as carbon, optimization of nitrogen source in YPD medium was also conducted based on OFAT approach by employing peptone concentrations of 0, 0.5, 1, 2, 3 and 4% (w/v), for high cell growth and biomass production, whilst maintaining the yeast extract and dextrose concentrations at 1% and 2% (w/v), respectively. In addition to dextrose, peptone is also a prominent component in YPD medium that serves as an organic nitrogen source because it is a soluble rich source of peptides, amino acids, vitamins, and trace elements; consequently, promoting growth, cell biomass production, and recombinant protein expression (Gandhi et al., 2015; Ramli et al., 2021).

Meyerozyma guilliermondii strain SO was able to grow in all tested concentrations of peptone, and the differences in cell growth and biomass production were apparent as in Fig. S2A and Fig. S2B, respectively. Based on both these figures, the growth on different peptone concentrations exhibited the peak of log phase at 24-h cultivation time, figuratively, with 3% (w/v) peptone concentration displaying the highest growth and production at $OD_{600} = 16.40$, and 10.95 g/L, respectively. In addition to qualitative analysis of growth curve, quantitative analysis of growth curve was performed through the calculation of exponential growth rate kinetics (Table 3). The highest specific growth rate (0.1861 1/h) and lowest doubling time (3.72 h) were both observed in 3% (w/v) pep- tone. Similar finding was reported by Li et al. (2017), in which, at 3% peptone concentration, the fermentation efficiency of *S. cerevisiae* was the highest at 74.6% with a yield of 17.9%. Furthermore, it was also reported that the production was only slightly higher than that at 3% peptone, when peptone concentration was adjusted at levels above 3% (Li et al., 2017). This is in line with the findings of this study, whereby at 4% peptone concentration, the cell growth increased slightly higher (at least 0.98×) at prolonged cultivation (30 h and onwards).

Since the DCW data is more prominent in this study; therefore, one-way analysis of variance (ANOVA) was performed for the DCW data, and tabulated in Table S2. The DCW at 36 h, 42 h and 48 h were significantly higher than 30 h for all peptone concentrations, except for 1, 2 and 3% (w/v), where certain time points shared the same subsets. However, increasing the incubation time by 1.2 times (30 h to 36 h), 1.4 times (30 h to 42 h), and 1.6 times (30 h to 48 h), did not guarantee the increase in cell biomass at similar or approximate multiplication factors. This indicated that prolonging the cultivation above 30 h was not time efficient; thus, unfavorable for yeast biomass production.

Besides, when compared to 24 h (1.25 times), the biomass production was at least 1.05× higher for all concentrations, with a maximum increase of only 1.15×, that too observed at 0.5% (w/v) peptone concentration. Since it did not meet the multiplication factor; therefore, it indicates that the biomass production did not increase at an acceptable level for 30 h, prompting 24 h as an optimum cultivation time.

At 24-h cultivation time, DCW at 3% (w/v) peptone was significantly higher than all other concentrations (Fig. 3). Therefore, maintaining the peptone concentration at 3% (w/v) is undeniably beneficial. However, when it comes to aiming for a cost-efficient and profitable production, 3% (w/v) peptone concentration is not suitable as the difference in biomass as compared to 2% (w/v) is scarce (1.15 g/L difference). This does not make up for the extra 1% (w/v) peptone that is to be placed in the culture media. Furthermore, at 24 h and 2% (w/v) peptone concentration, the biomass production reached 9.80 g/L; therefore, if the cultivation was run twice, the maximum biomass production can be increased to 19.60 g/L, leading to a time efficient, cost effective, large and profitable production.

Noteworthily, providing different concentrations of the carbon and nitrogen sources to *Meyerozyma guilliermondii* strain SO exerted an impact on the cell growth and biomass

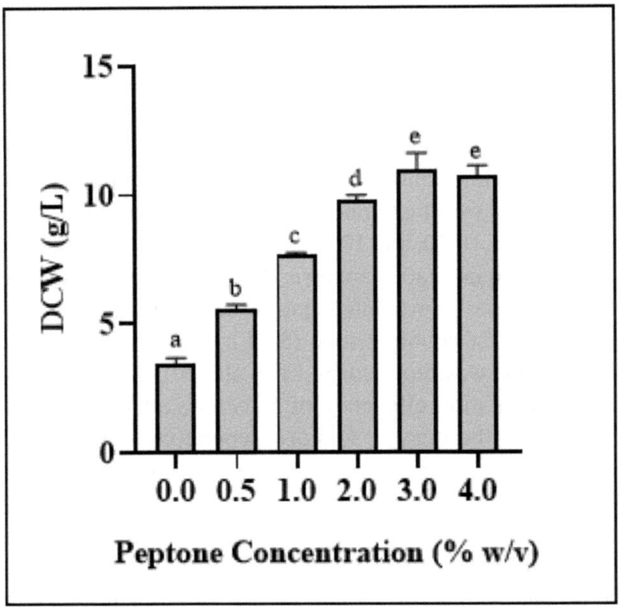

Fig. 3. Effect of different peptone concentrations (0–4% w/v) on *Meyerozyma guilliermondii* strain SO biomass production at 24-h cultivation time in 500 mL Erlenmeyer flasks. Data were presented as mean ± SD of three technical replicates (n = 3). Different letters indicate significant differences (p<0.05) by using Fisher's LSD and Duncan's MR tests.

Table 3. Exponential growth rate kinetics of *Meyerozyma guilliermondii* strain SO at various peptone concentrations.

Peptone concentrations	Kinetic parameters	
Specific growth rate, μmax (1/h)		Doubling time (h)
0.0%	0.1050	6.60
0.5%	0.1430	4.85
1.0%	0.1610	4.31
2.0%	0.1734	4.00
3.0%	0.1861	3.72
4.0%	0.1835	3.78

pro- duction. Therefore, the production medium must include all the necessary components in the right concentrations to provide an abundance of readily available nutrients for promoting the development and production of desired products at a large scale (Singh et al., 2017; Allikian et al., 2019; Ramli et al., 2021). In addition, the carbon to nitrogen

(C: N) ratio is a crucial consideration for biomass generation (Raita et al., 2022). However, the ratio is dependent on the type of yeast cells.

3.4 Scale up Biomass Production in 7.5-L to 30-L Stirred-Tank Bioreactor

In order to ensure a high degree of similarity in the fermentation conditions employed with both the 7.5-L and 30-L bioreactors, the impeller tip speed parameter was utilized as the primary criterion for process scale-up, while maintaining uniformity in all other fermentative conditions. Specifically, the impeller tip speed was sustained at 0.86 m/s, corresponding to agitation speeds of 300 and 200 revolutions per minute (rpm) in the 2-L and 10-L bioreactors, respectively. Furthermore, a constant culture pH of 7.0 was maintained throughout the entirety of the experiment. The effects of impeller tip speed on the cell biomass production of *Meyerozyma guilliermondii* strain SO in a 30-L stirred-tank bioreactor was investigated and presented in Fig. S3. Batch fermentation was conducted with constant impeller tip speeds of 0.56, 0.84 and 1.13 m/s corresponding to 134, 200 and 270 rpm, respectively. Based on the growth patterns, the growth was slow initially, with exponential growth initiating at 6 h of fermentation. This exponential growth continued until the culture reached the stationary phase after 24 h. The maximum cell biomass was achieved at impeller tip speed of 0.56 m/s, 0.84 m/s and 1.13 m/s were 6.0 g/L, 6.0 g/L and 5.1 g/L, respectively.

In general, elevated biomass production is correlated with greater cell growth rates. As a result, the growth rate kinetic parameters were computed and presented in Table 4. According to the data in Table 4, enhanced production was noted when maintaining a constant impeller tip speed of 0.84 m/s. This condition yielded the highest growth rate and specific growth rate of 0.25 g/h and 0.45 1/h, respectively, resulting in a peak biomass concentration of 6.0 g/L after a 24-h cultivation period. On the other hand, poor production was observed at impeller tip speed of 1.13 m/s. This may correspond to the high shear force exerted on the yeast cells by the impellers of the stirred-tank bioreactor. Despite possessing a rigid cell wall, myriad mechanical stress may negatively impact yeast cell proliferation and production as the cell wall may shear, causing cells to agglutinate with each other (Mishra et al., 2022).

As a result, impeller tip speed of 0.84 m/s is deemed the optimum speed for biomass production of *Meyerozyma guilliermondii* strain SO in a 30-L bioreactor due to uniform distribution that is permitted at this speed, allowing the yeast cells to uptake the nutrients and oxygen from the culture medium efficiently, without any stress. This proves that the strategy of maintaining a constant impeller tip speed value during scale up provides similar condition as in the 7.5-L bioreactor. The findings indicated that maintaining a consistent impeller tip speed resulted in statistically similar outcomes for maximum biomass, maximum specific growth rate, biomass productivity, and organic acid production in both the 7.5-L and 30-L stirred-tank bioreactors. Consequently, it can be concluded that the use of a constant impeller tip speed is suitable for scaling up the cultivation of *Meyerozyma guilliermondii* strain SO. Similarly, Stephenie et al. (2007) conducted a study investigating the impact of varying impeller tip speeds on bacterial growth. They reported an enhancement in growth when the impeller tip speed was set at 0.83 m/s, with higher impeller tip speeds leading to decreased growth rates.

Table 4. Growth rate kinetic parameters of *Meyerozyma guilliermondii* strain SO from batch fermentation in a 30-L stirred-tank bioreactor at different impeller tip speeds.

	Impeller tip speed (m/s)			
Kinetic parameters	7.5-L bioreactor	30-L bioreactor		
	0.84	**0.56**	**0.84**	**1.13**
Maximumbiomass (g/L)	12.0	6.0	6.0	5.1
Time taken to reach maximum biomass (h)	24	24	24	24
Growth rate (g/h)	0.24	0.25	0.25	0.21
Specific growth rate, μ_{max} (1/h)	0.43	0.08	0.45	0.33

In scale up process, the agitation is a major component required to be taken into consideration to maintain homogenous mixing, mass and heat transfer. Hence, an optimum hydrodynamic condition is imperative to be constituted as apart from improving the transfers of oxygen and nutrients, and enhancing mixing, agitation also creates shear forces, or turbulence, which negatively impacts cell physiology and rheology properties. Reproducibility in large bioreactors poses numerous challenges. Therefore, it is crucial to attain a satisfactory level of biomass at a small scale before embarking on the scale up to larger volumes. Various other factors can influence biomass production during the scale-up process. In this specific case, our investigation identified mixing and shear stress as potential factors affecting biomass production when transitioning to larger scales. Consequently, our approach of scaling up while maintaining a constant impeller tip speed proved effective in consistently achieving reproducible results across various scales of *Meyerozyma guilliermondii* strain SO fermentation.

4 Conclusion

Meyerozyma guilliermondii strain SO is of great significance for industrial production due to its less methanol toxicity and time conserving traits as an expression host. However, process optimization for the host and medium components is pre-requisite for enhanced productivity and high yield. Therefore, the medium compositions (YPD medium with 2% and 3% (w/v) dextrose and peptone, respectively) that were essential to the growth and production of *Meyerozyma guilliermondii* strain SO has been successfully optimized, followed by the impeller tip speed (0.84 m/s) in 30-L stirred-tank bioreactor. The highest biomass was achieved at 6.0 g/L at 24 h cultivation time, leading to an improved yield with a decrease in the cost of production; thus, making it industrially potential. Based on the findings of this scale-up study, maintaining a constant impeller tip speed is effective for upscaling the production of *Meyerozyma guilliermondii* to an industrial scale. However, further optimization is required for scale-down studies involving other parameters, such as temperature, pH, inoculum size, aeration rate, dissolved oxygen concentration, and other relevant factors. This optimization is necessary to enhance yield and productivity at smaller scales. Moreover, investigating the preservation techniques and its conditions can provide more insights on the future prototype development and commercialization.

Supplementary Material The following is the supplementary data to this article, Figure S1: Time course study of effect of different dextrose concentrations on *Meyerozyma guilliermondii* strain SO growth (A) and biomass production (B) in 500 mL Erlenmeyer flasks, Figure S2: Time course study of effect of different peptone concentrations on *Meyerozyma guilliermondii* strain SO growth (A) and biomass production (B) in 500 mL Erlenmeyer flasks, Figure S3: Biomass production and growth profile of *Meyerozyma guilliermondii* strain SO in 30-L stirred-tank bioreactor at (A) 0.56 m/s (B) 0.84 m/s (C) 1.13 m/s impeller tip speeds, Table S1. One-way Analysis of Variance (ANOVA) table of effect of different dextrose concentrations on biomass production of *Meyerozyma guilliermondii* strain SO in 500 mL Erlenmeyer flasks, Table S2. One-way Analysis of Variance (ANOVA) table of effect of different peptone concentrations on biomass production of *Meyerozyma guilliermondii* strain SO in 500 mL Erlenmeyer flasks.

Author Contributions Diveyan Mohan: Investigation, Methodology, Data curation, Formal analysis, Visualization, Software, Validation, Writing- Original draft, Writing-Review and editing. Che Haznie Ayu Che Hussian: Investigation, Methodology, Data curation, Formal analysis, Visualization, Software, Validation, Writing- Review and editing. Si Jie Lim: Methodology, Formal analysis, Writing- Review and editing. Mohammad Rizal Kapri: Methodology. Siti Nur Hazwani Oslan: Writing- review & editing. Adam Thean Chor Leow: Writing- Review and editing, Supervision. Abu Bakar Salleh: Writing- Review and editing, Supervision. Siti Nurbaya Oslan: Conceptualization, Writing- Review and editing, Supervision, Resources, Project administration. Funding acquisition.

Conflict of Interest and Disclosure The authors declare no conflict of interest.

Acknowledgments. This research was funded by the Ministry of Higher Education, Malaysia *via* Prototype Research Grant Scheme [PRGS/1/2021/STG02/UPM/02/2] which was awarded to the corresponding author (SNO). DM thanks Universiti Putra Malaysia for supporting him *via* Graduate Research Fellowship (GRF).

References

Abu, M.L., Mohammad, R., Oslan, S.N., Salleh, A.B.: The use of response surface methodology for enhanced production of a thermostable bacterial lipase in a novel yeast system. Prep. Biochem. Biotechnol. **51**(4), 350–360 (2020). https://doi.org/10.1080/10826068.2020.1818256

Abu, M.L., Nooh, H.M., Oslan, S.N., Salleh, A.B.: Optimization of physical conditions for the production of thermostable T1 lipase in *Pichia guilliermondii* strain SO using response surface methodology. BMC Biotechnol. **17**(1), 1–10 (2017). https://doi.org/10.1186/s12896-017-0397-7

Allikian, K., Edgar, R., Syed, R., Zhang, S.: Fundamentals of fermentation media. In: Berenjian, A. (ed.) Essentials in fermentation technology, pp. 41–84. Springer Cham (2019). https://doi.org/10.1007/978-3-030-16230-6_2

Bryan, A., Hart, C., Howell, A., Wise, M., Roberts, B.: Glucose concentrations effect on rate of fermentation in yeast. J. Undergrad. Biol. Lab. Invest. **1**(1), 1–4 (2018)

Canché-Collí, C., Barahona, F., Medina-Medina, L.A., Canto, A.: The effect of sugar concentration on growth of yeast associated to floral nectar and honey. Scientia Fungorum **52**, 1–12 (2021). https://doi.org/10.33885/sf.2021.52.1288

Ergün, B.G., Hüccetoğulları, D., Öztürk, S., Çelik, E., Çalık, P.: Established and upcoming yeast expression systems. In: Gasser, B., Mattanovich, D. (eds.) Recombinant Protein Production in Yeast (vol. 1923, pp. 1–74). Humana Press (2019). https://doi.org/10.1007/978-1-4939-9024-5_1

Fedorovych, D., Boretsky, Y., Bobak, Y., Prokopiv, T., Sybirny, A.: Putative ferroxidases in the flavinogenic yeast *Pichia guilliermondii* are regulated by iron acquisition. Cytol. Genet. **49**(5), 288–293 (2015). https://doi.org/10.3103/S0095452715050035

Gandhi, S., Salleh, A.B., Abd Rahman, R.N.Z.R., Chor Leow, T., Oslan, S.N.: Expression and characterization of *Geobacillus stearothermophilus* SR74 recombinant α-amylase in *Pichia pastoris*. Biomed. Res. Int. **1-9** (2015). https://doi.org/10.1155/2015/529059

Gomar-Alba, M., Morcillo-Parra, M.A., Olmo, M.: Response of yeast cells to high glucose involves molecular and physiological differences when compared to other osmostress conditions. FEMS Yeast Res. **15**(5), 1–14 (2015). https://doi.org/10.1093/fem-syr/fov039

Heitmann, M., Zannini, E., Arendt, E.: Impact of *Saccharomyces cerevisiae* metabolites produced during fermentation on bread quality parameters: A review. Crit. Rev. Food Sci. Nutr. **58**(7), 1152–1164 (2018). https://doi.org/10.1080/10408398.2016.1244153

Karbalaei, M., Rezaee, S.A., Farsiani, H.: *Pichia pastoris*: A highly successful expression system for optimal synthesis of heterologous proteins. J. Cell. Physiol. **235**(9), 5867–5881 (2020). https://doi.org/10.1002/jcp.29583

Klein, M., Swinnen, S., Thevelein, J.M., Nevoigt, E.: Glycerol metabolism and transport in yeast and fungi: established knowledge and ambiguities. Environ. Microbiol. **19**(3), 878–893 (2017). https://doi.org/10.1111/1462-2920.13617

Kokkonen, P., Beier, A., Mazurenko, S., Damborsky, J., Bednar, D., Prokop, Z.: Substrate inhibition by the blockage of product release and its control by tunnel engineering. RSC Chemical Biol. **2**(2), 645–655 (2021). https://doi.org/10.1039/D0CB00171F

Kumar, N., Aggarwal, N.K.: Process development for fungal cellulase production and enzymatic saccharification of *Parthenium hysterophorus* for bioethanol production. Waste Manage. Bull. **2**(1), 1–8 (2024). https://doi.org/10.1016/j.wmb.2023.11.004

Li, Z., Wang, D., Shi, Y.C.: Effects of nitrogen source on ethanol production in very high gravity fermentation of corn starch. J. Taiwan Inst. Chem. Eng. **70**, 229–235 (2017). https://doi.org/10.1016/j.jtice.2016.10.055

Ling, L.Y., Ithoi, I., Fong, M.Y.: Optimization for high-level expression in *Pichia pastoris* and purification of truncated and full length recombinant SAG2 of *Toxoplasma gondii* for diagnostic use. Southeast Asian J. Trop. Med. Public Health. **41**(3), 507–513 (2010)

Liu, Z., Tyo, K.E.J., Martínez, J.L., Petranovic, D., Nielsen, J.D., Nielsen, J.: Different expression systems for production of recombinant proteins in *Saccharomyces cerevisiae*. Biotechnol. Bioeng. **109**(5), 1259–1268 (2012). https://doi.org/10.1002/bit.24409

Mediboyina, M.K., et al.: Upscale fermenter design for lactic acid production from cheese whey permeate focusing on impeller selection and energy optimization. J. Food Sci. Technol. **59**(6), 2263–2273 (2022). https://doi.org/10.1007/s13197-021-05239-6

Mishra, R., Minc, N., Peter, M.: Cells under pressure: how yeast cells respond to mechanical forces. Trends Microbiol. **30**(5), 495–510 (2022). https://doi.org/10.1016/j.tim.2021.11.006

Nasir, N.S.M., Leow, C.T., Oslan, S.N.H., Salleh, A.B., Oslan, S.N.: Molecular expression of a recombinant thermostable bacterial amylase from *Geobacillus stearothermophilus* SR74 using methanol-free *Meyerozyma guilliermondii* strain SO yeast system. Bioresources. **15**(2), 3161–3172 (2020). https://doi.org/10.15376/biores.15.2.3161-3172

Olivares-Marin, I.K., González-Hernández, J.C., Regalado-Gonzalez, C., Madrigal-Pe- rez, L. A.: *Saccharomyces cerevisiae* exponential growth kinetics in batch culture to analyze respiratory and fermentative metabolism. J. Vis. Exp. **139**, 1–10 (2018). https://doi.org/10.3791/58192

Oslan, S.N., Salleh, A.B., Abd Rahman, R.N.Z.R., Basri, M., Chor, A.L.T.: Locally isolated yeasts from Malaysia: Identification, phylogenetic study and characterization. Acta Biochim. Pol. **59**(2), 225–229 (2012). https://doi.org/10.18388/abp.2012_2142

Oslan, S.N., Salleh, A.B., Abd Rahman, R.N.Z.R., Leow, T.C., Basri, M.: *Pichia pastoris* as a host to overexpress the thermostable T1 lipase from *Geobacillus zalihae*. GSTF J. BioSci. **3**(1) (2014). https://doi.org/10.5176/2251-2489_BioTech14.26

Oslan, S.N., Salleh, A.B., Abd Rahman, R.N.Z.R., Leow, T.C., Sukamat, H., Basri, M.: A newly isolated yeast as an expression host for recombinant lipase. Cell. Mol. Biol. Lett. **20**(2), 279–293 (2015). https://doi.org/10.1515/cmble-2015-0015

Oslan, S.N.H., Tan, J.S., Saad, M.Z., Halim, M., Mohamed, M.-S., Ariff, A.B.: Influence of amino acids and vitamins on the growth of *gdhA* derivative *Pasteurella multocida* B:2 for use as an animal vaccine. Bioprocess Biosyst. Eng. **42**, 355–365 (2019). https://doi.org/10.1007/s00449-018-2040-y

Oslan, S.N.H., et al: *Pasteurellosis* vaccine commercialization: Physiochemical factors for optimum production. PRO. **10**(7), 1248 (2022). https://doi.org/10.3390/pr10071248

Pallujam, A.D., Zali, S.A., Oslan, S.N.H., Salleh, A.B., Oslan, S.N.: Critical physical parameters for optimum recombinant protein production in yeast systems. Malaysian J. Biochem. Mol. Biol. **2**, 34–45 (2020) http://mjbmb.org/

Raita, S., Kusnere, Z., Spalvins, K., Blumberga, D.: Optimization of yeast cultivation factors for improved SCP production. Environ. Clim. Technol. **26**(1), 848–861 (2022). https://doi.org/10.2478/rtuect-2022-0064

Ramli, N.A., Abd Rahman, R.A., Md Illias, R.: Optimization of medium fermentation for isobutanol production in *Saccharomyces cerevisiae* using response surface methodology. Chem. Eng. Trans. **83**, 175–180 (2021). https://doi.org/10.3303/CET2183030

Singh, V., Haque, S., Niwas, R., Srivastava, A., Pasupuleti, M., Tripathi, C.K.M.: Strategies for fermentation medium optimization: An in-depth review. Front. Microbiol. **7**, 1–16 (2017). https://doi.org/10.3389/fmicb.2016.02087

Stephenie, W., Kabeir, B.M., Shuhaimi, M., Rosfarizan, M., Yazid, A.M.: Influence of pH and impeller tip speed on the cultivation of Bifidobacterium pseudocatenulatum G4 in a milk-based medium. Biotechnol. Bioproc. Eng. **12**(5), 475–483 (2007). https://doi.org/10.1007/BF02931343

Struyf, N., Van der Maelen, E., Hemdane, S., Verspreet, J., Verstrepen, K.J., Courtin, C.M.: Bread dough and baker's yeast: an uplifting synergy. Compr. Rev. Food Sci. Food Saf. **16**(5), 850–867 (2017). https://doi.org/10.1111/1541-4337.12282

Thamer, A.A., Issa Al haboubi, N.A.: Study the effect of different types impellers on the transfer coefficient in photobioreactor. IOP Conf. Series Mater. Sci. Eng. **928**(2), 022144 (2020). https://doi.org/10.1088/1757-899X/928/2/022144

Xia, J., Wang, G., Fan, M., Chen, M., Wang, Z., Zhuang, Y.: Understanding the scale-up of fermentation processes from the viewpoint of the flow field in bioreactors and the physiological response of strains. Chin. J. Chem. Eng. **30**, 178–184 (2021). https://doi.org/10.1016/j.cjche.2020.12.004

Xiberras, J., Klein, M., Nevoigt, E.: Glycerol as a substrate for *Saccharomyces cerevisiae* based bioprocesses – Knowledge gaps regarding the central carbon catabolism of this "non-fermentable" carbon source. Biotechnol. Adv. **37**(6), 1–15 (2019). https://doi.org/10.1016/j.biotechadv.2019.03.017

Zhang, Q., Wu, D., Lin, Y., Wang, X., Kong, H., Tanaka, S.: Substrate and product inhibition on yeast performance in ethanol fermentation. Energy Fuel. **29**(2), 1019–1027 (2015). https://doi.org/10.1021/ef502349v

Literature Review on QoS Prediction for Transportation Using Machine Learning

Dewanto Rosian Adhy, Mohd Fairuz Iskandar Othman(✉), and Najwan Khambari

Department of Computer Systems and Communication, Faculty of Information and Communication Technology (FTMK), Universiti Teknikal Malaysia Melaka (UTeM), Hang Tuah Jaya, 76100 Durian Tunggal, Melaka, Malaysia
p0131810010@student.utem.edu.my, {mohdfairuz,najwan}@utem.edu.my

Abstract. Quality-of-Service (QoS) Prediction supports the reliability of communication in the Internet of Vehicles (IoV). It forecasts whether future network parameters will remain above the safety threshold required by Intelligent Transportation Systems (ITS). This paper reviews machine-learning-based approaches for QoS Prediction, focusing on datasets, preprocessing, modeling, and evaluation metrics. Beyond existing surveys, it identifies unresolved challenges that hinder deployment—particularly real-time inference, data drift under dynamic vehicular conditions, and the computational constraints of edge devices. The analysis of recent supervised, unsupervised, and reinforcement-learning studies reveals that although machine learning improves accuracy, most models neglect non-stationary behavior, future-horizon fore-casting, and distributed computation. Addressing these issues is vital for adaptive, scalable, and low-latency prediction. This review therefore proposes a framework direction in which reinforcement learning, ensemble strategies, and edge-assisted continuous learning enable reliable and real-time.

Keywords: Machine learning · QoS · IoV · VANET · Intelligent Transportation Systems

1 Introduction

The rapid integration of communication technology into transportation has produced connected vehicles and emerging autonomous-driving features. Comfort and safety now depend on continuous, high-quality connectivity. Cellular networks evolving from 3G to 5G are expected to support such communication, yet fluctuating band-width, coverage gaps, and heterogeneous devices still cause service degradation. These failures can interrupt infotainment, delay safety-critical decisions, and jeopardize autonomous-driving functions.

Predicting network quality before vehicles enter a region is therefore essential. A QoS Prediction tool estimates future throughput, latency, jitter, and packet loss, enabling systems to judge whether communication performance meets application requirements.

M. A. b. Jamlos et al. (Eds.): ICSET 2025, LNNS 1897, pp. 80–85, 2026.
https://doi.org/10.1007/978-3-032-20722-7_9

Machine-learning methods are well suited for this task because they infer complex non-linear relationships from historical measurements. However, the variability of vehicular mobility and network evolution introduces unique modeling challenges. Early QoS-related studies in transportation focused on communication technologies such as VANET and IoV, emphasizing latency, throughput, and packet loss as key performance parameters. As ITS evolved, researchers adopted machine learning to model nonlinear network behavior and improve prediction accuracy. Various domains—including IoT, web services, and edge computing—have explored QoS Prediction, but IoV presents unique challenges due to mobility, dynamic topology, and time-varying environments.

Existing approaches can be grouped into four functions:

1. QoS Prediction – estimates future performance based on historical data,
2. Outage Probability – calculates the chance of QoS falling below threshold,
3. Service Selection – chooses services based on quality estimates,
4. Resource Allocation – optimizes network resources to maintain QoS.

QoS Prediction and Outage Probability are most aligned with transportation needs, as both assess future network feasibility. Most solutions use supervised learning (e.g., Random Forest, Neural Networks, LSTM), while some incorporate Kalman Filter or Reinforcement Learning to improve temporal modeling or adaptability.

However, several gaps persist:

1. Many studies ignore non-stationary behavior and data drift.
2. Most predictions are offline, not real-time.
3. Future-horizon (multi-step) forecasting is rarely addressed.
4. Edge computing integration is limited, causing latency and scalability issues.
5. Overfitting, sparsity, and missing data appear but are not deeply mitigated.

These limitations reveal the need for more adaptive, scalable, and latency-aware QoS Prediction frameworks.

2 Method

This review adopts a structured approach to identify and analyze machine-learning-based QoS Prediction studies relevant to IoV and ITS.

Articles were retrieved using Publish or Perish and filtered from IEEE Xplore, Scopus, ScienceDirect, ACM, and Google Scholar. Keywords: QoS Prediction, IoV, ITS, Machine Learning, Outage Probability, Edge Computing.

Inclusion criteria:

- Published 2019–2024 (to capture recent ML and 5G/6G developments)
- Focus on communication network QoS
- Use machine learning or related intelligent techniques
- English-language, peer-reviewed sources

This yielded 524 articles → filtered to 26 relevant → 12 core articles on QoS Prediction & Outage in IoV for in-depth analysis.

Each selected article was analyzed based on:

- Target Function: QoS Prediction, Outage Probability, Service Selection, Re-source Allocation
- Domain: IoV/ITS, IoT, Edge Computing, Web Services
- Dataset: Source (real, simulation, public), features (throughput, RSSI, latency, speed, location), size, stationarity
- Preprocessing: Cleaning, normalization, feature engineering, handling missing data/outliers
- Model Type: Neural Network (DNN, LSTM, BPNN), Random Forest, Reinforcement Learning, Hybrid or Ensemble
- Prediction Horizon: Single-step vs. multi-step (future-horizon)
- Performance Metrics: Accuracy, RMSE, MAPE, R^2, latency
- Computational Aspects: Feasibility, overfitting, sparsity, edge deployment

To answer the research objectives, three dimensions were emphasized:

- Dataset Challenges: Non-stationary behavior, drift, sparsity, missing data
- Model Capability: Ability to handle temporal dependencies, nonlinear patterns, and multivariable prediction
- Deployment Considerations: Real-time inference, computational efficiency, ability to run at the edge

This structured methodology ensures that technical novelty, reproducibility, and implementation feasibility are consistently evaluated.

3 Results and Discussion

From 26 relevant articles, 12 core studies focused specifically on QoS Prediction or Outage Probability in IoV/ITS using machine learning. These works were analyzed based on models, datasets, parameters, and deployment constraints.

Most studies targeted QoS Prediction (14) and Outage Probability (4), confirming alignment with forecasting future network quality. Domains were dominated by IoV (15) and IoT (4), with limited use of edge computing (3) despite its importance for latency-sensitive predictions. Neural Networks and their variants accounted for the majority (14), followed by LSTM (7), Random Forest, and one Reinforcement Learning approach. Neural Net-works were preferred due to their nonlinear modeling and integration potential (e.g., Kalman Filter, SSA, ICS, ensemble layers). LSTM models better captured temporal dependencies, but few addressed future-horizon prediction.

Datasets ranged from simulation-based to real-world measurements (4G/5G, Wi-Fi, V2X logs). While preprocessing (cleaning, normalization, feature engineering) was common, three major challenges remained insufficiently addressed:

- Data Drift/Non-stationarity: Most models trained offline without adaptation.
- Sparsity & Missing Values: Often acknowledged but rarely mitigated robustly.
- Overfitting Risks: Large feature sets and imbalanced data increased model in-stability.

QoS Prediction must estimate performance ahead of time for route planning and autonomous decisions. However:

- Most studies used single-step predictions.
- Few employed time-series architectures (LSTM, SES-LSTM) specifically for multi-step forecasting.
- Only 1–2 studies considered real-time inference or edge deployment.

Table 1. QoS Prediction for Internet of Vehicle

No	Model	Target Atribute	Learning Problem			Prediction	Data Capture	Network Tech.
			Drift	Sparsity	Overfitting			
1	Random Forest [1]	Throughput	r	r	r	r	Real Data Sampling	5G & 6G
2	Neural Network + Kalman Filter [2]	Availability	r	r	r	a	Simulation	NA
3	ICS-BP Neural Network [3]	Outage Probability	r	a	a	r	NA	4G & 5G
4	SES-LSTM [4]	Availability	r	a	a	a	Kaggle Dataset	4G
5	Neural Network [5]	Outage Probability	r	r	r	a	Real Data	WiFi
6	SSA-BPNN [6]	Outage Probability	r	a	a	r	NA	5G
7	DNN [7]	Throughput	r	a	a	a	Real Data	4G & 5G
8	Random Forest & MLP [8]	Throughput	a	r	r	a	Real Data	4G & 5G
9	Neural Network [9]	Throughput	r	r	r	r	NA	4G
10	RL [10]	Throughput	r	r	a	r	Simulation	5G
11	Neural Network [11]	Throughput	r	r	r	r	Real Data	4G
12	Neural Network [12]	Throughput	r	r	r	r	Simulation	4G

A consolidated comparison (from Table 1) confirms the research gap: accuracy alone is no longer sufficient—adaptability and latency-awareness are essential.

- Neural Networks dominate due to nonlinear power and integration capability.
- Preprocessing well-studied, but drift and sparsity are not fully solved.
- QoS Prediction is shifting from static tools to intelligent, adaptive systems.
- Real-time and edge deployment are critical but underexplored.
- Reinforcement Learning and Ensemble Methods offer promising adaptability but require

4 Conclusion

This review confirms that machine learning-based QoS Prediction is essential for reliable and safe operation of Internet of Vehicles and Intelligent Transportation Systems. Existing studies demonstrate strong progress in modeling throughput, latency, availability, and outage probability using Neural Networks, LSTM, and other learning-based approaches. However, this review identifies three critical gaps that must be addressed before real-world deployment is possible.

- First, real-time prediction must become a core capability, not an offline post-processing task. QoS estimation must support low-latency decision-making during vehicle movement.
- Second, data drift is unavoidable in dynamic vehicular environments. Most existing models assume static data distributions, but real-world networks evolve constantly. Continuous learning, adaptive models, or drift detection mechanisms are required to maintain accuracy.
- Third, edge computing is crucial to balance prediction speed, scalability, and resource constraints. Centralized architectures introduce latency and bandwidth overhead, making them unsuitable for time-sensitive ITS applications.

By emphasizing real-time inference, non-stationary data handling, and edge-enabled learning, this work provides a clear research agenda for next-generation QoS Prediction. Future models must support multivariable targets (throughput, latency, jitter, packet loss), operate over future horizons, and adapt to evolving network conditions without retraining from scratch. Integrating reinforcement learning, ensemble methods, and automated feature engineering at the network edge offers a promising pathway toward adaptive, scalable, and accurate QoS intelligence. QoS Prediction can therefore evolve from a supportive tool into a core intelligence layer that enables autonomous driving, resilient connectivity, and scalable ITS deployment.

Future QoS Prediction development should focus on:

- Multivariable forecasting: Throughput, latency, jitter, packet loss simultaneously
- Non-stationary data handling: Drift detection, continual learning, sliding windows
- Real-time execution: Low-latency inference, streaming data support
- Edge computing integration: Distributed computation, local model updates
- Reinforcement learning & ensembles: Improved adaptability and robustness
- Feature engineering automation: Handling complex, high-dimensional data
- Scalability and generalization: Support wider regions, heterogeneous networks

References

1. A. Krause, P. Schulz, A. Palaios, A. Kumar, and G. Fettweis, An Improved Data Collection Framework for Enabling ML-based QoS Prediction for Vehicular Communication, 2023. [Online]. https://www.ericsson.com/en/cases/2020/5g-connectedmobility
2. F. Bragato et al., Towards Decentralized Predictive Quality of Service in Next-Generation Vehicular Networks, Feb. 2023, [Online]. http://arxiv.org/abs/2302.11268
3. Xu, L., Wang, H., Gulliver, T.A.: Outage probability performance analysis and prediction for mobile IoV networks based on ICS-BP neural network. IEEE Internet Things J. **8**(5), 3524–3533 (2021). https://doi.org/10.1109/JIOT.2020.3023694
4. Alsaade, F.W., Hmoud Al-Adhaileh, M.: Cellular traffic prediction based on an intelligent model. Mob. Inf. Syst. **2021** (2021). https://doi.org/10.1155/2021/6050627
5. Suga, N., Yano, K., Webber, J., Hou, Y., Higashimori, T., Suzuki, Y.: Prediction of QoS outage probability for wireless communication in factory environments. In: Proceedings of the 2019 International Conference on Internet of Things, Embedded Systems and Communications (IINTEC 2019), pp. 124–129. IEEE, Piscataway (2019). https://doi.org/10.1109/IINTEC48298.2019.9112098
6. Li, Y., Zhang, Y., Tian, X., Liu, R.: Outage performance prediction of cooperative vehicle network based on sparrow search algorithm based on back-propagation neural network. IET Networks. **13**(1), 45–57 (2024). https://doi.org/10.1049/ntw2.12100
7. M. Boban, M. Giordani, and M. Zorzi, Predictive Quality of Service (PQoS): The Next Frontier for Fully Autonomous Systems, Sep. 2021, [Online]. http://arxiv.org/abs/2109.09376
8. Palaios, A., et al.: Machine learning for QoS prediction in vehicular communication: Challenges and solution approaches. IEEE Access. **11**, 92459–92477 (2023). https://doi.org/10.1109/ACCESS.2023.3303528
9. N. Sivapriya, R. Mohandas, and K. Kumar Vaigandla, A QoS perception routing protocol for MANETs based on machine learning, Int. J. Intell. Syst. Appl. Eng., vol. 2024, no. 1, pp. 733–745, 2023, [Online]. www.ijisae.org
10. Singh, M., Baranwal, G., Tripathi, A.K.: QoS-Aware selection of IoT-based service. Arab. J. Sci. Eng. **45**(12), 10033–10050 (2020). https://doi.org/10.1007/s13369-020-04601-8
11. Al-Thaedan, A., et al.: Downlink throughput prediction using machine learning models on 4G-LTE networks. Int. J. Inform. Technol. (Singapore). **15**(6), 2987–2993 (2023). https://doi.org/10.1007/s41870-023-01358-9
12. Kousaridas, A.: QoS prediction for 5G connected and automated driving. IEEE Commun. Mag. **59**(9) (2021). https://doi.org/10.1109/MCOM.110.2100042

Thermogravimetric Analysis (TGA) of Carbon Retention in EAF Slag Under Varied Carbon Dioxide and Humidity Cycling

Jason Ting Jing Cheng[1(✉)], Lee Foo Wei[1], Mo Kim Hung[2], and Yip Chun Chieh[1]

[1] Department of Civil Engineering, University Tunku Abdul Rahman, Sungai Long Campus, Jalan Sungai Long, Bandar Sungai Long Cheas, 43000 Kajang, Selangor, Malaysia
jasontingjingcheng@gmail.com

[2] Department of Civil Engineering, Sunway University Malaysia, 5, Jalan Universiti, Bandar Sunway, 47500 Petaling Jaya, Selangor, Malaysia

Abstract. This study examines carbon retention behavior in Electric Arc Furnace (EAF) slag through thermogravimetric analysis (TGA) under systematically varied environmental conditions. Nine experimental variants were tested, combining three CO_2 concentrations (10%, 20%, 30%) with three humidity levels (40%, 60%, 80%) at a constant temperature of 30 °C. Results demonstrate that carbon retention efficiency depends critically on the synergistic effects of these parameters. Optimal carbonation occurred at 20% CO_2 and 60% humidity, yielding an 8.2% mass gain from stable carbonate formation. Higher CO_2 levels (30%) at elevated humidity (80%) showed reduced effectiveness (5.1% mass gain), suggesting solubility limitations of reaction products. Conversely, low humidity conditions (40% RH) resulted in minimal carbonation (<3% mass gain) across all CO_2 concentrations. TGA derivative curves revealed distinct mass loss patterns between 200–800 °C, corresponding to different carbonate phases. The findings establish humidity as the dominant factor controlling reaction kinetics, while CO_2 concentration primarily determines long-term carbonate stability. These results provide essential operational parameters for deploying EAF slag in practical carbon capture systems, particularly in moderate humidity environments. The study further identifies threshold conditions beyond which carbon retention efficiency significantly deteriorates, offering crucial guidance for industrial applications.

Keywords: Carbon Retension · Carbon Dioxide sequestration · Carbonated EAF Slag · Thermogravimetric analysis

1 Introduction

Electric Arc Furnace (EAF) slag, a byproduct of steel production with annual global generation exceeding 400 million tons, has gained considerable attention for carbon sequestration applications due to its high calcium content and reactivity with CO_2 [1]. While previous investigations have demonstrated the carbonation potential of EAF slag under controlled conditions, these studies have typically examined individual parameters in isolation, neglecting the complex interactions between multiple environmental variables encountered in practical applications [2].

M. A. b. Jamlos et al. (Eds.): ICSET 2025, LNNS 1897, pp. 86–96, 2026.
https://doi.org/10.1007/978-3-032-20722-7_10

The present study was designed to address this knowledge gap through systematic investigation of carbon retention behavior in EAF slag under carefully controlled combinations of CO_2 concentration (10%, 20%, and 30%), relative humidity (40%, 60%, and 80%), and fixed temperature (30 °C). Thermogravimetric analysis (TGA) coupled with evolved gas analysis was employed to provide detailed insights into phase transformation kinetics and carbonate stability under these multifactorial conditions.

Several significant phenomena were observed. Maximum carbonate formation, corresponding to an 8.2% mass gain, was achieved at intermediate conditions of 20% CO_2 and 60% RH. At elevated conditions of 30% CO_2 and 80% RH, carbonate solubility limitations became apparent, resulting in reduced carbon retention efficiency with only 5.1% mass gain. Microstructural examination revealed that reaction kinetics were primarily governed by humidity through its influence on ion mobility, while long-term carbonate stability was determined mainly by CO_2 concentration.

These findings have important implications for industrial-scale carbon sequestration system design. The identification of optimal operational windows for maximum carbon retention efficiency enables more effective utilization of this abundant industrial byproduct in climate change mitigation strategies. Additionally, the established threshold conditions provide valuable guidance for maintaining performance in variable environments, representing a significant advancement in mineral carbonation technology development.

This investigation represents a meaningful contribution to understanding the complex relationships between environmental parameters in EAF slag carbonation, effectively bridging the gap between fundamental research and practical application. The results not only enhance scientific understanding of mineral carbonation processes but also provide essential data for engineering design and policy development in carbon capture and utilization technologies.

2 Materials and Methods

The materials employed in this investigation included Ordinary Portland Cement (OPC) and carbonated Electric Arc Furnace (EAF) slag, with the latter being procured from local steel manufacturing facilities. The EAF slag underwent an accelerated carbonation process under nine distinct environmental conditions, systematically combining three CO_2 concentrations (10%, 20%, and 30%) with three relative humidity levels (40%, 60%, and 80%) at a constant temperature of 30 °C, to evaluate its enhanced chemical reactivity and CO_2 sequestration potential. For concrete formulation, a consistent water-to-cement ratio of 0.55 was maintained with equal proportions of cement and sand, while the total material quantity was increased by 30% to compensate for processing losses. The absolute volume method was utilized for precise proportioning of constituent materials, with Tables 1 and 2 detailing the specific weights and mix proportions for both the control mixture (100% OPC) and the modified mixture incorporating 30% carbonated EAF slag as partial replacement for fine aggregates. The mixing procedure employed a mechanical mixer, commencing with dry blending of cement, EAF slag, and fine aggregates for five minutes to achieve homogeneity, followed by gradual water addition and continued mixing for an additional 2–3 min to ensure proper hydration.

After casting, the concrete specimens were cured under standard conditions prior to thermogravimetric analysis (TGA) to assess the carbon retention characteristics and thermal decomposition behavior of the incorporated carbonated EAF slag under the various pretreatment conditions.

Table 1. Weight of Raw Materials (Proportion mix with EAF slag, particle sizes between 0.8 mm–7 mm)

Type of Raw Materials	Proportion Ratio	Weight (kg)	Density (Kg/m^3)
OPC	1	3.878	1440
Sand	0.700	2.231	1560
Water	0.550	2.133	1000
EAF Slag	0.265	0.846	1380
Total	2.515	≈ 9.09	

Table 2. Nine distinct environmental conditions

Sample	Carbon Dioxide Concentration Percentage)	Temperature (Degree Celcius)	Humidity (Percentage)
1	10	30	40
2	10	30	60
3	10	30	80
4	20	30	40
5	20	30	60
6	20	30	80
7	30	30	40
8	30	30	60
9	30	30	80

2.1 EAF Slag

The Electric Arc Furnace (EAF) slag utilized in this investigation was obtained from Anjoo Steel Sdn Bhd, located in Malaysia. This industrial byproduct exhibits a chemical composition that contributes significantly to both cementitious properties and carbon sequestration capabilities. The material contains a substantial proportion of calcium oxide (CaO), accounting for roughly 40% of its composition, which serves as a crucial component in both cement hydration reactions and carbon dioxide mineralization. Approximately 30% of the slag consists of silicon dioxide (SiO_2), a compound known to enhance mechanical strength and long-term performance through pozzolanic activity. The presence of magnesium oxide (MgO) at about 8% helps maintain dimensional stability and promotes the formation of a dense microstructure. Iron oxide (Fe_2O_3) constitutes around 10% of the composition, improving resistance to chloride penetration and overall durability. The remaining portion, approximately 12%, includes various minor constituents such as aluminum oxide (Al_2O_3), manganese oxide (MnO), and sulfur trioxide (SO_3), all of which may affect the rate of hydration and the resulting mechanical characteristics of cement-based materials.

2.2 Fine Aggregate

The research employed natural river sand as fine aggregate, procured from local sources to ensure material accessibility and reduce environmental impact. Selected for its uniform quality and proven performance in construction applications, this sand type represents a conventional choice for concrete manufacturing. Prior to use, the material underwent comprehensive particle size distribution testing to verify compliance with ASTM C33 specifications, the established standard governing fine aggregate quality in concrete production.

The standardized testing protocol confirmed the material's adherence to strict requirements regarding particle gradation and permissible levels of contaminants. These specifications help control potentially harmful components - including clay particles, fine silts, and organic matter - that could compromise concrete performance by affecting hydration processes, mechanical strength development, and long-term durability. Furthermore, the optimized particle size distribution promotes efficient particle packing within the cementitious matrix, enhancing density while reducing undesirable void content. This rigorous quality assurance process using ASTM C33-certified river sand contributes to producing consistent, high-performance concrete specimens, thereby ensuring the experimental validity and reproducibility of the study's findings.

2.3 Water

The mixing water employed in this investigation was obtained from local Malaysian sources and carefully selected to meet the stringent requirements of ASTM C1602 for concrete production. This international standard establishes critical limits for potentially detrimental substances including chloride ions, sulfate compounds, alkaline materials, and organic contaminants - all of which could impair concrete's setting characteristics, compressive strength development, and service life.

In cementitious systems, water serves multiple essential functions beyond simple mixture fluidity. As the activating medium for cement hydration reactions, it facilitates the formation of strength-giving calcium silicate hydrate (C-S-H) gels while enabling proper workability during placement operations. The water phase also promotes uniform distribution of cement particles and fine aggregates throughout the mixture, ensuring microstructural homogeneity in the hardened composite.

2.4 Cement

The Ordinary Portland Cement (OPC) used in this study was procured from a certified Malaysian manufacturer, meeting the requirements of ASTM C150 and MS EN 197–1 standards for structural concrete applications. Chemical analysis showed the cement comprised 62.0% calcium oxide (CaO) as the primary component for forming calcium silicate hydrate (C-S-H) gel during hydration, 30.0% silicon dioxide (SiO_2) contributing to pozzolanic reactions and long-term strength development, 5.0% aluminum oxide (Al_2O_3) influencing early strength and workability, and 3.0% iron oxide (Fe_2O_3) enhancing sulfate resistance. The cement exhibited a Blaine fineness of 300 m^2/kg, providing optimal surface area for efficient hydration reactions, while its specific gravity of 3.15 facilitated accurate mix design calculations. These carefully balanced chemical and physical properties ensured proper workability during mixing, controlled setting characteristics, predictable strength development, and long-term durability

3 Methodology

The methodology for this study focused on the preparation and analysis of concrete cube specimens (150 mm × 150 mm × 150 mm) following ASTM C192 guidelines. The cube molds were constructed from rigid, non-absorbent materials and lightly coated with mineral oil prior to casting to facilitate proper demolding. Concrete mixtures were prepared using a mechanical mixer, with dry components including Ordinary Portland Cement (OPC) and fine aggregates initially blended for 3 min before the gradual addition of mixing water and further mixing for 2 min. For mixtures containing carbonated EAF slag, the material was pre-wetted to control water absorption during mixing. The fresh concrete was poured into molds in three equal layers, each compacted using a vibrating table to ensure proper consolidation, followed by surface finishing with a steel trowel. After casting, specimens were covered with damp burlap and plastic sheets for 24 ± 2 h at 23 ± 2 °C before demolding and subsequent curing in a controlled environment (23 ± 1 °C, $\geq 95\%$ relative humidity). Following the curing period, samples were prepared for thermogravimetric analysis (TGA) by extracting powdered material from the crushed cubes to evaluate carbon retention efficiency and hydration product formation under the different carbonation conditions of the EAF slag. The TGA testing protocol involved controlled heating of samples to assess mass loss characteristics and thermal decomposition behavior of the carbonated components. Quality control measures included monitoring ambient conditions throughout the process and preparing triplicate specimens for each mix design to ensure representative sampling. Fig. 1 presents a flowchart outlining the step-by-step methodology adopted in this study.

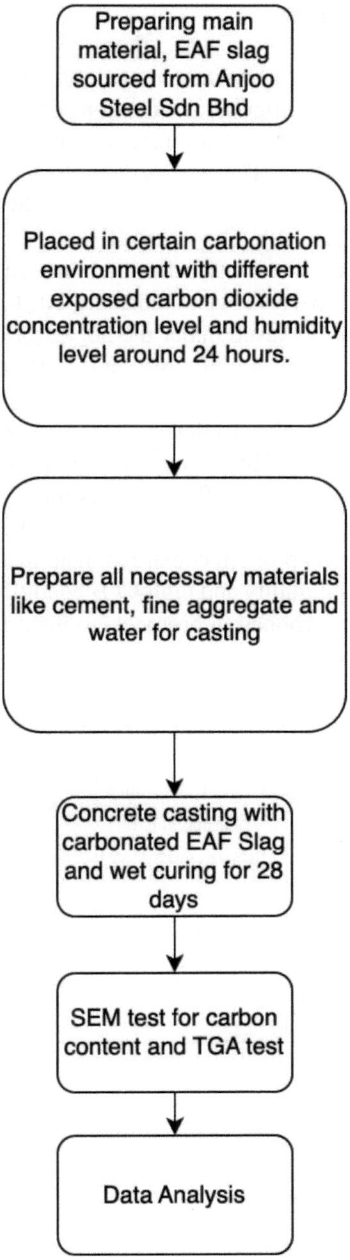

Fig. 1. Flow Chart

4 Results and Discussion

The data in Fig. 2 illustrates the carbon content variation in Electric Arc Furnace (EAF) slag and mortar mixtures under different carbonation conditions, characterized by varying CO_2 concentrations (10%, 20%, 30%), relative humidity levels (40%, 60%, 80%), and a fixed temperature of 30 °C. The results demonstrate a clear trend of increasing carbon content with higher CO_2 concentrations and humidity levels. For instance, under the 10_40_30 condition, Fig. 3 (10% CO_2, 40% RH, 30 °C), the carbon content in EAF slag rose from 12.21 wt% to 14.33 wt% after carbonation, while the mortar mixture exhibited a further increase to 16.41 wt% under the 20_40_30 condition, Fig. 4. The highest carbon retention was achieved under the 30_80_30 condition (30% CO_2, 80% RH, 30 °C), Fig. 5 with the mortar mixture reaching 19.92 wt% carbon content.

This progression highlights the synergistic effect of CO_2 concentration and humidity on carbonation efficiency, where elevated levels of both parameters significantly enhance carbon sequestration in EAF slag. The subsequent incorporation of carbonated slag into mortar mixtures further amplified this effect, suggesting continued carbonation during the hydration process. These findings underscore the potential of carbonated EAF slag as an effective material for CO_2 capture in construction applications, with optimal performance observed under high-humidity and high-CO_2 conditions. The data provides valuable insights for optimizing carbonation protocols to maximize environmental benefits in sustainable building practices.

Condition	Element Composition of Carbon in EAF Slags (wt %)	Elemental Composition of Carbon in mortar Mixed with Carbonated EAF slags (wt %)
10_40_30	12.21	14.33
20_40_30	13.98	16.41
30_40_30	14.87	17.46
10_60_30	13.72	16.11
20_60_30	14.73	17.29
30_60_30	15.68	18.41
10_80_30	14.86	17.45
20_80_30	15.34	18.01
30_80_30	16.97	19.92

Fig. 2. Comparison of Elemental Composition of Carbon of Slags and Mortar Mixed with Carbonated EAF Slags.

The thermogravimetric analysis (TGA) results reveal distinct thermal decomposition behaviors for carbonated EAF slag under varying CO_2 concentrations at constant 60% relative humidity and 30 °C temperature. For the 10% CO_2 condition (10_60_30),

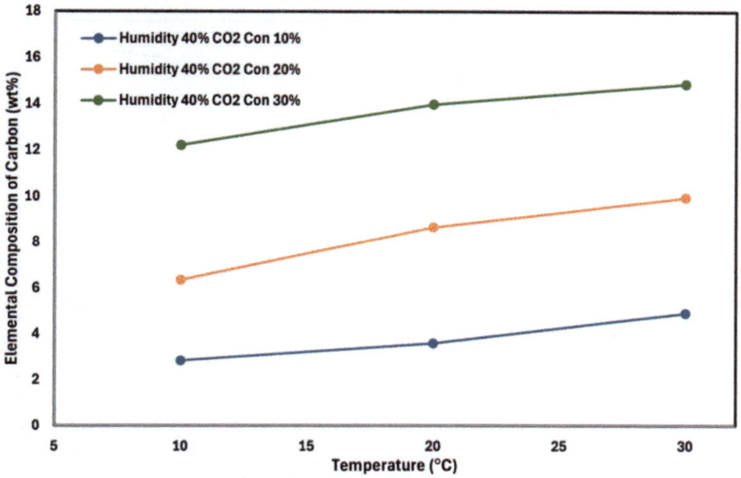

Fig. 3. Graph of Carbon Element Content versus Temperature under 10%, 20%, 30% CO2 Concentration at Relative Humidity of 40%.

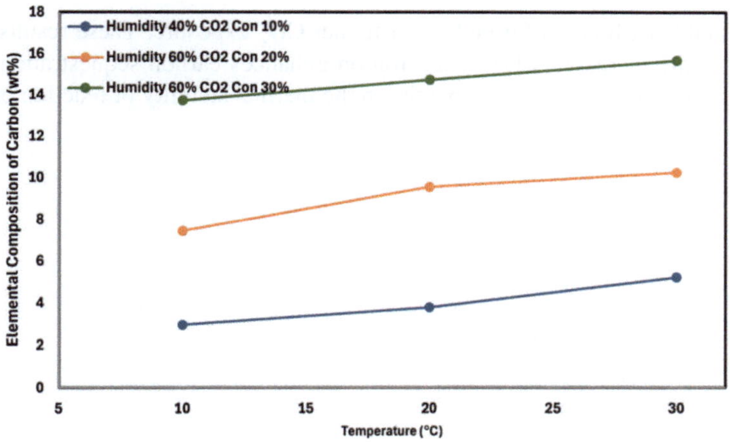

Fig. 4. Graph of Carbon Element Content versus Temperature under 10%, 20%, 30% CO2 Concentration at Relative Humidity of 60%.

Fig. 6, the sample exhibited a steady mass decrease until 700 °C, followed by rapid weight loss until 750 °C before stabilizing, consistent with the typical decomposition of calcium carbonate ($CaCO_3$) into calcium oxide (CaO) and CO_2. The 20% CO_2 sample (20_60_30), Fig. 7 showed a more complex pattern, with mass decreasing to 82.5% by 700 °C before a slight increase to 83–84%, likely due to concurrent oxidation of metallic impurities or intermediate carbonate reformation during heating. The 30% CO_2 condition (30_60_30), Fig. 8 demonstrated the most complex behavior, featuring initial mass fluctuations including a minor increase to 100.1% by 50 °C (possibly from surface re-adsorption) followed by a significant drop near 880 °C, indicating decomposition of

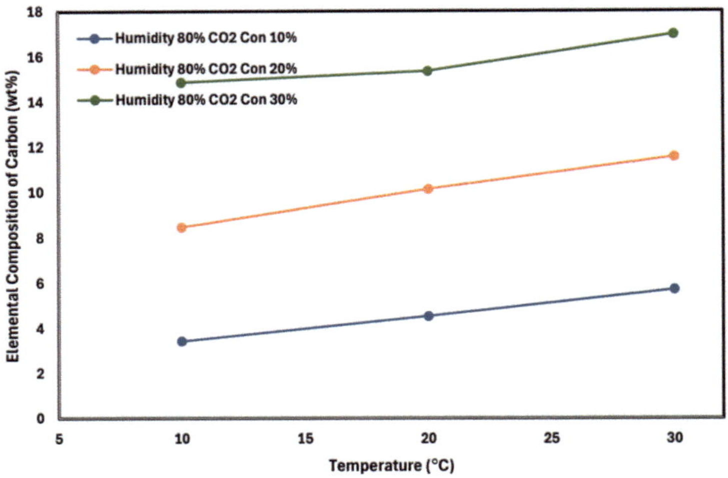

Fig. 5. Graph of Carbon Element Content versus Temperature under 10%, 20%, 30% CO2 Concentration at Relative Humidity of 80%.

thermally stable carbonates formed under higher CO_2 exposure. These results demonstrate that while increased CO_2 concentration enhances carbon sequestration in EAF slag, it also introduces greater complexity in the thermal stability of reaction products.

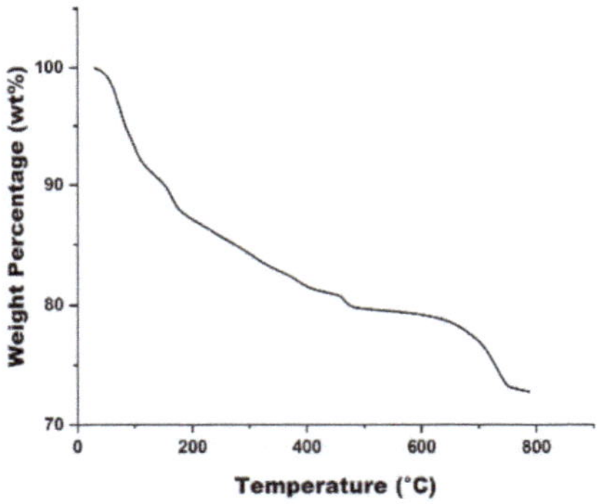

Fig. 6. TGA Gaph of Carbonated EAF Slag under 10_60_30

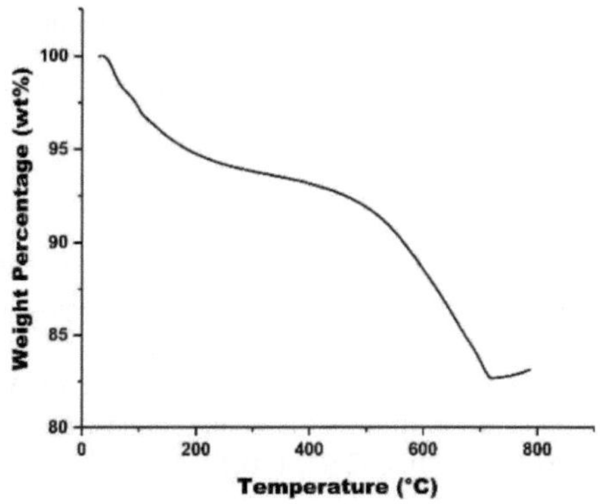

Fig. 7. TGA Gaph of Carbonated EAF Slag under 20_60_30

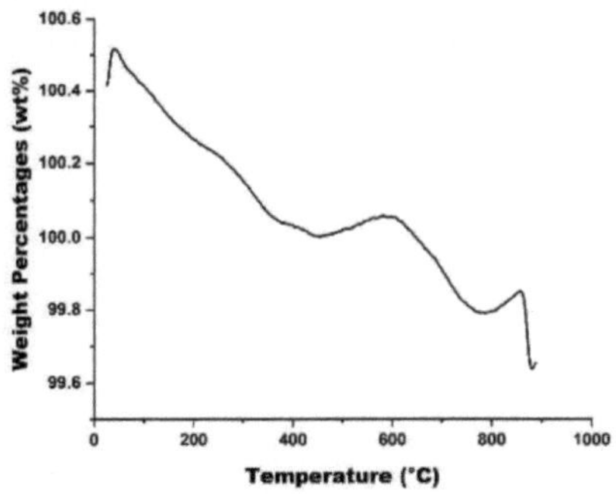

Fig. 8. TGA Gaph of Carbonated EAF Slag under 30_60_30

5 Conclusion

The thermogravimetric analysis of carbonated EAF slag under varying CO_2 concentrations (10%, 20%, and 30%) at 60% relative humidity and 30 °C revealed distinct thermal decomposition behaviors that correlate directly with carbonation conditions. The 10% CO_2 sample exhibited a straightforward mass loss pattern characteristic of calcium carbonate decomposition, while higher CO_2 concentrations introduced greater complexity in thermal stability, including mass fluctuations indicative of secondary reactions. Notably, the 20% CO_2 condition demonstrated an optimal balance between carbon sequestration efficiency and thermal stability, with minimal anomalous behavior compared to the 30% CO_2 sample. These findings highlight that while increased CO_2 exposure enhances carbon capture in EAF slag, it simultaneously affects the thermal stability of the resulting carbonation products. The results provide critical insights for industrial applications, suggesting that moderate CO_2 concentrations (around 20%) may offer the most favorable combination of carbon uptake and material performance when utilizing carbonated EAF slag in construction materials.

Acknowledgments. The authors would like to acknowledge Ann Joo Steel Berhad.

Funding. The work is supported by Universiti Tunku Abdul Rahman, Malaysia under the grant FRGS/1/2022/TK01/UTAR/02/2.

References

1. Biava, G., Depero, L.E., Bontempi, E.: Accelerated carbonation of steel slag and their valorisation in cement products: A review. Span. J. Soil Sci. **14** (2024). https://doi.org/10.3389/sjss.2024.12908
2. Omale, S.O., Choong, T.S.Y., Abdullah, L.C., Siajam, S.I., Yip, M.W.: Utilization of Malaysia EAF slags for effective application in direct aqueous sequestration of carbon dioxide under ambient temperature. Heliyon. **5**(10), e02602 (2019). https://doi.org/10.1016/j.heliyon.2019.e02602
3. Lee, H., et al.: Carbon dioxide capture and product characteristics using steel slag in a mineral carbonation plant. Processes. **11**, 1676 (2023). https://doi.org/10.3390/pr11061676
4. Wang, F., Dreisinger, D.: Acceleration of iron-rich olivine co2 mineral carbonation and utilization for simultaneous critical nickel and cobalt recovery. Minerals. **14**, 766 (2024). https://doi.org/10.3390/min14080766

Statistical Modeling and Resilience Management of Flood Mitigation Systems in Puncak Alam

Mak Jien Hui[1], Ong Wei Heng[1], Lim Ming Han[1], Lee Yee Ling[1], and Tan Ooi Kuan[2](✉)

[1] Centre of Climate Change, Disaster, Risk and Reduction (3CDRR), Department of Civil Engineering, Lee Kong Chian Faculty of Engineering and Science, Universiti Tunku Abdul Rahman, Sungai Long Campus, Cheras, Malaysia

[2] Centre of Climate Change, Disaster, Risk and Reduction (3CDRR), Department of Civil Engineering, Lee Kong Chian Faculty of Engineering and Science. and, Belt and Road Strategic Research Centre (BRSRC), Universiti Tunku Abdul Rahman, Sungai Long Campus, Cheras, Malaysia

oktan@utar.edu.my

Abstract. Recurrent flooding in Puncak Alam persists despite substantial government investment in mitigation infrastructure. This study examines how gender influences public evaluation of flood mitigation systems, assessing disparities in risk awareness, preparedness, and trust. Four variables risk perception, mitigation awareness, adaptive readiness, and trust in infrastructure were analyzed using survey data from 137 residents. SPSS was used for descriptive statistics and SmartPLS 4 for gender-stratified structural equation modeling. Results show significant gender differences: male respondents prioritized infrastructural trust and technical readiness, while females emphasized risk communication and community perception. Although adaptive readiness and risk perception were the most influential factors overall, their impact varied by gender. These findings underscore the need for gender-sensitive flood mitigation planning to address diverse community needs and improve system effectiveness in Puncak Alam.

Keywords: Resilience Management · Flood Mitigation · Gender · Perception

1 Background of the Study

Globalization and urbanization are irreversible worldwide phenomena, driven by rural-to-urban migration for better livelihoods and services. While promising economic advancement, unplanned urban expansion exacerbates environmental degradation, poverty, crime, and inequity, undermining quality of life. This interplay intensifies "grey footprints"—ecological and infrastructural strains from population growth [1]. By 2050, 70% of the global population will be urban, up from 54% in 2020. In Malaysia, this is evident in the Klang Valley (Selangor and Kuala Lumpur), home to about nine million people, reflecting uneven urban concentration [2].

Malaysia has a tropical equatorial climate with year-round heat, humidity, and high annual rainfall averaging 2,875 mm, ranking seventh globally for precipitation [3]. Its

M. A. b. Jamlos et al. (Eds.): ICSET 2025, LNNS 1897, pp. 97–104, 2026.
https://doi.org/10.1007/978-3-032-20722-7_11

climate is defined by two monsoons: Southwest (April–September) and Northeast (October–March). Anthropogenic activities like deforestation and industrial emissions disrupt these cycles, exacerbating climate variability and prolonging extreme weather. Rising precipitation, compounded by inadequate flood infrastructure, intensifies flood frequency and severity, highlighting the need for adaptive resilience.

Heavy monsoon rainfall significantly elevates flood risks with cascading economic and social impacts. Nationally, 2022 flood losses reached RM622 million in direct damages and RM6.1 billion in total economic losses, disproportionately affecting public infrastructure, housing, and agriculture [2].

In Puncak Alam, rapid urbanization and infrastructure gaps compound climate challenges. This study focuses on gender-disaggregated acceptance of smart flood mitigation systems, acknowledging that risk perception and technology trust often vary by gender. Analyzing how men and women assess strategies especially regarding risk communication, adaptive readiness, and infrastructure reliability reveals gendered priorities in resilience. These insights are vital for designing inclusive, gender-responsive policies that boost community buy-in and optimize mitigation systems, reducing socio-economic vulnerabilities. Thus, studying public perception of these systems in Malaysia remains essential.

2 Literature of the Study

Malaysia's high-income aspirations are challenged by climate resilience needs. Unplanned development and weak flood infrastructure in the Klang Valley heighten disaster vulnerability [5]. Urban flooding erodes lives, livelihoods, and public trust, hindering national progress [2].

Puncak Alam in Selangor experienced over 250 floods in 2021–2022, intensified by poor drainage [2], causing widespread losses. While adaptive strategies like smart flood mitigation are crucial, their success depends on public acceptance often neglected in top-down approaches. Community buy-in is vital in Puncak Alam, where risk awareness, trust, and technological readiness shape perceptions [6]. Gendered priorities further influence views: women stress preparedness, while men focus on infrastructure.

This study argues that improving acceptance requires addressing planning gaps and enabling gender-responsive engagement. Examining how gender shapes perceptions of mitigation efficacy can help align technical and community needs, advancing resilience and development.

Figure 1 presents the conceptual framework for analyzing gender-specific perceptions of smart flood mitigation systems in Puncak Alam. The model integrates gender as a key variable, mapping the influence of risk communication, adaptive readiness, and infrastructure trust on public acceptance. It uses structural equation modeling (SEM) to quantify gendered disparities in priorities, identifying pathways to more inclusive and effective policies.

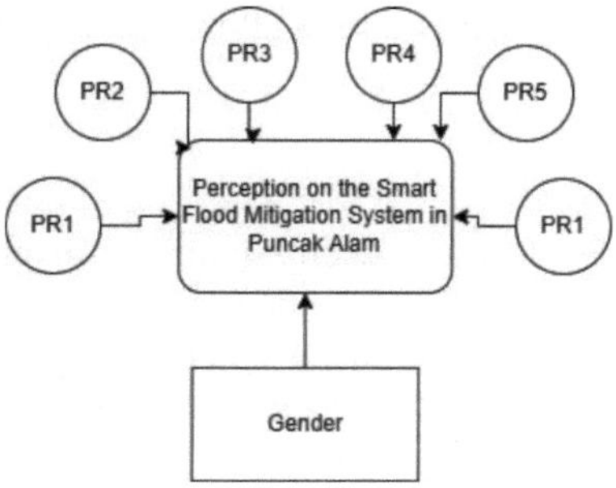

Fig. 1. Conceptual Framework

3 Methodology of the Study

This study uses a structured quantitative survey to assess public perception of smart flood mitigation systems. The questionnaire adapts Broadbent et al.'s [7] validated perception scale, known for its reliability (Cronbach's $\alpha > 0.85$). Originally developed for illness-related perceptions, its constructs perceived usefulness, ease of use, and trust have been repurposed to measure attitudes toward flood mitigation technology. The adapted instrument addresses flood risk awareness, adaptive readiness, and gender-specific priorities in Puncak Alam, enabling analysis of how demographic factors, particularly gender, shape system acceptance.

This study targeted residents of Puncak Alam and individuals who have visited the area. Given its status as a commercial zone with high population flow, including visitors in the sample was essential. Data were collected using snowball sampling to expand participation through respondent networks [8]. A pilot test ensured reliability, requiring a Cronbach's Alpha above 0.75 before full rollout. Out Of 147 initial respondents, 137 valid responses were retained—75 male (64.2%) and 62 female (35.8%). Responses were analyzed using BCa confidence interval tests for each question. Ratings were stratified by gender to enable comparative study.

I. Reliability Test

Cronbach's alpha (α) measures the internal reliability of multi-item scales, ranging from 0 to 1. Higher values indicate greater consistency among items in measuring a single construct [9]. Interpretive thresholds are: $\alpha \geq 0.95$ (excellent), 0.80–0.94 (high), 0.70–0.79 (acceptable), 0.60–0.69 (marginal), 0.50–0.59 (questionable), and < 0.50 (unacceptable for research).

II. Multicollinearity Test

Multicollinearity occurs when predictor variables in regression models are highly correlated, which can distort parameter estimates [11]. It often results from poorly specified or redundant variables. This study used the variance inflation factor (VIF) calculated in SPSS to diagnose multicollinearity across all independent variables, ensuring model stability.

III. Differences Test

Difference tests in SPSS detect statistically significant disparities between subgroups, such as gender. For comparing two independent groups (e.g., male vs. female perceptions), the Independent Samples t-test is used with normally distributed continuous data, reporting t- and p-values [12]. For ordinal data or non-normal distributions, the nonparametric Mann-Whitney U test is applied instead [13]. Both tests were conducted using SPSS's Compare Means module, with results including confidence intervals or effect sizes to assess practical significance beyond $p < 0.05$.

4 Results and Discussion

To ensure reliability, SPSS was used to conduct a reliability test verifying the collected data met established thresholds [14]. Internal consistency was assessed through Cronbach's alpha, yielding $\alpha = 0.725$ for the composite scale. All subconstructs also exceeded the 0.70 threshold, meeting social science research standards and confirming the instrument's suitability for further analysis.

A multicollinearity test was performed to identify problematic correlations among predictors [8]. As shown in Table 1, all Variance Inflation Factor (VIF) values fall within the acceptable range of 1–5, indicating no multicollinearity issues. This ensures model stability by preventing coefficient inflation and reducing estimator sensitivity.

Table 1. Multicollinearity Test

Model		Tolerance	VIF
1	(Constant)		
	P1 Smart Flood Mitigation has an impact on our daily lives.	0.493	2.030
	P2 Smart Flood Mitigation is applicable and suitable for our environment.	0.525	1.904
	P3 Smart Flood Mitigation can help address and solve flooding problems.	0.411	2.433
	P4 I am aware of and pay attention to issues related to Smart Flood Mitigation.	0.487	2.054
	P5 I have personally observed or received the advantages of Smart Flood Mitigation.	0.364	2.746
	P6 I am well-informed about how Smart Flood Mitigation works.	0.291	3.434

Kruskal-Wallis H test has been chosen to measure the gender differences that influence the perception towards the smart flood mitigation in Puncak Alam. The analysis of the data is illustrated in Table 2.

Table 2. Differences Test of Gender's Perception

		Sum of Squares	df	Mean Square	F	Sig.	Ranking
P1 Smart Flood Mitigation has an impact on our daily lives.	Between Groups	7.292	1	2.922	10.947	0.006	3
	Within Groups	51.363	135	0.368			
	Total	23.518	136				
P2 Smart Flood Mitigation is applicable and suitable for our environment.	Between Groups	5.248	1	4.118	11.769	0.007	4
	Within Groups	68.248	135	.608			
	Total	72.342	136				
P3 Smart Flood Mitigation can help address and solve flooding problems.	Between Groups	.001	1	0.001	12.891	013	5
	Within Groups	44.592	135	.440			
	Total	44.592	136				
P4 I am aware of and pay attention to issues related to Smart Flood Mitigation.	Between Groups	5.482	1	3.384	9.145	0.005	2
	Within Groups	69.474	135	0.415			
	Total	52.090	136				

(continued)

The study surveyed 137 individuals—75 men and 62 women living in or frequently traveling to Puncak Alam. Participants rated their agreement with six statements about the Smart Flood Mitigation System on a 1-to-5 Likert scale. Analysis revealed notable gender-based differences in perceptions:

(P5): Men and women differed significantly in perceived personal benefits from the system ($F = 11.460$, $p < 0.01$), suggesting men may have had greater direct exposure or involvement, leading to stronger recognition of its advantages. This disparity may reflect gender differences in roles, responsibilities, or information access.

Table 2. (*continued*)

		Sum of Squares	df	Mean Square	F	Sig.	Ranking
P5 I have personally observed or received the advantages of Smart Flood Mitigation.	Between Groups	6.080	1	6.080	11.460	0.001	1
	Within Groups	62.898	135	0.532			
	Total	53.827	136				
P6 I am well-informed about how Smart Flood Mitigation works.	Between Groups	3.384	1	3.384	0.002	0.937	6
	Within Groups	58.090	135	0.430			
	Total	61.474	136				

(P4): A significant gender difference existed in awareness and attention to system-related issues ($F = 9.145$, $p < 0.05$), indicating men and women may prioritize or engage with flood mitigation topics differently, influenced by varying interests, experiences, or societal roles.

(P1): Perceptions of the system's impact on daily life varied by gender ($F = 10.947$, $p < 0.05$), possibly due to differences in daily routines, responsibilities, or personal flood experiences, influencing its perceived relevance.

(P2): Men and women differed in their views on the system's applicability and suitability for the environment ($F = 11.769$, $p < 0.05$). This suggests that gender-based perspectives may influence opinions on how well the system fits within the local environmental context, potentially reflecting differing values or priorities regarding environmental interventions.

(P3): Gender differences emerged in beliefs about the system's ability to solve flooding problems ($F = 12.891$, $p < 0.05$), possibly reflecting varying levels of confidence or trust in technological solutions between genders, influenced by past experiences or societal narratives.

(P6): No significant gender difference was found in knowledge of how the system works ($F = 0.02$, $p = 0.937$), indicating similar levels of understanding between men and women and suggesting equal access to information or educational resources.

These findings align with research showing gender-based differences in flood risk perception. For example, a study in Ireland found men were more aware of well locations and past water testing than women, underscoring the need for gender-focused communication in flood risk management.

5 Conclusion and Recommendation

Based on data from flood-prone Puncak Alam, Malaysia, this study provides critical insights into public acceptance of smart flood mitigation systems. Gender differences emerged in perceptions of system effectiveness: while both genders acknowledged benefits from past experience, their confidence in the system's capacity to resolve floods varied. This underscores the need for inclusive public engagement.

Although general awareness and understanding of the system were high, readiness more than perception was identified as the key factor influencing public acceptance. The study's gender-imbalanced sample suggests future research should aim for balanced sampling to better examine behavioral differences and avoid bias.

From a policy perspective, authorities should implement targeted public education campaigns in Puncak Alam that not only raise awareness but also build readiness, especially among groups with lower trust or understanding. Analyzing acceptance through the lens of gender and behavioral readiness offers actionable insights for enhancing community resilience and advancing smart flood mitigation in Puncak Alam and other climate-vulnerable regions.

This research contributes to smart city and disaster management literature by positioning public perception and behavioral readiness as essential pillars for successful smart flood technology deployment.

Acknowledgments. This research was supported by the Ministry of Higher Education (MoHE) Malaysia through the Fundamental Research Grant Scheme project (Project Number: FRGS/1/2022/TK10/UTAR/03/1). The authors are also grateful to the Wijadi Resources Sdn. Bhd (Co. No. 647965-P) Industrial Grant Offer for Engineering Based Research (UTAR Vote No: 8269/0001), Mr. Chong Jia Chen and Dr. Ong Wei Heng (Postdoc) in contributing their support in this study.

References

1. Fleck, A.: Infographic: The World's Next Megacities. Statista Infographics, Hamburg (2023)
2. DOSM: Department of Statistics Malaysia. DOSM. https:Www.dosm.gov.my, Putrajaya (2022)
3. Nico: List of Top 10 Countries with Most Rainfall in the World-Basic Planet. Bionetwork Ltd, Basic Planet, UK (2022)
4. Fawzy, S., Osman, A.I., Doran, J., Rooney, D.W.: Strategies for ,mitigation of climate change: A Review. Environ. Chem. Lett. **18**(18), 2069–2094 (2020)
5. Adnan, N., Nordin, S.M., Bahruddin, M.A., Tareq, A.H.: A state-of-the-art review on facilitating sustainable agriculture through green fertilizer technology adoption: Assessing farmers behavior. Trends Food Sci. Technol. **86**, 439–452 (2019)
6. Broadbent, E., Wilkes, C., Koschwanez, H., Weinman, J., Norton, S., Petrie, K.J.: A systematic review and meta-analysis of the Brief Illness Perception Questionnaire. Psychol. Health. **30**(11), 1361–1385 (2015)
7. Sekaran, U., Bougie, R.: Research Methods for Business: A Skill-Building Approach, 8th edn. John Wiley & Sons, Inc, Hoboken, NJ (2020)
8. Tavakol, M., Dennick, R.: Making sense of Cronbach's Alpha. Int. J. Med. Educ. **2**(2), 53–55 (2011). https://doi.org/10.5116/ijme.4dfb.8dfd

9. Ghasemi, A., Zahediasl, S.: Normality tests for statistical analysis: A guide for non-statisticians. Int. J. Endocrinol. Metab. **10**(2), 486–489 (2012)
10. Uyanık, G.K., Güler, N.: A study on multiple linear regression analysis. Procedia. Soc. Behav. Sci. **106**(1), 234–240 (2013). https://doi.org/10.1016/j.sbspro.2013.12.027
11. Field, A.: Discovering Statistics Using IBM SPSS Statistics. SAGE Publications Ltd, London (2024)
12. Pallant, J.: SPSS Survival Manual, 7th edn. Routledge, London (2020)
13. Mishra, P., Pandey, C., Singh, U., Gupta, A.: Scales of measurement and presentation of statistical data. Ann. Card. Anaesth. **21**(4), 419–422 (2018)

Intelligent Analytical Techniques for Humidity Sensing, Natural Antioxidants and Urinalysis

Nurain Natasha Badarozaman, Muhammad Syahmi Khairul Azman, Nur Fatin Nadhirah Ab Aziz, and Siti Amira Othman(✉)

Faculty of Applied Sciences and Technology, Universiti Tun Hussein Onn Malaysia, Pagoh, Johor, Malaysia
sitiamira@uthm.edu.my

Abstract. Detection and sensing of humidity play a crucial role across various industries, including agriculture, building maintenance, environmental monitoring, and industrial process control. Uncontrolled humidity levels can cause structural degradation, reduced shelf life of products, and health and safety risks. With the rise of IoT-based and smart systems, there is growing demand for real-time, precise, and integrable humidity sensors. However, conventional sensors often suffer from limitations in response time, selectivity, and long-term stability, particularly in dynamic or harsh environments, creating a gap for the development of cost-effective, programmable, and high-performance alternatives. Similarly, the cosmetic industry relies heavily on synthetic antioxidants such as butylated hydroxytoluene (BHT) and butylated hydroxyanisole (BHA) to improve product stability and protect skin from oxidative damage by neutralizing free radicals. Despite their effectiveness, these synthetic compounds have been associated with adverse health effects including skin irritation, allergic reactions, and potential long-term toxicity, prompting interest in safer, natural antioxidant alternatives. In parallel, urinalysis remains a fundamental diagnostic method for assessing bodily health, as the urinary system plays a central role in eliminating waste and regulating blood composition. Visual examination of urine—considering aspects such as color, clarity, and concentration along with characterization of constituents, odor, and pH, provides valuable insights into metabolic and pathological conditions. This preliminary review highlights emerging trends and challenges in humidity sensing, the shift toward natural antioxidants in cosmetics, and the diagnostic potential of urinalysis, reflecting the interconnected importance of innovation and health monitoring in science and industry.

Keywords: Humidity · Antioxidants · Urinalysis · Analytical

1 Introduction

Biosensors are analytical devices that combine a biological sensing element and a physicochemical transducer. While initially used for the detection of biological analytes, biosensors presently adopt materials and mechanisms able to sense environment parameters, such as humidity. Polyaniline (PANI), a conducting polymer, has been in

M. A. b. Jamlos et al. (Eds.): ICSET 2025, LNNS 1897, pp. 105–111, 2026.
https://doi.org/10.1007/978-3-032-20722-7_12

great demand as a biosensing material due to its unique electrical properties, environmental robustness, low price, and ease of preparation [1]. In emeraldine salt form, PANI is well known for high proton conductivity and better interaction with water molecules. The characteristics of PANI make it a good promising contender for humidity sensors because the polymer's resistance changes immensely on adsorbing water. PANI is readily synthesized and processed into thin films that are easily applied as coatings on any number of substrates. Its flexibility in molecular structure and ability to be doped allow it to be tailored to the specific sensing requirements. Studies demonstrated that sensors based on PANI were capable of rapidly and sensitively monitoring changes in ambient humidity levels [2, 3]. Furthermore, its complementarity with flexible and miniaturized platforms enables it to be adopted in novel devices, from Internet of Things (IoT) devices to wearable electronics [4–6].

2 Biosensor Characterization: UV-Vis., FTIR, Four Point Probe, SEM

Characterization techniques are required to verify the quality, structure, and performance of PANI-based humidity sensors. Characterization techniques provide details on the chemical composition, morphology, and electrical properties of the sensor, enabling the researcher to fine-tune the sensor for requisite performance UV-V is spectroscopy involves measuring the absorption of ultraviolet and visible light by the sample to observe electronic transitions. It provides information on the conjugation of the polymer and the presence of specific oxidation states. For PANI, UV-V is spectra typically have characteristic peaks at ca. 320 nm (π–π^* transitions) and 600–650 nm (polaron transitions) (Nguyen et al., 2014). These peaks confirm the formation of the emeraldine salt form and doping level of PANI, which are critical parameters for sensor functionality. By comparing the UV-Vis spectra before and after exposure to humidity, the modification in the electronic structure of the polymer and its interaction with water molecules can also be studied by researchers Fourier-transform infrared (FTIR) spectroscopy determines the vibrational modes of chemical bonds in the sample by detecting the absorption of infrared radiation. It reveals the presence of specific functional groups and structural features in the polymer.

FTIR spectroscopy of PANI typically shows characteristic peaks for N–H stretching (~3400 cm^{-1}), C=N stretching of quinoid rings (~1580 cm^{-1}), and C=C stretching of benzenoid rings (~1500 cm^{-1}) [7]. These peaks provide confirmation of PANI synthesis and provide information about its chemical structure, including the interaction with water molecules that influence the humidity sensing mechanism [8].

Four-point probe technique involves flowing a constant current through the outer probes and detecting the voltage drop across the inner probes to calculate the sheet resistance of the material. This method eliminates errors caused by contact resistance, enabling electrical conductivity to be determined more accurately. In PANI-based sensors, four-point probe measurement is used to determine the sheet resistance of the thin film before and after exposure to humidity. The measurements allow calculation of conductivity changes and provide information on the performance of the sensor at different humidity conditions. The technique plays an important role in determining the reproducibility and sensitivity of the sensor [9].

Scanning Electron Microscopy (SEM) is a highly effective imaging technique on the basis of a focused high-energy electron beam to generate a variety of signals at the surface of solid samples. Signals comprise secondary electrons, backscattered electrons, and X-rays that are collected to create high-resolution images of the surface topography and chemistry of the sample. SEM provides morphological and structural information such as grain size, surface homogeneity, and porosity and is thus highly suitable for the characterization of thin films and nanostructured materials [1]. In the case of polyaniline (PANI) thin films, SEM is used to analyse the surface morphology, film thickness, and film homogeneity deposited on a substrate. Homogeneously formed porous film is essential for successful water molecule adsorption in humidity sensing. The porosity enhances the surface area with strong interaction between the polymer and ambient moisture, improving the sensor sensitivity. In addition, SEM imaging helps to verify the coverage and continuity of the PANI film as surface defects or uneven coatings will degrade electrical performance [10]. Samples are usually coated with a thin conductive film (such as gold) to prevent charging and enhance image quality for SEM analysis.

3 Dragon Fruit Peel as a Source of Bioactivity

Bioactive compounds are substances present in foods and plants that, upon consumption, can prevent lipid peroxidation owing to their antioxidant properties, modulate immune responses to antigens, and exhibit neuroprotective, anticancer, and anti-inflammatory effects, offering health advantages to consumers. To investigate the antioxidant properties of powdered dragon fruit peel for cosmetic applications, a systematic experimental approach will be adopted. Dragon fruit holds strong anti-oxidant potential as it contains polyphenols, carotenoids, flavonoids, tocopherol, and vitamin C. The polyphenols are mostly responsible for this activity due to their hydrogen-donor ability to scavenge the free radicals [11]. Not only that but fresh dragon fruit peels (Hylocereus spp.) will be collected from local markets or fruit processing centers. The peels will be thoroughly washed with distilled water to remove any surface impurities and then sliced into smaller pieces. These pieces will be dried using a hot air oven at a controlled temperature of 50–60 °C until they reach a constant weight. Once dried, the peels will be ground into a fine powder using a laboratory grinder and stored in airtight containers at room temperature to preserve their bioactive integrity until further use [12].

The powdered peel will then be subjected to solvent extraction to isolate antioxidant compounds. Ethanol or methanol (80%) will be used as the primary solvent due to their effectiveness in extracting phenolic and flavonoid compounds. The extraction process will involve maceration or Soxhlet extraction over a period of 24 to 48 h, followed by filtration and solvent removal using a rotary evaporator to obtain concentrated extracts. Preliminary phytochemical screening will be conducted on the extracts to qualitatively detect the presence of flavonoids, phenols, tannins, and other antioxidant-related compounds [13].

To evaluate the antioxidant potential of the dragon fruit peel extract, several in vitro assays will be performed. These include the DPPH (2,2-diphenyl-1-picrylhydrazyl) radical scavenging assay to measure free radical inhibition, the ABTS assay for assessing the ability to scavenge ABTS radicals, and the FRAP (ferric reducing antioxidant power) assay to evaluate the reducing potential of the extract. Additionally, total phenolic content (TPC) will be determined using the Folin-Ciocalteu method and expressed as mg gallic acid equivalent per gram of extract (mg GAE/g), while total flavonoid content (TFC) will be measured via the aluminum chloride colorimetric method and expressed as mg quercetin equivalent per gram of extract (mg QE/g). All assays will be conducted in triplicate, and absorbance readings will be taken using a UV-Vis spectrophotometer.

Dragon Fruit Characterization: FTIR, UV-Vis., SEM Fourier Transform Infrared (FTIR) Spectroscopy was used in this study to investigate the functional groups responsible for antioxidant properties in powdered dragon fruit (*Hylocereus* spp.) peel. FTIR is a non-destructive analytical technique that allows for the rapid identification of chemical bonds within organic compounds by measuring their vibrational transitions. This method is especially valuable for screening plant-based antioxidants, as it provides qualitative insight into the presence of hydroxyl (-OH), carbonyl (C=O), and aromatic ring structures—common functional groups found in phenolic compounds, flavonoids, and betalains, which are associated with antioxidant activity.

The powdered peel sample was air-dried, finely ground, and prepared either by direct ATR (attenuated total reflectance) or by KBr pellet method depending on sample consistency. Spectral data were collected across the range of 4000–400 cm^{-1} using an FTIR spectrometer. The key absorption peaks were interpreted to identify characteristic antioxidant-related structures. For instance, broad peaks around 3200–3600 cm^{-1} typically correspond to O-H stretching in phenolic groups, while peaks near 1600 cm^{-1} may indicate C=C stretching from aromatic rings. These observations help confirm the presence of bioactive antioxidant components within the peel matrix.

FTIR has been widely applied in previous studies to assess antioxidant potential in various plant species due to its speed, reproducibility, and minimal sample preparation requirements [14]. In the context of this study, FTIR provided a chemical fingerprint of dragon fruit peel, supporting the hypothesis that it contains antioxidant-active compounds. This analytical step served as a foundation for further evaluation of antioxidant activity using UV-V is spectroscopy and DPPH assay. The antioxidant activity of the powdered dragon fruit peel extract will be assessed using UV-V is spectroscopy and the DPPH (2,2-diphenyl-1-picrylhydrazyl) assay because of their effectiveness and precision in measuring free radical scavenging capacity [15]. The DPPH assay will consist of creating a methanolic solution of DPPH, generally at a concentration of 0.1 mM. Different concentrations of the dragon fruit peel extract will be combined with the DPPH solution in test tubes or microplates and incubated in the dark at room temperature for 30 min to facilitate the reaction [16]. The decrease of the DPPH radical, shown by a shift from purple to yellow, signifies the sample's antioxidant properties [17, 18].

4 Urinalysis

Urine was varied in appearance, depending principally upon a body's level of hydration, as well as other factors. Normal urine was a transparent solution ranging from colourless to amber but is usually a pale yellow. There is some indication about the various appearance of urine colour; no colour indicates that a lot drinking of water or over-hydration, a cloudy or white urine may be due to the vaginal discharge, dehydration or due to kidneys stones, Researchers prove that a transparent yellow urine is a sign of normal or hydrated person, a dark yellow or brown urine due to certain medicines or bilirubin from liver, green urine typically caused by food colours but it is also related to some diseases such as diabetes, reddish or pinkish urine may also due to certain food colours or some problems such as prostate problem [19–21].

Urine turbidity affected by some handling and transportation conditions. If a urine specimen is unpreserved, it can become cloudy from bacterial overgrowth. Refrigeration lead to faster precipitation of amorphous urates or phosphates resulting in cloudiness. This condition became slower when the specimen was brought to room temperature. The collection time and storage conditions of the specimen were reviewed to determine that the cloudiness has been caused by storage conditions. As for abnormal colouration, a lack of clarity leads to test result inaccuracies. Clarity and concentration of urine were related to its colour. The darker the colour of urine, the lower the clarity and the higher the concentration. For the brighter, the clarity and concentration were vice versa. Urinalysis by urea showed that the concentrations of urea was very high in the urine. When urea was diluted with water, it decayed into ammonia and bicarbonate as chemical equation shown below which associated with the increase in pH value [22, 23].

Instruments to Measure Urinary pH The determination of urine pH is using three instruments which is pH meter, pH paper and reagent strip (dipstick). For exact pH values, the pH meter should be used [24, 25]. However, since this instrument is rather complex, it is not used very often in urinalysis due to time limitations. The pH of urine can be determined using indicator paper such as Nitra zine paper. The tip of the paper is dipped into the specimen, or a drop may be placed on the paper. The resulting colour is compared with the standard chart supplied with the paper. Nitra zine paper has a range of 4.5 to 7.5 [26]. The colour varies from yellow at 4.5 to blue at 7.5.

For reagent strips (dipstick). Some multiple reagent strips include a test region with the indicator's methyl red and bromothymol blue. This combination of indicators gives a pH range from 5.0 to 8.5 [27]. The resulting colours range from orange to blue. Care should be taken to follow the directions supplied by the manufacturer. Excessive immersion time will wash the chemicals out of the test regions. This can affect the results of the readings on one or all the test regions. Studies have shown that dipstick measurement of pH are inherently inaccurate when compared to those taken using pH meter, so when an accurate measurement of urine pH is critical for clinical decision making, for example when evaluating cases of urolithiasis, the measurement should be made with a pH meter [28, 29]. The pH determination of a specimen will be reported as the numerical value obtained, or the relative degree of acidity or alkalinity will depend upon the procedure used.

5 Conclusion

Humidity detection, antioxidant development, and urinalysis each play pivotal roles in enhancing safety, health, and efficiency across diverse sectors. The growing need for real-time, reliable, and integrable humidity sensors reflects the increasing adoption of smart and IoT-enabled systems, while the limitations of conventional sensors drive the pursuit of high-performance, cost-effective alternatives. In the cosmetic industry, the shift from synthetic antioxidants toward natural, safer compounds addresses both consumer health concerns and product stability. Meanwhile, urinalysis continues to serve as a vital diagnostic tool, offering non-invasive insights into metabolic and pathological conditions. Collectively, these fields underscore the importance of innovation, safety, and sustainability, highlighting ongoing efforts to meet the evolving demands of technology, health, and the environment.

Acknowledgement. The authors would like to thank the Faculty of Applied Sciences and Technology for facilities provided that make the research possible.

References

1. Singh, P., Shukla, S.K.: Structural, optical and electrical properties of ZnO thin films grown by sol–gel spin coating method. Surf. Interfaces. **18**, 100410 (2020)
2. Alhartomy, O.A., Almessiere, M.A., Ahamad, T., Khan, M.: A new facile synthesis of ultra fine magnesium oxide nanowires and their application in chemical sensing. Sens. Actuators B Chem. **221**, 1120–1127 (2015)
3. Rahman, M.M., Ahammad, A.J.S., Jin, J.H., Ahn, S.J., Lee, J.J.: A comprehensive review of glucose biosensors based on nanostructured metal-oxides. Mater. Today Commun. **33**, 104634 (2022)
4. Nguyen, H.D., et al.: pH sensitivity of emeraldine salt polyaniline and poly(vinyl butyral) blend. Adv. Nat. Sci. Nanosci. Nanotechnol. **5**, 045001 (2014)
5. Grotthuss, C.J.T.: Memoir on the mechanism of electrochemical decomposition. Ann. Chim. **58**, 54–74 (1806)
6. Li, X., Wang, Y., Liu, Y., Chen, Q.: Preparation and properties of poly(vinyl alcohol)/cellulose nanocrystal nanocomposite films with enhanced mechanical strength and barrier performance. J. Appl. Polym. Sci. **136**(27), 47902 (2019)
7. Krishna, D.N.G., Philip, J.: Review on surface-characterization applications of X-ray photoelectron spectroscopy (XPS): Recent developments and challenges. Appl. Surf. Sci. Adv. **12**, 100332 (2022)
8. Li, T., Zhang, J., Wang, D.: A novel electrochemical sensor based on poly(3,4-ethylenedioxythiophene) film modified electrode for the detection of dopamine. Sens. Actuators B Chem. **160**, 542–548 (2011)
9. Radhakrishnan, S., Sethulekshmi, A.: Studies on the thermal and mechanical properties of polylactic acid/polyethylene glycol blends. Int. J. Polym. Anal. Charact. **19**(5), 345–356 (2014)
10. Zhu, L., Wang, C., Xie, Z.: Microstructural characterization and mechanical properties of Ti–Nb–Zr alloys processed by thermomechanical treatment. Mater. Charact. **169**, 110601 (2020)
11. Chatterjee, D., et al.: CO_2 to dimethyl ether (DME): Structural and functional insights of hybrid catalysts. Hybrid Adv. **4**(3), 100185 (2024)
12. Som, A.M., Ahmat, N., Abdul Hamid, H.A., Azizuddin, N.: A comparative study on foliage and peels of *Hylocereus undatus* (white dragon fruit) regarding their antioxidant activity and phenolic content. Heliyon. **5**(2), e01244 (2019)
13. Pandel, R., Poljšak, B., Godic, A., Dahmane, R.: Skin photoaging and the role of antioxidants in its prevention. ISRN Dermatol. **2013**, 930164 (2013)

14. Pilarski, R., et al.: Antioxidant activity of ethanolic and aqueous extracts from *Uncaria tomentosa* (Willd.) DC. J. Herb. Med. **3**(3), 149–156 (2013)
15. Thaipong, K., et al.: Comparison of ABTS, DPPH, FRAP, and ORAC assays for estimating antioxidant activity from guava fruit extracts. J. Food Compos. Anal. **19**(6–7), 669–675 (2006)
16. Sasidharan, S., et al.: Extraction, isolation and characterization of bioactive compounds from plants' extracts. Afr. J. Tradit. Complement. Altern. Med. **8**(1), 1–10 (2010)
17. Machado, M., Silva, S., Costa, E.M.: Byproducts as a sustainable source of cosmetic ingredients. Appl. Sci. **14**(22), 10241 (2024)
18. Bassey, E.J., Cheng, J.-H., Sun, D.-W.: Comparative elucidation of bioactive and antioxidant properties of red dragon fruit peel as affected by electromagnetic and conventional drying approaches. Food Chem. **439**, 138118 (2023)
19. Andrade, R.C., et al.: Valorisation of fruit by-products as sustainable sources of bioactive compounds: Advances, challenges, and future perspectives. Trends Food Sci. Technol. **153**, 104721 (2024)
20. Ali, A., Zhang, N., Santos, R.M.: Mineral characterization using scanning electron microscopy (SEM): A review of the fundamentals, advancements, and research directions. Appl. Sci. **13**(23), 12600 (2023)
21. Udert, K.M., Larsen, T.A., Biebow, M., Gujer, W.: Urea hydrolysis and precipitation dynamics in a urine-collecting system. Water Res. **37**(11), 2571–2582 (2003)
22. Ratkalkar, V.N., Kleinman, J.G.: Mechanisms of stone formation. Curr. Mol. Med. **9**(2), 187–193 (2011)
23. Langley, S., Fry, C.: The role of the bladder in urinary continence. Br. J. Urol. **79**(1), 8–11 (1997)
24. Frassetto, L., Kohlstadt, I.: Diets alkaline and acid: Do they matter? Am. Fam. Physician. **84**(11), 1234–1242 (2011)
25. Heiserman, L.D.: Clinical Laboratory Principles & Practices: Urinalysis. F.A. Davis Company, Philadelphia (2015)
26. Pratumvinit, B., et al.: Urinary stone analysis in Thailand: A study of 1,500 cases. Clin. Chim. Acta. **426**, 46–50 (2013)
27. Villiers, E., Blackwood, L.: BSAVA Manual of Canine and Feline Clinical Pathology, 2nd edn. BSAVA, Gloucester (2005)
28. Trautvetter, U., et al.: Effect of calcium phosphate and vitamin D_3 supplementation on bone remodelling and metabolism of calcium, phosphorus, magnesium and iron. Nutr. J. **13**(6), 1–9 (2014)
29. Isabell, I.K., et al.: Phosphorus recovery from sewage sludge ash by thermochemical treatment. Water Res. **95**, 361–370 (2016)

Evaluation of Palm Oil and Fly Ash Suspensions for Sustainable Machining of Aluminum: A Taguchi Approach

Ainusyafiqah Shaaroni[1], Pay Jun Liew[1(✉)], Raja Izamshah Raja Abdullah[1], and Jingsi Wang[2]

[1] Fakulti Teknologi dan Kejuruteraan Industri dan Pembuatan, Universiti Teknikal Malaysia Melaka, Hang Tuah Jaya, 76100 Durian Tunggal, Melaka, Malaysia
payjun@utem.edu.my

[2] Marine Engineering College, Dalian Maritime University, 1 Linghai Road, Ganjingzi District, Dalian 116026, People's Republic of China

Abstract. This study aimed to investigate the effects of machining parameters and type of cutting on the machining performance of Al6061-T6 under Minimum Quantity Lubrication (MQL). A Taguchi L18 orthogonal array was employed to analyze the influence of type of cutting fluid, cutting speed, feed rate, and depth of cut on surface roughness and cutting force. The results showed that cutting speed emerged as the most influential factor for both surface roughness and cutting force. Surface finish was primarily affected by cutting speed and depth of cut, while cutting force was most influenced by cutting speed and type of cutting fluid. Compared to pure palm oil, the fly ash suspension reduced surface roughness and reduced cutting force by up to 13% and 12% respectively, owing to its superior tribological properties. The optimal combination for minimizing surface roughness was achieved at 3500 rpm, 600 mm/min feed rate, and 0.55 mm depth of cut, whereas the minimum cutting force occurred at 2500 rpm, 400 mm/min, and 0.55 mm. Overall, the results highlight the effectiveness of fly ash as a lubricant additive, demonstrating its potential to enhance machining performance and offering a sustainable solution for clean manufacturing.

Keywords: Cutting fluid · Minimum quantity lubricant · Fly ash · Taguchi Method · Milling · Aluminium alloy

1 Introduction

Aluminum alloys such as AL6061-T6 are widely used in aerospace, automotive, and structural applications due to their high strength-to-weight ratio, corrosion resistance, and good machinability [1]. However, machining challenges persist, particularly in maintaining surface finish and tool life because of built-up edge (BUE) formation, especially under dry or high-speed conditions [2]. Cutting speed, feed rate, and depth of cut are key parameters influencing surface roughness, cutting force, and tool wear. Bushan [3]

M. A. b. Jamlos et al. (Eds.): ICSET 2025, LNNS 1897, pp. 112–119, 2026.
https://doi.org/10.1007/978-3-032-20722-7_13

reported reduced tool life at higher cutting speeds when machining AA7075/SiC composites, while Imran et al. [4] found that increased cutting speed improved surface finish. Ikhries and Al-Shawabkeh [5] identified depth of cut as the dominant factor affecting MRR and surface roughness in milling 7075-T6, and El-Deen [6] showed that all cutting parameters significantly influenced machining performance using a Taguchi–grey relational approach.

The Taguchi method is widely used in machining research to optimize multiple parameters efficiently, as its orthogonal arrays allow systematic evaluation with fewer trials than conventional DOE approaches [7]. It is particularly effective in MQL studies, where interactions among cutting speed, feed rate, depth of cut, and lubricant type must be considered. Using S/N ratios, Taguchi analysis helps identify robust parameter settings and assess process variability [8].

Cutting fluid selection also plays an important role in machining efficiency. Conventional cutting fluid poses environmental concerns, leading to increasing interest in biodegradable vegetable oils [9]. Their performance can be enhanced by nanoparticles which improve lubrication and thermal behavior [10]. Tiwari et al. [11] reported significant improvements using nano-enhanced vegetable oils. Fly ash, a waste product rich in Al_2O_3 and SiO_2, offers similar tribological benefits but remains underexplored in biodegradable cutting fluids [12]. Therefore, this study investigates fly ash–enhanced vegetable oil and machining parameters on surface roughness and cutting force in milling aluminum alloy using a Taguchi L18 design.

2 Materials and Methods

This study examines the effects of cutting parameters and cutting fluid type on the milling performance of aluminum alloy. Experiments were conducted on a HAAS CNC machine using an uncoated carbide insert (APET103504PDFR-S) and Al6061-T6 workpiece (100 × 45 × 30 mm. All runs were performed under MQL with the nozzle directed at the cutting zone. Surface roughness and cutting force were measured using a Mitutoyo tester and a Kistler dynamometer. The experimental setup is shown in Fig. 1. Fly ash from Jimah Power Plant was sieved to 20 μm and had an average size of ~13 μm. A 1.2 wt% suspension (0.6 wt% fly ash +0.6 wt% SLS) was prepared using a two-step ultrasonication method, with total solids calculated using Eq. (1) [13]. The experiments were designed in Minitab using a Taguchi L18 orthogonal array (Table 1), with parameter ranges set according to Sumitomo recommendations for aluminum milling. All tests were performed under a constant MQL flow rate of 1.5 ml/min and a tool travel of 5000 mm. Each L18 trial was conducted once, and a "smaller-the-better" S/N ratio was applied to both responses.

$$\text{wt\%} = (\text{Wn} + \text{W})/(\text{Wn} + \text{Ws} + \text{Wf}) \times 100 \quad (1)$$

Where; Wn = weight of fly ash powder
Ws = weight of surfactant (SLS)
Wf = weight of palm oil

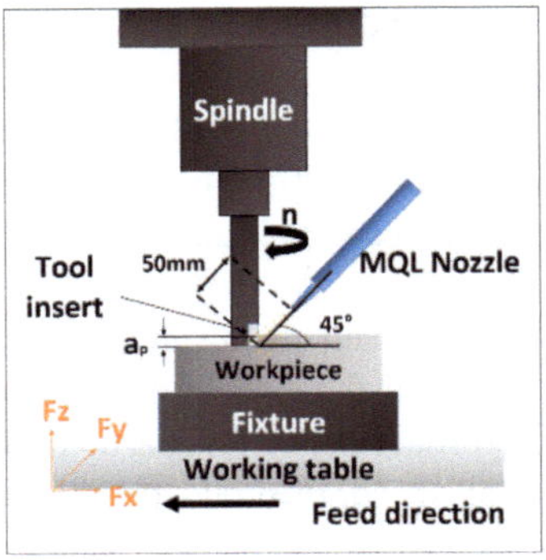

Fig. 1. Schematic Diagram of Experimental Set-up

Table 1. Cutting Parameters and Their Levels

Factor	Level 1	Level 2	Level 3
Type of cutting fluid (A)	Palm oil	Fly ash suspension	–
Cutting speed, rpm (B)	1500	2500	3500
Feed rate, mm/min (C)	400	600	800
Depth of cut, mm (D)	0.40	0.55	0.70

3 Results and Discussion

Table 2 showing the measured responses and S/N ratios for each run.

Table 2. Experimental Results and Their S/N Ratio

Run	Type of cutting fluid	Cutting speed	Feed rate	Depth of cut	Ra, µm	SNRA of Ra	Fc, N	SNRA of Fc
1	palm oil	1500	400	0.4	1.02	−0.172	106.04	−40.509
2	palm oil	1500	600	0.55	0.47	6.558	104.19	−40.356
3	palm oil	1500	800	0.7	0.89	1.012	98.65	−39.882
4	palm oil	2500	400	0.4	0.27	11.373	102.3	−40.198
5	palm oil	2500	600	0.55	0.30	10.458	100.64	−40.055
6	palm oil	2500	800	0.7	0.83	1.618	97.93	−39.819
7	palm oil	3500	400	0.55	0.19	14.425	101.04	−40.090
8	palm oil	3500	600	0.7	0.57	4.883	109.61	−40.797
9	palm oil	3500	800	0.4	0.67	3.479	111.35	−40.934

(continued)

Table 2. *(continued)*

Run	Type of cutting fluid	Cutting speed	Feed rate	Depth of cut	Ra, μm	SNRA of Ra	Fc, N	SNRA of Fc
10	fly ash suspension	1500	400	0.7	1.01	−0.086	99.2	−39.930
11	fly ash suspension	1500	600	0.4	0.75	2.499	91.64	−39.242
12	fly ash suspension	1500	800	0.55	0.52	5.680	89.26	−39.013
13	fly ash suspension	2500	400	0.55	0.43	7.331	59.41	−35.477
14	fly ash suspension	2500	600	0.7	0.55	5.193	68.61	−36.727
15	fly ash suspension	2500	800	0.4	0.39	8.179	68.24	−36.681
16	fly ash suspension	3500	400	0.7	0.47	6.558	98.55	−39.873
17	fly ash suspension	3500	600	0.34	0.17	15.391	132.74	−42.460
18	fly ash suspension	3500	800	0.55	0.24	12.396	115.98	−41.288

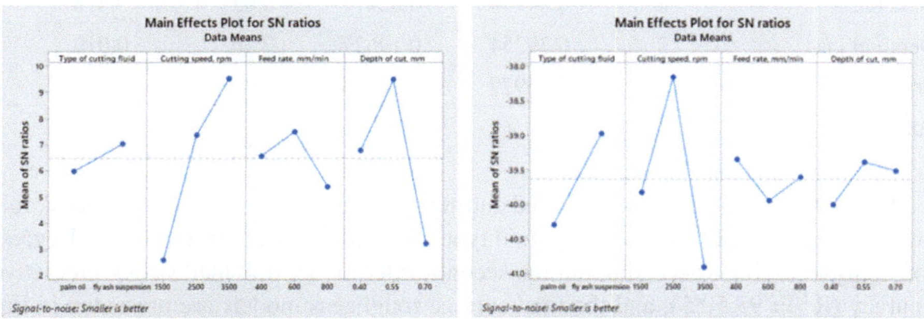

(a) (b)

Fig. 2. Main Effects Plot for S/N Ratios (a)Surface Roughness (b) Cutting Forces

3.1 Surface Roughness

Surface roughness is a key indicator of machining quality. Table 3 presents the S/N response table, while Fig. 2(a) shows the main effects plot. Cutting speed had the strongest influence on surface roughness, followed by depth of cut, feed rate, and cutting fluid type. The optimal condition was achieved using the fly ash suspension at 3500 rpm,

600 mm/min, and 0.55 mm. Increasing cutting speed reduced surface roughness due to lower built-up edge formation, improved shear action, and reduced tool–material adhesion. The selected feed rate and depth of cut likely balance material removal and tool stability, minimizing rubbing, vibration, and chatter. Medium feed rates reduce chip thickness and ploughing, while very low feed rates may promote BUE formation. The fly ash suspension further improved Ra values through micro-polishing and a protective tribological film [14].

Table 3. Response Table for S/N Ratios of Surface Roughness

Level	Type of cutting fluid	Cutting speed	Feed rate	Depth of cut
1	5.959	2.582	6.571	6.791
2	7.015	7.358	7.497	9.474
3		9.522	5.394	3.196
Delta	1.056	6.940	2.103	6.278
Rank	4	1	3	2

Table 4. Analysis of Variance of Surface Roughness

Source	DF	Adj SS	Adj MS	F-Value	P-Value
Type of cutting fluid	1	0.02569	0.02569	0.99	0.343
Cutting speed	2	0.51701	0.25851	9.95	0.004
Feed rate	2	0.04954	0.02477	0.95	0.418
Depth of cut	2	0.39254	0.19627	7.56	0.010
Error	10	0.25979	0.02598		
Total	17				

ANOVA results (Table 4) confirmed cutting speed as the most significant factor, followed by depth of cut, while cutting fluid type and feed rate were insignificant. Regression analysis (Table 5) showed that the second-order model provided strong predictive accuracy ($R^2 = 93.50\%$), and the final surface roughness models are presented in Eq. (2) and Eq. (3).

Table 5. Regression Statistic of Surface Roughness

Regression statistic	1st order model	2nd order model
R2	46.57%	93.50%
R2 (adj)	30.13%	84.21%
Standard error	0.22616	0.107532

For palm oil,

$$\text{Ra (mm)} = 4.834\text{--}0.001378\text{B--}0.00483\text{C--}5.68\text{D} + 0.000003\text{C2} + 4.94\text{ D2} + 0.000001\text{BC} + 0.00044\text{BD} \quad (2)$$

For fly ash suspension,

$$\text{Ra (mm)} = 5.312\text{--}0.001378\text{B--}0.00575\text{C--}5.68\text{D} + 0.000003\text{C2} + 4.94\text{ D2} + 0.000001\text{BC} + 0.00044\text{BD} \quad (3)$$

3.2 Cutting Force

Table 6. Response Table for S/N Ratios of Cutting Force

Level	Type of cutting fluid	Cutting speed	Feed rate	Depth of cut
1	−40.55	−39.82	−39.35	−40.00
2	−38.72	−38.16	−39.94	−39.38
3		−40.91	−39.60	−39.50
Delta	1.33	2.75	0.59	0.62
Rank	2	1	4	3

Cutting force reflects the energy required for material removal and the mechanical load on the tool. As shown in Table 6 and Fig. 2 (b), cutting speed was the most dominant factor, followed by cutting fluid type, depth of cut, and feed rate. The cutting force reached its minimum at 2500 rpm, indicating an optimal balance between shearing efficiency and thermal stability. Higher resistance at 1500 rpm resulted from insufficient shearing and BUE formation, while the slight increase at 3500 rpm suggests added friction and vibration at high speeds. The fly ash suspension also reduced cutting force by forming a protective, low-friction tribological layer on the tool surface [15]. Feed rate and depth of cut contributed modestly, with lower feed rates reducing chip thickness and contact stress. The lowest cutting force occurred at 2500 rpm, 400 mm/min, and 0.55 mm. ANOVA (Table 7) identified cutting speed as the only statistically significant factor ($p = 0.015$). Regression results (Table 8) confirmed this, with the second-order model showing excellent fit; the predictive equations are given in Eq. (4) and Eq. (5).

Table 7. Analysis of Variance of Cutting Force

Source	DF	Adj SS	Adj MS	F-Value	P-Value
Type of cutting fluid	1	649.6	649.63	3.45	0.093
Cutting speed	2	2472.9	1236.47	6.56	0.015

(*continued*)

Table 7. (*continued*)

Source	DF	Adj SS	Adj MS	F-Value	P-Value
Feed rate	2	142.7	71.37	0.38	0.694
Depth of cut	2	185.2	92.60	0.49	0.626
Error	10	1883.9	188.39		
Total	17				

Table 8. Regression Statistic of Cutting Force

Regression statistic	1st order model	2nd order model
R2	25.07%	98.76%
R2 (adj)	2.01%	97.97%
Standard error	17.5353	14.0976

For palm oil,

$$Fc\ (N) = -0.0542B + 0.460C + 103.5D + 0.000017B2\text{--}0.000385C2\text{--}0.0466BD \quad (4)$$

For fly ash suspension,

$$Fc\ (N) = -11.62\text{--}0.0542B + 0.460C + 103.5D + 0.000017B2\text{--}0.000385C2\text{--}0.0466BD \quad (5)$$

4 Conclusion

This study evaluated the effects of machining parameters and cutting fluid type on the milling of AL6061-T6 under MQL. Taguchi S/N analysis showed that cutting speed and depth of cut mainly affected surface roughness, while cutting speed and cutting fluid type significantly influenced cutting force. The fly ash suspension gave the best results, achieving the lowest surface roughness at 3500 rpm, 600 mm/min, and 0.55 mm, and the lowest cutting force at 2500 rpm, 400 mm/min, and 0.55 mm. Overall, it improved surface roughness by up to 13% and reduced cutting force by 12% compared to pure palm oil, demonstrating its potential as an effective and sustainable machining fluid.

Acknowledgments. This study is funded by Ministry of Higher Education (MOHE) of Malaysia through the Fundamental Research Grant Scheme (FRGS), No: FRGS/1/2023/TK10/UTEM/02/6. The authors also would like to thank Universiti Teknikal Malaysia Melaka (UTeM) for all the support.

References

1. Duan, Z., et al.: Milling force model for aviation aluminum alloy: Academic insight and perspective analysis. Chin. J. Mech. Eng. **34**(1), 18 (2021)
2. Namlu, R.H., Yılmaz, O.D., Lotfisadigh, B., Kılıç, S.E.: An experimental study on surface quality of Al6061-T6 in ultrasonic vibration-assisted milling with minimum quantity lubrication. Procedia CIRP. **108**, 311–316 (2022)
3. Bhushan, R.K.: Impact of SiC particle size and weight% on tool life while machining of AA7075/SiC composite. J. Alloys Metall. Syst. **3**, 100016 (2023)
4. Imran, M., Shuangfu, S., Yuzhu, B., Yuming, W., Raheel, N.: Optimising subsurface integrity and surface quality in mild steel turning: A multi-objective approach to tool wear and machining parameters. J. Mater. Res. Technol. **35**, 3440–3462 (2025)
5. Ikhries, I.I., Al-Shawabkeh, A.F.: Novel methods for optimizing CNC aluminum alloy machining parameters in polymer mold cavities. Int. J. Lightweight. Mater. Manuf. **7**(4), 507–519 (2024)
6. El-Deen, A.B.: Multi optimization in slot milling of CFRP composites through grey relational-based Taguchi analysis. Niger. J. Technol. Dev. **21**(1), 11–18 (2024)
7. Alaba, E.S. et al.: Evaluation of palm kernel oil as cutting lubricant in turning AISI 1039 steel using Taguchi-grey relational analysis optimization technique. Adv. Ind. Manuf. Eng. **6**, 100115 (2023)
8. Prasad, J., Sonwani, R.K.: Optimize chemical milling of aluminium alloys to achieve minimum surface roughness in aerospace and defense industry. J. Indian Chem. Soc. **102**(1), 101537 (2025)
9. Luo, X., Wu, S., Wang, D., Yun, Y., An, Q., Li, C.: Sustainable development of cutting fluids: the comprehensive review of vegetable oil. J. Clean. Prod. **473**, 143544 (2024)
10. Kumar, G., Ghosh, S., Rao, P.V.: A sustainable approach for the preparation and stability of mono and hybrid nanofluids in the milling of Nickel based superalloy. J. Clean. Prod. **467**, 143015 (2024)
11. Tiwari, S., Amarnath, M., Gupta, M.K.: Synthesis, characterization, and application of Al2O3/coconut oil-based nanofluids in sustainable machining of AISI 1040 steel. J. Mol. Liq. **386**, 122465 (2023)
12. Kanti, P., Korada, V.S., Ramachandra, C.G., Sai, S.T.: P.H.V.: Experimental study on density and thermal conductivity properties of Indian coal fly ash water-based nanofluid. International Journal of Ambient Energy. **43**(1), 2557–2562 (2022)
13. Peng, R., Shen, J., Tang, X., Zhao, L., Gao, J.: Performances of a tailored vegetable oil-based graphene nanofluid in the MQL internal cooling milling. J. Manuf. Process. **134**, 814–831 (2025)
14. Liew, P.J., Tay, H.S., Yap, C.Y., Othman, I.S., Raja, R.I., Tunggal, D.: Tribological behaviour of palm oil mixed with fly ash microparticles as a bio-based lubricant for manufacturing processes. J. Tribol. **31**, 73–83 (2021)
15. Chu, A., et al.: Nanofluids minimal quantity lubrication machining: From mechanisms to application. Lubricants. **11**(10), 422 (2023)

Finite Element Analysis of Steel Fiber Reinforced Concrete Beams Using the CDP Model for Rigid Pavement Applications

Mohammad Nasir Mohamad Taher[1](✉), Mohamad Yusri Aman[1], Hazirah Bujang[2], and Sinatu Sadiah Shapie[3]

[1] Smart Driving Research Center, Faculty of Civil Engineering and Built Environment, Universiti Tun Hussein Onn Malaysia, 86400 Parit Raja, Johor, Malaysia
nasirt@uthm.edu.my

[2] Department of Civil Engineering, Centre for Diploma Studies, Universiti Tun Hussein Onn Malaysia, 84600 Panchor, Johor, Malaysia

[3] Department of Civil Engineering, Polytechnic Port Dickson, KM 14, Jalan Pantai, 71050 Si Rusa, Port Dickson, Negeri Sembilan, Malaysia

Abstract. In this study, the Concrete Damage Plasticity (CDP) model is employed to simulate the flexural behavior of steel fiber-reinforced concrete (SFRC) beams under three-point bending loads. The Finite Element Method (FEM) in Abaqus is used to analyze crack propagation, stress distribution, and failure mechanisms for 100 mm × 100 mm × 400 mm beams with 0%, 1%, 3%, and 5% steel fiber content. The study reveals that the increase in fiber content further enhances crack resistance, reduces brittle failure, and improves energy absorption. Compared to other proportions, beams with 5% steel fiber show the highest ductility and very little crack formation. The present study provides sound guidelines for optimizing steel fiber content in enhancing the concrete performance for structural applications.

Keywords: Steel Fiber Reinforced Concrete (SFRC) · Concrete Damage Plasticity (CDP) · Finite Element Method (FEM) · Three-Point Bending Test · ABAQUS

1 Introduction

Concrete serves as a most popular and widely used construction material due to high compressive strength, durability, and cost-effectiveness. But it is brittle, and low tensile strength fails or even cracks under the application of flexural loads. Conventional constitutive approaches simplified concrete's nonlinear characteristics too easily, so that such changes could hardly be used for a realistic description of damage and failure mechanisms. Therefore, advanced models, like the Concrete Damage Plasticity (CDP) model, were developed for just this purpose, incorporating nonlinear behaviour from cracking or damage evolution which improves accuracy, as validated by Lee et al., [1]. Recent studies have also discussed modifications into the CDP model to make it more accurate for simulation regarding the behaviour of SFRC as proposed by Tran et al., [2].

M. A. b. Jamlos et al. (Eds.): ICSET 2025, LNNS 1897, pp. 120–126, 2026.
https://doi.org/10.1007/978-3-032-20722-7_14

Steel fiber reinforced concrete (SFRC) is one of the most effective ways to improve concrete ductility, toughness, and post-cracking behavior. Steel fibers inclusion controls crack propagation & enhances flexural and compressive strength. Its mechanical performance is changed side as per increasing percentage of steel fibers in the mix. ABAQUS CAE is widely considered a cutting-edge mobilization useful to solve complex engineering problems where analytical methods cannot easily be utilized, as discussed by Khennane, [3]. Recently, further FEM proved the efficacy of ABAQUS as to its ability to simulate fracture behavior in SFRC, as shown in studies conducted by Shams et al., [4]. ABAQUS, specifically the ABAQUS/CAE platform, is a widely used commercial finite element analysis (FEA) software developed by Dassault Systemes. It provides an integrated environment for pre-processing, modeling, simulation, and post-processing of structural and material behavior. ABAQUS/CAE (Complete Abaqus Environment) enables users to construct detailed 3D geometries, assign nonlinear material models, generate meshes, apply loads and boundary conditions, and visualize stress, strain, and damage responses. Its ability to capture crack initiation, stiffness degradation, and progressive failure makes it especially suitable for research in rigid pavement engineering and structural concrete applications.

A volume fraction of steel fibers of 1% can enhance compressive strength up to 10% and flexural strength up to 80%, as reported by Harki et al., [5]. However, excessive fiber content may impair workability and lead to fiber balling. Research has suggested that the optimum fiber percentage is a function of the mix design and intended use. Guan et al., (2023) show that ductility and service life of SFRC are considerably governed by fiber distribution and orientation, affecting control on cracks and durability [6].

SFRC is growing in popularity for the construction of rigid pavements because of enhanced fatigue resistance and long-term durability. CDP model is particularly efficient to simulate SFRC, since its prediction of progressive damage mechanisms like crack initiation and propagation under different loading is proved, according to Hafezolghorani et al., [7]. Unlike simpler models, CDP accounts for both tensile and compressive damage evolution which enhances its reliability in predicting the structural performance of SFRC.

The present study focuses on the flexural behavior of SFRC beams containing various steel fiber volume fractions (0%, 1%, 3%, and 5%) using a nonlinear finite element approach. The beams are analyzed under three-point bending conditions, and a mesh sensitivity study is performed to ensure numerical reliability. The findings support the development of optimized SFRC mixtures for structural applications, particularly in rigid pavement systems.

This study aims to evaluate how different steel fiber volume fractions influence the flexural response, crack propagation, and energy absorption capacity of SFRC beams using the CDP model. It is hypothesized that increasing steel fiber content will significantly enhance ductility, reduce crack severity, and delay brittle failure under three-point bending loads.

2 Methodology

For modeling numerically through the CDP model under the Finite Element Method (FEM) behavior of concrete beams under three-point bending, the methodology concerning different percentages of steel fibers was discussed in this sect. (0%, 1%, 3%, and 5%). The study emphasizes the necessity of accurate data collection in order to simulate the beam response under flexural loads reflecting the role of steel fibers. Different steel fiber percentages were used to see their influence on the mechanical performance pertaining to the experimental data.

FEM was activated on the mechanical properties obtained from experimental investigations, and numerical analyses were done in ABAQUS software as presented by Fares & Bakir [8]. The numerical modeling phase included some considerations that were crucial for obtaining reliable results for Finite Element Analysis (FEA): these included geometrical design, material properties, meshing strategies, loading conditions, and boundary conditions. In addition, nonlinear finite element analysis has prominently been adopted to study SFRC beams especially in determining their composite layering and the propagation of cracks in them as done by Fares & Bakir [9].

The experimental validation of the results obtained via FEM was thereafter done in order to show how accurately SFRC is predicted in view of its failure modes, deformations, and flexural performance. The grades of progressive damage mechanisms simulated by this model-also amenable to crack initiation and propagation-also make it a suitable model for analysis of SFRC beams under flexural loading as proved by Fares & Bakir [9].

2.1 Idealization in Modeling

In accordance with previous experimental findings, the beam and loading setup were replicated in this stage. An SFRC beam was modeled 100 mm × 100 mm × 400 mm with a span of 400 mm. Considering the experimental model, the supports and loading indenter were defined as rigid bodies. A concentrated load of 1000 kN was applied at the mid-span to simulate the three-point bending test. Figure 1 illustrates the modelled beam.

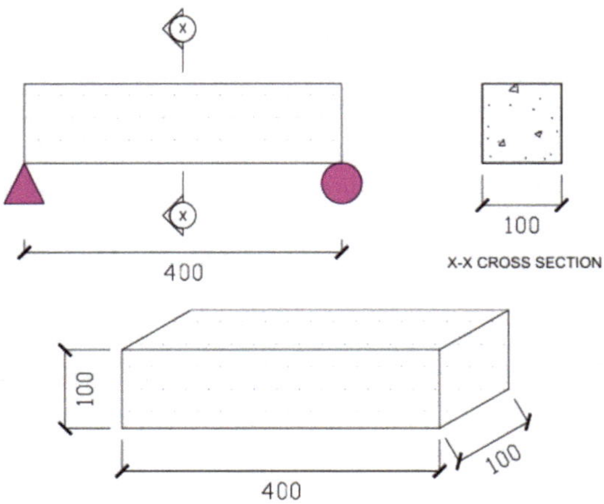

Fig. 1. The measurements of a Steel fiber reinforced concrete beam

3 Results and Discussion

3.1 Mesh Sensitivity Analysis

In Abaqus, mesh sensitivity analysis is one of the key factors regarding finite element simulations, especially when assessing the influence of element size on the accuracy of results in SFRC beams under three-point bending. As mentioned in Abbas *et al.*, [14], a mesh sensitivity analysis was carried out to select the best mesh size to be used and to avoid mesh dependency. Thorough validation of this process guarantees consistency and reliability in the simulation of results irrespective of any mesh refinement. Larger mesh sizes yield a more accurate result as they account for minute stress, strain, and crack propagation obfuscations.

Mesh sensitivity analysis was conducted for SFRC beams with 3% steel fibers to find the suitable mesh size for the current study. Five mesh sizes were tested which were 30 mm, 25 mm, 20 mm, 15 mm, and 10 mm as shown in Figure. 8 and Table 5. The findings indicate that mesh sizes of 25 mm, 20 mm, 15 mm, and 10 mm produce comparable results, with a 20 mm mesh size deemed optimal for all specimen models in the tensile simulation. This selection ensures a balance between accuracy and computational efficiency, making it suitable for further analysis of SFRC beams under three-point bending loads.

3.2 Development of SFRC Beam Models Under Three-Point Bending Using CDP Equations

This section focuses on analyzing the structural response of SFRC beams under three-point bending by integrating the CDP model with the FEM. The primary objective is to develop CDP models that accurately represent the mechanical behavior of SFRC beams when subjected to bending loads. The CDP constitutive equations play a crucial role in simulating crack initiation and propagation within the concrete material. These equations define the material response under uniaxial compression and tension, aligning with the stress-strain behavior of concrete was also conducted by Halmandge & Harsoor, [15].

The CDP model treats concrete as a nonlinear, inelastic material, simulating crack development and material degradation under bending forces. This model introduces a damage factor, which starts at 0 (undamaged state) and increases as cracks form and propagate. The crack formation process is influenced by Kent and Park's., [16]. The equations dynamically track strain evolution under applied stress, allowing for a progressive softening response, which accurately represents the loss of stiffness and strength in cracked regions. By employing the CDP model, the simulation captures the realistic deformation behavior of SFRC beams, even beyond the initial cracking stage [17] (Fig. 2).

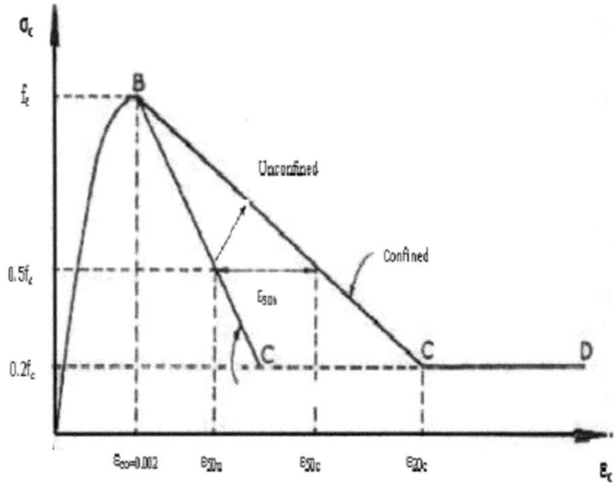

Fig. 2. Kent and Park model for confined and unconfined concrete [17]

4 Conclusions

In conclusion, the simulation of three-point-bending tests on beams made of SFRC using the CDP model in ABAQUS was performed successfully, confirming findings of previous investigations. This numerical modeling has proved to be a predominant tool in understanding the damage evolution, crack propagation, and structural behaviour of SFRC beams under flexural loading conditions. Findings contribute to the growing acknowledgment of the ABAQUS package, where application of the CDP model has come out as a prominent model for nonlinearly simulating the behavior of SFRC, as previously established by Youssef-Miaari *et al.,* [18].

Higher steel fiber content is related to greater improvements in structural integrity, reduced crack severity, and enhanced post-cracking strength. That is what Heylemelecot's., [19] report has substantiated. The analysis has shown extensive flexural cracks and brittle failure for the 0% steel fiber beam, while progressive reductions in crack propagation and damage accumulation were observed, respectively, for the 1%, 3%, and 5% fiber-reinforced beams. The best result was obtained by the 5% steel fiber beam because it has very few cracks and higher energy absorption, showing that structural failure can be minimized through the fiber-reinforcement of structures.

While the numerical results indicate that a 5% steel fiber content provides the highest crack resistance and energy absorption, such a high dosage is rarely practical for real pavement applications. Fiber contents above 2–3% typically present significant challenges in terms of mix workability, uniform fiber dispersion, potential balling, and increased mixing energy. These issues can compromise quality control during batching and placement, particularly for large-scale pavement projects. Moreover, the cost of steel fibers rises almost linearly with dosage, making a 5% volume fraction economically prohibitive for most rigid pavement designs. Therefore, although the numerical model shows superior performance at 5% fiber content, a more realistic engineering compromise would likely fall in the 1–2% range, where improvements in ductility and crack control can be achieved without severe penalties in workability or project cost.

This highlights the importance of balancing analytical performance with constructability and economic feasibility when recommending fiber dosages for pavement engineering practice.

Acknowledgments. This research was funded by a grant from Universiti Tun Hussein Onn Malaysia (UTHM) through Tier 1 (Vot Q471)

Disclosure of Interests The authors declare that there are no financial or non-financial competing interests that could have influenced the work reported in this manuscript.

References

1. Lee, S., Abolmaali, A., Shin, K., Lee, H.: ABAQUS modeling for post-tensioned reinforced concrete beams. J. Build. Eng. **30**, 101273 (2020). https://doi.org/10.1016/j.jobe.2020.101273
2. Tran, T.T., Pham, T.M., Tran, D.T., Ha, N.S., Hao, H.: Modified plastic damage model for steel fiber reinforced concrete. Struct. Concr. **26**(1), 13–42 (2024). https://doi.org/10.1002/suco.202300640
3. Khennane, A.: Introduction to Finite Element Analysis Using MATLAB and Abaqus, 1st edn. CRC Press, Boca Raton (2013). https://doi.org/10.1201/b15042
4. Shams, M.A. et al.: Fracture analysis of steel fibre-reinforced concrete using finite element method modeling. Front. Mater. **11**, 1355351 (2024). https://doi.org/10.3389/fmats.2024.1355351
5. Harki, B.Q.K.A., Jawahery, M.S.A., Abdulmawjoud, A.A.: Hybrid steel fiber of rigid pavements: A 3D finite element and parametric analysis. Coatings. **12**(10), 1478 (2022). https://doi.org/10.3390/coatings12101478
6. Guan, J.L.L., Too, Y.Y., Teh, B.J.W., Kum, Y.J.: Characterisation of steel fiber reinforced concrete: Ductility and service life. Construction. **3**(2), 300–307 (2023). https://doi.org/10.15282/construction.v3i2.10071
7. Hafezolghorani, M., Hejazi, F., Vaghei, R., Jaafar, M.S.B., Karimzade, K.: Simplified damage Plasticity model for concrete. Struct. Eng. Int. **27**(1), 68–78 (2017). https://doi.org/10.2749/101686616x1081
8. Fares, A.M., Bakir, B.B.: Parametric study on the flexural behavior of steel fiber reinforced concrete beams utilizing nonlinear finite element analysis. Structures. **65**, 106688 (2024a). https://doi.org/10.1016/j.istruc.2024.106688
9. Fares, A.M., Bakir, B.B.: Nonlinear finite element analysis of layered steel fiber reinforced concrete beams. Comput. Struct. **307**, 107637 (2024b). https://doi.org/10.1016/j.compstruc.2024.107637
10. Mokhatar, A., Abdullah, J.: Computational analysis of reinforced concrete slabs subjected to impact loads. Int. J. Integr. Eng. **4**(2), 29–36 (2012) https://publisher.uthm.edu.my/ojs/index.php/ijie/article/view/322
11. Omar, R., et al.: Investigation of mesh size effect on dynamic behaviour of an assembled structure with bolted joints using finite element method. Int. J. Automot. Mech. Eng. **15**(3), 5695–5708 (2018). https://doi.org/10.15282/ijame.15.3.2018.22.0437
12. Melesse, G., Behailu, T., Kassa, H.K.: Finite element analysis of a reinforced concrete dapped-end beam under the effects of impact velocity and dapped-end beam cross-section geometry. Adv. Mater. Sci. Eng. **2023**, 1–13 (2023). https://doi.org/10.1155/2023/5869552
13. Hou, H.Z., Nordin, N.N.: Finite element analysis on the steel fiber-reinforced concrete beams: A systematic review. Int. J. Infrastruct. Res. Manag. **10**(1), 1–15 (2022) https://iukl.edu.my/rmc/wp-content/uploads/sites/4/2022/06/1.-IJIRM-Vol.10_1_Hou-Zhicheng.pdf

14. Abbas, A. (2021). A simplified finite element model for assessing steel fibre reinforced concrete structural performance. School of Architecture, Computing and Engineering, University of East London, 1–36. https://repository.uel.ac.uk/download/ca0911483375d94ac466313c52f5512025193ffc835b9986fc748f2832b34fcc/2469198/Abbas%20-%20A%20simplified%20FE%20analysis%20model%20for%20assessing%20SFRC%20-20ACCEPTED.pdf
15. Halmandge, G., Harsoor, R.: Response of simply supported concrete beams reinforced with steel and polypropylene fibers under the influence of low velocity impact. Preprints. (2023). https://doi.org/10.20944/preprints202312.0944.v1
16. Kent, D.C., Park, R.: Flexural members with confined concrete. J. Struct. Div. ASCE. **97**(7), 1969–1990 (1971). https://doi.org/10.1061/jsdeag.0002957
17. Karasin, A., Gunaslan, S.E., Öncü, M.E.: Models for confined concrete columns with fiber composites. Int. J. Adv. Res. Eng. Technol. **5**(12), 55–63 (2015) https://www.researchgate.net/publication/276202656_MODELS_FOR_CONFINED_CONCRETE_COLUMNS_WITHFIBER_COMPOSITES
18. Youssef-Miaari, Y.M. (2020). Numerical modeling of the mechanical behavior of a SFRC beam reinforced with FRCM composites. In School of Civil, Environmental and Land Management Engineering. https://www.politesi.polimi.it/retrieve/c007aa41-648c-49a8-8f76-869f88ea690f/YOUSSEF-MIAARI-918473.pdf
19. Heylemelecot, T.M.: Effect of hooked steel fibers on shear strength of beams without stirrups. Abyssinia J. Eng. Comput. **3**(2), 14–26 (2023) https://abjol.org.et/index.php/ajec/article/view/897
20. Kytinou, V.K., Chalioris, C.E., Karayannis, C.G., Elenas, A.: Effect of steel fibers on the hysteretic performance of concrete beams with steel reinforcement—tests and analysis. Materials. **13**(13), 2923 (2020). https://doi.org/10.3390/ma13132923

Fire Resistance and Thermal Characteristics of Palm Empty Fruit Bunches-Fiberglass (PEFB-FG) Composites as Insulation Material in HVAC Applications

Mohamad Fikri Mohamad Yunus[1,3], Noraini Marsi[1,2(✉)], Nik Normunira Mat Hassan[4], Abdul Mutalib Leman[1], Abdul Murad Zainal Abidin[3], Zamil Hisham Abdul Rashid[1,3], and Tengku Muhammad Ibrahim Salahuddin[1,3]

[1] Faculty of Engineering Technology, Universiti Tun Hussein Onn Malaysia, Pagoh Campus KM 1, Jln Panchor, 86400 Pagoh, Johor, Malaysia
mnoraini@uthm.edu.my

[2] Advanced Manufacturing and Material Centre (AMMC), Institute of Integrated Engineering, Universiti Tun Hussein Onn Malaysia, Parit Raja, 86400 Batu Pahat, Johor, Malaysia

[3] Centre of Excellent for Engineering and Technology (CREaTE), Jabatan Kerja Raya Malaysia, Jalan Kemus, Simpang Ampat, 78000 Alor Gajah, Melaka, Malaysia

[4] Faculty of Mechanical Engineering, Universiti Teknologi MARA, 40450 Shah Alam, Selangor, Malaysia

Abstract. Conventional HVAC insulation materials often lack a balance between thermal performance and fire safety. This study investigates an innovative palm empty fruit bunch–fibreglass (PEFB–FG) composite designed to enhance thermal resistance, fire retardancy, and mechanical strength. We fabricated six composite samples (50 mm thickness) with varying lamination ratios using 8 mm needle-punched PEFB mats. Sample F achieved the highest tensile strength (180.35 kPa) due to optimized layer composition. Under ASTM D3841 fire resistance testing, Sample F retained 58.25% of its structural integrity after 90 s. Additionally, it demonstrated superior thermal performance, maintaining a thermal resistance of 35.38% after 15 min of exposure. These findings confirm the potential of PEFB–FG composites as high-performance, sustainable insulation for HVAC applications.

Keywords: fire resistance · thermal · PEFB · fiberglass · insulation

1 Introduction

Insulation material is critical for the safety and energy efficiency of HVAC systems. While conventional materials like foam boards and mineral wool are standard, they often compromise between thermal performance and fire resistance [1]. The urgency for better materials is reflected in the global HVAC insulation market, which was valued at

M. A. b. Jamlos et al. (Eds.): ICSET 2025, LNNS 1897, pp. 127–133, 2026.
https://doi.org/10.1007/978-3-032-20722-7_15

USD 14.6 billion in 2024 and is projected to reach USD 20.3 billion by 2032 (CAGR 4.5%). This growth is driven by stricter energy conservation standards and fire safety regulations in commercial buildings.

However, traditional materials face limitations. For instance, standard glass wool (thermal conductivity 0.035–0.045 W/m·K) often lacks the fire retardancy required for high-temperature ducting. To address this, researchers are investigating hybrid composites that combine natural fibres with synthetic reinforcements [3, 4]. Palm Empty Fruit Bunch (PEFB), a renewable by-product of the palm oil industry, offers a sustainable solution [5]. With Malaysia alone producing over 23 million tons of palm biomass in 2023, PEFB represents a significant, underutilized resource for the circular economy.

While PEFB has inherent insulating properties, it requires reinforcement to meet industrial fire standards [6]. Fibreglass (FG), known for maintaining stability at temperatures > 500 °C, is an ideal partner [8]. Hybridizing PEFB with FG can improve fire resistance by 30–40% while maintaining sustainability [9]. This study develops a laminated PEFB-FG composite, aiming to optimize the balance between thermal conductivity, fire resistance, and tensile strength for HVAC applications.

2 Methodology

The fabrication process for the non-woven PEFB mats is illustrated in Fig. 1. Raw PEFB fibres were processed through opening, blowing, and carding to align the fibers and remove contaminants, followed by needle-punching to form 8 mm thick non-woven mats. These mats were then laminated with fibreglass (FG) sheets to create composite samples with a total thickness of 50 mm.

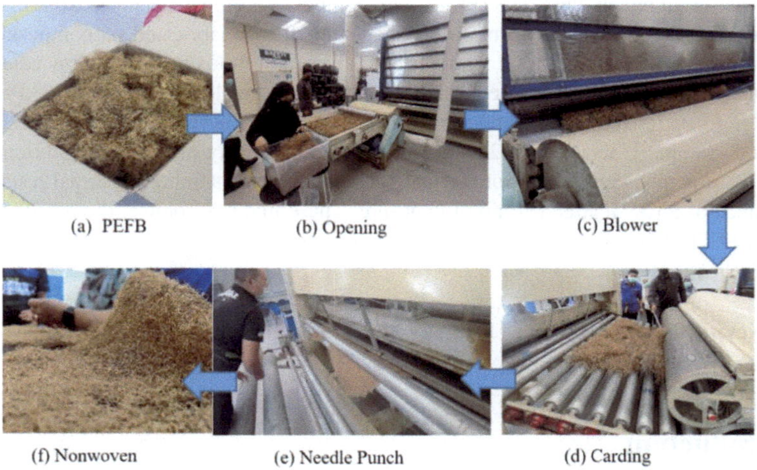

Fig.1: Fabrication process of laminated layer of Palm Empty Fruit Bunches-Fiberglass (PEFB-FG) Composites

To evaluate the effect of layering, six distinct sample configurations (Samples A–F) were fabricated by varying the stacking sequence and thickness ratios of the PEFB and

FG layers. The specific layer compositions are detailed in Table 1, and the resulting cross-sections are visualized in Fig. 2.

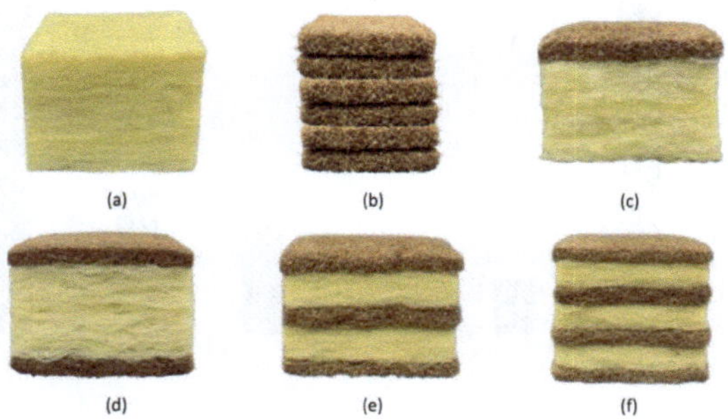

Fig. 2. Cross-sectional views of the six laminated PEFB-FG composite samples (A–F).

Table 1: Laminated layer Palm Empty Fruit Bunches-Fiberglass (PEFB-FG)

Sample name	Ratio of layer (mm)						
	Layer 1	Layer 2	Layer 3	Layer 4	Layer 5	Layer 6	Layer 7
	PEFB	FG	PEFB	FG	PEFB	FG	PEFB
Sample A	0	50	-	-	-	-	-
Sample B	25	0	25	-	-	-	-
Sample C	8	42	-	-	-	-	-
Sample D	8	34	8	-	-	-	-
Sample E	8	13	8	13	8	-	-
Sample F	8	6	8	6	8	6	8

3 Result and Discussion

3.1 Tensile Strength Analysis

Figure 3 illustrates the tensile strength of the composites, which ranged from 25.55 kPa (Sample B) to 180.35 kPa (Sample F). Sample F achieved the highest strength, attributed to its optimized alternating layer structure. This confirms that distributing FG layers enhances interfacial adhesion and stress transfer, as detailed by Li et al. [11]. Conversely, Sample B exhibited the lowest strength, consistent with findings that excessive natural fiber loading without adequate synthetic reinforcement compromises mechanical integrity [12].

Intermediate samples (C, D, E) showed a progressive increase in strength, validating the role of symmetrical layering in improving load distribution [13, 14]. The overall trend indicates that while pure PEFB lacks structural rigidity, the hybridization with FG significantly enhances tensile performance, making Sample F comparable to established synthetic composites.

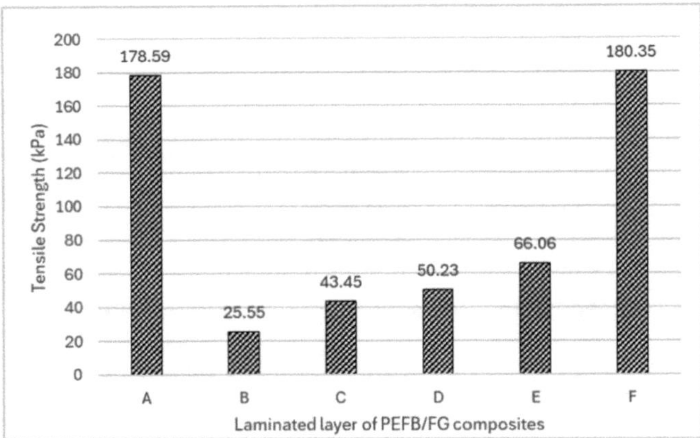

Fig. 3. Tensile strength of PEFB/FG composite samples.

3.2 Fire Resistance Test

Figure 4 presents the fire resistance rates under direct flame exposure over 90 s. Sample F demonstrated the highest overall performance, retaining 58.25% resistance after 90 s. This superiority is attributed to its intricate layering (8:6:8:6:8:6:8), where multiple FG interfaces act as repeated thermal barriers to delay heat transfer [16, 17]. Conversely, the PEFB-dominant Sample B exhibited the lowest resistance (51.15% at 90s), confirming the impact of natural fiber flammability [15]. Notably, the hybrid Sample F slightly outperformed the pure fiberglass Sample A (57.10% at 90s), validating that the alternating structure enhances thermal stability beyond that of the individual constituents [18–20].

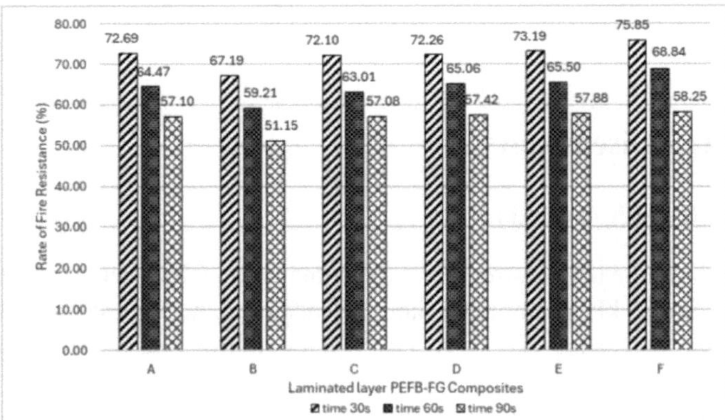

Fig. 4. Rate of fire resistance for PEFB/FG composites.

3.3 Rate of Thermal Characteristic Analysis

Figure 5 illustrates the thermal resistance rates over a 15-min exposure duration. Sample F demonstrated the highest stability, maintaining a resistance rate between 34.15% and 35.38%. This superior performance is attributed to the multiple FG interfaces acting as thermal barriers, which delay heat transfer [21]. Conversely, the PEFB-dominant Sample B recorded the lowest stability (dropping to 11.20%), reflecting the rapid thermal degradation inherent to organic fibers [22]. Interestingly, hybrid samples (D, E, and F) consistently outperformed the pure fiberglass Sample A (~25%), confirming that strategic layering creates heat dissipation pathways that are more effective than monolithic materials [23, 24].

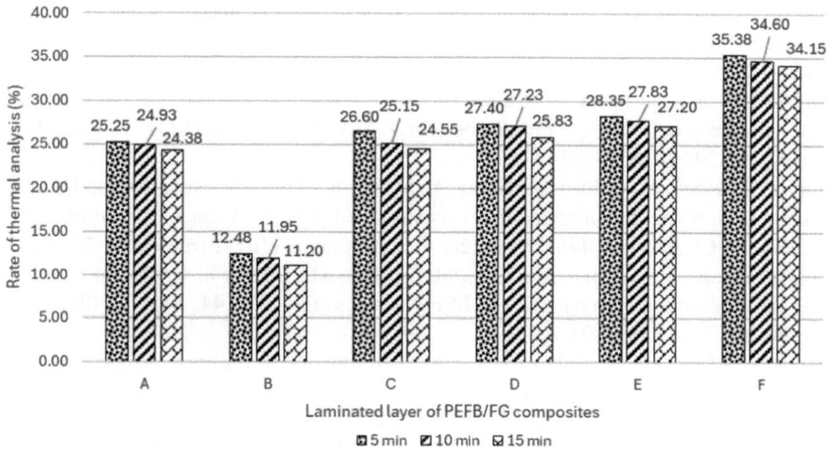

Fig. 5. Rate of thermal resistance for PEFB/FG composites over time.

4 Conclusion

This study validates the feasibility of Palm Empty Fruit Bunch–Fibreglass (PEFB–FG) laminated composites as sustainable, high-performance HVAC insulation. Among the tested configurations, the seven-layer alternating structure (Sample F) consistently demonstrated superior performance. It achieved the highest tensile strength (180.35 kPa) and effectively delayed thermal degradation, retaining over 34% thermal resistance after 15 min of fire exposure. These results confirm that intricate layering acts as a potent fire barrier, mitigating the inherent combustibility of natural PEFB fibers. Economically, utilizing abundant PEFB by-products offers a potential cost reduction of 20–30% compared to traditional fiberglass-only systems. Future studies should prioritize real-world performance testing, particularly regarding acoustic absorption, long-term aging, and field installation challenges in operating HVAC duct networks.

5 Disclosure of Interests.

The authors have no competing interests to declare that are relevant to the content of this article.

Acknowledgments. The authors gratefully acknowledge the Malaysian Public Works Department (JKR), Centre of Excellent for Engineering and Technology (CREaTE) and Universiti Tun

Hussein Onn Malaysia (UTHM) for their technical support and resources. Appreciation is also extended to all colleagues and contributors whose assistance was invaluable to the successful completion of this research.

References

1. Sepehri, A., Pavlak, G.S.: Evaluating optimal control of active insulation and HVAC systems in residential buildings. Energy Build. **281**, 1–10 (2023). https://doi.org/10.1016/j.enbuild.2022.112728
2. Ali, A., Issa, A., Elshaer, A.: A comprehensive review and recent trends in thermal insulation materials for energy conservation in buildings. Sustainability **16**(20), 1–14 (2024). https://doi.org/10.3390/su16208782
3. Horvat, B., Knez, N., Hribar, U., Konig, J., Music, B.: Thermal insulation and flammability of composite waste polyurethane foam encapsulated in geopolymer for sustainable building envelope. J. Clean. Prod. **446**, 28–36 (2024). https://doi.org/10.1016/j.jclepro.2024.141387
4. Gnanachelvam, S., Ariyanayagam, A., Mehendran, M.: Effects of insulation materials and their location on the fire resistance of LSF walls. J. Build. Eng. **44**, 104–115 (2021). https://doi.org/10.1016/j.jobe.2021.103323
5. Harahap, M.I.P., Othman, I., Dinata. N.A.M.: Investigation on the nature safety rule violations in high-rise constructions projects. J. Adv. Res. Appl. Sci. Eng. Technol. **41**(2), 164–179 (2024). https://doi.org/10.37934/araset.41.2.164179
6. Deraman, R., et al.: Production of roof board insulation using agricultural wastes towards sustainable building material. J. Adv. Res. Fluid Mech. Therm. Sci. **99**(1), 66–89 (2022). https://doi.org/10.37934/arfmts.99.1.6689
7. Marin-Calvo, N., Gonzalez-Serrud, S., James-Rivas, A.: Thermal insulation material produced from recycled materials for building applications: cellulose and rice husk-based material. Front. Built Environ. **9**, 1–13 (2023). https://doi.org/10.3389/fbuil.2023.1271317
8. Aheran, D.G., et al.: Fungal colonization of fiberglass insulation in the air distribution system of a multi-story office building: VOC production ans possible relationship to a sick building syndrome. J. Ind. Microbiol. **16**, 280–285 (1996). https://doi.org/10.1007/BF01570035
9. Marsi, N., Kien, C.L., Rani, J.A., Main, N.M., Jamir, M.R.M., Shaari, M.F.: Preparation and performance test of PEFB reinforced box waste coated superhydrophobic coating for shoe sole application. Int. J. Integrat. Eng. **12**(5), 210–216 (2020). https://doi.org/10.30880/ijie.2020.12.05.026
10. Rus, A.Z.M., Salim, N.S.M., Alzomor, A., Marsi, N., Junoh, M., Zamdin, A.Z.R.B.: Plastics in thermal insulation. Encycl. Mater. Plast. Polym. **1–4**, 273–284 (2022). https://doi.org/10.1016/B978-0-12-820352-1.00087-0
11. Al-hassany, M.O.A., Al-Dulaimy, A., Al-Sammarraie A., Ali, A.F.: Effects of fiberglass form on the tensile and bending characteristics of epoxy composite materials. AIMS Mater. Sci. **7**(5), 583–595 (2020). https://www.aimspress.com/aimspress-data/aimsmates/2020/5/PDF/matersci-07-05-583.pdf
12. Xuan, Y.Y., et al.: Cure behaviour and tensile properties of pineapple leaf fibre reinforced natural rubber composites. J. Adv. Res. Appl. Mech. **115**(1), 88–97 (2024). https://doi.org/10.37934/aram.115.1.8897
13. Hamdan, H., et al.: Layering effect on mechanical, thermal & physical properties carbon fibre reinforced polyphenylene sulfide. J. Adv. Res. Fluid Mech. Thermal Sci. **98**(2), 128–145 (2022). https://doi.org/10.37934/arfmts.98.2.128145

14. Yaozu, H., et al.: Synergistic enhancement of strength and toughness of fiber-reinforced composites by constructiing biomimetic intermittent porous structure. Compos. Part A: Appl. Sci. Manuf. **185**, 1–18 (2024). https://doi.org/10.1016/j.compositesa.2024.108335
15. Kim, N.K., Dutta, S., Bhattacharyya, D.: A review of flammability of natural fibre reinforced polymeric composites. Compos. Sci. Technol. **162**, 64–78 (2018). https://doi.org/10.1016/j.compscitech.2018.04.016
16. Andrew, J., Sain, M., Ramakrishna, S., Jawaid, M., Dhakal, H.N.: Environmentally friendly fire retardant natural fibre composites: a review. Int. Mater. Rev. **69**(5–6), 1–16 (2024). https://doi.org/10.1177/09506608241266302
17. Venkatesan, K., Ramanathan, K., Vijayanandh, R., Selvaraj, S., Raj Kumar, G., Senthil Kumar, M.: Comparative structural analysis of advanced multi-layer composite materials. Mater. Today Proc. **27**(3), 2673–2687 (2020). https://doi.org/10.1016/j.matpr.2019.11.247
18. Dalfi, H.K., et al.: Enhancing the fire-resistance performance of composite laminates via multi-scale hybridisation: a review. J. Ind. Text. **54**, 1–24 (2024). https://doi.org/10.1177/15280837241226988
19. Gang, L., Fang, Q., Zhi, W., Xuhai, X., Yanying, X.: Experimental study of thermal and fire reaction properties of glass fiber/bismaleimide composites for aeronautic application. Polymers **15**(10), 1–17 (2023). https://doi.org/10.3390/polym15102275
20. Ayuni, E.L.F., et al.: The effect of insulation thickness on heat transfer characteristics and flammability in tube mesoscale combustors. J. Adv. Fluid Mech. Thermal Sci. **116**(2), 157–171 (2024). https://doi.org/10.37934/arfmts.116.2.157171
21. Neto, J.S.S., Queiroz, H.F.M., Aguiar, R.A.A., Banea, M.D.: A review on the thermal characterisation of natural and hybrid fiber composites. Polymers **13**(24), 4425–4436 (2021). https://doi.org/10.3390/polym13244425
22. Amal, N., Eman, N.: Effect of fiber orientation on the mechanical properties of multi layers laminate nanocomposites. Heliyon **6**(1), 1–14 (2020). https://doi.org/10.1016/j.heliyon.2020.e03167
23. Grace, J., Wilder, A.T.: Complex thermal expansion of an epoxy-fiberglass laminate. In: IEEE Electrical Insulation Conference (EIC), pp. 209–211 (2022). https://doi.org/10.1109/EIC51169.2022.9833160
24. Desgrosseilliers, L., Groulx, D., White, M.A.: Heat conduction in laminate multilayer bodies with applied finite heat source. Int. J. Thermal Sci. **72**, 47–59 (2013). https://doi.org/10.1016/j.ijthermalsci.2013.05.001

Valorization of Malaysian Reclaimed Tin Mining Sand for Sustainable Glass-Ceramic Tiles: Influence of Quartz Size on Properties

Syarifah Aminah Ismail(✉), Mohd Idham Mustaffar, Mohd Hakim Ibrahim, and Nabihah Othman

Mineral Research Centre, Mineral and Geoscience Department Malaysia, Jalan Sultan Azlan Shah, 31400 Ipoh, Perak, Malaysia
syarifahaminah@jmg.gov.my

Abstract. This investigation assesses the function of reclaimed tin-mining sand from Malaysia as a practical substitute raw material for glass-ceramic tiles, ultimately striving to lessen environmental harm and energy usage. Conventional tiles are fired at approximately 1200 °C utilizing pristine materials, whereas this research amalgamates mining sand (0.075–4.75 mm) with soda-lime glass and a foaming agent, sintering at temperatures ranging from 850 to 1000 °C. Characterization techniques (XRD, MOR, density, water absorption) indicated that an optimal composition of 49% sand and 32% glass attained a MOR of 15.20 MPa, a density of 1.92 g/cm^3, a water absorption rate of 12.31%, and a shrinkage of 0.42%, conforming to ISO 10545 standards. Finer quartz (~106 µm) enhanced sintering and performance metrics. This research elucidates that mining by products can yield high-performance, eco-efficient tiles at reduced temperatures, underscoring the potential for sustainable, industrial-scale applications. Subsequent investigations should delve into surface modification and the feasibility of large-scale production.

Keywords: Glass ceramic tiles · Optimization · Mechanical strength · Quartz particle size · Reclaimed tin mine sand residue · sintering · Sustainable Materials

1 Introduction

The rapid depletion of natural resources and growing demand for sustainable construction materials highlight the need for resource recovery and circular economy practices. The ceramics industry is energy intensive and relies heavily on primary raw materials, raising concerns under global climate agreements. In Malaysia, tin mining generates large amounts of leftover sand, which, despite being an environmental issue, is rich in silica and offers potential as a secondary raw material for glass-ceramic production.

Quartz, a major component of these by-products, plays a key role in densification, phase formation, and sintering behavior [1]. Previous research shows that raw material particle size strongly affects sintering temperature, mechanical strength, and energy consumption. Studies from 2020–2025 (Andreola et al., 2021; Silva et al., 2023; Ziejewska

M. A. b. Jamlos et al. (Eds.): ICSET 2025, LNNS 1897, pp. 134–140, 2026.
https://doi.org/10.1007/978-3-032-20722-7_16

et al., 2023) found that fine waste-derived particles (fly ash, stone dust, slag, glass cullet) improve densification, reduce firing temperatures, and enhance mechanical strength due to better diffusion and particle packing [2–4]. Similarly, Yusuf et al. (2022) reported that using fine glass waste can reduce firing energy by up to 20% [5].

The effect of quartz particle size from Malaysian reclaimed tin-mining sand on tile sintering and strength is not well studied. This research examines how particle size influences the physical and mechanical properties of recycled glass tiles, aiming to produce high performance, ISO-compliant tiles at lower temperatures and support sustainable ceramic manufacturing.

2 Materials and Methods

The study was carried out in three main stages, as shown in Fig. 1. First, the by-products were analyzed to assess their suitability as raw materials. Next, a sustainable ceramic tile was designed and produced from these materials. Finally, various tests were conducted to evaluate the finished product.

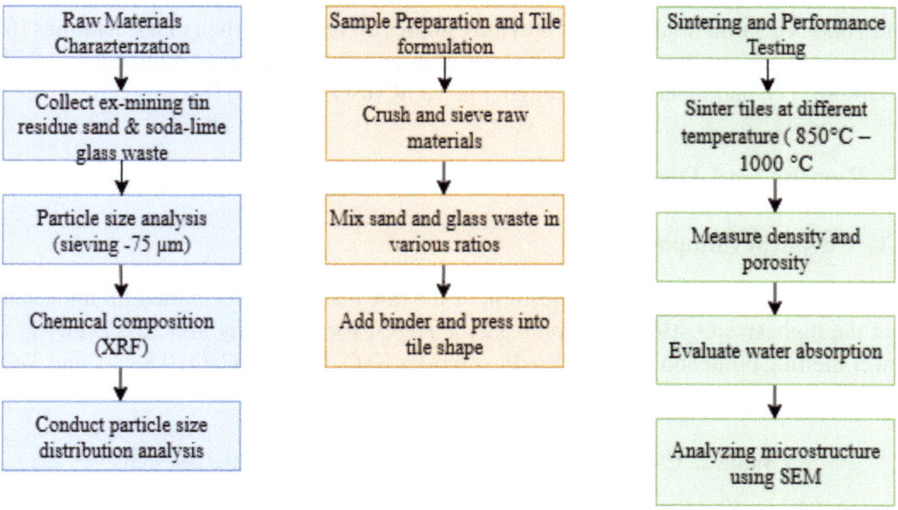

Fig. 1. Schematic of the process for making lightweight glass-ceramic tiles from ex-mining tin sand and soda-lime glass waste. The process involves three stages: raw material analysis, sample preparation, and sintering with performance testing.

2.1 Materials

Lightweight glass ceramic tiles were synthesized utilizing ex-mining tin residue sand as the primary raw material, incorporating soda-lime glass waste as a flux, along with bentonite serving as a binder/filler and a foaming agent to create porosity.

2.2 Sample Preparation

Quartz particles of 75, 106, and 150 μm were tested for their effects on sintering and mechanical properties. Finer particles improve densification and lower sintering temperature, while coarser particles can reduce strength. Combining different sizes (<75 μm and 75–150 μm) helps achieve higher density and lower energy use. Tin mining sand and soda-lime glass were ground to 75 μm. Five MOR tests were done per particle size, and the materials were mixed with a foaming agent and bentonite (Table 1). Chemical composition was measured using a Shimadzu XRF-1700 spectrometer.

Table 1. Formulation details of lightweight tile mixtures

Material	Ex-tin residue sand	Soda lime waste	Bentonite	Foaming agent
Composition (wt.%)	49	32	18	1.0

Specimens were sintered from 850 °C, 900 °C, 950 °C and 1000 °C with staged heating and 30-min holds, then cooled over 24 h. The range (850 °C, 900 °C, 950 °C and 1000 °C, 5 min hold) follows previous studies on recycled silica glass-ceramics [6]. Properties were measured by water absorption and bulk density (Archimedes, ASTM C373–88) [7], and mechanical strength via MOR (ISO 10545-4) [8].

3 Results and Discussion

3.1 Chemical Composition

Table 2 shows the chemical composition of the raw materials. Reclaimed tin mine sand had the highest SiO_2 (90.02%), plus 5.21% Al_2O_3 for durability and 1.63% Fe_2O_3 to lower melting point and darken tiles. Flux oxides K_2O (0.89%), CaO (0.88%), and TiO_2 (0.47%) promote densification by forming a glassy phase [9].

Table 2. Chemical composition of materials utilized in the research

Oxide	Composition of Raw Materials (weight %)			
	Reclaimed Tin Mine Sand	Soda lime waste	Binder	Foaming Agent
SiO_2	90.02	74.87	40.32	0.33
Na_2O	0.03	4.85	3.85	0.04
CaO	0.88	8.71	5.57	70.13
MgO	0.24	2.24	2.43	0.61
Al_2O_3	5.21	0.82	0.44	0.05
K_2O	0.89	0.59	0.13	0.03
Fe_2O_3	1.63	1.04	1.03	0.13

(continued)

Table 2. *(continued)*

Oxide	Composition of Raw Materials (weight %)			
	Reclaimed Tin Mine Sand	Soda lime waste	Binder	Foaming Agent
TiO_2	0.47	0.21	0.28	0.02
LOI	1.4	6.67	45.95	28.65

3.2 Temperature-Dependent Densification and Mechanical Properties

As sintering temperature rises from 850 °C, 900 °C, 950 °C and 1000 °C, tiles densify, showing higher bulk density and lower water absorption from pore closure and vitrification. MOR increases to 15.20 MPa at 1000 °C due to microstructural consolidation, while linear shrinkage rises with enhanced viscous flow [10] (Fig. 2).

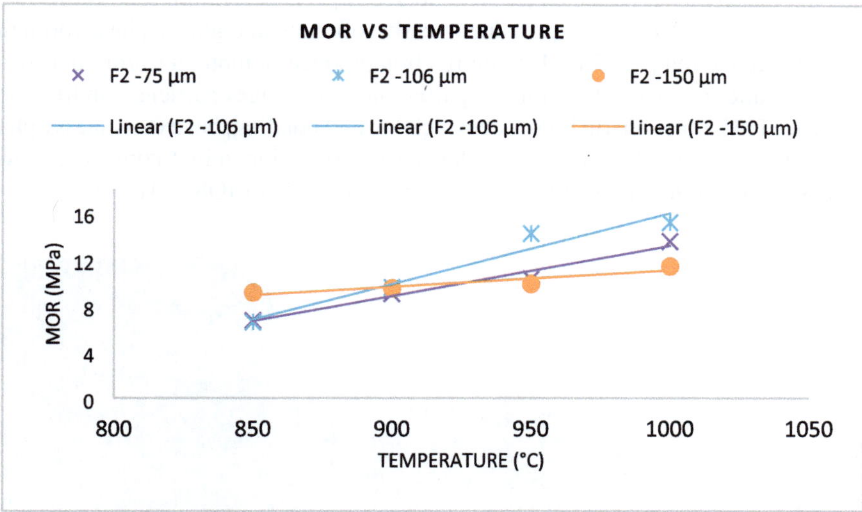

Fig. 2. Densification characteristics of glass ceramic tiles after sintering at the temperature in range 850 to 1000 °C

At 1000 °C, 106 µm quartz particles reached a MOR of 15.20 MPa, exceeding ISO 10545-4 standards. Strength is highly temperature-dependent ($R^2 = 0.97$). The 75 µm fraction had slightly lower MOR ($R^2 = 0.95$), while 150 µm was less consistent ($R^2 = 0.87$), showing 106 µm as the best size for densification and performance.

3.3 Microstructural Balance and Mechanical Performance Trade offs

Although shrinkage increases, bulk density stays stable, indicating balanced pore structures. Particle size significantly affects density: medium quartz 1.92 g/cm^3, fine

1.84 g/cm^3, and coarse 1.94 g/cm^3 (Table 3). Using 106 μm particles at 950–1000 °C raises shrinkage (~0.42%) while lowering water absorption (12.31%).

Table 3. Correlation of Particle Size with Ceramic Properties using Coefficient of Determination (R^2) at temperature 1000 °C

Particle Size (μm)	Water Absorption (%)	R^2 (WA)	Bulk Density (g/cm^3)	R^2 (BD)	Linear Shrinkage (%)	R^2 (LS)
75	13.06	0.8179	1.84	0.0032	−0.44	0.7784
106	12.31	0.8494	1.92	0.5331	−0.42	0.7294
150	9.42	0.0515	1.94	0.4016	−0.31	0.8257

Particle size strongly affects sintered tile properties. Finer fractions (75 μm and 106 μm) improve water absorption ($R^2 \approx 0.82$–0.85) and shrinkage ($R^2 \approx 0.73$–0.83), while bulk density is less affected ($R^2 = 0.00$–0.53) [11]. The 75 μm sand forms many surface cavities ($\approx$19.45–$\gtrsim$180 μm) due to early softening and glassy-phase formation from Fe_2O_3 reduction [12]. The 106 μm fraction produces a more uniform microstructure with smaller pores ($\approx$48.37 μm) as partial melting bridges particles, limiting pore coalescence [13]. Finer quartz boosts densification and bonding, but excess glassy phase can cause microcracks. The 106 μm size balances vitrification, minimizing pore growth and cracking, resulting in lower water absorption and higher MOR [14] (Fig. 3).

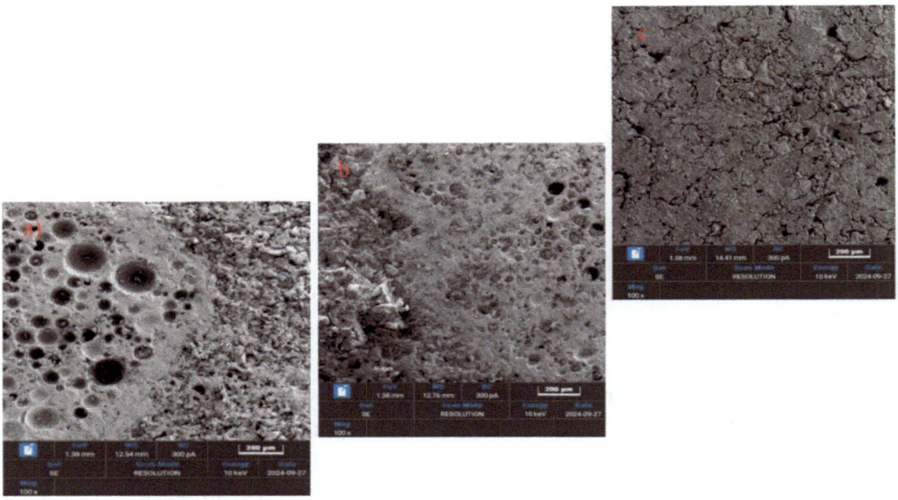

Fig. 3. SEM-EDX characterization of samples at 100x magnifycation for varying particle sizes: a) 75 μm, b) 106 μm, and c) 150 μm

4 Sustainability and Industrial Relevance

Reducing the firing temperature from 1200 °C to 1000 °C saves 15–20% energy (≈0.49–0.65 GJ/t) and lowers costs. For 10,000 t/year, this saves 4,875–6,500 GJ, RM 230,000–305,000 (≈USD 49,000–65,000), and 273–364 t CO_2. Using 49 wt% reclaimed mine sand repurposes ~4,900 t of waste annually, reducing landfill by 4.9–24.5%, demonstrating financial, environmental, and circular-economy benefits.

5 Conclusion

Glass-ceramic tiles from recycled tin-mining sand showed optimal properties at 1000 °C, meeting ISO 10545 standards. Finer particles improved sintering, though temperature most strongly affected strength. The process is energy efficient, cost-effective, and sustainable. Future work should explore higher temperatures, durability, and Life Cycle Assessment to enhance performance and environmental impact.

Acknowledgments. The authors sincerely thank the Senior Director and the support team of Malaysia's Mineral Research Centre for their guidance and assistance throughout this study.

Disclosure of Interests The authors declare no financial or personal conflicts of interest related to this research.

References

1. Kobayashi, Y., Maekawa, H., Okuno, W., Mizuno, T., Sugimura, T., Negawa, A.: Effect of quartz grain size on elastic and thermal expansion properties of porcelain. J. Ceram. Soc. Jpn. **131**(8), 451–457 (2023). https://doi.org/10.2109/jcersj2.23050
2. Andreola, F., Lancellotti, I., Pozzi, P., Barbieri, L.: Eco-compatible construction materials containing ceramic sludge and packaging glass cullet. Appl. Sci. **11**(8), 3545 (2021). https://doi.org/10.3390/app11083545
3. Silva, L.S., Amario, M., Stolz, C.M., Figueiredo, K.V., Haddad, A.N.: A comprehensive review of stone dust in concrete: mechanical behavior, durability, and environmental performance. Buildings. **13**(7), 1856 (2023)
4. Ziejewska, C., Grela, A., Hebda, M.: Influence of waste glass particle size on the physico-mechanical properties and porosity of foamed geopolymer composites based on coal fly ash. Materials. **16**(5), 2044 (2023)
5. Yusuf, M.O., Saleh, T.A., Suleiman, H.T., Gidado, S.S., Abdullahi, M.A.: Microstructural characteristics, modeling of mechanical strength and thermal performance of industrial waste glass blended concrete. Appl. Sci. **12**(17), 8600 (2022). https://doi.org/10.3390/app12178600
6. Cetin, S.: Production of sintered glass-ceramic composites from low-cost materials. Ceram. Int. **49**(13), 22386–22392 (2023). https://doi.org/10.1016/j.ceramint.2023.04.068
7. ASTM C373-88: Standard Test Method for Water Absorption, Bulk Density, Apparent Porosity, and Apparent Specific Gravity of Fired Whiteware Products, Ceramic Tiles, and Glass Tiles. ASTM International, West Conshohocken (2008). https://doi.org/10.1520/C0373-88R06

8. International Organization for Standardization: ISO 13006. Ceramic Tiles — Definitions, Classification, Characteristics and Marking. International Organization for Standardization, Geneva (2018)
9. Qian, C., Li, Y., Liu, Z., Wang, Q., Zhang, J.: The effect of particle size distribution on the microstructure and properties of Al_2O_3 ceramics formed by stereolithography. Ceram. Int. **48**(15), 21600–21609 (2022). https://doi.org/10.1016/j.ceramint.2022.04.133
10. Ji, Y., Li, E., Zhu, G., Wang, R., Sha, Q.: Preparation and performance of ceramic tiles with steel slag and waste clay bricks. Materials. **17**(8), 1755 (2024). https://doi.org/10.3390/ma17081755
11. Ochen, W., D'ujanga, F.M., Oruru, B.: Effect of quartz particle size on sintering behavior and flexural strength of porcelain tiles made from raw materials in Uganda. Adv. Mater. **8**(1), 33–40 (2019)
12. Kim, Y., Park, H.: A value-added synthetic process utilizing mining wastes and industrial byproducts for wear-resistant glass ceramics. ACS Sustain. Chem. Eng. **8**(5), 2196–2204 (2020). https://doi.org/10.1021/acssuschemeng.9b05884
13. Zhang, N., Xie, F., Wang, W., Lu, D.: Study on performance and mechanism of foamed ceramics based on lead fuming furnace slag. Ceram. Int. **50**(2), 3307–3318 (2024). https://doi.org/10.1016/j.ceramint.2023.11.072
14. Suárez González, J., Lopez Boadella, I., López Gayarre, F., López-Colina Pérez, C., Serrano López, M., Stochino, F.: Use of mining waste to produce ultra-high-performance fibre-reinforced concrete. Materials. **13**(11), 2457 (2020). https://doi.org/10.3390/ma13112457

Seismic Disaster Response of Low-Rise Buildings with Passive Viscous Dampers Under Pseudo-Dynamic Loading

Chun Chieh Yip[1(✉)], Jing Ying Wong[2], Foo Wei Lee[1], Yann Ling Goh[1], Yen Sin Lim[1], and Yue Yue Zhao[1]

[1] Department of Civil Engineering, Lee Kong Chian Faculty of Engineering and Science, Universiti Tunku Abdul Rahman, 43000 Kajang, Malaysia
yipcc@utar.edu.my

[2] Department of Civil Engineering Programme, Faculty of Science and Engineering, University of Nottingham, Malaysia, 43500 Semenyih, Selangor, Malaysia

Abstract. Earthquakes are among the most destructive natural disasters, causing fatalities and destruction globally. Malaysia is considered a region with a low seismicity profile. However, earthquakes in neighbouring countries such as Indonesia and the Philippines have induced tremors that affected buildings in Malaysia. Three brands viscous damper systems have been introduced and are considered effective in enhancing the seismic resistance of buildings. The main focus of this research is to assess the inter-storey drift performance and mode shape of a scaled 1:8 reinforced concrete building resembling critical parts of a high school building's with and without dampers under dynamic loads test. The model specifications were derived using Similitude Theory and Buckingham's Pi Theorem to establish the correlation between the prototype and the downscaled model. The failure modes and damage mechanisms of the scaled model under different levels of earthquake intensity revealed that inter-storey drift was significantly large when the ground experienced substantial movement with low aggressiveness. The structure exhibited a higher translational mode of oscillation, transitioning from single curvature to double curvature, as ground acceleration intensified. The key finding is the rooftop displacements of the scaled model were reduced by 45%, 63%, and 34% under different levels of ground motion when equipped with APIDO, SKK, and ESPADA viscous dampers, respectively. Hence, the novelty of this research lies in identifying that the APIDO viscous damping system is most effective in absorbing large displacements, while the SKK viscous damping system is more suitable for mitigating lateral displacements generated under intense ground vibration.

Keywords: Seismic · Disaster Response · Low-Rise Buildings · Viscous Dampers · Pseudo-Dynamic Loading

1 Introduction

Earthquakes are among the most destructive natural disasters in human history, causing the deaths of hundreds of thousands of people and the destruction of property worth billions of dollars. The deadliest earthquake ever recorded occurred in Shaanxi, China,

M. A. b. Jamlos et al. (Eds.): ICSET 2025, LNNS 1897, pp. 141–147, 2026.
https://doi.org/10.1007/978-3-032-20722-7_17

on January 23, 1556, with a magnitude of 8, resulting in approximately 830,000 deaths [1, 2]. Malaysia, located on the Sunda Plate within the Eurasian Plate and surrounded by the Philippine Plate and the Indo-Australian Plate, is considered a region with a low seismicity profile. However, regional earthquakes originating from Sumatra's active fault lines and subduction zones have caused seismic events in Malaysia. In addition, the eastward movement of the Sunda Plate at a rate of about 10 mm/year relative to the Eurasian Plate also contributes to local earthquakes in Malaysia [3].

From 1922 to 2020, at least 59 earthquake events in West Malaysia were recorded by international and national seismological centres, including a local earthquake caused by a strike-slip fault along the Bukit Tinggi Fault Zone [4]. Therefore, seismic performance must be improved in structural design to minimize earthquake response and ultimately reduce the devastating impacts of earthquakes [5]. Energy dissipation devices or system, such as viscous dampers, should be utilized to protect structures against sudden shocks and destructive vibrations.

In conventional buildings, earthquake energy is dissipated through the yielding of construction materials, such as the formation of plastic hinges at beams and columns [6]. However, this mechanism can lead to structural failure if the plastic deformation capacity of the structural members within the system is insufficient [7]. Viscous dampers, on the other hand, absorb earthquake energy as compressible hydraulic fluid flows through small orifices in an enclosed cylinder system [8]. As a result, the ability of the structure to withstand seismic activity increases, since viscous damper systems effectively reduce the impact of earthquakes [9].

This study examines the structural behaviour and damage mechanisms of a scaled 1:8 reinforced concrete structure accordance to Buckingham and Similitude Theorem under pseudo-dynamic loading [10]. Scaled model are commonly been use as prototype and yield reasonable result with the correct customization of the scaled factors agreed by researchers [11]. The laboratory experiment conducted can serve as guidance or reference for improving displacement response as well as enhancing the seismic resistance of structures during earthquakes. The study of viscous damping systems is also crucial for improving building performance, as seismic effects can be mitigated by installing viscous dampers system.

The aim of this research is to investigate the inter-storey drift performance of a scaled reinforced concrete building with and without dampers. The specific objectives include fabrication of a downscaled one-bay, three-storey reinforced concrete model based on Buckingham's Pi Theorem and Similitude Law. Assessment of the building's inter-storey drift with and without equipped with APIDO, SKK, and ESPADA viscous dampers and comparison of the mode shapes of the building.

2 Methodology

2.1 Building Specification

The dimension of the structural model was scaled down 1:8 from actual building bracket in red as shown in Fig. 1. The total height of the 3-storey structure is 1.5 m while each storey is 0.5 m. The footings bolted on the shaking table have a dimension of 175 mm in width and length, while the footing depth is 50 mm. The dimension of the scaled beam is

31 mm in width and 75 mm in depth. The column in the model is a square column with a dimension of 40 mm. The slab is scaled to 16 mm in thickness, 830 mm in width and length for the whole structure. According to the dimension of scaled structural elements, the total volume of concrete in the structural model is around 0.072 m^3, equivalent to 170 kg of mass with concrete Grade 30 and bar schedule of 3 mm ∅ as main bar and 1 mm ∅ as shear links. Since the shaking table in the Advanced Geotechnical Laboratory is capable of sustaining 3000 kg, the shaking table test is able to be performed using this model. Hence, the dimension of the downscaled 1 bay 3-storey reinforced concrete model is acceptable. Total of four LVDTs and accelerometer were installed on each of the model as shown in the Fig. 1.

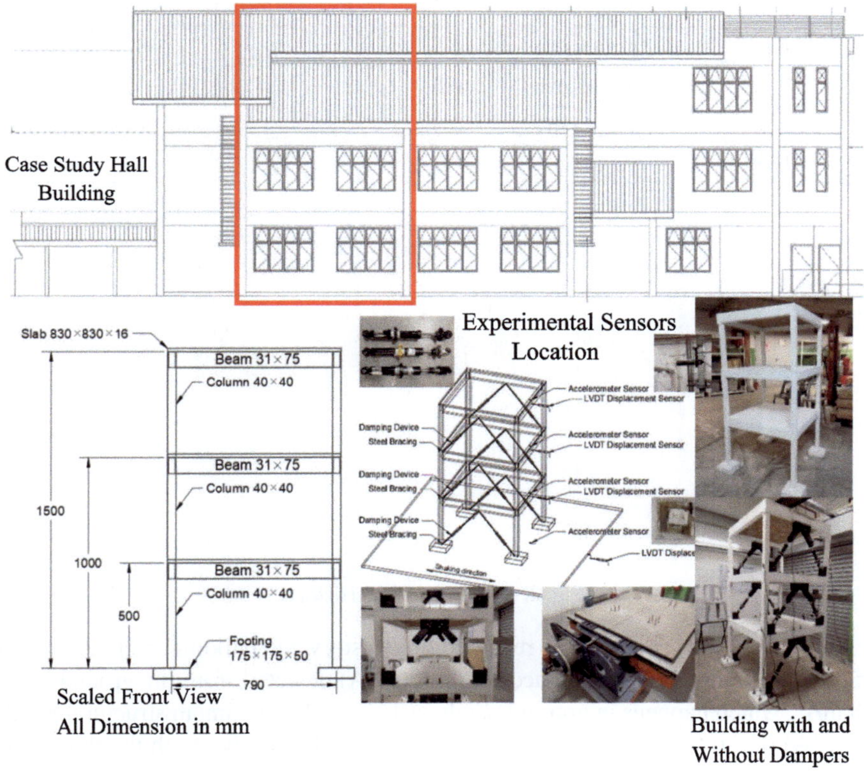

Fig. 1. Building Specification.

2.2 Input Ground Motions

In this study, ten levels of earthquake simulation were selected to conduct the shaking table test. The earthquake simulation was carried out randomly without predetermination of peak ground acceleration. Each level has a varying intensity in terms of displacement and aggressiveness of shaking, as shown in Table 1. The earthquake simulation gets intense when the shaking is more aggressive with a shorter displacement. Unit displacement (referring to motor's rotation) and frequency of each level were determined and served as the input in the operating shaking table.

Table 1. Intensity of Each Level of Earthquake Simulation.

Level	Intensity		Shaking Table Inputs	
	Displacement	Aggressiveness	Unit Displacement (mm/mm)	Frequency (Hz)
1	Large	Very low	3	1
2	Large	Low	3	2
3	Moderate	Low	1.5	3
4	Moderate	Low	1.9	3
5	Moderate	Aggressive	1	5
6	Short	Aggressive	0.7	7
7	Short	Aggressive	0.6	8
8	Short	Aggressive	0.6	10
9	Short	Very high	0.3	15
10	Short	Very high	0.5	15

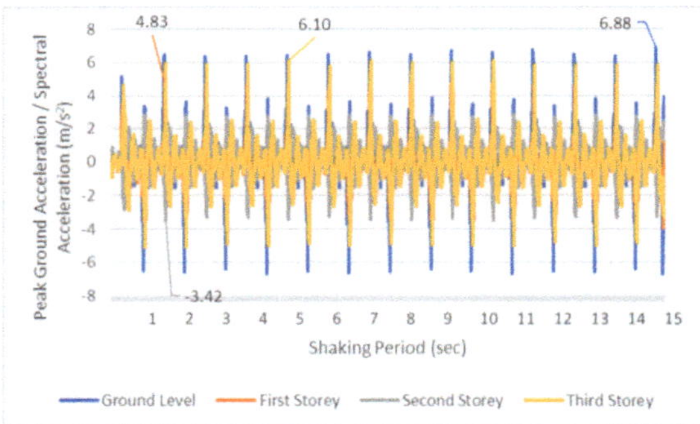

Fig. 2. Recorded Time History

Figure 2 show a sample of the recorded time history and building inter storey acceleration. Maximum Peak ground acceleration (PGA) detected from the ground level is 6.88 m/s^2, which happens in approximately at 14.7 s. The maximum structural acceleration (SA) of the first storey was recorded concurrently together with the third storey, where both happened at 1.5 s of 4.83 m/s^2, while for the second storey is -3.42 m/s^2. Lastly, maximum SA at the third storey has the value of 6.10 m/s^2, which happens approximately at 4.8 s.

3 Results and Discussion

3.1 Bare Frame Structure (Without Damper)

Table 2 shows the inter-storey drift calculated from the maximum displacement at each floor for each shaking level. The results for Level 1 and Level 2 show that the maximum displacement at Level 2 is larger than at Level 1. Although the unit displacement is consistent at 3 mm/mm for both levels, the shaking frequency at Level 2 is 1 Hz higher, causing more significant motion.

At Level 4, the structure's overall displacement is greater than at Level 3, despite the shaking frequency being the same (3 Hz). This is due to the larger unit displacement (0.4 mm/mm) at Level 4. Similar results are observed at Levels 9 and 10, where an increase in either shaking frequency or unit displacement leads to greater maximum displacement.

Level 5 simulates more aggressive seismic activity, where maximum displacement decreases significantly compared to previous levels. The building's behaviour shifts from large lateral movement to more intense vibration.

The highest inter-storey drift of 5.16 mm occurs at Level 4. At these levels, rooftop displacement is about 10 mm greater than ground displacement. This indicates that the higher floors experience the largest displacement during moderate shaking, which could lead to significant damage or collapse.

Table 2. Inter-Storey Drift for Bare Frame Structure at All Levels.

Level	Maximum displacement (mm)				Inter-storey drift (mm)		
	GF	1F	2F	3F	GF - 1F	1F - 2F	2F - 3F
1	11.28	11.93	12.63	12.27	0.65	0.70	-0.36
2	11.76	13.21	15.28	15.16	1.45	2.08	-0.13
3	7.91	12.60	16.31	17.53	4.69	3.71	1.22
4	8.52	13.68	17.46	18.26	5.16	3.77	0.80
5	3.85	2.59	5.96	7.10	-1.26	3.37	1.13
6	2.43	2.05	2.98	3.71	-0.38	0.94	0.73
7	2.52	2.47	3.27	3.58	-0.06	0.80	0.31
8	2.61	2.62	2.99	3.45	0.01	0.36	0.47
9	1.16	1.40	0.55	1.10	0.25	-0.86	0.56
10	1.54	2.00	1.96	2.36	0.46	-0.04	0.40

The mode shape of the bare frame structure is shown in Fig. 3. From Level 1 to Level 4, the structure exhibits a single curvature shape, with translational motion in one direction. At Levels 5 to 8, the mode shape changes to double curvature, indicating movement in two directions. At Levels 9 and 10, the mode shape becomes double S curvature, where the second floor has lower displacement than the first and third floors.

3.2 Structure (With APIDO, SKK & ESPADA Dampers)

Figure 4 shows the structural mode shapes for the APIDO, SKK, and ESPADA dampers. At Level 1, there was no noticeable reduction in lateral displacement for the damped structures. However, the bare frame structure had a single curvature mode shape, while all damped structures had a preferred double curvature shape.

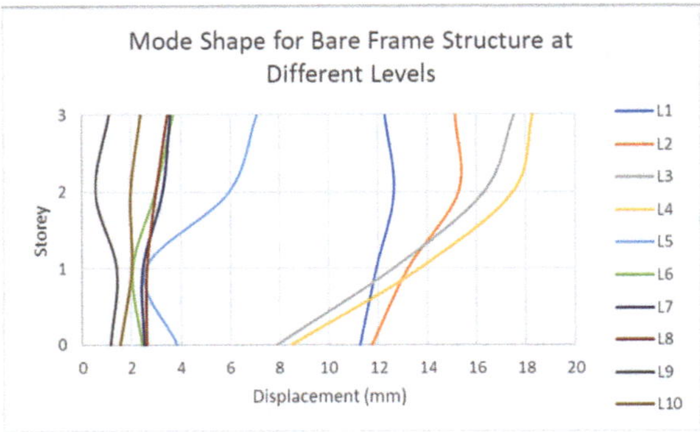

Fig. 3. Mode Shape for Bare Frame Structure at Different Levels

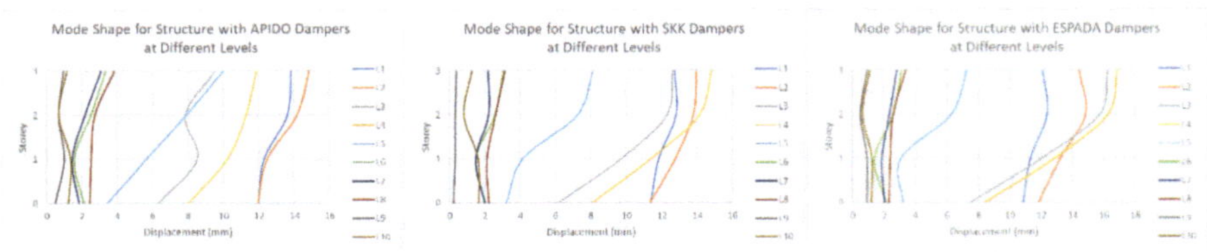

Fig. 4. Comparison of Structural Mode Shape of different damper

Table 3 shows the summary reduction in percentage for rooftop displacement of each damper at all levels in the shaking table test. The results show general improvements in the reduction of rooftop displacement of the structure throughout most of the earthquake simulation.

Table 3. Rooftop Displacement and Reduction Percentage of Each Damper at All Levels.

Level	Intensity		Rooftop displacement (mm)				Reduction percentage (%)		
	Displacement	Aggressiveness	Bare frame	APIDO	SKK	ESPADA	APIDO	SKK	ESPADA
1	Large	Very low	12.27	13.82	12.73	12.12	-12.61	-3.72	1.21
2	Large	Low	15.16	14.86	13.97	14.45	1.98	7.83	4.69
3	Moderate	Low	17.53	9.57	12.61	16.25	45.43	28.08	7.32
4	Moderate	Low	18.26	11.88	14.85	16.86	34.91	18.66	7.68
5	Moderate	Aggressive	7.10	10.01	8.15	7.28	-41.05	-14.81	-2.60
6	Short	Aggressive	3.71	3.36	3.05	3.16	9.46	17.91	14.85
7	Short	Aggressive	3.58	3.09	2.18	2.83	13.58	39.11	20.78
8	Short	Aggressive	3.45	3.82	3.15	3.43	-10.49	8.91	0.79
9	Short	Very high	1.10	0.97	0.40	0.92	12.47	63.44	16.27
10	Short	Very high	1.66	1.14	1.27	1.10	31.16	23.71	33.70

The end results show that the SKK damper achieved the highest rooftop displacement reduction, up to 63.44% at Level 9. The APIDO damper provided the second-best improvement, reducing displacement by 45.43% at Level 3, while the ESPADA damper performed best at Level 10 with a 33.70% reduction.

4 Conclusion

Inter-Storey Drift where LVDTs were installed to measure lateral movements of the structure. Tests with three viscous dampers (APIDO, SKK, and ESPADA) showed that dampers significantly reduced displacement. Rooftop displacements were reduced by up to 63% (SKK), 45% (APIDO), and 34% (ESPADA). Overall, the study confirms that viscous dampers improve seismic performance by reducing inter-storey drift and promoting more stable mode shapes, making structures safer against earthquake.

Acknowledgments. Author is grateful this research was supported by UTAR RF (IPSR/RMC/UTARRF/2022-C1/Y05) and External Grant from YI FENG CONSTRUCTION SDN. BHD.

Disclosure of Interests The authors have no competing interests to declare that are relevant to the content of this article.

References

1. Zhang, Y., Xu, Y., Du, P., Li, W., Chen, L., Tian, Q.: The volume calculation method of rock collapses and loess landslides triggered by the 1556 AD Huaxian M 8½ earthquake in Shaanxi Province, China. Earthq. Res. China. **34**(4), 599–616 (2020)
2. Yip, C.C., Marsono, A.K., Wong, J.Y., Lee, S.C.: Seismic performance of scaled IBS block column for static nonlinear monotonic pushover experimental analysis. J. Teknol. (Sci. Eng.). **80**(1), 1–9 (2018)
3. Lee, J.Y., Chow, T.L., Chong, F.L., Cheah, Y.B., Lau, T.L.: Development of seismic microzonation maps for low to moderate seismicity regions using microtremor observation data. Res. Square. **PPR577919**, 1–31 (2022)
4. Nazaruddin, D.A., Duerrast, H.: Intraplate earthquake occurrence and distribution in Peninsular Malaysia over the past 100 years. SN Appl. Sci. **3**(7), 693 (2021)
5. Yip, C.C., Wong, J.Y., Ling, L., Goh, Y.L., Ong, H.E.: Dynamic response of scaled structure with one liquid tuned mass damper. Case Stud. Constr. Mater. **14**, e00512 (2021)
6. Lee, T.L., Yip, C.C., Ling, L., Kwong, K.Z., Wong, J.Y., Ean, L.W., 2025, Correlation of sustainable material in pozzolanic composite fly ash concrete. IOP Conf. Ser.: Earth Environ. Sci. 1467(1), 012020, Scopus
7. Bayrak, B., Kaplan, G., Akarsu, O., Aydin, A.C.: The plastic hinge length prediction of RC members by using ANN. Sādhanā – Acad. Proc. Eng. Sci. **48**(3), 114 (2023)
8. Yip, C.C., Wong, J.Y., Amran, M., Fediuk, R., Vatin, N.I.: Reliability analysis of reinforced concrete structure with shock absorber damper under pseudo-dynamic loads. Materials. **15**(7), 2688 (2022)
9. Yip, C.C., Wong, J.Y., Lim, S.S., Ooi, J.S.: Pseudo dynamic loads energy dissipation in mechanical shock absorber seismic damper. E3S Web Conf. **347**, 03012 (2022)
10. Chong, W.K., Wong, J.Y., Yip, C.C., Ling, L., Amran, M.: Evaluating viscous damping for seismic energy dissipation in downscaled concrete structures under consistent earthquake loadings. Struct. **78**, 109313 (2025)
11. Yip, C.C., Marsono, A.K.: Structural seismic performance of reinforced concrete block system for two storeys safe house. J. Teknol. (Sci. Eng.). **78**(2), 83–97 (2016)

Automation of IBS Scoring Through a BIM Revit–Dynamo Framework

Jing Ying Wong[1](✉), Chun Chieh Yip[2], Kai Yang Tiew[1], Yu Hoe Tang[1], and Zhi Xian Chew[1]

[1] Department of Civil Engineering, Faculty of Science and Engineering, University of Nottingham Malaysia, 43500 Semenyih, Selangor, Malaysia
JingYing.Wong@nottingham.edu.my

[2] Department of Civil Engineering, Lee Kong Chian Faculty of Engineering and Science, Universiti Tunku Abdul Rahman, Cheras, 43000 Kajang, Selangor, Malaysia

Abstract. The Malaysian construction industry is experiencing rapid digital transformation through national initiatives such as the National Construction Policy (NCP) and the Construction Industry Transformation Programme (CITP). At the forefront of this transformation are the Industrialised Building System (IBS) and Building Information Modelling (BIM), which aim to improve productivity, sustainability, and overall project quality. This research introduces a BIM Revit–Dynamo framework developed to automate the IBS scoring process for structural and wall system components, using Dynamo's visual programming within Autodesk Revit. A prototype was created and tested, demonstrating full accuracy compared to manual scoring methods. Industry experts confirmed the framework's potential to enhance efficiency, accuracy, and IBS adoption, while highlighting the importance of further customization for wider application. The study underscores the role of BIM-driven automation in advancing Malaysia's construction digitalisation, providing streamlined workflows, reduced risks, and increased competitiveness to support national goals and Construction 4.0 strategies.

Keywords: Building Information Modelling (BIM) · Industrialised Building System (IBS) · Automation

1 Introduction

Malaysia's construction sector is rapidly digitalising through the National Construction Policy 2030 [1] which promotes the adoption of digital technologies, innovation, and automation to achieve sustainable infrastructure and address global challenges, alongside the Construction Industry Transformation Programme (CITP) [2] and the Construction 4.0 Strategic Plan [3], which provide a roadmap for embracing the Fourth Industrial Revolution (IR 4.0) to enhance industry productivity and competitiveness. Central to this agenda are Industrialised Building System (IBS) and Building Information Modelling (BIM), promoted to enhance productivity, sustainability, and quality. Government support, including the IBS Score [4] and levy exemptions, has encouraged wider adoption.

M. A. b. Jamlos et al. (Eds.): ICSET 2025, LNNS 1897, pp. 148–153, 2026.
https://doi.org/10.1007/978-3-032-20722-7_18

Key initiatives include the IBS Roadmap (2011–2015) [5], the Construction 4.0 Strategic Plan (2021–2025) [3], and the upcoming BIM mandate effective July 1, 2025, requiring BIM for projects above RM10 million across public and private sectors [6] tends to be seen as a cutting-edge technology and processes [7]. The Construction Industry Development Board (CIDB) has been instrumental in advancing these policies through initiatives such as the Malaysian BIM Steering Committee and the IBS Manufacturer & Product Registration Programme, aligning Malaysia's construction industry with Industry 4.0 and international best practices.

To evaluate adoption, the Construction Industry Development Board (CIDB) introduced the IBS Scoring System [4] which quantifies the extent of IBS application in projects, with a minimum score of 70 points mandated for government projects. Complementing this, the Quality Assessment System in Construction (QLASSIC) [8], based on Construction Industry Standard (CIS 7), provides a structured evaluation of workmanship quality across structural, architectural, and M&E trades. Together, BIM, IBS scoring, and QLASSIC establish a comprehensive framework that integrates policy direction, technical standards, and performance benchmarking, driving Malaysia towards a more innovative, sustainable, and globally competitive construction sector.

The assessment of Industrialised Building System (IBS) adoption in construction projects currently relies on manual IBS Scoring, which is labour-intensive, prone to inconsistency, and inefficient for complex projects. Although Building Information Modelling (BIM) provides a data-rich platform, its potential for automating IBS scoring has not been fully realised. For instance, BIM has been integrated with Generative Design (GD) to automate design exploration and optimisation, demonstrating its capability to support automation in construction processes [9]. This gap creates challenges in ensuring accuracy, compliance, and timely decision-making, thereby limiting the effectiveness of policy-driven IBS implementation. Therefore, there is a need to develop an automated BIM-based framework, using Revit and Dynamo, to streamline IBS scoring and enhance standardisation in Malaysia's construction industry.

2 Methodology

2.1 Dynamo Script Development for IBS Scoring

Dynamo is an open-source visual programming tool that seamlessly integrates with Revit, providing a node-based environment for automating and enhancing design and modelling tasks within BIM platforms. Building on this capability, the methodology developed in this research emphasizes two key IBS score components: structural systems and wall systems. A prototype was developed, as shown in Fig. 1, utilizing Dynamo scripts to automate the scoring process. The categories of IBS elements were defined in accordance with CIS 18:2018 (CIDB, 2018), each assigned an appropriate IBS factor. For instance, floor slabs were categorized and scored based on their construction method, with the assignment of IBS Factor categories demonstrated in Fig. 2. Similar classifications were also applied to other elements such as roofs and walls, ensuring consistency with industry standards. This systematic categorization provided the foundation for accurate automation of IBS scoring within the Dynamo–Revit environment.

Innovacia Sdn. Bhd

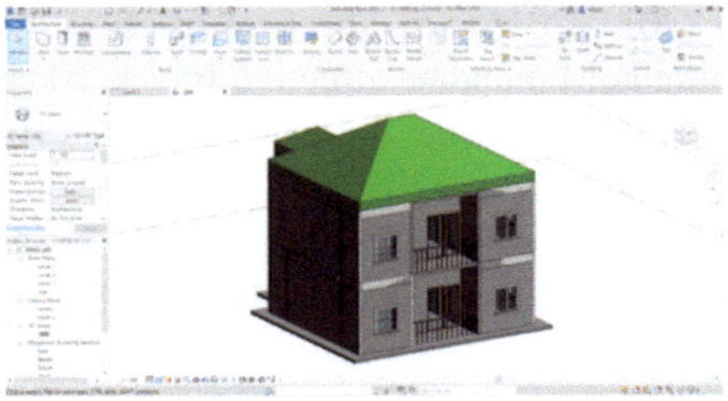

Fig. 1. A Revit model developed by

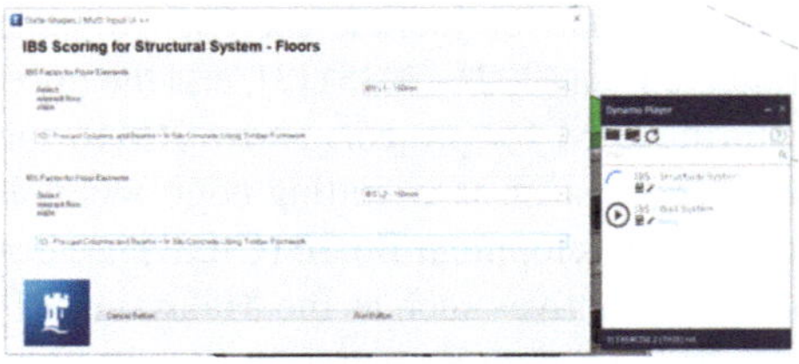

Fig. 2. Assignment of IBS Factor categories for floor slabs

2.2 Sample Dynamo Script for Structural Systems

Figure 3 illustrates the design script workflow for structural systems in Dynamo, while Figure 4 presents a sample user interface for floor slabs within the structural system. The workflow begins with a user interface form that prompts the user to specify the number of element "Types" to which IBS Factors will be assigned. The List of Repeated Item node processes this input to determine how many times the subsequent script will run, enabling multiple elements to be assigned to IBS Factors efficiently. To support systematic selection of elements by "Type," the workflow employs the All Elements of Type node, which retrieves all element types for floor slabs and roofs. These are then displayed in a dropdown menu using the DropDown Data node by Data-Shapes.

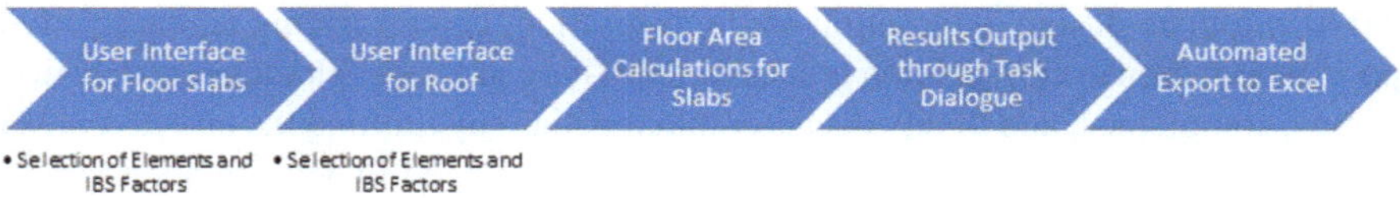

Fig. 3. Design script workflow for structural systems in Dynamo

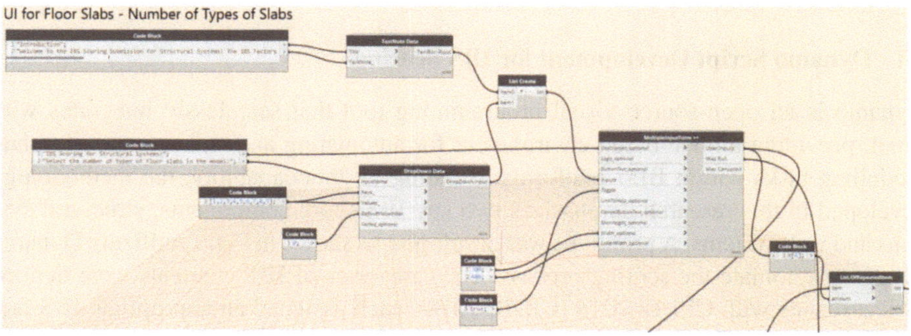

Fig. 4. Sample of design script - Floor slab user interface for structural system

Figure 5 shows the user interface in Revit when the Dynamo script named "IBS – Structural System" is executed. Users can select their desired value to specify the number of types of elements and proceed to the next user interface form.

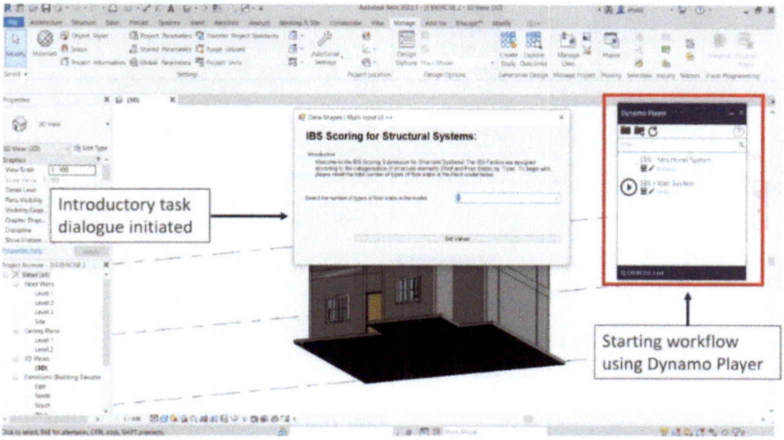

Fig. 5. User interface (UI) for structural system in Revit

2.3 Qualitative Survey—Semi-structured Interview

A semi-structured interview was held with four AEC professionals to evaluate the BIM-based IBS Scoring framework. After a brief presentation on its background, objectives, and Revit demonstration, respondents were asked targeted questions to provide detailed insights.

Part 1: Overview of BIM and IBS

- Challenges in adopting BIM and IBS.
- Current status of practice and alignment with government goals.
- Trends in construction technologies.
- Government incentives supporting BIM and IBS.
- Expectations for BIM's future role in IBS and smart construction.

Part 2: Feedback on IBS-BIM Framework

- Benefits of BIM automation for IBS applications.
- User-friendliness and potential adoption of Revit API.
- Ways to encourage industry uptake.
- Value of early IBS scoring in improving design and achieving GBI compliance.

3 Results and Discussion

3.1 Verification of Revit API Output

After executing the script in Dynamo Player and assigning IBS Factors to each element type for structural and wall systems, the Revit API generated the results. For the structural system, the IBS Score was 33.55 out of 50, as shown in Fig. 6. The breakdown of calculations was automatically exported to Excel, detailing element type, area, IBS factor, coverage, and corresponding scores (Figure 7). To ensure accuracy, manual calculations based on Revit floor plans were performed, and the results confirmed consistency with the automated output. All results were identical to the manual calculations and fully accurate.

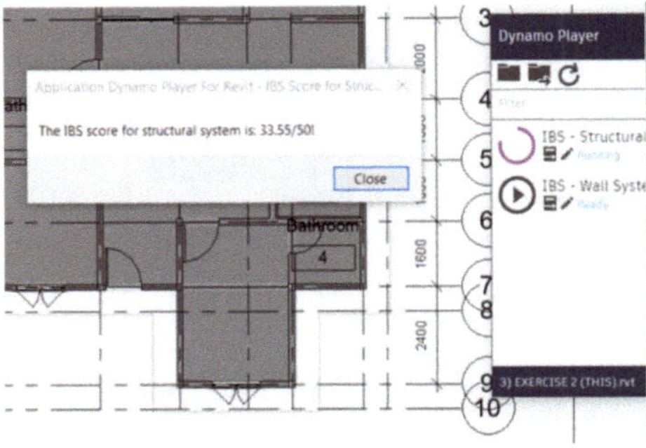

Fig. 6. Output of case study results for structural system through task dialogue

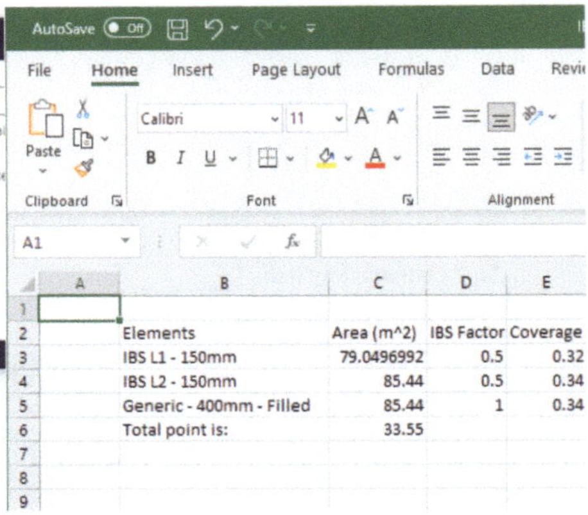

Fig. 7. Automated export of results for structural system to Excel spreadsheet

3.2 Experts' Feedback Analysis

Part 1: Overview of BIM and IBS

Experts identified key challenges in adopting BIM and IBS in Malaysia, including issues of demand, productivity, costs, and OSH. Prefabrication technologies, such as PPVC in Singapore, demonstrate how high demand can accelerate adoption (Respondent 1), while in Malaysia uptake remains slower. Nonetheless, BIM and IBS are gaining momentum, supported by government mandates for public and large-scale private projects. This aligns with the Construction 4.0 agenda, which promotes digitalisation and automation. Experts agreed that stronger collaboration between government, contractors, and consultants is essential, with further policy incentives needed to strengthen adoption. Looking ahead, BIM is expected to play a central role in embedding IBS scoring and compliance checks within smart construction workflows.

Part 2: Feedback on IBS-BIM Framework

Respondents highlighted the framework's novelty in automating IBS scoring through Dynamo, reducing manual effort and ensuring consistency with CIS 18:2018 standards. Respondent 2 noted that while Singapore has implemented prefabricated prefinished volumetric construction (PPVC) with a scoring system since 2007, the use of Dynamo to integrate scoring processes with building systems, as demonstrated in Malaysia, is still uncommon there. Feedback was unanimously positive: all agreed that the framework enhances design workflows and holds strong potential for industry use; three out of three found the Revit API application user-friendly; and all four agreed Dynamo could see wider industry uptake. Early IBS scoring was also valued for supporting sustainable design and GBI compliance.

Two considerations were raised for successful adoption: (i) further refinement to suit company-specific needs, and (ii) stronger government support to drive industry-wide implementation. Overall, the feedback validates the framework's relevance while emphasising the importance of aligning it with industry readiness and national policy directions.

4 Conclusion

This research presents a BIM–IBS integration framework that automates IBS scoring using Revit API and Dynamo, developed in response to industry needs and validated through case studies. The framework proved accurate, reliable, and consistent with Malaysia's Construction 4.0 strategies. Feedback from industry experts confirmed its potential to improve efficiency, accuracy, and IBS adoption, while also stressing the importance of customization and stronger policy support. By digitizing and streamlining a traditional manual process, the framework minimizes errors, saves time, and ensures compliance with national standards. Ultimately, the research demonstrates the practicality of BIM-based automation in advancing Malaysia's construction digitalization agenda, paving the way for smarter workflows, enhanced competitiveness, and future innovation in construction technology.

Acknowledgements. The research is supported by Innovacia Sdn Bhd and PLYTEC Group for the valuable training session on IBS score submission and Revit model.

References

1. Ministry of Works Malaysia: National Construction Policy 2030 (NCP 2030). Ministry of Works Malaysia, Putrajaya (2021)
2. CIDB Malaysia: Construction Industry Transformation Programme (CITP) 2016–2020. Ministry of Works Malaysia, Putrajaya (2015)
3. CIDB Malaysia: Construction 4.0 Strategic Plan 2021–2025. CIDB Malaysia, Putrajaya (2021)
4. CIDB Malaysia: IBS Score Manual (Edition 2018). CIDB Malaysia, Kuala Lumpur (2018)
5. CIDB Malaysia: IBS Roadmap 2011–2015. CIDB Malaysia, Kuala Lumpur (2010)
6. Ministry of Finance Malaysia: Pekeliling Perbendaharaan Malaysia PK 1.15: Pelaksanaan Building Information Modelling (BIM). Effective 1 July 2025. Ministry of Finance Malaysia, Putrajaya (2025)
7. Wong, J.Y., et al.: BIM-VR framework for building information modelling in engineering education. Int. J. Interact. Mob. Technol. **14**(6), 15–39 (2020)
8. CIDB Malaysia: Quality Assessment System in Construction (QLASSIC). CIDB Malaysia, Kuala Lumpur (n.d.)
9. Chew, Z.X., Wong, J.Y., Tang, Y.H., Yip, C.C., Maul, T.: Generative design in the built environment. Autom. Constr. **166**, 105775 (2024)

An Analysis of the Run Sum Coefficient of Variation Chart for Accurate Monitoring of Engineering Systems

Wai Chung Yeong[1(✉)], Sok Li Lim[2], Azizul Hakim bin Khairil Aznan[2], Peh Sang Ng[3], and Khai Wah Khaw[4]

[1] Centre for Big Data and Blockchain Technologies, CoE for Advanced Cloud, Multimedia University, Persiaran Multimedia, 63100 Cyberjaya, Selangor, Malaysia
wcyeong@mmu.edu.my

[2] School of Accounting and Finance, Faculty of Business & Law, Taylor's University, 1, Jln Taylors, Subang Jaya 47500, Selangor, Malaysia

[3] Department of Physical and Mathematical Science, Faculty of Science, Universiti Tunku Abdul Rahman, 31900 Perak, Malaysia

[4] School of Management, Universiti Sains Malaysia, 11800 Penang, Malaysia

Abstract. Control charts are powerful tools for assisting engineering practitioners in process monitoring. During their implementation, performance measures that can accurately represent the chart's actual performance is essential. One of the performance measures that is frequently used is the average run length (*ARL*). However, it is only accurate for a symmetrical run length distribution, and leads to misleading results when it is skewed. The run sum chart has so far been assessed solely based on its *ARL*, without any studies on its run length distribution. To address this gap, this paper examines the complete run length distribution of the chart. From the numerical findings in this paper, it was observed that the in-control median run length and out-of-control median run length for small shift sizes is smaller than its corresponding *ARL*, indicating a skewed run length. The numerical analysis conducted in this paper also reveals that 60% to 70% of the actual run lengths are shorter than its *ARL*, demonstrating that the *ARL* should not be used as the sole performance measure. This paper also studies the entire run length distribution, allowing practitioners to identify the range and uncertainties in the possible run length values. The study demonstrates that performance evaluations should be accompanied with its run length distribution and not based on the *ARL* alone. This ensures practitioners avoid misleading conclusions regarding false alarms and the detection of small and moderate shifts.

Keywords: Average Run Length · Coefficient of Variation · Median Run Length · Percentiles of the Run Length · Run Sum Chart

1 Introduction

In most engineering applications, consistency of process outputs is critical, as deviations from the norm indicates the presence of unusual causes that needs to be removed. Control charts are useful in this aspect by detecting shifts in statistical parameters. A chart has

M. A. b. Jamlos et al. (Eds.): ICSET 2025, LNNS 1897, pp. 154–160, 2026.
https://doi.org/10.1007/978-3-032-20722-7_19

good performance if it does not require a large number of samples until the detection of out-of-control conditions, while at the same time avoiding frequent false alarms.

The run sum chart proposed by [1] improves the detection of shifts in the coefficient of variation (γ). It divides its charting region into k equal regions above and below the centerline (CL), then assigs positive and negative scores to these regions, with potentially larger magnitude scores associated with regions farther from the CL. Samples that fall in these regions are assigned these scores, and the scores are accumulated as the sampling process continues, with the accumulated scores being reverted to zero when a sample is on opposite sides of the CL. The accumulated scores are then compared with a predetermined threshold. Once the accumulated value exceeds this threshold, out-of-control signals are generated. The chart outperforms the chart by [2] under all conditions, while outperforming the chart by [3] for large shift sizes. The chart was later extended to multivariate processes by [4], and its performance was improved by varying its charting parameters by [5–8].

The existing run sum γ chart is evaluated through its average run length (ARL), which consists of the in-control and out-of-control ARL (ARL_0 and ARL_1). The former looks at the average number of samples until the chart signals a false alarm, while the ARL_1 measures the average number of samples until the out-of-control condition is detected. However, these measures are only accurate for symmetrically distributed run length; otherwise, the ARL provides an inaccurate representation of the chart's performance. In a practical context, correct interpretation of the chart's performance is essential for practitioners. This includes an accurate assessment on the frequency of false alarms and the number of samples until the detection of out-of-control shifts. If false alarms frequently occur earlier than indicated by the ARL_0, or if the number of samples required to detect the shift is frequently different from the ARL_1, practitioners may lose confidence in the chart. Hence, if the run length is skewed, the ARL should not be used as the sole performance measure.

A past study on the variable sample size (VSS) chart by [9] found that its run length distribution is skewed, and false alarms frequently occurs before the ARL_0; while out-of-control signals for small shift sizes also occurs before its ARL_1. Similar studies on other charts also leads to the same conclusions [10–15]. These studies demonstrates that evaluation based on the ARL is inaccurate for these charts. However, no prior work has studied the run length distribution of the run sum γ chart despite its increasing use, and the chart is only evaluated based on its ARL. Given the significant impact of run length skewness on the accuracy of ARL-based evaluations, this paper aims to study the entire run length distribution of the run sum γ chart and assess the suitability of using the ARL as its performance measure.

The following section describes the operations of the run sum γ chart and presents the formula for its ARL along with the cumulative distribution function (cdf) of its run length. Then, Sect. 3 analyzes its run length distribution. Conclusions are then presented in the final section.

2 Run Sum γ Chart

To operate this chart, at each sampling time t, n independent and normal variables $\{X_1, \ldots, X_n\}$ is collected, and its sample γ ($\hat{\gamma}$) is obtained as $\hat{\gamma} = \frac{S}{\overline{X}}$, where $\overline{X}$ and S are the sample mean and standard deviation, respectively.

The run sum γ chart divides the chart into k segments above and below the CL, through its lower control limits (LCL_ts) and upper control limits (UCL_ts), where $t = 1, 2, \ldots, k-1$, as illustrated in Fig. 1. For example, when $UCL_1 < \hat{\gamma} < UCL_2$, the sample belongs to region R_{+1}, and the sample is assigned a score of S_2. The LCL_ts and UCL_ts in Fig. 1 are obtained as follows [1]:

$$LCL_t = \max\left[0, \mu_0(\hat{\gamma}) - \left(\frac{3t}{k-1}\right)K\sigma_0(\hat{\gamma})\right], \tag{1}$$

$$UCL_t = \mu_0(\hat{\gamma}) + \left(\frac{3t}{k-1}\right)K\sigma_0(\hat{\gamma}), \tag{2}$$

where K is the charting parameter that needs to be determined, while $\mu_0(\hat{\gamma})$ and $\sigma_0(\hat{\gamma})$ are defined as:

$$\mu_0(\hat{\gamma}) = \gamma_0\left[1 + \frac{1}{n}\left(\gamma_0^2 - \frac{1}{4}\right) + \frac{1}{n^2}\left(3\gamma_0^4 - \frac{\gamma_0^2}{4} - \frac{7}{32}\right) + \frac{1}{n^3}\left(15\gamma_0^6 - \frac{3\gamma_0^4}{4} - \frac{7\gamma_0^2}{32} - \frac{19}{128}\right)\right] \tag{3}$$

$$\sigma_0(\hat{\gamma}) = \gamma_0\sqrt{\frac{1}{n}\left(\gamma_0^2 + \frac{1}{2}\right) + \frac{1}{n^2}\left(8\gamma_0^4 + \gamma_0^2 + \frac{3}{8}\right) + \frac{1}{n^3}\left(69\gamma_0^6 + \frac{7\gamma_0^4}{2} + \frac{3\gamma_0^2}{4} + \frac{3}{16}\right)}, \tag{4}$$

with γ_0 denoting its in-control γ. Note that $CL = \mu_0(\hat{\gamma})$.

The probabilities $p_{+(t-1)}$ and $p_{-(t-1)}$ in Fig. 1 is computed as:

$$p_{+(t-1)} = F_{\hat{\gamma}}(\mathrm{UCL}_t|n, \gamma) - F_{\hat{\gamma}}(\mathrm{UCL}_{t-1}|n, \gamma) \tag{5}$$

$$p_{-(t-1)} = F_{\hat{\gamma}}(\mathrm{LCL}_{t-1}|n, \gamma) - F_{\hat{\gamma}}(\mathrm{LCL}_t|n, \gamma), \tag{6}$$

where from [3], $F_{\hat{\gamma}}(x|n, \gamma) = 1 - F_t\left(\frac{\sqrt{n}}{x}\middle|n-1, \frac{\sqrt{n}}{\gamma}\right)$, with $F_t(\cdot)$ representing the cdf of the non-central t distribution with $n-1$ degrees of freedom and non-centrality parameter $\frac{\sqrt{n}}{\gamma}$.

Scores for successive samples are then accumulated through $U_i = U_{i-1} + S(\hat{\gamma}_i)$ for $\hat{\gamma}_i \geq CL$ and $L_i = L_{i-1} + S(\hat{\gamma}_i)$ for $\hat{\gamma}_i \leq CL$, where $S(\hat{\gamma}_i)$ is the score for the ith sample. Note that U_i and L_i resets to zero when $\hat{\gamma}$ falls on opposite sides of CL. The process is out-of-control if $U_i \geq S_k$ or $L_i \geq -S_k$.

The ARL is evaluated through a Markov chain, where all possible combinations of (L_i, U_i) less than $-S_k$ or S_k are distinct in-control transient states, alternatively they are absorbing states [1]. has shown that the ARL is obtained as follows:

$$ARL = \boldsymbol{q}^T(\boldsymbol{I} - \boldsymbol{Q})^{-1}\boldsymbol{1} \tag{7}$$

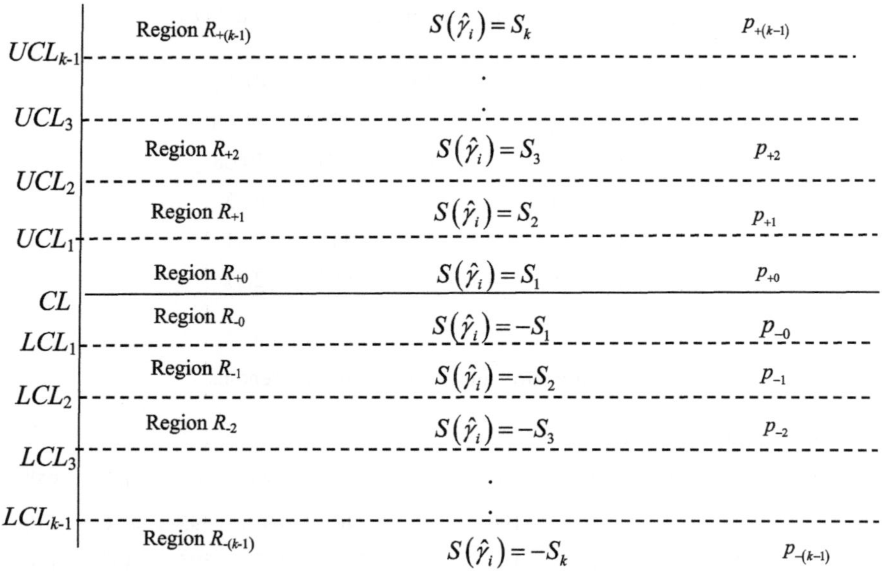

Fig. 1. An illustration of the run sum γ chart.

where $\boldsymbol{q}$ and $\boldsymbol{Q}$ are respectively its initial probability vector and transition probability matrix between its in-control states, available in [1], while $\boldsymbol{I}$ and $\boldsymbol{1}$ are its identity matrix and vector of ones, respectively.

In this paper, the run length distribution is studied through its cdf, as follows [7]:

$$F_{RL}(l) = 1 - \mathbf{q}^{\mathrm{T}}\left(\mathbf{Q}^{l-1}\right)\mathbf{1}. \tag{8}$$

The $(100\theta)^{\text{th}}$ percentile, l_θ, can be obtained by searching for integer values that satisfy

$F_{RL}(l_\theta - 1) \leq \theta$ and $F_{RL}(l_\theta) \geq \theta$, (9)where $0 < \theta < 1$. For example, to obtain the 50th percentile, i.e. the median run length (*MRL*), we let $\theta = 0.5$.

3 Run Length Analysis

This section analyzes the percentiles of the run length distribution by implementing the optimal charting parameters from [1] for $n = 5$, $\gamma_0 = 0.05$ and $\tau \in \{1.10, 1.25, 1.50, 1.75, 2.00\}$, where τ is the shift size. The optimal charting parameters and its ARL_1 are presented in Table 1. The charting parameters in Table 1 will be adopted to compute its fifth to 95th in-control and out-of-control percentiles of the run lengths, as shown in Table 2. The percentiles in Table 2 are verified through 10,000 simulation trials, where each trial is made up of one implementation of the run sum γ chart. Percentiles are identified through the percentage of trials that produces out-of-control signals prior to that sample. For example, a 30th in-control percentile of 133 indicates that for 30% of the trials, a false alarm occurs by the 133th sample.

Table 1. Existing optimal charting parameters and ARL_1.

τ	S_1	S_2	S_3	S_4	K	ARL_1
1.10	0	1	5	7	1.1734	81.66
1.25	0	1	4	6	1.1749	19.68
1.50	0	1	4	6	1.1749	6.13
1.75	0	1	5	8	1.1183	3.41
2.00	0	0	1	2	1.0975	2.37

Table 2. In-control and out-of-control percentiles.

τ	In-control Percentiles										
	5th	10th	20th	30th	40th	50th	60th	70th	80th	90th	95th
1.10	21	41	84	133	190	257	340	446	595	850	1106
1.25	21	41	84	133	190	257	339	445	595	850	1105
1.50	21	41	84	133	190	257	339	445	595	850	1105
1.75	20	40	84	133	190	257	339	445	595	850	1106
2.00	20	40	83	133	190	257	339	445	595	851	1107
τ	**Out-of-control Percentiles**										
1.10	6	10	20	30	43	57	75	98	130	185	241
1.25	3	4	6	8	11	14	18	23	31	43	55
1.50	1	2	3	3	4	5	6	7	9	12	15
1.75	1	1	1	2	2	3	3	4	5	7	8
2.00	1	1	1	1	2	2	2	3	3	4	6

From Table 2, an in-control *MRL* (MRL_0) of 257 is obtained, although from [1], the ARL_0 is 370.4. Hence, for 50% of the time, a false alarm occurs significantly earlier than suggested by the ARL_0. Large discrepancies between its MRL_0 and ARL_0 shows an in-control run length distribution that is right-skewed, making evaluation based solely on the ARL_0 inappropriate. Furthermore, through the percentiles, a better understanding of the variation in the run length is possible. In fact, the range for the fifth to 95th in-control percentiles are quite wide, from 20 to 1107, showing uncertainty in the actual run length. For the middle 40%, i.e. between its 30th and 70th percentile, the range is between 133 and 446. Hence, users should not be surprised if false alarms frequently happen before the ARL_0.

The out-of-control percentiles of small and moderate τ shows a similar trend. For example, for $\tau = 1.10$, an obvious difference is observed between the ARL_1 of 81.66 and out-of-control *MRL* (MRL_1) of 57, showing a right-skewed out-of-control run length. The range between its fifth to 95th percentile is also quite wide, and the range between the 30th and 70th percentile for $\tau = 1.10$ is 30 to 98. Hence, the actual out-of-control run length for small shift sizes is quite uncertain, showing the inappropriateness of evaluating the chart through its ARL_1.

By contrast, for large τ, a small difference is observed between the ARL_1 and MRL_1. For example, for $\tau = 2.00$, the ARL_1 is 2.37, while the MRL_1 is 2, showing a symmetrical run length distribution. Furthermore, variation in the percentiles decreases as τ increases, showing more consistency as τ increases. For example, for $\tau = 1.10$, a difference between the fifth and 95th percentile of 235 is observed, while for $\tau = 2.00$, the corresponding difference is only 5.

4 Conclusion

The existing run sum γ chart has been evaluated exclusively through its *ARL*; however, the *ARL* is not a good performance measure for skewed run lengths. The key numerical finding in this paper show that the in-control run length and out-of-control run length for small and moderate shifts are skewed to the right and exhibit large variability. Specifically, 60% to 70% of the run lengths differ substantially from the *ARL*, resulting in a loss of confidence towards the chart if evaluation is based solely on its *ARL*. Conversely, out-of-control run lengths of large shifts are quite symmetrical. As a conclusion, the ARL_0 and ARL_1 for small and moderate shifts should not be used as the only performance measure as its actual run length may deviate significantly from the *ARL*. Instead, its entire run length distribution should be analyzed to obtain an accurate understanding of the chart's actual performance. The main limitation of this paper is the chart only considers univariate quality characteristics; future work can look at analyzing the chart's performance for multivariate quality characteristics.

References

1. Teoh, W.L., Khoo, M.B.C., Castagliola, P., Yeong, W.C., Teh, S.Y.: Run-sum control charts for monitoring the coefficient of variation. Eur. J. Oper. Res. **257**(1), 144–158 (2017)
2. Kang, C.W., Lee, M.S., Seong, Y.J., Hawkins, D.M.: A control chart for the coefficient of variation. J. Qual. Technol. **39**(2), 151–158 (2007)
3. Castagliola, P., Celano, G., Psarakis, S.: Monitoring the coefficient of variation using EWMA charts. J. Qual. Technol. **43**(3), 249–265 (2011)
4. Lim, A.J.X., Khoo, M.B.C., Teoh, W.L., Haq, A.: Run sum chart for monitoring multivariate coefficient of variation. Comput. Ind. Eng. **109**, 84–95 (2017)
5. Yeong, W.C., Lim, S.L., Khoo, M.B.C., Ng, P.S., Chong, Z.L.: A variable sampling interval run sum chart for the coefficient of variation. J. Stat. Comput. Simul. **92**(15), 3150–3166 (2022)
6. Yeong, W.C., Tan, Y.Y., Lim, S.L., Khaw, K.W., Khoo, M.B.C.: Variable sample size and sampling interval (VSSI) and variable parameters (VP) run sum charts for the coefficient of variation chart. Qual. Reliab. Eng. Int. **38**(4), 1869–1885 (2022)
7. Yeong, W.C., Tan, Y.Y., Lim, S.L., Khaw, K.W., Khoo, M.B.C.: Variable sample size and sampling interval (VSSI) and variable parameters (VP) run sum charts for the coefficient of variation chart. Qual. Technol. Quant. Manag. **21**(2), 177–199 (2024)
8. Cui, D., Khoo, M.B.C., Saha, S., Chong, Z.L., Chua, S.: A variable sampling interval run sum chart for monitoring multivariate coefficient of variation. PLoS One. **20**(9), e0330936 (2025)

9. Karimi, H., Khaw, K.W., Chew, X.Y., Chong, Z.L., Lim, S.T.: Variable sample size control charts for monitoring the multivariate coefficient of variation based on median run length and expected median run length. J. Sci. Technol. **15**(2), 54–80 (2023)
10. Teoh, W.L., Chong, J.K., Khoo, M.B.C., Castagliola, P., Yeong, W.C.: Optimal designs of the variable sample size $\overline{X}$ chart based on median run length and expected median run length. Qual. Reliab. Eng. Int. **33**(1), 121–134 (2017)
11. Lim, S.L., Yeong, W.C., Khoo, M.B.C., Chong, Z.L., Khaw, K.W.: An alternative design for the variable sample size coefficient of variation chart based on median run length and expected median run length. Int. J. Ind. Eng. **26**(2), 199–220 (2019)
12. Teoh, W.L., Lim, J.Y., Khoo, M.B.C., Chong, Z.L., Yeong, W.C.: Optimal designs of EWMA charts for monitoring the coefficient of variation based on median run length and expected median run length. J. Test. Eval. **47**(1), 459–479 (2019)
13. Lee, M.H., Lim, V.Y.C., Chew, X.Y., Lau, M.F., Yakub, S., Then, P.H.H.: Design of the synthetic multivariate coefficient of variation chart based on the median run length. Adv. Math. **9**(9), 7397–7406 (2020)
14. Yeong, W.C., Lim, S.L., Khoo, M.B.C., Khaw, K.W., Ng, P.S.: An optimal design of the synthetic coefficient of variation chart based on the median run length. Int. J. Reliab. Qual. Saf. Eng. **28**(3), 2150018 (2021)
15. Yeong, W.C., Lee, P.Y., Lim, S.L., Ng, P.S., Khaw, K.W.: Optimal designs of the side sensitive synthetic chart for the coefficient of variation based on the median run length and expected median run length. PLoS One. **16**(7), e0255366 (2021)

Enhancing Tint Film Quality with DMAIC and Vision Systems

Azizol Heirey Bahari(✉) and Zamzuri Hamedon

Faculty of Manufacturing and Mechatronics Engineering Technology, Universiti Malaysia Pahang, 26600 Pekan, Pahang, Malaysia
pfo21002@adab.umpsa.edu.my

Abstract. This article argues that the manufacturing industry always aims at increasing the quality and effectiveness of its production. This research focuses on evaluating the cut and trim mechanisms of vehicle window films as well as the quality of the final product. Thus, the key goal was to increase process performance and product quality by applying Six Sigma DMAIC and AI vision technologies. The implemented strategy included the DMAIC process outlined by Define, Measure, Analyze, Improve, and Control with additional support of AI vision technologies for the assessment of quality. Hypothesized relation was supported where there was a 13% reduction on cycle time, work output increased by 3%, the machine downtime for minor breakdowns reduced by 27% and the dust related product rejections reduced by 44%. Considering these conclusions, it is possible to suggest significant increases in production quality and effectiveness, and to maintain that the combined use of AI and the Six Sigma DMAIC system promotes the development of more efficient and effective processes in productivity and quality management. The approach used in the study also offers the example of how the company can scale its solutions and recommendations to other similar organizations that may experience similar issues, with ideas on areas of future research and application thrown in the mix.

Keywords: Quality System · DMAIC · AI · Operational Cost · Vision System

1 Introduction

The global market landscape, characterized by heightened competition and increasing customer demands for superior quality, requires businesses to improve their quality systems continuously. At the same time, pressures from stakeholders, economic uncertainties, and fluctuating market dynamics necessitate effective cost-management strategies. Quality systems are the lifeline of any enterprise that values customer satisfaction and continuous improvement. A well-designed quality system contributes significantly to a company's efficiency, competitiveness, and profitability.

Regarding operational costs, the traditional quality systems often entail significant expenditure in resource allocation, equipment maintenance, human labor, and error rectification. Further, these systems may lack the agility and flexibility required to respond

M. A. b. Jamlos et al. (Eds.): ICSET 2025, LNNS 1897, pp. 161–168, 2026.
https://doi.org/10.1007/978-3-032-20722-7_20

to changes promptly, leading to potential delays, inefficiencies, and increased costs. Considering these challenges, there is a pressing need to transform how quality systems are managed. An innovative approach that blends technology and quality management principles could solve these predicaments. This is where Artificial Intelligence (AI) comes into the picture.

The core hypothesis of this article is that implementing AI can significantly enhance quality systems, thereby leading to substantial reductions in operational costs. In recent years, AI has made remarkable strides, moving from the realm of science fiction to practical, real-world applications. The disruptive potential of AI lies in its ability to learn, reason, adapt, and perform tasks that traditionally require human intelligence.

The review study is to provide a comprehensive guide for businesses seeking to harness the power of AI to revolutionize their quality systems and drive down op-erational costs. Accordingly, Dahmen et al. (2023) [1] in their study showed that AI could significantly transform quality control by automating procedures, decreasing mistake rates, enhancing productivity, and eventually lowering expenses, which is why more and more companies are recognizing AI's potential advantages and im-plementing AI-based quality systems to make their operations more efficient and improve productivity. This journey will be explored through an understanding of quality systems, an analysis of AI and its implications for quality control, case studies of successful.

Bahari and Hamedon (2022) [2] claim that the DMAIC method improves processes through data. DMAIC improves cutting process of window tint film quality. First, define project objectives and client deliverables such as cutting precision, uniformity, and minimizing waste material. Next, the procedure is measured, and relevant data is collected for comparison. The data is then analyzed to identify the key issues and opportunities for development (Cláudio et al., 2023) [3]. Solutions like improved procedures or equipment enhancements address the key is-sues. Finally, the upgraded process is monitored and modified to ensure quality improvement.

The Control phase guarantees the long-term sustainability of improvements by implementing control systems. DMAIC offers a methodical approach to enhancing quality in corporate processes. Dewi et al. (2022) [4] and Cláudio et al. (2023) [5] have observed that implementing Six Sigma DMAIC reduces variance and errors in production, resulting in cost savings and improved productivity. Jou et al. (2022) [6] and Narottam et al. (2020) [7] have found that Six Sigma DMAIC improves manu-facturing processes by reducing waste and improving quality. This is achieved through identifying quality issues, measuring dimensions, and detecting deviations in real time using AI Vision Systems (Weiss et al., 2024) [8]. Table 1 below highlights the DMAIC phases.

Quality and defect management in auto window tinting production are crucial. Automotive manufacturing relies on quality control to eliminate faults and issues. This is necessary for excellent engineering, safety, and comfort for the end user. Defect management in auto window tint manufacturing involves detecting, assessing, and fixing production defects, Soundararajan & Reddy, 2019) [9]. Hoffmann and Reich (2023) [10] emphasize that AI Vision Systems can improve quality and defect management through real-time monitoring and feedback to discover and correct deviations from the set standards. As such, real-time (Weiss et al., 2024) [8] and precise inspection (Sundaram & Zeid, 2023 [11]; Wang et al., 2023) [12] can be enabled, and quality assurance can be strengthened and improved by combining Six Sigma DMAIC with AI Vision Systems.

Table 1. DMAIC phases

Phase	Details
Define (D)	Set project goals and objectives Improve quality and productivity
Measure (M)	Measure the defect where they occur
Analyze (A)	Evaluate data/information for trends, patterns and root causes
Improve (I)	Develop, implement and evaluate the Solution targeted
Control (C)	Problems have been cleared, and the method is improving

2 Methodology

This study explores the challenges and benefits of integrating AI into quality systems. It focuses on its potential to reduce operational costs and enhance quality using Company A as a case study, which applies Six Sigma DMAIC methodology and machine vision systems to address defects in the window tint film-cutting process. Defects have been identified in the final product of the window tint film during the cutting process, the cutting process illustration scheme is mentioned in Fig. 2. In this diagram process by machine efficiency was also mentioned and a comparative analysis between manual & machine also highlighted in the Fig. 2. The faults are identified throughout the Cutting Process by visual inspection conducted during the in-process checking activity. The quality inspector performed a comprehensive visual examination of the final product following the completion of the cutting process. The imperfections include dust particles, line marks, discoloration, fur fibres, air bubbles, scratches, dirt, adhesive residue, stains, straw-like lines, and other flaws. Many defects in the product are irreparable and will result in waste, necessitating the implementation of a scrap operation. The company has incurred significant financial losses due to a substantial increase in rejected products after the cutting process and frequent line stoppages that have negatively impacted on production efficiency.

2.1 Measure

Performance metrics are crucial to measure the phase of DMAIC method, especially for a project that aims to reduce automotive window film-cutting mistakes. In this study the number and frequency of defects in each roll of film was measured and calculated the proportion of defective cuts out of the total number of cuts, and monitored the rate of customer returns due to cutting defects, which directly affects customer satisfaction. The main goal was to align these measures with Six Sigma's process variation reduction and quality improvement goals. Using eq. 1 suggested for the appropriate solution of defect reduction. And through this equation a maximum of 3.4 defects per million possibilities (DPMO) obtainment is desired with designed study as mentioned in Table 2.

$$\text{DPMO} = \left(\frac{\textit{Number of Defects}}{\textit{Total Opportunities for a Defect}}\right) \times 1{,}000{,}000$$

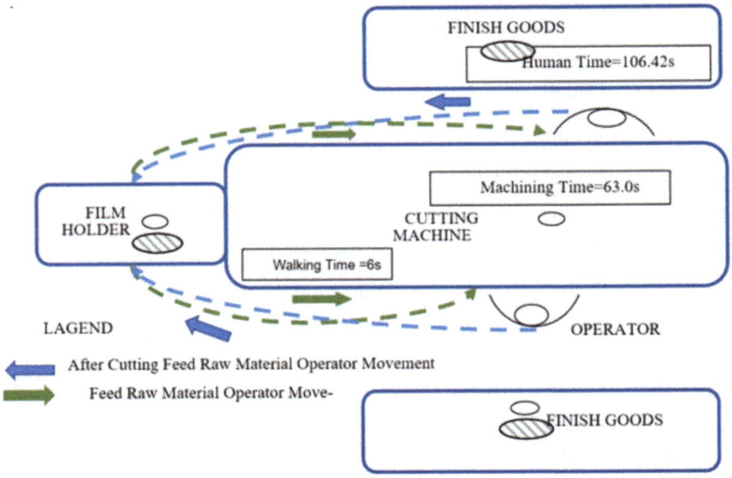

Fig. 1. Cutting Process Diagram

3 Result and Discussion

3.1 Pareto Chart Analysis

The Pareto charts were used to identify the most frequent defects and Control charts to monitor process stability, as mentioned in Fig. 3 and Table 2 was used to identify the quality defects in the production line in Company A for the occurrence of defects by pareto chart analysis. This analysis was conducted a factor analysis to understand the variables that influence the cutting process and the occurrence of defects.

Based on accurate data collection methods and measurements it was confirmed that the data has been recorded and examined correctly, without errors or omissions, sample size is 26000pcs for one month production. In the root cause analysis stage, a fish-borne diagram.

In the Improve phase of Six Sigma DMAIC, the goal is to generate, choose, and evaluate solutions that address the root causes of the errors found in Analyze. Authors developed a Prioritization Matrix to compare solutions to reduce vehicle window film-cutting errors. A systematic approach to the matrix construction included effect assessment, cost estimation, feasibility study, and business objectives.

3.2 Implementation of AI Vision Systems

AI Vision Systems in industrial applications are best implemented with the Axiomtek AI Starter Kit. Its comprehensive features are developed for AI and machine vision jobs, making it strategic. The Axiomtek AI Starter Kit fulfills visual AI algorithm requirements. The kit includes the IPC962–511-FL, an Intel® Core™ processor-powered AI edge platform for high-performance GPU/VPU computation, crucial for complex AI algorithms. This integrated AIS uses TensorFlow Framework, Intel® Edge Software Hub for Industrial, and Intel® OpenVINO™.

Table 2. Defect Specification

INSPECTION ITEM	SKETCH	SPECIFICATION
Dust		i. Dust Max size Ø0.6mm. ii. In the area of 1 side window glass, Max size Ø0.6mm x 3 pieces.
Dirt	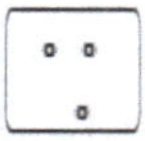	i. Dirt Max size Ø 0.6mm. ii. In the area of 1 side window glass, Max size Ø0.6mm x 3 pieces.
Furry	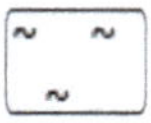	i. Fury Max size 3.0mm. ii. In the area of 1 side window glass, Max size Ø3.0mm x 3 pieces.
Straw Line / Finger	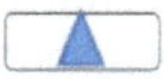	i. In the area of 1 side window glass. No straw line/finger
Folded	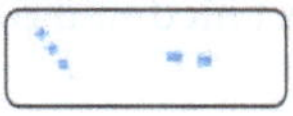	i. Folded Max size 3.0mm. ii. In the area of 1 side window glass, Max size Ø3.0mm x 3 pieces. Inline area not accepted.
Pin Hole	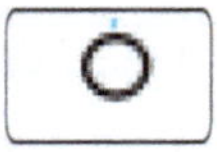	i. In the area of 1 side window glass. No pinhole
Bubble		i. In the area of 1 side window glass. No Air Bubble ii. In the area of 1 side window glass. No Water Bubble
Scratch		i. In the area of 1 side window glass. No deep scratch. ii. Minor scratch (cannot feel with nail) In area 100mm x 100mm, Max size 50mm
Cutting Edge	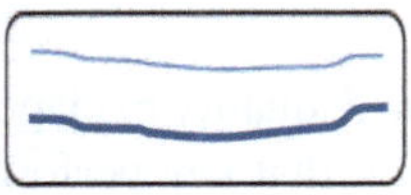	i. In the area of 1 side window glass. There is no cutting edge.

The Axiomtek AI Starter Kit can increase car window tint manufacturing, cutting quality, and efficiency. The AI Starter Kit, high-resolution camera, and AI Suite allow window tint film inspection after cutting. This allows quick repairs, reducing waste and

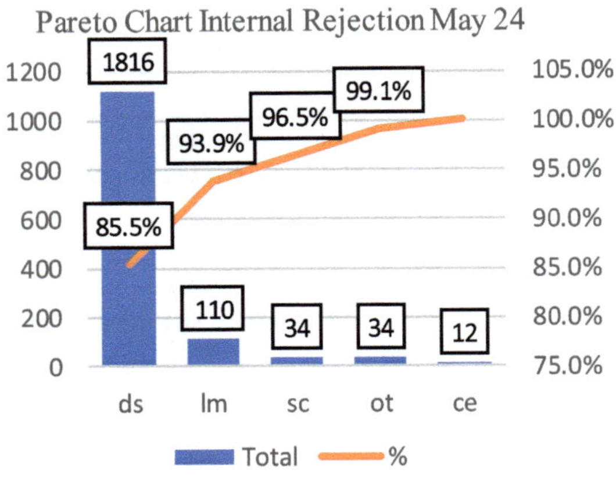

ds	Dust
lm	Line Mark
sc	Scratch
ot	Others
ce	Cutting Edge
fg	Finger
dt	Dirt
hp	Hot Problem
bb	Buble
fr	Fur
fd	Folded

Fig. 2. Pareto Chart Internal Rejection (Company A)

increasing production. The AI Starter Kit can be integrated into cutting machinery control systems. By providing real-time cutting quality input, the technology may automatically adjust cutting blades or process settings to maintain high standards. The Vision System Tooling's camera and software functions are detailed in the specs as shown in Fig. 3.

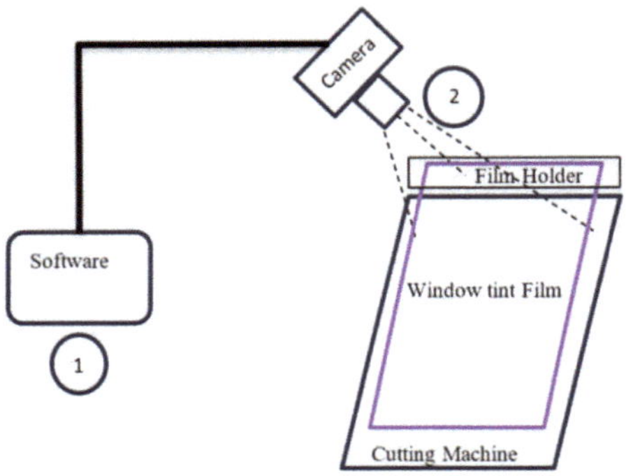

Fig. 3. Basic Camera Concept

3.3 Six Sigma DMAIC Analysis

In order to enhance the cutting process, waste & quality management DMAIC technique was studied. The results highlighted in Fig. 4 that key performance metrics improved after integrating an AI vision system using Six Sigma DMAIC.

The AI vision system and DMAIC technique improved the cutting process, improving performance, waste, and quality in vehicle window film production. AI in quality inspection may have improved fault identification and flaw eradication. In addition, DMAIC provided a systematic framework for analyzing and improving the process.

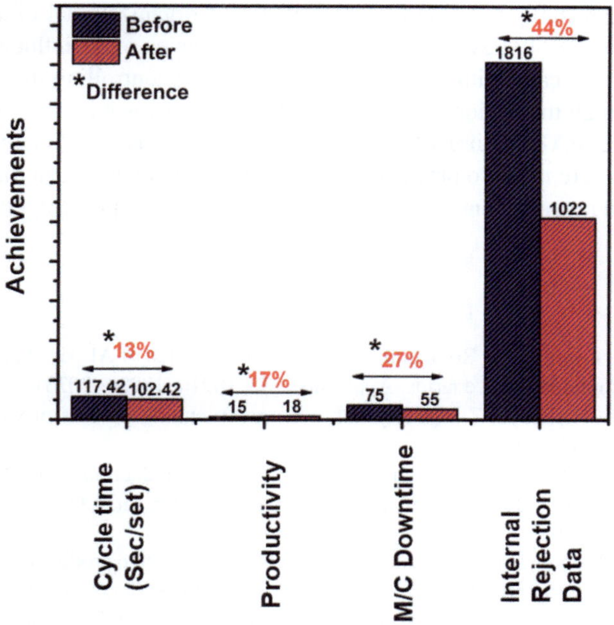

Fig. 4. Final Result after improvement implementation.

4 Conclusion

The highest number of rejections, thus, emerged as the key process constraint and mainly attributed to the faults in the cutting operation. Thus, to have a complete understanding of this question, a Pareto analysis was made to determine the repeated mistakes. Further research was conducted to analyze the causes of emergence of these defects. The outcome of this analysis was the definition of improvement endeavors that were to be presented to the management and then high. For improvement of the car window film cutting operation, Six Sigma DMAIC method and the AI vision technology were employed on the project. To implement the working effectively, the steps of defining the process, measuring it, analyzing it, improving it as well as controlling the process were followed. The project delivered an impressive performance improvement as follows cycle time reduced by 13%, productivity increased by 3%, the machine downtime which is due to brief interruption reduced by 27% and internal rejections due to dust were reduced by 44%. This applied to the discussion where the focus was made on how the AI vision system can help in the detection and avoidance of quality deviations. This was accomplished through the fact that the structured analysis and improvement process of the DMAIC framework played a critical role in these enhancements. These are findings demonstrate that AI can improve quality assurance, which is critical in maintaining quality in goods and services. The usage of AI vision system and the DMAIC approach implies an element of capacity improvement. This method can be applied to many other industries and other levels of quality inspection such as high technology products like electronics, to industries that produces drugs like pharmaceuticals, to food products like food and beverage industries. It would be important for any industry that employs the use of big data techniques towards improvement of organizational efficiency and its products. The future studies could focus on the tendencies of AI influencing the labor environment or the capabilities of AI in identifying and controlling the quality-related problems. Through the exploratory analysis, the study proves it is easy to introduce AI, and Six Sigma DMAIC optimizes production processes, thus inspiring other industries. The conclusions are made to promote advancements in knowledge on the relationships between AI and management to provide better quality and improve operations.

References

1. Dahmen, U., Osterloh, T., Roßmann, J.: Structured validation of AI-based systems by virtual testing in simulated test scenarios. Appl. Intell. **53**, 18910–18924 (2023)
2. Bahari, A.H., Hamedon, Z.: Quality improvement in cutting process of window tint using DMAIC approach (2022)
3. Condé, G.C.P., Oprime, P.C., Pimenta, M.L., et al.: Defect reduction using DMAIC and Lean Six Sigma: a case study in a manufacturing car parts supplier. Int. J. Qual. Reliab. Manag. **40**, 2184–2204 (2023)
4. Dewi, S.K., Widodo, R.D.S.A., Lukman, M.: Reducing defective products using six sigma for production process improvement. In: AIP Conference Proceedings. AIP Publishing (2022)
5. Claudio, A.B.: A statistical analysis for infrastructure investment. Rev. Bras. Transp. **3**, 20–63 (2023)
6. Jou, Y.-T., Silitonga, R.M., Lin, M.-C., et al.: Application of Six Sigma methodology in an automotive manufacturing company: a case study. Sustainability **14**, 14497 (2022)
7. Narottam, Mathiyazhagan, K., Sharma, V.: Defect reduction in manufacturing industry using lean Six Sigma approach. In: Recent Advances in Mechanical Engineering: Select Proceedings of NCAME 2019. Springer, pp 19–30 (2020)
8. Weiss, E., Caplan, S., Horn, K., Sharabi, M.: Real-time defect detection in electronic components during assembly through deep learning. Electronics **13**, 1551 (2024)
9. Soundararajan, K.: Cost-reduction and quality improvement using DMAIC in the SMEs. Int. J. Product. Perform. Manag. **68**, 1528–1540 (2019)
10. Yousif, I., Burns, L., El Kalach, F., Harik, R.: Leveraging computer vision towards high-efficiency autonomous industrial facilities. J. Intell. Manuf. **36**(7), 2983–3008 (2024)
11. Hoffmann, R., Reich, C.: A systematic literature review on artificial intelligence and explainable artificial intelligence for visual quality assurance in manufacturing. Electronics **12**, 4572 (2023)
12. Wang, W., Wang, P., Zhang, H., et al.: A real-time defect detection strategy for additive manufacturing processes based on deep learning and machine vision (2023)

Electrodeposition of ZnO Nanorod Films as Efficient Photocatalysts for Solar Degradation of Malachite Green

Chuang-Jyhe Boh[1], Jin-Chung Sin[1,2](✉), Sze-Mun Lam[1,2](✉), Yit-Thai Ong[1], Wei-Ming Yeoh[1], and Honghu Zeng[2]

[1] Faculty of Engineering and Green Technology, Universiti Tunku Abdul Rahman, Jalan University Bandar Barat, 31900 Kampar, Perak, Malaysia
bohchuangjyhe@gmail.com, {sinjc,lamsm}@utar.edu.my
[2] College of Environmental Science and Engineering, Guilin University of Technology, Guilin 541004, China

Abstract. The discharge of synthetic dyes from textile industries into aquatic ecosystems can cause severe environmental concerns owing to their toxic, carcinogenic and persistent nature. Among various remediation strategies, photocatalytic treatment has emerged as a highly effective and sustainable approach for the elimination dye pollutants from water. This study introduced ZnO photocatalytic film synthesized via a seed layer-assisted electrodeposition method, forming a unique nanorods structure which aimed to achieve good photodegradation of malachite green (MG) under simulated solar light. Characterization techniques like X-ray diffraction (XRD), energy dispersive X-ray spectroscopy (EDX), field-emission scanning electron microscope (FESEM) and transient photocurrent response (TPR) were utilized to analyze the physiochemical and photoelectrochemical properties of the as-synthesized ZnO film. The 1 mM-ZnO and 5 mM-ZnO film were able to degrade 47.9% and 49.9% MG in 360 min, respectively, which was significantly higher than the photoactivity obtained by photolysis itself. This confirmed that MG removal was performed at photocatalytic conditions, demonstrating the potential of ZnO films as efficient photocatalysts for wastewater treatment. Although the preliminary study highlighted the potential of ZnO films for MG degradation, further optimization is still ongoing to enhance its charge separation and long-term stability to maximize efficiency.

Keywords: Photocatalysis · ZnO thin film · Electrodeposition · Malachite green · Degradation

1 Introduction

In recent decades, rapid urbanization and industrial growth have led to the uncontrolled release of persistent organic pollutants (POPs). Among them, dye pollutants posing severe toxicity to aquatic organisms and plants [1]. Malachite green (MG) is a metallic green triphenylmethane dye and owing to the presence of benzene rings, its molecules

M. A. b. Jamlos et al. (Eds.): ICSET 2025, LNNS 1897, pp. 169–174, 2026.
https://doi.org/10.1007/978-3-032-20722-7_21

exhibit high reactivity, leading to harmful effects that threaten human health [2]. Conventional dye removal methods are often limited by drawbacks such as significant energy demands and production of toxic secondary wastes. At present, semiconductor based photocatalysis is regarded as one of the most effective and eco-friendly techniques for MG elimination [3, 4].

ZnO has emerged as a particularly attractive material and has been extensively investigated because of its strong redox potential and good photoactivity, thereby improving their participation in degradation processes [5]. Moreover, the ZnO possesses a wide bandgap of ~3.37 eV, which enables strong ultraviolet absorption and efficient electron-hole pair generation for photocatalysis [6]. Photocatalysts were typically utilized in powdered form, but when dispersed as a slurry, they encounter drawbacks such difficulty in upscaling and recovering nanoparticles, which significantly restrict their practical use [7, 8]. In contrast, thin film photocatalysts offer several advantages including uniform light exposure, adaptability on various substrates and ease of scale up [7, 9].

In this research, ZnO photocatalytic film with nanorods morphology was fabricated through electrodeposition method. X-ray diffraction (XRD), energy dispersive X-ray spectroscopy (EDX), field-emission scanning electron microscope (FESEM) and transient photocurrent response (TPR) were employed to characterize the synthesized photocatalysts films. The photocatalytic activity of ZnO films was assessed with MG dye degradation under simulated solar light. To the best of our knowledge, previous studies on ZnO photocatalysts for MG photodegradation [10–12], have predominantly utilized ZnO powders, whereas investigations into the effect of electrodeposition electrolyte concentration on the performance of ZnO films remain limited.

2 Materials and Methods

Zinc acetate dihydrate ($(CH_3COO)_2Zn{\bullet}2H_2O$, assay ≥99%) and sodium hydroxide (NaOH, assay ≥99%) were obtained from Chem Soln. Ethanol (C_2H_5OH, assay ≥99%), acetone (C_3H_6O assay ≥99.5%) and sodium sulphate (Na_2SO_4, assay ≥99.84%) were procured from Bendosen. Zinc nitrate hexahydrate ($(Zn\ (NO_3)2{\bullet}6H_2O$, assay ≥98.5%) was purchased from SYSTERM. Hydrochloric acid (HCl, assay ~37%) was acquired from Merck. Fluorine-doped tin oxide (FTO) glass (75 mm × 25 mm) was obtained from Liaoning Yike Precision.

The ZnO film was fabricated through a two-step process involving spray coating followed by electrochemical deposition. ZnO seed solution was prepared by dissolving 0.8780 g of $(CH_3COO)_2Zn{\bullet}2H_2O$ and 0.08 g of NaOH in 200 mL C_2H_5OH. The solution was then mixed and stirred at 70 °C for 2 h. After that, the seed solution was spray-coated onto FTO glass and placed into muffle furnace at 300 °C for an hour. ZnO nanorods were grown on ZnO-seeded FTO glass via electrodeposition in a two-electrode system, consisting of a zinc plate anode and seeded FTO glass cathode in 200 mL of 5 mM $Zn\ (NO_3)_2{\bullet}6H_2O$ solution at 80 °C, 1.1 V, 0.01 A for 2 h. After that, the sample was rinsed with C_2H_5OH and H_2O, and calcined in a muffle furnace at 400 °C for an hour. The ZnO synthesized at 1 mM and 5 mM $Zn\ (NO_3)_2{\bullet}6H_2O$ concentrations were designated as 1 mM-ZnO and 5 mM-ZnO, respectively.

The crystalline structure of the as-synthesized samples was examined using a Philips PW1820 X-ray diffractometer, operating over a scanning range of 20–80° with Cu Kα radiation source ($\lambda = 0.15406$ nm). EDX mapping and FESEM imaging were performed using a JEOL JSM-6701F microscope equipped with an Oxford X-Max 20 EDX detector. Photoelectrochemical tests were performed in 0.5 M Na_2SO_4 electrolyte using a three-electrode configuration on Gamry Interface 1000 T electrochemical workstation. The Ag/AgCl electrode, platinum wire and the ZnO film were employed as the reference, counter and working electrode respectively.

The MG dye degradation performance of as-fabricated ZnO films was carried out using 150 W LED lamp as simulated solar light source. The light intensity was measured at 5 cm from the lamp, where the lux meter recorded 56,983 lx. The ZnO film was immersed into 100 mL of 3 ppm MG solution. Prior the reaction, the solution with ZnO film was placed in dark conditions for 30 min to achieve adsorption-desorption equilibrium. During the reaction, 2 mL of dye solution withdrawn, and its absorbance at 617 nm was recorded using UV-vis spectrophotometer (Model V-730).

3 Results and Discussion

3.1 Structural and Compositional Analysis

Fig. 1(a) reveals the XRD patterns of 1 mM-ZnO and 5 mM-ZnO. The XRD patterns exhibited distinct diffraction peaks at 2θ value of approximately 31°, 34°, 36°, 47°, 56°, 63°, 66°, 67° and 69° which can be indexed to the (100), (002) (101), (102), (110), (103), (200), (112) and (201) planes. The reflections were found to be consistent with the standard reference patterns (JCPDS No. 89–0510), confirming the formation of crystalline ZnO with a hexagonal wurtzite structure [13, 14]. The prominent (002) peak at 34° exhibited the highest intensity in all samples, indicating a preferred orientation along the *c*-axis [15]. Further verification on the ZnO formation has been revealed by EDX analysis as shown in Fig. 1(b). Moreover, EDX mapping confirmed the uniform distribution of Zn and O element throughout 5 mM-ZnO.

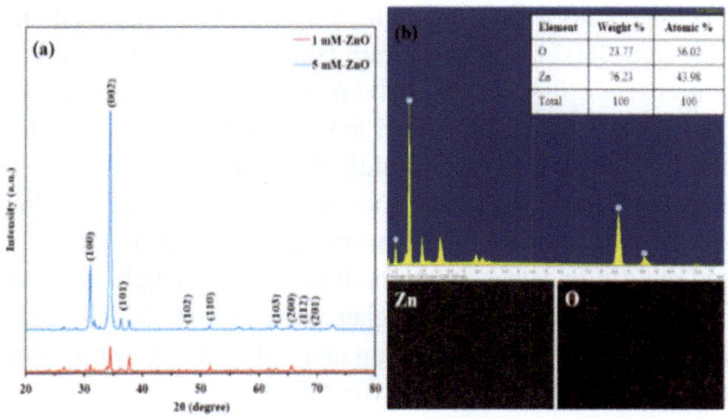

Fig. 1. (a) XRD patterns of 1 mM-ZnO and 5 mM-ZnO. (b) EDX analysis of 5 mM-ZnO.

3.2 Morphological Analysis

FESEM is an advanced characterization technique employed to visualize the surface morphology of samples. All the samples were examined directly on the FTO glass substrate. As illustrated in Fig. 2(a), a homogeneous and densely packed distribution of ZnO seed layer has been formed on FTO glass surface. In Fig. 2(b) 1 mM-ZnO sample showed a uniform distribution of ZnO nanorods grown vertically from the FTO surface, with rod length of approximately at 0.631 μm. In contrast, the 5 mM-ZnO sample shown in Fig. 2(c) displayed well-defined rod-like nanostructures that were interconnected into dense flower-like morphology, with rod length of about 1.006 μm.

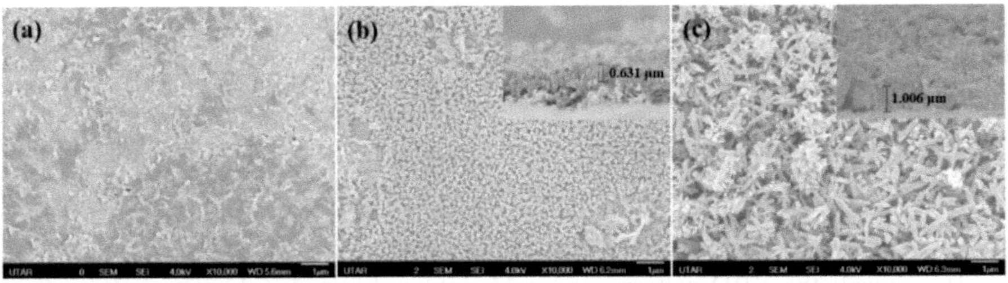

Fig. 2. FESEM images of (a) ZnO seed (top view), (b) 1 mM-ZnO (top and side view) and (c) 5 mM-ZnO (top and side view)

3.3 Dye Degradation and Mechanism

The photodegradation of MG dye over as-synthesized ZnO samples were revealed in Fig. 3(a). Photolysis of MG dye only showed 11.7% degradation. For photocatalytic tests, both 5 mM-ZnO and 1 mM-ZnO films exhibited comparable degradation efficiencies of 49.9% and 47.9% within 360 min, respectively. However, the 5 mM-ZnO film performed slightly better. These results revealed that the removal of MG occurs mainly through the photocatalytic process. Besides that, the rate constants were determined by fitting the obtained data with pseudo-first-order kinetic model as shown in Fig. 3(b) [12, 16]. The kinetic rate constant of 1 mM-ZnO and 5 mM-ZnO were 0.0022 and 0.0024 min^{-1}. The rate constant of 5 mM-ZnO was slightly better than 1 mM-ZnO. From transient photocurrent response plot in Fig. 3(c), the presence of photocurrent reading indicated the generation of charge carriers under illumination. Under light irradiation, the photocurrent density produced rapidly, and dropped immediately once the light was switched off. The 5 mM-ZnO sample exhibited slightly higher photocurrent density than 1 mM-ZnO, indicating a more efficient charge separation and reduced electron-hole recombination, which accounts for its enhanced photoactivity [17, 18].

When ZnO was irradiated with photon of energy higher than its bandgap, electron-hole pairs were generated. The photoexcited charge carriers subsequently migrate to the surface, where they participate in a series of redox reactions. The photogenerated holes (h^+) act as strong oxidizing agents, reacting with surface-adsorbed water molecules or hydroxide ions to produce hydroxyl radicals. In parallel, the photogenerated electrons

(e^-) reduced molecular oxygen to form superoxide radicals. These highly reactive oxygen species (ROS) possessed high oxidative potential and effectively degraded adsorbed MG molecules into less complex intermediates, eventually mineralizing them into H_2O and CO_2 [19, 20].

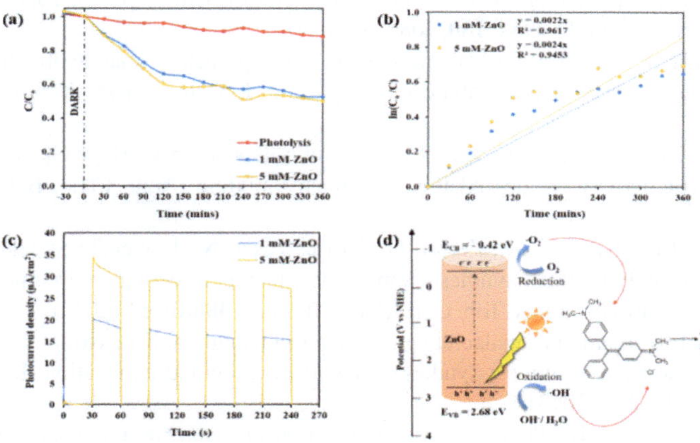

Fig. 3. (a) MG degradation curves and (b) its corresponding kinetic plots (Experimental conditions: [MG] = 3 mg/L; volume of MG = 100 mL and reaction time = 360 min). (c) TPR of as-synthesized ZnO film and (d) Degradation mechanism of MG over prepared ZnO.

4 Conclusion

In conclusion, ZnO films with a nanorod structure was successfully synthesized via an electrodeposition method. Characterization analyses confirmed crystal structure, morphology and charge separation efficiency of samples. Photocatalytic evaluation demonstrated that as-synthesized ZnO films exhibited remarkable performance in MG dye degradation under simulated solar light. Among the prepared samples, the 5 mM-ZnO achieved a slightly better performance than 1 mM-ZnO. However, further optimization was still required to maximize dye degradation efficiency.

References

1. Ma, D., et al.: Enhancing photocatalytic degradation of rhodamine B with visible-light-driven HCl-assisted Bi_2O_3 photocatalysts: Activity, mechanism, and pathways. Mater. Sci. Semicond. Process. **181**, 108672 (2024)
2. Yao, D., Xie, X., Liang, X., Lu, S., Lai, H.: Photocatalytic degradation of malachite green by titanium dioxide/covalent organic framework composite: characterization, performance and mechanism. Chem. Open. **13**(6), e202300209 (2024) 2016
3. Amigun, A.T., Adekola, F.A., Tijani, J.O., Mustapha, S.: Photocatalytic degradation of malachite green dye using nitrogen/sodium/iron-TiO_2 nanocatalysts. Results Chem. **4**, 100480 (2022)

4. Michel, C., Martinez, A.: Photocatalytic performance of β-Ga_2O_3 microcubes towards efficient degradation of malachite green. Ceram. Int. **48**(7), 9746–9752 (2022)
5. Gupta, A., Likozar, B., Jaidka, S.: A review on photocatalytic seawater splitting with efficient and selective catalysts for hydrogen evolution reaction. Renew. Sust. Energ. Rev. **208**, 115074 (2025)
6. Ahmad, I., et al.: Recent progress in ZnO-based heterostructured photocatalysts: A review. Mater. Sci. Semicond. Process. **180**, 108578 (2024)
7. Chinnakonda, S., Naresh, N.: A scalable and thin film approach for solar hydrogen generation: A review on enhanced photocatalytic water splitting. J. Mater. Chem. A. **9**(3), 1353–1371 (2021)
8. Zargazi, M., Entezari, M.: Anodic electrophoretic deposition of Bi_2WO_6 thin film: high photocatalytic activity for degradation of a binary mixture. Appl. Catal. B Environ. **242**, 507–517 (2019)
9. Šuligoj, A., Cerc Korošec, R., Žerjav, G., Novak Tušar, N., Lavrenčič Štangar, U.: Solar-driven photocatalytic films: synthesis approaches, factors affecting environmental activity, and characterization features. Top. Curr. Chem. (Cham). **380**(6), 51 (2022)
10. Alanzi, A.: Investigation of malachite green removal using graphene oxide-zinc oxide composite from aqueous solutions: synthesis, characterization and application. Desalin. Water Treat. **293**, 243–252 (2023)
11. Elkady, M., Hassan, H.: Photocatalytic degradation of malachite green dye from aqueous solution using environmentally compatible Ag/ZnO polymeric nanofibers. Polymers. **13**(13), 2033 (2021)
12. Meena, P., Poswal, K., Surela, A.: Facile synthesis of ZnO nanoparticles for the effective photodegradation of malachite green dye in aqueous solution. Water Environ. J. **36**(3), 513–524 (2022)
13. Tang, C., Chen, C., Zhang, H., Zhang, J., Li, Z.: Enhancement of degradation for nitrogen doped zinc oxide to degrade methylene blue. Physica B: Condens. Matter. **583**(43), 412029 (2020)
14. Guo, W., Chen, K., Zhang, H.: Synthesis and characterization of ZnO hexagonal sheets wrapped MoS_2 sphere for triethylamine gas sensing application. Ceram. Int. **50**(23), 50890–50905 (2024)
15. Nizar, B., Lanjnef, M., Chaste, J., Chtourou, R., Herth, E.: Highly C-oriented (002) plane ZnO nanowires synthesis. RSC Adv. **13**(22), 15077–15085 (2023)
16. Pavithra, M., Jessie Raj, M., Jeno Blair, M., Jeno Baptis, M., Melaine, N.: Solar-photocatalytic and bactericidal efficacy of self-assembled Co:ZnO hybrid flower-rod like nanostructures in dye wastewater remediation. Ceram. Int. **51**(25), 43893–43902 (2025)
17. Hezam, A., et al.: Direcr Z-scheme Cs_2O-Bi_2O_3-ZnO heterostructures as efficient sunlight-driven photocatalysts. ACS Omega. **3**(9), 12260 (2018)
18. Balu, S., Venkatesvaran, H., Lan, K., Yang, C.-K., T.: Synthesis of highly efficient (0D/1D) Z-scheme CdS-NPs@ZnO-NRs visible-light-driven photo(electro)catalyst for PEC Oxygen evolution reaction and removal of tetracycline. Catalysts. **12**(12), 1601 (2022)
19. Dodoo-Arhin, D., Asiedu, T., Agyei-Tuffour, B., Nyankson, E., Obada, D., Mwabora, J.M.: Photocatalytic degradation of Rhodamine B dyes using zinc oxide nanoparticles. Mater. Today Proc. **38**, 809–815 (2021)
20. Gebrye, A., Mehari, B., Atlabachew, M.: Photocatalytic performance of ZnO immobilized on bentonite for catalytic degradation of malachite green under sunlight irradiation. Hybrid Adv. **11**, 100542 (2025)

Role of HMTA in the Electrochemically Synthesized ZnO Nanorod Arrays for Photocatalysis Application

Kuan-Chee Low[1], Sze-Mun Lam[1(✉)], Jin-Chung Sin[1], Yit-Thai Ong[1], Wei-Ming Yeoh[1], and Honghu Zeng[2]

[1] Faculty of Engineering and Green Technology, Universiti Tunku Abdul Rahman, Jalan Universiti, Bandar Barat, 31900 Kampar, Perak, Malaysia
lowkuanchee@gmail.com, lamsm@utar.edu.my

[2] College of Environmental Science and Engineering, Guilin University of Technology, Guilin 541004, China

Abstract. In this study, a ZnO nanorod array (ZnO NRA) was successfully fabricated via an electrochemical deposition method using hexamethylenetetramine (HMTA) as a surfactant. The as-prepared photocatalysts were characterized in terms of morphology and structure using X-ray diffraction (XRD), field-emission scanning electron microscopy (FESEM) and transient photocurrent response (TPR). A possible formation mechanism of the nanorod array structure was explained in terms of the crystal nucleation and crystal growth direction. In comparison to unmodified-ZnO, substantial photocatalytic degradation of hexamethylenetetramine-modified ZnO (HMTA–ZnO) was observed towards sunset yellow (SY) dye under natural sunlight irradiation. The profound photocatalytic degradation of SY attributed to the one-dimensional highly order rod-like nanostructures of fabricated HMTA-ZnO with excellent crystallinity and electrochemical characteristics which might improve the charge carrier segregation and might resulted in the production of a massive number of active radicals.

Keywords: ZnO nanorod array · Electrodeposition · Surfactant · Dye degradation

1 Introduction

In recent decades, industrial effluents have become a major source of water pollution, as they are often discharged directly into natural water bodies. These effluents commonly contain toxic substances such as dyes, organic compounds, and heavy metals [1]. Among them, azo dyes have raised significant environmental and health concerns. Sunset yellow (SY), an azo dye, is widely used in confectionery, pharmaceuticals, and beverages, and is recognized as a persistent organic pollutant due to its chemical stability. Its molecular structure consists of phenyl and naphthyl rings linked by an azo functional group, along with sulfonate substituents. Excessive consumption of SY has been associated with harmful effects including renal failure, liver damage, and carcinogenic risks [1]. The

M. A. b. Jamlos et al. (Eds.): ICSET 2025, LNNS 1897, pp. 175–180, 2026.
https://doi.org/10.1007/978-3-032-20722-7_22

effective treatment of azo dye–contaminated wastewater is therefore crucial not only for environmental protection but also for attaining Sustainable Development Goal 6 (Clean Water and Sanitation), which emphasizes the need for clean and safe water resources worldwide. Developing sustainable and cost-effective treatment technologies capable of addressing industrial dye pollution aligns directly with this global agenda. Various treatment strategies have been developed, including adsorption, coagulation, advanced oxidation processes (AOPs), and photocatalysis. Among these, photocatalysis stands out as particularly promising as a highly efficient, eco-friendly, and cost-effective approach that utilizes solar energy for dye degradation.

Zinc oxide (ZnO) is a wide-bandgap semiconductor (3.37 eV) widely employed as photocatalyst due to its low cost, non-toxicity, chemical stability, and unique intrinsic properties. Among various ZnO nanostructures, one-dimensional (1D) nanorods are particularly attractive due to their high surface area, efficient charge transport, and short diffusion paths [2]. Several techniques have been employed to prepare ZnO thin films, including spray pyrolysis, sol–gel processing, hydrothermal methods, and electrodeposition. Among these, electrodeposition is particularly suitable for synthesizing ZnO nanorods due to its simplicity, low cost, mild temperature and pressure requirements, and versatility [2]. In this approach, hexamethylenetetramine (HMTA) is commonly employed as a surfactant and chemical additive to attain the rod-like growth of ZnO structures, predominantly when $Zn(NO_3)_2$ is used as the precursor [3]. This method also enables the direct formation of highly crystalline films immediately after deposition [4].

Herein, ZnO NRA photocatalyst was fabricated using an electrochemical deposition method in HMTA aqueous solution. The synthesized photocatalyst was examined using X-ray diffraction (XRD) and field-emission scanning electron microscopy (FESEM) to determine the physical and chemical characteristics. The photocatalytic performance of hexamethylenetetramine-modified ZnO (HMTA–ZnO) photocatalyst was evaluated by photodegradation of SY under sunlight irradiation. Moreover, the possible growth mechanism of ZnO is proposed.

2 Experimental Procedure

All chemicals and reagents employed in this study were of analytical grade and utilized directly without further purification.

2.1 Material Synthesis

Electrochemical deposition was employed to form a ZnO photocatalyst on an FTO glass substrate. First, ZnO seed layer solution was prepared by dissolving 20 mM of $(CH_3COO)_2Zn{\cdot}2H_2O$ and 10 mM NaOH in ethanol, followed by stirring at 70 °C for 2 h. The seed solution was then sprayed onto FTO glass placed on a hot plate at 100 °C by using a spray gun and subsequently calcined at 300 °C for 1 h in a muffle furnace. After that, a two-electrode system was employed, with the seeded FTO as the cathode (−) and a zinc plate as the anode (+). A potential of 1.1 V and a current of 0.01 A were applied for 2 h during the electrodeposition in a 5 mM $Zn(NO_3)_2{\cdot}6H_2O$ aqueous solution at 80 °C. In the modified synthesis, an equimolar concentration of HMTA was added into the electrodeposition solution. Finally, the deposited samples were rinsed with ethanol and distilled water, then calcined at 400 °C for 1 h in a muffle furnace. The samples without and with HMTA were denoted as unmodified-ZnO and hexamethylenetetramine-modified ZnO (HMTA–ZnO), respectively.

2.2 Characterization

The morphology and the crystal structure of the prepared photocatalysts was characterized through field emission scanning electron microscopy (FESEM, Jeol JSM-6701F) and X-ray diffraction (XRD, Shimadzu-6000) respectively. The Gamry Interface 1000 T potentiostat electrochemical workstation was utilized to carry the Transient photocurrent response (TPR) test. A 105 W compact fluorescent lamp was used as the light source, the sample-coated fluorine-doped tin oxide (FTO) glass as the working electrode, an Ag/AgCl electrode as the reference electrode and a platinum electrode as the counter electrode.

2.3 Photocatalytic Degradation Experiment

The photocatalytic activity of the as-synthesized catalysts was evaluated based on the degradation of sunset yellow (SY) solution. Photocatalyst-coated FTO substrate was immersed in a 100 mL beaker containing 100 mL of 5×10^{-3} mM SY solution and kept in the dark for 30 min to establish adsorption–desorption equilibrium. Photodegradation experiments were conducted on sunny days between 11:00 a.m. and 2:00 p.m. The initial and final concentrations of SY were determined by measuring the absorbance of the supernatant at the maximum wavelength of 480 nm using a UV–Vis spectrophotometer.

3 Results and Discussion

3.1 Characterization of HMTA-ZnO Photocatalyst

Fig. 1(a–b) illustrates the morphological differences between unmodified-ZnO and HMTA-ZnO. As shown in Fig. 1(a), the unmodified-ZnO exhibited a nanoflower-like structure having average particle size of 500–520 nm. In contrast, Fig. 1(b) reveals that the presence of HMTA as a surfactant promoted the formation of well-aligned ZnO nanorods with a diameter in the range of 490–540 nm, which were slightly smaller and more uniform as compared to that of unmodified-ZnO. The FESEM images clearly revealed the morphological transformation from irregular, larger ZnO particles to more uniform and well-defined nanostructures in the presence of HMTA, confirming its crucial role in directing nucleation and growth of ZnO crystals. Fig. 1(c) presents the XRD spectra of synthesized photocatalyst. The XRD diffraction peaks of both unmodified-ZnO and HMTA-ZnO at 2θ values of 31.9, 34.4, 36.3, 47.7, 56.5, 63.0, 66.7, 68.1 and 72.7° are correlated with (100), (002), (101), (102), (110), (103), (200), (112) and (004) planes of hexagonal wurtzite ZnO structure (JCPDS card no. 36–1451), respectively. Compared to unmodified-ZnO, the sharper and more intense peaks in the HMTA-ZnO sample indicated improved crystallinity and it was consistent with various reported studies [4].

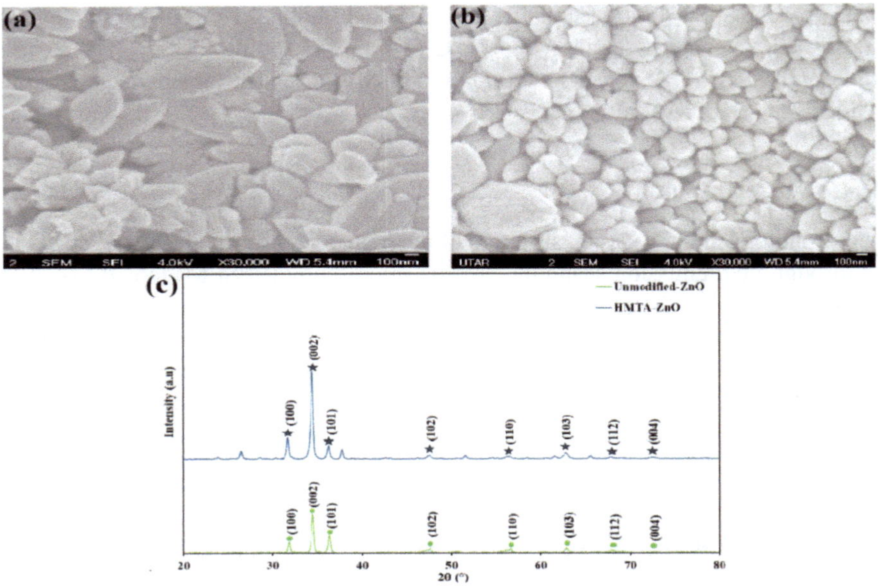

Fig. 1. FESEM images of (a) unmodified-ZnO and (b) HMTA-ZnO; (c) XRD spectra of unmodified-ZnO and HMTA-ZnO.

3.2 Proposed Mechanism of ZnO Growth

The electrodeposition growth mechanism of ZnO nanorods is illustrated in Fig. 2. Initially, a ZnO seed layer is deposited onto the FTO substrate, providing heterogeneous nucleation sites with lower activation energy that facilitate crystal growth [3, 5]. In the second step, nanorods grow over this layer, with morphology determined by the relative supply of OH^- and Zn^{2+} ions. When Zn^{2+} diffusion is slower than OH^- generation, the growth occurs mainly along the longitudinal axis, whereas comparable rates promote both longitudinal and lateral growth, producing thicker nanorods [3]. Hexamethylenetetramine (HMTA) is commonly employed as a chemical agent to promote the rod-like growth of ZnO structures with $Zn(NO_3)_2$ precursor. During the synthesis process, HMTA decomposes into ammonia and formaldehyde, which act as additional sources of OH^- ions [5]. The decomposition pathways can be represented as follows:

$$(CH_2)_6N_4 + 6H_2O \rightarrow 4NH_3 + HCHO \tag{1}$$

$$NH_3 + H_2 \rightarrow NH_4^+ + OH^- \tag{2}$$

Moreover, HMTA binds to the non-polar facets of ZnO, suppressing lateral growth and promoting vertical elongation, which leads to well-aligned nanorods with uniform structures [5].

3.3 Effect of HMTA on Photocatalytic Activity of ZnO

The addition of HMTA during ZnO synthesis played a crucial role in enhancing the photocatalytic degradation efficiency of SY. As shown in Fig. 3(a), ZnO without HMTA

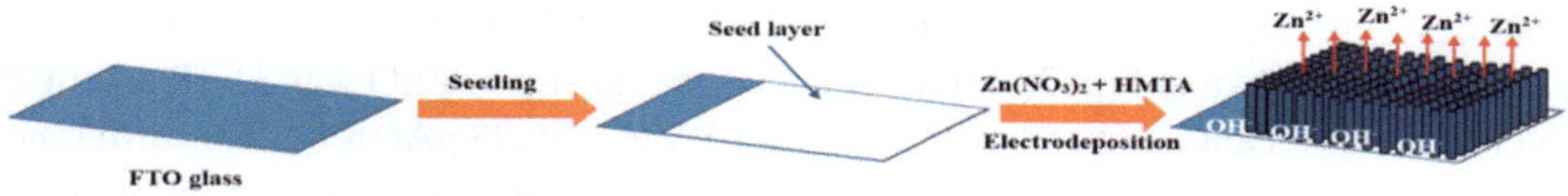

Fig. 2. Schematic view of the growth mechanism of ZnO NRA by electrodeposition from $Zn(NO_3)_2$ and HMTA solution.

modification achieved 44% degradation of SY after 3 h of sunlight irradiation. When HMTA was introduced, the efficiency increased slightly to 54%. This enhancement can be attributed to the role of HMTA in modifying the growth of ZnO into well-aligned rod-shaped structures with uniform size distribution and improved crystallinity. This structural enhancement suppressed the recombination of photo-generated charge carriers, consequently improving the efficiency of the photocatalytic reaction.

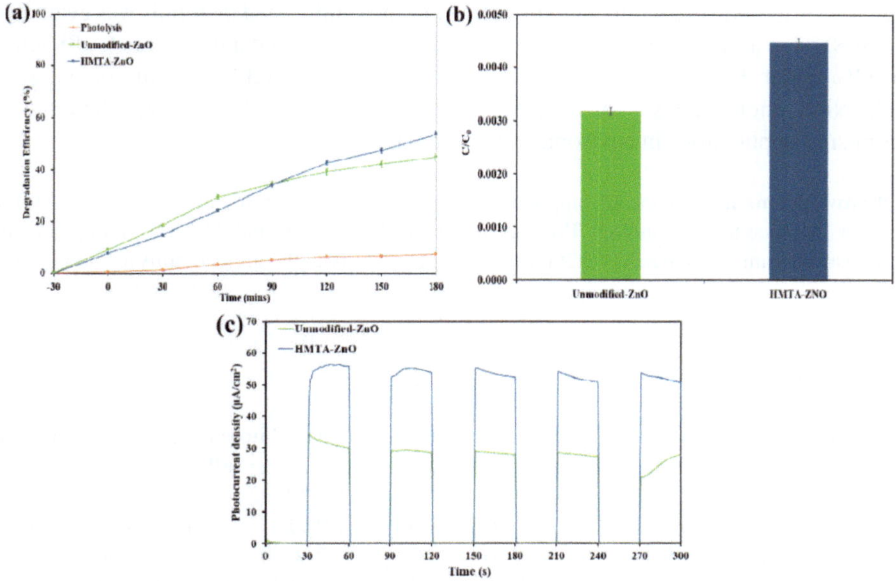

Fig. 3. (a) SY degradation of photolysis, unmodified-ZnO and HMTA-ZnO; (b) Pseudo-first-order degradation rate constant of SY and (c) TPR spectra.

Figure 3(b) displays the kinetic curves for SY degradation under sunlight irradiation. The photodegradation process followed a pseudo-first-order kinetic model, as expressed by Eq. (3)

$$-\ln(Ct/C0) = kt \tag{3}$$

where C_t and C_0 referred to the concentrations of SY (ppm) at particular irradiation times t (min) and the initial concentration, respectively, while k denoted the reaction rate constant. For the HMTA-ZnO photocatalyst, the calculated k value was 0.00447 min^{-1}, which was 1.41 times higher than ZnO without HMTA. The photoinduced charge carrier dynamics were further investigated through transient photocurrent response (TPR) measurements. As shown in Fig. 3(c), the photocurrent density of ZnO was 31.18 μAcm^{-2}, which was significantly lower than that of HMTA-ZnO (56.28 μAcm^{-2}). This difference indicated a higher recombination rate of photogenerated charge carriers in unmodified-ZnO compared to HMTA-ZnO. These results confirmed that the separation and migration of photogenerated charge carriers were more efficient in the ZnO photocatalyst-loaded with HMTA.

4 Conclusion

In summary, ZnO NRA photocatalysts were successfully synthesized using the electrodeposition method. Their crystal structures, morphologies, optical properties, and electrical properties were systematically characterized through XRD, FESEM and TPR analyses. In addition, the growth mechanism of ZnO nanorod arrays via electrodeposition was proposed. The photocatalytic activity of the synthesized materials was evaluated using SY dye as a model pollutant, where ZnO NRA achieved a degradation efficiency of 54% under 3 h of sunlight irradiation. These findings highlighted the potential of ZnO NRA photocatalysts for advancing research in photocatalytic degradation with promising applications in environmental pollutant remediation.

Acknowledgements. Financial supports by the Petronas Research Sdn. Bhd. (8151/0002), Guangxi Science and Technology Program (Guike AD25069074) and ASEAN Young Talented Scientist Program of Guangxi (DT2200002995 and DT2200002996) are gratefully acknowledged.

References

1. Zhang, S., et al.: Designing, characterization, and evaluation of chitosan-zinc selenide nanoparticles for visible-light-induced degradation of tartrazine and sunset yellow dyes. Environ. Res. **213**, 113722 (2022). https://doi.org/10.1016/j.envres.2022.113722
2. Yong, Z.J., Lam, S.M., Sin, J.C., Zeng, H., Mohamed, A.R., Jaffari, Z.H.: Boosting sunlight-powered photocatalytic fuel cell with S-scheme Bi_2WO_6/ZnO nanorod array composite photoanode. Inorg. Chem. Commun. **143**, 109826 (2022). https://doi.org/10.1016/j.inoche.2022.109826
3. Ait Hssi, A., et al.: Electrodeposition of oriented ZnO nanorods by two-steps potentiostatic electrolysis: Effect of seed layer time. Solid State Sci. **104**, 106207 (2020). https://doi.org/10.1016/j.solidstatesciences.2020.106207
4. Gboglo, A.D., Baneto, M., Ako, O., Haris, M., Senthilkuma, M.: Role of hexamethylenetetramine/zinc nitrate hexahydrate molar ratio in controlling structural, morphological and optical properties of ZnO thin films synthesized by CBD. AIMS Mater. Sci. **12**, 893–908 (2025). https://doi.org/10.3934/matersci.2025039
5. Biswal, H.J., Srivastava, T., Vundavilli, P.R., Gupta, A.: Facile fabrication of hydrophobic ZnO nanostructured nickel microtubes through pulse electrodeposition as promising photocatalyst for wastewater remediation. J. Manuf. Process. **75**, 538–551 (2022). https://doi.org/10.1016/j.jmapro.2022.01.001

Predictive Analytics for NDUM Foundation Admissions via Naïve Bayes Algorithm

Syahaneim Marzukhi(✉), Ammar Hafiy Muhammad Habib, Nor Fatimah Awang, Sharifah Aishah Syed Ali, and Nor Azliana Akmal Jamaludin

National Defense University Malaysia (NDUM), Kem Sg. Besi, Kuala Lumpur, Malaysia
syahaneim@upnm.edu.my

Abstract. Student selection process of the National Defense University of Malaysia (NDUM) Foundation Program leveraging the Naïve Bayes algorithm, aims to automate and optimize applicant evaluation by identifying the most qualified applicants. A dataset comprising 768 instances from historical data of previous applicants and 16 attributes was processed and analyzed using RapidMiner software. The model achieved an accuracy of 98.04%, precision of 84.88%, and recall of 97.33%, demonstrating high predictive reliability. Further, the results were integrated into a system with visual and tabular output components to support academic decision-making and include the data export feature for administrative use. The results suggest that the application of machine learning technique, particularly Naïve Bayes algorithm, provides a scalable and data-driven approach to improving fairness, transparency, and efficiency in the student admission process. Future work may involve extending the feature set and applying other algorithms to enhance generalizability and robustness.

Keywords: optimize applicant evaluation · data-driven approach · Naïve Bayes

1 Introduction

National Defense University of Malaysia (NDUM) was established by the Malaysian Government on 21 June 2006 to produce military and civilian graduates for national defense needs. Thus, the selection process for the foundation students at NDUM is crucial as the institution is responsible for producing future military officers. Therefore, NDUM has established stringent criteria including physical fitness and academic achievement to select the most suitable applicants. These two factors are considered vital, as a military career demands both exceptional physical endurance and strong intellectual capabilities. Physical fitness is important as the students are required to undergo rigorous military training that demands extraordinary physical resilience. In addition, academic achievement is also a key determinant in order to comprehend military theories and apply this knowledge in practical settings.

Thus, this work is proposed to enhance foundation program admissions at NDUM using Naïve Bayes predictive modelling that can facilitate the selection of suitable applicants based on a certain key criterion. This model aims to address several challenges and issues during the current selection process as follows:

M. A. b. Jamlos et al. (Eds.): ICSET 2025, LNNS 1897, pp. 181–187, 2026.
https://doi.org/10.1007/978-3-032-20722-7_23

1. Selection Based on Historical Data.

Often the number of applicants exceeds the available quota. A major issue lies related to the lack of comprehensive data analysis and the limited utilization of historical data in the current system to ensure that only the most qualified applicants are selected.

2. Physical Fitness versus Academic Achievement as a Primary Selection Criterion.

The applicants are required to undergo thorough medical and physical screenings to ensure they meet the high standards of physical endurance necessary for military training. A key challenge in the selection process involves differentiating the applicants with strong academic features and meet the criteria for physical fitness.

There is a critical need to model and analyze the interaction between physical fitness and academic achievement as interdependent features within a predictive model. Capturing these multimodal relationships can improve the model's ability to identify the best applicants and enhance the overall accuracy and fairness of the selection process. Therefore, this work aims to enhance the accuracy and efficiency of the selection process through a predictive model that can identify suitable applicants for the NDUM Foundation Program. By leveraging historical admissions data, the work employs the Naïve Bayes algorithm; a probabilistic machine learning approach known for its simplicity and effectiveness in high-dimensional datasets to develop the predictive model. Finally, results from the predictive model are integrated into the selection system to support real-time decision-making.

The paper is structured as follows. Sect. 2 provides the related background of the work under study. Sect. 3 presents the methodology for the work, Sect. 4 discusses findings and results, and Sect. 5 contains concluding remarks.

2 Research Background

2.1 Related Works

Based on previous study, there are numerous works that have functional similarities to "Predictive Analytics for NDUM Foundation Admissions via Naïve Bayes Algorithm". Though these studies vary in implementation, each presents distinct advantages and disadvantages that provide critical insights into the enhancement of the proposed work. Various research such as in [1, 2, 4, 6–8, 10, 11], applying data mining techniques, in particular Naive Bayes for analyzing the students' performance. In [2], the study proposes a framework employing the Naïve Bayes classifier to predict the academic performance of first-year Computer Science students at UNISZA, Malaysia. Utilizing data from six academic years, the model considers variables such as race, gender, family income, institutional entry mode, and GPA. Data Mining tool (i.e. WEKA) is used to build a prediction model. The data were collected from 6-year period intakes from July 2006/2007 until July 2011/2012. The findings indicate that factors like family income and gender significantly influence academic performance, demonstrating the classifier's effectiveness in educational data mining.

Similar study in [2, 10, 11], the researcher applies the Naïve Bayes algorithm to predict students' final grades. Those studies focus on developing a methodological platform using the Naïve Bayes classifier to evaluate and enhance student academic performance. The study reports that the algorithm's potential in forecasting academic performance and aiding in the enhancement of teaching and learning processes.

There are also studies on Student Dropout Prediction Using Naive Bayes Classification [4]. The authors of [4], this study investigates the application of the Naïve Bayes algorithm to predict first year students from undergraduate at the institution who at risk of dropping out. High dropout rates can significantly impact the overall quality of education and diminish students' future opportunities. The model analyzes historical data of students who have either dropped out or successfully completed their studies, estimating the likelihood of dropout based on variables such as aca-demic performance, attendance records, participation in co-curricular activities, and socio-economic background. One of the key strengths of the Naïve Bayes algorithm lies in its efficiency in handling large datasets and generating rapid predictions, making it particularly suitable for real-time educational decision-making.

Yet, the authors of [8], explores the use of the Naïve Bayes classifier to predict new student admissions. This study aims to predict the number of students that will register for the next coming semester. The data was obtained by distributing survey randomly to 200 students who were about to enter high school. By analyzing factors such as extracurricular activities, cost, and distance, the model achieved a high precision rate, demonstrating its applicability in admission processes. The conclusion is that the extracurricular, cost and distance criteria need attention and improvement.

Thus, this work aims to develop a predictive model using the Naïve Bayes algorithm to support the foundation student selection at NDUM, with the goal of enhancing accuracy and efficiency in applicant screening and selection. For this work, Naïve Bayes algorithm is used to identify and analyze relevant features (e.g., academic performance, co-curricular involvement, physical fitness) that influence the student selection for the NDUM Foundation program. The proposed model will be evaluated using standard metrics such as accuracy, precision and recall. This approach offers the decision maker with a solution to provide recommendations by integrating the predictive model into the NDUM admission process for improvement.

2.2 Machine Learning for Predictions

Machine Learning (ML) technique has become an essential tool in predictive analytics, offering significant potential in the education field for tasks such as student assessment, academic performance forecasting, and admission decision support [5, 10]. By leveraging historical and multidimensional data, ML can detect patterns and generate accurate predictions that inform strategic interventions and institutional planning. In the context of student selection, the previous student selection processes rely on manual screening and subjective judgment, that can lead to inconsistencies and inefficiencies. Thus, to address these limitations, ML techniques are increasingly being adopted to enhance the objectivity and reliability of applicant evaluations [4]. By identifying underlying patterns in historical student data using Naïve Bayes, the educational institutions can leverage this model to make more informed, objective, and data-driven decisions rather than subjective judgment. The model will analyze applicants based on various factors, including academic performance, co-curricular involvement, physical fitness to predict their suitability for admission.

Naïve Bayes algorithm is a probabilistic ML algorithm based on Bayes' Theorem. The algorithm is designed to predict the likelihood of a data instance belonging to a particular class based on observed feature values. The algorithm is termed 'naïve' due to its fundamental assumption of conditional independence among features; meaning that each attribute is independently contribute to the final classification outcome. The Naïve Bayes process generally involves three primary steps. First, the algorithm computes the prior probability of each class without regard to feature values. Second, the algorithm estimates the likelihood of each feature given a class label, capturing how attribute distributions relate to specific classes. Finally, Bayes' Theorem applies to calculate the posterior probability of each class for a given instance, selecting the class with the highest posterior as the predicted outcome.

The principle of Naïve Bayes is discussed as follows. Given a set of features $X=\{x1, x2, \ldots, xn\}$ and a class label C, the Naïve Bayes algorithm estimates the posterior probability $P(C|X)$ using Bayes' Theorem:

$$P(C|X) = \frac{P(X|C) \cdot P(C)}{P(X)} \tag{1}$$

Since P(X) is constant across all classes for a given input, prediction is typically based on maximizing the numerator. The 'naïve' assumption refers to treating all features xi as conditionally independent given the class C.

For example, the Naïve Bayes algorithm can be trained on historical student data to predict the students into two meaningful categories, either as accepted applicants for "Pass" or rejected applicants for "Fail" based on predetermined features (e.g. academic performance, co-curricular involvement, physical fitness). An illustrative computation is presented as following using the first row of the training dataset to demonstrate the process of determining selection eligibility criteria prior to the prediction. Given a new student with the following attributes: {CGPA = Medium, Co-Curricular = Active, BMI = Good}, Fig. 1 shows the Naïve Bayes calculation either the student will "Pass" or "Fail" to be selected. Since $P(Yes|X)>P(No|X)$, the model predicts "Pass".

For "Yes":

$$P(Yes \mid X) = P(Yes) \cdot P(CGPA = Medium \mid Yes) \cdot P(CoCurricular = Active \mid Yes) \cdot P(BMI = Good \mid Yes)$$

$$= 0.5 \cdot \frac{1}{3} \cdot \frac{2}{3} \cdot \frac{2}{3} = 0.5 \cdot 0.296 = 0.148$$

For "No":

$$P(No \mid X) = 0.5 \cdot \frac{1}{3} \cdot \frac{1}{3} \cdot 0 = 0$$

Fig. 1. Illustrative computation using Naïve Bayes.

3 Methodology—CRISP-DM

Cross Industry Standard Process for Data Mining (CRISP-DM) is among the most widely used framework for solving data science problems and is used for this study. In this paper [10, 11], the author discussed how this CRISP-DM Framework is translating business problems into data mining tasks. CRISP-DM consists of six phases: (1) business understanding, (2) data understanding, (3) data preparation, (4) modelling, (5) evaluation and (6) deployment and is described as the following [10].

In Phase 1, the primary objective is to improve the fairness and operational efficiency of the NDUM Foundation student selection process. Thus, the development of predictive models is required in identifying applicants with the highest likelihood of success. Consequently, the academic and admissions staff must understand the business process to define a measurable success criterion and establish data-driven selection policies aligned with institutional goals. Phase 2 involves collecting historical data on applicants, such as academic performance, co-curricular involvement, physical fitness and final selection decisions. Exploratory data analysis is conducted to assess data quality, detect outliers, and identify patterns or correlations between attributes and outcomes. Phase 3 focuses on cleaning, transforming, and organizing the data for modeling. This includes handling missing values, encoding categorical variables, and scaling numerical features. At this stage, irrelevant or redundant features may be removed to improve model performance. Phase 4, Naïve Bayes algorithm is also applied to train the models. Hyper-parameter tuning and cross-validation are used to optimize model performance and prevent overfitting. Phase 5 assesses the effectiveness of each model using metrics such as accuracy, precision and recall. This ensures that the model aligns with institutional objectives, particularly in avoiding bias or unfair exclusion from the applicant groups. Finally, in Phase 6, the best-performing model is integrated into the proposed selection system to support real-time decision-making. This involves developing a web-based system that provides a list of successful applicants.

4 Results and Discussion

For this study, the dataset was past applicants' data. The results are derived from the implementation of the Naïve Bayes algorithm to the NDUM Foundation student dataset and visualized using the web-based system. First, the Naïve Bayes algorithm as trained and tested using RapidMiner Studio 9.8.01 on the dataset comprising 768 instances with 16 attributes, including 5 standard attributes and 10 special attributes. The special attributes encompass a diverse range of information such as Identification Number, Gender, Academic Score, Merit Score, Body Mass Index (BMI), Name, Program Name, Disability Status, and Offer Status. The model performance was assessed using standard classification metrics: accuracy, precision, and recall (sensitivity). Secondly, the prediction results were integrated into the web-based system, where a list of successful applicants was presented and visualized through graphical representations to effectively communicate the analytical outcomes.

The Naïve Bayes algorithm achieved accuracy of 98.04% demonstrating its ability to correctly classify the instances in the dataset. The precision is 84.88%, indicates that most applicants predicted as "Pass/Lulus" were indeed true positives, thereby reflecting the algorithm's reliability in reducing false positive predictions. The recall is 97.33%, signifying the algorithm's strong capability in identifying actual positive cases; correctly recognizing applicants who should be accepted. Findings show that Academic Score, Merit Score and Body Mass Index (BMI) are considered most important criteria for NDUM student selection. The results indicate that the Naïve Bayes is a high-performing algorithm in the context of student selection. Its high recall suggests strong suitability for admission scenarios minimizing false negatives; failing to select successful applicants.

Next, the results are uploaded to the web-based system and displayed in graphical forms. Fig. 2 illustrates the stacked bar chart, where the *x-axis* represents the foundation program and the *y-axis* represents the count of applicants; it categorizes the student into two classes (i.e. "Pass/Lulus" in blue color vs. "Fail/Gagal" in pink color) across three foundation academic streams (i.e. Medicine, Engineering and Technology and Foundation in Management and Strategy).

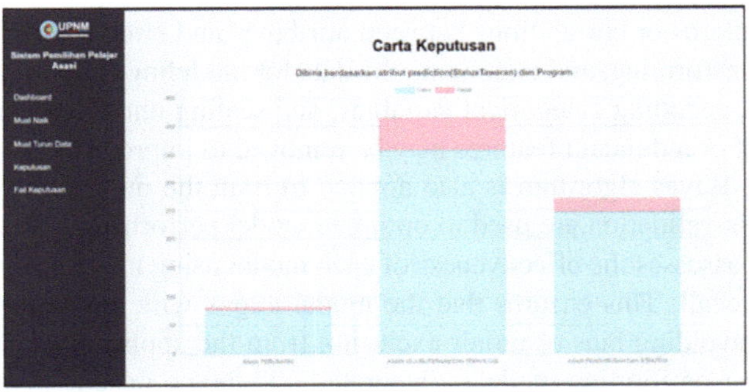

Fig. 2. Graphical Results in Web-based System.

5 Conclusion

This study successfully developed and evaluated a predictive modeling system for the NDUM Foundation student selection process using the Naïve Bayes algorithm. The model achieved strong performance metrics, including an accuracy of 98.04%, a precision of 84.88%, and a recall of 97.33%, confirming its effectiveness in correctly identifying applicants who should be accepted (i.e. "Pass/Lulus"). Naïve Bayes algorithm accelerates and automates the screening and prioritization process by generating a simple, transparent probabilistic score for each applicant; thereby reducing manual workload, shortening processing time, and enhancing consistency in decision-making. The system also incorporated multiple interfaces to enhance interpretability and usability for the decision-makers. These included a stacked bar chart for visualizing selection outcomes across academic programs and a tabular interface for reviewing individual applicant predictions. The inclusion of an export functionality for downloading results in Excel format supports data archival and integration into institutional workflows.

The integration of ML into the student admission process offers clear advantages in terms of efficiency, transparency, and data-driven decision-making. The findings suggest that the Naïve Bayes algorithm is a viable and high-performing tool for automating applicant evaluation, reducing human bias, and supporting scalable admissions management. Future work may explore the integration of ensemble learning methods to further improve model accuracy. In addition, expanding the dataset and including more diverse features such as interview scores or psychometric data could provide deeper insights and improve the model's generalizability.

References

1. Adrian, R., Perdana, M.A.J.S., Asroni, A., Riyadi, S.: Applying the naïve bayes algorithm to predict the student final grade. Emerg. Inf. Sci. Technol. **1**(1) (2022)
2. Aziz, A.A., Ismail, N.H., Ahmad, F., Hassan, H.: A framework for students' academic performance analysis using naïve bayes classifier. J. Teknol. (Sci. & Eng.). **75**(3), 13–19 (2015)
3. Patel, D., Savaj, K., Malani, P., Patel, J., Trivedi, H.: Unlocking enigmatic pathways: empowering student dropout analysis with machine learning and energizing holistic investigation. In: International Conference for Convergence in Technology 2024 (I2CT 2024), pp. 1–7 (2024)
4. Almahadeen, L., Akkaya, M., Sari, A.: Mining Student Data Using Crisp-DM Mod-el. Comput. Sci. Inf. Syst. **15**(2), 305–316 (2017)
5. Marbouti, F., Diefes-Dux, H.A., Madhavan, K.: Models for early prediction of at-risk students in a course using standards-based grading. Comput. Educ. **103**, 1–15 (2016)
6. Nakhipova, V., et al.: Use of the naïve bayes classifier algorithm in machine learning for student performance prediction. Int. J. Inf. Educ. Technol. **14**(1), 92–98 (2024)
7. Romero, C., Ventura, S.: Educational data mining and learning analytics: An updated survey. Wiley Interdiscip. Rev.: Data Min. Knowl. Discov. **10**(3), e1355 (2020)
8. Suwayudhi, S., Irawan, E., Damanik, B.E.: Classification techniques in predicting new student admission using the naïve bayes method. JOMLAI: J. Mach. Learn. Artif. Intell. **1**(3), 251–256 (2022)
9. Sweeney, M., Lester, J., Rangwala, H.: Next-term student performance prediction: A recommender systems approach. J. Educ. Data Min. **8**(1), 22–51 (2016)
10. Kotu, V., Deshpande, B.: Data Science: Concepts and Practice. Morgan Kaufman (2019)
11. Wiemer, H., Drowatzky, L.: Ihlenfeldt: Data mining methodology for engineering applications (DMME)—A holistic extension to the CRISP-DM model. Appl. Sci. **9**, 2407 (2019)

Optimization of Magnetic Field Distribution via Finite Element Magnetic Analysis of Magnetorheological Elastomer Based Engine Mount

Rusdi Mat Song[1(✉)], M. S. M. Shamsuddin[1], and S. A. Mazlan[2]

[1] Universiti Pertahanan Nasional Malaysia, Kuala Lumpur 57000, Malaysia
rusdi@upnm.edu.my
[2] Malaysia-Japan International Institute of Technology, Universiti Teknologi Malaysia, Kuala Lumpur 54100, Malaysia
lncs@springer.com

Abstract. The class of smart materials known as magnetorheological (MR) materials includes those that can be dramatically changed under the effect of an external stimulus, such as a change in stress, pH level, moisture content, electric fields, or, in the case of magnetorheological elastomers (MREs), magnetic fields. In order for the MRE to be fully induced, the magnetic field must be circulated sufficiently throughout its effective area and thus the device must be designed to optimize the magnetic flux density generated by an electromagnetic coil. Therefore, design analysis by Finite Element Method Magnetic or FEMM software is being applied to analyze several types of engine mount design which incorporates the placement of the electromagnetic coil and the application of full cover. The results demonstrated that the placement of an external electromagnetic coil with 3584 total turns with the application of full cover manage to provide a magnetic flux density distribution of up to 0.31 Tesla (T) at the MRE effective area without changing the original design of the existing industrial mount. The results had demonstrated that such of the device utilizing MR material must be properly designed to gain maximum effect of the material under external stimulus.

Keywords: Magnetorheological elastomer · engine mount · FEMM design

1 Introduction

1.1 Magnetorheological Elastomer

Magnetorheological elastomer or MRE is considered as a smart material in which its mechanical and rheological properties can respond towards the external stimuli such as magnetic fields [1]. Categorized as composite material in a solid state, it consists of a mixture between magnetic particles and elastomeric matrix [2]. Due to its capability of being controllable in both mechanical and rheological, this material had been widely applied in device instrumentation such as the engine mount [3]. MRE engine mount

M. A. b. Jamlos et al. (Eds.): ICSET 2025, LNNS 1897, pp. 188–194, 2026.
https://doi.org/10.1007/978-3-032-20722-7_24

is a device in which the elastomeric rubber part is composed of dispersed micron-size magnetizable particles suspended in its matrix carrier, making it magnetorheological based thus giving effect on its storage and loss modulus with the presence of magnetic field [4]. The strength of the magnetic field regulates the material's shear modulus, which in turn influences the material's stiffness and damping [5]. Mostafa et al. reported that the required strength of the magnetic field is range up to 250 kA/m (equivalent to 0.3 Tesla) [6]. Moreover, for the MRE to be fully induced, the magnetic field must be circulated evenly and uniformed throughout its effective area [7] and thus the device must be designed to optimize the magnetic field distribution generated by an electromagnetic coil and at the same time avoid magnetic flux leakage as much as possible.

Therefore, prior to any fabrication of the MRE-based device, thorough analysis should be carried out in term of optimizing the magnetic field distribution at the MRE affective area for a maximum stimulation. This is being carried out with correct design combination which consists of device material selection and the determination of electromagnetic coil geometry. Several research highlight the importance of ferromagnetic materials selection to ensure a closed property of the magnetic circuit and increase the magnetic flux intensity through the MRE [8]. This is because ferromagnetic materials exhibit a high magnetic permeability, which can efficiently absorb and redirect electromagnetic fields [9]. This property makes them highly effective in shielding applications.

This work aims to analyze several designs of proposed MRE-based engine mount in the initial stage of product design and thus proceed with the fabrication process itself. The original design of the device is based on a conventional engine mount retrieved from local industry. Several material combinations will be analyzed in terms of magnetic field circulation and distribution using a magnetic software called Finite Element Method Magnetic or FEMM. The effects of magnetic field density distribution on several material combinations (ferromagnetic and paramagnetic) were properly simulated.

2 Methodology

2.1 Conceptual Design

The process of constructing the entire device structure begins with identifying a suitable rubber-based engine mount to be used as the main component. The process then continued with designing a suitable electromagnetic coil to suit with the engine mount to supply the magnetic field throughout the engine mount. Several engine mounts had been collected to study the commonly used dimension together with its cross section. Figure 1(a) shows a cross section of an engine mount's internal form obtained from local market and automotive workshops. It consists of rubber compound between the outer casing and center core. The diameter of the engine mounts mostly in the range of 60 to 70 mm.

2.2 Finite Element Method Magnetic (FEMM)

Once the selection of the engine mount has been finalized, the next process involves the design of the electromagnetic coil. It is important to measure the dimension of the mount prior in designing the electromagnetic coil to determine the geometrical size of the coil bobbin. This is because the coil bobbin would function as a base for the engine mount placement as well as for the fabrication of the electromagnetic coil. Subsequently, the fabrication of the electromagnetic coil needs to consider the width and the length of the coil area, which would determine the total number of copper wire turns required in providing sufficient magnetic field throughout the device structure. Therefore, considering all these requirements, several designs concept needed to be carried out in searching for suitable configurations and Finite Element Method Magnetics (FEMM) software was used to assist with several initial designs' outcomes.

Electromagnetic distribution is the most important criteria to be considered when designing the MRE engine mount because it is the source of external stimuli. Hence, most of the magnetic circulation must be guided towards the MRE area, which requires an optimal design of the device structure. Therefore, in this work, the FEMM software was used to analyze the distribution of magnetic flux throughout the device structure and to measure the magnetic field strength at the effective area of the MRE engine mount. The first step conducted in FEMM simulation is to construct a two-dimensional cross section of the device structure, as shown in Fig. 1(b).

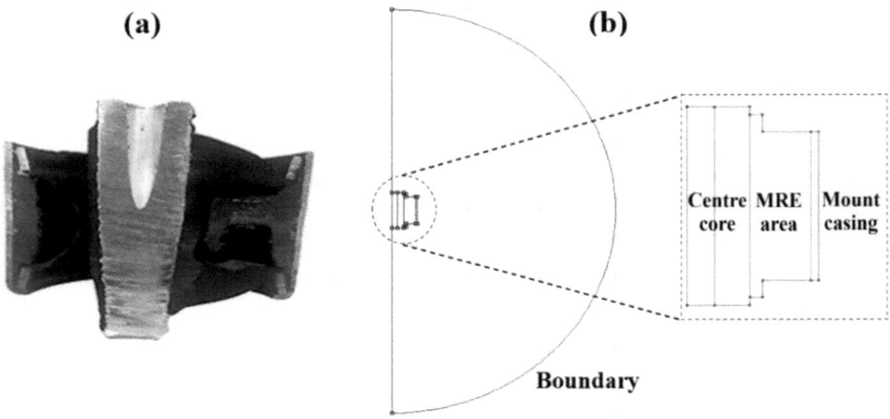

Fig. 1. (a) Cross section of a conventional engine mount (b) Boundary condition and the two-dimensional axisymmetric diagram of the engine mount used in FEMM

The diagrams of the engine mount are then labelled with associate sections, which are the center core, the MRE area and the engine mount casing. They are then assigned with appropriate types of material for the simulations analysis. After the two-dimensional MRE mount was modelled, the second step for FEMM simulation was the creation of the boundary condition using a mixed boundary condition type. This type of boundary condition consisted of an interface of a high permeability metal. Thus, the magnetic field generated by the coil could be modelled in an unbound space, while modelling a finite region of that space. The purpose of applying boundary condition was to assign the properties of line segment or arc segment that served as boundaries of the solution domain. The coverage area of the boundary was set to four times greater than the size of the modelled device structure.

The material and the circuit properties were then altered to produce an optimum value of magnetic flux density at the MRE area of the engine mount. FEMM provided a built-in material library, which enabled the use of any available material properties according to the design. The magnetic and non-magnetic together with air properties were assigned to each part in the model. For the design, the properties for air, SWG copper wire, MRE compound and 6061-T6 aluminum were selected.

3 Results and Discussion

Several design simulations had been conducted in selection for suitable device design configuration, which included electromagnetic coil placement either internally or externally. The summary of the design simulations is tabulated in the following Table 1. From the tabulated simulation preliminary findings, it was observed that the design with internal electromagnetic coil placement as in number 1 produced a scattered distribution of magnetic flux towards the outside area of the device and thus, a closed loop magnetic circulation could not be produced with such design. Further, the measured magnetic flux density at the MRE effective area yield a low value which was lower than 0.1 Tesla (T). To add, the design with internal coil placement had a result in low utilization of MRE compound volume, which already being allocated for the placement of the internal coil. Such condition was a disadvantage in the application of load bearing device.

Next, in the design simulation of number 2 and 3 with the external electromagnetic coil placement, it could be observed that the magnetic flux density at the MRE effective area differs compared to the previous design simulation, denoted by the coverage of the density plot color. Measurement of magnetic flux density at the effective area is 0.15 T, higher than previous design. It could be seen as well from the design that the distribution of magnetic flux scattered outside of the device structure and has not produced a closed loop magnetic circulation. Further in design simulation of number 4, partial cover has been introduced to control the magnetic flux distribution, thus resulting in slight increase of magnetic flux density with 0.18 T. However, there was still magnetic flux leakage, scattering outside of the device.

Therefore, in the final simulation design of number 5, the device structure has been improvised by adding full cover. The application of full cover improves in the circulation of magnetic field circuits originating from the coil through the MRE engine mount and forms a closed loop magnetic circulation as shown in Fig. 2. As such, much of the magnetic flux leakage could be prevented to yield a measurement of magnetic flux density of up to 0.31 T. Based on the tabulated simulation analysis and also in Fig. 3, device structure with the simulation design number 5 is expected to produce uniform and even distribution of magnetic flux density and the highest values of the magnetic flux density at the effective area of the MRE engine mount.

In this study, after referring to the materials' availability in the industry, the selected material for the full cover was mild steel with property compatible with AISI 1020. Since the coil bobbin must be made from non-magnetic material, aluminum 6061-T6 type was chosen for the material of the coil bobbin. The copper wire type 24SWG was chosen.

Table 1. Simulation design comparisons for electromagnetic coil.

No. 1	No. 2	No. 3
Internal coil	External coil: Plastic bobbin	External coil: Aluminum bobbin
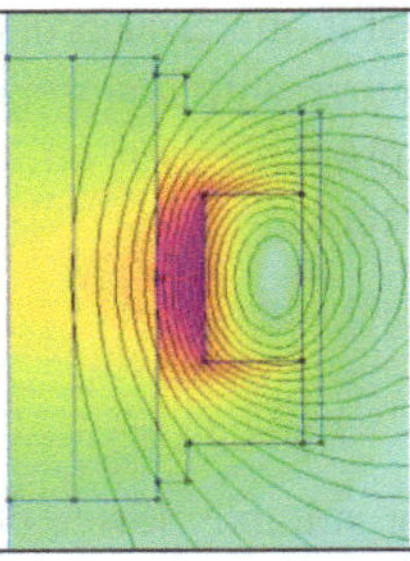	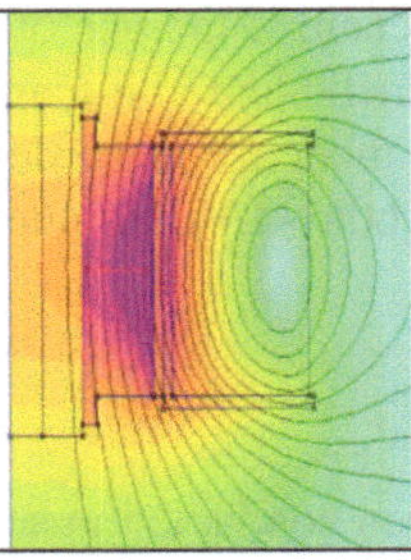	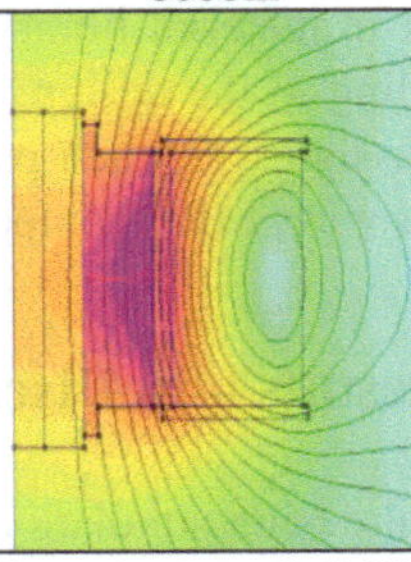
• Magnetic strength 0.07T. • Low utilization of elastomeric volume due to internal coil placement.	• Magnetic strength 0.15T. • Most of the magnetic fields scatter outside the effective region.	• Magnetic strength 0.15T. • Most of the magnetic fields scatter outside the effective region.

No. 4	No. 5
External coil: Aluminum bobbin with partial cover	External coil: Aluminum bobbin with full cover
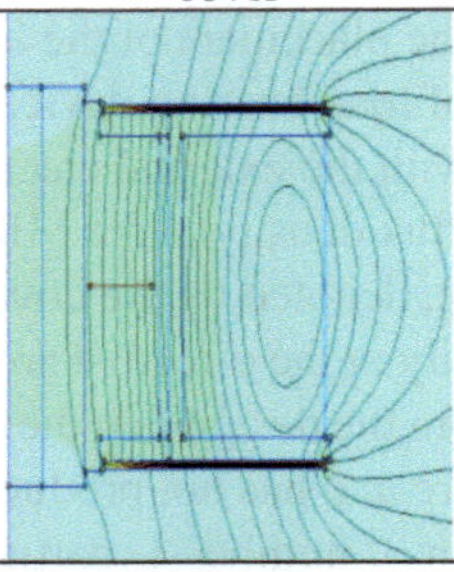	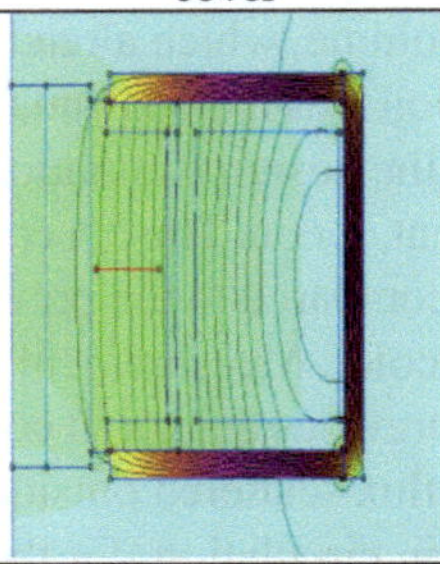
• Magnetic strength of 0.18T. • Most of the magnetic fields are scattered outside the effective region.	• High magnetic strength of 0.31T. • Most of the magnetic field circulation focus inside the device structure thus avoiding flux leakage. • Even distribution of magnetic flux at the effective area.

The total turns for the electromagnetic coil were calculated based on the area of the coil bobbin. The selection of coil was based on the highest intensity of magnetic field, H. Therefore, after careful consideration on the design of various segments starting from the coil bobbin material selection towards the full cover of the device structure using FEMM, the results yield an entire structure of MRE engine mount with high magnetic flux density distribution. Details of material arrangement are shown in Table 2.

Table 2. Part list of the device structure

Item	Part	Material Type	Material	Type
1	Coil bobbin	Non-magnetic	Aluminum	6061
2	Engine mount casing	Non-magnetic	Aluminum	6061
3	Top and bottom plates	Magnetic	Low carbon steel	1020
4	Cylindrical housing	Magnetic	Low carbon steel	1020
5	Coil wire	Non-magnetic	Copper	24SWG

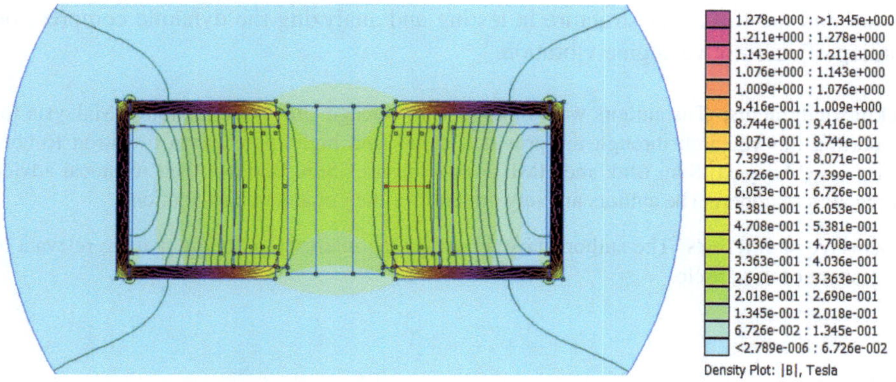

Fig. 2. Schematic of FEMM analysis showing minimum magnetic leakage with the application of full cover

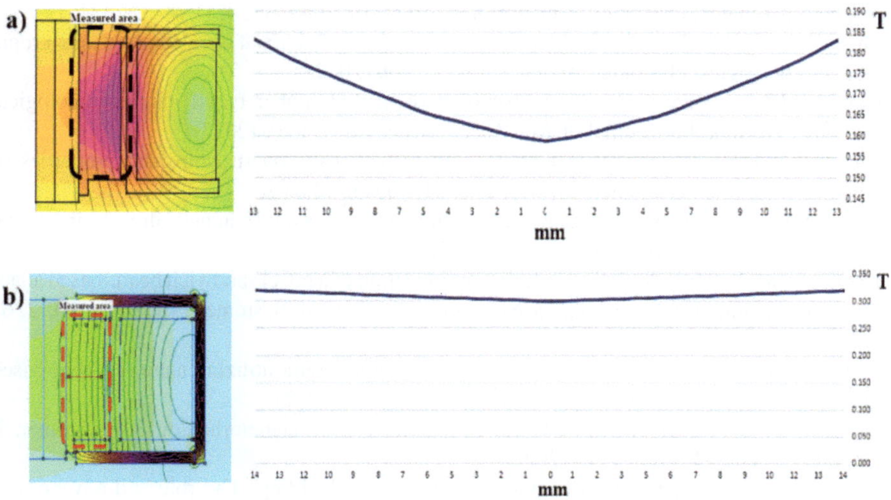

Fig. 3. Uniformity and strength of magnetic distribution based on: a) initial device design and b) final device design

4 Conclusion

In this work, MRE based engine mount with a full cover structure has been developed. Simulation processes were carried out by FEMM to ensure the effectiveness of the magnetic field distribution throughout the entire structure including suitable material selection and combination. The combination of high permeability (magnetic material) and low permeability material (non-magnetic material) influenced the magnetic distribution throughout the device structure. Based on Table 2, design arrangement as in number 5 capable of providing the highest magnetic field strength of 0.31 T in the MRE region with minimum magnetic flux leakage. This arrangement consisted of materials combination as shown in Table 2. Future works will focus on the fabrication of the MRE engine mount and the full cover structure in testing and analyzing the dynamic compression loading of automotive engine vibration.

Acknowledgments. The authors would like to acknowledge Universiti Teknologi Malaysia for supporting this research through the provided grants and high appreciation to Titron Rubber Industries (Malaysia) Sdn. Bhd. and HML Auto Industries Sdn. Bhd. for their technical advice and facilities support. The authors are very grateful for their eloquent contributions.

Disclosure of Interests The authors have no competing interests to declare that are relevant to the content of this article.

References

1. Rashid, R.Z.A.: Effects of silica on mechanical and rheological properties of EPDM-based magnetorheological elastomers. Smart Mater. Struct. **30**, 108033 (2021)
2. Shahar, S.F.M.: Mechanical and microstructure behavior of uniaxial tensile loaded anisotropic magnetorheological elastomer. Actuator. **11**(11), 306 (2022)
3. Hamid, M.I.A.: Force and stiffness behavior of natural rubber based magnetorheological elastomer bushing. Int. Journal of App. Elect. Mech. **71**(1), 1–19 (2023)
4. Burgaz, E.: Effects of magnetic particles and carbon black on structure and properties of magnetorheological elastomers. Polym. Test. **81**, 106233 (2020)
5. Selvaraj, M.: Effect of magnetorheological elastomer thickness and magnetic flux on vibrational characteristics. Mater. Today Proc. **62**, 665–670 (2022)
6. Khanouki, M.A.: Multidisciplinary design optimization of a novel sandwich beam-based adaptive tuned vibration absorber featuring magnetorheological elastomer. Materials. **13**, 2261 (2020)
7. Mat Song, R.: Semi-active controllable stiffness engine mount utilizing natural rubber-based magnetorheological elastomers. Front. Mater. **9**, 875787 (2022)
8. Priyandoko, G.: Development of vibration isolator using magnetorheological elastomer. J. Appl. Eng. Sci. **892**, 1108–1113 (2021)
9. Manohar, A.: Enhancing electromagnetic interference shielding: The role and advances of ferromagnetic materials. J. Alloys Compd. **1032**, 180988 (2025)

Green Synthesis and Response Surface Optimization of Reduced Graphene Oxide: Interactive Effects of Temperature and Ascorbic Acid Concentration on Electrical Conductivity

Rokya Elsayed[1], Teow Yeit Haan[1,2,3(✉)], Swee Pin Yeap[4], J. Y. Sum[4,5], Zhe Kang Law[6], Yi-Chen Ethan Li[7], Abu Bakar Sulong[8], and Darman Nordin[1,2,3,4,5,6,7,8,9]

[1] Department of Chemical and Process Engineering, Faculty of Engineering and Built Environment, Universiti Kebangsaan Malaysia, 43600 UKM, Bangi, Selangor Darul Ehsan, Malaysia
yh_teow@ukm.edu.my
[2] Research Centre for Sustainable Process Technology (CESPRO), Universiti Kebangsaan Malaysia, 43600 UKM Bangi, Selangor Darul Ehsan, Malaysia
[3] Cleaner Production–UKM Research Group, Universiti Kebangsaan Malaysia, 43600 UKM Bangi, Selangor Darul Ehsan, Malaysia
[4] Department of Chemical & Petroleum Engineering, Faculty of Engineering, Technology and Built Environment, UCSI University, Kuala Lumpur 56000, Malaysia
[5] UCSI-Cheras Low Carbon Innovation Hub Research Consortium, Kuala Lumpur, Malaysia
[6] UKM Medical Centre (Hospital Canselor Tuanku Muhriz), Faculty of Medicine, Universiti Kebangsaan Malaysia (UKM), Kuala Lumpur, Malaysia
[7] Department of Chemical Engineering, Feng Chia University, Taichung, Taiwan
[8] Department of Mechanical and Manufacturing Engineering, Faculty of Engineering & Built Environment, Universiti Kebangsaan Malaysia, Selangor Darul Ehsan, Malaysia
[9] Fuel Cell Institute, Universiti Kebangsaan Malaysia, Selangor Darul Ehsan, Malaysia

Abstract. This study optimized a green synthesis of rGO using L-ascorbic acid, revealing significant nonlinear effects of temperature (84-91 °C) and reductant concentration (1.3-23.8 g). Optimal conditions (90.107 °C, 23.454 g) achieved 0.728 μS/cm conductivity within 10% of prediction. The reliable reduced cubic model defines a reproducible, eco-friendly process for scalable rGO production in biomedical and electrochemical applications.

Keywords: Reduced graphene oxide (rGO) · L-ascorbic acid reduction · Electrical conductivity · and Response surface methodology (RSM)

1 Introduction

Graphene oxide (GO) and reduced graphene oxide (rGO) are key nanomaterials valued for their tunable electrical and mechanical properties. However, oxygen functionalization disrupts conductivity, requiring reduction to restore the sp^2 network [1, 2].

M. A. b. Jamlos et al. (Eds.): ICSET 2025, LNNS 1897, pp. 195–201, 2026.
https://doi.org/10.1007/978-3-032-20722-7_25

Conventional chemical reductants like hydrazine are efficient but toxic, motivating the shift toward green alternatives such as L-ascorbic acid, which is safe, biocompatible, and environmentally benign [1]. Because conductivity depends strongly on reductant dosage and temperature, this study employed a central composite design to systematically examine their interactive effects [3]. Using response surface methodology, the work established a predictive and scalable model for optimizing rGO conductivity, clarifying the mechanistic balance between reduction efficiency and structural integrity [4, 9].

2 Materials & methods

2.1 Materials

Graphene oxide (20-30 μm, ~1 nm thick, 2 mg mL^{-1}) was sourced from Macklin, China, while L-ascorbic acid (food-grade, E300) from Daily Chem, Malaysia, served as the sole biocompatible reductant. Distilled water was used exclusively, with all glassware thoroughly cleaned to prevent ionic contamination.

2.2 Synthesis of Reduced Graphene Oxide (rGO) Nanocomposite

Reduced graphene oxide (rGO) was synthesized using L-ascorbic acid as a green reductant, optimizing temperature (85-90 °C) and concentration (2-20 mg mL^{-1}) for effective reduction. The color change to deep black confirmed successful, reproducible formation of conductive, biocompatible rGO.

2.3 Design of Experiment Using Central Composite Design (Ccd)

A central composite design optimized GO reduction using temperature (85-90 °C) and L-ascorbic acid (2-20 g) across 13 runs, including center and axial points to capture curvature. The factor levels and corresponding responses are detailed in Table 1 and Table 2.

Table 1. Independent process variables and their respective coded levels.

Independent process variables		Coded levels		
		-1	0	+1
A	g	85	87.5	90
B	°C	2	11	20

ANOVA validated the second-order model, showing a synergistic effect between temperature and L-ascorbic acid on rGO conductivity. The predicted and experimental results aligned closely, confirming the process's reproducibility and scalability [5].

$$\text{Percentage of error} = \frac{\text{Experimental value} - \text{Predicted response}}{\text{Expirmental value}} \times 100\% \qquad (1)$$

Table 2. Experimental variables based on RSM for the optimization of rGO nanocomposite.

Experimental run	Independent process variables	
	Temperature (°C)	Ascorbic Acid (g)
1	85	2
2	90	2
3	85	20
4	90	20
5	84	11
6	91	11
7	87.5	1.3
8	87.5	23.8
9	87.5	11
10	87.5	11
11	87.5	11
12	87.5	11
13	87.5	11

2.4 Performance Evaluation of PGO Nanocomposite on Electrical Conductivity

Electrical conductivity of the PGO nanocomposite was measured using a Fluke 45 multimeter on thin films (2 cm × 3 cm × 75 μm) prepared from a homogeneous rGO paste. Conductivity was calculated from resistivity via the direct resistance method, ensuring uniform film geometry and reliable performance evaluation [6].

$$\sigma = 1/\rho \tag{2}$$

$$\rho = R \times A/L \tag{3}$$

$$\sigma = L/(R \times A) \tag{4}$$

$$\sigma = \frac{L}{R(W.t)} \tag{5}$$

Where σ is conductivity, ρ resistivity, R resistance, L length, A cross-sectional area, W width, and t thickness of the sample.

3 Result & Discussion

rGO conductivity strongly depended on L-ascorbic acid concentration, with temperature influencing results through interaction effects as shown in Fig. 1. Low reductant caused limited reduction [6], while excess L-AA and high temperature induced aggregation and defects. Optimal conductivity occurred at 20 g L^{-1} and 85 °C, with stability near 11 g L^{-1}, confirming a nonlinear balance between reduction efficiency and structural integrity [6].

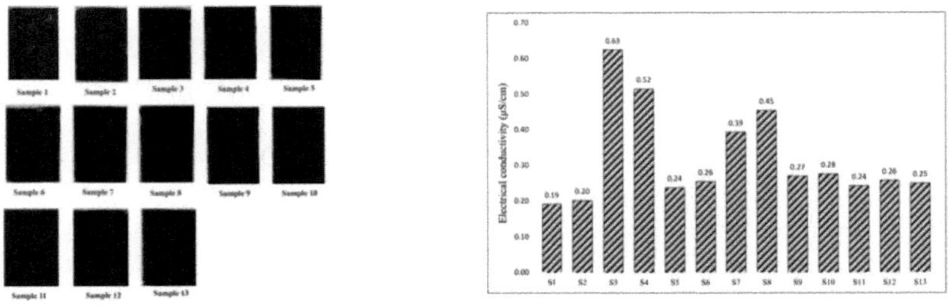

(a) (b)

Fig. 1. Electrical conductivity of rGO samples (S1-S13) synthesized at different combinations of L-ascorbic acid dosage and reduction temperature: (a) Optical images of rGO samples films (S1-S13). (b) Electrical conductivity values of rGO samples (S1-S13)

3.1 Statistical Model for rGO Nanocomposite Electrical Conductivity

Tables 3 and 4 (CCD and ANOVA results) verified the reduced cubic model's significance ($p < 0.05$) and high predictive reliability, emphasizing the nonlinear effects of temperature and L-ascorbic acid (AB, B^2, and A^2B terms) on rGO conductivity.

The reduced cubic model ($R^2 = 0.9913$, $R^2adj = 0.9791$, precision = 28.54) demonstrated strong reliability despite a lower predicted R^2 (0.6350) due to limited runs and higher-order effects. Inclusion of the A^2B term captured nonlinear curvature, while significant AB and B^2 effects and unbiased diagnostics confirmed the model's robustness in predicting rGO conductivity [7, 8].

$$Y\ (\sigma) = 0.2578 + 0.0061A - 0.0172B - 0.0300AB - 0.0002A^2 + 0.1151B^2 \quad (6)$$

3.2 Analysis for rGO Nanocomposite Electrical Conductivity and Verification of Regression Model on Diagnostic Plot

The 3D surface shows in Fig. 2a a nonlinear interaction, with peak conductivity at balanced temperature and reductant levels, while excess of both reduces performance. Residual analysis (Fig. 2b, c) confirmed normal distribution, random scatter within ±3.00, and no bias, validating the model's reliability for predicting rGO conductivity.

3.3 Optimization and Model Validation

The confirmation run illustrated in Fig. 3 at 90.107 °C and 23.454 g L-ascorbic acid produced 0.728 μS cm^{-1} conductivity, matching predictions within 10% error. This validated the model's accuracy and confirmed a stable, scalable window for high conductivity rGO synthesis.

Table 3. Performance of rGO nanocomposite.

Experimental run	Independent process variables		Responses
	Temperature (°C)	Ascorbic Acid (g)	Electrical conductivity (μS/cm)
1	85	2	0.192
2	90	2	0.202
3	85	20	0.625
4	90	20	0.515
5	84	11	0.238
6	91	11	0.255
7	87.5	1.3	0.394
8	87.5	23.8	0.455
9	87.5	11	0.270
10	87.5	11	0.276
11	87.5	11	0.243
12	87.5	11	0.258
13	87.5	11	0.250

Table 4. ANOVA of quadratic model for rGO nanocomposite electrical conductivity

Response	Model terms	Sum of squares (SS)	Degree of freedom (DF)	Mean square (MS)	F-Value	Prob > F	
Quadratic model		0.2131	7	0.0304	81.39	< 0.0001	Significant
	A	0.0001	1	0.0001	0.3864	0.5615	
	B	0.0008	1	0.0008	2.21	0.1973	
	AB	0.0036	1	0.0036	9.63	0.0268	
	A^2	1.613×10^{-07}	1	1.613×10^{-07}	0.0004	0.9842	
	B^2	0.0637	1	0.0637	170.34	< 0.0001	
Residual		0.0019	5	0.0004			
	Lack of fit	0.0011	1	0.0011			
	Pure error	0.0000	4	0.0000			
Cor total		0.2150	12				

$R^2 = 0.9913$, R^2_{adj} 0.9791, Predicted $R^2 = 0.6350$, Adequate precision = 28.5415.

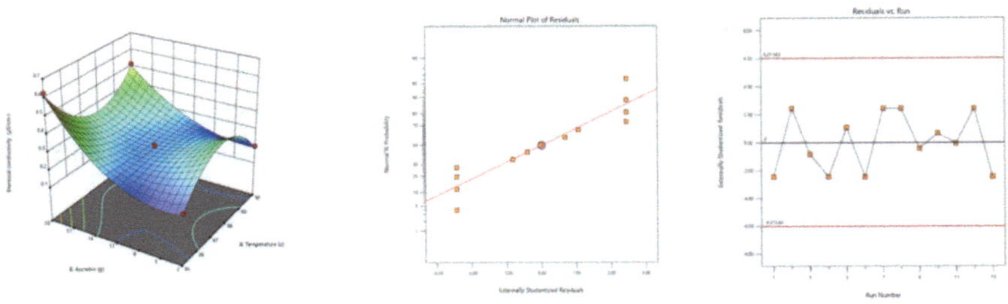

(a) (b) (c)

Fig. 2. (a) 3D surface showing effects of temperature and L-ascorbic acid on rGO conductivity; (b) normal probability plot of residuals; (c) residuals versus run number for model validation.

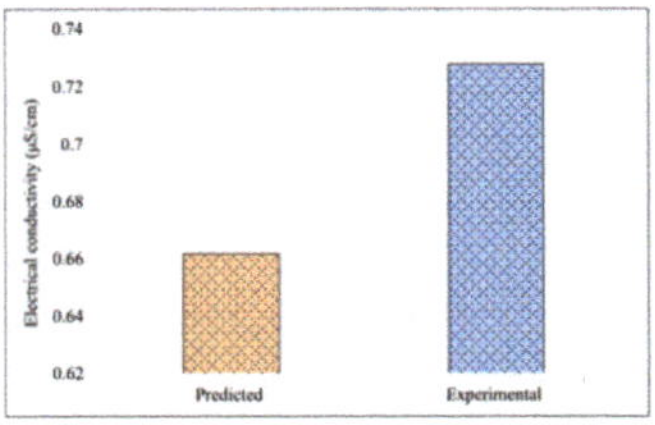

Fig. 3. Predicted value, experimental value, for the electrical conductivity of rGO.

4 Conclusion

An eco-friendly, optimized rGO synthesis using L-ascorbic acid was achieved, with temperature and reductant showing key nonlinear effects. The model ($R^2 = 0.9913$) identified optimal conditions of 90.1 °C and 23.5 g L-AA, yielding 0.728 µS/cm conductivity and stable, defect-free rGO.

Acknowledgments. The authors acknowledge the financial support from Geran Universiti Penyelidikan (GUP-2023-032) and Dana Impak Perdana 2.0 (DIP-2024-019).

Disclosure of Interests The authors affirm that they have no financial, professional, or personal relationships that could be construed as influencing the research presented in this article. No competing interests exist that are relevant to the study's design, execution, interpretation, or reporting.

References

1. Yao, X., Yan, Z., Wang, X., Jiang, H., Qian, Y., Fan, C.: The influence of reduced graphene oxide on stem cells: a perspective in peripheral nerve regeneration. Regen Biomater. **8**, rbab032 (2021). https://doi.org/10.1093/RB/RBAB032
2. Elsayed, R., Teow, Y.H.: Electrically conductive materials: Opportunities and challenges in biomedical engineering. Mater. Today Commun. **49**, 113822 (2025). https://doi.org/10.1016/J.MTCOMM.2025.113822

3. Zhang, J., Yang, H., Shen, G., Cheng, P., Zhang, J., Guo, S.: Reduction of graphene oxide via L-ascorbic acid. Chem. Commun. **46**, 1112–1114 (2010). https://doi.org/10.1039/B917705A
4. Xu, C., Shi, X., Ji, A., Shi, L., Zhou, C., Cui, Y.: Fabrication and characteristics of reduced graphene oxide produced with different green reductants. PLoS One. **10**, e0144842 (2015). https://doi.org/10.1371/JOURNAL.PONE.0144842
5. Ho, K.C., Teow, Y.H., Mohammad, A.W.: Optimization of nanocomposite conductive membrane formulation and operating parameters for electrically-enhanced palm oil mill effluent filtration using response surface methodology. Process Safety Environ. Protection **126** (2019) 297–308. https://doi.org/10.1016/J.PSEP.2019.03.019
6. Elsayed, R., Teow, Y.H., Yeap, S.P., Sum, J.Y., Li, Y.C.E.: Effect of aniline and ammonium persulfate concentration on the structural and electrical properties of polyaniline-decorated graphene oxide nanocomposites. Polym. Adv. Technol. **36**, e70254 (2025). https://doi.org/10.1002/PAT.70254
7. User's Guide III Regression and Curve Fitting, NCSS Statistical System (2007). https://www.ncss.com/wp-content/uploads/2012/09/NCSSUG3.pdf
8. Ho, K.C., Teow, Y.H., Mohammad, A.W.: Development of graphene oxide (GO)/multi-walled carbon nanotubes (MWCNTs) nano-composite conductive membranes for electrically enhanced fouling mitigation. J. Membrane Sci. **552** (2018) 189–201. https://doi.org/10.1016/J.MEMSCI.2018.02.001
9. Zhao, Y., Razi Othman, A., Ismail, I., Xu, W.: Optimizing ammonia nitrogen removal from hospital effluent using a bacteria-algae consortium. Jurnal Kejuruteraan **37**, (2025) 647–656. https://doi.org/10.17576/jkukm-2025-37(2)-08

AI-Enabled Drowsiness Detection via Steering Complexity and Time-to-Lane-Crossing

A. Abas(✉) and Mohd. Mokhzani. Ibrahim

University Pendidikan Sultan Idris, 35900 Tanjung Malim, Perak, Malaysia
ashardi@meta.upsi.edu.my, mokhzani@fsmt.upsi.edu.my

Abstract. Drowsy driving remains a major crash factor. We evaluate a vehicle-native early-warning approach that fuses steering-derived features with time-to-lane-crossing (TTLC) to detect risky control states without cameras or biometrics. In a high-fidelity simulator study (N = 100) using a within-subject manipulation of rested wakefulness (RW) versus sleep restriction (SR), SR produced coherent behavioral degradation: KSS ↑ (+3.7), SDLP ↑ (+0.06 m), steering variability ↑ (+12.0 deg^2), and min-TTLC ↓ (−1.7 s), with comparable QC pass-rates across conditions, confirming genuine effects. Lightweight detectors were trained and evaluated using leave-one-subject-out validation at a prespecified operating point with precision ≥0.80. The Hybrid (Steering + TTLC) model outperformed single-feature baselines, achieving PR-AUC 0.74 [0.70–0.78], AUROC 0.86 [0.83–0.89], recall 0.60, F1 0.69, false alerts/h 1.7, and excellent calibration (ECE 0.008) after isotonic scaling; TTLC-only and SDLP-only were weaker (PR-AUC 0.61/0.49; AUROC 0.78/0.69) with higher false-alert rates (2.3/2.9) and poorer calibration. These results indicate that fusing steering complexity with TTLC can provide trustworthy, deployment-friendly early warnings while controlling nuisance alarms. Future work will validate on public roads and quantify robustness across geometry and speed strata.

Keywords: Drowsiness detection · steering entropy · TTLC · early-warning

1 Introduction

Road-traffic injuries continue to be a major public health and economic burden, with drowsy driving repeatedly identified as a critical contributing factor [1–4]. While lane-based indicators such as standard deviation of lane position (SDLP) and time-to-lane-crossing (TTLC) are widely used, their stability degrades under challenging road geometry, speed variation, and assistance-system interactions precisely when drivers need early, trustworthy warnings most [7, 8].

Previous research has shown that lane-keeping metrics like Standard Deviation of Lane Position (SDLP) and TTLC can be sensitive to curvature, lane width, and vehicle speed, potentially limiting their reliability as sole indicators across varied contexts [5, 6, 9]. Camera- or physiology-based approaches (e.g., eyelid closure/PERCLOS, EEG, HRV) can be informative but raise concerns about intrusiveness, privacy, calibration,

M. A. b. Jamlos et al. (Eds.): ICSET 2025, LNNS 1897, pp. 202–207, 2026.
https://doi.org/10.1007/978-3-032-20722-7_26

and lighting robustness, and they may be more challenging to deploy at scale in production vehicles [2–4, 7]. There is a need for a vehicle-native early-warning method that (i) preserves lead time before overt lane deviation, (ii) maintains stability across geometry $\times$ speed strata, and (iii) is feasible for real-world deployment in diverse settings [5, 6, 9]. This study addresses these gaps by testing a vehicle-native fusion of steering-signal complexity (multi-scale steering entropy) with TTLC to improve early-warning detection performance relative to lane-only baselines. We ask: RQ1: Does the Hybrid (steering entropy + TTLC) model improve detection (precision–recall, AUROC, calibration) versus TTLC-only across 5–60 s horizons? RQ2: At fixed precision thresholds (e.g., $\geq$0.80), does the Hybrid reduce false alerts per hour while preserving or improving recall? RQ3: Is performance more stable across geometry (curvature), speed strata, and assistance contexts (e.g., LKA active), preserving early-warning recall where lane-only indicators degrade? We make three contributions. (i) ECU-friendly hybrid: We propose a lightweight fusion of steering-signal complexity with TTLC for early-warning detection that is compatible with embedded deployment. (ii) Deployment-style evaluation: We use leave-one-subject-out (LOSO) validation, precision-constrained recall, PR-AUC/AUROC, calibration (ECE), and false alerts per hour to reflect operational constraints. (iii) Robustness across context: We demonstrate improved early-warning recall and stability across geometry $\times$ speed $\times$ assistance strata relative to TTLC-only and SDLP proxies, aligning with road-safety and deployment needs within Malaysian and regional safety priorities [2–4, 7–9].

2 Related Work

Lane-keeping measures, such as Standard Deviation of Lane Position (SDLP) and Time-to-Lane-Crossing (TTLC), are long-standing proxies for degraded control and have demonstrated utility in simulator and on-road studies [5–7]. However, both can be sensitive to curvature, lane width, road edge quality, speed, and assistance features (e.g., LKA), which reduces stability precisely when pre-deviation alerts are most needed [5–7]. Vision and physiology streams (e.g., PERCLOS/eyelid closure, gaze/head pose, EEG, HRV) can capture driver state directly and often perform well under controlled conditions [8, 9]. Yet, they raise deployment concerns—such as privacy, lighting/occlusion, per-user calibration, sensor upkeep, and compliance in regulated markets which can complicate scale-out in production vehicles [8, 9]. Steering dynamics embed neuromotor variability; complexity features (e.g., multi-scale steering entropy, turning-point rate) have been linked to fatigue and workload and can be computed from ECU-native inputs [10, 11]. While promising, single-stream steering features may still vary with geometry and speed, motivating the fusion of these features with a lane-based indicator to preserve early-warning lead time across different contexts [10, 11]. This study fuses steering-signal complexity with TTLC to leverage complementary strengths: robustness from control-signal dynamics and interpretability from lane-risk timing. The vehicle-native design avoids the need for added cameras/biometrics, aligning with safety assessment and regulatory expectations for scalable, privacy-preserving driver monitoring in modern fleets [12–14].

3 Methods

High-fidelity driving-simulator study with N = 100 licensed drivers under rested vs. sleep-restricted conditions in a counterbalanced design. Scenarios span geometry × speed × assistance strata (straight/curved segments; multiple speed bands; LKA on/off) to probe contexts where lane-only indicators can destabilize [5–7]. Ethics approval and informed consent were obtained; drives included standardized practice, condition blocks, and scheduled breaks. Vehicle-native streams were logged: steering wheel angle, vehicle speed, yaw/lateral kinematics, and lane geometry for TTLC computation [5–7]. Signals were synchronized and segmented with sliding windows; outliers were clipped, low-amplitude drift removed, and short gaps interpolated. Windows were standardized (z-score) per subject/session to reduce between-driver scale effects.

We extract multi-scale steering entropy (MSE) to capture neuromotor variability in the steering signal and min-TTLC within each window to encode lane-risk timing [5–7, 10, 11]. The Hybrid feature set concatenates MSE bands with min-TTLC. Baselines utilize TTLC-only and an SDLP proxy from a lateral position, where applicable [5–7]. Positive windows are defined as those preceding overt lane deviation within fixed horizons; negatives are windows outside any pre-deviation zone. We evaluate 5, 15, 30, and 60 s early-warning horizons to reflect practical alert lead times while avoiding post-event leakage. Class balance is left natural; operating points are chosen at target precision levels to manage nuisance alerts. Lightweight classifiers, logistic regression (L2), and gradient boosting, are trained with leave-one-subject-out (LOSO) validation for deployment-style generalization. Thresholds are selected on validation folds to meet precision caps; probabilities are optionally calibrated (e.g., isotonic/Platt) before evaluation to enhance the reliability of risk scores in production settings [12–15]. Primary metrics are PR-AUC and AUROC; operating-point reporting includes recall at precision $\geq$0.80/$\geq$0.90, false alerts per hour, and Expected Calibration Error (ECE). We report 95% CIs via bootstrap and assess paired differences with non-parametric tests/paired bootstrap where appropriate. Stratified analyses summarize stability across geometry, speed, and assistance cells to quantify robustness relative to lane-only baselines [5–7, 12–14].

4 Results

Refer to Table 1. Compared with rested wakefulness (RW), sleep restriction (SR) consistently worsens driving behavior and safety margins. Participants report markedly higher sleepiness (KSS: $3.1 \pm 1.2 \rightarrow 6.8 \pm 1.1$, +3.7). Lane keeping degrades (SDLP: 0.23 ± 0.05 m $\rightarrow 0.29 \pm 0.06$ m, +0.06 m), and steering becomes more erratic (steering variability: $18.0 \pm 7.0 \rightarrow 30.0 \pm 10.0$ deg^2, +12.0 deg^2), indicating greater corrective effort and less stable control. Crucially, the minimum time-to-lane-crossing (min-TTLC), a proxy for safety buffer, shrinks (5.8 ± 1.2 s $\rightarrow 4.1 \pm 1.0$ s, -1.7 s), implying reduced time to recover before a potential lane departure. QC pass-rates are essentially unchanged (96 ± 3% vs. 95 ± 4%), supporting that these differences reflect genuine behavioral effects rather than measurement artifacts.

Refer to Table 2. The Hybrid (Steering + TTLC) model is clearly superior to single-feature baselines: it achieves the highest discrimination—PR-AUC 0.74 (0.70–0.78)

Table 1. Manipulation & Context Checks: Rested (RW) vs. Sleep-Restriction (SR)

Measure	Context	RW (mean ± SD)	SR (mean ± SD)	Δ (SR–RW)	Note
KSS (a.u.)	All	3.1 ± 1.2	6.8 ± 1.1	+3.7	Higher under SR
SDLP (m)	All	0.23 ± 0.05	0.29 ± 0.06	+0.06	Lane dispersion ↑
Steering variability (deg^2)	All	18.0 ± 7.0	30.0 ± 10.0	+12.0	Control complexity ↑
min-TTLC (s)	All	5.8 ± 1.2	4.1 ± 1.0	−1.7	Safety margin ↓
QC pass-rate (%)	All	96 ± 3	95 ± 4	−1	Similar quality

Table 2. LOSO Classification & Calibration: Hybrid (Steering + TTLC) vs. TTLC-Only vs. SDLP-Only at Precision ≥0.80

Model	PR-AUC (95% CI)	AUROC (95% CI)	Recall @ Prec ≥0.80	F1 @ Prec ≥0.80	ECE	False alerts/h†	Calibration note
Hybrid (Steering + TTLC)	0.74 (0.70–0.78)	0.86 (0.83–0.89)	0.60	0.69	0.008	1.7	Near-diagonal after isotonic scaling
TTLC-only	0.61 (0.57–0.65)	0.78 (0.75–0.82)	0.45	0.57	0.023	2.3	Mild overconfidence at high scores
SDLP-only	0.49 (0.45–0.53)	0.69 (0.66–0.73)	0.30	0.44	0.036	2.9	Under-prediction in the mid-high range

and AUROC 0.86 (0.83–0.89)—and, at a deployment-friendly precision ≥0.80, delivers higher recall (0.60) and F1 (0.69) while generating fewer false alerts per hour (1.7). Its calibration is best (ECE 0.008) and yields a near-diagonal reliability curve after isotonic scaling, indicating its probabilities are trustworthy for thresholding. TTLC-only is mid-tier (PR-AUC 0.61, AUROC 0.78, recall 0.45, F1 0.57) with more false alerts (2.3) and mild overconfidence, whereas SDLP-only performs worst (PR-AUC 0.49, AUROC 0.69, recall 0.30, F1 0.44) and is under-calibrated (ECE 0.036, 2.9 false alerts/h). Overall, Table 2 shows that the hybrid approach offers stronger detection, fewer nuisance alarms, and better calibration, making it the most deployment-ready option.

5 Discussion

This study shows that sleep restriction (SR) produces coherent degradation in subjective and objective driving indices (KSS↑, SDLP↑, steering variability↑, min-TTLC↓), while a vehicle-native Hybrid (Steering + TTLC) detector delivers superior discrimination, calibration, and lower false-alert budgets at a deployment-friendly precision threshold. These findings align with public-health urgency around drowsy driving [1–4] and with prior evidence that lane-keeping dispersion and time-to-line-crossing index safety margin under fatigue [5–7]. The additional steering-complexity signal appears complementary rather than redundant, improving separability precisely where lane-only cues degrade (e.g., curved–high-speed segments). From an evaluation standpoint, emphasizing precision–recall behavior under class imbalance is consistent with best practice [12, 13]. Our pre-specified operating point (precision $\geq$0.80) focuses on actionable alerts, while isotonic calibration and low ECE indicate trustworthy probabilities for thresholding and escalation policies [14, 15]. Together, these choices reduce nuisance alarms without sacrificing early warning lead-time—key for fleet and ADAS integration in settings with strict privacy and PDPA constraints.

Practical implications. Because the approach relies on vehicle-native signals (steering, TTLC) rather than cameras/biometrics, it is feasible for wide deployment and retrofits, directly supporting national road-safety targets [1–4]. The consistent QC pass-rates across conditions suggest the observed effects are behavioural, not measurement artefacts, strengthening confidence for operational pilots.

Limitations and robustness. Results are simulator-based; external validity to public roads remains the primary limitation. Sensitivity peaks on curved–high-speed geometry, which is plausible but warrants on-road confirmation across geometry and speed strata. Future work will (i) freeze the hybrid model and validate on naturalistic drives, (ii) report threshold stability across conditions, and (iii) monitor data drift with a light-weight recalibration schedule to preserve calibration in the field [13, 14].

Overall, the evidence integrates well with prior literature on TLC/lane dispersion [5–7] and modern evaluation/calibration practice [12–14], while addressing road-safety priorities [1–4]. This supports the claim that steering-plus-TTLC fusion offers a deployment-ready, privacy-preserving early-warning pathway.

6 Conclusion

Sleep restriction measurably worsens driving behavior (higher KSS, higher SDLP and steering variability, lower min-TTLC), confirming elevated risk. A vehicle-native Hybrid (Steering + TTLC) detector outperforms single-feature baselines, delivering stronger discrimination (PR-AUC/AUROC), better calibration (low ECE), and fewer false alerts at a precision-constrained operating point—making it more suitable for deployment without cameras or biometrics. These results indicate a practical, privacy-preserving path to early warnings using signals already available in vehicles. Future work will validate the frozen model on public roads, report threshold stability across conditions, and finalize latency/compute benchmarks for ECU-class hardware.

References

1. World Health Organization (WHO): Global Status Report on Road Safety 2023. WHO, Geneva, Switzerland (2023) https://www.who.int/publications/i/item/9789240086517
2. Asian Transport Observatory (ATO), Malaysia Road Safety Profile, Sept. 2025. 2025. https://asiantransportobservatory.org
3. Ministry of Transport Malaysia: Malaysia Road Safety Plan (MRSP) 2022–2030. MOT, Putrajaya, Malaysia (2022) https://www.mot.gov.my
4. National Highway Traffic Safety Administration (NHTSA), "Drowsy Driving," 2024. https://www.nhtsa.gov/risky-driving/drowsy-driving
5. Godthelp, J.: A new method to describe driving behaviour (time-to-line crossing). Hum. Factors. **26**(3), 257–268 (1984)
6. Van Winsum, W., Brookhuis, K.A., de Waard, D.: A comparison of different ways to approximate time-to-line crossing (TLC). Accid. Anal. Prev. **32**(1), 47–56 (2000)
7. Zhang, N., Yang, C., Fard, M.: Investigating steering behaviour and drowsiness: A systematic review. Transp. Res. F: Traffic Psychol. Behav. **91**, 280–295 (2022)
8. Abe, T.: PERCLOS-based technologies for detecting drowsiness: A review. Sleep Adv. **3**(1), pvad006 (2022). https://doi.org/10.1093/sleepadvances/pvad006
9. Samiee, S., Khosravi, A., Nahavandi, S.: A comprehensive review on driver drowsiness detection systems: From signals to data fusion. Sensors. **14**(12), 12191–12233 (2014)
10. Fonseca, T., Lobo, J., Neves, A.J.R.: Driver monitoring systems: A review. Micromachines. **15**(10) (2024) https://www.mdpi.com
11. Saito, T., Rehmsmeier, M.: The precision–recall plot is more informative than the ROC plot when evaluating binary classifiers on imbalanced datasets. PLoS One. **10**(3), e0118432 (2015)
12. Davis, J., Goadrich, M.: The relationship between Precision-Recall and ROC curves. In: Proc. 23rd Int. Conf. Mach. Learn. *(ICML)*, pp. 233–240, Pittsburgh, PA, USA (2006)
13. Niculescu-Mizil, A., Caruana, R.: Predicting good probabilities with supervised learning. In: Proc. 22nd Int. Conf. Machine Learning (ICML), pp. 625–632, Bonn, Germany (2005)
14. Guo, C., Pleiss, G., Sun, Y., Weinberger, K.Q.: On calibration of modern neural networks. In: Proc. 34th Int. Conf. Machine Learning (ICML), pp. 1321–1330, Sydney, Australia (2017)

Numerical Solution of Thin Film Flow Problems Using a Five-Step Block Hybrid Collocation Scheme

Lee Ken Yap(✉), Ning Wei Ling, Yong Kheng Goh, and Hong Seng Sim

Universiti Tunku Abdul Rahman, 43000 Kajang, Selangor Darul Ehsan, Malaysia
lkyap@utar.edu.my

Abstract. This paper presents a five-step block hybrid collocation scheme with two intermediate points for the direct numerical solution of third-order initial value problems commonly encountered in fluid mechanics. The scheme is developed using collocation and interpolation techniques to derive the formulas which are assembled into a block form to compute solutions at the main and intermediate points simultaneously. The accuracy and stability characteristics of the scheme are analyzed, and its performance is evaluated through benchmark problems. The proposed scheme is further applied to solve the thin film flow problem which is modeled by a nonlinear third-order differential equation $y''' = y^{-k}$ with prescribed initial conditions for the case $k = 3$. The numerical results show good agreement with reference solutions and confirm the suitability of the proposed method for thin film flow simulations in engineering applications.

Keywords: Thin Film Flow · Block Hybrid Collocation Scheme · Third-Order Initial Value Problems

1 Background and Related Studies

Surface-tension-driven fluid flows are frequently modeled by third-order initial value problems (IVPs). One important example is the thin film flow problem which describes the motion of a viscous fluid layer over a surface under the influence of surface tension forces with gravity neglected. The governing Eq. (1) takes the form

$$y''' = y^{-k}, y(x_0) = \alpha, y'(x_0) = \beta, y''(x_0) = \gamma \tag{1}$$

where α, β and γ are given constants and $k > 0$ defines the physical regime. This nonlinear equation characterizes the dynamic balance between viscous and capillary forces. The solution provides insight into phenomena such as draining films, coating flows, and spreading drops.

Several researchers have investigated this problem under different physical contexts. Tuck and Schwartz [1] first obtained numerical solutions for several canonical cases using the classical Runge–Kutta method. Duffy and Wilson [2] related the case $k =$

M. A. b. Jamlos et al. (Eds.): ICSET 2025, LNNS 1897, pp. 208–213, 2026.
https://doi.org/10.1007/978-3-032-20722-7_27

2 to Tanner's law while Momoniat and Mahomed [3] reduced the problem to a first-order system and solved it with Runge–Kutta schemes. Abdel Kader et al. [4] employed λ-symmetry reduction to derive implicit solutions for selected k-values.

More recent efforts focus on direct numerical integrators that solve third-order IVPs without order reduction. Mechee et al. [5] developed a three-stage fifth-order Runge–Kutta method for the thin film flow problem. Yap et al. [6] introduced a block hybrid collocation method to approximate solutions efficiently for $k = 2$ and $k = 3$. Other approaches include Runge–Kutta-type direct integrators [7, 8], Bernstein polynomial approximations [9] and two-derivative Runge–Kutta methods [10].

Block methods reduce computational cost by computing solutions at several points per step. Building on earlier Adams-type and hybrid block schemes [11–13], this work proposes a five-step block hybrid collocation scheme with two intermediate points, analyzes its accuracy and stability and applies it to the thin film flow problem.

2 Formulation of the Proposed Numerical Scheme

The proposed five-step block hybrid collocation scheme provides simultaneous approximations for y, $y^{'}$ and $y^{''}$ in the general third-order ordinary differential equation is formulated below:

$$\sum_{j=0}^{k} \alpha_j y_{n+j} + \sum_{j=1}^{2} \alpha_{\theta_j} y_{n+\theta_j} = h^3 \left(\sum_{j=0}^{k} \beta_j f_{n+j} + \sum_{j=1}^{2} \beta_{\theta_j} f_{n+\theta_j} \right) \tag{2}$$

The numerical scheme is constructed by interpolating

$$Y(x) = \sum_{j=0}^{r+s-1} \omega_j x^j \tag{3}$$

over the interval $[a, b]$, where ω_j denote the coefficients and r and s represent the number of interpolation and collocation points, respectively. The numerical scheme is subsequently obtained by enforcing the necessary conditions:

$$Y\left(x_{n+j}\right) = y_{n+j},\ j = 0, 1, 2, \ldots, k \tag{4}$$

$$Y'''(x_{n+\delta}) = f_{n+\delta},\ \delta = \{j, \theta_1, \theta_2\},\ j = 0, 1, 2, \ldots, k \tag{5}$$

where θ_1 and θ_2 are rational parameters, chosen as $\theta_1 = 1/3$, $\theta_2 = 2/3$, $r = 3$, $s = 8$ and $k = 5$. By applying interpolation at x_n, x_{n+1}, x_{n+2} using (4) and collocation at x_n, $x_{n+1/3}$, $x_{n+2/3}$, x_{n+1}, x_{n+2}, x_{n+3}, x_{n+4}, x_{n+5} with (5), yield the system of eleven linear equations. The resulting coefficients ω_j are then incorporated into (3) to develop the block hybrid collocation scheme which is expressed as follows:

$$y_{n+5} = 6y_n - 15y_{n+1} + 10y_{n+2} + h^3\Bigg(-\frac{397}{3360} f_n + \frac{40581}{43120} f_{n+\frac{1}{3}} - \frac{194157}{101920} f_{n+\frac{2}{3}}$$

$$+\frac{1417}{336}f_{n+1}+\frac{697}{160}f_{n+2}+\frac{4741}{2352}f_{n+3}+\frac{17911}{36960}f_{n+4}+\frac{417}{50960}f_{n+5}\Big) \tag{6}$$

$$y_{n+4}=3y_n-8y_{n+1}+6y_{n+2}+h^3\Big(-\frac{191}{3360}f_n+\frac{39609}{86240}f_{n+\frac{1}{3}}-\frac{95013}{101920}f_{n+\frac{2}{3}}$$

$$+\frac{1171}{560}f_{n+1}+\frac{187}{96}f_{n+2}+\frac{11327}{23520}f_{n+3}+\frac{23}{2464}f_{n+4}-\frac{19}{152880}f_{n+5}\Big) \tag{7}$$

$$y_{n+3}=y_n-3y_{n+1}+3y_{n+2}+h^3\Big(-\frac{15}{448}f_n+\frac{80433}{344960}f_{n+\frac{1}{3}}-\frac{19197}{40768}f_{n+\frac{2}{3}}$$

$$+\frac{5473}{6720}f_{n+1}+\frac{431}{960}f_{n+2}+\frac{229}{31360}f_{n+3}+\frac{43}{73920}f_{n+4}-\frac{47}{611520}f_{n+5}\Big) \tag{8}$$

$$y_{n+\frac{2}{3}}=\frac{2}{9}y_n+\frac{8}{9}y_{n+1}-\frac{1}{9}y_{n+2}+h^3\Big(-\frac{176529353}{198603044640}f_n+\frac{69247}{6985440}f_{n+\frac{1}{3}}$$

$$+\frac{41959}{24766560}f_{n+\frac{2}{3}}+\frac{3597607}{99202320}f_{n+1}+\frac{75641}{28343520}f_{n+2}-\frac{436127}{1388832480}f_{n+3}$$

$$+\frac{105863}{2182451040}f_{n+4}-\frac{38029}{9027411120}f_{n+5}\Big) \tag{9}$$

$$y_{n+\frac{1}{3}}=\frac{5}{9}y_n+\frac{5}{9}y_{n+1}-\frac{1}{9}y_{n+2}+h^3\Big(-\frac{325421}{396809280}f_n+\frac{265127}{16765056}f_{n+\frac{1}{3}}$$

$$+\frac{136187}{16511040}f_{n+\frac{2}{3}}+\frac{2862509}{79361856}f_{n+1}+\frac{30503}{11337408}f_{n+2}-\frac{353989}{1111065984}f_{n+3}$$

$$+\frac{215311}{4364902080}f_{n+4}-\frac{154879}{36109644480}f_{n+5}\Big) \tag{10}$$

The formulas for the first- and second-derivative methods are obtained by substituting the computed coefficients ω_j into $Y'(x) = \sum_{j=0}^{r+s-1} j\omega_j x^{j-1}$ and $Y''(x) = \sum_{j=0}^{r+s-1} j(j-1)\omega_j x^{j-2}$, respectively. Their explicit expressions are omitted here for brevity.

3 Convergence Properties

The proposed five-step block hybrid collocation scheme is consistent with order $p = 6 > 1$. The scheme is zero stable since its characteristic polynomial $p(z) = z^6(z-1)$ has roots with $|z| \leq 1$. Together, these properties guarantee the convergence of the proposed numerical scheme [14].

4 Benchmark Problems and Accuracy Assessment

The following benchmark problems are selected to evaluate the accuracy of the proposed numerical scheme. These problems are commonly used in the literature because they contain nonlinearities of different types and have known exact solutions for error analysis.

Benchmark Problem 1. A third-order nonlinear model [12] with a rational forcing term and exponential feedback: $y''' = -2e^{-3y} + 4(1+x)^{-3}$, $y(0) = 0$, $y'(0) = 1$, $y''(0) = -1$, $x \in [0, 4]$ with the exact solution $y(x) = (e^{-2x} - 1)x^2 + 3$.

Benchmark Problem 2. A third-order nonlinear model [5] with quadratic feedback and periodic forcing: $y''' = (\cos x - 1)\cos x + y^2 - 1$, $y(0) = y''(0) = 0$, $y'(0) = 1$, $x \in [0, 10]$ with exact solution $y(x) = \sin x$.

For performance evaluation, the proposed five-step block hybrid collocation scheme (5-BHCS) is compared with the three-point block hybrid collocation method (3-BHCM) [6], the four-point block hybrid collocation method (4-BHCM) [12], the hybrid linear multistep method (H-LMM) [15] and the hybrid collocation method (HC-M) [16]. The abbreviations for these methods are used in Tables 1 and 2 to report the maximum errors at different step sizes for the benchmark problems.

Table 1. Maximum absolute error across step size (h) for Benchmark Problem 1.

h	5-BHCS	3-BHCM	4-BHCM	H-LMM	HC-M
0.1	2.16 (−7)	1.19 (−6)	5.18 (−7)	1.24 (−5)	1.29 (−2)
0.05	2.52 (−9)	1.87 (−8)	5.01 (−9)	3.01 (−7)	1.14 (−3)
0.025	1.40 (−11)	3.31 (−10)	6.42 (−11)	9.16 (−9)	1.22 (−4)
0.0125	6.12 (−14)	4.98 (−12)	4.91 (−13)	2.56 (−10)	1.37 (−5)
0.00625	2.48 (−16)	7.93 (−14)	3.55 (−15)	7.87 (−12)	1.63 (−6)

Table 2. Maximum absolute error across step size (h) for Benchmark Problem 2.

h	5-BHCS	3-BHCM	4-BHCM	H-LMM	HC-M
0.1	1.09 (−8)	1.34 (−6)	8.71 (−8)	6.98 (−5)	5.69 (−2)
0.05	4.84 (−11)	1.97 (−8)	7.47 (−10)	1.79 (−6)	6.45 (−3)
0.025	2.01 (−13)	3.17 (−10)	6.66 (−12)	5.39 (−8)	8.12 (−4)
0.0125	8.10 (−16)	4.87 (−12)	5.12 (−14)	1.60 (−9)	9.87 (−5)
0.00625	5.77 (−19)	7.66 (−14)	3.84 (−16)	4.95 (−11)	1.24 (−5)

The results in Tables 1 and 2 indicate that the proposed 5-BHCS consistently yields smaller maximum errors than 3-BHCM, 4-BHCM, H-LMM and HC-M across all step sizes. The results confirm the reliability of the proposed scheme and demonstrate its suitability for modeling thin film flow problems.

5 Numerical Values for Thin Film Flow

The proposed five-step block hybrid collocation scheme is implemented to solve the thin film flow problem [8] for $k = 3$ with $y(0) = y'(0) = y''(0) = 1$ using a step size $h = 0.00625$ over the interval [0,1]. As presented in Table 3, the numerical values produced by 5-BHCS are consistently closer to the expected solution compared with 3-BHCM, 4-BHCM, H-LMM and HC-M. The proposed scheme demonstrates its superior accuracy. In addition, block formulation allows simultaneous computation at multiple points while offering improved computational efficiency over one-step schemes.

Table 3. Numerical Approximation of y for Thin Film Flow ($k = 3$), $h = 0.00625$.

x	5-BHCS	3-BHCM
0.2	1.221155142395689	1.221155142395689
0.4	1.488105284219075	1.488105284219076
0.6	1.804262548146545	1.804262548146548
0.8	2.171522798126401	2.171522798126406
1.0	2.590958259113169	2.590958259113176
x	4-BHCM	H-LMM
0.2	1.221155142395689	1.221155142395704
0.4	1.488105284219075	1.488105284219160
0.6	1.804262548146545	1.804262548146777
0.8	2.171522798126401	2.171522798126867
1.0	2.590958259113168	2.590958259113962

x	HC-M
0.2	1.221155142636966
0.4	1.488105285837896
0.6	1.804262552925606
0.8	2.171522808203834
1.0	2.590958276814426

6 Concluding Remarks

A five-step block hybrid collocation scheme (5-BHCS) has been developed for solving third-order ODEs directly. The scheme satisfies consistency and zero stability and thus ensures convergence. Numerical experiments demonstrate that it produces more accurate results than 3-BHCM, 4-BHCM, H-LMM and HC-M. A key advantage of the proposed scheme is its block structure which computes at five points per block enhancing computational efficiency and reducing computation time. The successful application to the thin film flow problem for $k = 3$ confirms its effectiveness and practical relevance.

References

1. Tuck, E.O., Schwartz, L.W.: A numerical and asymptotic study of some third-order ordinary differential equations relevant to draining and coating flow. SIAM Rev. **32**(3), 453–469 (1990)
2. Duffy, B.R., Wilson, S.K.: A third-order differential equation arising in thin-film flows and relevant to tanner's law. Appl. Math. Lett. **10**(3), 63–68 (1997)
3. Momoniat, E., Mahomed, F.M.: Symmetry reduction and numerical solution of a third-order ODE from thin film flow. Math. Comput. Appl. **15**(4), 709–719 (2010)
4. Abdel Kader, A.H., Abdel Latif, M.S., Nour, H.M.: Exact solutions of a third-order ODE from thin film flow using symmetry method. Int. J. Non-Linear Mech. **55**, 147–152 (2013)
5. Mechee, M., Senu, N., Ismail, F., Nikouravan, B., Siri, Z.: A three-stage fifth-order Runge-Kutta method for directly solving special third-order differential equation with application to thin film flow problem. Math. Probl. Eng. **2013**., Article ID 795397, 1–7 (2013)
6. Yap, L.K., Ismail, F., Senu, N.: An accurate block hybrid collocation method for third order ordinary differential equations. J. Appl. Math. **2014**., Article ID 549597(4), 1–9 (2014)
7. Fawzi, F.A., Senu, N., Ismail, F., Majid, Z.A.: A new integrator of Runge-Kutta type for directly solving general third-order ODEs with application to thin film flow problem. Appl. Math. Inf. Sci. **12**(4), 775–784 (2018)
8. Fawzi, F.A., Senu, N., Ismail, F., Majid, Z.A.: An efficient of direct integrator of Runge-Kutta type method for solving $y''' = f(x, y, y')$ with application to thin film problem. Appl. Math. Inf. Sci. **120**(1), 27–50 (2018)
9. Khataybeh, S.N., Hashim, I., Alshbool, M.: Solving directly third-order ODEs using operational matrices of Bernstein polynomials method with applications to fluid flow equations. J. King Saud Univ. Sci. **31**, 822–826 (2019)
10. Lee, K.C., Senu, N., Ahmadian, A., Ibrahim, S.N.I., Baleanu, D.: Numerical Study of Third-order Ordinary Differential Equations using A New Class of Two Derivative Runge-Kutta Type Methods. Alex. Eng. J. **59**, 2449–2467 (2020)
11. Mehrkanoon, S.: A direct variable step block multistep method for solving general third-order ODEs. Numer. Algorithms. **57**(1), 53–66 (2011)
12. Yap, L.K., Ismail, F.: Four point block hybrid collocation method for direct solution of third order ordinary differential equations. In: AIP Conference Proceedings 1974, vol. 020055, pp. 1–9. AIP Publishing, Pahang, Malaysia (2018)
13. Ramos, H., Rufai, M.A.: A two-step hybrid block method with fourth derivatives for solving third-order boundary value problems. J. Comput. Appl. Math. **404**, 1–21 (2022)
14. Henrici, P.: Discrete Variable Methods in Ordinary Differential Equations. John Wiley & Sons, New York (1962)
15. Mohammed, U., Adeniyi, R.B.: A three step implicit hybrid linear multistep method for the solution of third order ordinary differential equations. Gen. Math. Notes. **5**, 62–74 (2014)
16. Awoyemi, D.O., Idowu, O.M.: A class of hybrid collocation methods for third-order ordinary differential equations. Int. J. Comput. Math. **82**(10), 1287–1293 (2005)

Eco-Engineered Reduced TiO_2 via TiH_2–HCl Processing for Enhanced Photocatalytic Degradation of Paracetamol in Aqueous Phase

Zul Adlan Mohd Hir[1(✉)], Nur Aisyah Umirah Zakaria[1], Hamizah Mokhtar[2], Hartini Ahmad Rafaie[3], Nur Hashimah Alias[4], Ahmad Zamani Ab Halim[5], and Fariza Aina Abd Manan[1]

[1] Faculty of Applied Sciences, Universiti Teknologi MARA Pahang, 26400 Bandar Tun Abdul Razak Jengka, Pahang, Malaysia
zuladlan@uitm.edu.my

[2] Faculty of Civil Engineering, Universiti Teknologi MARA Pahang, 26400 Bandar Tun Abdul Razak Jengka, Pahang, Malaysia

[3] Centre of Foundation Studies, Universiti Teknologi MARA, Selangor Branch, Dengkil Campus, 43800, Dengkil, Selangor, Malaysia

[4] Faculty of Chemical Engineering, Universiti Teknologi MARA, 40450 Shah Alam, Selangor, Malaysia

[5] Faculty of Industrial Science and Technology, Universiti Malaysia Pahang Al-Sultan Abdullah, Lebuh Persiaran Tun Khalil Yaakob, 26300 Kuantan, Pahang, Malaysia

Abstract. The present study explores an innovative approach to tackling the pressing environmental issue of pharmaceutical residues in water bodies, focusing on the photodegradation of paracetamol (PARA), an analgesic and antipyretic drug. By utilizing a reduced-TiO_2 photocatalyst under UVC irradiation, this study aims to enhance the efficiency of PARA from aqueous solutions. The photocatalysts were synthesized via eco-engineered and a facile TiH_2-HCl processing method with varying mass ratios of TiO_2 to TiH_2 by 2:0.5, 2:1, and 2:1.5. They were characterized by using SEM-EDX and FTIR to confirm its morphological, compositions, and functional groups. The photodegradation of PARA by reduce-TiO_2 (2:0.5) was succeeded under normal conditions with ~65% removal, and the kinetic followed pseudo-first order model (5.2×10^{-3} min^{-1}), significantly surpassing the performance of pure TiO_2. This finding showed that the photocatalyst exhibits promising potential for degrading pharmaceutical residues under environmentally appropriate conditions.

Keywords: Paracetamol · photodegradation · reduced TiO_2 · titanium hydride · water

1 Introduction

Photocatalytic water treatment is gaining popularity due to its environmental benefits and potential to address future water pollution issues. Titanium dioxide (TiO_2) has been extensively researched as promising semiconductor oxide photocatalyst due to its non-toxicity, inexpensive, outstanding physical and chemical stability, and broad-spectrum

M. A. b. Jamlos et al. (Eds.): ICSET 2025, LNNS 1897, pp. 214–219, 2026.
https://doi.org/10.1007/978-3-032-20722-7_28

efficiency against a variety of contaminants (El Mchaouri et al., 2025). However, the TiO_2 photocatalyst is restricted by the large band gap of about 3.0–3.2 eV, and quick recombination of the photogenerated charge carriers. Therefore, many strategies have been developed to overcome the constraints to obtain high photocatalytic efficiency.

Recent studies show that the construction of defects on TiO_2 surfaces could improve light absorption by increasing oxygen vacancies and surface Ti^{3+} ions (Meroni et al., 2023). This modification has resulted in a higher number of active sites that enhance photocatalytic performance. In this view, oxygen vacancies act as shallow electron donors in reduced TiO_2 ($rTiO_{2-x}$), which are recognized as electron-hole trap sites and can help separate photogenerated charge carriers effectively. Previously, several conventional synthesis routes have been introduced such as calcination under H_2 atmosphere, plasma treatment, and vacuum activation (Chen et al., 2011; An et al., 2016; Hir et al., 2018). However, the methods use high pressure and temperature (T > 500 °C), the extensive heat treatment route, the modern and high-end facilities, and the potential negative effects on morphological properties, these methods are difficult to control and require multiple steps, severely limiting their practical application. Thus, constructing a facile and relatively inexpensive technique to achieve stable $rTiO_2$ remains a worthy challenge.

In comparison to the previously adopted procedure, herein, we describe a straightforward, low-temperature thermal reduction process that produces $rTiO_2$ in a comparatively short amount of time using titanium hydride (TiH_2), as potent reducing agent. This chemical method diminishes the possibility of an explosion during the synthesis process because it is stable in both air and hydrolysis. Therefore, the present work has continued focuses on the surface morphological, compositional characteristics, and photocatalytic performance of $rTiO_2$ under UVC light with a very low intensity (9 W, $\lambda = 254$ nm) irradiation over paracetamol (PARA) degradation in aqueous phase.

2 Materials & Method

2.1 Materials

Commercial titanium dioxide (TiO_2, 99%) and titanium hydride (TiH_2, 98%) were supplied by Sigma-Aldrich (Selangor, Malaysia), respectively. Hydrochloric acid (HCl, 37%) and isopropanol (C_3H_8O, 99.9%) were obtained from HmbG Chemicals (Selangor, Malaysia). Paracetamol (PARA) was purchased from a local pharmacy in Malaysia. The chemicals were utilized as procured, without any refinement. Deionized water was used throughout the photocatalyst preparation and photocatalytic testing.

2.2 Synthesis of Reduced-TiO_2 Photocatalyst and Characterization

Initially, reduced-TiO_2 ($rTiO_2$) photocatalyst was prepared with different mass ratios of pure TiO_2 to TiH_2 (2:0.5, 2:1.0, and 2:1.5) through simple mixing process. The samples were grinded for 30 min prior to the calcination in a muffle furnace. The calcination was carried out at 300 °C for 50 min. Next, the dried powder was let to cool at room temperature. The light grey $rTiO_2$ was obtained and carefully washed with 1.0 M HCl solution, isopropanol, and deionized water three times each, to remove the unreacted TiH_2 and catalytic activation. The process was continued by drying the powder in the oven at 80 °C overnight and labelled as $rTiO_2$ accordingly. Pure TiO_2 was also used to examine photocatalytic activity for comparison purposes. The surface morphology and composition of the synthesized $rTiO_2$ were analyzed by using Scanning electron microscopy attached with energy dispersive X-ray analyser (SEM-EDX, TESTAN VEGA3).

2.3 Photodegradation Setting

First, 0.2 g of $rTiO_2$ was dispersed in a beaker containing the paracetamol solution (10 mg/L, 100 mL). The PARA solution was stirred in the dark for 30 min to achieve the adsorption-desorption equilibrium. Then, the solution was irradiated under UVC lamp (9 W, $\lambda = 254$ nm) for 180 min with continuous stirring. About 5 mL of the samples were drawn in every 30 min intervals, and the concentration of the collected samples were analyzed using a UV-vis spectrophotometer (PerkinElmer Lambda 35) at λ_{max} = 243 nm. The percentage of photodegradation (%) was evaluated using the equation below (Sazman et al., 2024):

$$\%\text{Photodegradation} = (\text{Co} - \text{Ct})/\text{Co} \times 100 \quad (1)$$

The Langmuir–Hinshelwood (L–H) model was used to assess the kinetic of PARA removal by the prepared photocatalysts. The following linear form can be used to express the pseudo first-order:

$$\ln\left(\frac{C_o}{C_t}\right) = k_1 t \quad (2)$$

where C_o and C_t are the concentration of PARA at initial and "t min", respectively and k_1 is the rate constant (min^{-1}).

3 Results and Discussion

3.1 Morphological and Compositions Analyses

Figure 1 shows the surface morphological and composition of pure TiO_2 and $rTiO_2$ photocatalysts. It can be seen that the pure TiO_2 appear smooth surface without any clustering or voids (Fig. 1(a)). Meanwhile, the observations for all the $rTiO_2$ exhibited almost similar surface structure with slightly denser particles and a rougher surface, and greyish colour as evidenced by the visible colour changes after reduction reaction. The images depicted as slightly different distribution with that of pure TiO_2 as they have undergone TiH_2 reduction and HCl treatment (Fig. 1(b)-(d)). In consideration of that, there was a great homogeneous dispersion for the mixture of pure TiO_2 and $rTiO_2$. This finding is in a good agreement with previously reported study (Zhang et al., 2017). The mapping analysis for rTiO2 (0.5) is presented in Fig. 1(e)(i)-(iii) also exhibited high amount of titanium (Ti) and oxygen (O) elements indicated by the red and green colour. The presence of insignificant amounts of carbon (C) was observed due to some carbonaceous impurities accumulate on the surface of the samples during the process of reduction such as CO_2 gaseous in air (Fang et al., 2014). Due to the minimal amount, the C element was negligible in this study.

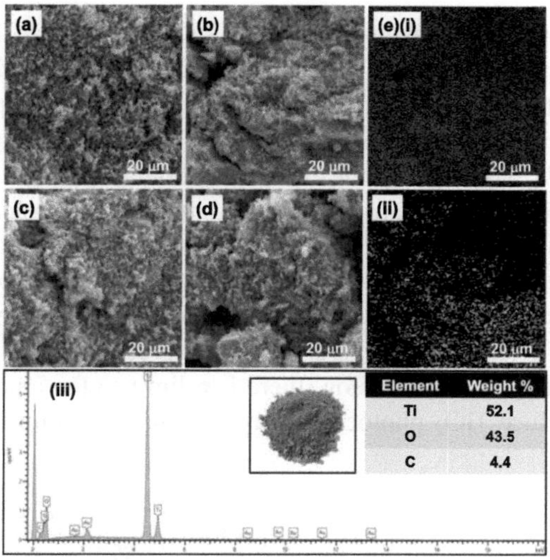

Fig. 1. SEM images of (a) pure TiO_2, (b) $rTiO_2$ (0.5), (c) $rTiO_2$ (1.0), (d) $rTiO_2$ (1.5) at 3 kx magnification; and (e)(i–iii) elemental mapping images of $rTiO_2$ (0.5) (inset: image of $rTiO_2$ (0.5) powder).

3.2 Photodegradation Evaluation

Figure 2 depicts the results of photodegradation testing and kinetic studies of PARA degradation in aqueous solutions using pure TiO_2 and $rTiO_2$ photocatalysts, with varying mass ratios. It was observed that PARA was stable under UVC irradiation, since no significant degradation was seen throughout the photolysis process (5%). Pure TiO_2 demonstrated good degradation activity against PARA, of about 54%. When 0.5 g of TiH_2 was mixed to TiO_2 ($rTiO_2$ (0.5)), the percentage of PARA degradation rose to a maximum of almost 65%. It shows that, $rTiO_2$ (0.5) expressed greater efficiency with 1.2 times better than that of pure TiO_2. As the TiH_2 content increased further up to 1.5 g, the percentage of photodegradation progressively declined down to 57% (Fig. 2(a)). A considerable amount of PARA was degraded using the photocatalysts, suggesting that photocatalysis was the main reason for PARA removal. Theoretically, when the photocatalyst is added into the solution and exposed to UVC light, it starts to absorb photon energy from the light and generates electron-hole pairs (e_{cb}^-/h_{vb}^+). These charge carriers drive a series of redox reactions at the catalysts' surface that further produce reactive radical species (RRS), including superoxide radicals ($\bullet O_2^-$) and hydroxyl radicals ($\bullet OH$), which are responsible for the oxidative degradation of pollutants. These powerful radicals attack the molecule and break down into other small organic molecules, water and carbon dioxide. For the pseudo first-order kinetics, the experimental data exhibited linearity and a high correlation coefficient for the best $rTiO_2$ (0.5) photocatalyst ($R^2 = 0.9879$) (Fig. 2(b)). The best rate constant for k_1 was found to be 5.2×10^{-3} min^{-1}.

The optimum ratio between TiH_2 and TiO_2 played a crucial role in the construction of high performance photocatalyst. Accordingly, the high percentage of photodegradation was due to the presence of bulk oxygen vacancies and Ti^{3+} ion in TiO_2 surface defect (SD), hence resulted in a higher number of active sites that enhance the concentration of electron-hole pairs which then generates the RRS (Fang et al., 2017). In addition, the generation of oxygen vacancies and Ti^{3+} has lower the level of band gap energy, which then reduces the excitation energy of photogenerated electrons to jump from valence band to the conduction band, results in a highly effective photocatalyst (Fig. 2(c); Jaafar et al., 2015; Hir et al., 2018). Nonetheless, the excessive amount of the reducing agents has led to excessive defect formation where it introduced more oxygen vacancies and Ti^{3+} ion centers. The continuous formation of Ti^{3+} stimulates localized electronic states under the conduction band minimum, which then causes the quick recombination of electron-hole pairs on TiO_2 surfaces that reduced the generation of RRS, and thus diminishing the photocatalytic activity. This finding is in a good agreement with previously reported study (Ping et al., 2022). In conclusion, there is a limit to the amount of TiH_2 as the reducing agent to reduce TiO_2 photocatalyst by which the optimum ratio was found to be 4:1 of TiO_2:TiH_2.

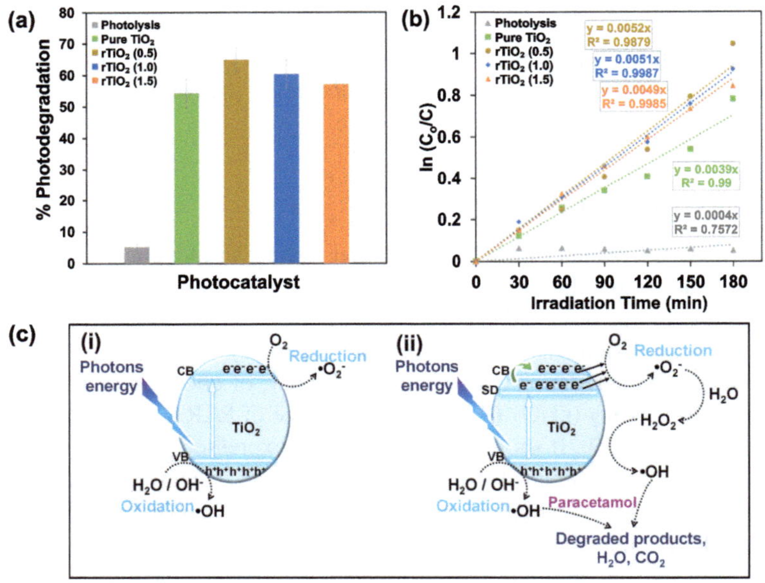

Fig. 2. (a) Degradation percentage, (b) pseudo first-order kinetics, and (c) mechanism for photodegradation of PARA in aqueous solution (i) before, and (ii) after TiH_2 reduction and HCl treatment. [PARA] = 10 mg/L, pH = 6.8, photocatalyst = 0.2 g.

4 Conclusion

Reduced TiO_2 synthesized by an eco-engineered and facile reduction process using TiH_2 as reductant has been successfully carried out. In addition, HCl treatment helps to eliminate the unreacted TiH_2 and activate the catalysts' surface through defect formations. The presence of surface defects induced by Ti^{3+} ions or oxygen vacancies, led to the improvement of photodegradation of PARA in aqueous phase up to 65%. The kinetic model that best suited PARA degradation was pseudo first-order with highest value for rate constant (5.2×10^{-3} min^{-1}). The $rTiO_2$ photocatalyst demonstrated good potential, making it suitable alternative photocatalyst in photocatalytic remediation processes.

Acknowledgement. This work was supported by the Ministry of Higher Education, Malaysia (FRGS/1/2024/STG05/UITM/02/12) and facilities by UiTM Pahang Branch.

References

An, H., et al.: Advanced nanoporous TiO2 photocatalysts by hydrogen plasma for efficient solar-light photocatalytic application. Nat. Publ. Group. Sci. Reports. **6**, 29683 (2016). https://doi.org/10.1038/srep29683

Chen, X., Liu, L., Yu, P.Y., Mao, S.S.: Increasing solar absorption for photocatalysis with black hydrogenated titanium dioxide nanocrystals. Science. **80-.). 331**, 746–750 (2011). https://doi.org/10.1126/science.1200448

El Mchaouri, M., et al.: Engineering TiO2 photocatalysts for enhanced visible-light activity in wastewater treatment applications. Tetrahedron Green Chem. **6**, 100084 (2025). https://doi.org/10.1016/j.tgchem.2025.100084

Fang, W., Xing, M., Zhang, J.: Modifications on reduced titanium dioxide photocatalysts: A review. J Photochem Photobiol C: Photochem Rev. **32**, 21–39 (2017). https://doi.org/10.1016/j.jphotochemrev.2017.05.003

Hir, Z.A.M., Abdullah, A.H., Zainal, Z., Lim, H.N.: Visible light-active hybrid film photocatalyst of polyethersulfone–reduced TiO2: photocatalytic response and radical trapping investigation. J. Mater. Sci. **53**, 13264–13279 (2018). https://doi.org/10.1007/s10853-018-2570-3

Jaafar, N.F., Jalil, A.A., Triwahyono, S., Shamsuddin, N.: New insights into self-modification of mesoporous titania nanoparticles for enhanced photoactivity: effect of microwave power density on formation of oxygen vacancies and Ti3+ defects. RSC Adv. **5**, 90991–91000 (2015). https://doi.org/10.1039/C5RA15120A

Meroni, D., et al.: Oxygen vacancies in the Spotlight: On the engineering of intrinsic defects in highly defective TiO2 photocatalysts. J. Photochem. Photobiol. A Chem. **444**, 114916 (2023). https://doi.org/10.1016/j.jphotochem.2023.114916

Ping, X., Zhang, Y., Zhao, Q., Lu, Y., Hao, L.: Flexible TiO2 nanograss array film decorated with oxygen vacancies introduced by facile chemical reduction and their photocatalytic activity. Environ. Technol. Innov. **25**, 102185 (2022). https://doi.org/10.1016/j.eti.2021.102185

Sazman, N.S.N., Hir, Z.A.M., Daud, S., Rafaie, H.A.: Charge transport properties and photostability of microcrystalline cellulose derived from Gigantochloa scortechinii-supported ZnO for enhanced acetaminophen degradation. Ind. Crop. Prod. **222**, 119558 (2024). https://doi.org/10.1016/j.indcrop.2024.119558

Zhang, M., Pei, Q., Chen, W., Liu, L., He, T., Chen, P.: Room temperature synthesis of reduced TiO2 and its application as a support for catalytic hydrogenation. RSC Adv. **7**, 4306–4311 (2017). https://doi.org/10.1039/C6RA26667C

Thermally Driven Electrical Response of Cobalt(ii) Spin Crossover/Graphene Oxide Composite Devices

N. A. A. M. Amin[1(✉)], S. M. Said[2], N. M. J. N. Ibrahim[2], M. M. I. M. Hasnan[2], and S. V. U. Sharma[3]

[1] Faculty of Information Sciences and Engineering, Management and Science University, 40100 Shah Alam, Selangor, Malaysia
nuraishah_aminah@msu.edu.my

[2] Department of Electrical Engineering, Faculty of Engineering, Universiti Malaya, 50603 Kuala Lumpur, Malaysia

[3] Institutional of Aeronautical Engineering, Dundigal Road, Dundigal, Hyderabad, Telangana 500043, India

Abstract. Spin-crossover (SCO) materials, which reversibly switch between low-spin and high-spin states under external stimuli such as temperature, offer promising functionality for sensors, switches, and memory devices. In this study, three SCO/graphene oxide (GO) composite devices were fabricated using $[Co_2(CH_3COO)_4(LCn)_2]$ (n = 8, 10, 12) SCO complexes. The thermally induced electrical behaviour of these devices was studied as a function of alkyl chain length, voltage bias, and the presence of GO. Temperature-dependent measurements revealed increased conductance during heating and decreased conductance upon cooling, correlating with spin-state transitions. At 0 V bias, large hysteresis (~20 K) was observed in the SCO/GO devices with the longest alkyl chain length (Co-LC12/GO) while at 1 V bias, large hysteresis with strong memory effect was evident in the same device. Longer alkyl chains increase spacing and disorder between SCO complexes, leading to reduced electron mobility and enhanced charge trapping. The presence of GO further contributes to charge trapping and induces dipole alignment under thermal and electrical stimuli. These findings highlight the potential of SCO material and molecular design strategies for applications ranging from thermal sensors to switching and memory applications.

Keywords: Spin crossover (SCO) · graphene oxide · memory · thermal sensors · carbon alkyl chain length

1 Introduction

The spin-crossover (SCO) phenomenon involves a reversible transition between low-spin (LS) and high-spin (HS) electronic configurations in response to stimuli such as temperature, pressure, magnetic fields, or light-induced spin-state trapping (LIESST) [1, 2]. This behaviour is commonly observed in octahedral complexes of first-row transition metal ions (e.g., Mn, Fe, and Co) with d^4-d^7 valence electronic configurations [3].

M. A. b. Jamlos et al. (Eds.): ICSET 2025, LNNS 1897, pp. 220–230, 2026.
https://doi.org/10.1007/978-3-032-20722-7_29

A key feature of these SCO complexes is their ability to undergo reversible spin state transitions between high-spin (HS) and low-spin (LS) states, a process that is accompanied by significant changes in molecular geometry and other properties. This spin transition behaviour is central to their potential use in electronic devices, particularly where thermal hysteresis which is critical for molecular bistability is required for applications like thermal sensors, switches, and memory devices.

The electrical conductivity of these materials is closely linked to their spin state, with longer metal-ligand bonds (M-L) in the HS state resulting in lower conductivity, and shorter bonds in the LS state leading to higher conductivity due to phonon contributions [4, 5]. However, the naturally insulting property of these complexes may pose challenges for optimal detection based on spin state changes. To overcome this, GO has been incorporated as a conductive matrix in SCO composites, enhancing electron transport and overall conductivity. For example, multifunctional hybrid $Fe(Htrz)_2(trz)$ (trz = triazole) SCO nanoparticles/graphene composites [6] have been shown to exhibit both thermal and optical addressability of the spin state through electrical detection.

In our previous work [7], a series of cobalt(II) SCO complexes, $[Co_2(CH_3COO)_4(LCn)_2]$ (L = ligand, n = 8, 10, 12), were investigated to evaluate the effect of alkyl chain length on their thermally induced spin transition and charge transport properties in both solution and thin film forms. The findings showed that variations in ligand length significantly affect molecular packing and spin transition cooperativity, providing a structural–functional basis for the development of composite thin film devices.

The present work extends the investigation of thin film composites of these SCO complexes with graphene oxide (GO) to determine the optimal alkyl chain length that balances thermal stability, SCO behaviour, and electrical performance. The incorporation of GO is expected to enhance the electrical response through interfacial coupling, which facilitates charge transfer and modulation during the spin transition. The effects of alkyl chain length and voltage bias on the structural and electrical characteristics of the SCO/GO films are examined to optimise their performance for low-power thermal sensing, switching, and memory device applications.

2 Methodology

2.1 Preparation of SCO Cantilever

A layer of silver (Ag) as conductive layer was pre-printed on a flexible Polyethylene terephthalate (PET) substrate and cut into three strips strip of dimension 7 mm x 1 mm each. Prior to deposition, solution of 0.005 M of SCO complexes was prepared by dissolving 0.0215 g of Co-LC8, 0.0117 g of Co-LC10, and 0.0129 g of $Co(CH_3CO_2)_2$ LC12 (Co-LC12) fine powders in 1 mL of chloroform to form solution with 0.005 M concentration, respectively. Each solution was pipetted and dropped onto each cut strip to forms bilayers of SCO/Ag. The change in the position of the cantilevers was observed under Olympus BX51 Microscope.

2.2 Thin Film Device Fabrication

The SCO/GO composites were prepared by adding commercially obtained GO in water dispersion into the solution Co-LC8, Co-LC10 and Co-LC12 in mol ratio 1:1. Dimethyl-formamide (DMF) was added to aid with the dispersion of GO in the SCO solution. The mixture was then deposited on a clean interdigitated Indium-Tin-Oxide (ITO) glass substrates with 12 μm gaps size before further annealed at at 45 °C-50 °C to ensure strong adhesion between the thin film and substrate. The dimension of the interdigitated ITO substrate is schematically drawn as in Fig. 1.

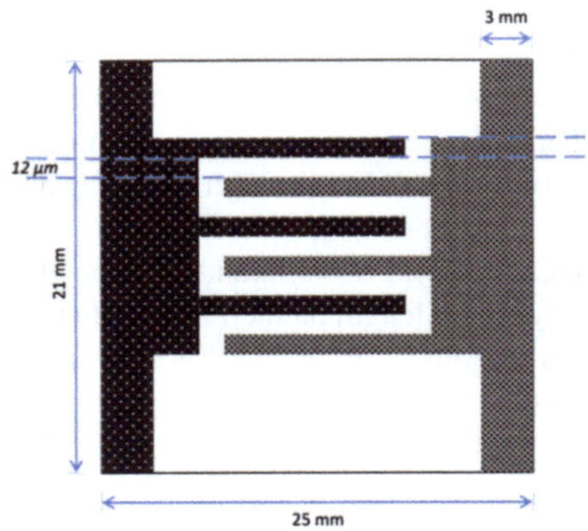

Fig. 1. Schematic of interdigitated Indium-Tin-Oxide (ITO) substrate

2.3 Fabrication of Cantilevers

A layer of silver (Ag) as conductive layer was pre-printed on a flexible Polyethylene terephthalate (PET) substrate and cut into three strips strip of dimension 7 mm x 1 mm each. Prior to deposition, solution of 0.005 M of SCO complexes was prepared by dissolving 0.0215 g of Co-LC8, 0.0117 g of Co-LC10, and 0.0129 g of Co-LC12 fine powders in 1 mL of chloroform to form solution with 0.005 M concentration, respectively. Each solution was pipetted and dropped onto each cut strip to forms bilayers of SCO/Ag.

2.4 Voltage Response Measurement

The voltage responses were measured using Agilent - 34461A 61.2 Digital Multimeter connected to Wheatstone bridge configuration (Fig. 2). Voltage bias was supplied using a Keithley 2230G-30-6 triple Channel DC Power Supply. A heating plate is placed directly below the substrate, while the temperature readings were observed using infrared thermal camera. Resistors of 220 KΩ each and the SCO substrate were connected in a Wheatstone bridge configuration to obtain the changes in output voltage (V_g) in response to the resistance variation in the SCO/GO system. Temperature variation between room temperature (RT) to 85 °C was applied onto the substrate. The voltage response, Vg (mV) as a function of temperature (°C) was measured under bias conditions of Vb = 0 V and Vb = 1 V during the heating cycle. The results were recorded as an average of three heating and cooling cycles. In all measurements, open circuits potentials at room temperature were removed to zero out all output values.

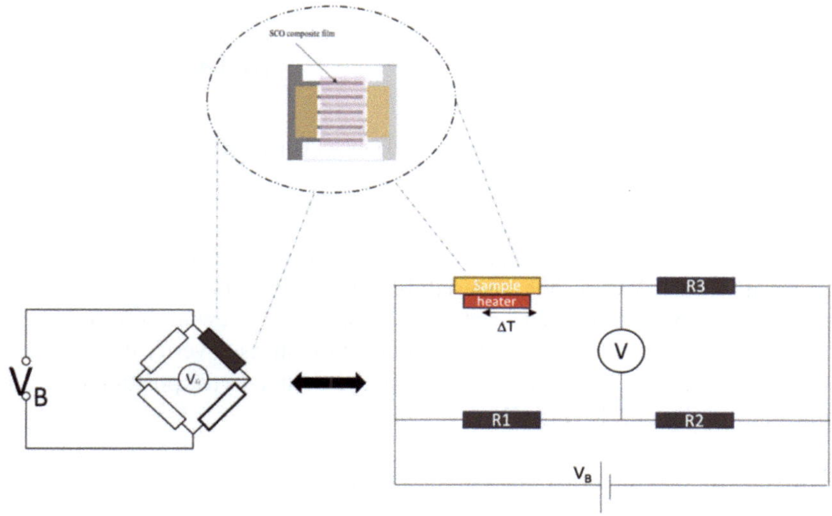

Fig. 2. Schematic of Wheatstone bridge configuration used to probe the voltage response corresponding to resistance variation.

3 Results and Discussions

3.1 Spin Transition and Strain-Stress Response

From magnetic properties of the pure complexes Co-LC8, Co-LC10, and Co-LC12 [7], these complexes were found to possessed high-spin (HS) Co(II) atoms, resulting in weak Co-N and Co-O bonds attributed to the extended metal-to-ligand bonds. The spin transition behaviour obtained from our previous work [7] displayed a gradual shift from predominantly high-spin (HS) at 25 °C to a higher proportion of low-spin (LS) at 70 °C, returning to the HS state upon cooling. This reversible spin transition indicated the dynamic nature of these complexes. Typically, the transition from LS to HS or vice versa involves changes in the metal-to-ligand (M-L) bond lengths. The HS state is associated with longer M-L bonds, while the LS state is linked to shorter M-L bonds (Fig. 3). Previous reports on a family of Co(II) complexes have documented a length expansion ranging from 0.06 Å to 0.21 Å during the transition from LS to HS [8, 9].

Expansion of molecules due to spin transition will also translate to macrospic observation of samples expansion upon application of temperature which induced spin state change. This results in a change in the stress-strain response of the sample, in other

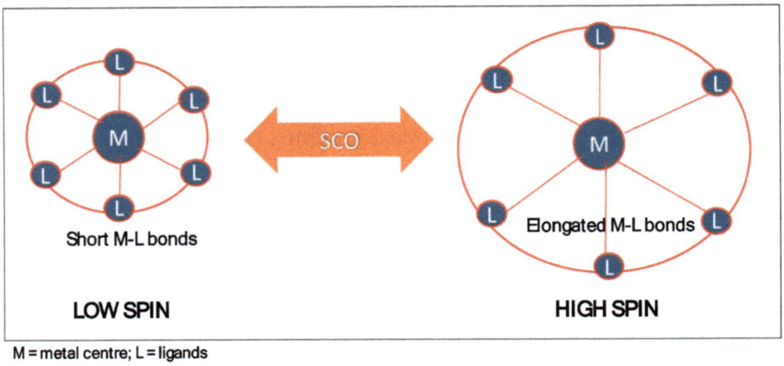

Fig. 3. Illustration of mechanism of shortening and elongation of metal-to-ligands (M-L) bonds during SCO process.

words, a piezoelectric effect upon application of temperature. Elongation and compression of the molecular level of the material during spin transition in the SCO complex can be further be proven through mechanical actuation. Following this, a straightforward experiment was conducted using SCO-embedded strips to verify the presence of a stress-strain response in the SCO, by observing the change in the position of the strips along heating-cooling cycle (Fig. 4).

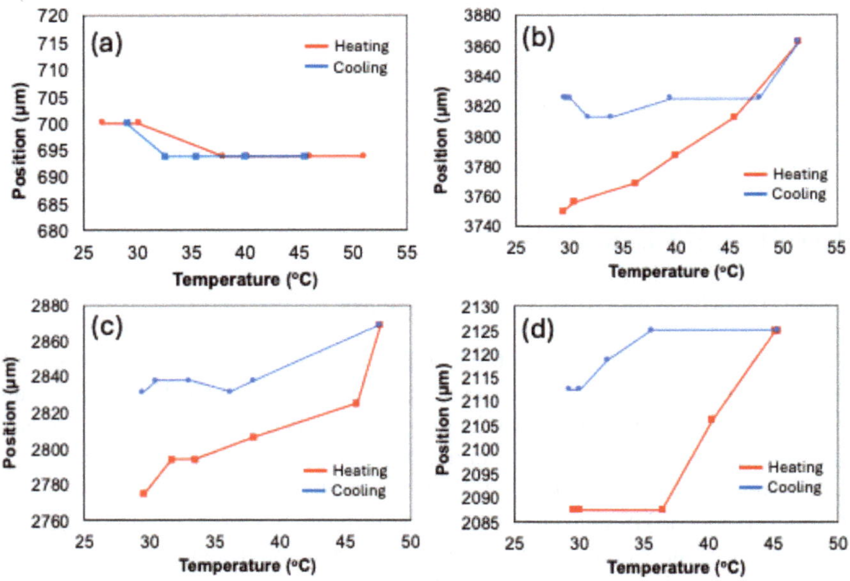

Fig. 4. Position changes upon heating cooling for (a) Plain Ag strip; (b) Co-LC8; (c) Co-LC10; and (d) Co-LC12 strips

Compared to the plain Ag strip (Fig. 4a), all SCO strips exhibited deflection upon heating up to 55 °C, except for the Co-LC12 strip (Fig. 4d). In this case, weak heat transfer from the heating element to the strip end prevented the temperature from exceeding

50 °C. The measured deflection distances between 25–45 °C were 1.87%, 1.80%, and 1.90% for the Co-LC8, Co-LC10, and Co-LC12 strips, respectively. The thermal stress–strain response that causes strip deflection may be associated with the spin transition behaviour of the complexes. Based on temperature-dependent UV–vis absorbance data [7], the spin transition magnitudes were 1.21%, 0.93%, and 1.90% for Co-LC8, Co-LC10, and Co-LC12 thin films, respectively, indicating a strong correlation between stress and strain during the spin transition. Similar observation was reported in several studies. For example, spin transition exhibited in SCO complex Fe(trz)(H-trz)$_2$ dispersed in a poly(methylmethacrylate) (PMMA) matrix [10] resulted in significant isotropic expansion of the composite material. This expansion causes strong deformation of the bimorph at the temperature of the spin transition. Such behaviour highlights the potential for developing electromechanical actuating devices, including soft actuators for artificial muscles, piezoelectric energy harvesting (PEH) systems, and micro-electromechanical systems (MEMS) [11–13] .

3.2 Voltage Response Measurement

The cooperative spin-crossover (SCO) behaviour can induce conductive bistability in which translates into bistable voltage outputs. To investigate these characteristics, voltage response measurements on these systems were conducted. The effect of carbon alkyl chain / ligand length and bias voltages (0 V and 1 V) on the SCO/GO composites under varying temperatures were assessed. Plots of voltage response vs. temperature for Co-LC8/GO, Co-LC10/GO and Co-LC12/GO under 0 V bias are shown in Fig. 5.

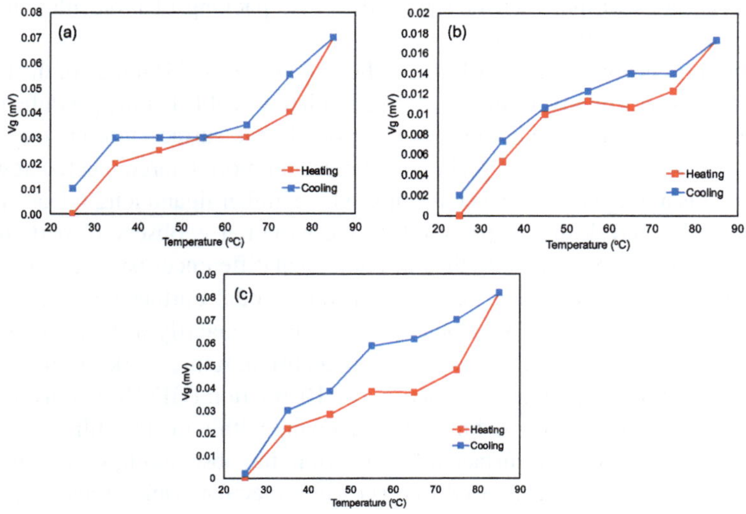

Fig. 5. Voltage response vs. temperature for (a) Co-LC8/GO; Co-LC10/GO and (c) Co-LC12/GO under 0 V bias.

Voltage response observations showed clear dependence on spin state transitions. At 0 V, the responses were governed purely by thermally induced processes. During heating, voltage increased and decreased upon cooling, corresponding to conductance changes driven by thermal spin transitions and related molecular structural variations. The stress or strain from M–L bond elongation and contraction in HS or LS states significantly influences charge carrier hopping distance and alters dc conductivity. This modulation of hopping distance is closely linked to changes in the vibrational density of states, affecting overall charge transport mechanisms [4].

In a system subjected to thermal perturbation, the HS-to-LS transition (resulting in shorter M–L bonds) enhances phonon modulation and shifts charge transfer from hopping to tunnelling, leading to higher conductivity. These effects have been widely studied in spin-state-dependent electrical junctions [4, 14–16]. Rotaru et al. [4, 17] reported higher conductivity in the LS state due to increased phonon frequencies, while conductivity decreased in the HS state. Shalabaeva et al. [18], reported that the conductance of the ITO/SCO/Al junction decreases significantly during the LS-to-HS transition, which is consistent with the findings of this study. Therefore, it is inferred that the low voltage level (low conductance) corresponds to the HS state, while a higher voltage level (high conductance) reflects a greater proportion of LS-state SCO molecules within the devices.

A wider hysteresis window of about 20 K was observed in the Co-LC12/GO device compared to Co-LC8/GO, attributed to differences in alkyl chain length. Shorter chains promote better lateral stacking of SCO complexes that enhance electron mobility and charge transport [19]. Longer chains reduce mobility and cause charge trapping as shown by DFT calculations from Ma et al. [20]. It is hypothesised that the extension of the alkyl carbon chain leads to increased spacing and partial ordering between SCO complexes, which in turn lowers electron mobility. Additionally, longer chains introduce more disorder and defects, disrupting charge transport. The increased steric hindrance from the longer alkyl chains further interferes with the close packing of molecules, impeding efficient charge transport pathways.

The Field Emission Scanning Electron Microscope (FESEM) image of the GO-only film (Fig. 6a) shows a relatively smooth surface with few visible features or defects. Upon incorporation of Co(II) SCO complexes (Fig. 6b–d), the surface morphology appears more textured with formation of grain boundaries and more interconnected aggregated structures. This aggregation may act as traps, where higher ligand length leads to larger grain boundaries and deeper traps. While some variation is observed, particularly in the Co-LC12/GO sample (Fig. 6d), the morphological differences between Co-LC8/GO (Fig. 6b) and Co-LC10/GO (Fig. 6c) are relatively subtle. Further investigation using Atomic Force Microscopy (AFM) would be beneficial to quantify surface roughness and topographical variations at higher resolution. In addition, future works to include X-ray Diffraction (XRD) or Grazing-Incidence X-ray Diffraction (GIXRD) analysis will be useful to evaluate crystallinity and molecular packing within the thin films.

Despite the limitations in surface differentiation, the observed hysteresis behaviour in Co-LC12/GO may still relate to differences in molecular packing and trap density. As temperature increases, charge carriers tend to be either easily trapped or released, depending on the thermal activation energies, which is outside of the scope of this paper. It is expected that charge carriers become easily trapped in longer ligand length, thus explaining the wider hysteresis and stronger memory effect in the Co-LC12/GO.

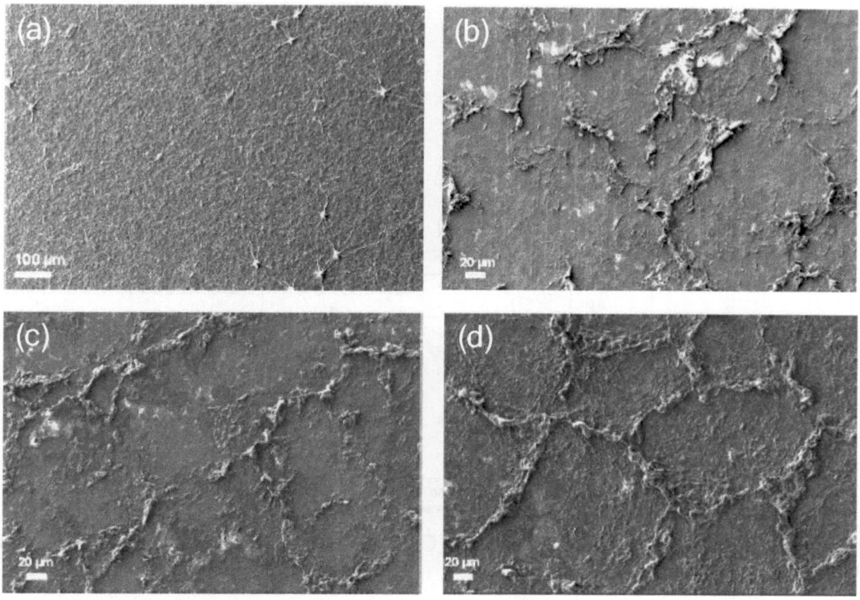

Fig. 6. FESEM images showing surface morphologies of (a) GO; (b) Co-LC8/GO; (c) Co-LC10/GO; and (d) Co-LC12/GO.

Under a 1 V bias, the change in conductance level corresponding to the spin transition remains observable (Fig. 7), with magnitude of significantly higher than that for 0 V. Although it seemed like Co-CL10/GO (Fig. 7b) showed large hysteresis, the response sensitivity was the lowest (4.7 μV/K) compared to Co-LC8/GO (92.3 μV/K) (Fig. 7a) and Co-LC12/GO (94.5 μV/K) (Fig. 7c) due to weak cooperativeness as explained in earlier discussion. Notably, a significant hysteresis and strong memory effect were evident at higher temperatures and even more pronounced in the Co-LC12/GO compared to other samples. The voltage levels reached a maximum voltage at around 45-50oC, and then remained until further heating. The voltage response might have reached a saturation point where further temperature increases do not significantly affect the voltage, but the intrinsic properties of the material (like increased carrier density or decreased scattering) continue to reduce the resistance, thereby increasing the current.

When a bias voltage is applied to the systems, it induces polarisation, facilitating the movement of charge carriers. This effect is more pronounced in shorter alkyl chains due to the reduced distance between conductive pathways, allowing easier charge carrier movement. Longer alkyl chain can exhibit fastener effect due to strong intermolecular interaction, thereby enhancing the spin transition [21]. It was observed that in Co-LC12/GO sample, the voltage increased from roughly 0.5 mV at 20 °C and reached maximum voltage level at lower temperature of about 6.0 mV at 50 °C, showing a more cooperative transition than that of Co-LC8/GO (60 °C) (Note: The Co-CL10/GO is under exception due to non-cooperativeness).

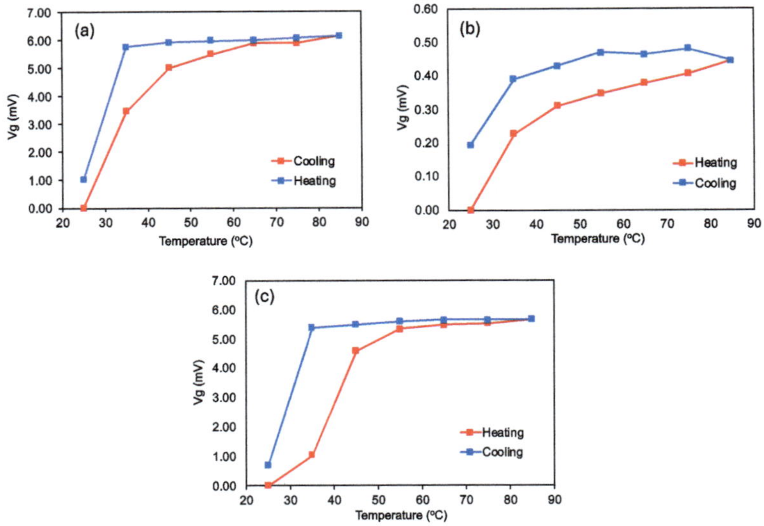

Fig. 7. Voltage response vs. temperature for (a) Co-LC8/GO; Co-LC10/GO and (c) Co-LC12/GO under 1 V bias.

4 Conclusion

Three devices containing thin film composites of Co(II)-SCO/GO were fabricated and investigated for their cooperative SCO behaviour and conductive bistability. These devices, featuring varying carbon alkyl chain lengths (8, 10, and 12-carbon) and subjected to different voltage biases, revealed significant findings:

1. The complexes exhibit a reversible HS to LS transition with corresponding changes in M-L bond lengths, resulting in measurable stress-strain responses or piezoelectricity upon heating.
2. The combination of SCO complexes with GO enhanced electrical properties via interfacial coupling. Temperature-dependent voltage and current analyses showed increased conductance upon heating and reduced conductance during cooling. This behaviour arises from conductance and charge transport changes driven by thermally induced spin transitions, where the HS-to-LS transition enhances phonon modulation and charge transfer, increasing conductivity.
3. At 0 V bias, Co-LC12/GO exhibited large hysteresis due to the longer alkyl chains, which reduce charge mobility and promote charge trapping. The longer alkyl chains increase spacing and disorder between SCO complexes, reducing electron mobility due to steric hindrance and defects. FESEM images confirmed that longer alkyl chains result in rougher, more textured surfaces with grain boundaries and interconnected structures, acting as charge traps.
4. Under a 1 V bias, the combination of alkyl chain length, polarisation effects in both GO and SCO moieties, and spin state transition upon temperature changes contributed to enhanced voltage response levels, charge trapping, wide hysteresis, and strong memory effects, particularly in Co-LC12/GO.

The study provides a pathway for molecular design through addition of guest moieties and optimisation of alkyl chain length for high-performing thin film devices, suitable for low-powered thermal sensors, switching, and memory applications.

Acknowledgments. We are grateful for the financial support by the Ministry of Higher Education Malaysia via the Fundamental Research Grant Scheme (FRGS/1/2020/TK0/UM/02/20) and Management & Science University via the MSU Research Grant Scheme (MCSG-008-022025).

Disclosure of Interests The authors have no competing interests to declare that are relevant to the content of this article.

References

1. Gütlich, P., Goodwin, H.A.: Spin crossover—An overall perspective. In: Spin Crossover in Transition Metal Compounds I, pp. 1–47 (2004)
2. Gaspar, A.B., Seredyuk, M., Gutlich, P.: Spin crossover in iron(II) complexes: Recent advances. J. Mol. Struct. **924**, 9–19 (2009)
3. Gaspar, A.B., Seredyuk, M., Gutlich, R.: Spin crossover in metallomesogens. Coord. Chem. Rev. **253**, 2399–2413 (2009)
4. Rotaru, A., Molnár, G., Salmon, L., Demont, P., Bousseksou, A.: Spin state dependence of electrical conductivity of spin crossover materials. Chem. Commun. **48**, 4163–4165 (2012)
5. Rosseinsky, D.R., Tonge, J.S.: Electron transfer in solids. Temperature dependence of dielectric relaxation and conductivity in mixed-valence potassium manganate–permanganate. J. Chem. Soc., Faraday Trans. 1: Physical Chemistry in Condensed Phases. **78**, 3595–3603 (1982)
6. Dayen, J.-F., et al.: Room temperature optoelectronic devices operating with spin crossover nanoparticles. Mater. Horiz. **8**, 2310–2315 (2021)
7. Amin, N.A.A.M., Said, S.M., Ibrahim, N.M.J.N., Hasnan, M.M.I.M., Salleh, M.F.M., Afifi, A.M.: Comparative analysis between solution-phase and thin films of cobalt (II) spin crossover (SCO) complexes with 8, 10, 12-carbon alkyl chains based on structural, optical and electrical properties. J. Mater. Sci. Mater. Electron. **35**, 1930 (2024)
8. Ondo, A., Ishida, T.: Cobalt(II) Terpyridin-4′-yl Nitroxide Complex as an exchange-coupled spin-crossover material. Crystals. **8**, 155 (2018)
9. Krivokapic, I., et al.: Spin-crossover in cobalt (II) imine complexes. Coord. Chem. Rev. **251**, 364–378 (2007)
10. Gural'skiy, I.Y.A., et al.: Spin crossover composite materials for electrothermomechanical actuators. J. Mater. Chem. C. **2**, 2949–2955 (2014)
11. Zahid, S.F.B.M., Al-Humairi, S.N.S., Han, G.W.: Innovative piezoelectric energy harvesting design for footstep-driven power generation: Analysis and performance evaluation. In: 2024 IEEE 12th Conference on Systems, Process & Control (ICSPC), pp. 378–383. IEEE, Malacca, Malaysia (2024)
12. Piedrahita-Bello, M.: Fabrication of spin crossover nanocomposites and devices for electromechanical applications. Coordination Chemistry. Doctoral dissertation, Université Paul Sabatier-Toulouse III (2020)
13. Yashout, M.M.S.B., AI-Humairi, S.N.S., Abdullah, M.I., Zulkipli, M.A.: Magic mat: Design a smart power generation system using piezoelectric mechanisms. In: 2021 IEEE 9th Conference on Systems, Process and Control (ICSPC 2021), pp. 103–107. IEEE (2021)
14. Senthil Kumar, K., Ruben, M.: Emerging trends in spin crossover (SCO) based functional materials and devices. Coord. Chem. Rev. **346**, 176–205 (2017)

15. Burzurí, E., et al.: Spin-state dependent conductance switching in single molecule-graphene junctions. Nanoscale. **10**, 7905–7911 (2018)
16. Villalva, J., et al.: Spin-state-dependent electrical conductivity in single-walled carbon nanotubes encapsulating spin-crossover molecules. Nat. Commun. **12**, 1578 (2021)
17. Rotaru, A., et al.: Nano-electromanipulation of spin crossover nanorods: Towards switchable nanoelectronic devices. Adv. Mater. **25**, 1745–1749 (2013)
18. Shalabaeva, V., et al.: Room temperature current modulation in large area electronic junctions of spin crossover thin films. Appl. Phys. Lett. **112**, 013301 (2018)
19. Welford, A., et al.: Influence of alkyl side-chain type and length on the thin film microstructure and OFET performance of naphthalene diimide-based organic semiconductors. Org. Electron. **75**, 105378 (2019)
20. Ma, Z., Geng, H., Wang, D., Shuai, Z.: Influence of alkyl side-chain length on the carrier mobility in organic semiconductors: Herringbone: vs. pi-pi stacking. J. Mater. Chem. C. **4**, 4546–4555 (2016)
21. Komatsu, Y., et al.: Spin-crossover behaviors based on intermolecular interactions for Cobalt(II) complexes with long Alkyl chains. Eur. J. Inorg. Chem. **2012**, 2769–2775 (2012)

Virtual Reality as a Methodological Framework for Daylight Visual Comfort Assessment

Athira Azmi(✉), Maszura Abd Ghafar, Sarah Abdulkareem Salih, and Annisa Ummihusna

Department of Architecture, Faculty of Design and Architecture, Universiti Putra Malaysia, Serdang, Malaysia
athira.azmi@upm.edu.my

Abstract. Virtual reality (VR) has emerged as a transformative tool for evaluating daylight visual comfort in buildings, offering immersive and interactive simulations that extend beyond the capabilities of traditional assessment methods. This paper examines the methodological applications of VR in indoor daylight visual comfort research, with a focus on experimental design, immersive virtual environment (IVE) settings, data collection strategies, and measurement tools. Drawing on recent studies, we identify key trends in the use of VR in daylight visual comfort evaluations in buildings. Our analysis highlights VR's ability to replicate real-world daylight scenarios and elicit authentic user feedback such as their emotional, cognitive, and physiological responses, supporting its use as a pre-occupancy evaluation tool in design processes. However, the findings reveal persistent challenges, including reliance on limited scope of participants and building contexts simulated in VR. To address these gaps, we propose a framework for future VR-based experiments that integrates user diversity, sense of presence validation, and multi-modal response measurement. This framework not only advances methodological rigor in VR research but also enhances its potential for informing sustainable and user-centered building design. By positioning VR as both a scientific and practical tool, this paper contributes to the growing discourse on digital innovation in architecture and environmental psychology, paving the way for more resilient, energy-efficient, and occupant-focused indoor environments.

Keywords: Virtual Reality (VR) · Daylight Visual Comfort · Immersive Virtual Environment (IVE) · Pre-Occupancy Evaluation · Indoor Environmental Quality (IEQ)

1 Introduction

Indoor environmental quality (IEQ) is a central concern in architectural design, with daylight visual comfort playing a critical role in well-being, spatial perception, productivity, and energy efficiency [1]. Assessing occupant comfort in the early design phase is a process known as Pre-Occupancy Evaluation (POE), is essential for creating resilient, user-centered buildings. Traditional POE methods often fall short, relying on 2D visualizations or physical mock-ups. This underscores the need for immersive,

M. A. b. Jamlos et al. (Eds.): ICSET 2025, LNNS 1897, pp. 231–238, 2026.
https://doi.org/10.1007/978-3-032-20722-7_30

human-centered approaches to assess daylight comfort in the early design stage. VR offers such potential, enabling users to experience and evaluate design alternatives in IVEs. By replicating daylight conditions and capturing emotional, cognitive, and physiological responses, VR provides richer insights and supports pre-occupancy evaluation, making it increasingly valuable in architectural research and practice.

However, the application of VR in daylight research is currently constrained by methodological inconsistencies that limit experimental rigor and generalizability. This paper reviews existing VR methodologies, identifies emerging trends and challenges, and proposes a framework that emphasizes user diversity, presence validation, and multimodal response measurement. By positioning VR not merely as a visualization tool but as a formal methodological foundation, this study ultimately contributes to advancing daylight research by offering more rigorous, user-centered approaches essential for sustainable building design.

2 Methodology

This paper adopted a systematic literature review (SLR) using the Preferred Reporting Items for Systematic Reviews and Meta-Analyses (PRISMA) guidelines. This guideline divides the process into four main steps: identification, screening, eligibility, and data extraction and analysis [2]. Figure 1 depicts the SLR procedure conducted for this paper. During screening, only English-language, peer-reviewed research articles published between 2019 and 2024 were considered, focusing on VR applications for daylight visual comfort or lighting-related assessments, while reviews, conference papers, and book chapters were excluded. In the eligibility phase, studies were retained only if they clearly described the indoor and virtual environments, specified the VR equipment used, and reported user data collected after the VR experience. This process resulted in 17 articles selected for data extraction and analysis.

3 Results

The SLR findings are illustrated in Table 1. Most studies employed within-subjects designs and purposive sampling, though reliance on student participants highlights the need for broader user diversity to improve generalizability. The existing body of VR-daylight research suffers from a critical lack of methodological rigor stemming from three interconnected flaws. First, it exhibits significant sampling homogeneity, with 71% of studies relying exclusively on student populations, thus severely restricting the ecological validity and generalizability of findings to diverse building occupants. Second, 11 out of 17 studies exhibit a methodological reliance on subjective measures, depending exclusively on biased questionnaires. This is compounded by the core failure of Presence Validation: despite immersion being fundamental to VR's validity, only six studies reported using standardized scales (e.g., IPQ, SUS). This results in a pervasive issue of Presence Ambiguity, transforming the core assumption that the IVE is a valid representation of reality into an untested hypothesis, thereby restricting the rigor and reliability of the current body of literature

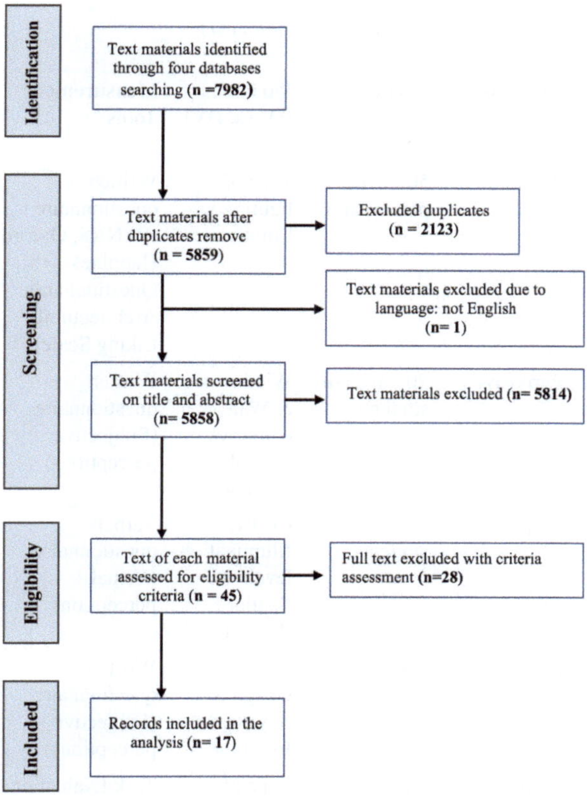

Fig. 1. Flowchart of the SLR procedure using the PRISMA method.

Table 1. Methodological Overview

ID	Experiment Design Type	Sampling Method	Participants	Variables (IV vs. DV)	Measurement Tools	Sense of Presence Measurement
[4]	Within Subjects	Purposive	40 university students	WWR & Window orientation vs. Cognition & Energy consumption	Task Evaluation (Attention & Memory)	IPQ
[5]		Not mentioned	20 university students	Time of day vs. Lighting Preference	Written questionnaire (System Usability)	–

(*continued*)

Table 1. (*continued*)

ID	Experiment Design Type	Sampling Method	Participants	Variables (IV vs. DV)	Measurement Tools	Sense of Presence Measurement
[6]		Random	50 public participants	Type of lighting vs. Emotion	Written questionnaire (PANAS, Oxford Happiness Questionnaire, Architectural Liking Scale)	–
[7]		Purposive	20 university students	Window size & Wall colour vs. Spatial Perception	Written questionnaire (Subjective perceptions)	–
[8]		Purposive	36 university students	CCT & Illumination levels vs. Spatial Perception	Verbal questionnaire (Visual perception)	IPQ
[9]		Random	406 public participants	Window design & Sky type vs. Emotion	Written questionnaire (Subjective perceptions)	–
[10]		Purposive	200 university students	Type of Lighting & Wall colour vs. Cognition	Task Evaluation (Attention & Memory)	SUS
[11]		Convenience	25 university students	CCT & Illumination levels vs. Emotion & Physiology	Written questionnaire (Self-Assessment Manikin)	IPQ
[12]		Random	33 public participants	WWR, Window design, No. of window vs. Visual sensation & Lighting preference	Verbal questionnaire (Visual perception)	–
[13]		Random	61 school students	Light colour vs. Emotion & Spatial Perception	Written questionnaire (Subjective perceptions)	SUS

(*continued*)

Table 1. (*continued*)

ID	Experiment Design Type	Sampling Method	Participants	Variables (IV vs. DV)	Measurement Tools	Sense of Presence Measurement
[14]		Purposive	142 university students	Illuminance level vs. Physiology & Cognition	Task Evaluation (Memory) & Physiological data (EEG, HRV, EDA)	SUS
[1]	Between Subjects	Purposive	120 university students	Window orientation & Time of Day vs. Cognition, Lighting Preference, Productivity & Energy consumption	Verbal questionnaire (Lighting preference)	–
[15]		Purposive	57 public participants	Light colour vs. Physiology & Spatial perception	Verbal questionnaire (Visual perception) & Physiological data (EDA, BVP, HR)	–
[16]	Within subjects & Between subjects	Purposive	150 university students	Window size & Sky type vs. Emotion & Spatial perception	Written questionnaire (Subjective perceptions)	–
[17]		Purposive	23 university students	Wall colours vs. Visual sensation & Productivity	Written questionnaire (Visual sensation)	–
[18]		Purposive	41 university students	Window size & Sky type vs. Physiology & Spatial perception	Written questionnaire (Subjective perceptions) & Physiological data (HRV, EEG)	–
[19]		Random	256 public participants	Window design & Sky type vs. Emotion & Physiology	Written questionnaire (Subjective perceptions) & Physiological data (EDA, BVP, HR)	–

Note:

IPQ: Igroup Presence Questionnaire, SUS: Slater-Usoh-Steed Questionnaire, EEG: Electroencephalography, EDA: Electrodermal activity, BVP: blood volume pulse data, HR: heart rate

4 Methodological Framework for Future VR-based Daylight Research

This review proposes a comprehensive methodological framework, detailed in Fig. 2, structured around three essential components designed to address current rigor deficits of the reviewed literature: User Diversity, Presence Validation, and Multi-Modal Response Measurement. By moving beyond convenience sampling to include varied participant profiles, the framework significantly enhances the generalizability of findings. Crucially, the mandatory application of standardized presence scales ensures that all measured responses reliably reflect a validated sense of immersion within the IVE. Furthermore, the framework mandates the fusion of subjective feedback with objective physiological (e.g., heart rate, skin conductance) and cognitive data. This integrated approach elevates VR from a simple visualization platform to a formal methodological foundation for daylight research, thereby strengthening experimental reproducibility and supporting more reliable, user-centered building design.

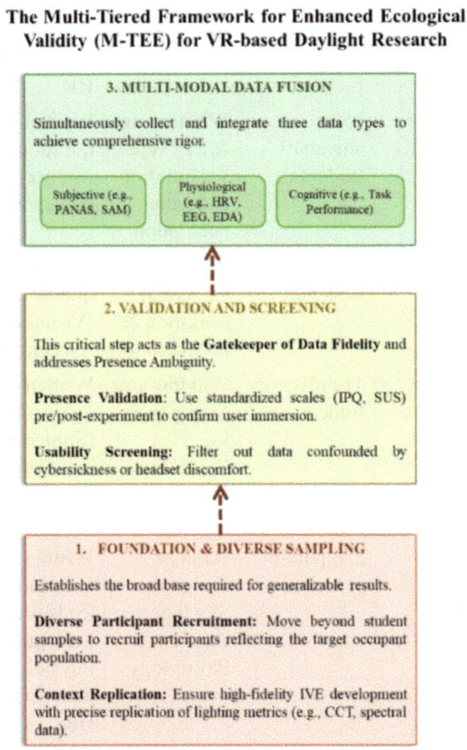

Fig. 2. The Multi-Tiered Framework for Enhanced Ecological Validity (M-TEE) for VR-based Daylight Research

5 Conclusion

This review highlighted the growing methodological importance of VR for daylight visual comfort assessment, revealing established research streams in building openings and lighting properties, primarily focused on office settings. Despite this potential, analysis of 17 studies revealed gap in VR-based research methodology, specifically concerning participant sampling and validation. To address these deficits, we propose a new framework emphasizing user diversity, presence validation, and multi-modal response measurement to guide future experiments. Crucially, the immediate next step is the empirical validation of this framework through a two-part future study, correlating multi-modal data from the controlled IVE with measurements from its real-world physical twin. By positioning VR beyond a visualization tool, this study underscores its value as a rigorous, reproducible, and practical approach for advancing daylight comfort research and informing energy-efficient, occupant-focused environments.

References

1. Carneiro, J.P., Aryal, A., Becerik-Gerber, B.: Understanding the influence of orientation, time-of-day and blind use on user's lighting choices and energy consumption using immersive virtual environments. Adv. Build. Energy Res. **15**, 603–629 (2021). https://doi.org/10.1080/17512549.2019.1639075
2. Page, M.J., et al.: The PRISMA 2020 statement: An updated guideline for reporting systematic reviews. BMJ. **372**, n71 (2021). https://doi.org/10.1136/bmj.n71
3. Latini, A., Di Loreto, S., Di Giuseppe, E., D'Orazio, M., Di Perna, C.: Crossed effect of acoustics on thermal comfort and productivity in workplaces: a case study in virtual reality. J. Archit. Eng. **29**(2) (2023). https://doi.org/10.1061/JAEIED.AEENG-1533
4. Yeom, S., Kim, H., Hong, T., Lee, M.: Determining the optimal window size of office buildings considering the workers' task performance and the building's energy consumption. Build. Environ. **177**, 106872 (2020). https://doi.org/10.1016/j.buildenv.2020.106872
5. Hegazy, M., Yasufuku, K., Abe, H.: Evaluating and visualizing perceptual impressions of daylighting in immersive virtual environments. J. Asian Archit. Build. Eng. **20**(3), 1–17 (2020). https://doi.org/10.1080/13467581.2020.1800477
6. Kayhan, A.M., Şahin, A., Erkan, İ.: The effect of types of light on people's mood using a church as an example in the virtual reality. Ment. Health Relig. Cult. **24**(5), 504–518 (2021). https://doi.org/10.1080/13674676.2020.1850665
7. Moscoso, C., Nazari, M., Matusiak, B.S.: Stereoscopic images and virtual reality techniques in daylighting research: A method-comparison study. Build. Environ. **214**, 108962 (2022). https://doi.org/10.1016/j.buildenv.2022.108962
8. Ma, J.H., Lee, J.K., Cha, S.H.: Effects of lighting CCT and illuminance on visual perception and task performance in immersive virtual environments. Build. Environ. **209**, 108678 (2022). https://doi.org/10.1016/j.buildenv.2021.108678
9. Chamilothori, K., Wienold, J., Moscoso, C., Matusiak, B., Andersen, M.: Regional differences in the perception of daylit scenes across europe using virtual reality. Part II: effects of façade and daylight pattern geometry. LEUKOS— J. Illum. Eng. Soc. North Am. **18**(3), 316–340 (2022). https://doi.org/10.1080/15502724.2021.1999257
10. Nolé Fajardo, M.L., Higuera-Trujillo, J.L., Llinares, C.: Lighting, colour and geometry: Which has the greatest influence on students' cognitive processes? Front. Archit. Res. **12**(4), 575–586 (2023). https://doi.org/10.1016/j.foar.2023.02.003

11. Mostafavi, A., Cruz-Garza, J., Kalantari, S.: Enhancing lighting design through the investigation of illuminance and correlated color Temperature's effects on brain activity: An EEG-VR approach. J. Build. Eng. **75**, 106776 (2023). https://doi.org/10.1016/j.jobe.2023.106776
12. Mirdamadi, M.S., Zomorodian, Z.S., Tahsildoost, M.: Evaluation of occupants' visual perception in day lit scenes: A virtual reality experiment. J. Daylighting. **10**, 45–59 (2023). https://doi.org/10.15627/jd.2023.4
13. Gómez Sirvent, J.L., Fernández-Sotos, D., Sánchez-Reolid, R., de la Rosa López, F., Fernández-Sotos, A., Fernández-Caballero, A.: Pre-occupancy evaluation of a virtual music school classroom: Influence of color and type of lighting on music performers. Build. Environ. **246**(3), 110989 (2023). https://doi.org/10.1016/j.buildenv.2023.110989
14. Castilla, N., Higuera-Trujillo, J.L., Llinares, C.: The effects of illuminance on students′ memory. A neuroarchitecture study. Build. Environ. **228**, 109833 (2023). https://doi.org/10.1016/j.buildenv.2022.109833
15. Chinazzo, G., Chamilothori, K., Wienold, J., Andersen, M.: Temperature–color interaction: subjective indoor environmental perception and physiological responses in virtual reality. Hum. Factors. **63**(3), 474–502 (2020). https://doi.org/10.1177/0018720819892383
16. Moscoso, C., Chamilothori, K., Wienold, J., Andersen, M., Matusiak, B.: Window size effects on subjective impressions of daylit spaces: indoor studies at high latitudes using virtual reality. LEUKOS—J. Illum. Eng. Soc. North Am. **17**, 242–264 (2021). https://doi.org/10.1080/15502724.2020.1726183
17. Latini, A., Di Giuseppe, E., D'Orazio, M., Di Perna, C.: Exploring the use of immersive virtual reality to assess occupants' productivity and comfort in workplaces: An experimental study on the role of walls colour. Energ. Buildings. **253**, 111508 (2021). https://doi.org/10.1016/j.enbuild.2021.111508
18. Kong, Z., et al.: Subjective and physiological responses towards interior natural lightscape: Influences of aperture design, window size and sky condition. Buildings. **12**(10), 1612 (2022). https://doi.org/10.3390/buildings12101612
19. Chamilothori, K., Wienold, J., Moscoso, C., Matusiak, B., Andersen, M.: Subjective and physiological responses towards daylit spaces with contemporary façade patterns in virtual reality: Influence of sky type, space function, and latitude. J. Environ. Psychol. **82**, 101839 (2022). https://doi.org/10.1016/j.jenvp.2022.101839

Laboratory Evaluation of Microbial Biofilm-Induced Carbonate Precipitation for Rock Integrity and CO_2 Storage

Farah Adilla Shahidan[1(✉)], Nur Asyraf Md Akhir[1], Fairolniza Mohd Shariff[2], Wan Mohamed Nazeem Wan Mohamed Den[2], and Afif Izwan Abd Hamid[1,3]

[1] Petroleum Engineering Department, Universiti Teknologi PETRONAS, Bandar Seri Iskandar 32610, Perak, Malaysia
farah.shahidan@utp.edu.my

[2] Enzyme and Microbial Technology Research Centre (EMTech), Faculty of Biotechnology and Biomolecular Sciences, Universiti Putra Malaysia, Serdang 43400, Selangor, Malaysia

[3] Decarbonisation and Renewable Energy Research Centre, TNB Research Sendirian Berhad, Kajang 43000, Selangor, Malaysia

Abstract. Microbially induced carbonate precipitation (MICP) is a sustainable biotechnology that leverages microbial metabolism to precipitate calcium carbonate minerals. Traditionally used in bio-cementation and soil stabilization, MICP enhances strength, reduces permeability, and offers an eco-friendly alternative to chemical grouting. Recent advances, particularly biofilm development, have broadened its potential for CO_2 sequestration and subsurface stability. This study investigates the growth and precipitation capacity of *Sporosarcina pasteurii* (ATCC 11859™) and *Bacillus* sp. Strain 42. Urease assays confirmed strong enzymatic activity in both strains. Growth analysis showed *S. pasteurii* achieved rapid proliferation with peak OD of 1.42 at 16 h, while *Bacillus* sp. displayed delayed growth, reaching OD 1.07 at 24 h. Pre-biomineralization analysis showed *Sporosarcina Pasteurii* have strong ability to produce $CaCO_3$ through ureolysis, with the PR of 10.88%. Overall, *S. pasteurii* demonstrated greater potential for rapid $CaCO_3$ deposition, supporting MICP's dual role in formation strengthening and CO_2 storage.

Keywords: Microbial Carbonate Precipitation · Biofilm · Biocementation · CO_2 Storage · Geomechanical Integrity

1 Introduction

Microbially induced carbonate precipitation (MICP) is increasingly recognized as a sustainable biotechnology that leverages microbial metabolism to induce calcium carbonate ($CaCO_3$) formation, offering advantages over conventional soil improvement and chemical grouting methods [1, 2]. In geotechnical engineering, MICP enhances soil strength, controls permeability, and improves geomechanical properties, providing an eco-friendly pathway for soil stabilization and ground improvement.

M. A. b. Jamlos et al. (Eds.): ICSET 2025, LNNS 1897, pp. 239–247, 2026.
https://doi.org/10.1007/978-3-032-20722-7_31

Beyond geotechnical applications, MICP has gained attention for subsurface CO_2 sequestration. Geological storage remains a key strategy for mitigating greenhouse gas emissions, but challenges of long-term containment and rock integrity persist. Engineered microbial biofilms present a novel solution, serving as abundant nucleation sites for carbonate mineralization and improving sealing performance in porous rocks [3, 4].

At its core, MICP relies on the metabolic activity of microorganisms, predominantly ureolytic bacteria, to create an alkaline environment conducive to calcium carbonate precipitation. The most common pathway involves urea hydrolysis, catalyzed by the enzyme urease, which is produced by bacteria such as Sporosarcina pasteurii. This enzymatic reaction breaks down urea into ammonium (NH_4^+) and carbonate (CO_3^{2-}) ions [5]. The chemical reactions involved in urea hydrolysis, carbonate formation, and calcium carbonate precipitation are presented in Eqs. (1), (2) and (3), respectively:

$$CO(NH_2)_2 + 2H_2O \rightarrow 2NH_3 + H_2CO_3\ 2NH_3 + 2H_2O \rightleftharpoons 2NH_4^+ + 2OH^- \quad (1)$$

$$H_2CO_3 \rightleftharpoons HCO_3^- + H^+\ HCO_3^- + H^+ \rightleftharpoons CO_3^{2-} + 2H^+ \quad (2)$$

$$Ca^{2+} + CO_3^{2-} \rightarrow CaCO_3(s) \quad (3)$$

The increase in pH due to the production of hydroxide ions (OH^-) and the generation of carbonate ions (CO_3^{2-}) from carbonic acid (H_2CO_3) shifts the chemical equilibrium, leading to supersaturation of calcium carbonate in the presence of calcium ions (Ca^{2+}). The bacterial cell surfaces, often negatively charged, act as nucleation sites, attracting Ca^{2+} ions and facilitating the heterogeneous precipitation of $CaCO_3$ crystals [6]. This biomineralization process effectively binds soil particles, fills pores, and forms a solid matrix, thereby enhancing the mechanical properties of the treated material.

Recent studies highlight MICP's versatility in contaminated soil remediation, monument restoration, self-healing concrete, erosion control, and dust suppression [7–10]. It can also immobilize heavy metals and radionuclides through co-precipitation [9]. Biofilm-mediated MICP further improves carbonate nucleation, as extracellular polymeric substances (EPS) trap Ca^{2+} and CO_3^{2-}, yielding denser precipitation and enhancing long-term geomechanical stability of CO_2 reservoirs [11, 12].

Despite these advances, gaps remain in understanding how environmental conditions influence bacterial activity, the efficiency of biofilms in different rock types, and the durability of treated formations. This study therefore aims to evaluate bacterial growth under varying subsurface-relevant conditions and investigate the sealing efficacy of biofilms in carbonate, sandstone, and shale, advancing MICP applications for geotechnical engineering and CO_2 storage.

2 Materials and Methods

2.1 Materials

Core specimens of Indiana Limestone (carbonate), Berea Sandstone, and shale were sourced from Polygon Synergy Ventures Sdn. Bhd., Malaysia. These lithologies, selected for their distinct mineralogy and structure, allowed comparative evaluation of microbial colonization, carbonate precipitation, and sealing efficiency. Two ureolytic strains, *S. pasteurii* (ATCC 11859™) and *Bacillus* sp. Strain 42, were obtained from EMTech lab culture collection. Bacteria were cultivated in GM-A medium containing yeast extract, urea, ammonium sulfate, and sodium acetate, with concentrations detailed in Table 1.

Table 1. Nutrients composition for to bacteria cultivation.

Culture medium	Ingredients				Initial pH
GM-A (deionized water)	Yeast extract	Urea	Ammonium sulphate	Sodium acetate	7.00
Concentration (g/L)	15	40	10	4.1	

2.2 Methods

A urease active strain of *S. pasteurii* (ATCC 11859™) was cultivated in Luria Broth at 37 °C for 24 h, subcultured on tryptone soy broth with 4% urea until pure colonies were obtained, and maintained at 4 °C for up to one month following Omoregie et al. [8]. The growth medium (GM A; Table 1) was prepared in 1 L Schott bottles with deionized water, sterilized by autoclaving, and supplemented with filter sterilized urea to prevent thermal decomposition. The pH was adjusted to 7.0 before inoculation, where 12.5 mL of overnight culture was transferred into 250 mL flasks containing 112.5 mL medium and incubated at 32 °C for 72 h with shaking at 130 rpm [13, 14].

Urease activity was verified using a phenol red broth containing yeast extract, phosphates, and urea, which was filtered sterilized, stored at 4–8 °C, and exhibited a color change upon positive activity [15]. Biocement treatment was carried out on sand columns via the surface percolation method. After 92 h of treatment, the columns were cured for 3 weeks at room temperature before analyzing their respective surface strength, $CaCO_3$ content, and pH of effluents.

3 Results and Discussion

3.1 Detection of Urease Activity

A qualitative urease assay was conducted to assess the ureolytic activity of *S. pasteurii* (Sp) and *Bacillus* sp. Strain 42 (Bs) using urea broth with phenol red as a pH indicator. Urease hydrolyzes urea to ammonia and CO_2, raising pH and shifting the broth color from yellow orange to fuchsia. The results are presented in Table 2.

Table 2. Identification of the urea presence.

Bacteria	Sample Name	Urease Test
S. pasteurii	Sp_1	Positive
	Sp_2	Positive
	Sp_3	Positive
Bacillus sp. Strain 42	Bs_1	Positive
	Bs_2	Positive
	Bs_3	Positive

As shown in Table 2, all replicates of Sp (Sp_1–3) and Bs (Bs_1–3) consistently exhibited positive urease activity, with distinct color transitions confirming urea hydrolysis. The positive control (+ve) produced the same fuchsia change, validating the assay, while the negative control (−ve) retained its yellow orange color. These results confirm that both strains possess strong ureolytic activity, highlighting their suitability for microbial induced carbonate precipitation and soil stabilization applications [15].

3.2 Growth Performance of the Bacteria

Growth profiles (Fig. 1) showed distinct dynamics between *S. pasteurii* (Sp) and *Bacillus* sp. Strain 42 (Bs). *S. pasteurii* exhibited rapid growth with a short lag phase, reaching peak OD 1.42 at 16 h before gradually declining due to nutrient depletion and by product accumulation. This strong adaptation reflects its ability to thrive in nutrient rich media with calcium support for biomineralization [16, 17]. In contrast, Bs showed delayed and lower growth, consistent with earlier findings on growth limitations of related strains [18].

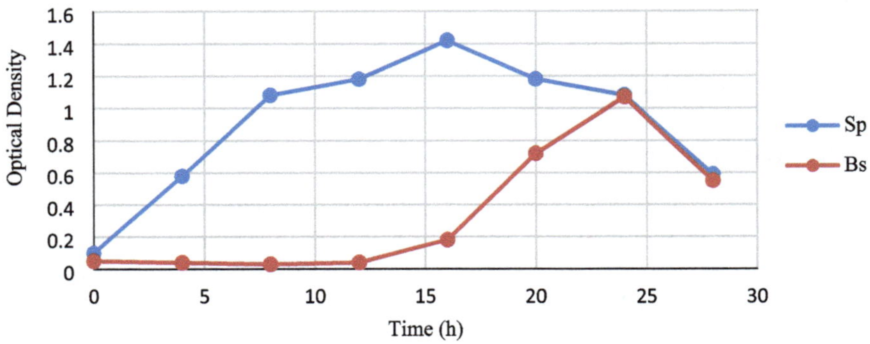

Fig. 1. Growth rate of the bacteria.

3.3 Analysis of pH Evolution in Bacteria Culture

Based on the provided line graph (Fig. 2), it shows that *Sporosarcina pasteurii* demonstrates a significantly higher and more rapid increase in pH compared to the Bacillus strain. Initially, at the one-hour mark, the Sp culture already showed a higher pH of approximately 7.0, while the Bs culture was at about 6.7. Although both cultures experienced a slight dip in pH at the two-hour incubation period, the pH for Sp began a steady and steep climb, reaching a final pH of approximately 8.0 by the end of the nine hours. This difference in performance highlights the superior alkalizing capability of *Sporosarcina pasteurii*, which is consistent with its well-documented ability to perform efficient ureolysis, a metabolic process that produces ammonia and raises the pH of its environment [19, 20]. The generated ammonia, along with carbon dioxide from urea hydrolysis, leads to an increase in pH and provides carbonate ions, which are crucial for subsequent calcium carbonate precipitation in applications like biocementation. The rapid increase in pH by *S. pasteurii* is a direct consequence of its urease activity, which quickly converts urea into ammonium, causing the culture medium's pH to rise significantly [21]. In contrast, the Bs strain showed a much more modest and fluctuating pH trend, dipping to a low of 6.35 before gradually rising to a peak of around 6.95 at the sixth hour and finishing at approximately 7.1. While some *Bacillus* species can exhibit pH changes depending on their metabolic activities and environmental conditions, their capacity for alkalization, particularly through ureolysis, is generally not as pronounced or rapid as that of *S. pasteurii* [22]. This result underscores why *Sporosarcina pasteurii* is a key organism in applications where a rapid and substantial increase in environmental pH is required.

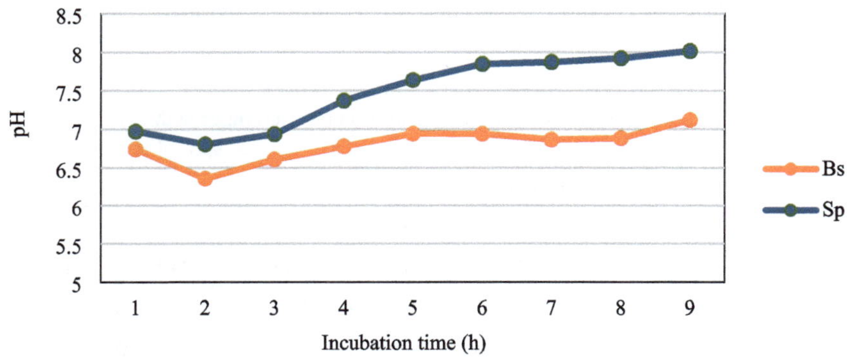

Fig. 2. Growth rate of the bacteria.

3.4 Optimization of Bacteria Growth Analysis

Tables 3, 4 and 5 present the growth effect of *Sporosarcina pasteurii* and *Bacillus sp.* Strain 42 at varying temperature, pH, and salt (NaCl) concentration, respectively. Table 3 shows that *S. pasteurii* grew best at moderate temperatures (40–45 °C), confirming it is mesophilic and not able to survive above 50 °C. This finding is in agreement with Lapierre et al. [23]. Based on Tables 4 and 5, *S. pasteurii* also thrived in alkaline conditions up to pH 10 and tolerated up to 5% NaCl, showing it prefers moderate salt levels and neutral to high pH. This matches its known role in ureolysis and calcite precipitation under alkaline environments. *Bacillus sp.* Strain 42, on the other hand, tolerated higher temperatures (40–60 °C), showing thermophilic traits useful for high-temperature applications. It also grew in a wider pH range (8–11) and handled up to 5% NaCl but not higher concentrations. These results suggest *Bacillus sp.* Strain 42 has stronger temperature and pH tolerance than *S. pasteurii*, making it potentially useful for biotechnological processes in more extreme conditions.

Table 3. Effect of temperature on bacteria growth.

Bacteria	Temperature (°C)				
	40	45	50	55	60
S. pasteurii	Positive	Positive	Negative	Negative	Negative
Bacillus sp. Strain 42	Positive	Positive	Positive	Positive	Positive

Table 4. Effect of pH on bacteria growth.

Bacteria	pH				
	8	9	10	11	12
S. pasteurii	Positive	Positive	Positive	Negative	Negative
Bacillus sp. Strain 42	Positive	Positive	Positive	Positive	Negative

Table 5. Effect of NaCl concentration on bacteria growth.

Bacteria	NaCl Concentrations (%)					
	5		10		15	
	24 h	48 h	24 h	48 h	24 h	48 h
S. pasteurii	Positive	Positive	Negative	Negative	Negative	Negative
Bacillus sp. Strain 42	Positive	Positive	Negative	Negative	Negative	Negative

3.5 Pre-Biomineralization Test

Results in Table 6 show the MICP efficiency of *Sporosarcina pasteurii* and *Bacillus sp.* Strain 42. Both showed strong ability to produce $CaCO_3$ through ureolysis, where urea is broken down, increasing pH and triggering $CaCO_3$ precipitation. *S. pasteurii* had an average precipitation weight of 1.4630 g and a precipitation rate (PR%) of 10.88%, while *Bacillus sp.* Strain 42 showed almost identical results (1.4620 g and 10.59%).

This similarity suggests *Bacillus sp.* Strain 42 can be as effective as *S. pasteurii* for MICP. Since it also tolerates higher temperatures and pH, it could be useful in harsher conditions where *S. pasteurii* might not thrive [24]. Overall, this highlights the potential of using different bacterial strains not just *S. pasteurii* for flexible and efficient MICP applications in various industries.

Table 6. Pre-Biomineralization Test.

Sample	Sp_1	Sp_2	Sp_3	Bs_1	Bs_2	Bs_3
Precipitation weight (g)	1.2773	1.7310	1.4509	1.6609	1.3440	1.3110
Final weight (g)	13.4163	13.8700	13.5899	13.7999	13.4830	13.4500
PR (%)	9.5	12.48	10.68	12.04	9.97	9.75

4 Conclusion

This study provides clear evidence that *Sporosarcina pasteurii* and *Bacillus sp.* Strain 42 exhibit strong ureolytic activity, enabling efficient urea hydrolysis and subsequent calcium carbonate precipitation. Comparative growth analysis revealed that *S. pasteurii* demonstrated a higher proliferation rate, more rapid pH elevation, and faster calcite formation, highlighting its suitability for applications requiring accelerated sealing and strengthening. In contrast, *Bacillus sp.* Strain 42 exhibited slower yet stable growth, indicating its potential utility in environments where gradual and sustained mineralization is advantageous, particularly under elevated temperature and alkaline conditions. The mineralization performance varied across lithologies, with carbonate cores exhibiting the most extensive $CaCO_3$ precipitation, followed by sandstone, while shale showed minimal deposition due to its low permeability. These findings underscore the dual functionality of MICP in biocementation and CO_2 sequestration, offering a sustainable and environmentally compatible approach to enhance rock integrity and subsurface sealing. Future research should focus on optimizing bacterial delivery mechanisms, improving cell survivability in subsurface environments, scaling the process to reservoir-level applications, and integrating advanced monitoring strategies to evaluate long-term durability and environmental impacts.

Acknowledgement. The authors would like to express their sincere gratitude to Universiti Teknologi PETRONAS and Universiti Putra Malaysia for providing the laboratory facilities and technical support. The financial assistance provided by Yayasan UTP under project no. 015LC0-464 is also gratefully acknowledged.

References

1. Zhang, X., Wang, H., Wang, Y., Wang, J., Cao, J., Zhang, G.: Improved methods, properties, applications and prospects of microbial induced carbonate precipitation (MICP) treated soil: A review. Biogeotechnics. **3**(1), 100123 (2024)

2. Wang, Y., Konstantinou, C., Tang, S., Chen, H.: Applications of microbial-induced carbonate precipitation: A state-of-the-art review. Biogeotechnics. **1**(1), 100008 (2023)
3. Li, F., et al.: A genetically engineered composite biofilm for microbial induced calcium carbonate precipitation by synergic effect of urease, protein adhesive and xanthan gum. J. Environ. Chem. Eng. **10**(5), 108431 (2022)
4. Mitchell, A.C., Dideriksen, K., Spangler, L.H., Cunningham, A.B., Gerlach, R.: Microbially enhanced carbon capture and storage by mineral-trapping and solubility-trapping. Environ. Sci. Technol. **44**(13), 5270–5276 (2010)
5. Sjoblom Kurt, J.: Coarse-grained molecular dynamics approach to simulating clay behavior. J. Geotech. Geoenviron. **142**(2), 06015013 (2016)
6. Krajewska, B.: Urease-aided calcium carbonate mineralization for engineering applications: A review. J. Adv. Res. **13**, 59–67 (2018)
7. Almajed, A., Lateef, M.A., Moghal, A.A., Lemboye, K.: State-of-the-art review of the applicability and challenges of microbial-induced calcite precipitation (MICP) and enzyme-induced calcite precipitation (EICP) techniques for geotechnical and geoenvironmental applications. Crystals. **11**(4), 370 (2021)
8. Omoregie, A.I., et al.: Bio-based solutions for concrete infrastructure: A review of microbial-induced carbonate precipitation in crack healing. Buildings. **15**(7), 1052 (2025)
9. Zeng, Y., et al.: Mechanism of microbiologically induced calcite precipitation for cadmium mineralization. Sci. Total Environ. **852**, 158465 (2022)
10. Liu, Y., Hu, K., Pan, M., Dong, W., Wang, X., Zhu, X.: Research and application of green technology based on microbially induced carbonate precipitation (MICP) in mining: A review. Sustainability. **17**(17), 7587 (2025)
11. Robles-Fernández, A., Areias, C., Daffonchio, D., Vahrenkamp, V.C., Sánchez-Román, M.: The role of microorganisms in the nucleation of carbonates, environmental implications and applications. Minerals. **12**(12), 1562 (2022)
12. Fang, C., Achal, V.: Enhancing carbon neutrality: A perspective on the role of microbially induced carbonate precipitation (MICP). Biogeotechnics. **2**(2), 100083 (2024)
13. Omoregie, A.I., Khoshdelnezamiha, G., Senian, N., Ong, D.E.L., Nissom, P.M.: Experimental optimisation of various cultural conditions on urease activity for isolated *Sporosarcina pasteurii* strains and evaluation of their biocement potentials. Ecol. Eng. **109**, 65–75 (2017)
14. Omoregie, S.I., Palombo, E.A., Ong, D.E.L., Nissom, P.M.: Biocementation of sand by *Sporosarcina pasteurii* strain and technical-grade cementation reagents through surface percolation treatment method. Constr. Build. Mater. **228**, 116828 (2019)
15. Urease Test Protocol, A.S.f. Microbiology. https://asm.org/getattachment/ac4fe214-106d-407c-b6c6-e3bb49ac6ffb/urease-test-protocol-3223.pdf (2016).
16. Nature: Microbial makers help humans to build tough stuff. Nature. **591**(7849), 180 (2021)
17. Lapierre, F.M., Schmid, J., Ederer, B., Ihling, N., Büchs, J., Huber, R.: Revealing nutritional requirements of MICP-relevant Sporosarcina pasteurii DSM33 for growth improvement in chemically defined and complex media. Sci. Rep. **10**(1), 22448 (2020)
18. Dong, Y., et al.: Optimization of growth conditions and biological cementation effect of Sporosarcina pasteurii. Constr. Build. Mater. **395**, 132288 (2023)
19. Lauchnor, E.G., et al.: Whole cell kinetics of ureolysis by *Sporosarcina pasteurii*. J. Appl. Microbiol. **118**(6), 1321–1330 (2015)
20. Ma, L., et al.: Beneficial factors for biomineralization by ureolytic bacterium *Sporosarcina pasteurii*. Microb. Cell Factories. **19**(1), 1–12 (2020)
21. Wu, Y., et al.: Biomineralization induced by cells of *Sporosarcina pasteurii*. Front. Microbiol. **12**, 786048 (2021)
22. Lapierre, F.M., et al.: Feeding strategies for *Sporosarcina pasteurii* cultivation for microbially induced calcium carbonate precipitation. Biotechnol. Bioeng. (2024)

23. Lapierre, F.M., et al.: Revealing nutritional requirements of MICP-relevant *Sporosarcina pasteurii* strains for enhanced biomineralization. Sci. Rep. **10**(1), 1–13 (2020)
24. DeJong, J.T., et al.: Biogeochemical processes and geotechnical applications of microbially induced calcite precipitation. Environ. Sci. Technol. **44**(15), 5732–5740 (2010)

Performances of SIR, SEIR and SEIRS Models in Predicting Contagious Disease: Case Study for Covid-19 Cases in Malaysia

Noor Amalina Nisa Ariffin[1(✉)], Nabihah Azmi Shah[2], Ainnill Munirah Kahar[2], and Nurin Athirah Muslihat[2]

[1] Faculty of Computer and Mathematical Sciences, Universiti Teknologi MARA Pahang, 26400 Bandar Tun Abdul Razak Jengka, Pahang, Malaysia
amalinanisa@uitm.edu.my

[2] Faculty of Computer and Mathematical Sciences, Universiti Teknologi MARA Seremban, 70300 Seremban, Negeri Sembilan, Malaysia

Abstract. Coronavirus disease (COVID-19) is an infectious disease caused by the SARS-CoV-2 virus. Anyone can be infected by the disease and become severely ill or face death at any age. The best way to prevent and slow down transmission is to be well informed about the disease and how the virus spreads. What makes the situation complicated is the ambiguity about the duration and ultimate spread of the pandemic. It is critical for the governments, healthcare systems, and economic sectors to have an estimate future case of this pandemic. This study simulated the infectious count for the same duration to assess the predictive capability of the Susceptible-Infectious-Recovered (SIR) model, Susceptible-Exposed-Infected-Recovered (SEIR) model and Susceptible-Exposed-Infected-Recovered-Susceptible (SEIRS) model in predicting the pattern of COVID-19 cases in Malaysia. Different mathematical approaches are used in order to evaluate the performance of each model in predicting the number of contagious diseases. This study aims to suggest the best model that can represent COVID-19 cases in Malaysia. In this study, the epidemic in Malaysia have been simulated by using those models and all models have been solved by using fourth-five order Runge-Kutta (RK45) method via MATLAB. Then, the performance of each model has been compared with the data of COVID-19 cases in Malaysia.

Keywords: COVID-19 · SIR model · SEIR model · SEIRS model · Runge-kutta method

1 Introduction

The news of an infectious disease, COVID-19, spread globally in 2019 left everyone in shock. In December 2019, the first case of Coronavirus COVID-19 was discovered in Wuhan, China (COVID-19 and Your Health, 2020). On February 4, 2020, first Malaysian was quarantined at the Sungai Buloh Hospital in Selangor. At that time, infected people experience moderate to mild respiratory illness and will recover without the need for

M. A. b. Jamlos et al. (Eds.): ICSET 2025, LNNS 1897, pp. 248–254, 2026.
https://doi.org/10.1007/978-3-032-20722-7_32

special treatment, according to the WHO. Following that, Malaysians were astounded to learn of their country's first citizen, who was confirmed by COVID-19 on February 4, 2020. The high rate of infection spread, and the significant number of deaths caused by Malaysia's COVID-19 are of great concern to researchers, governments, and all people. Mathematical based approach are the best representation of outbreaks for understanding the behaviour of an infectious disease when it enters a community. The simple SIR model is the most well-known mathematical model for describing an infectious disease. Researchers all over the world later developed new models based on this standard SIR model, such as the SEIR and SEIRS models. The SEIR model, on the other hand, divides the population into four groups based on how well they are currently doing in terms of disease. To allow recovered people to return to a susceptible state in this scenario, the SEIRS model can be applied. SEIRS describes them as susceptible (S), exposed (E), infectious (I), recovered (R), and susceptible (S). Generally, in order to see the clear picture of COVID-19 cases in Malaysia, all mathematical models above will be solved via numerical methods to provide a precise prediction and to be relevant to reality. This development of mathematical models of an infectious disease can also assist in decision-making by making projections about important issues.

2 Methodology

The spread of a virus, the number of infected people, the duration of an epidemic to occur and estimation of other epidemiological parameters such as reproduction number are all aspects that these models attempt to predict. SIR, SEIR and SEIRS models can demonstrate how different public health initiatives may influence the course of epidemics. In this section, SIR, SEIR and SEIRS models have been presented along with their equations and parameters (Figs. 1, 2 and 3).

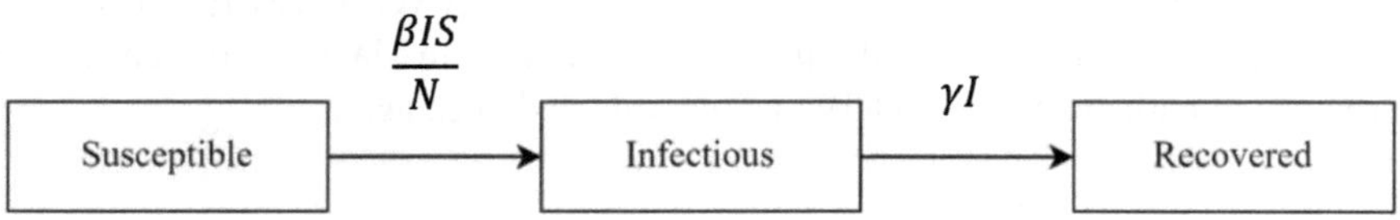

Fig. 1. Component of SIR Model

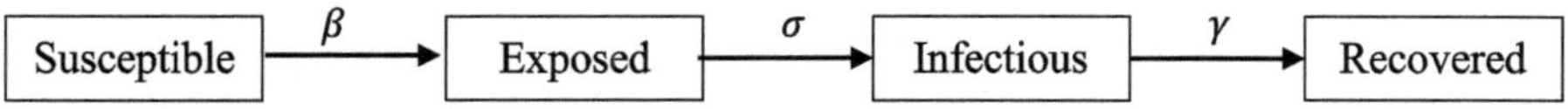

Fig. 2. Components of SEIR Model

Where β is contact-infection rate, σ is the rate at which an exposed individual becomes infective, γ is the recovery rate and ξ is the rate at which immune status is degraded and recovered persons become susceptible. Fig. 4 below represents the proposed method for developing the performances of SIR, SEIR and SEIRS models in predicting the COVID-19 cases in Malaysia (Figs. 5, 6, 7, 8, 9 and 10).

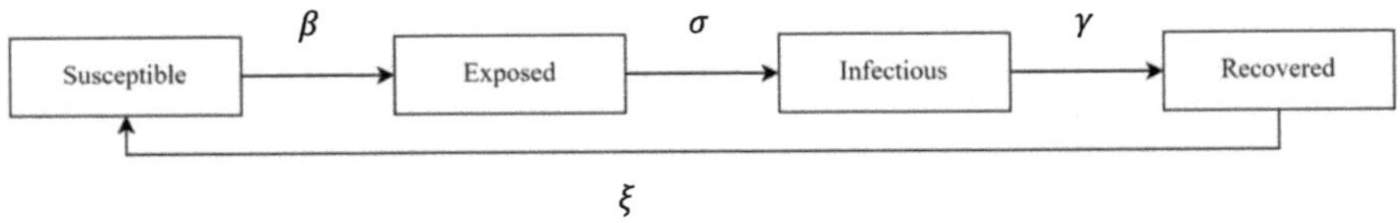

Fig. 3. Components of SEIRS Model

SIR, SEIR and SEIRS models will be solved by using Runge-kutta 45 (RK45) method via MATLAB software and the solutions obtained will be compared to the actual data.

3 Results and Discussion

Considering the comparison of solution between each model with real data and the performances of each model in predicting COVID-19 cases in Malaysia, SEIRS model is the most suitable model to be used in predicting COVID-19 cases in Malaysia. As had been previously stated, SEIRS model is taking reinfected or susceptible again into consideration for contagious disease. By observing the situation of COVID-19 cases in Malaysia in Year 2025, the probability for those who have recovered to be reinfected is high. This is due to the fact that COVID-19 virus is a new discovery where the virus is still under the observation of scientists around the world especially in finding the cure to this disease. Furthermore, the National COVID-19 Immunisation Programme (*Program Imunisasi COVID-19 Kebangsaan*) is begin implemented by Malaysian government in early 2021 by phase. Hence, the immunity rate of Malaysians towards COVID-19 virus was low at that point in time. A full COVID-19 vaccination rate of 49.9% was achieved in Malaysia as of 12th January 2023. However, there is no confirmation regarding whoever that had been infected with COVID-19 previously will not get infected once again even after being vaccinated. These supporting evidence certainly verified SEIRS model to be the most appropriate model in predicting COVID-19 cases in Malaysia to such an extent that recovered patients of COVID-19 may get infected once again.

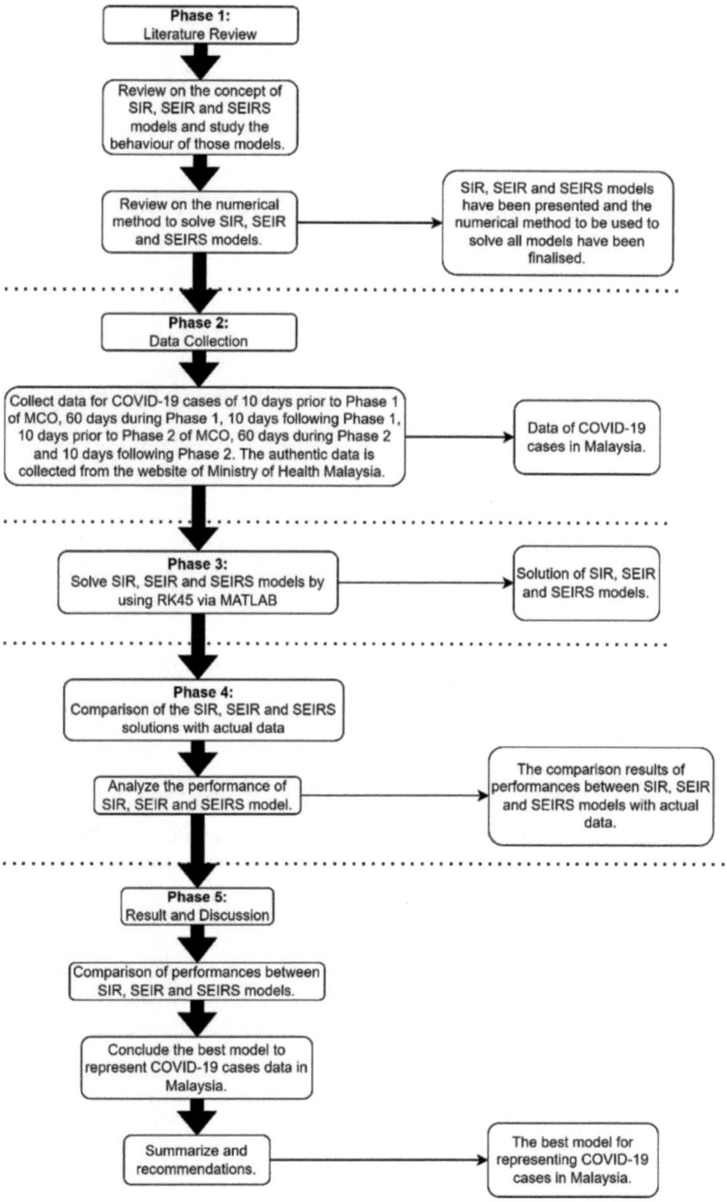

Fig. 4. Flowchart of Research Methodology

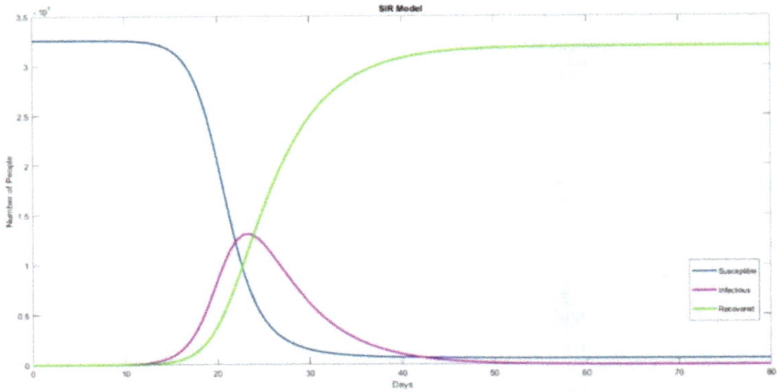

Fig. 5. SIR model for Malaysia's Covid-19 cases in Year 2025

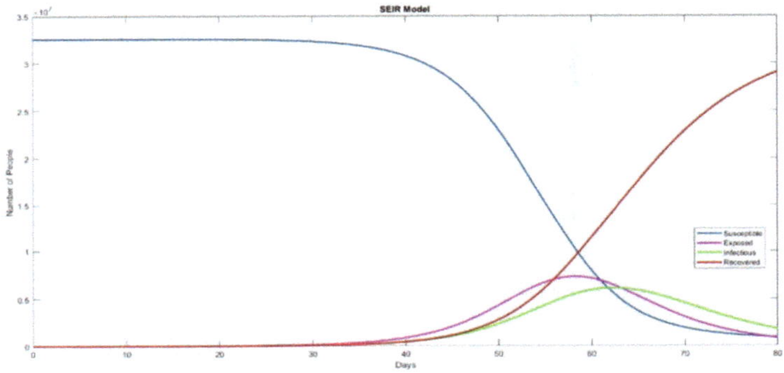

Fig. 6. SEIR model for Malaysia's Covid-19 cases in Year 2025

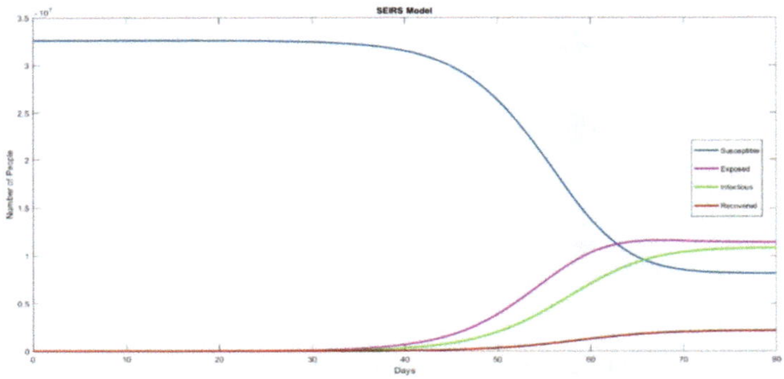

Fig. 7. SEIRS model for Malaysia's Covid-19 cases in Year 2025

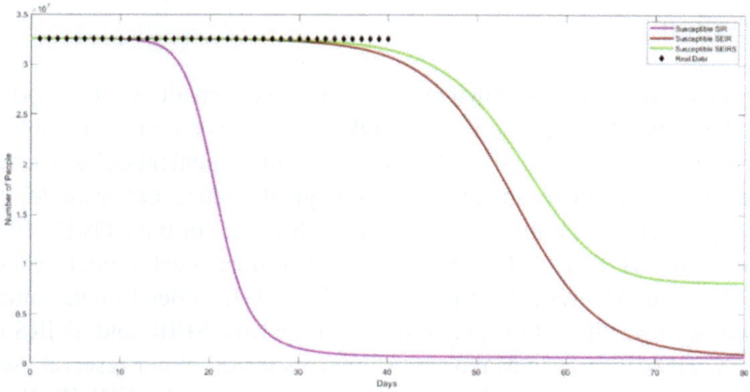

Fig. 8. The Comparison of Susceptible Level Between SIR, SEIR and SEIRS Models in Year 2025

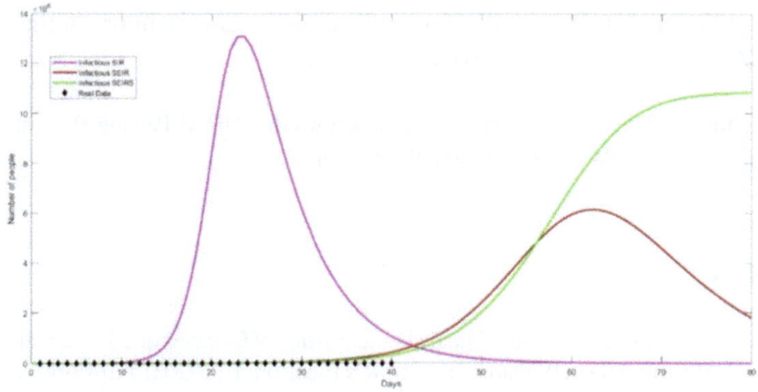

Fig. 9. The Comparison of Infectious Level Between SIR, SEIR and SEIRS Models in Year 2025

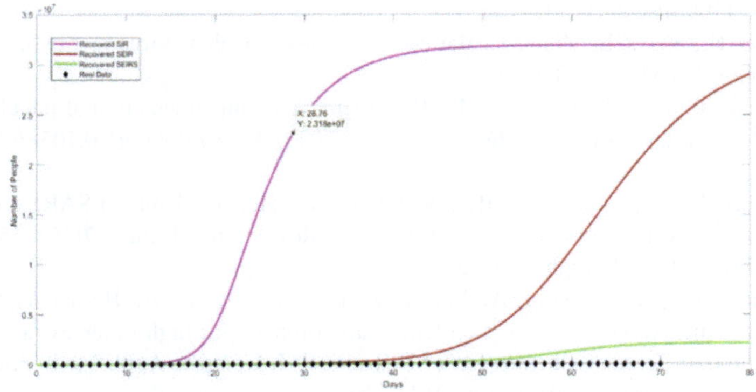

Fig. 10. SEIRS model for Malaysia's Covid-19 cases in Year 2025

4 Conclusion

This unexpected pandemic has drawn many researchers from all over the world to investigate how the COVID-19 pattern works. COVID-19 cases are unpredictable, with ups and downs despite numerous studies. The SIR epidemiological model is one of the most ancient and has far-reaching implications in biological science. Other models, however, can be used to observe the pattern for contagious disease. For the COVID-19 outbreak, we used three models: SIR, SEIR, and SEIRS. To compare each model, real data from COVID-19 cases in Malaysia for the year 2025 used. In order to determine the best model to use, we investigated the performance of the SIR, SEIR, and SEIRS models in predicting COVID-19 cases in Malaysia. At the conclusion of our research, we demonstrated that the SEIRS model is the best model to represent the COVID-19 pattern in Malaysia because its pattern is the closest to the real data. Furthermore, the situation of COVID-19 cases in Malaysia fit the SEIRS model's assumptions the best. This research work may be valuable for the Malaysian government in making decisions such as Movement Control Order (MCO) and also for medical organizations in predicting the future impact of COVID-19 virus on Malaysia or the world.

Acknowledgement. The authors would like to acknowledge UiTM Pahang Branch for support and encouragement towards the completion of this projects.

References

Chen, N., et al.: Epidemiological and clinical characteristics of 99 cases of 2019-novel coronavirus (2019-nCoV) Pneumonia in Wuhan, China. SSRN Electron. J. (2020a). https://doi.org/10.2139/ssrn.3523861

Chen, Y.C., Lu, P.E., Chang, C.S., Liu, T.H.: A time-dependent SIR model for COVID-19 with undetectable infected persons. IEEE Trans Netw Sci Eng. **7**(4), 3279–3294 (2020b). https://doi.org/10.1109/tnse.2020.3024723

Fan, Y., Zhao, K., Shi, Z.L., Zhou, P.: Bat coronaviruses in China. Viruses. **11**(3), 210 (2019). https://doi.org/10.3390/v11030210

Giordano, G., et al.: Modelling the COVID-19 epidemic and implementation of population-wide interventions in Italy. Nat. Med. **26**(6), 855–860 (2020). https://doi.org/10.1038/s41591-020-0883-7

Godio, A., Pace, F., Vergnano, A.: SEIR modeling of the italian epidemic of SARS-CoV-2 using computational swarm intelligence. Int. J. Environ. Res. Public Health. **17**(10), 3535 (2020). https://doi.org/10.3390/ijerph17103535

Martín-Vaquero, J., Queiruga-Dios, A., Martín Del Rey, A., Encinas, A., Hernández Guillén, J., Rodríguez Sánchez, G.: Variable step length algorithms with high-order extrapolated non-standard finite difference schemes for a SEIR model. J. Comput. Appl. Math. **330**, 848–854 (2018). https://doi.org/10.1016/j.cam.2017.03.031

Mathematical model. In Wikipedia. https://en.wikipedia.org/wiki/Mathematical_model (2022, April 24)

Mohajan, H.: Mathematical analysis of sir model for COVID-19 transmission. J. Innov. Med. Res. **1**(2), 1–18 (2022) https://mpra.ub.uni-muenchen.de/114390/

Embedded Control Electrostatic Double Layer Capacitor-Battery Hybrid Energy Storage System

Noah Lee[1], Chen Hon Nee[1](✉), Seong Shan Yap[2], Guang Liang Ong[1], and Kwong Keong Tham[3]

[1] Faculty of Artificial Intelligence and Engineering, Multimedia University, Cyberjaya, Malaysia
chnee@mmu.edu.my
[2] Department of Physics, Xiamen University Malaysia, Sepang, Selangor, Malaysia
[3] Faculty of Computer Science and Technology, First City University College, Petaling Jaya, Selangor, Malaysia

Abstract. The increasing demand for high-performance and sustainable energy storage solutions has motivated the integration of electrostatic double layer capacitors (EDLC) lithium-based batteries and lead acid batteries through advanced embedded control systems. This study presents a full active topology for hybrid energy storage, designed to synergize the fast charge–discharge dynamics of EDLCs with the high energy density of lithium batteries. The embedded control strategy enables adaptive power flow management, ensuring optimal charge distribution, load balancing, and extended lifetime of the lithium battery by mitigating peak current stresses. The embedded control hybrid energy storage system is tested to acquire the voltage, current data and perform the necessary energy redirection for energy outflow from storage system, functioned in full active topology when subject to constant load.

Keywords: Hybrid energy storage system · embedded control · electrostatic double layer capacitor · supercapacitor · lithium battery · lead acid battery

1 Introduction

An energy storage system is crucial to the power generation system and especially important to the renewable energy system [1, 2]. For example, solar or wind energy is intermittent in power generation, thus an efficient energy-storage system is required to provide uninterrupted energy supply. Currently battery is the most common energy storage device. However, in a battery only energy storage system, battery capacity is often oversized to meet load demand, thus incurred high overhead cost. In a hybrid energy storage system, another energy storage device is used with batteries to meet energy and power density required by the load [3]. One of the best candidates is the electrostatic double layer capacitor, also known as supercapacitor [4].

M. A. b. Jamlos et al. (Eds.): ICSET 2025, LNNS 1897, pp. 255–261, 2026.
https://doi.org/10.1007/978-3-032-20722-7_33

The integration of lithium batteries and supercapacitor has been proposed for electric and hybrid vehicle in an early work [5]. The test has been conducted on compact, mid-sized, full-sized, small SUV, mid-sized SUV, and delivery truck. Each of the energy storage device is set to supplied 50% of the power during peak power conditions. From the test results, it is found that the efficiency of the supercapacitors during peak power conditions is approximately 10–20% higher than the battery for all vehicles. A similar passive scheme was also used in another work [6].

It is noted that an active hybrid energy storage system would has higher efficiency under high load condition over passive one because energy is actively drawn from EDLC that has a lower internal resistance than batteries. The sizing and matching of EDLC and battery are also crucial factors, which can enhance system and cost efficiency, with lower the reliance on the battery [7–11]. In this work, an embedded control electrostatic double layer capacitor-battery hybrid energy storage system is investigated. The system consists of an electrostatic double layer capacitor (EDLC), a lithium polymer battery and a lead acid battery. The energy delivering to load and charging of each energy system is controlled to meet the load demand which maximizing the system life, safety with minimum oversizing.

2 Methodology

The specification of the EDLC, lithium polymer battery and the lead acid battery are shown in Tables 1, 2 and 3. We design the system to function as a 12 V module, which is compatible with 6 EDLC cell in series and 3 lithium polymers batteries in series. Lead acid in series serves as the backup for the system.

Table 1. EDLC Specification

Specification	Data
Voltage of 6 EDLC cell in series	$2.7\ \text{V} \times 6 = 16.2\ \text{V}$
Maximum discharge current	180 A
Cycle life	500000 cycles

The hybrid energy storage system is designed to work in full active mode where the embedded system has full control of: from which energy system is/are flowing to/from according to the load characteristics. Full active control allows us to actively control the particular energy system the energy to be drawn from. The strategy enables minimal oversizing while meeting the power and energy demand, and maximizing system operation life span. Thus, the main thread is programmed for system control and the second is used for battery parameters data acquisition concurrently, such as voltage and current measurement.

To ensure safety and best performance, the limit of each EDLC and battery are considered in the designed. The total voltage of 6 EDLCs will be 16.2 V, but since the voltage for other batteries are 12 V in the system, the EDLC will only be charged until

Table 2. Lithium ion battery

Specification	Data
Maximum voltage of 3 lithium battery cell in series	4.2 V × 3 = 12.6 V
Minimum voltage of 3 lithium battery cell in series	3.2 V × 3 = 9.6 V
Capacity	2.2 AH
Maximum discharge current	55 A
Cycle life	300–500 cycles

Table 3. Lead-acid battery

Specification	Data
Capacity	4.5 AH
Cycle life	300–500 cycles

12 V. The best and safety limit such as voltage, current limit, charging and discharging rate of the lithium polymer batteries will be obtained for the charging and discharging measurement. There is no overcharging problem for the lead acid battery hence the maximum voltage while charging will be 14 V. Excess charge energy is redirected to the lead-acid battery to prevent overvoltage (Fig. 1).

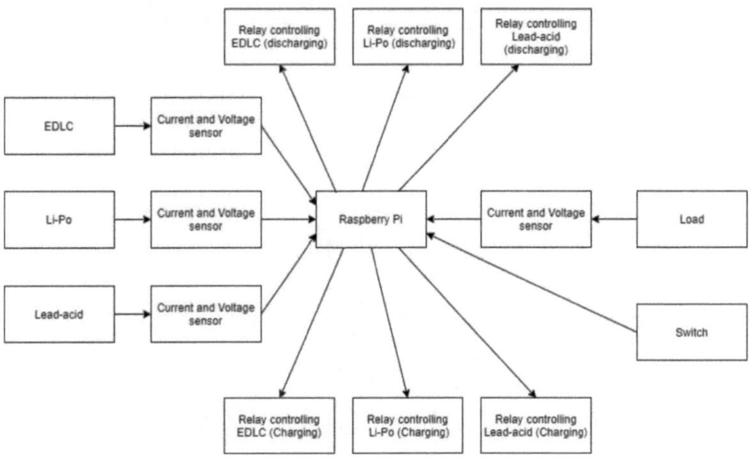

Fig. 1. Block Diagram of Embedded System.

3 Results and Discussion

The charging and discharging data of the EDLC were recorded at constant current of 0.2 A, 0.7 A and 1.5 A as shown in Fig. 2 (a) and (b). This value is used in the embedded control to determine state of charge (SOC) for EDLC. A linear relationship between voltage and time when it is charged at constant current allows an accurate SOC calculations.

The charging data of the lithium polymer battery is recorded at constant current of 1.0 A, 1.5 A and 2.2 A. The values obtained are used in the embedded control to determine state of charge for the lithium polymer battery as the life-cycle of lithium polymer is much lower than that of EDLC. A "knee" is observed at 11.75 V which translate to 3.92 V per cell. Care must be taken on the characteristics when determining SOC for LiPo cell. For discharging, the "knee" is observed at 11.5 V which translate to 3.83 V per cell. The rest of the curve are relative linear (Fig. 3).

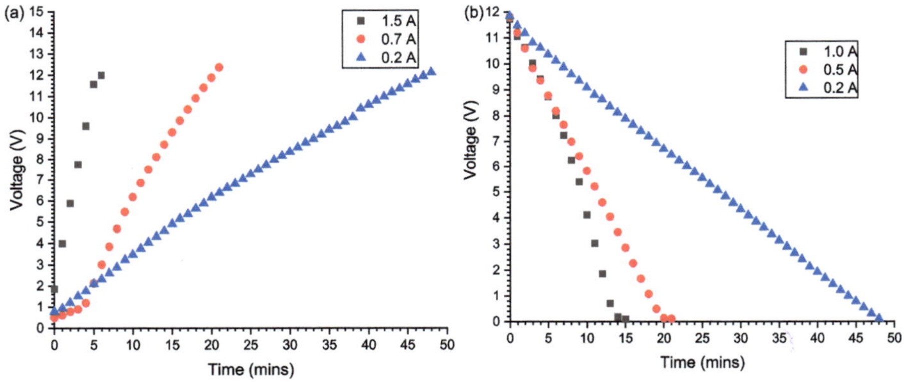

Fig. 2. EDLC (a) charging and (b) discharging at the respective constant current.

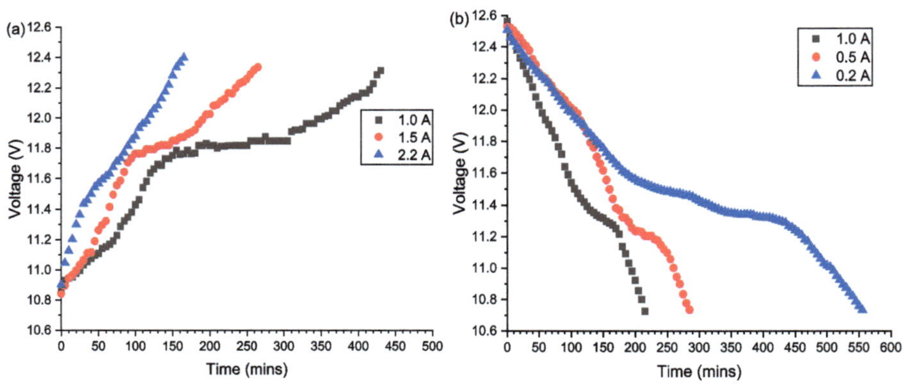

Fig. 3. Lithium polymer battery (a) charging and (b) discharging at the constant current.

The charging and discharging data of the lead acid were recorded at constant current of 1.5 A, 1.7 A and 2.0 A as shown in Fig. 4. This value is used in the embedded control to determine state of charge of the lead acid to synchronized with the other energy storage device for the system. With the SOC of each EDLC and battery determined, the embedded control hybrid energy storage system is constructed based on the strict safety of each of the energy storage device and also optimized for performance. The embedded control hybrid energy storage system is subjected to constant load testing.

Figure 5(a) shows the voltage characteristics when a 12 V, 55 W resistive load is drawing approximately 4.6 A from the embedded control hybrid energy storage system. The constant load discharging curve show the switching process between the EDLC and Li-Po when the constant load is applied. The lithium polymer battery is always within the safety limit. The system is configured to draw power from EDLC for the first 4 s, then switch the load to the lithium polymer battery while the lead acid battery function as backup to the first two system. The rationale is that EDLC has much lower internal resistance, thus be able to deliver current at a lower loss at the start, once the peak current gradually stabilized, the lithium polymer battery will graduate take over the load as EDLC energy dwindle of.

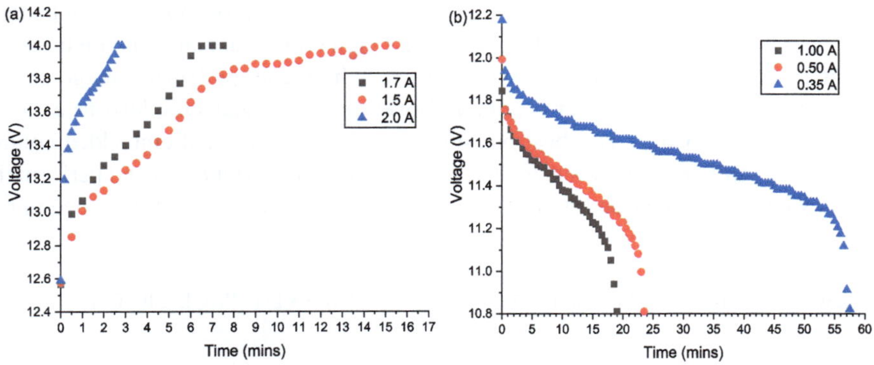

Fig. 4. Lead-acid battery (a) charging and (b) discharging characteristics at constant current.

The discharging current characteristics of the embedded control hybrid energy storage system under the same constant resistive load is shown in Fig. 5(b). EDLC is delivering above ~ 5.2 A from t = 0 s, while the current of the lithium polymer battery remained low. After t = 5 s, the embedded control redirected the current delivery from the EDLC to the lithium polymer battery, while load is drawing the same constant current. The current drawn from lead acid remain zero because the SOC of lithium polymer battery has yet to be reached for the current delivery to be redirected to lead acid battery. The working voltage range each energy-storage device is set within 10–12 V. This is to ensure that the supply is sufficient for the load. When the voltage level of any of the energy-storage device is above or below the range, the embedded control will switch the discharging or charging supply to other energy-storage devices following the predetermined order. The lead-acid will be the last device to receive or supply power due to its robustness against over-charging or over-discharging.

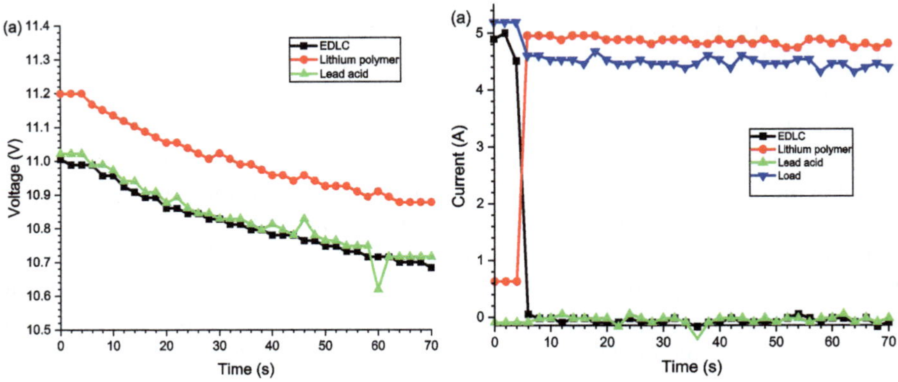

Fig. 5. The (a) voltage and (b) current discharging characteristics for the embedded control hybrid energy storage system at constant load.

4 Conclusion

An embedded control hybrid energy storage system has been constructed using 3 types of energy storage device, namely EDLC, lithium polymer battery and lead acid battery. EDLC is superior in cycle life, able to withstand high current demand and high instantaneous power delivery. Lithium polymer provides high energy density, stable energy delivery but is sensitive to over-charge and over-discharge that may lead to a safety hazard. Lead acid battery is robust but slow in charging. The final embedded control hybrid energy storage system is able to acquire the voltage, current data and perform the necessary energy redirection for energy outflow from storage system, functioned in full active topology when subject to constant load.

Acknowledgement. We acknowledge the support from TM R&D (MMUE/240106).

References

1. Tan, K.M., Babu, T.S., Ramachandaramurthy, V.K., Kasinathan, P., Solanki, S.G., Raveendran, S.K.: Empowering smart grid: A comprehensive review of energy storage technology and application with renewable energy integration. J. Energy Storage. **39**, 102591 (2021)
2. Sun, J., Lin, W., Hong, M., Loparo, K.A.: Voltage regulation of DC-microgrid with PV and battery. IEEE Trans, Smart Grid. **11**, 4662–4675 (2020)
3. Sahari, M., et al.: Investigation of hybrid energy storage system based on battery and supercapacitor. AIP Conf. Proc. **2319**(1), 070002 (2021)
4. Sharma, P.K., Arora, A., Tripathi, S.K.: Review of supercapacitors: Materials and devices. J. Energy Storage. **21**(January), 801–825 (2019)
5. Burke, H.Z., Andrew, Z.L.: Present and future applications of supercapacitors in electric and hybrid vehicles. In: 2014 IEEE International Electric Vehicle Conference (IEVC), pp. 1–8. IEEE, Florence, Italy (2014)
6. Chuan, Y., Mi, C., Zhang, M.: Comparative study of a passive hybrid energy storage system using lithium ion battery and ultracapacitor. World Electr. Veh. J. **5**(1), 83–90 (2012)

7. Sengupta, A., Pereira, T., Liserre, M.: Discharge ratio optimization of supercapacitor storage for size-critical applications. IEEE Trans. on Ind. Appl. **61**, 7825 (2025)
8. Lee, N., et al.: Capacity sizing of embedded control battery–supercapacitor hybrid energy storage system. Energies. **15**(10), 3783 (2022)
9. Choi, M.E., Kim, S.W., Seo, S.W.: Energy management optimization in a battery/supercapacitor hybrid energy storage system. IEEE Trans. Smart Grid. **3**, 463–472 (2012)
10. Zhang, P., Zheng, Y., Wang, H., Jin-Ming, W., Zhang, Z., Wen, W.: A battery-supercapacitor hybrid energy storage device that directly uses seawater or saltwater lake water. Mater. Today Adv. **24**, 100535 (2024)
11. Papageorgiou, P., Oureilidis, K., Tsakiri, A., Christoforidis, G.: A modified decentralized droop control method to eliminate battery short-term operation in a hybrid supercapacitor/battery energy storage system. Energies. **16**, 2858 (2023)

Hybrid-Based Algorithm for the Optimal Location and Sizing of Multiple DGs and DSTATCOMs

Bazilah Ismail[1,2(✉)], Noor Izzri Abdul Wahab[2], Mohammad Lutfi Othman[2], and Mohd Amran Mohd Radzi[2]

[1] School of Technology and Engineering, Peninsula College Malaysian, The Ship Campus, Bandar Cassia, Penang, Malaysia
bazilahismail87@gmail.com

[2] Advanced Lightning and Power Energy Research (ALPER), Department of Electrical and Electronics Engineering, Faculty of Engineering, Universiti Putra Malaysia, Serdang, Selangor, Malaysia
{izzri,lutfi,amranmr}@upm.edu.my

Abstract. An efficient and modernized power grid has become essential nowadays. Excessive power losses, poor voltage profiles and voltage stability remain the major challenges, particularly at distribution level. Integrating distributed generation (DG) and distribution synchronous static compensator (DSTATCOM) has been shown in many studies to alleviate these issues, with further improvements achieved when they are optimally located at the vulnerable area or bus. This study presents a hybrid analytical-metaheuristic framework for solving the optimal location and sizing of DGs and DSTATCOMs. A combined voltage stability and power loss index (CI) was formulated as the analytical approach for predetermined candidate buses and next, the optimal location and sizing for both DGs and DSTATCOMs were utilized using non-dominated sorting genetic algorithm II. The proposed framework was tested on IEEE 69 bus test system. The results had demonstrated the efficiency and competitiveness of proposed hybrid-based framework in solving complex optimization problems.

Keywords: combined index (CI) · power loss · voltage profile · voltage stability index · hybrid-based algorithm and optimal location and sizing

1 Introduction

Efficiency, reliability and resilience are the among the important keys for an electrical grid and this is known as grid modernization. This involves the integration of advanced technologies, smart grid systems and renewable energy resources that enable two-way communication between the power utilities and end-users [1]. Hence, the role of power utilities is crucial in ensuring that secure, resilient and efficient electrical power is delivered to the consumer under all circumstances.

To meet this advancement, both sufficient real and reactive power must be managed effectively to ensure system's stability and reliability. The increasing integration

M. A. b. Jamlos et al. (Eds.): ICSET 2025, LNNS 1897, pp. 262–274, 2026.
https://doi.org/10.1007/978-3-032-20722-7_34

of renewable energy-based distributed generations (RE-DGs) introduces variability and uncertainty, making reactive power compensation (RPC) essential [2]. RPC helps to mitigate voltage fluctuations and power imbalances in grid contributed by RE-DGs. It also help to preserve power quality and maintain voltage stability [3–7].

The optimal number, size and location of DGs and RPCs to be installed in the network must be determined as part of modernizing the electrical power grid. The location of DGs and RPCs can be installed at any potential location, however the network's performance varies for different network characteristics at both transmission and distribution levels. For maximizing the overall performance and providing an effective control to the network, a proper location, sizing and number are essential. Economic feasibility, security, reliability, quality of supply and availability are among the main perspectives for selecting of location, size and number of DGs and RPCs. These factors have motivated the researchers to investigate optimal placement frameworks, focusing optimization techniques, with the most common objectives according to [8–10], which are to enhance the voltage stability, minimize the power losses and reduce the bus voltage deviation.

Yuvaraj et al. in [11] investigated the optimal location and sizing of both DG and distributed static synchronous compensator (DSTATCOM) in various IEEE radial distribution test system (RDS); 12-bus, 34-bus and 69-bus. The authors proposed a hybrid-based technique utilizing two indices, loss sensitivity factor (LSF) and voltage stability index (VSI) to pre-determine the candidate buses for DG and DSTATCOM placement. In the next stage, cuckoo search algorithm (CSA) was developed to solve the optimization problem for minimizing total power loss by focusing on the selected buses identified in the pre-determined stage. The authors then extended the similar framework in [12], with different algorithm, lightning search algorithm (LSA) and three objective, power loss, total voltage deviation and VSI, were derived in the objective function. The study was tested on two IEEE radial distribution test systems; 33-bus and 69-bus with various loading scenarios.

A similar hybrid-based framework was adopted in [13], where the potential locations of both DG and DSTATCOM were obtained using LSF and the final optimal location and size of DG and DSTATCOM were determined using Dward Mongoose Optimization (DMO) method, utilizing the same there objective as in [13]. The study was conducted on IEEE 30 bus test system. In [14], the author proposed a multi-objective particle swarm optimization (MOPSO) for solving multi-objective problems. Three objectives were considered in this study which includes minimizing the real power losses, enhancing bus voltage profile and reducing branch current. The optimal location and sizing of DGs and shunt capacitors were simultaneously obtained and the proposed framework was tested on IEEE 33 RDS test system. The findings indicate almost 95% of reduction in network's losses with all bus voltage measured above 0.975 pu.

In this paper, an analytical-metaheuristic hybrid-based algorithm was developed, by combining the Power Loss Sensitivity Index (PLS) and Voltage Stability Index (BVSI), forming a Combined Index (CI) to pre-determine the candidate buses for both DGs and DSTATCOMs. In the next stage, the optimal location and sizing were obtained using Non-dominated Sorting Genetic Algorithm II (NSGA-II). The proposed CI-NSGA-II algorithm aims to minimize the power losses, reduce bus voltage deviation and enhance the voltage stability. The study was conducted on the IEEE 69-bus RDS test system and results were compared with other algorithms reported in literature to validate the effectiveness of the proposed framework.

2 Methodology

2.1 Candidates Locations of DG(s) and/or STATCOM(s)

The formulation of the index-based algorithm for determining the candidate buses for DG(s) and/or STATCOM(s) are briefly discussed in this section.

2.1.1 Line Voltage Stability Index

Line voltage stability indices (VSIs) are one of the common and effective tools which have been utilized in many studies in identifying the critical lines or buses that operate close to their stability limits. The formulation of line VSIs are based on the power transmission concept of two bus models. The new line voltage stability index (BVSI) proposed by Ismail et al. [15] is utilized in this study to determine the candidate locations of DG(s) and/or STATCOM(s) from the voltage stability perspective. BVSI index is defined as in (1),

$$BVSI = \frac{4R_{ij}P_jZ_{ij}^2}{\left(V_i\left(R_{ij}\cos\delta + X_{ij}\sin\delta\right)\right)^2} \tag{1}$$

Where, i, j are the sending and receiving end bus, P_j is the real power at bus j, R_{ij}, X_{ij} and Z_{ij} are the line resistance, reactance and impedance connected between bus i and bus j, V_i is the bus voltage magnitude at bus i and δ is the angle difference between bus i and bus j.

2.1.2 Power Loss Sensitivity Index

Power loss index (PLS) is extensively adopted by various study related to optimal placement of DG and capacitor. The formulation of PLS is developed by taking the partial derivative of the real power losses in line with respect to the reactive power injection at the receiving bus (j). The PLS is computed as given in [16],

$$PLS = \frac{\partial P_{loss}}{\partial Q_j} = \frac{2Q_jR_{ij}}{V_j^2} \tag{2}$$

Where, P_{loss} is the power loss in line connected between bus i and j, Q_j is the reactive power at bus j and V_j is the bus voltage magnitude at bus j.

2.1.3 Combined Voltage Stability – Power Loss Index

The combined voltage stability and power loss index, which is known as *CI* is proposed in this study to determine the candidate bus for DG(s) and/or STATCOM(s) from both of voltage stability and power loss perspective. *CI* is obtained after calculating both BVSI and PLS in every line and CI is computed as in (3),

$$CI = w_1 \times BVSI_{ij} + w_2 \times PLS \tag{3}$$

Where w_1 and w_2 are the weighting factors for both BVSI and PLS and the sum of these weights is unity. These weighting factors are utilized to adjust the relative contribution between the two indices. In this work, w_2 is set 0.7 and w_2 is set 0.3 to prioritize the real power loss contribution, as it represents a more significant concern than voltage stability in radial configuration network.

2.2 Optimal Location and Sizing of DG(s) and/or STATCOM(s)

This section discusses a complete framework of the proposed multi-objective optimization algorithm which include the formulation of the objective functions, constraints and a complete proposed hybrid-based algorithm framework. The proposed algorithm is designed to minimize real power loss, reduce total bus voltage deviation and total bus voltage stability index.

2.2.1 Objective Function

The first objective function is derived to minimize the total real power loss (T_{RPL}) in the distribution line as follows,

$$T_{RPL} = \min \sum_{j=1}^{nline} G_{ij}\left(V_i^2 + V_j^2 - 2V_iV_j\cos\delta_{ij}\right) \tag{4}$$

Where, i, j are the sending and receiving buses, G_{ij} is the line conductance, V_i, V_j are the sending and receiving end bus voltage, δ_{ij} is the phase angle between the sending and receiving end buses, respectively. RPL is determined for each distribution line and T_{RPL} is obtained by summing all RPL values across the lines.

Next, the second objective function is formulated to minimize the total bus voltage deviation (T_{BVD}), where BVD is calculated by taking the difference between the reference voltage and the recorded bus voltage at a given time. Then, T_{BVD} is calculated by summing all the BVD obtained in every bus as below,

$$T_{BVD} = \min \sum_{i=1}^{nbus} \left|V_{ref} - V_i\right| \tag{5}$$

Where, *nbus* is the number of bus and V_{ref} is the voltage reference (typically set as 1 pu).

Lastly, the third objective is set to reduce the total lines voltage stability index (T_{BVSI}). The BVSI in every line is initially obtained and next, T_{BVSI} is calculated by summing all line BVSI for every case as follows,

$$T_{BVSI} = \min \sum_{j=1}^{nl} \frac{4R_{ij}P_jZ_{ij}^2}{\left(V_i\left(R_{ij}\cos\delta_{ij} + X_{ij}\sin\delta_{ij}\right)\right)^2} \tag{6}$$

Where, R_{ij}, X_{ij} and Z_{ij} are the line resistive, reactance and impedance and P_j is the real power at the receiving end bus.

2.2.2 The Equal and Inequality Constraints

The following set of equality and inequality constraints must be satisfied when solving the optimal location and sizing of DG(s) and/or DSTATCOM(s),

$$P_{Gi} - P_{Li} = \sum_{i=1}^{nl} \sum_{j=1}^{nl} V_i V_j \left(G_{ij} \cos \delta + B_{ij} sin\delta \right) \tag{7}$$

$$Q_{Gi} - Q_{Li} = \sum_{i=1}^{nl} \sum_{j=1}^{nl} V_i V_j \left(G_{ij} \sin \delta - B_{ij} cos\delta \right) \tag{8}$$

Where, P_{Gi} and Q_{Gi} are the real and reactive power generation at bus i, P_{Li} and Q_{Li} are the real and reactive power load at bus i and B_{ij} is the susceptance of line ij.

The bus voltage magnitude limit, DG real power generation limit, and DSTATCOM reactive power limit are the inequality constraints which are presented as below,

The bus voltage magnitude limit:

$$V_i^{min} \le V_i \le V_i^{max}, \quad i = 1, \ldots, N_{bus} \tag{9}$$

The DG real power generations limit,

$$P_{DG}^{min} \le P_{DG} \le P_{DG}^{max}, \quad i = 1, \ldots, N_{DG} \tag{10}$$

The reactive power allowable values of DSTATCOM,

$$Q_{DSTAT}^{min} \le Q_{DSTAT} \le Q_{DSTAT}^{max} \tag{12}$$

2.2.3 Non-dominated Sorting Genatic Algorithm II

Deb et al. in [17] introduced non-dominated sorting genetic algorithm 2 (NSGA II). The crowding distance proposed by the author capable of achieving a uniform spread of Pareto optimal fronts, eliminating the need for an addition fitness sharing parameter. Under this technique, the two solutions are randomly selected using the crowded comparison operator and crowding distance is obtained for every solution. The solution with a better non-domination rank is preferred and if both solutions are located at the same pareto front, then the solution with higher crowding distance value is preferred.

The pre-defined NSGA II parameters adopted in this work include a populations (nPoP) of 100, a maximum number of iterations (MaxIt) of 100, a crossover probability (pCrossover) of 0.85 and a mutation probability (pMutation) of 0.35 and mutation rate (MR) is 0.02. A complete flowchart of the proposed CI-NSGA II hybrid-based approach is presented in Fig. 1.

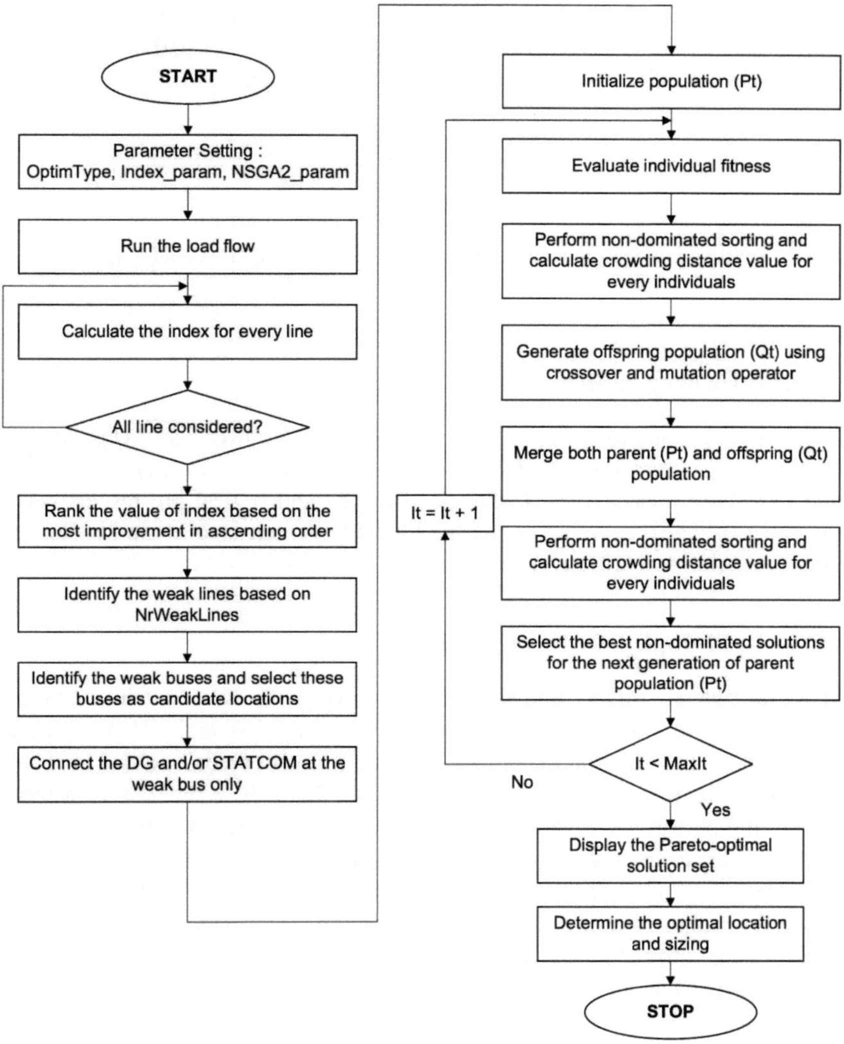

Fig. 1. The complete flowchart of the proposed CI-NSGA-II framework.

3 Result and Analysis

This section presents the results of proposed technique, where the effectiveness of proposed method is adopted in IEEE 69-bus RDS. The IEEE 69-bus RDS comprises 69 bus, 68 load buses and 68 lines with a total real and reactive power load of 3.175 MW and 2.3 MVAR. The test system depicts a modern power distribution systems which operates at a voltage of 12.66 kV. The complete modelling of the test system was done using DigSILENT PowerFactory environment, and the complete script of proposed algorithm was developed in DigSILENT programming language (DPL). Table I presents the performance analysis of the test system with various cases studies listed below and a brief discussion of the results are provided in the next sub-section:

Case I	:	Uncompensated
Case II	:	Multiple DGs integration
Case III	:	Multiple DSTATCOMs integration
Case IV	:	Multiple DGs and DSTATCOMs integration

3.1 Case I: Uncompensated

Under a base case condition, both of the real and reactive power losses (P_{loss} and Q_{loss}) are recorded at 224.948 kW and 102.141 kVAR. The minimum (V_{min}) and average bus voltage (V_{ave}) are measured at 0.9092 pu (Bus 65) and 0.9734 pu, respectively, and the total bust voltage deviation (T_{BVD}) and total line voltage stability index (T_{BVSI}) are recorded at 1.8559 and 1.2133, respectively.

3.2 Case II: Multiple DGs Integration

In this case, Bus 61, Bus 17 and Bus 11 appear as the optimal location for multiple DG integration with optimal size of 1721.99 kW, 364.66 kW and 571.06 kW, respectively. A significant reduction in both P_{loss} and Q_{loss} is observed, recorded at 69.41 kW and 34.94 kVAR, respectively, corresponding to a 65 ~ 70% reduction. In terms of voltage profile performance, both V_{min} and V_{ave} are increased to 0.9793 pu (Bus 65) and 0.9936 pu, respectively. This results in substantial reduction in T_{BVD}, measured at 0.4425, indicating 75.9% reduction. T_{BVSI} is also slightly reduced to 1.1318 with 6.7% reduction. Overall, the findings clearly shows that the performance of system is significantly improved with multiple DGs integrations, except for voltage stability, where only marginal improvement is observed.

3.3 Case III: Multiple DSTATCOMs Integration

For this case, the overall system performance is only marginally improved compared to multiple DGs case. Multiple DSTATCOMs are optimally located at Bus 61, Bus 11 and Bus 21 with optimal size of 1232.8 kW, 408.5 kW and 231.8 kW, respectively. P_{loss} and Q_{loss} are reduced to 35.65 kW and 33.81 kVAR, respectively, showcasing a reduction of 33 ~ 36%. The voltage profile in slightly improved, with both V_{min} and V_{ave} are increased to 0.9314 pu (Bus 65) and 0.9800 pu, respectively, resulting T_{BVD}, measured at 1.3778, indicating 24.95% reduction. However, the voltage stability greatly improved, with T_{BVSI} recorded at 0.1563 and representing a 87.12% improvement. The clearly shows the need of DTSTATCOMs, which is well-known for its effectiveness in improving the voltage stability.

3.4 Case IV: Multiple DGs and STATCOMs Integration

In the final case, the integration of multiple DGs and DSTATCOMs demonstrates more comprehensive improvement in all aspects. DGs are optimally placed at Bus 61, Bus 17, and Bus 11 with capacities of 1701.98 kW, 398.1 kW, and 503.1 kW, respectively, while DSTATCOMs are allocated at Bus 61, Bus 17, and Bus 11 with optimal sizes of 1198.78 kW, 291.01 kW, and 301.1 kW. As a result, both P_{loss} and Q_{loss} are substantially reduced to 4.35 kW and 6.78 kVAR, showcasing a reduction of 93 ~ 98%. The voltage profile is further enhanced, with both V_{min} and V_{ave} are improved to 0.9943 pu (Bus 50) and 0.9996 pu, respectively, which in turn reduces TBVD to 0.0487, corresponding to an 97.35% reduction. Most notably, T_{BVSI} improves remarkably to 0.1539, representing a 87.316% enhancement. These outcomes clearly confirm the effectiveness of integrating both DGs and DTSTATCOMs integrations into the network, in providing the optimal solutions.

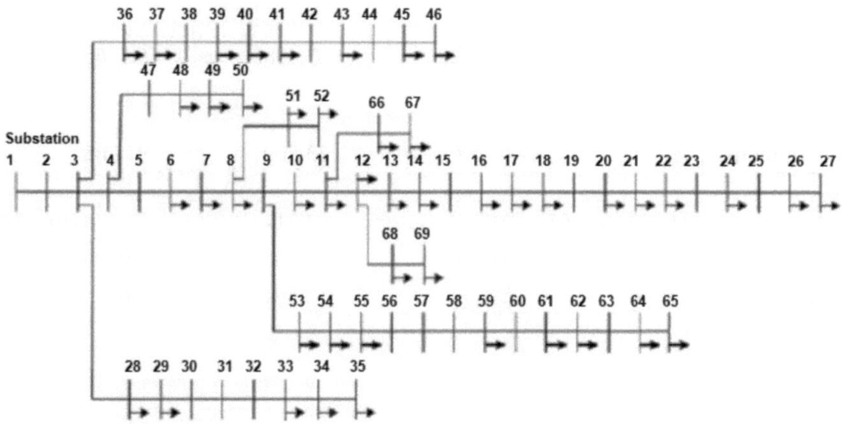

Fig. 2. Single line diagram of IEEE 69.bus RDS.

3.5 Comparative Study

Table II presents the comparative study between the proposed hybrid CI-NSGA-II algorithm with other algorithms in literature. It is observed that the proposed algorithm outperforms other algorithms across all cases. In multiple DGs integration (Case I), the proposed algorithm achieves better loss reduction and voltage profile improvement than other algorithms; Multi-objective Grey Wolf Optimizer (MOGWO), Adaptive Fuzzy Clonal Particle Quantum Algorithm (AFCPQA) and NSGA-II. In Case II, with multiple DSTATCOMs integration, CI-NSGA-II provides superior voltage stability improvement compared to combined LSF and Cuckoo Search Algorithm (LSF-CSA) and combined Firefly Algorithm-Sequential Coordinated Algorithm Control-PSO (FA-SCAC-PSO). It also delivers best overall performance, surpassing CSA and FA-SCAC-PSO in Case III. This findings confirm effectiveness and competitiveness of CI-NSGA-II in solving various optimization problems.

Table 1. The Performance Analysis of The Proposed Technique Under Various Case

	Uncompensated	Multiple Integration of		
		DGs	DSTATCOMs	DGs and DSTATCOMs
DG(s) size in kW (location)	–	Bus 61(1721.99) Bus 17 (364.66) Bus 11 (571.06)	–	Bus 61 (1701.98) Bus 17 (398.1) Bus 11 (503.1)
DSTATCOM(s) size in kVAr (location)	–	–	Bus 61 (1232.8) Bus 11 (408.5) Bus 21 (231.8)	Bus 61 (1198.78) Bus 17 (291.01) Bus 11 (301.1)
Real power loss (P_{loss}, kW)	224.948	69.4121	144.7608	4.3549
% reduction in P_{loss}	–	69.143	35.647	98.064
Reactive power loss (Q_{loss}, kVAr)	102.1407	34.9428	67.6085	6.7766
% reduction in Q_{loss}	–	65.790	33.809	93.365
V_{min} (pu)	0.9092 (Bus 65)	0.9793 (Bus 65)	0.9314 (Bus 65)	0.9943 (Bus 50)
V_{ave} (pu)	0.9734	0.9936	0.9800	0.9996
Total bus voltage deviation (*TBVD*)	1.8359	0.4425	1.3778	0.0487
% reduction in *TBVD*	–	75.897	24.952	97.347
Total line voltage stability index (*TBVSI*)	1.2133	1.1318	0.1563	0.1539
% reduction in *TBVSI*	–	6.717	87.118	87.316
Line with highest index ($BVSI_{max}$)	0.24941(line 56–57)	0.23421(line 56–57)	0.0192 (line 48–49)	0.01919 (line 45–46)
% reduction in $BVSI_{max}$	–	6.094	92.302	92.306
Computation Time (min)	–	2.633333	3.000000	3.483333

Table 2. The Performance Analysis of The Proposed Technique Under Various Case

	Algorithms	DG/RPC Location and Size (MW/MVAR)	P_{loss} (MW)	*TBVD*	*TBVSI*
Case I Uncompensated	-	-	224.948	1.8359	1.2133
Case II Multiple DGs	**CI- NSGA2**	Bus 61(1721.99), Bus 17(364.66), Bus 11(571.06)	69.4121	0.4425	1.1318
	MOGWO [18]	Bus 61(2123.5), Bus 20(743.4), Bus 11(713.5)	85.5703	0.2197	1.1145
	AFCPOA [19]	Bus 15(736.6), Bus 61(546.8), Bus 61(1000)	74.4198	0.4847	1.1396
	NSGA-II [20]	Bus 61(1497.7), Bus 64(374.8), Bus 17(567.8)	71.5321	0.4224	1.1327
Case III Multiple RPCs	**CI- NSGA2**	Bus 61(1232.8), Bus 11(408.5), Bus 21(231.8)	144.7608	1.3778	0.1563
	LSF-CSA [21]	Bus 61 (1170), Bus 11 (350), Bus 18 (230)	145.066	1.3761	0.1601
	FA-SCAC-PSO [22]	Bus 61 (1236.7), Bus 20 (235.5), Bus 11 (386.8)	144.7745	1.3807	0.1577
Case IV Multiple DGs-RPCs	**CI- NSGA2**	Bus 61 (1701.98), Bus 17 (398.1), Bus 11 (503.1) Bus 61 (1198.78), Bus 17 (291.01), Bus 11 (301.1)	4.3549	0.0487	0.1539

(*continued*)

Table 2. (*continued*)

	Algorithms	DG/RPC Location and Size (MW/MVAR)	P_{loss} (MW)	*TBVD*	*TBVSI*
	CSA [23]	Bus 17 (490), Bus 61 (1400), Bus 63 (250) Bus 25 (270), Bus 61 (980), Bus 63 (200)	8.2386	1.4226	0.2202
	FA-SCAC-PSO [22]	Bus 17 (355.9), Bus 61 (1753.9), Bus 69 (407.0) Bus 11 (400.7), Bus 17 (267.3), Bus 61 (1158.8)	5.1608	0.0356	0.1592

4 Conclusion

This study presented a hybrid analytical-metaheuristic framework utilizing a combination BVSI and PLS indices, forming CI, to initially obtain the candidate buses for both DGs and DSTATCOMs location. Next, in the next stage, NSGA-II was developed to obtain the optimum result based on three objectives set; reducing real power loss, improving voltage profile and enhancing the voltage stability. Based on the findings, the combination of multiple DGs and DSTATCOMs integration yielded the most substantial improvements, achieving up to 93–98% power loss reduction, near unity voltage profile as well as remarkable enhancement in voltage stability indices. A comparative study further validated the superiority of the proposed hybrid-based framework over the other exiting optimization algorithms in literature, proving its capability of delivering more optimal, stable and reliable solutions across various power system optimization problems

Acknowledgments. The authors sincerely acknowledge Universiti Putra Malaysia for providing the financial support and necessary facilities to support this work.

References

1. Khalid, M.: Smart grids and renewable energy systems: Perspectives and grid integration challenges. Energ. Strat. Rev. **51**, 101299 (2024)

2. Mohammed, A., Sakr, E.K., Abo-Adma, M., Elazab, R.: A comprehensive review of advancements and challenges in reactive power planning for microgrids. Energy Inform. **7**(63), 1–27 (2024)
3. Astapov, V., Shabbir, N., Rosin, A., Kutt, L., Maask, V., Tiismus, H.: Review of technical solutions addressing voltage and operational challenges in a distribution grid with high penetration of intermittent RES. Energy Rep. **14**, 1738–1760 (2025)
4. Adegoke, S.A., Sun, Y.: Power system optimization approach to mitigate voltage instability issues: A review. Cogent Eng. **10**(1), 2153416 (2022)
5. Ismail, B., et al.: The optimal location and sizing of single and multiple statcom using analytical approaches under high loading occasion. J. Eng. Sci. Technol. **17**(6), 3973–3988 (2022)
6. Ismail, B., Wahab, N.I.A., Othman, M.L., Radzi, M.A.M., Kahar, N.H.A., Awalin, L.J.: Optimal placement and sizing of STATCOM(s) for power system security improvement using MOPSO. In: 2022 IEEE International Conference on Power and Energy (PECon), pp. 304–309. IEEE, Langkawi, Kedah, Malaysia (2022)
7. Ismail, B., Abdul Wahab, N.I., Othman, M.L., Mohd Radzi, M.A., Vijyakumar, K.N.: Multiple DGs and DSTATCOMs allocation in radial distribution network using hybrid BVSI-CPVI MOPSO. In: 2023 International Conference on Engineering Technology and Technopreneurship (ICE2T), Kuala Lumpur, pp. 57–62. IEEE, Malaysia (2023)
8. Ismail, B., Abdul Wahab, N.I., Othman, M.L., Radzi, M.A.M., Naidu Vijyakumar, K., Mat Naain, M.N.: A comprehensive review on optimal location and sizing of reactive power compensation using hybrid-based approaches for power loss reduction, voltage stability improvement, voltage profile enhancement and loadability enhancement. IEEE Access. **8**, 222733–222765 (2020)
9. Al Ahmad, A., Sirjani, R.: Optimal placement and sizing of multitype FACTS devices in power systems using Metaheuristic optimisation techniques: An updated review. Ain Shams Eng. J. **11**(3), 611–628 (2020)
10. Zhang, D., Shafiullah, G.M., Das, C.K., Wong, K.W.: A systematic review of optimal planning and deployment of distributed generation and energy storage systems in power networks. J. Energy Storage. **56**(part A), 105937 (2022)
11. Yuvaraj, T., Ravi, K., Devabalaji, K.R.: Optimal allocation of DG and DSTATCOM in radial distribution system using cuckoo search optimization algorithm. Model. Simul. Eng. **2017**(1), 2857926 (2017)
12. Yuvaraj, T., Ravi, K.: Multi-objective simultaneous placement of DG and DSTATCOM using novel lightning search algorithm. J. Appl. Res. Technol. **15**(5), 477–491 (2017)
13. Raj, A.F., Saravanan, A.G.: An optimization approach for optimal location & size of DSTATCOM and DG. Appl. Energy. **336**, 120797 (2023)
14. Patel, P., et al.: Multi-objective particle swarm optimization algorithm-based method for optimal placement and sizing of distributed generations and shunt capacitors in a radial distribution network. Results Eng. **27**, 106514 (2025)
15. Ismail, B., Wahab, N.I.A., Othman, M.L., Radzi, M.A.M., Naidu Vijyakumar, K., Rahmat, M.K.: New line voltage stability index (BVSI) for voltage stability assessment in power system: the comparative studies. IEEE Access. **10**, 103906–103931 (2022). https://doi.org/10.1109/ACCESS.2022.3204792
16. El-Fergany, A.A., Abdelaziz, A.Y.: Capacitor allocations in radial distribution networks using cuckoo search algorithm. IET Gener. Transm. Distrib. **8**(2), 223–232 (2014)
17. Deb, K., Pratap, A., Agarwal, S., Meyarivan, T.: A fast and elitist multiobjective genetic algorithm: NSGA-II. IEEE Trans. Evol. Comput. **6**(2), 182 (2002)
18. Bendriss, B., Sayah, S., Hamouda, A.: Efficient multi-objective optimization approach for solving optimal DG placement and sizing problem in distribution systems. J. Eng. Res. **13**(4) (2025 in press)

19. Sunil, A., Venkaiah, C.: Multi-objective adaptive fuzzy campus placement based optimization algorithm for optimal integration of DERs and DSTATCOMs. J. Energy Storage. **75**, 109682 (2024)
20. Prasad, K.R.K.V., Kollu, R., Ramkumar, A., Ramesh, A.: A multi-objective strategy for optimal DG and capacitors placement to improve technical, economic, and environmental benefits. Int. J. Electr. Power Energy Syst. **165**, 110491 (2025)
21. Rukmani, D.K., et al.: A new approach to optimal location and sizing of DSTATCOM in radial distribution networks using bio-inspired cuckoo search algorithm. Energies. **13**(18), 4615 (2020)
22. Zellagui, M., Lasmari, A., Settoul, S., El-Sehiemy, R.A., El-Bayeh, C.Z., Chenni, R.: Simultaneous allocation of photovoltaic DG and DSTATCOM for techno-economic and environmental benefits in electrical distribution systems at different loading. Int. Trans. Electr. Energy Syst. **31**, e12992 (2021)
23. Salimon, S.A., Omofuma, O.I., Akinrogunde, O.O., Thomas, T.G., Edwin, T.E.: Optimal allocation of shunt capacitors in radial distribution networks using Constriction-Factor Particle Swarm Optimization and its techno-economic analysis. Frankl. Open. **7**, 100093 (2024)

Critical Parameter Identification in Furniture Strength Testing Using Probabilistic Data Matching Approach

Muhammad Firdhaus Che Hassan[1,2](✉), Shahruddin Mahzan[1](✉), Abdullah Al Rashid Ab Hamid[2], Ahmad Anwar Safwan Sidek[2], Mohd Hasni Chumiran[2], Muhammad Firdaus Md Rawi[2], and Mohd Zarir Yusoff[3]

[1] Functional Composite Structure (FCS) Focus Group, Faculty of Mechanical and Manufacturing Engineering, Universiti Tun Hussein Onn Malaysia (UTHM), Parit Raja, Batu Pahat 86400, Johor, Malaysia
muhammadfirdhauschehassan@gmail.com, sharudin@uthm.edu.my

[2] Creative Research and Innovation for Technology (CREATE) Focus Group, Faculty of Engineering Technology, Universiti Tun Hussein Onn Malaysia (UTHM), Pagoh Campus, Panchor 84600, Johor, Malaysia

[3] Human Capital Development (HCD) Focus Group, Faculty of Technology Management and Busniess, Universiti Tun Hussein Onn Malaysia (UTHM), Parit Raja, Batu Pahat 86400, Johor, Malaysia

Abstract. Ensuring the structural safety and durability of school chairs is critical for both manufacturers and regulatory compliance. This research applies a probabilistic data matching approach to integrate the Strength Design of Furniture theory with EN 1729–2:2019 testing requirements. Using phonetic and fuzzy matching methods, critical test parameters were identified: (1) Vertical Loads on Stretchers linked with Static Load of Foot Rail and Foot Rail Durability, and (2) Vertical Seat Loads linked with Safety Requirements, Stability in all orientations, and Seat/Back Load and Durability tests. These matched results exceeded threshold similarity levels, enabling their application in simulation-based pre-testing of wooden school furniture. The approach supports optimized design, reduced physical testing cycles, and consistent compliance with international safety standards, offering value to furniture manufacturers, testing laboratories, and policymakers.

Keywords: Probabilistic Data Matching · School Furniture Testing · Standards Compliance in Furniture Design · Furniture Strength Design

1 Introduction

The process of identifying and establishing associations between records by comparing and connecting data from various sources is known as data matching, often referred to as entity resolution and record linkage. It looks for similarities between two entities and a technique to prevent duplicated material. It assesses if and how data points and records

M. A. b. Jamlos et al. (Eds.): ICSET 2025, LNNS 1897, pp. 275–287, 2026.
https://doi.org/10.1007/978-3-032-20722-7_35

correspond using statistical techniques, machine learning, and occasionally manual verification. Data matching enables you to conduct more accurate and exact searches and conduct more sophisticated and dependable data analysis when working with big volumes of data. Comparing data, identifying trends, and flagging abnormalities are all made possible by data matching. The following is a simplified version of the process: Data standardization, choose characteristics that are unlikely to alter, sort the data into chunks, using probability to match, give matches a value, to determine the total weight, and summarize. The target is to continuously improve the data matching algorithms through experience in order to achieve better outcomes. To put it briefly, data matching improves accuracy, efficiency, and compliance in a variety of settings and sectors.

Despite their long-standing concern for the strength of the furniture designed, no methodology has ever been developed that enables a designer to examine the strength requirements of a particular piece of furniture and then determine, using scientific principles, the size of the members and joints required to meet these requirements. The goal of Strength Design of Furniture theory is to present and expand on the ideas and tenets of strength design as they relate to furniture [1].

According to the Standards of Malaysia Act 1996 [Act 549], the Minister of Investment, Trade, and Industry has authorized the Malaysian Standard (MS), a national standard created by the Standards Development Committee (SDC) inside the Malaysian Standards Development System. The Technical Committee on Furniture, which is under the jurisdiction of the Industry Standards Committee on Mechanical Engineering, suggested that the EN Standard be adopted as a Malaysian standard for the European Committee for Standardization (CEN) produced EN 1729–2:2012 + A1:2015, Furniture - Chairs and tables for educational institutions - Part 2: Safety standards and test procedures. Nonetheless, the following is relevant for the purposes of this Malaysian Standard (MS), which are; that "this European Standard" should actually read "this Malaysian Standard" in the original text; the comma, if it is used as a decimal symbol, should read as a point; and the Equivalent Malaysian Standards should be used in place of references to European Standards. The said standard is the MS EN 1729–2:2019 (Furniture – Chairs and tables for educational institutions – Part 2: Safety requirements and test methods) [2].

This research is attempting to identify critical parameters to be used for a school furniture testing in simulation mode by data matching from the Strength Design of Furniture theory and the MS EN 1729–2:2019 (Furniture – Chairs and tables for educational institutions – Part 2: Safety requirements and test methods) standards.

2 Literature Review

The idea of similarity is the foundation of data matching; for example, two features, A and B, from two distinct datasets are matched if they are similar. For many applications, data matching is a first step rather than an end in itself. Pre-processing, candidate selection, criteria combination and choice, and assessment are the five processes that make up the matching process [3]. According to [4], similarity functions are used as the primary resource in a number of data matching applications, such as data integration, data deduplication, data cleaning, similarity querying, and similarity joining, to determine whether two distinct data instances reflect the same real-world item. Errors in data arrived at a data warehouse from outside sources are found and removed via data matching

The interdisciplinarity of product development and furniture mechanical property testing is the current trend in the furniture testing field. The manufacturing and mechanical testing procedures for chairs, tables, and joints made using CNC machines are explained. Critical construction displacement and deformation are analyzed digitally. Together with furniture mechanical testing, the numerical simulation predicts the most stressed parts, enabling us to improve the furniture's quality. The test findings allow for the optimization of strength attributes and offer fresh information that may be utilized during the development phase [5].

The history, present state, and potential future developments of furniture product engineering, strength design, and performance testing are narratively described in the article by [6]. Product engineering is discussed both generally and in relation to furniture. The key component of the piece is the furniture's strength design. The generic application of reliability concepts to furniture and their integration into performance testing standards are both illustrated. Enhancing furniture goods' safety and durability as well as anticipating failure or unforeseen issues are the main goals of reliability and performance testing. In addition to providing manufacturers and consumers with relevant expected performance information, testing and assessment are necessary to provide safe and dependable furniture. Along with recommendations for applying strength design to enhance furniture building, the history of the field's evolution and present state of development are covered. Finally, a comprehensive process for creating high-strength furniture in light of recent advancements in technology is described.

The critical parameters found in this article will be used to do simulation testing for furniture, focusing on the strength side, withstanding any weight applied on for different movements during a certain period of time. An example of simulation furniture testing can be seen in article [8], where a flat wooden seat used in Colombian schools was redesigned using ManneQuin Pro and AutoCAD to model anatomy-based Medium Density Fibreboard (MDF) seat curves for each size, aligning to ischial tuberosity position for maximum contact and comfort. By using Force-sensing array (FSA) mats with 256 sensors to be placed on the seat surface, the measurement of buttock/leg pressure distribution was simulated. Another example from article [9], shows a method for physical design based on ISO 15537 guidelines for anthropometric testing, combined with defined target population, determined sample size, set design limits, analyzed data, and finally solve design problems. By using the IDEO methodology (observation/research, brainstorming, prototyping). An adjustable furniture with separate desk and chair for greater adaptability was simulated.

As stated in article [7], without a national standard, schools will continue to use furniture that is physically unsuitable, leading to long-term health and educational consequences, with the current furniture appears to be designed without a specific anthropometric data, it will lead to systemic mismatches across school levels. If a national standard is formed, it will guide manufacturers to design the school furniture according to a specific country-oriented data, in this case in Nigeria.

3 Research Methodology

For this particular research, there will be a few steps taken to accomplish the objective. First and foremost, the Data Matching Process must be adhered to by fulfilling each process with the necessary procedure to finally come out with matched critical parameters for the school furniture testing in simulation mode from the Strength Design of Furniture theory and the MS EN 1729–2:2019 (Furniture – Chairs and tables for educational institutions – Part 2: Safety requirements and test methods) standards.

3.1 Data Matching Process

The Data Matching Process can be simplified as follows:

- Define Scope – in this process, there needs determination of specific goals. There is a total of 17 tests for MS EN 1729–2:2019 (Furniture – Chairs and tables for educational institutions – Part 2: Safety requirements and test methods) standards & 9 tests for the Strength Design of Furniture theory for a Light/Medium/Heavy Duty Institutional Side Chair. The number of tests will be determined for the next step based on the official school chair specification in Malaysian government school from the Malaysian Education Ministry (KPM) as shown in Fig. 1 as an example.

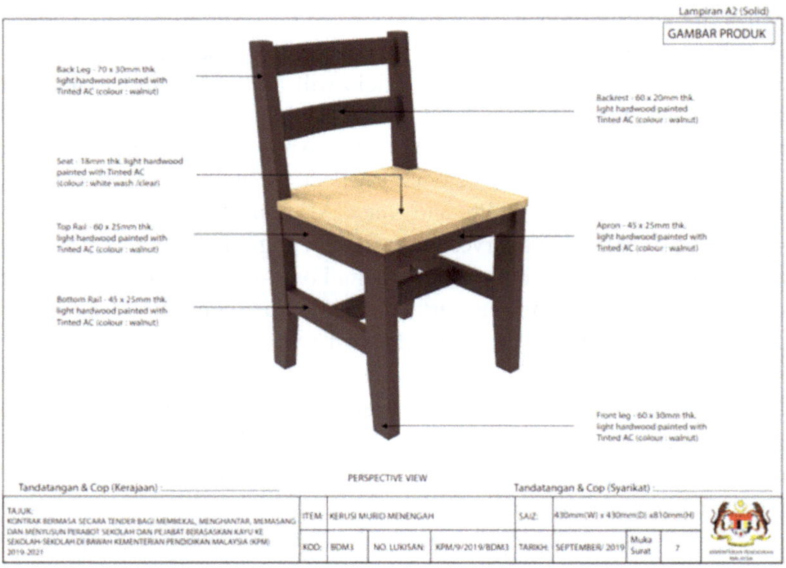

Fig. 1. *Example of a school chair specification*

- Data Preparation – in this procedure, there will be a clean and normalize data formatting process, whereby each data for matching process are defined, as shown in Appendix A for Strength Design of Furniture theory and Appendix B for MS EN

1729–2:2019 (Furniture – Chairs and tables for educational institutions – Part 2: Safety requirements and test methods) standards.

- Choose Algorithm – to select the types of data matching tools, which are primarily categorised as probabilistic and deterministic.

The tools for Probabilistic Data Matching statistical techniques used by probabilistic data matching tools to ascertain the probability that two records pertain to the same entity. They operate by comparing several fields in the records and giving each one a similarity score. The two records' total similarity score is then used to determine whether they should be linked in a probabilistic manner.

Most probabilistic tools available now use machine learning algorithms to produce even more precise outcomes. Important features are automatically found and extracted from data using regression and natural language processing algorithms. These features are then utilized in the similarity scoring procedure. Usually, under this type of tools use fuzzy and phonetic matching method to identify similar but distinct instances recognized by this entity resolution technique. Typographical errors, missing data, and spelling variants are a few examples.

The benefits of this kind of tools are; any kind of data, even unstructured data, can be matched using probabilistic record matching methods; due to manual data entry, real-world data is frequently unorganized and badly maintained; it can deal with a wide range of spelling and coding variations and exceptions that would be difficult to address with a strict set of rules or even a comprehensive dictionary; and compared to deterministic data matching techniques, they are typically more accurate.

However, there are its drawbacks such as; because more parameters need to be set, they can be more challenging to configure and tune; to train the machine learning models used for feature extraction and similarity scoring, they need a sizable amount of data; since they rely on probabilistic techniques, a large portion of the process takes place in the background, making some of their findings challenging to understand.

Then, the tools for Deterministic Data Matching are conversely, matching records using rule-based techniques. In other words, they use a system to compare various fields in the records and search for exact matches; if two fields match, the records are connected. Deterministic tools are typically far simpler to set up than their probabilistic equivalents since they employ a preset set of rules. It might, however, miss some relationships that don't fit the parameters, which could make them less accurate. When dealing with highly organized and well-defined data, such as when connecting records from two distinct databases that share a schema, deterministic data matching methods are frequently employed.

As stated before, the benefits compared to probabilistic data connection technologies, they are far simpler to set up and adjust, it can be applied to unstructured data as well as any other kind of data, and it don't require training data for machine learning models, thus you don't need a lot of data to use them.

The drawbacks; it might overlook some relationships that don't satisfy the rules, they may be less accurate than probabilistic data matching techniques; limited to comparing fields in a specific manner, they are frequently less adaptable than probabilistic data matching methods; and since it rely on deterministic approaches, a large portion of the process takes place behind the scenes, making some of their conclusions hard to decipher. As deterministic matching, whereby exact and numeric matching method are usually placed, is sometimes excessively rigid, probabilistic data matching is the most popular.

- Review Results – where it is manually verified to avoid false positives/negatives. This is to make sure the record and matching record really does comply with each other to form a master record. This can be verified by referring to the match definitions defined beforehand, especially when setting the fuzzy levels to make sure the higher percentage testing method score are used for record matching purposes. In Fig. 2, an example of match results is to have shown all the information explained previously.

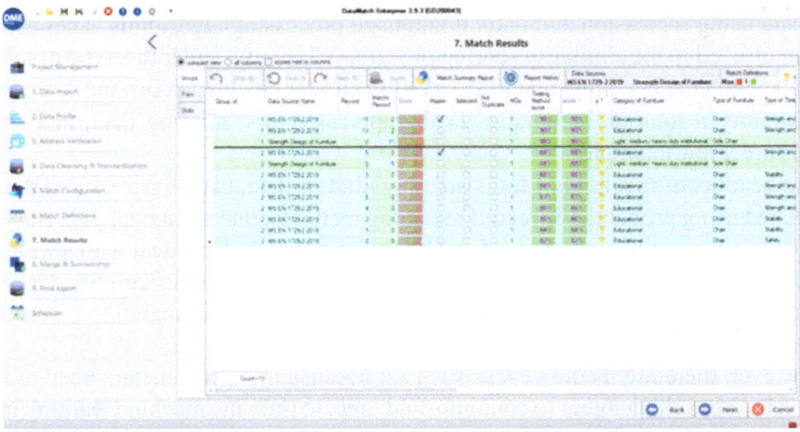

Fig. 2. Example of match results from the DataMatch Enterprise 3.9.3 software

- Merge, Purge, Set Master – this is to create comprehensive master records by merging duplicates, when finally, the matched data can be used.

All these can be done manually or by software for more accurate data matching process. The software used for this particular research is the DataMatch Enterprise 3.9.3, where it can apply both the probabilistic and deterministic data matching method with data imported from what was gathered. The flowchart of the Data Matching Process is shown in Fig. 3.

4 Result and Discussion

Match Results, the process or Merge, Purge, Set Master can be done manually or by software to create master records, thus publishing the matched data.

The data to be imported are the defined MS EN 1729–2:2019 (Furniture – Chairs and tables for educational institutions – Part 2: Safety requirements and test methods)

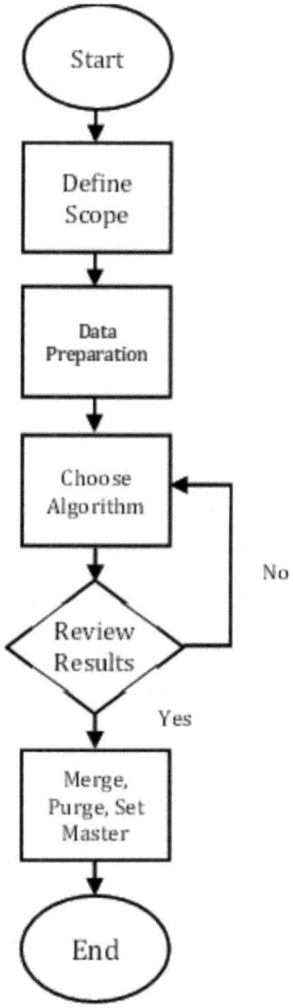

Fig. 3. The Data Matching Process flowchart

standards and Strength Design of Furniture theory for a Light/Medium/Heavy Duty Institutional Side Chair, which the process can be seen in Fig. 4 above.

From the 17 tests for MS EN 1729–2:2019 (Furniture – Chairs and tables for educational institutions – Part 2: Safety requirements and test methods) standards and 9 tests for the Strength Design of Furniture theory for a Light/Medium/Heavy Duty Institutional Side Chair, it was determined that only 14 and 7 tests respectively will be used for this data matching process, as it fits the type of furniture which is a school chair. Refer Appendix A and Appendix B for the defined sets of data. This was proven from the Data Profile, whereby when profile generated, the data are 100% filled, as shown below in Figs. 5 and 6.

After that, the next procedure is the Match Configuration procedure, which confirms the selection of sets of data, in which this case between the set of data from the Strength Design of Furniture theory for a Light/Medium/Heavy Duty Institutional Side Chair and

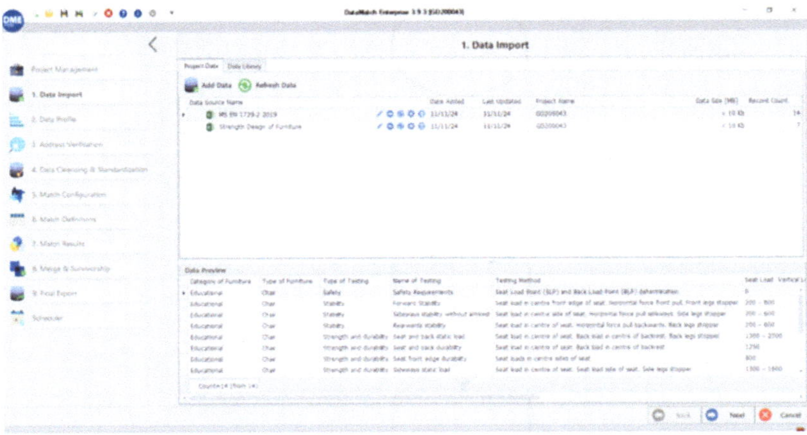

Fig. 4. *Data import procedure*

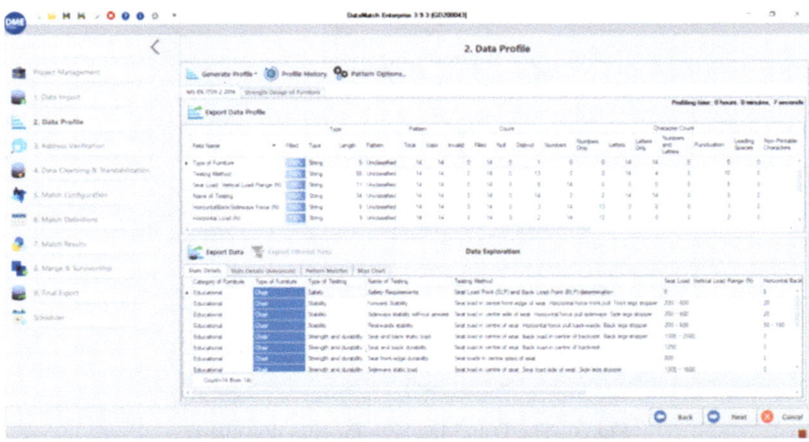

Fig. 5. *Data profile procedure*

the MS EN 1729–2:2019 (Furniture – Chairs and tables for educational institutions – Part 2: Safety requirements and test methods) standards (Fig. 7).

Finally, when the chosen Match Definitions parameter to be used for the data matching process is set, as in Fig. 8, the Match Results are published. From the Match Definitions procedure, the method for probabilistic was chosen which are phonetic and fuzzy, due to the chosen parameter, Testing Method are of words and descriptively displayed.

Match Results procedure shows all the details needed to be used for the next step of creating a master record which are the record and matching record linkage, master data, and testing method score (Fig. 9). There are 2 master records that can be published. The first are the Vertical Loads on Stretchers test from the Strength Design of Furniture theory for a Light/Medium/Heavy Duty Institutional Side Chair, that is matched with the Static Load of Foot Rail, and the Foot Rail Durability tests from the MS EN 1729–2:2019 (Furniture – Chairs and tables for educational institutions – Part 2: Safety requirements

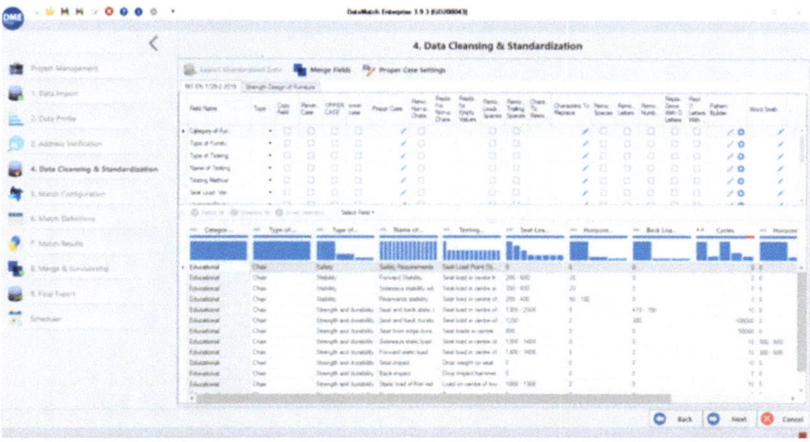

Fig. 6. Data cleansing and standardization procedure

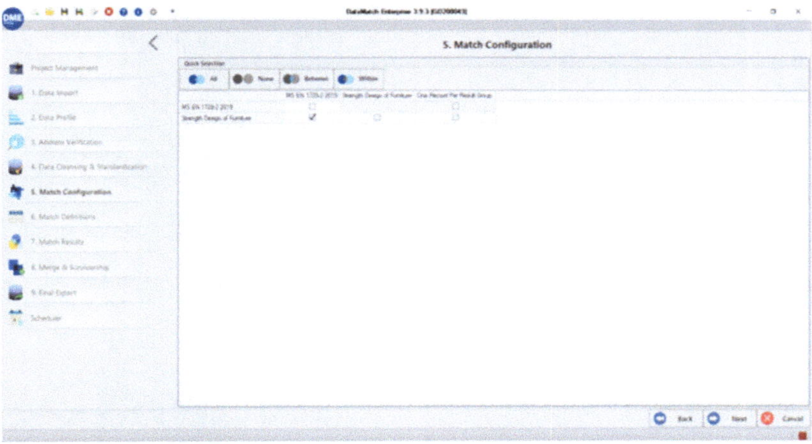

Fig. 7. Match configuration procedure

and test methods) standards. The second are the Vertical Seat Loads test from the Strength Design of Furniture theory for a Light/Medium/Heavy Duty Institutional Side Chair, that is matched with the Safety Requirements, Forward Stability,

Rearwards Stability, Sideways Stability without Armrest, Forward Static Load, Seat and Back Static Load, Sideways Static Load, and Seat and Back Durability tests from the MS EN 1729–2:2019 (Furniture – Chairs and tables for educational institutions – Part 2: Safety requirements and test methods) standards. The master records for both sets of data are shown in Table 1.

In the Match Summary Report, the Matches Between Sources (Fig. 10) indicate that 10 tests out 14 (71%) from the MS EN 1729–2:2019 (Furniture – Chairs and tables for educational institutions – Part 2: Safety requirements and test methods) standards were of matched with 2 out of 7 tests (29%) from the Strength Design of Furniture theory

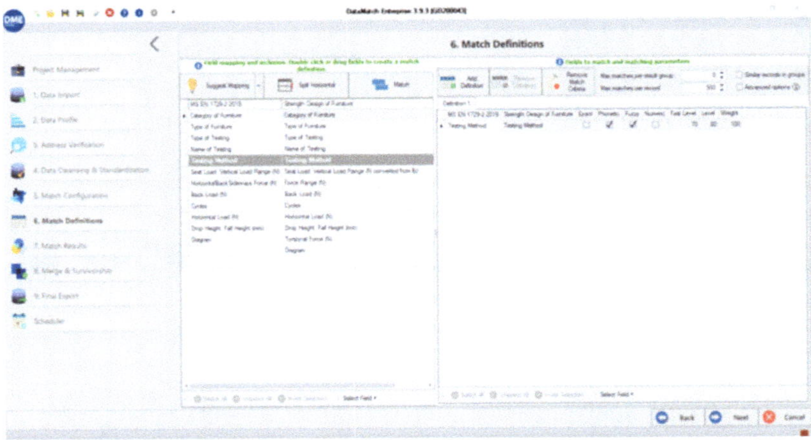

Fig. 8. Match definitions procedure

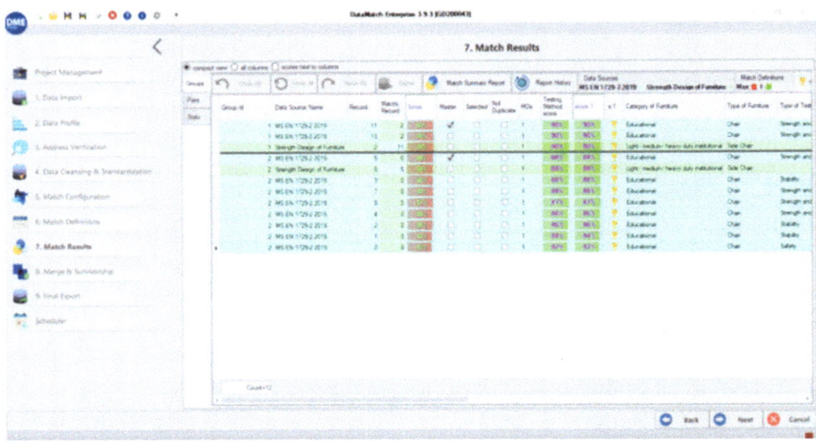

Fig. 9. Match results procedure

for a Light/Medium/Heavy Duty Institutional Side Chair. This constituents towards the proportion of matches found between these sets of data, which stands at 57% (Fig. 11).

By score, the range for the matches varies but with the least being 85%. In this range, there are 2 matches found. Next, in the score range of 90%, there are 7 matches, which is the highest out of all within the predetermine set of data matching process. Finally, in the 95% score range there are 3 matches in this highest set found. All are shown in Fig. 12. This shows that the matches found are within the high end of percentage of score range, for more than half of the data defined beforehand.

In the final process, the master record was published due to the overlap of records or matching of data. The number of matching group suggested in the software is two (Fig. 13), whereby as stated before, 10 tests from the MS EN 1729–2:2019 (Furniture – Chairs and tables for educational institutions – Part 2: Safety requirements and

Table 1. Master records of matched critical parameters

Set of Imported Data	Master Record 1	Master Record 2
Strength Design of Furniture theory for a Light/Medium/Heavy Duty Institutional Side Chair	Vertical Loads on Stretchers	Vertical Seat Loads
MS EN 1729–2:2019 (Furniture – Chairs and tables for educational institutions – Part 2: Safety requirements and test methods) standards	Static Load of Foot Rail Foot Rail Durability	Safety Requirements Forward Stability Rearwards Stability Sideways Stability without Armrest Forward Static Load Seat and Back Static Load Sideways Static Load Seat and Back Durability

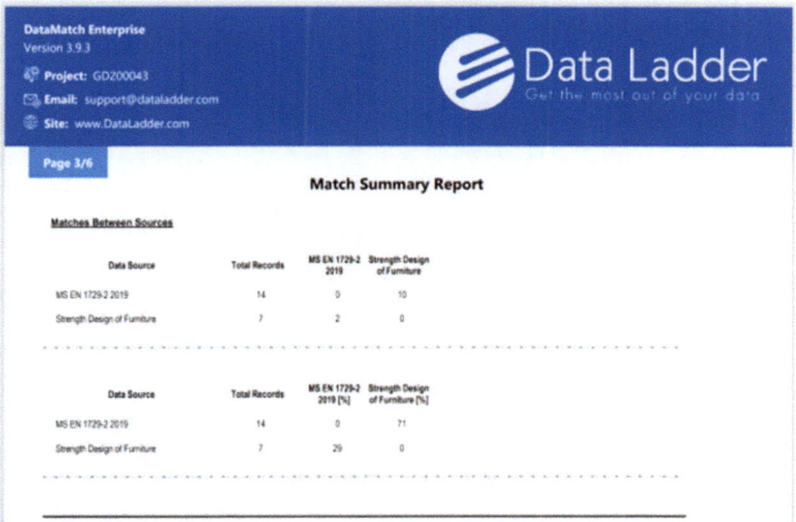

Fig. 10. Matches between sources

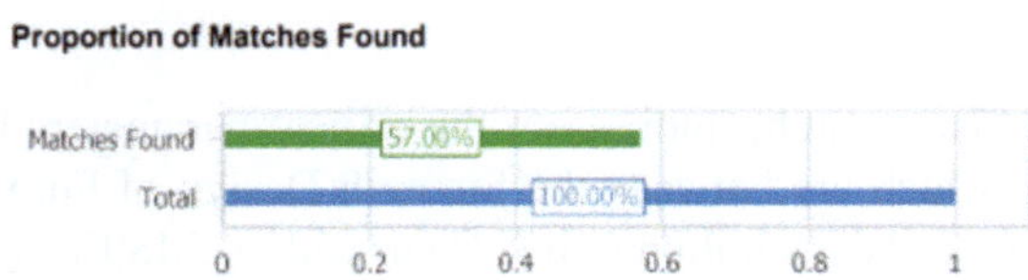

Fig. 11. Proportion of matches found

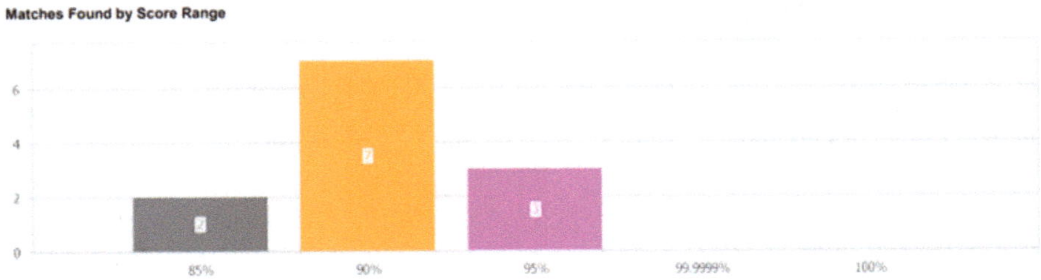

Fig. 12. Matches found by score range

test methods) standards and 2 tests from the Strength Design of Furniture theory for a Light/Medium/Heavy Duty Institutional Side Chair make up the two master records.

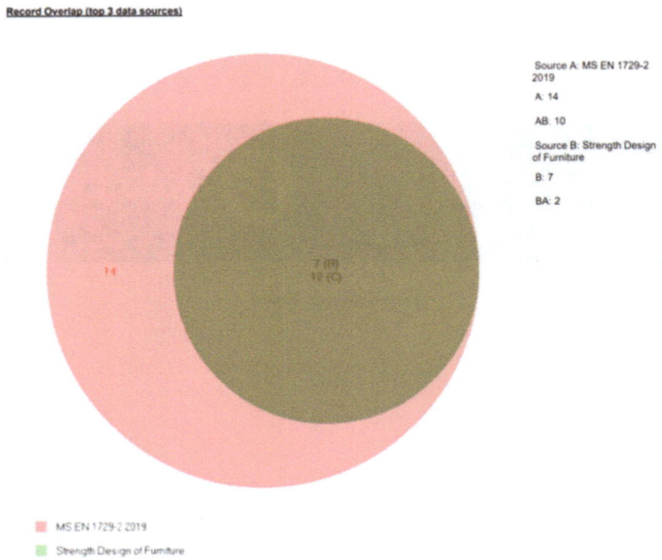

Fig. 13. Record overlap

5 Conclusion

To conclude from this research, finding out the critical parameters by data matching the two set of data, which are between the Strength Design of Furniture theory for a Light/Medium/Heavy Duty Institutional Side Chair and the MS EN 1729–2:2019 (Furniture – Chairs and tables for educational institutions – Part 2: Safety requirements and test methods) standards, was done using the data matching process. This was completed by using the probabilistic tool type for which the phonetic and fuzzy method were set for during the process.

The master records of matched set of data from the Strength Design of Furniture theory for a Light/Medium/Heavy Duty Institutional Side Chair and the MS EN 1729–2:2019 (Furniture – Chairs and tables for educational institutions – Part 2: Safety requirements and test methods) standards are first; the Vertical Loads on Stretchers test from the Strength Design of Furniture theory for a Light/Medium/Heavy Duty Institutional Side Chair, that is matched with the Static Load of Foot Rail, and the Foot Rail Durability tests from the MS EN 1729–2:2019 (Furniture – Chairs and tables for educational institutions – Part 2: Safety requirements and test methods) standards; and second the Vertical Seat Loads test from the Strength Design of Furniture theory for a Light/Medium/Heavy Duty Institutional Side Chair, that is matched with the Safety Requirements, Forward Stability, Rearwards Stability, Sideways Stability without Armrest, Forward Static Load, Seat and Back Static Load, Sideways Static Load, and Seat and Back Durability tests from the MS EN 1729–2:2019 (Furniture – Chairs and tables for educational institutions – Part 2: Safety requirements and test methods) standards.

The master records will be used in the simulation furniture testing element of a United Furniture Strength Design method in the future research of a guideline provisioning of a school chair in Malaysia.

References

1. Eckelman, C.A.: Textbook of Product Engineering and Strength Design of Furniture, 1st edn. Purdue University, Indiana, USA (2003)
2. Department of Standards Malaysia: Malaysian Standard. MS EN 1729–2:2019 Furniture – Chairs and tables for educational institutions – Part 2: Safety requirements and test methods. Department of Standards Malaysia, Cyberjaya, Selangor, Malaysia (2019)
3. Olteanu-Raimond, A.-M., Mustiere, S., Ruas, A.: Knowledge formalization for vector data matching using belief theory. J. Spat. Inf. Sci. **10**, 21–46 (2015)
4. dos Santos, J.B., Heuser, C.A., Orengo, V.M., Wives, L.K.: Automatic threshold estimation for data matching applications. Inf. Sci. **181**(13), 2685–2699 (2011)
5. Simek, M.: Furniture testing for higher competitiveness, better quality and design. In: 10th Annual International Scientific Conference on More Wood, Better Management, Increasing Effectiveness: Starting Points and Perspective WoodEMA 2017, pp. 165–170. WoodEMA, Prague, Czechia (2017)
6. Erdil, Y.Z., Haviarova, E., Eckelman, C.A.: Product engineering and performance testing in relation to strength design of furniture. Wood Fiber Sci. **36**(3), 411–416 (2004)
7. Fidelis, O.P., Ogunlade, B.: Anthropometric perspective to classroom furniture ergonomics and the need for standards in Nigerian schools. Work. **72**(1), 279–289 (2022)
8. Rincón, O., Bernal, M.L., Salazar, J.J., Zea, C.R.: Relationship between seat surface shape and pressure distribution in school seat models. Work. **66**(1), 161–171 (2020)
9. Velasco, T.W.L., Cristina Camacho, C., C. C.: Latin American schoolchildren anthropometry: Study of the anthropometric differences of the rural and urban zones in Cotopaxi, Ecuador. In: 20th Congress of the International Ergonomics Association (IEA 2018), pp. 104–110. Springer, Cham, Florence, Italy (2018)

Mathematical Insight into Dynamic Roguing Strategies for Controlling Rice Tungro Disease

Abdul Rahman Mohd Kasim[1(✉)], Adeshina Taofeeq Adeosun[1], Laila Amera Aziz[1], Zulkhibri Ismail[1], Livia Owen[2], and Siti Hanani Mat Yasin[1]

[1] Centre for Mathematical Sciences, Universiti Malaysia Pahang Al-Sultan Abdullah (UMPSA), Gambang, Malaysia
rahmanmohd@umpsa.edu.my

[2] Center for Mathematics and Society, Department of Mathematics, Parahyangan Catholic University, Bandung 40141, Indonesia

Abstract. Rice Tungro Disease (RTD) is a viral complex that the green leafhopper (Nephotettix virescens) spreads. It is still a big problem for rice production in Southeast Asia, especially in Malaysia and Indonesia. If not properly managed, the disease can cause a lot of crop losses and put regional food security at risk. Roguing, the selective removal of infected plants, is a widely employed control strategy to mitigate RTD transmission; however, its efficacy under diverse biological and environmental conditions has not been adequately assessed. This study introduces a mathematical framework employing nonlinear ordinary differential equations to characterise the interactions between plant and vector populations facilitating RTD transmission. A new way to use the Spectral Weighted Residual Method (SWRM) to solve the governing equations is presented. This method is a simple but strong alternative to traditional analytical methods. The model is further expanded to include dynamic roguing strategies, with simulations performed to assess the impact of roguing intensity and frequency on disease suppression. To check the stability of a system and find the important thresholds for good management, the sensitivity analysis, optimal control, and bifurcation analysis is used. The findings provide a mathematical comprehension of the role of adaptive roguing in diminishing RTD prevalence and improving resource allocation for disease management. This study demonstrates the application of advanced mathematics in practical agricultural management. It also shows how modelling can help people make decisions that will lower the number of rice viral diseases and make food more secure in areas that are at risk.

Keywords: Rice Tungro Disease · Mathematical Modeling · Roguing · Spectral Weighted Residual Method

1 Introduction

Rice is a key part of food security all over the world. More than half of the people in the world eat it every day, and it gives them more than one-fifth of their calories [1]. The area where rice is grown has grown from 1.2 million km^2 in 1960 to 1.7 million km^2 in

M. A. b. Jamlos et al. (Eds.): ICSET 2025, LNNS 1897, pp. 288–297, 2026.
https://doi.org/10.1007/978-3-032-20722-7_36

2022. This has caused production to go up from 157 million tonnes to 520 million tonnes [2].As the population grows, rice demand is expected to rise by 6% by 2030. This makes it even harder to make sure that production is sustainable [3]. However, biotic stresses, especially viral infections, have a big impact on rice productivity, causing annual yield losses of more than 40%. Rice Tungro Disease (RTD) is one of the most serious threats in South and Southeast Asia. It costs the world economy about USD 109 million a year [4]. RTD is still causing problems with production in Indonesia and Malaysia. In South Sulawesi, where there have been severe outbreaks, researchers are testing the durability of resistance in several rice cultivars [5]. RTD results from a simultaneous infection by Rice Tungro Bacilliform Virus (RTBV) and Rice Tungro Spherical Virus (RTSV). RTBV is a double-stranded DNA virus that belongs to the Caulimoviridae family. Its particles are 30–35 nm wide and 100–300 nm long [6–8]. RTSV, on the other hand, is a polyhedral single-stranded RNA virus in the Sequiviridae family. Its particles are about 30 nm in size [9]. The green leafhopper Nephotettix virescens is the only way for the viruses to spread, and it does so by feeding on plants. Plants that have the disease show orange-yellow leaf discolouration, stunted growth, and less tillering within 6 to 15 days [10]. Traditional control methods, like using insecticides, biological control, and roguing (getting rid of infected plants), don't always work the same way in different agroecological conditions. Consequently, mathematical modelling has emerged as a crucial methodology for investigating RTD epidemiology and control. Blas et al. [11] created a model for how N. virescens spreads tungro, and Blas and David [12] built on that model to look at roguing strategies. Anggriani et al. [13] integrated the effects of insecticides, whereas Suryaningrat et al. [14] utilised optimal control analysis to amalgamate chemical and biological interventions. Maryati et al. [15] conducted a more in-depth analysis of the stability of RTD transmission throughout various stages of rice growth, emphasising the significance of plant developmental dynamics. Unlike previous RTD models that examined either roguing or biological control separately, this study integrates both strategies within a unified mathematical framework. To our knowledge, no prior work has combined dynamic roguing and biological control using the multi-domain Spectral Weighted Residual Method (SWRM). The study thus contributes a novel modeling framework and a robust numerical technique for optimizing RTD management strategies. The model shows that roguing works best on small to medium-sized rice farms, where finding and removing infected plants quickly can greatly slow the spread of RTD. Moderate and frequent roguing reduces infection levels without causing a lot of yield loss. This fits with what people have seen in the fields in Indonesia and the Philippines, where timely removal of sick plants cut outbreaks by up to 70% [4, 5]. These results show that adaptive roguing could be a cost-effective and long-term way to control diseases that helps keep rice safe in the area. The findings also show that roguing works better when used with biological control methods, which is a more holistic and eco-friendly way to stop the spread of RTD.

2 Model Formulation

Mathematical modelling is an essential tool for analysing the transmission and control of Rice Tungro Disease (RTD), a significant viral threat to rice cultivation in Asia. Rice plants, insect vectors, and biological control agents all need to work together for

RTD to spread. Because of this, field-based assessments are usually limited by money, time, and changes in the environment. Researchers can analyse these interactions using a mathematical framework to evaluate the efficacy of control methods such as roguing (the removal of diseased plants) and biological predation, and to predict long-term outcomes under varying conditions. This model explains how Tungro disease spreads in rice fields by looking at healthy rice plants (P_0), infected rice plants (P_1), healthy insect vectors (V_0), infected insect vectors (V_1), and natural predators (B) such as ladybugs. Healthy plants become infected when bitten by infected insects, and healthy insects become infected when feeding on diseased plants. Once infected, plants and insects cannot recover. To reduce the disease, two strategies are considered: removing diseased plants (rogueing), represented by (ρ) with $\rho = 0$ indicating no removal and $\rho = 1$ representing perfect removal, while biological control effectiveness is described by the parameters β_1 and β_2.

$$
\begin{aligned}
\frac{dP_0}{dt} &= r\left(K - N_p\right) - \frac{\alpha P_0 V_1}{N_P} - q_0 P_0, \\
\frac{dP_1}{dt} &= \frac{\alpha(1-\rho)P_0 V_1}{N_P} - q_1(1-\rho)P_1 - \rho P_1, \\
\frac{dV_0}{dt} &= bN_V\left(1 - \frac{N_V}{V}\right) - \frac{aV_0 P_1}{N_P} - \mu V_0 - \beta_1 B V_0, \\
\frac{dV_1}{dt} &= \frac{aV_0 P_1}{N_P} - \mu V_1 - \beta_1 B V_1, \\
\frac{dB}{dt} &= r_1 B\left(1 - \frac{B}{N_B}\right) + \beta_2 B(V_0 + V_1) - \gamma B,
\end{aligned} \tag{1}
$$

In simple terms, these equations track how the numbers of healthy and infected plants, insects, and predators change over time, and help us test how removing diseased plants and using natural predators can slow down the spread of Tungro disease. To better understand the interactions among host plants, vectors, and biological control strategies, a compartmental model was developed. The schematic representation is presented in Fig. 1, where the arrows indicate the possible transmission pathways and control interventions, including roguing and biological agents.

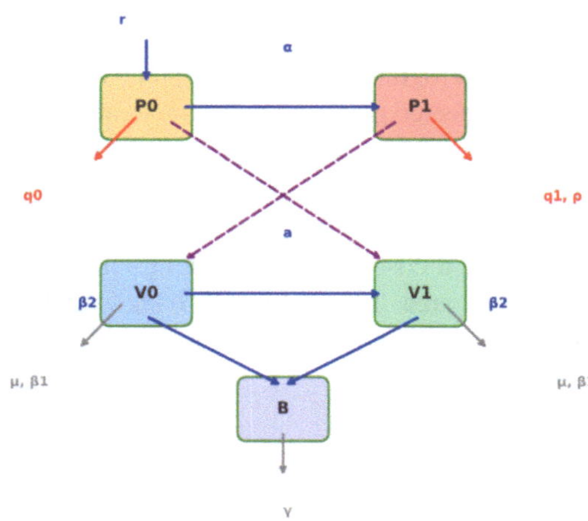

Fig. 1. Compartmental diagram of the Tungro disease transmission model with roguing and biological control interventions.

Table 1. Assigned values for parameters and initial conditions

Parameter	Value	References	Parameter	Value	References
$V_0(0)$	0	Blass and David [12], Amelia et al. [16]	ρ	0.40	Blass and David [12], Amelia et al. [16]
$V_1(0)$	4000	Blass and David [12], Amelia et al. [16]	q_1	0.009	Blass and David [12], Amelia et al. [16]
$P_0(0)$	20000	Blass and David [12], Amelia et al. [16]	b	0.033	Blass and David [12], Amelia et al. [16]
$P_1(0)$	0	Blass and David [12], Amelia et al. [16]	V	5000	Blass and David [12], Amelia et al. [16]
$B(0)$	10	Assumed value	a	0.996	Blass and David [12], Amelia et al. [16]
r	0.001	Blass and David [12], Amelia et al. [16]	μ	0.022	Assumed value
K	30000	Blass and David [12], Amelia et al. [16]	r_1	0.01	Assumed value
α	0.035	Blass and David [12], Amelia et al. [16]	N_B	80	Assumed value
q_0	0.008	Blass and David [12], Amelia et al. [16]	β_1	0.0002	Assumed value
γ	0.01	Assumed value	β_2	0.0001	Assumed value

Equation (1), the parameter P_0 and V_0 represent susceptible plants and vectors, while P_1 and V_1 denote the infected populations. The total plant and vector populations are defined as $N_P = P_0 + P_1$ and $N_V = V_0 + V_1$, respectively. The term $r(K - N_p)$ models replanting, where r is the replanting rate and K is the plant carrying capacity. The parameters include b (vector birth rate), α (transmission rate), and a (average acquisition time of the virus by vectors). Vector mortality or emigration is given by μ, while q_0 and q_1 represent plant loss rates and ρ denotes the roguing rate. The vector population is limited by capacity V. Biological agents are represented by B, with growth rate r_1, capacity N_B, predation rate β_1, growth contribution β_2, and mortality rate γ. Parameter values are summarized in Table 1.

3 Solution Procedures

The multi-domain spectral weighted residual method (MSWRM) is used to solve the system of differential equations that describes how rice, vectors, and biological agents interact in the numerical simulation. The time interval [0,T] is partitioned into distinct subdomains, and within each subinterval, the state variables $[P_0, V_0, P_1, V_1, B]$ are estimated utilising shifted Chebyshev polynomials with indeterminate coefficients. Next, collocation points are set up in both the global and local domains to turn the system into residual equations. These equations are then solved iteratively with continuity conditions between subintervals. We set initial conditions for healthy and infected plants, healthy and infected vectors, and biological agents. Then, we use Mathematica's FindRoot function to find the unknown polynomial coefficients. This method lets you get a good idea of how the model changes over time and lets you compare it to standard methods like Runge–Kutta to make sure it's correct.

3.1 Implementation MSWRM

In order to solve Eq. (1), an approximate solution is assumed for P_0, P_1, V_0, V_1, and B as:

$$\begin{aligned} P_0 \approx P_0^{(n)} &= \textstyle\sum_{i=0}^{N_t} a_i^{(n)} T_i\left(\frac{2t}{T}-1\right), \\ P_1 \approx P_1^{(n)} &= \textstyle\sum_{i=0}^{N_t} b_i^{(n)} T_i\left(\frac{2t}{T}-1\right), \\ V_0 \approx V_0^{(n)} &= \textstyle\sum_{i=0}^{N_t} c_i^{(n)} T_i\left(\frac{2t}{T}-1\right), \\ V_1 \approx V_1^{(n)} &= \textstyle\sum_{i=0}^{N_t} d_i^{(n)} T_i\left(\frac{2t}{T}-1\right), \\ B \approx B^{(n)} &= \textstyle\sum_{i=0}^{N_t} e_i^{(n)} T_i\left(\frac{2t}{T}-1\right), \end{aligned} \tag{2}$$

where $a_i^{(n)}$, $b_i^{(n)}$, $c_i^{(n)}$, $d_i^{(n)}$, and $e_i^{(n)}$ are unknown coefficients to be determined. The functions $T_i\left(\frac{2t}{T}-1\right)$ are shifted Chebyshev polynomials from [−1, 1] to [0,T]. At $n = 1$, Eq. (1) becomes

$$\begin{aligned} \frac{dP_0^{(1)}}{dt} - r(K - N_P) + \frac{\alpha P_0^{(1)} V_1^{(1)}}{N_P} + q_0 P_0^{(1)} &= 0, \\ \frac{dP_1}{dt} - \frac{\alpha(1-\rho) P_0^{(1)} V_1^{(1)}}{N_P} + q_1(1-\rho) P_1^{(1)} + \rho P_1^{(1)} &= 0, \\ \frac{dV_0^{(1)}}{dt} - bN_V^{(1)}\left(1 - \frac{N_V^{(1)}}{V}\right) + \frac{aV_0^{(1)} P_1^{(1)}}{N_P} + \mu V_0^{(1)} + \beta_1 B V_0^{(1)} &= 0, \\ \frac{dV_1^{(1)}}{dt} - \frac{aV_0^{(1)} P_1^{(1)}}{N_P} + \mu V_1^{(1)} + \beta_1 B V_1^{(1)} &= 0, \\ \frac{dB^{(1)}}{dt} - r_1 B^{(1)}\left(1 - \frac{B^{(1)}}{N_B}\right) - \beta_2 B^{(1)}\left(V_0^{(1)} - V_1^{(1)}\right) + \gamma B^{(1)}, \end{aligned} \tag{3}$$

subjected to initial conditions

$$P_0^{(1)}(0) = 20000, \quad P_1^{(1)}(0) = 0, \quad V_0^{(1)}(0) = 0, \quad V_1^{(1)}(0) = 4000, \quad B^{(1)}(0) = 10. \tag{4}$$

Likewise, for $n > 1$, Eq. (1) becomes

$$\begin{aligned} \frac{dP_0^{(n)}}{dt} - r(K - N_P) + \frac{\alpha P_0^{(n)} V_1^{(n)}}{N_P} + q_0 P_0^{(n)} &= 0, \\ \frac{dP_1}{dt} - \frac{\alpha(1-\rho) P_0^{(n)} V_1^{(n)}}{N_P} + q_1(1-\rho) P_1^{(n)} + \rho P_1^{(n)} &= 0, \\ \frac{dV_0^{(n)}}{dt} - bN_V^{(n)}\left(1 - \frac{N_V^{(n)}}{V}\right) + \frac{aV_0^{(n)} P_1^{(n)}}{N_P} + \mu V_0^{(n)} + \beta_1 B V_0^{(n)} &= 0, \\ \frac{dV_1^{(n)}}{dt} - \frac{aV_0^{(n)} P_1^{(n)}}{N_P} + \mu V_1^{(n)} + \beta_1 B V_1^{(n)} &= 0, \\ \frac{dB^{(n)}}{dt} - r_1 B^{(n)}\left(1 - \frac{B^{(n)}}{N_B}\right) - \beta_2 B^{(n)}\left(V_0^{(n)} - V_1^{(n)}\right) + \gamma B^{(n)} &= 0, \end{aligned} \tag{5}$$

with initial conditions:

$$P_0^{(n)}(t_{n-1}) = P_0^{(n-1)}(t_{n-1}),\quad P_1^{(n)}(t_{n-1}) = P_1^{(n-1)}(t_{n-1}),\quad V_0^{(n)}(t_{n-1}) = V_0^{(n-1)}(t_{n-1}),$$
$$V_1^{(n)}(t_{n-1}) = V_1^{(n-1)}(t_{n-1}),\quad B^{(n)}(t_{n-1}) = B^{(n-1)}(t_{n-1}). \tag{6}$$

Equations (3), (4), (5) and (6) are evaluated at collocation points $t_j^n = t_{n-1} + \frac{j(t_n - t_{n-1})}{N_t}$, for $j = 0, 1, 2, \cdots, N_t$, and $n = 1, 2, 3, \cdots, Q$ to obtain the following residual equations.

$$\begin{aligned}
\mathrm{Res}\left[P_0^{(n)}\left(t_j^n\right)\right] &= \frac{dP_0^{(n)}}{dt} - r(K - N_P) + \frac{\alpha P_0^{(n)} V_1^{(n)}}{N_P} + q_0 P_0^{(n)} = 0,\\
\mathrm{Res}\left[P_1^{(n)}\left(t_j^n\right)\right] &= \frac{dP_1}{dt} - \frac{\alpha(1-\rho)P_0^{(n)} V_1^{(n)}}{N_P} + q_1(1-\rho)P_1^{(n)} + \rho P_1^{(n)} = 0,\\
\mathrm{Res}\left[V_0^{(n)}\left(t_j^n\right)\right] &= \frac{dV_0^{(n)}}{dt} - bN_V^{(n)}\left(1 - \frac{N_V^{(n)}}{V}\right) + \frac{aV_0^{(n)} P_1^{(n)}}{N_P} + \mu V_0^{(n)} + \beta_1 B V_0^{(n)} = 0,\\
\mathrm{Res}\left[V_1^{(n)}\left(t_j^n\right)\right] &= \frac{dV_1^{(n)}}{dt} - \frac{aV_0^{(n)} P_1^{(n)}}{N_P} + \mu V_1^{(n)} + \beta_1 B V_1^{(n)} = 0,\\
\mathrm{Res}\left[B^{(n)}\left(t_j^n\right)\right] &= \frac{dB^{(n)}}{dt} - r_1 B^{(n)}\left(1 - \frac{B^{(n)}}{N_B}\right) - \beta_2 B^{(n)}\left(V_0^{(n)} - V_1^{(n)}\right) + \gamma B^{(n)},
\end{aligned} \tag{7}$$

for $j = 0, 1, 2, \cdots, N_t$ and $n = 1, 2, 3, \cdots, Q$,

with

$$\begin{aligned}
Res\left[P_0^{(n)}(t_{n-1})\right] &= \begin{cases} P_0^{(n)}(t_{n-1}) - 20000, & \text{if } n = 1,\\ P_0^{(n)}(t_{n-1}) - P_0^{(n-1)}(t_{n-1}), & \text{if } n > 1, \end{cases}\\
Res\left[P_1^{(n)}(t_{n-1})\right] &= \begin{cases} P_1^{(n)}(t_{n-1}), & \text{if } n = 1,\\ P_1^{(n)}(t_{n-1}) - P_1^{(n-1)}(t_{n-1}), & \text{if } n > 1, \end{cases}\\
Res\left[V_0^{(n)}(t_{n-1})\right] &= \begin{cases} V_0^{(n)}(t_{n-1}), & \text{if } n = 1,\\ V_0^{(n)}(t_{n-1}) - V_0^{(n-1)}(t_{n-1}), & \text{if } n > 1, \end{cases}\\
Res\left[V_1^{(n)}(t_{n-1})\right] &= \begin{cases} V_1^{(n)}(t_{n-1}) - 4000, & \text{if } n = 1,\\ V_1^{(n)}(t_n - t_{n-1}) - V_1^{(n-1)}(t_n - t_{n-1}), & \text{if } n > 1, \end{cases}\\
Res\left[B^{(n)}(t_{n-1})\right] &= \begin{cases} B^n(t_{n-1}) - 10, & \text{if } n = 1,\\ B^n(t_{n-1}) - B^{n-1}(t_{n-1}), & \text{if } n > 1. \end{cases}
\end{aligned} \tag{8}$$

Equations (7) and (8) are solved by employing FindRoot in Wolfram Mathematica to obtain the values of unknown constant coefficients, $a_i^{(n)}$, $b_i^{(n)}$, $c_i^{(n)}$, $d_i^{(n)}$, and $e_i^{(n)}$, where and Q are chosen as $N_t = 4$ and $Q = 301$.

Table 2. Absolute error computation for P_0 as time (t) increases

t	0	50	100	150	200
MSWRM	20000	12926.9394594999	9047.8499205230	6574.4345933572	4997.3153518870
RK4	20000	12926.9394612417	9047.8499198568	6574.4345941192	4997.3153540933
Absolute error	0	1.74×10^{-6}	6.66×10^{-7}	7.62×10^{-7}	2.20×10^{-6}

Table 3. Absolute error computation for P_1 as time (t) increases

t	0	10	20	30	50
MSWRM	0	78.3377753166	5.2415600543	0.2326478573	0.0011847167
RK4	0	78.3377627373	5.2415717757	0.2326478760	0.0011847109
Absolute error	0	1.26×10^{-5}	1.17×10^{-5}	1.87×10^{-8}	5.89×10^{-9}

4 Results and Discussion

The numerical experiments were carried out to assess both the validity of the proposed spectral weighted–residual scheme and the epidemiological insights obtained from the rice–vector–biological agent model. The initial conditions were set as $V_0(0) = 0$, $V_1(0) = 4000$, $P_0(0) = 20000$, $P_1(0) = 0$, $B(0) = 10$. Unless otherwise specified, the main parameter values used were $a = 0.996$, infection rate, $\rho = 0.4$, roguing rate, which are consistent with previous modelling studies on rice tungro disease. To establish the accuracy of the numerical solver, simulations over time interval [0,200] were compared with the classical fourth-order Runge–Kutta method. Figure 2 illustrates the trajectories of healthy (P_0) and infected (P_1) rice plants obtained from both approaches, which show excellent agreement. We compute the absolute error between the two solutions (see Tables 2 and 3) and achieve a minimum error magnitude on the order of 10^{-5}. This confirms that the MSWRM provides reliable approximations and is appropriate for analysing the nonlinear disease dynamics. The epidemiological relevance of the model is highlighted in Fig. 2, which demonstrates the impact of increasing the roguing rate (ρ) on the infected rice plant population. For fixed $\mu = 0.022$ and $\beta_1 = 0.0002$, The simulations show that as roguing gets stronger, the number of infected plants goes down a lot. This shows that roguing is a good way to get rid of sources of inoculum and stop the virus from spreading through vectors. This result shows that roguing is an important way to control tungro. Figures 2 and 3 are the focus here because of space limits, but the other results also give useful information. These encompass the impacts of roguing on healthy plants and vector populations, in addition to the function of biological predation in diminishing both healthy and infected leafhopper densities. These new findings support the idea that integrated management, which combines roguing with biological control, is a strong way to fight rice tungro disease. In summary, these findings demonstrate that the proposed framework is both numerically sound and biologically significant, effectively representing the dynamics of tungro transmission and offering a quantitative foundation for assessing control strategies. This study employed a spectral weighted-residual methodology to analyse a rice-vector-biological agent model for Rice Tungro Disease (RTD). When compared to the standard fourth order Runge-Kutta method (Fig. 2), the proposed solver gave very accurate estimates of plant populations. This shows that it can be used for nonlinear epidemic modelling. The effectiveness of roguing as a control measure was subsequently demonstrated (Fig. 3), showing that a higher roguing rate significantly reduced the number of infected rice plants. These results demonstrate two significant conclusions: firstly, that the suggested numerical scheme is both robust and efficient for simulating intricate plant–virus–vector interactions; and secondly, that roguing serves as a practical and effective measure to reduce the spread of RTD. The results collectively validate that the model offers both a robust numerical framework and a biologically relevant instrument for assessing integrated management strategies.

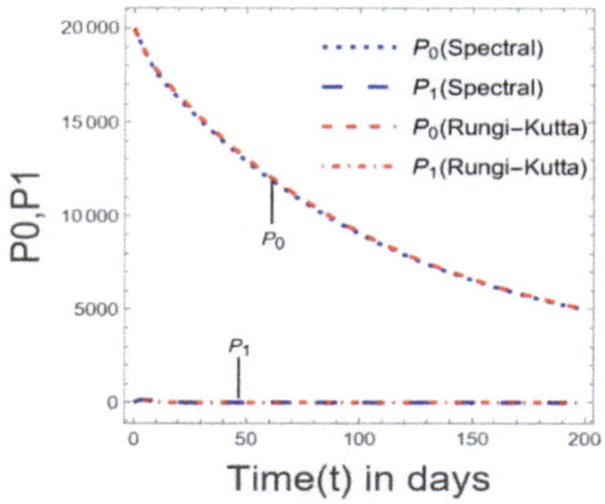

Fig. 2. Dynamic behaviour of healthy and infected rice plant.

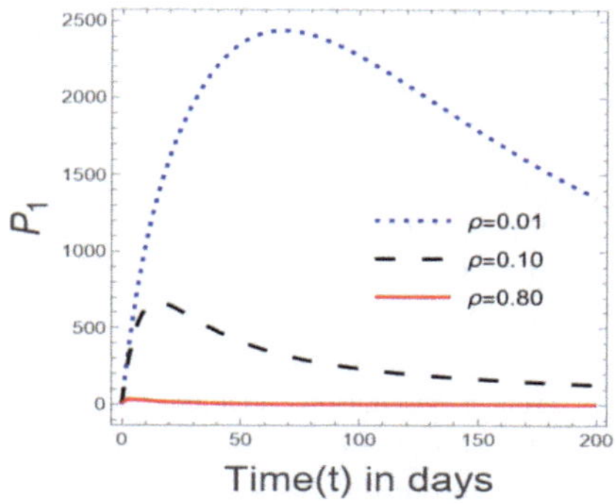

Fig. 3. Impact of roguing on infected rice plant.

While the method and findings are promising, detailed analyses of equilibrium stability, threshold conditions, and parameter sensitivity were beyond the scope of this manuscript. These components will be addressed in future work to strengthen model-based decision support for RTD control policies.

5 Conclusion

This research utilised a spectral weighted-residual approach to examine a rice-vector-biological agent model for Rice Tungro Disease (RTD). Validation against the standard fourth-order Runge–Kutta method (Fig. 2) showed that the proposed solver gives very accurate estimates of plant populations, which shows that it can be trusted for nonlinear epidemic modelling. The efficacy of roguing as a control measure was subsequently illustrated (Fig. 3), demonstrating that an increased roguing rate significantly diminished the incidence of infected rice plants. These results demonstrate two significant conclusions: firstly, that the suggested numerical scheme is both robust and efficient for simulating intricate plant–virus–vector interactions; and secondly, that roguing serves as a practical and effective measure to reduce the spread of RTD. The results collectively validate that the model offers both a thorough numerical framework and a biologically relevant instrument for assessing integrated management strategies.

Acknowledgement. Financial support (UIC251504 (RDU252703)) and Tabung Persidangan Dalam Negara (TPDN) was provided by UMPSA. The writers would like to extend their gratitude to Parahyangan Catholic University for their valuable advice and assistance.

Nomenclature

P_0	healthy rice plants
P_1	infected rice plants
V_0	healthy vectors
V_1	infected vectors
B	natural predators
ρ	roguing rate
β_1	predation rate
β_2	predator growth rate
r	replanting rate
K	plant population capacity
N_p	total population of plants
N_v	total population of vectors
V	vector population capacity
N_B	predator population capacity
γ	predator death rate
α	transmission rate from infected vectors to the health plants
a	transmission rate from infected plants to the health vectors
q_0, q_1	plant loss rate due to crop period or loss of biomass
b	birth rate of vectors
μ	death rate or immigration of vectors
r_1	birth rate of predators

References

1. Yuan, S., et al.: Sustainable intensification for a larger global rice bowl. Nat. Commun. **12**(1), 1–11 (2021)
2. Liu, Y., et al.: Rice paddy soils are a quantitatively important carbon store according to a global synthesis. Commun. Earth Environ. **2**, 1–9 (2021)
3. Uwizeye, A., et al.: Nitrogen emissions along global livestock supply chains. Nat. Food. **1**(7), 437–446 (2020)
4. Pangga, I.B., Sta. Cruz, F.C.: Rice tungro disease. In: Khan, M.S., Ahmad, J. (eds.) Viral Diseases of Field and Horticultural Crops, pp. 81–86. Elsevier, Amsterdam (2024)
5. Rosida, N., Kuswinanti, T., Amin, N., Nasruddin, A.: Resistance durability of several rice cultivars against rice tungro disease in South Sulawesi, Indonesia. Pak. J. Phytopathol. **32**(2), 225–233 (2020)
6. Qu, R., et al.: Characterization of the genome of rice tungro bacilliform virus: Comparison with commelina yellow mottle virus and caulimoviruses. Virology. **185**(1), 354–364 (1991)
7. Bao, Y., Hull, R.: Mapping the 5′-Terminus of rice Tungro bacilliform viral Genomic RNA. Virology. **197**(1), 445–448 (1993)
8. Laco, G.S., Beachy, R.N.: Rice tungro bacilliform virus encodes reverse transcriptase, DNA polymerase, and ribonuclease H activities. Proc. Natl. Acad. Sci. USA. **91**(7), 2654–2658 (1994)
9. Dey, S.R., Das, R., De, M.: A brief review on present status of rice Tungro disease: Types of viruses, vectors, occurrence, symptoms, control and resistant rice varieties. Int. J. Adv. Life Sci. Res. **7**(3), 15–23 (2024)

10. Bigirimana, V.D., Hua, G.K., Nyamangyoku, O.I., Höfte, M.: Rice sheath rot: An emerging ubiquitous destructive disease complex. Front. Plant Sci. **6**, 1–16 (2015)
11. Blas, N.T., Addawe, J.M., David, G.: A mathematical model of transmission of rice tungro disease by *Nephotettix virescens*. AIP Conf. Proceeding. **1787**, 080004 (2016)
12. Blas, N., David, G.: Dynamical roguing model for controlling the spread of tungro virus via *Nephotettix virescens* in a rice field. J. Phys. Conf. Ser. **893**, 1–5 (2017)
13. Anggriani, N., Yusuf, M., Supriatna, A.K.: The effect of insecticide on the vector of rice Tungro disease: Insight from a mathematical model. Information. **20**(9A), 6197–6206 (2017)
14. Suryaningrat, W., Anggriani, N., Supriatna, A.K., Istifadah, N.: The optimal control of rice tungro disease with insecticide and biological agent. AIP Conf. Proceeding. **2264**, 1–12 (2020)
15. Maryati, A., Anggriani, N., Carnia, E.: Stability analysis of tungro disease spread model in rice plant using matrix method. BAREKENG: J. Ilmu Mat. dan Terap. **16**(1), 217–228 (2022)
16. Amelia, R., Anggriani, N., Supriatna, A.K., Istifadah, N.: Analysis and optimal control of the tungro virus disease spread model in rice plants by considering the characteristics of the virus, roguing, and pesticides. Mathematics. **11**(5), 1–14 (2023)

Finite Element Analysis of Particle Shape and Impact Angle Effects on Steam Turbine Blade Erosion

Wan Fathul Hakim W. Zamri[1(✉)], Nazar Julasrin[2], Azhari Shamsudeen[1], and Muhamad Faiz Md Din[3]

[1] Department of Mechanical & Manufacturing Engineering, Universiti Kebangsaan Malaysia, 43000 Bangi, Selangor, Malaysia
wfathul.hakim@ukm.edu.my

[2] Taisek Lamick Asia (Malaysia) Sendirian Berhad, 43000 Kajang, Malaysia

[3] Faculty of Engineering, The National Defence University of Malaysia, Kuala Lumpur, Malaysia

Abstract. The turbine blade is a critical component of a steam turbine, enabling the conversion of thermal energy from steam into mechanical energy for electricity generation. However, turbine blades are highly vulnerable to erosive wear due to the high-velocity steam flow that often carries abrasive particles such as dust and sand. This study investigates the influence of impact angle and particle shape on erosion behavior by developing a finite element model in Abaqus/CAE. Two im-pact angles (30° and 90°) and two particle geometries (spherical and angular) were examined to assess their effects on erosion resistance, with Plastic Equivalent Strain (PEEQ) used as the primary indicator of material degradation. The results show that spherical particles consistently produced higher PEEQ values than angular particles at both shallow and perpendicular angles, indicating greater plastic strain absorption but lower actual material removal due to rolling and sliding interactions. In contrast, angular particles, despite generating lower PEEQ values, caused more localized surface damage through sharp-edge penetration and interlocking effects, leading to more severe erosion. These findings highlight the importance of particle shape and impact angles in determining turbine blade erosion mechanisms, providing valuable insights for improving material design, erosion resistance, and operational durability in steam turbines.

Keywords: Erosion · Turbine Blade · Impact Particles

1 Introduction

Steam turbines remain a fundamental driving force in modern turbomachinery, primarily used for electricity generation due to their ability to efficiently convert thermal energy from steam into mechanical power [1]. The stator blades accelerate and direct high-temperature, high-pressure steam onto the rotating blades, transmitting torque to the rotor and ultimately driving the generator [2].

M. A. b. Jamlos et al. (Eds.): ICSET 2025, LNNS 1897, pp. 298–304, 2026.
https://doi.org/10.1007/978-3-032-20722-7_37

Steam turbines are vital in electricity generation due to their ability to convert thermal energy from high-pressure steam into mechanical energy. The turbine blade plays a crucial role in this process, as it transfers energy from the steam to the rotor [3]. How-ever, turbine blades operate in extreme environments, where high-velocity steam often contains entrained water droplets and abrasive solid particles such as dust and sand [4]. The continuous impact of these particles on the blade surface leads to erosive wear, causing material loss, surface deformation, and increased roughness [5]. Such damage reduces the effective surface area for steam interaction, disrupts flow patterns, and increases turbulence, ultimately decreasing power output and overall turbine efficiency. In addition, erosion accelerates maintenance requirements, raising operational costs due to frequent repairs and blade replacements [6].

Although numerous experimental studies have explored turbine blade erosion, they are generally constrained by high costs, long testing durations, and difficulty in isolating the combined influence of multiple operating factors [7–10]. Moreover, much of the existing work tends to focus on individual parameters such as either particle velocity, impact angle, or particle shape without considering their combined effects. Computational modeling, particularly through finite element analysis (FEA), offers a cost-effective and systematic approach to overcoming these limitations [11, 12]. However, current computational research remains limited in providing an integrated analysis of the two most influential factors: impact angle and particle shape. This represents a critical knowledge gap, as understanding the interaction of these variables is essential for developing predictive models of erosion and for guiding more durable turbine blade designs.

To address the limitations of previous studies, this research develops a finite element model using Abaqus/CAE to systematically investigate the erosion behavior of steam turbine blades under particle impact. Unlike prior works that typically focused on single parameters or relied heavily on experimental testing, this study simultaneously examines the combined influence of impact angle (30° and 90°) and particle shape (spherical and angular) on erosion response. By employing Plastic Equivalent Strain (PEEQ) as an indicator of surface damage, the model provides new insights into how particle geometry and impact orientation govern erosion mechanisms [13–15]. The outcomes of this study not only fill a critical gap in understanding the relationship between particle characteristics and erosion resistance but also offer valuable guidance for optimizing turbine blade design, enhancing erosion-resistant strategies, and improving the durability and efficiency of steam turbine systems. While other parameters such as particle velocity, size, material properties, and temperature also play significant roles in erosion phenomena, the present study focuses exclusively on the effects of impact angle and particle geometry. This limitation allows for a more con-trolled investigation and clearer interpretation of how these two primary factors influence erosion behavior in steam turbine blades.

2 Methodology

Using Abaqus/CAE, a finite element model of the steam turbine blade subjected to particle impact was developed, as illustrated in Fig. 1. Two types of impacting particles were considered in this analysis: a spherical (rounded) particle and an angular particle.

The corresponding input parameters employed in the simulation are summarized in Table 1. To accurately represent erosion behavior, the turbine blade was modeled as a deformable body, allowing material deformation upon contact with the impacting particles. This approach reflects the progressive material loss characteristic of erosive wear. The turbine blade material was selected as Nimonic alloy, which is widely used in high-temperature turbine applications due to its excellent mechanical and thermal properties. In this study, the particle velocity is set to 500 m/s.

Table 1. Material properties of turbine blade

Part/Material Properties	Dimension/ Values
Turbine blade plate (mm)	200 × 200 × 50
Impact particle radius (mm)	10
Young's Modulus (GPa)	210
Poisson ratio	0.3
Density (g/cm3)	8.40

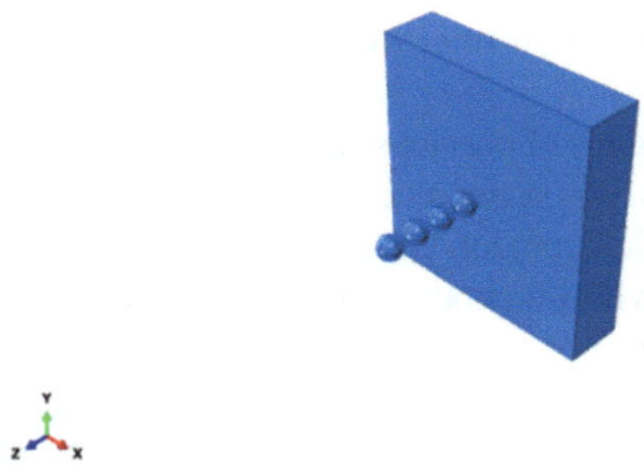

Fig. 1. Model for a steam turbine blade with the impacting particle.

A mesh validation test was conducted to determine the optimal mesh size for the simulation. This validation ensures that the selected meshing provides accurate and reliable results. The initial mesh size of 20 mm was applied and subsequently refined in increments of 5 mm for each test. It was observed that smaller mesh sizes produced a smoother mesh and improved result accuracy. Based on this validation, a mesh size of 20 mm was adopted for the simulation, as it provided a sufficiently smooth mesh with results deemed accurate and computationally efficient.

3 Results and Discussion

The shape of impactor particles plays a critical role in determining erosion behavior, as it directly influences the mechanism of material removal from the turbine blade surface. Spherical particles, with their smooth surfaces and curved edges, generally produce lower erosive effects because they tend to roll or slide upon impact. This rolling action reduces the effective cutting and ploughing force applied to the surface, thereby minimizing material loss. In contrast, angular particles, characterized by sharp edges and irregular geometry, exert higher erosive forces. Their edges penetrate the target surface, causing localized stress concentrations, micro-cutting, and dis-placement of surface material, which accelerate erosion.

In erosion studies of steam turbine blades, PEEQ is a critical indicator of material damage severity. Erosion of turbine blades (whether by solid particles like dust/ash or liquid droplets in wet steam) is fundamentally a surface degradation process caused by repeated high-velocity impacts [6]. Each impact can induce plastic deformation in the blade material; over time this can lead to material removal (wear) through mechanisms like micro-cutting, fatigue cracking, or splashing off of material. PEEQ is directly relevant because it measures the plastic strain energy absorbed by the material during these impacts, which correlates with the damage imparted. Regions experiencing higher PEEQ in the simulation are those where the material has undergone significant plastic flow. These are the zones where one would expect dents, microscopic pits, or eventual erosion to occur. In essence, the higher the PEEQ from an impact, the more severe the local deformation, and thus the higher the likelihood that material will be removed either immediately or after repeated loading. There is generally a direct relationship between increasing PEEQ values and the severity of erosion or wear in ductile materials. As PEEQ increases, it signifies more severe plastic deformation, which usually corresponds to more severe wear. Small PEEQ values indicate that the material experienced only light yielding (or none at all), while large PEEQ values indicate heavy plastic straining that can be a precursor to material removal. In finite element simulations (such as in Abaqus), the Plastic Equivalent Strain (PEEQ) is essentially the accumulated equivalent plastic strain, often derived from the second invariant of the plastic strain tensor.

In this study, two impact angles 30° and 90° were analyzed to assess erosion performance under different conditions. The results highlight the significant influence of particle shape and impact angle on erosion behavior of steam turbine blade materials (Table 2). At an impact angle of 30°, which generally corresponds to maximum erosion conditions due to cutting and ploughing mechanisms, the spherical particle produced a higher PEEQ value (0.5015) compared to the angular particle (0.3992). Similarly, at a perpendicular impact angle of 90°, which typically results in lower erosion severity dominated by deformation rather than cutting, the spherical particle again exhibited a higher PEEQ (0.4639) relative to the angular particle (0.2946). These findings may appear counterintuitive since angular particles, with their sharp edges, are often associated with higher erosion severity due to localized stress concentrations and cutting action. However, the higher PEEQ values observed for spherical particles indicate that the target material absorbs more plastic strain energy when impacted by spherical particles. This suggests that spherical particles induce greater overall deformation, but the deformation is distributed more uniformly across the surface, resulting in less material removal. In other words, the material undergoes more plastic strain (higher PEEQ) yet remains more resistant to actual erosive wear because the smooth, curved geometry of spherical particles promotes rolling or sliding interactions rather than aggressive cutting.

Table 2. The significant influence of particle shape and impact angle on erosion behavior of steam turbine blade materials

Angle	Rounded Shape	Angular Shape
30°	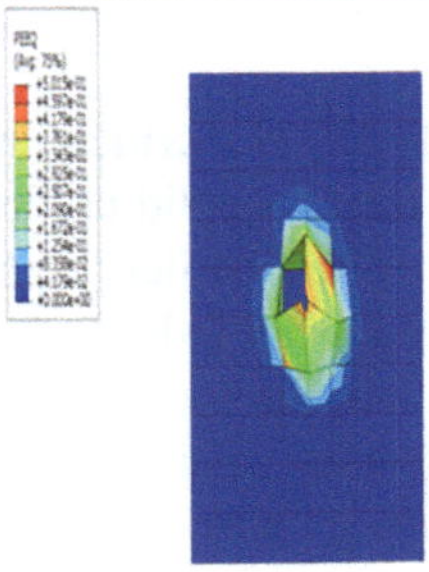	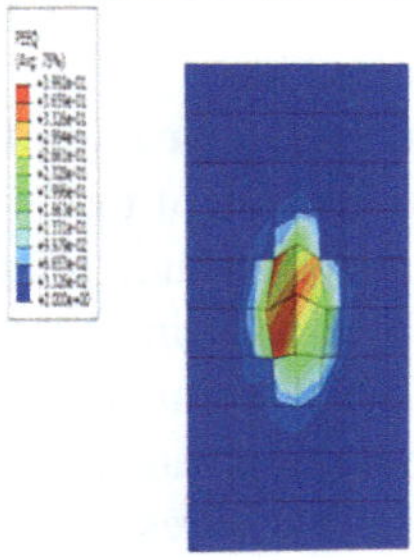
90°	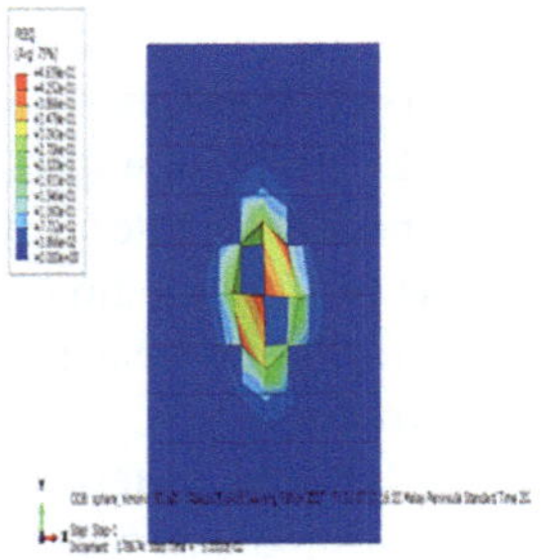	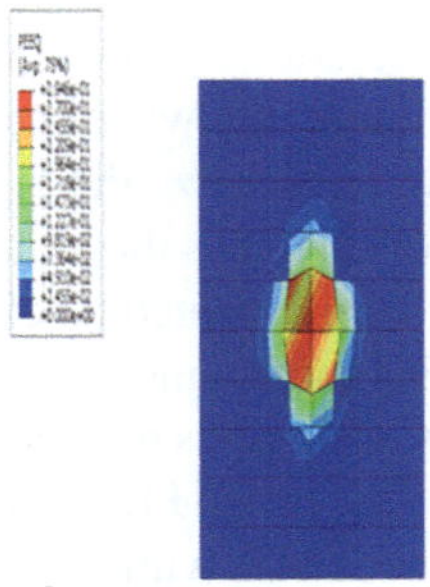

Conversely, angular particles, despite producing lower PEEQ values, tend to generate localized damage through sharp-edge penetration and interlocking effects. These mechanisms can dislodge surface material more efficiently, resulting in greater erosion rates even though the cumulative plastic strain is lower. The importance of this finding lies in distinguishing between plastic strain accumulation (PEEQ) and erosion wear severity. A higher PEEQ does not necessarily equate to higher erosion rates; rather, it reflects the material's ability to absorb deformation without immediate material loss. Thus, spherical particles demonstrate superior erosion resistance, as they induce higher plastic strain but less effective cutting, whereas angular particles, with lower PEEQ, cause more severe erosive wear due to their sharper geometry.

4 Conclusion

This study demonstrated the significant influence of impact angle and particle shape on the erosion behavior of steam turbine blade materials using finite element modeling in Abaqus/CAE. The results revealed that spherical particles consistently produced higher PEEQ values than angular particles at both shallow (30°) and perpendicular (90°) impact angles, indicating greater plastic strain absorption but reduced material removal due to rolling and sliding interactions. In contrast, angular particles, despite showing lower PEEQ values, generated more localized damage through sharp-edge penetration and interlocking effects, leading to more severe erosion. Overall, the findings highlight that spherical particles exhibit superior erosion resistance compared to angular particles, offering valuable insight into erosion mechanisms and contributing to improved strategies for turbine blade design and durability

Acknowledgments. The authors would like to express their gratitude and thanks to Ministry of Higher Education Malaysia and Universiti Kebangsaan Malaysia for funding this research under the Fundamental Researchers Grant Scheme (FRGS/1/2022/TK09/UKM/02/31).

References

1. Zainuddin, N.S., Zamri, W.F.H.W., Omar, M.Z., bin Md Din, M.F.: Comprehensive insight into the failure mechanisms, modes, and material selection of steam turbine blades. Journal of Current. Sci. Technol. **14**(3), 47–47 (2024)
2. Pauzi, A.A., Yunus, S.M., Husin, S., Ghazali, M.J., Zamri, W.F.H.W.: Wear characteristics of a combustion liner for power generation gas turbine. Int. J. Mater. Prod. Technol. **57**(1–3), 230–239 (2018)
3. Ramadhan, D.F., Rusirawan, D.: Evaluation turbine design and materials steam power plant: literature review. J. Rekayasa Energi dan Mek. **5**(1), 8 (2025)
4. Di, J., et al.: Experimental investigation on the effect of solid particle erosion on the water droplet erosion damages for blade materials. Sci. Rep. **15**(1), 18183 (2025)
5. Zainuddin, N.S., Zamri, W.F.H.W., Omar, M.Z., Din, M.F.B.M., Pauzi, A.A.B.: Unveiling the crucial factors and coating mitigation of solid particle erosion in steam turbine blade failures: A review. Rev. Adv. Mater. Sci. **64**(1), 20250089 (2025)
6. Azman, A.A.S., Ghazali, M.J., Zamri, W.F.H.W., Pauzi, A.A.: A review on coating and finite element analysis for optimization gas turbine. In: Proceedings of the Malaysian International Tribology Conference 2020, pp. 52–60. Malaysian Tribology Society, Langkawi, Kedah (2020)
7. Zhang, Z., Yang, B., Zhang, D., Xie, Y.: Experimental investigation on the water droplet erosion characteristics of blade materials for steam turbine. Proc. Inst. Mech. Eng. C J. Mech. Eng. Sci. **235**(20), 5103–5115 (2021)
8. Li, D., Jiang, P., Sun, F., Yuan, X., Zhang, J., Cao, X.: Water-droplet erosion behavior of high-velocity oxygen-fuel-sprayed coatings for steam turbine blades. Corros. Rev. **40**(1), 39–49 (2022)
9. Suang, N.J., Zamri, W.F.H.W., Din, M.F.M., Shamsudeen, A., Mohamed, I.F., Ariffin, A.K.: Effects of layer addition on hardness of a multi-layered TiAlN and TiN coating. Solid State Technol. **64**(2), 7339–7351 (2021)
10. Wang, Z., Liu, T., Luo, R.: Study on water erosion and preventive measures of last stage blade of steam turbine. J. Phys.: Conf. Ser. **2076**(1), 012071 (2021)
11. Avevor, J., Aikins, S.A., Enyejo, L.A.: Optimizing gas and steam turbine performance through predictive maintenance and thermal optimization for sustainable and cost-effective power generation. Int. J. Innov. Sci. Res. Technol. **10**(3), 994–1010 (2025)
12. Yau, L.C., Zamri, W.F.H.W., Mohamed, I.F., Kamal, A., Ariffin, M.F.M.D.: An elasto-plastic ball bearing contact with bilinear hardening. Int. J. Innov. Technol. Explor. Eng. **9**(4), 438–445 (2020)
13. Li, M., You, J.H.: Structural impact of creep in tungsten Monoblock divertor target at 20 MW/m2. Nucl. Mater. Energy. **14**, 1–7 (2018)

14. Zhang, Z., Ren, Y., He, G., Yang, Z.: Enhancing erosion resistance of ceramic/metal multilayer coating by crack inhibition design. J. Vac. Sci. Technol. A. **42**(5), 053412 (2024)
15. Li, W., et al.: Cavitation erosion prevention using laser shock peening: development of a predictive evaluation system. Materials. **16**(14), 5096 (2023)
16. Author, F.: Article title. Journal. **2**(5), 99–110 (2016)

AISprayAndWait: An AI-Enhanced Routing Protocol for Delay Tolerant Networks

Halikul Bin Lenando[1(✉)], Sanjay Charles Albert[2], Inson Din[2], and Chong Eng Tan[1]

[1] Institute Social Informatics and Technological Innovations, University Malaysia Sarawak, 94300 Kota Samarahan, Malaysia
{cool,cetan}@unimas.my

[2] Faculty of Computer Science and Information Technology, University Malaysia Sarawak, 94300 Kota Samarahan, Malaysia
21010271@siswa.unimas.my, inson@unimas.my

Abstract. Delay Tolerant Networks (DTNs) operate in intermittent network connectivity and dynamic topology. This poses a challenge to achieving efficient data exchange. Traditional routing protocols such as Epidemic, Spray and Wait, and PRoPHET aim for high delivery but suffer from high overhead. This paper introduces AISprayAndWait, a lightweight AI-enhanced routing protocol that builds upon the Spray and Wait mechanism by integrating intelligent forwarding decisions. The protocol evaluates four contextual features—encounter frequency, recency of contact, buffer availability, and a bias term—to compute a forwarding score using a sigmoid function. The simulation results demonstrate that AISprayAndWait significantly improves delivery probability and reduces overhead compared to existing protocols. These findings show that lightweight AI integration is possible in optimizing DTN routing strategies.

Keywords: Delay Tolerant Networks (DTNs) AISprayAndWait · Artificial Intelligence (AI) · Opportunistic routing

1 Introduction

Delay Tolerant Networks (DTNs) [1] operate in intermittent connectivity with dynamic topology. These networks rely on the store-carry-forward paradigm. Traditional routing protocols such as Epidemic [2], Spray and Wait [3, 4], and PRoPHET [5] have been used to address these challenges but suffer from trade-offs between delivery probability, latency, and overhead. The advancements in AI-driven networking have shown potential in optimizing data dissemination in DTNs. Techniques like ML-Fresh [6] and RL4DTN [7] demonstrate how machine learning enhances routing decisions using AI. Similarly, frameworks like Edge MLOps [8] and TinyML [9] utilize intelligent processing on limited resources to improve latency and energy efficiency. However the integration of lightweight AI mechanisms into routing protocols remains unexplored. This paper proposing an AI-enhanced routing protocol that intelligently selects relay nodes based on various parameters or features.

M. A. b. Jamlos et al. (Eds.): ICSET 2025, LNNS 1897, pp. 305–312, 2026.
https://doi.org/10.1007/978-3-032-20722-7_38

The remainder of this paper is organized as follows. Section 2 reviews related works on data dissemination in DTNs and AI-driven routing strategies, highlighting the strengths and limitations of existing approaches. Section 3 introduces the AISprayAndWait protocol, including details of its architecture, scoring mechanism, and forwarding algorithm. Section 4 describes the simulation setup using the ONE simulator. Section 5 presents the discussion and comparison between AISprayAndWait, Epidemic, Spray and Wait, and PRoPHET across multiple performance indicators. Finally, Sect. 6 concludes the paper by highlighting the findings and outlining the next directions of this research.

2 Related Works

2.1 Data Dissemination Approaches in Opportunistic Network

Ensuring efficient data dissemination is challenging due to unstable and unpredictable network conditions. In general, the traditional data dissemination approaches can be categorized into three major categories, i.e., Publish/Subscribe and social-based, Context-Aware Forwarding, and Replication, coding, and incentive-based.

Publish/Subscribe methods are models decoupling senders and receivers [10]. It is a one-to-many communication. Whereas social-based dissemination always leverages social metrics (e.g., centrality, similarity, closeness, etc.) [11, 12] to choose relay nodes to improve the likelihood of encountering subscribers. Context-aware methods use node attributes (energy, buffer, mobility, social ties) [13] to adapt forwarding decisions. They often use fuzzy logic or utility functions to optimize delivery and resource use. Replication-based (e.g., epidemic, two-hop) [14] and coding-based (e.g., network coding, erasure coding) [15] techniques increase delivery probability but can cause high overhead due to the lack of a decision-making approach. Therefore, many emerging routing approaches are improving this type of routing using context-based, social-based, and, more recently, integrating artificial intelligence.

2.2 AI-Driven Data Dissemination in IoT Environment.

The integration of Artificial Intelligence (AI) into Internet of Things (IoT) enabling intelligent decision-making and automation possible across various disciplines [16]. Khadam and his colleagues conducted a systematic mapping study the roles of AI in IoT and identify application that the most used in that domains. As AI-enabled IoT systems continue to expand, ensuring reliable and low-latency performance in data processing becomes crucial. Oikonomou and his colleagues pointed out that significant processing power is necessary to maintain system reliability [17].

While Edge Computing focuses on reducing latency and processing data closer to the source, it requires a well-structured architecture and components to support a wide range of applications and services [8]. Raj and his colleagues proposed a framework to facilitate machine learning at the edge for AIoT applications. The aim is to enable continuous delivery, deployment, and monitoring of machine learning models through Edge MLOps.

Integrating Artificial General Intelligence (AGI) to achieve more adaptive and context-aware IoT systems is the next step in the evolution of IoT environments. AGI has garnered significant interest across scientific, commercial, and societal domains [18]. However, adapting AGI to IoT environments presents a major challenge, as IoT systems are limited in resources.

Garg, Dixit, and Sethi (2022) introduced ML-Fresh, which integrates machine learning techniques to establish optimal communication paths between nodes. The use of machine learning assists the protocol by predicting node behavior and connectivity patterns, allowing dynamic adjustment of forwarding decisions and minimizing transmission redundancy. In their experiment, ML-Fresh outperformed existing Opportunistic Network frameworks in various metrics such as delivery ratio, latency, and throughput. Similarly, Visca and Baliosian (2022) proposed the RL4DTN approach, which leverages Q-learning to mitigate unpredictable situations in opportunistic networks. A temporal graph is used to enable nodes to learn connectivity patterns and temporal evolution for better forwarding decisions. They found that integrating Q-learning features enhances the algorithm's overall performance. This indirectly shows that reinforcement learning has potential in optimizing performance in stochastic Opportunistic Network environments [7].

Another study explored collaborative learning in AI to enhance data dissemination in Opportunistic Networks. Galliera, Venable, Bassani, and Suri (2024, 19] introduced a graph-based Multi-Agent Reinforcement Learning (MARL) framework to disseminate information effectively through rapid and reliable communication. Using this agent-based approach, the system can anticipate network changes while maintaining effective information exchange. Their experimental results show that the proposed method achieves better network coverage across dynamic graphs compared to the Deep Graph Network (DGN) and the Multipoint Relay (MPR) heuristic [20].

3 AISprayAndWait (AISaW) Routing Protocol

The AISprayAndWait integrates a lightweight AI-based decision-making mechanism to improve the Spray-and-Wait approach. There are four key features: encounter frequency, recency of contact, buffer availability, and a bias term used to assist the AI-based forwarding process. Each feature is assigned a weight, and then the score is calculated to determine whether a message should be forwarded to a neighbor or not. The forwarding decision is controlled by a threshold parameter $\theta \in [0,1]$ to ensure that only neighbors with sufficiently high scores are selected for relaying the message. The learning rate η is not actively updated in order to keep the algorithm lightweight.

3.1 Forwarding Function

The Algorithm 1 is designed to select the best current neighbors for forwarding message intelligently. In the case, the score more than the predefined threshold, it then add the connection into the can candidate connection list. Otherwise, the message will be broadcasted to all available connection at particular period of time.

Algorithm 1 Forwarding Decision

Input: Set of connections $\mathcal{C}$, message set M, forwarding threshold θ
Output: Set of selected forwarding connections
$C \leftarrow \emptyset$ ▷ Initialize candidate connection set
for each connection $c \in \mathcal{C}$ **do**
    $n \leftarrow \text{peer}(c)$
    $S(n) \leftarrow \max_{m \in M} s(n, m)$ ▷ Compute best score for neighbor
    if $S(n) \geq \theta$ **then**
        $C \leftarrow C \cup \{c\}$ ▷ Add to candidate set
    end if
end for
if $C \neq \emptyset$ **then**
    Forward messages in M to connections in C
else
    Forward messages in M to all available connections in $\mathcal{C}$
end if

3.2 Scoring Function

The scoring function is used to measure whether it is appropriate to forward a message to neighbour node or not as shown in Algorithm 2.

Algorithm 2 Scoring Function

Input: Neighbor node n, message m, weight vector $\mathbf{w} = [w_1, w_2, w_3, w_4]$
Output: Score $s(n, m) \in [0, 1]$
if $n = \text{destination}(m)$ **then**
    return 1
else
    Compute feature vector $\mathbf{f}(n) = [f_1(n), f_2(n), f_3(n), f_4(n)]$
    Compute linear combination: $z \leftarrow \sum_{i=1}^{4} w_i \cdot f_i(n)$
    Compute score: $s(n, m) \leftarrow \frac{1}{1+e^{-z}}$
    return $s(n, m)$
end if

4 Simulation Setup

This study is conducted based Opportunistic Network Environment (ONE) [39] simulator. The goal is to assess the performance of the proposed routing protocol, AISaW, against the Epidemic, Spray and Wait, and PRoPHET algorithms. The simulator settings are presented in Table 1.

Table 1. Simulator parameter settings

Simulation time	:	43200 s
Interface	:	Bluetooth interface
Transmit speed	:	250 k (2 Mbps)
Protocols	:	Epidemic, Spray and Wait, ProPHET, AISprayAndWait
Event Interval	:	25 to 35 s
Message size	:	0 to 126 k

(continued)

Table 1. *(continued)*

Simulation time	:	43200 s
No. of nodes	:	40
Mobility Model	:	ShortestPathMapBasedMovement

5 Result and Analysis

Figure 1 shows the comparison of routing protocols based on average buffer time. The higher buffer time in the AISprayAndWait approach indicates that messages are retained longer due to the selective forwarding process based on AI scoring. This might lead to increased latency and congestion.

Based on Fig. 2, the AISprayAndWait protocol outperforms the other approaches with the lowest average hop count (2.45). This shows that AISprayAndWait optimizes node selection by minimizing unnecessary transmissions. In contrast, the Epidemic protocol suffers from the highest hop count because of its broadcasting nature. This leads to higher overhead and resource consumption.

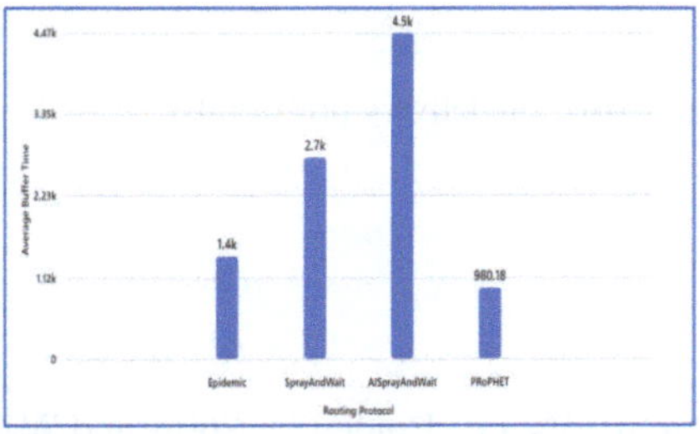

Fig. 1. A Comparison of routing protocols based on average buffer time

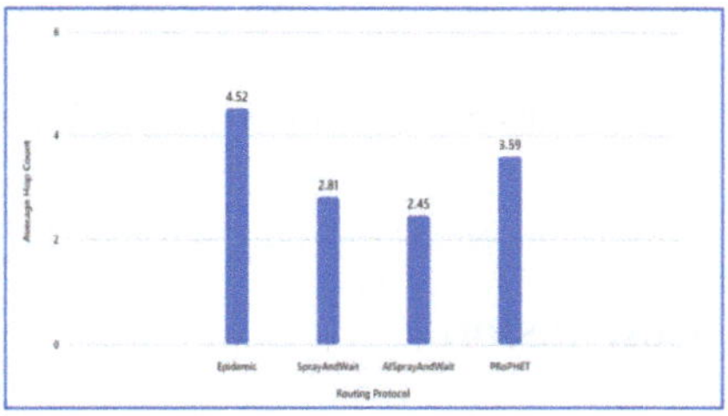

Fig. 2. Comparison of routing protocols based on average Hop Count

In terms of average latency, as shown in Fig. 3, SprayAndWait has the lowest value at 2.9 k ms, which indicates better control and more efficient message forwarding decisions. AISprayAndWait shows slightly higher latency than SprayAndWait. The artificial intelligence improves decision-making in selecting potential relays.

Based on the overhead ratio metric (Fig. 4), AISprayAndWait demonstrates the most efficient performance, followed by SprayAndWait. Both have shown an effective way of selecting neighbors, which ultimately limits message duplication. This comparison highlights that AISprayAndWait is the most resource-efficient protocol, making it ideal for environments that require low energy consumption.

Figure 5 shows the comparison of routing protocols based on delivery probability. As expected, AISprayAndWait achieves the highest delivery probability due to its intelligent feature of selecting and adapting to the dynamicity of node mobility. SprayAndWait also performs closely, as it maintains good control over replication and resource consumption.

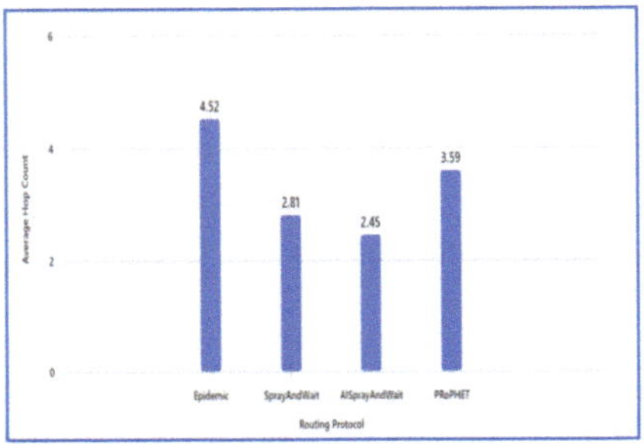

Fig. 3. Comparison of routing protocols based on average Latency

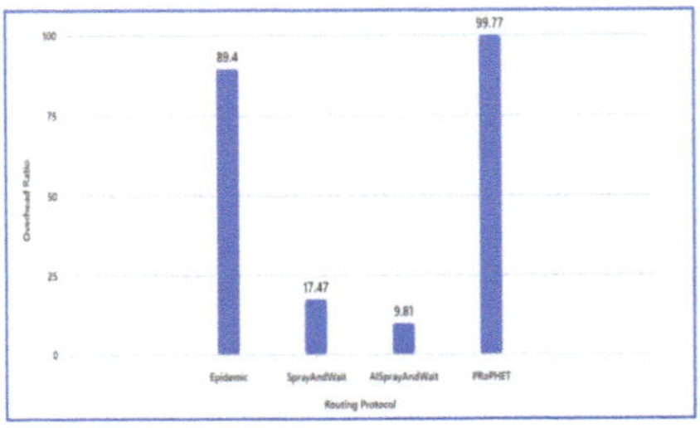

Fig. 4. Comparison of routing protocols based on overhead ratio

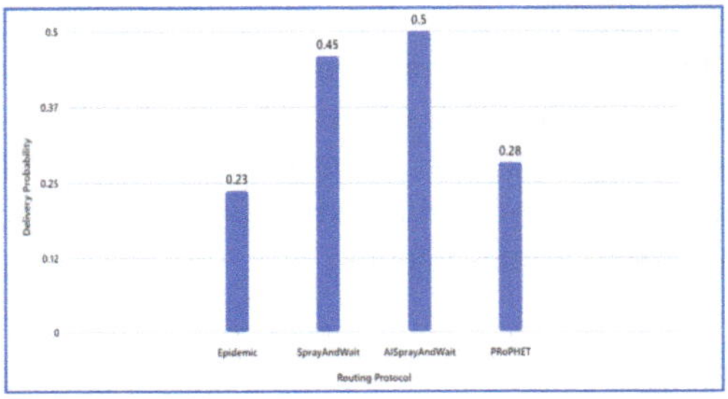

Fig. 5. Comparison of routing protocols based on delivery probability

6 Conclusion

In this paper, we present three popular routing protocols, i.e., Epidemic, Spray and Wait, and PRoPHET, in comparison with our new routing protocol called AISprayAndWait. Inspired by Spray and Wait, which is effective in controlling message overhead, we introduced lightweight Artificial Intelligence into Spray and Wait to make it more efficient in message forwarding. The presented results show that Artificial Intelligence improves and optimizes the forwarding selection strategy; however, it introduces higher latency as messages spend more time in the buffer. We will investigate further how to reduce latency using various Artificial Intelligence techniques. Overall, this study achieves its goal of exploiting Artificial Intelligence to improve the forwarding selection strategy in the Spray and Wait routing protocol.

References

1. Jain, S., Fall, K., Patra, R.: Routing in a delay tolerant network. In: Proceedings of the 2004 conference on Applications, technologies, architectures, and protocols for computer communications, in SIGCOMM '04, pp. 145–158. Association for Computing Machinery, New York, NY, USA (Aug. 2004). https://doi.org/10.1145/1015467.1015484

2. Garg, P., Kumar, H., Johari, R., Gupta, P., Bhatia, R.: Enhanced epidemic routing protocol in delay tolerant networks. In: 2018 5th International Conference on Signal Processing and Integrated Networks (SPIN), pp. 396–401. IEEE, Noida, India (2018). https://doi.org/10.1109/SPIN.2018.8474132
3. Cui, J., Cao, S., Chang, Y., Wu, L., Liu, D., Yang, Y.: An adaptive spray and wait routing algorithm based on quality of node in delay tolerant network. IEEE Access. **7**, 35274–35286 (2019). https://doi.org/10.1109/ACCESS.2019.2904750
4. Spyropoulos, T., Psounis, K., Raghavendra, C.S.: Spray and wait: An efficient routing scheme for intermittently connected mobile networks. In: Proceedings of the 2005 ACM SIGCOMM workshop on Delay-tolerant networking, in WDTN '05, pp. 252–259. Association for Computing Machinery, New York, NY, USA (Aug. 2005). https://doi.org/10.1145/1080139.1080143
5. Pathak, S., Gondaliya, N., Raja, N.: A survey on PROPHET based routing protocol in delay tolerant network. In: 2017 International Conference on Emerging Trends & Innovation in ICT (ICEI), pp. 110–115. IEEE, Pune, India (2017). https://doi.org/10.1109/ETIICT.2017.7977020
6. Garg, P., Dixit, A., Sethi, P.: ML-fresh: Novel routing protocol in opportunistic networks using machine learning. Social Science Research Network, Rochester, NY. https://papers.ssrn.com/abstract=3925374 (2021). Accessed 12 Oct 2025
7. Visca, J., Baliosian, J.: rl4dtn: Q-learning for opportunistic networks. Future Internet. **14**(12), 12 (2022). https://doi.org/10.3390/fi14120348
8. Raj, E., Westerlund, M., Espinosa-Leal, L.: Reliable fleet analytics for edge IoT solutions. arXiv. (Jan. 12, 2021). https://doi.org/10.48550/arXiv.2101.04414
9. Medeiros, T., et al.: TinyML custom ai algorithms for low-power iot data compression: A bridge monitoring case study. In: 2023 IEEE International Workshop on Metrology for Industry 4.0 & IoT (MetroInd4.0&IoT), pp. 54–59. IEEE, Piscataway, NJ (2023). https://doi.org/10.1109/MetroInd4.0IoT57462.2023.10180152
10. Li, L., Gaddah, A., Kunz, T.: Mobility support in a tactical P2P publish/subscribe overlay. In: MILCOM 2008–2008 IEEE Military Communications Conference, pp. 1–7. IEEE, Piscataway, NJ (2008). https://doi.org/10.1109/MILCOM.2008.4753344
11. Lenando, H., Alrfaay, M.: EpSoc: Social-based epidemic-based routing protocol in opportunistic mobile social network. Mob. Inf. Syst. **2018**, 1–12 (2018) https://onlinelibrary.wiley.com/doi/full/10.1155/2018/6462826. Accessed 14 Oct 2025
12. Ciobanu, R.-I., Marin, R.-C., Dobre, C., Cristea, V.: Interest-awareness in data dissemination for opportunistic networks. Ad Hoc Netw. **25**, 330–345 (2015). https://doi.org/10.1016/j.adhoc.2014.07.004
13. Sreenivasa, B.R., et al.: Social context-aware macroscopic routing scheme for opportunistic network. Trans. Emerg. Telecommun. Technol. **34**(12), e4844 (2023). https://doi.org/10.1002/ett.4844
14. Balasubramanian, A., Levine, B.N., Venkataramani, A.: Replication routing in dtns: a resource allocation approach. IEEEACM Trans. Netw. **18**(2), 596–609 (2010). https://doi.org/10.1109/TNET.2009.2036365
15. Feldmann, M., Walter, F., de Cola, T., Liva, G.: DTN coding. In: Silva, A.P., Burleigh, S., Obraczka, K. (eds.) Delay and Disruption Tolerant Networks: Interplanetary and Earth-Bound — Architecture, Protocols, and Applications, 1st edn, pp. 121–146. CRC Press, Boca Raton, FL (2018)
16. Khadam, U., Davidsson, P., Spalazzese, R.: Exploring the role of artificial intelligence in internet of things systems: A systematic mapping study. Sensors. **24**(20), 6511 (2024). https://doi.org/10.3390/s24206511

17. Oikonomou, P., Karanika, A., Anagnostopoulos, C., Kolomvatsos, K.: On the use of intelligent models towards meeting the challenges of the edge mesh. ACM Comput. Surv. **54**(6), 125:1–125:42 (2021). https://doi.org/10.1145/3456630
18. Dou, F., et al.: Towards artificial general intelligence (AGI) in the internet of things (IoT): Opportunities and challenges." arXiv. (Sep. 14, 2023). https://doi.org/10.48550/arXiv.2309.07438
19. Galliera, R., Venable, K.B., Bassani, M., Suri, N.: Collaborative information dissemination with graph-based multi-agent reinforcement learning. In: Freeman, R., Mattei, N. (eds.) Algorithmic Decision Theory, pp. 160–173. Springer Nature Switzerland, Cham (2025). https://doi.org/10.1007/978-3-031-73903-3_11
20. Galliera, R., Venable, K.B., Suri, N., Bassani, M.: Cooperative information dissemination with graph-based multi-agent reinforcement learning. In: RLC 2024 Workshop CoCoMARL: Coordination and Cooperation for Multi-Agent Reinforcement Learning Methods. OpenReview, Amherst, MA (2024) https://openreview.net/forum?id=R6Z14qDXH0 Accessed 14 Oct2025

Intuitionistic Fuzzy AHP for Robust Authentication Methods Selection in Information Security Context

Bee Wah Loo[1(✉)], Pei Ling Tan[1], Jeng Young Liew[2], and Choon Keat Low[1]

[1] Tunku Abdul Rahman University of Management and Technology, Setapak, 53300 Kuala Lumpur, Malaysia
loobw@tarc.edu.my

[2] Universiti Malaysia Kelantan, Jeli Campus, Jeli 17600, Kelantan, Malaysia

Abstract. The selection of an appropriate authentication method in cybersecurity has become increasingly complex due to the wide variety of available options. This study proposes a systematic mathematical framework that aids organizations in making informed decisions by evaluating and ranking authentication methods based on their security and operational criteria. A key contribution of this work is the application of the intuitionistic fuzzy analytic hierarchy process (IF-AHP) to the decision-making process for information security authentication, a method that has been relatively underutilized in this domain. Traditional decision-making approaches, such as the analytic hierarchy process (AHP), do not adequately account for uncertainty and vagueness in the evaluation process. By integrating intuitionistic fuzzy set theory with AHP, the proposed model addresses these limitations, offering more reliable outcomes, particularly in scenarios involving multiple alternatives. This approach is essential as the increasing number of authentication methods often leads to greater decision-maker confusion, necessitating the inclusion of both uncertainty and hesitancy factors. A case study demonstrates the practical application of the framework, revealing that fingerprint-based authentication ranks highest, followed by certificate-based authentication. In descending order of preference, the remaining methods are SMS-based authentication, multifactor authentication, and password authentication. These findings underscore the growing preference for biometric and certificate-based authentication solutions in contemporary information security practices.

Keywords: Information Security Authentication · Intuitionistic Fuzzy AHP (IF-AHP) · Multi-Criteria Decision-Making (MCDM)

1 Introduction

In today's interconnected digital landscape, information security is essential for protecting organizational assets and maintaining operational continuity. As cyber threats become more sophisticated and frequent [1, 2], authentication remains a fundamental defense mechanism for controlling access to sensitive resources.

M. A. b. Jamlos et al. (Eds.): ICSET 2025, LNNS 1897, pp. 313–318, 2026.
https://doi.org/10.1007/978-3-032-20722-7_39

Existing authentication methods range from traditional passwords and PINs to advanced approaches such as biometrics, smart cards, and multi-factor authentication (MFA) [3, 4]. Each method differs in security level, usability, and implementation cost, making the selection a complex multi-criteria decision problem [5]. Multi-criteria decision-making (MCDM) techniques, particularly the analytic hierarchy process (AHP) proposed by Saaty [6], are widely used to structure and prioritize such evaluations. However, AHP's reliance on precise judgments limits its performance in uncertain situations where hesitation or incomplete information exists [7].

To address this limitation, this study integrates intuitionistic fuzzy sets (IFS) into AHP. Introduced by Atanassov [8], IFS incorporates membership, non-membership, and hesitation degrees, better capturing ambiguity in expert evaluations. The proposed IF-AHP framework enhances decision flexibility and supports realistic selection of authentication methods under uncertainty. This research aims to (1) develop a mathematical model to assist organizations in selecting authentication mechanisms aligned with their security and operational requirements, and (2) improve MCDM techniques by addressing uncertainty and hesitation in expert evaluations.

2 Related Work

The increasing complexity of cyber threats has heightened the need for systematic and transparent decision-making in cybersecurity. Traditional qualitative assessments often lack consistency, leading to the adoption of multi-criteria decision-making (MCDM) techniques that structure and prioritize risks, controls, and strategies. Among these, the AHP [6] is widely used for its ability to decompose complex problems and derive quantifiable priorities from expert judgments.

AHP has been applied to various cybersecurity contexts. Awang et al. [9] developed an AHP-based framework to prioritize cybersecurity management guidelines in higher education, improving consensus among experts and reducing bias. Petrova [10] extended AHP to assess confidentiality, integrity, and availability (CIA) but noted that crisp judgments fail to capture uncertainty, suggesting fuzzy extensions. Similarly, Moreira et al. [11] combined AHP with sensitivity analysis to prioritize controls under the NIST Cybersecurity Framework, emphasizing the need for uncertainty-aware models.

Building on these efforts, Loo et al. [12] integrated AHP with evolutionary game theory (EGT) to study defender–attacker dynamics, later expanding the model to include more strategies [13]. These studies confirm AHP's value for structuring cybersecurity decisions such as guideline prioritization and control selection. However, reliance on crisp judgments limits its capacity to represent vagueness in expert opinions. To mitigate this, fuzzy AHP incorporates linguistic or interval preferences, improving realism under uncertainty [14].

Fuzzy AHP has also been applied across domains. Saakshi and Bhardwaj [15] used it for agroforestry suitability mapping, while Kumar, Singh, and Chaurasia [16] applied it to secure medical-image transmission. Yet, fuzzy AHP cannot explicitly model hesitation between agreement and disagreement. Intuitionistic fuzzy AHP (IF-AHP) overcomes this by including membership, non-membership, and hesitation degrees [17]. Despite such progress, IF-AHP remains underexplored in cybersecurity. Authentication-method selection exemplifies such complexity, requiring a balance among security, usability, and cost under uncertainty. Motivated by this gap, this study applies IF-AHP to improve authentication decision-making in cybersecurity.

3 Methodology

This study applied IF-AHP to support decision-making in selecting authentication methods within cybersecurity contexts. IF-AHP enhances the classical AHP framework by integrating intuitionistic fuzzy sets (IFS), which represent not only membership (μ_{ij}) and non-membership (ν_{ij}) but also hesitation (π_{ij}). This enables more realistic modeling of uncertainty and imprecise expert judgments.

The process begins by defining the goal (optimal authentication selection), criteria (security, usability, cost, reliability, and scalability), and alternatives (password, SMS, certificate-based, multi-factor, fingerprint). Experts perform pairwise comparisons using the traditional AHP scale [18], later converted into intuitionistic fuzzy values. Following Xu and Liao [18], an IFS is defined as

$$A = \{(x, \mu_A(x), \nu_A(x))|x \in X\} \tag{1}$$

where X is a non-empty set with elements x, $\mu_A(x)$, $\nu_A(x) \in [0, 1]$, and hesitation degree is $\pi_A(x) = 1 - \mu_A(x) - \nu_A(x)$ with $0 \leq \mu_A(x) + \nu_A(x) \leq 1$.

An intuitionistic fuzzy preference relation (IFPR) on set X is denoted by a matrix $\overline{A} = (a_{ij})_{m\times m}$, where $a_{ij} = (\mu_{ij}, \nu_{ij})$. Aggregation of two intuitionistic fuzzy values, $a_{ij} = (\mu_{ij}, \nu_{ij})$ and $a_{st} = (\mu_{st}, \nu_{st})$ in matrix $\overline{A}$, is expressed as:

$$a_{ij} \oplus a_{st} = \left(\mu_{ij} + \mu_{st} - \mu_{ij}\mu_{st}, \nu_{ij}\nu_{st}\right) \tag{2}$$

$$a_{ij} \bigotimes a_{st} = \left(\mu_{ij}\mu_{st}, \left(\nu_{ij+}\nu_{st}\right) - \nu_{ij}\nu_{st}\right). \tag{3}$$

The IFPR satisfies max-min transitivity property if $a_{ij} \geq min\ \{a_{it}, a_{tj}\}$, for $i, j, t = 1, 2, \cdots m$.

A consistency check is then performed to validate the expert judgments. If the consistency ratio (CR) exceeds 0.10 [19], revisions are required. Using IF-AHP, inconsistency is corrected through a three-cases algorithm [18]. For a 5 × 5 comparison matrix, Case 1 maintains the repair values for elements a_{12}, a_{23} a_{34} and a_{45}. In Case 2, the repair values for elements a_{21}, a_{31}, a_{32}, a_{41}, a_{42} a_{43}, a_{51}, a_{52}, a_{53} and a_{54} are defined as $\overline{a}_{ij} = \left(\overline{\mu}_{ij}, \overline{\nu}_{ij}\right) = \left(\nu_{ij}, \mu_{ij}\right)$. In Case 3, repair values for remaining off-diagonal elements using Eqs. (4) and (5), while the diagonal elements are fixed at (0.5,0.5).

$$\overline{\mu}_{ij} = \frac{\sqrt[j-i-1]{\prod_{t=i+1}^{k-1} \mu_{it}\mu_{tj}}}{\sqrt[j-i-1]{\prod_{t=i+1}^{j-1} \mu_{it}\mu_{tj}} + \sqrt[j-i-1]{\prod_{t=i+1}^{j-1}(1 - \mu_{it})(1 - \mu_{tj})}} \tag{4}$$

$$\overline{\nu}_{ij} = \frac{\sqrt[j-i-1]{\prod_{t=i+1}^{j-1} \nu_{it}\nu_{tj}}}{\sqrt[j-i-1]{\prod_{t=i+1}^{j-1} \nu_{it}\nu_{tj}} + \sqrt[j-i-1]{\prod_{t=i+1}^{j-1}(1 - \nu_{it})(1 - \nu_{tj})}} \tag{5}$$

Following this, Eq. (6) is applied to calculate the difference, d, between the repaired and the original IFPR.

$$d = \frac{1}{2(m-1)(m-2)} \sum_{i=1}^{m} \sum_{j=1}^{m} \left(\left| \overline{\mu}_{ij} - \mu_{ij} \right| + \left| \overline{v}_{ij} - v_{ij} \right| + \left| \overline{\pi}_{ij} - \pi_{ij} \right| \right) \tag{6}$$

The pairwise judgments are considered consistent if $d < 0.10$. Finally, the relative weights, w_i, of each criterion and alternative for the authentication methods are calculated using Eq. (7), and the priority ranking, $\rho(w)$, is established using Eq. (8), providing a robust decision basis for selecting the most suitable authentication method under uncertainty.

$$w_i = \left[\frac{\sum_{j=1}^{m} \mu_{ij}}{\sum_{i=1}^{m} \sum_{j=1}^{m} \left(1 - v_{ij}\right)}, \frac{\sum_{j=1}^{m} \left(1 - v_{ij}\right)}{\sum_{i=1}^{m} \sum_{j=1}^{m} \left(1 - \mu_{ij}\right)} \right] \tag{7}$$

$$\rho(w) = 0.5(1 + \pi_w)(1 - \mu_w) \tag{8}$$

4 Results and Discussion

A case study was conducted with eight experts from academia and industry to validate the proposed IF-AHP framework for selecting the optimal authentication method. The experts performed pairwise comparisons across five criteria, which are cost (T_1), usability (T_2), reliability (T_3), security strength (T_4) and scalability (T_5), as well as five alternatives: passwords (M_1), SMS (M_2), certificate-based (M_3), multi-factor (M_4) and fingerprints (M_5). Initial IFPR matrices showed minor inconsistencies, which were corrected using the proposed repair algorithm, significantly improving consistency. Table 1 presents both the original and repaired IFPR values for the criteria.

Applying Eqs. (7) and (8), the final priority scores, $\rho(w)$, for the authentication alternatives were: $\rho\left(w_{M_1}\right) = 0.4670$, $\rho\left(w_{M_2}\right) = 0.4903$, $\rho\left(w_{M_3}\right) = 0.4953$, $\rho\left(w_{M_4}\right) = 0.4843$ and $\rho\left(w_{M_5}\right) = 0.5117$, yielding the ranking $M_5 \succ M_3 \succ M_2 \succ M_4 \succ M_1$. Fingerprint authentication ranked highest, followed by certificate-based, SMS, multi-factor, and password-based methods. The ranking aligns with current trends favoring biometric and certificate-based solutions for their superior security and usability, while traditional methods remain less secure. The results validate the IF-AHP framework's ability to handle uncertainty and hesitation in expert assessments, producing rankings consistent with current cybersecurity practices.

Table 1. Original and repaired IFPR of comparison judgement of criteria.

Criteria	T_1	T_2	T_3	T_4	T_5
Original T_1	(0.50, 0.50)	(0.34, 0.54)	(0.36, 0.51)	(0.37, 0.50)	(0.37, 0.50)
Repaired T_1	(0.50, 0.50)	(0.34, 0.54)	(0.62, 0.19)	(0.53, 0.23)	(0.52, 0.23)
Original T_2	(0.54, 0.34)	(0.50, 0.50)	(0.76, 0.17)	(0.74, 0.18)	(0.74, 0.18)
Repaired T_2	(0.54, 0.34)	(0.50, 0.50)	(0.76, 0.17)	(0.83, 0.06)	(0.83, 0.06)
Original T_3	(0.51, 0.36)	(0.17, 0.76)	(0.50, 0.50)	(0.61, 0.25)	(0.66, 0.22)
Repaired T_3	(0.19, 0.62)	(0.17, 0.76)	(0.50, 0.50)	(0.61, 0.25)	(0.68, 0.10)
Original T_4	(0.50, 0.37)	(0.18, 0.74)	(0.25, 0.61)	(0.50, 0.50)	(0.58, 0.26)
Repaired T_4	(0.23, 0.53)	(0.06, 0.83)	(0.25, 0.61)	(0.50, 0.50)	(0.58, 0.26)
Original T_5	(0.50, 0.37)	(0.18, 0.74)	(0.22, 0.66)	(0.26, 0.58)	(0.50, 0.50)
Repaired T_5	(0.23, 0.52)	(0.06, 0.83)	(0.10, 0.68)	(0.26, 0.58)	(0.50, 0.50)

5 Conclusion

This study proposed an IF-AHP-based framework for selecting authentication methods suited to organizational cybersecurity needs. By integrating intuitionistic fuzzy sets and a consistency-repair algorithm, the model captures uncertainty and hesitation in expert evaluations, improving decision reliability.

The findings identified fingerprint-based authentication as the top-ranked method, followed by certificate-based, SMS, multi-factor, and password-based options. These results highlight the growing adoption of advanced authentication mechanisms due to their security and usability advantages. Future research should extend this framework by integrating dynamic MCDM or adaptive modeling techniques to better capture evolving cybersecurity decision environments.

References

1. Dharavath, K., Talukdar, F.A., Laskar, R.H.: Study on biometric authentication systems, challenges and future trends: A review. In: 2013 IEEE International Conference on Computational Intelligence and Computing Research, pp. 1–7. IEEE, Madurai, India (2013)
2. Bishop, M.: Introduction to Computer Security. Addison-Wesley Professional, Boston, MA (2004)
3. Josang, A., AlZomai, M., Suriadi, S.: Usability and privacy in identity management architectures. In: ACSW Frontiers, 2007 Proceedings of 5th Australasian Symposium on Grid Computing and E-Research, 5th Australasian Information Security Workshop (Privacy Enhancing Technologies), and Australasian Workshop on Health Knowledge Management and Discover, pp. 143–152. Australian Computer Society, Ballarat, VIC (2007)
4. NIST Special Publication 800-63B: Digital identity guidelines: Authentication and Lifecycle Management. National Institute of Standards and Technology, Gaithersburg, MD (2020)
5. Maček, D., Magdalenić, I., Ređep, N.B.: A systematic literature review on the application of multicriteria decision making methods for information security risk assessment. Int. J. Saf. Secur. Eng. **10**(2), 161–174 (2020)

6. Golden, B.L., Wasil, E.A., Harker, P.T.: The analytic hierarchy process. Appl. Stud. **2**(1), 1–273 (1989)
7. Mikhailov, L., Tsvetinov, P.: Evaluation of services using a fuzzy analytic hierarchy process. Appl. Soft Comput. **5**(1), 23–33 (2004)
8. Atanassov, K.: Intuitionistic fuzzy sets. Int. J. Bioautomation. **20**, S1–S6 (2016)
9. Awang, N., Samy, G.N., Hassan, N.H.: Prioritizing cybersecurity management guidelines using analytical hierarchy process (AHP) decision technique. Open Int. J. Inform. **10**(Special Issue 1), 1–10 (2022)
10. Petrova, V.: A cybersecurity risk assessment. Industry 4.0. **6**(1), 37–40 (2021)
11. Moreira, R.S., Moraes, R.O., Fernandes, C.A.: Cybersecurity risk assessment through analytic hierarchy process integrating multicriteria and sensitivity analysis. J. Inf. Secur. Appl. **83**, 103745 (2025)
12. Loo, B.W., Tan, P.L., Tey, S.K., Chin, W.Y.: Integrated decision support system of AHP and EGT for information security control analysis. J. Pharm. Negat. Results. **13**(10), 342–350 (2023)
13. Loo, B.W., Tan, P.L., Tey, S.K., Chin, W.Y.: Authentication methods selection in information security through hybrid AHP and EGT. J. Adv. Res. Appl. Sci. Eng. Technol. **50**(2), 171–185 (2025)
14. Nadeem, M.: Analyze quantum security in software design using fuzzy-AHP. Int. J. Inf. Technol. **16**(7), 4341–4352 (2024)
15. Saakshi, C., Bhardwaj, D.R.: Suitability mapping of agroforestry in Shimla district, Himachal Pradesh, India, using analytic hierarchy process (AHP) and fuzzy analytic hierarchy process (Fuzzy AHP). Agrofor. Syst. **99**, 1194 (2025)
16. Kumar, S., Singh, S., Chaurasia, P.K.: Enhancing security of medical image transmission: an innovative fuzzy-AHP approach. In: Proceedings of the Fifth International Conference on Trends in Computational and Cognitive Engineering, TCCE 2023, vol. 961, pp. 471–483. Springer, Singapore (2024)
17. Zhou, F., Chen, T.-Y.: A synergetic intuitionistic fuzzy model combining AHP, entropy, and ELECTRE for data fabric solution selection. Artif. Intell. Rev. **58**(2), 1–32 (2025)
18. Xu, Z., Liao, H.: Intuitionistic fuzzy analytic hierarchy process. IEEE trans. Fuzzy Syst. **22**(4), 749–761 (2014)
19. Saaty, T.L.: How to make a decision: The analytic hierarchy process. Eur. J. Oper. Res. **48**(1), 9–26 (1990)

An Ontology-Based Digital Platform for Malaysia Smart Tourism Initiatives

Ummi Maisarah Izani(✉) and Nur Syadhila Che Lah

Department of Computing, Universiti Teknologi PETRONAS (UTP), Bandar Seri Iskandar, Malaysia
{ummi_21000213,nursyadhila.chelah}@utp.edu.my

Abstract. Malaysia's tourism sector, characterized by cultural diversity and diverse offerings, faces persistent data fragmentation and lack of semantic integration. This study proposes an ontology-driven platform that unifies and contextualizes tourism information through a structured knowledge representation framework. The ontology integrates key domains—Destinations, Restaurants, and Health Tourism—enabling semantic search, intelligent recommendations, and contextual awareness. Evaluation through SPARQL queries achieved 100% success in answering all predefined Competency Questions (CQs), confirming the model's ability to support complex, multi-domain reasoning such as activity-based destination retrieval and cuisine-based restaurant filtering. The validated OWL ontology establishes a foundational model for Malaysia's Smart Tourism.

Keywords: Ontology-Driven platform · Smart Tourism · Semantic Integration · Knowledge Graph · Competency Questions

1 Introduction

Tourism is a vital pillar of Malaysia's economy, with 29 million international arrivals in 2024 [16]. However, the vast data generated from diverse sources like travel websites and social media remains fragmented, limiting its utility. Traditional keyword-based search and relational databases fail to provide personalized, context-aware results as they cannot semantically link traveler preferences, activities, and destination nuances [1]. This fragmentation across platforms like TripAdvisor and government portals creates isolated data silos [4], worsened by limited stakeholder collaboration [5]. Fragmentation challenges in tourism governance have been widely recognized [12].

Ontology, a method for structured knowledge representation, enables semantic search, intelligent recommendations, and data inference [2], supported by knowledge-graph principles for linking heterogeneous data [17]. Unlike static databases, ontologies define entities and their relationships for dynamic information retrieval [3], as shown in recent Semantic Web–driven smart-tourism systems [18]. However, existing ontology-based solutions in tourism often suffer from a narrow domain focus, lack of requirement-driven design using formal Competency Questions (CQs), and inadequate operational validation via semantic querying [6, 7].

M. A. b. Jamlos et al. (Eds.): ICSET 2025, LNNS 1897, pp. 319–325, 2026.
https://doi.org/10.1007/978-3-032-20722-7_40

To bridge these gaps, this study develops a comprehensive CQ-driven OWL ontology for Malaysia, integrating three tourism domains into a coherent knowledge graph populated with public data and validated through SPARQL queries. The framework establishes a unified foundation for Malaysia's Smart Tourism ecosystem, supporting integrated, cross-domain reasoning.

2 Literature Review

Ontology supports semantic integration and reasoning beyond keyword search [10] and underpins large-scale knowledge graphs [17]. Defined as a "formal, explicit specification of a shared conceptualization" [13], ontology facilitates intelligent tourism services. Foundational guidance for ontology development is provided by Noy and McGuinness [15]. Nevertheless, existing approaches exhibit methodological gaps summarized in Table 1.

Table 1. Comparative Analysis of Tourism Ontologies.

Study	Methodological Approach	Domain Coverage	Key Limitations
Ho et al. [5] (TOML)	Machine Learning Integration	Attractions, Activities	Not CQ-driven; limited to recommendations; lacks SPARQL validation.
Zhang et al. [6]	Thematic Taxonomy	Broad Themes, Attractions	Theoretical; lacks operational validation
She et al. [3]	Property Graph (Neo4j)	POIs, User Profiles	Non-standard semantics; evaluation not CQ-centric.
Gkevreki et al. [7] (HTH)	OWL Application	Health Tourism (Single Domain)	Domain-specific; not integrated with general tourism.
This Work	CQ-driven OWL Ontology	Destinations, Restaurants, Health Tourism	Unified, requirement-driven, SPARQL-validated.

Prior studies reveal a trade-off between specialization and semantic rigor. Ho et al. [5] improve recommendation accuracy yet omit CQs, reducing traceability. Zhang et al. [6] provide taxonomies without SPARQL validation, while She et al. [3] prioritizes performance at the expense of OWL interoperability. Gkevreki et al. [7] remain limited to health tourism. The Malaysian health tourism context is discussed by Abd Manaf et al. [8]. These efforts show isolated progress but lack a unified, empirically validated ontology for cross-domain reasoning [22, 23]. This study fills that gap through a CQ-driven OWL ontology evaluated via SPARQL for practical deployment.

3 Methodology

This study adopted the Cross-Industry Standard Process for Data Mining (CRISP-DM) methodology [9], adapted for ontology engineering following recent ontology-integration extensions [19]. The iterative process focused on: (1) defining requirements via Competency Questions, (2) developing the OWL ontology, and (3) evaluating it through SPARQL queries. The iterative process is summarized in Table 2.

Table 2. CRISP-DM Application Stages.

Phase	Description
Business Understanding	Identified fragmented tourism data as the core issue and set goals for semantic integration
Data Understanding	Collected tourism-related datasets from TripAdvisor, Kaggle, and MHTC
Data Preparation	Cleaned and reclassified data into ontology formats
Ontology Modelling	Constructed classes and properties in Protégé using OWL.
Evaluation	Executed SPARQL queries to verify ontology completeness and consistency

3.1 Requirement Elicitation via Competency Questions (CQs)

The project scope was defined through a set of CQs natural language questions the ontology must be capable of answering [14, 22]. This requirement-driven approach ensures the resulting model fulfills practical information retrieval tasks for stakeholders. The core CQs, derived from the target domains, are detailed in Table 3:

Table 3. Defined Competency Questions (CQs).

CQ	Domain	Question	Primary Objective
CQ1	Destinations	What tourist destinations fall under Activities and where are they located?	Enable activity-based filtering for personalized trip planning
CQ2	Restaurants	Which restaurants and ratings are available by cuisine type and location?	Support personalized dining recommendations by preference and geography.
CQ3	Health Tourism	What health-tourism facilities exist, by type and location?	Facilitate medical travel planning through clear service mapping.

3.2 Ontology Modelling in OWL

The ontology was implemented in Protégé using OWL, with the class hierarchy structured around three core domains (Fig. 1; Table 4). Semantic relationships were defined through formally specified object and data properties (Table 5) and supported by standard description-logic frameworks [21]. The ontology was populated using curated datasets from restaurants (2,378 entries), destinations (173), and health-tourism facilities (272). Overall, the knowledge base comprises 33 classes, 4 properties (1 object, 3 data), 2,823 individuals, and approximately 8,500 RDF triples—demonstrating scalability consistent with tourism knowledge-graph models [20].

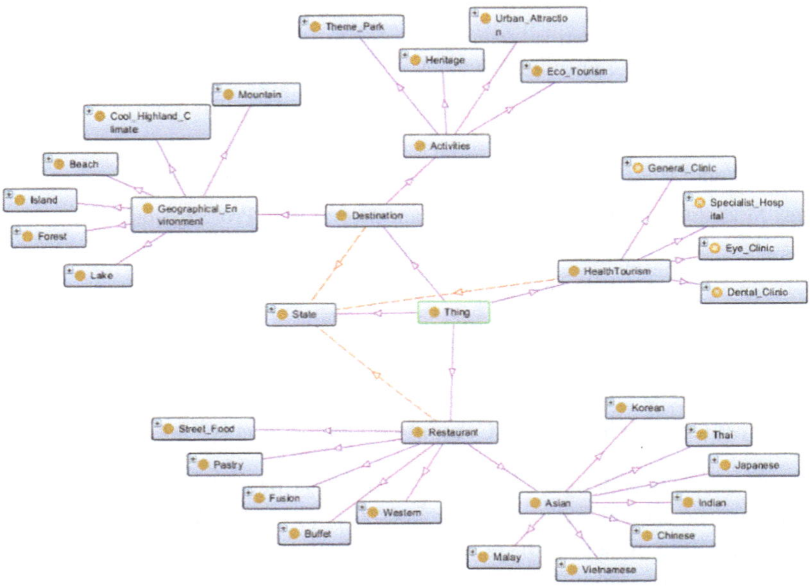

Fig. 1 Ontology Class Hierarchy in Protégé.

Table 4. Class Hierarchy.

Domain	Subclasses / Types
Destination	Geographical_Environment: (Mountain, Forest, Island, Beach, Lake, Cool_Highland_Climate) Activities: (Eco_Tourism, Heritage, Theme_Park)
Restaurant	Asian, Fusion, Western, Middle_Eastern, Buffet, Pastry, Street_Food
Health Tourism	Specialist_Hospital, General_Clinic, Dental_Clinic, Eye_Clinic

Table 5. Object and Data Properties.

Property Name	Type	Domain	Range	Description
hasLocation	Object Property	Destination, Restaurant, HealthTourism	State	Links entities to locations
hasRating	Data Property	Restaurant	xsd:float	Stores user ratings.
hasCategory	Data Property	Restaurant, Destination	xsd:string	Defines entity type.
offerService	Data Property	HealthTourism	xsd:string	Describes medical service.

3.3 Evaluation Through SPARQL Query Execution

SPARQL queries corresponding to each CQ tested the ontology's reasoning capacity. CQ1 retrieved activity-based destinations through rdfs: subClassOf hierarchy traversal; CQ2 linked restaurant subclasses with cuisine and rating; CQ3 used transitive paths (rdfs:subClassOf+) to list health facilities by type and location. This query-driven evaluation validated completeness and logical consistency.

4 Evaluation and Findings

SPARQL-based evaluation confirmed semantic reasoning and integration, as illustrated by the CQ1 query in Fig. 2 retrieving Activities destinations.

```
SPARQL query:
PREFIX tourism: <http://www.semanticweb.org/tourism#>
PREFIX rdf: <http://www.w3.org/1999/02/22-rdf-syntax-ns#>
PREFIX rdfs: <http://www.w3.org/2000/01/rdf-schema#>

SELECT ?destination ?category ?location
WHERE {
   ?destination rdf:type ?category .
   ?category rdfs:subClassOf tourism:Activities .  # Ensure it checks the correct subclass
   ?destination tourism:hasLocation ?location .
}
```

Fig. 2 SPARQL query for CQ1: Retrieving destinations under the Activities subclass.

The query for CQ1 ("What tourist destinations fall under Activities and where are they located?") retrieved activity-based destinations including Taman Negara and Sunway Lagoon. The SPARQL query leveraged rdfs:subClassOf to traverse the Activities hierarchy, demonstrating proper implementation of subclass relationships. This structured semantic retrieval surpasses traditional keyword-based approaches, aligning with Zhang et al. [6] on hierarchical categorization for tourism information systems. Similar patterns validated restaurant ratings (CQ2) and health facilities (CQ3), confirming multi-domain reasoning and practical integration aligned with established SPARQL validation frameworks.

5 Conclusion

This study addresses data fragmentation in Malaysia's tourism sector through a unified ontology framework. Developed via a requirement-driven methodology with formal Competency Questions, the OWL ontology integrates three core domains—Destinations, Restaurants, and Health Tourism—into a knowledge base of 2,823 individuals across 33 classes. Rigorous SPARQL evaluation achieved 100% success in semantic retrieval, confirming robust reasoning for hierarchical traversal and multi-criteria filtering. A key challenge involved harmonizing heterogeneous datasets, underscoring the need for standardized, machine-readable tourism data. The framework directly supports MyDIGITAL 2030 by enabling semantic integration across tourism datasets and advancing national goals for open-data interoperability, smart-city readiness, and AI-enabled digital ecosystems under the Smart Tourism Framework. Tourism data spaces represent an emerging direction for smart destinations [11]. Future work will extend the ontology to accommodation and transport, incorporate real-time data, and include stakeholder usability testing.

References

1. Sastriani, R., Baizal, Z.A., Kusumo, D.S.: Ontology-based semantic search on tourism information search system. Indones. J. Comput. (Indo-JC). **5**(1), 103–114 (2020)
2. Camacho, V.T., Cruz, J.: Ontology-based context-aware recommender system application for tourism. arXiv:2301.00768 (2023). https://doi.org/10.48550/arXiv.2301.00768
3. She, X., Ding, L., Zhang, Y.: AK tourism: A property graph ontology-based tourism recommender system. In: Proceedings of the Knowledge Management International Conference (KMICe) 2018, pp. 1–6. Universiti Utara Malaysia, Sintok (2018)
4. Mohamad, U.H., Ahmad, M.N., Zakaria, A.M.U.: Ontologies application in the sharing-economy domain: A systematic review. Online Inf. Rev. **46**(4), 807–825 (2022)
5. Dung, H.Q., Le, L.T.Q., Nguyen, H.H.T., Truong, T.Q., Nguyen-Dinh, C.-H.: A novel ontology framework supporting model-based tourism recommender. IAES Int. J. Artif. Intell. **10**(1), 70–79 (2021)
6. Zhang, P., Wang, J., Li, R.: Tourism-type ontology framework for tourism-type classification, naming and knowledge organization. Heliyon. **9**(4), e15192 (2023)
7. Gkevreki, S., Fiska, V., Nikolopoulos, S., Kompatsiaris, I.: Enhancing sustainability in health tourism through an ontology-based booking application for personalized packages. Sustainability. **16**(15), 6505 (2024)
8. Abd Manaf, N.H., Hussin, H., Jahn Kassim, P.N., Alavi, R., Dahari, Z.: Country perspective on medical tourism: The Malaysian experience. Leadersh. Health Serv. **28**(1), 43–56 (2015)
9. Azeroual, O., Nacheva, R., Nikiforova, A., Störl, U.: A CRISP-DM and predictive analytics framework for enhanced decision-making in research information management systems. Informatica. **49**(18), 1–18 (2025)
10. Hrankai, R., Mak, B.: Bridging the affordance-actualization gap in user preferences for AI-assisted trip planning. J. Travel Res. **64**(1), 1–16 (2025)

11. Ordóñez Martínez, M.D.: Tourism data spaces in smart tourism destinations: Conceptual and methodological approach. SHS Web Conf. **187**, 03033 (2024)
12. Maimaitiaili, Y.: Integrated tourism: A holistic approach to resolving fragmentation challenges in tourism governance. SHS Web Conf. **187**, 03033 (2024)
13. Gruber, T.R.: A translation approach to portable ontology specifications. Knowl. Acquis. **5**(2), 199–220 (1993)
14. Zhang, X., Zhao, C., Wang, X.: A survey on knowledge representation in materials science and engineering: An ontological perspective. Comput. Ind. **73**, 8–22 (2015)
15. Noy, N.F., McGuinness, D.L.: Ontology Development 101: A Guide to Creating Your First Ontology. Stanford Knowledge Systems Laboratory, Stanford (2001)
16. Malaysia, T.: Visitor Performance to Malaysia, January–December 2024 (Infographic). Tourism Malaysia, Putrajaya (2025)
17. Hogan, A., et al.: Knowledge graphs. ACM Comput. Surv. **54**(4), 1–37 (2021)
18. Nguyen, L.T.H., Nieto, A.A., Linh, G.T.R.: A semantic web-based approach for context-aware smart tourism applications. IEEE Access. **9**, 123255–123274 (2021)
19. Gill, M.S., Westermann, T., Steindl, G., Gehlhoff, F., Fay, A.: Integrating ontology design with the CRISP-DM in the context of cyber-physical systems maintenance. arXiv:2407.06930 (2024). https://doi.org/10.48550/arXiv.2407.06930
20. Yao, L., Lin, C., Zhou, X., Li, J., Ma, X.: An ontology-driven knowledge graph for tourism information management. J. Biomed. Semantics. **15**(1), 12 (2024)
21. Motik, B., Shearer, R., Horrocks, I.: Hypertableau reasoning for description logics. J. Artif. Intell. Res. **36**, 165–228 (2009)
22. de Oliveira, F.L.G., et al.: An ontology engineering methodology based on competency questions and conceptual graphs. In: Proceedings of the 13th International Conference on Semantic Systems (SEMANTICS 2017), pp. 25–32. ACM, New York, NY (2017)
23. Poveda-Villalón, M., et al.: LOT: An industrial-oriented ontology engineering framework. Eng. Appl. Artif. Intell. **111**, 104755 (2022)

Probabilistic Risk Assessment of Pumped Hydro Energy Storage Projects Using Monte Carlo Simulation

Chai Hoon Lee(✉)

TNB Research Sdn. Bhd., 43000 Kajang, Malaysia
chaihoon.lee@tnb.com.my

Abstract. Uncertainty analysis is well established in solar and wind techno-economics, yet equivalent studies for pumped hydro energy storage (PHES) remain scarce despite its central role in enabling long-duration storage and grid flexibility. Existing probabilistic frameworks are often proprietary or limited to conventional hydropower contexts, offering little transparency for early-stage feasibility studies. This paper addresses that gap by developing a reproducible, spreadsheet-native Monte Carlo framework that quantifies probabilistic ranges of levelized cost of energy (LCOE) and net present value (NPV) for PHES projects. The approach defines stochastic input distributions for cost, tariff, discount rate, and performance parameters, then performs thousands of random trials to generate probabilistic indicators such as P10, P50, and P90 outcomes. Tornado charts and exceedance curves are used to visualize dominant risk drivers and the likelihood of meeting target returns. Results demonstrate how probabilistic evaluation reveals both upside and downside risk that deterministic analyses overlook. The framework offers a transparent, adaptable decision-support tool suitable for feasibility screening of pumped hydro and other renewable-energy projects under uncertainty.

Keywords: Pumped hydro energy storage · Monte Carlo simulation · Probabilistic technoeconomic assessment

1 Introduction

Energy-transition pathways worldwide increasingly rely on large-scale energy storage to balance variable renewable-energy generation. Among available technologies, pumped hydro energy storage (PHES) remains the most mature and cost-effective option for long-duration storage, providing frequency regulation, load shifting, and reserve capacity. In Malaysia and Southeast Asia, PHES is recognized under national decarbonization strategies as a key enabler of grid flexibility and cross-state energy exchange [1].

Despite its maturity, PHES development faces significant investment uncertainty arising from hydrological variability, evolving electricity tariffs, and long construction lead times. Conventional feasibility analyses commonly adopt deterministic assumptions for these variables, generating single-point results that mask the probability of adverse or favorable outcomes. Monte Carlo simulation [2] offers a natural way to express such

M. A. b. Jamlos et al. (Eds.): ICSET 2025, LNNS 1897, pp. 326–331, 2026.
https://doi.org/10.1007/978-3-032-20722-7_41

uncertainty by repeatedly sampling input distributions and computing corresponding project outcomes. Although Monte Carlo methods are widely used in renewable-energy techno-economic assessments, including solar, wind, hybrid storage systems, and micro-hydro [3, 4], published examples specific to PHES remain limited. Existing studies of hydropower risk primarily examine plant availability or construction cost uncertainty [5, 6], while storage focused analyses typically isolate financial parameters without coupling them to hydrological performance. Furthermore, most implementations rely on proprietary software such as @Risk, Crystal Ball, or MATLAB, which can be costly or opaque to practitioners conducting early-stage feasibility screening.

Consequently, a methodological gap exists for a transparent, reproducible framework that integrates hydrological, financial, and market-price uncertainty for PHES evaluation. This study develops such a framework entirely within Microsoft Excel, enabling probabilistic computation of LCOE and NPV through Monte Carlo simulation. The paper presents the modelling workflow, applies it to a simplified per-kilowatt PHES representation, and compares probabilistic percentile outputs (P10–P50–P90) against deterministic results to demonstrate its usefulness in risk-informed project appraisal. PHES also intersects with water-resource management, where probabilistic inflow analysis is well established [7], linking energy storage planning to broader energy–water-nexus considerations.

2 Methodology

The modelling framework follows five main steps: (1) define probability distributions for input variables, (2) perform random sampling, (3) compute project outcomes, (4) aggregate statistical results, and (5) visualize risk metrics. The structure reflects typical stages of feasibility analysis and allows transparent reproduction within standard spreadsheet environments.

Each uncertain parameter was represented by a probability distribution based on engineering judgment and available benchmark data. The capital cost (C_{capex}) per kilowatt was modelled as a triangular distribution (min 1100, mode 1300, max 1700 USD kW^{-1}) to reflect expert-estimated ranges. The plant capacity factor (CF) followed a PERT distribution bounded between 0.22 and 0.33 with a mode of 0.27. The operation-and-maintenance cost ($C_{O\&M}$) was assumed log-normal with logarithmic mean ln(30) and standard deviation 0.35 USD kW^{-1} $yr.^{-1}$. The discount rate (r) was drawn from a uniform distribution (0.06–0.10) to represent financing uncertainty. The electricity tariff (Pt) used a triangular distribution (min 0.07, mode 0.10, max 0.14 USD kWh^{-1}) representing tariff-band variability. Plant efficiency (η) followed a PERT distribution (min 0.80, mode 0.88, max 0.92). All variables were treated as independent; correlation effects are reserved for future work.

A 20-year financial horizon was selected to match common hydropower concession and loan tenors [5]. The annuity factor (AF), which converts a uniform annual series into its present value over the 20-year horizon, was included to represent discounted cash-flow effects. Each simulation trial was executed on a per-kilowatt basis using the following relationships:

$$AF = \frac{1 - (1 + r)^{-20}}{r} \tag{1}$$

$$E = 8760 \times CF \times \eta \tag{2}$$

$$LCOE = \frac{C_{capex} \cdot \frac{r(1+r)^{20}}{(1+r)^{20}-1} + C_{O\&M}}{8760 \times CF \times \eta} \tag{3}$$

$$\mathrm{NPV} = -C_{capex} + [(P_t \times 8760 \times CF \times \eta) - C_{O\&M}] \times$$

$$\left(1 - (1+r)^{(-20)}\right)/\mathrm{r} \tag{4}$$

Random sampling was implemented using built-in spreadsheet random-number functions consistent with inverse-transform methods. Two thousand iterations were executed, which ensured stable mean and standard-deviation convergence. Rank correlations between input variables and each output metric were computed using Spearman's method to generate tornado charts showing the relative influence of each input.

3 Case Illustration

The model was applied to a simplified per-kilowatt PHES representation using the parameter ranges defined above. Two thousand Monte Carlo trials produced probabilistic distributions for both LCOE (USD kWh^{-1}) and NPV (USD kW^{-1}).

Table 1 Comparison between deterministic and Monte Carlo (probabilistic) results.

Metric	Method	P10	P50	P90	Remarks
LCOE (USD kWh^{-1})	Deterministic	–	0.082	–	Single-point estimate
LCOE (USD kWh^{-1})	Monte Carlo	0.069	0.081	0.098	Probabilistic distribution
NPV (USD kW^{-1})	Deterministic	–	+441	–	Single-point estimate
NPV (USD kW^{-1})	Monte Carlo	−17	+423	+927	Captures downside and upside

Sensitivity results showed that capacity factor (−0.52) and O&M cost (+0.47) had the largest correlations with LCOE, while tariff (+0.66) and capacity factor (+0.41) dominated NPV outcomes (Fig. 1). Discount rate and CAPEX had moderate influence, whereas efficiency had smaller effects ($<\pm$ 0.2).

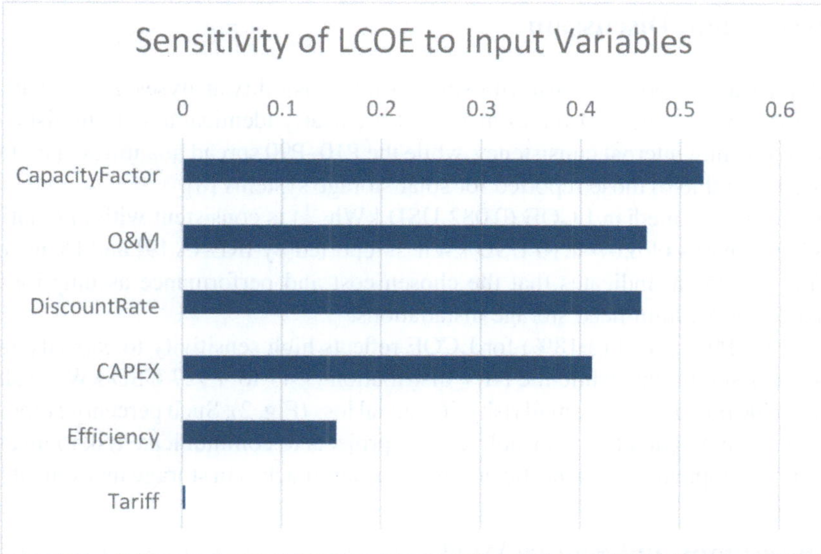

Fig. 1 Tornado chart (Spearman rank correlation) for LCOE.

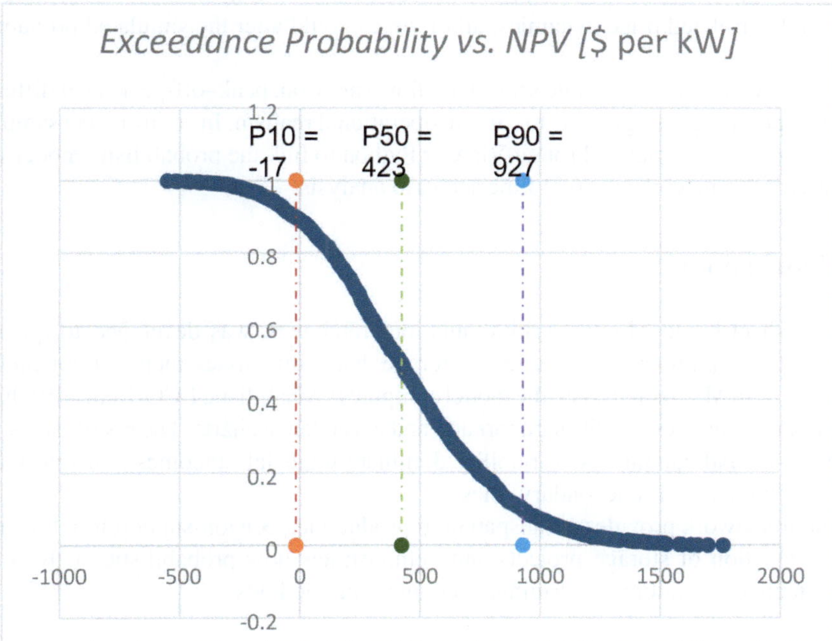

Fig. 2 Exceedance curve of NPV showing P10–P50–P90 probabilities.

4 Results and Discussion

The Monte Carlo approach transforms single-value feasibility analyses into probability-based assessments. The P50 values in Table 1 are nearly identical to deterministic estimates, confirming internal consistency, while the P10–P90 spread quantifies uncertainty ranges comparable to those reported for solar storage systems [8].

The simulated median LCOE (0.082 USD kWh^{-1}) is consistent with international PHES benchmarks of 0.07–0.10 USD kWh^{-1} reported by IRENA [9] and Deane et al. [6]. This agreement indicates that the chosen cost and performance assumptions are reasonable for medium-head storage installations.

The P10–P90 spread ($\pm 18\%$) for LCOE reflects high sensitivity to capacity-factor and O&M assumptions, while the NPV distribution (-17 to $+927$ USD kW^{-1}) shows higher upside potential but a small risk of financial loss (Fig. 2). Such percentile reporting has been recommended for renewable-energy projects to communicate risk to investors [8, 10], and complements stochastic optimization approaches in storage investment [11].

5 Limitations and Future Work

The present analysis assumes independent variables and constant annual averages. In practice, pumped hydro operation is influenced by temporal hydrology, market fluctuations, and correlated price dynamics, all of which could alter the simulated probability spreads.

Future work will incorporate seasonal inflow variation, peak–off-peak tariff differentials, and pumping-energy costs to capture operational realism. In addition, the sampling process may be automated through VBA or Python to link the probabilistic model with digital-twin frameworks for real-time scenario analysis.

6 Conclusions

A spreadsheet-based Monte Carlo simulation framework was developed to quantify uncertainty in pumped hydro energy storage feasibility assessments. Implemented entirely within Microsoft Excel, the model computes probabilistic LCOE and NPV distributions and visualizes risk through tornado and exceedance charts. The results show that energy-price and capital-cost variability dominate financial outcomes, whereas O&M cost and efficiency play secondary roles.

The framework provides a transparent, reproducible decision-support tool for early-stage evaluation of storage projects and demonstrates how probabilistic analysis can complement conventional deterministic engineering methods.

References

1. Omar, R., Isa, N.M., Rahman, H.A.: Hydropower development in Malaysia: Challenges and opportunities. Energ. Strat. Rev. **38**, 100763 (2021)
2. Metropolis, N., Ulam, S.: The Monte Carlo method. J. Amer. Stat. Assoc. **44**(247), 335–341 (1949)

3. Zhu, B., Li, X., Wang, Y.: Probabilistic assessment of hydropower projects under climate uncertainty. Renew. Energy. **175**, 471–482 (2021)
4. Louie, H., Pearce, J.: Techno-economic analysis of micro-hydro systems with Monte Carlo simulation. Energy Sustain. Dev. **31**, 111–119 (2016)
5. Deeney, P., Deeney, P., Cummins, M., Heintz, K., Pryce, M.T.: Economic evaluation of energy-storage technologies using Monte Carlo simulation. Energy Rep. **7**, 892–903 (2021)
6. Deane, J.P., Ó Gallachóir, B.P., McKeogh, E.J.: Techno-economic review of pumped hydro energy storage. Renew. Sust. Energ. Rev. **78**, 129–148 (2017)
7. International Energy Agency (IEA): Hydropower Technology Brief: Risk Assessment Methods for Hydropower. International Energy Agency, Paris (2022)
8. Fan, Y.V., Lee, C.T., Lim, J.S., Klemeš, J.J., Le, P.: Monte Carlo simulation for techno-economic uncertainty in renewable-energy systems. J. Cleaner Production. **379**, 134832 (2022)
9. Ralon, P., Taylor, M., Ilas, A., Diaz-Bone, H., Kairies, K.: Electricity Storage and Renewables: Costs and Markets to 2030. IRENA, Abu Dhabi (2017)
10. Agyekum, E.B., Amjad, F., Ansah, M.N.S.: Review of renewable-energy investment risk assessment models. Renew. Sust. Energ. Rev. **125**, 109795 (2020)
11. Liu, Y., He, X., Wang, S.: Uncertainty analysis in energy-storage investment using stochastic optimisation. Appl. Energy. **269**, 115082 (2020)

Selecting an Optimal ESP32 Module for Smart Lighting System using ESP-NOW

Muhammad Izat Firdaus Roslan[1], Aimi Syamimi Ab Ghafar[1](✉), Norezmi Jamal[1], Sree Sharvindren Subramaniam[1], Mohd Nurul Al-Hafiz Sha'abani[1], and Mohd Noh Sarip[2]

[1] Department of Electrical Engineering Technology, Faculty of Engineering Technology, Universiti Tun Hussein Onn Malaysia, Hab Pendidikan Tinggi Pagoh, KM 1, Jalan Panchor, 86400 Panchor, Johor, Malaysia
aimi@uthm.edu.my

[2] Makmal Penyelidikan Elektrik, Pusat Kecemerlangan Kejuruteraan dan Teknologi JKR (CREaTE), Jalan Kemus, Simpang Ampat, 78000 Alor Gajah, Melaka, Malaysia

Abstract. The advancement of smart building technology has increased the demand for compact, energy-efficient, and reliable wireless sensor nodes for real-time automation. This study focuses on developing and evaluating an IoT-enabled smart lighting system using ESP32-based microcontrollers and PIR motion sensors. Two modules, Seeed Studio XIAO ESP32C6 and Arduino Nano ESP32, are compared in terms of Received Signal Strength Indicator (RSSI), latency, and power consumption to determine their suitability for smart lighting applications. The SR501 PIR sensor exhibited superior detection performance, providing the longest and most consistent sensing range. Experimental results revealed that the Seeed Studio XIAO ESP32C6 outperformed the Arduino Nano ESP32, offering stronger wireless connectivity, lower mean latency (6.33 ± 0.93 ms), and low power consumption during both idle and motion-activated operations. These findings identified the Seeed Studio XIAO ESP32C6 as a more efficient and reliable option for IoT-based smart lighting, contributing to enhanced energy efficiency, faster system response, and improved scalability in smart building environments.

Keywords: smart lighting · ESP-NOW · wireless communication · PIR sensor · RSSI · Latency

1 Introduction

As cities expand and sustainability becomes increasingly important, energy-efficient lighting has become a priority in modern urban development. Smart lighting integrated with Internet of Things (IoT) technology enables automated control, reduced energy consumption, and improved operational flexibility through sensor-driven and networked lighting strategies [1, 2].

Smart lighting systems typically combine microcontroller-based nodes with motion or ambient-light sensors to adjust illumination according to occupancy and environmental conditions. Studies show that IoT-enabled dimming and motion-based control can

M. A. b. Jamlos et al. (Eds.): ICSET 2025, LNNS 1897, pp. 332–338, 2026.
https://doi.org/10.1007/978-3-032-20722-7_42

achieve substantial energy savings where it is reported to be between 17% and 60% compared to fixed-output lighting, depending on deployment scenarios [2, 3]. Low-cost IoT microcontrollers such as ESP-class devices further support adaptive dimming, wireless control, and simplified maintenance [1, 3].

Despite these advancements, many lighting installations continue to rely on grid-supplied electricity with limited automation, particularly in regions where fossil fuels dominate power generation [4]. Smart lighting technologies that enhance efficiency, controllability, and reliability can therefore contribute significantly to reducing operational cost and environmental impact [1, 5]. However, maintaining consistent performance across diverse environmental and sensor conditions remains a challenge for large-scale implementations [3, 5].

Wireless communication is another critical component of smart-lighting performance. Distributed nodes must maintain stable links to support reliable switching, dimming, and monitoring. Existing studies highlight challenges such as latency, packet loss, and limited scalability in IoT and WSN-based lighting systems [3, 6]. Sensor characteristics which include sensitivity, detection distance, and false-trigger behaviour, also influence user comfort and overall energy-saving potential [6].

This study addresses these challenges by developing and evaluating an IoT-enabled smart lighting system using microcontroller-based nodes and motion sensors. Multiple PIR sensor models are assessed for detection performance, while ESP32-class modules are compared in terms of wireless signal strength, latency, and power consumption. Communication is implemented using a lightweight protocol (ESP-NOW), and the findings identify the most suitable configuration for efficient and scalable smart-lighting applications.

2 Methodology

2.1 Design and Testing Block Diagram

Figure 1 shows the system process begins with motion detection using a PIR sensor as the primary input. When motion is detected, the ESP32 microcontroller records the timestamp and transmits motion data to the smart switch controller via the low-latency, energy-efficient ESP-NOW protocol. The controller receives the data, activates an LED indicator, and logs performance metrics such as RSSI and latency.

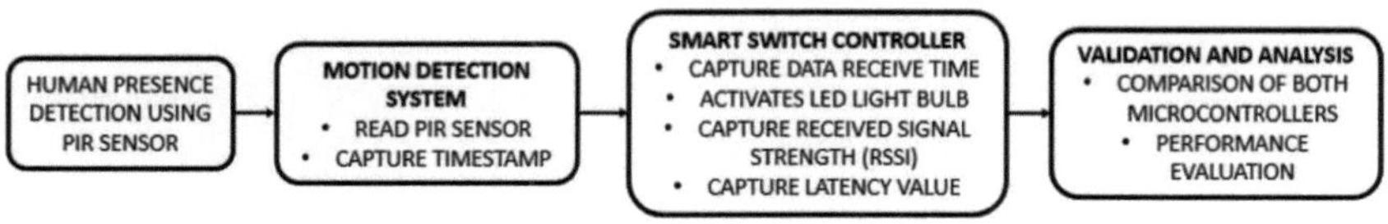

Fig. 1. Design and testing block diagram

2.2 System Block Diagram

Figure 2 shows the block diagram of this project, consisting of several interconnected components. The motion detection system is equipped with PIR motion sensor connected to ESP32-based microcontroller. 3 types of PIR motion sensors are tested in this project. This system is also powered by 5 V mini solar system. On the other hand, the smart switch system is made up of an ESP32-based microcontroller, a relay module and a light bulb.

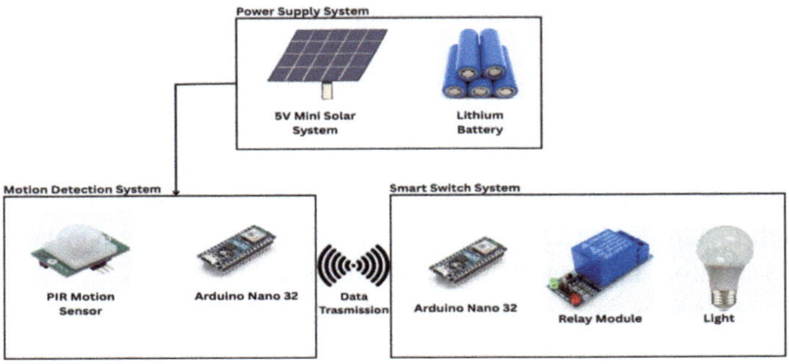

Fig. 2. Proposed System Block Diagram

Figure 3 illustrates the system flowcharts for the IoT-based smart lighting system. In Figure 3(a), the motion detection system begins by initializing the PIR sensor and ESP32 controller. The PIR sensor continuously monitors for movement. When motion is detected, the occupancy variable is updated to "1" and this information is sent wirelessly using ESP-NOW to the controller node. A one-minute delay is introduced before resuming detection to reduce redundant transmissions and save energy.

In Fig. 3(b), the smart switch controller receives the occupancy data. If occupancy equals "1," the controller turns on the lamp and maintains it for ten minutes before re-evaluating the status. If no motion is detected, the lamp remains off to conserve power.

This integrated process ensures responsive, energy-efficient lighting control through real-time IoT communication between motion detection system and smart switch controller.

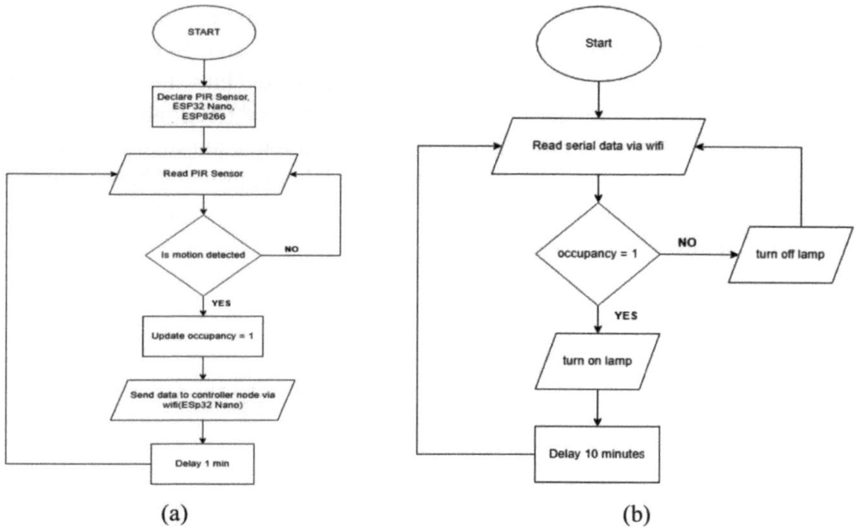

Fig. 3 System flowcharts (a) Motion detection system (b) Smart switch controller

3 Results and Discussions

3.1 Detection Capability Analysis of Different PIR Sensors

Three PIR sensors which are AM312, SR505, and SR501 were tested to evaluate motion detection performance. Both AM312 and SR505 detected motion up to 390 cm, while SR501 achieved the longest and most consistent range of 690 cm. SR501 demonstrated superior sensitivity and reliability, making it the best choice for efficient, responsive smart lighting control.

Table 1 Result of PIR Sensor

PIR Sensor Model	Detection Range (cm)
AM312	390
SR505	390
SR501	690

3.2 Received Signal Strength Indicator (RSSI) Performance Analysis

Reliable wireless communication is crucial for coordinating the sensor node and actuator in the smart lighting system. The RSSI is measured over various distances setup to assess signal strength and stability. Experiments were conducted in an open outdoor area with clear line-of-sight (LOS) between the transmitter and receiver. No obstacles or major interference sources were present. As shown in Fig. 4, the Seeed Studio demonstrated better RSSI performance compared to the Arduino Nano, primarily due to differences in antenna design, board layout, and RF front-end sensitivity. Although both devices were tested without external antennas, the Seeed Studio's optimized built-in antenna, coupled with the ESP32-C6's higher receiver sensitivity and improved RF front-end [7], yielded stronger signal strength and more stable wireless communication than the ESP32-S3-based Arduino Nano. This demonstrates that the Seeed Studio provides better long-range performance and reliability, making it more suitable for stable communication link in smart lighting applications.

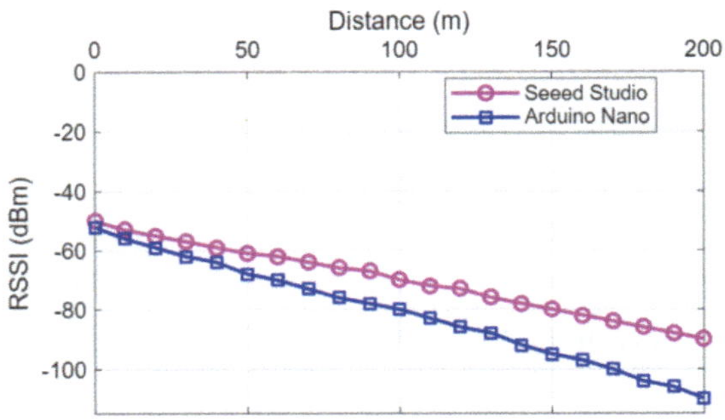

Fig. 4 RSSI Performance Analysis for Seeed Studio XIAO ESP32C6 and Arduino Nano ESP32

3.3 Latency Performance Analysis

Latency measurements were recorded for both modules over 21 distance points ranging from 0 m to 200 m, in an open outdoor area with clear LOS (no obstacle). Based on Fig. 5, the Seeed Studio XIAO ESP32C6 recorded a lower mean latency (6.33 ± 0.93 ms) compared to the Arduino Nano (10.48 ± 1.01 ms). A paired, two-tailed t-test at a 0.05 significance level indicated a statistically significant difference between the average latency of the Arduino Nano and Seeed Studio modules ($t(20) = 13.88$, $p < 0.001$), confirming that the Seeed Studio achieved consistently lower latency.

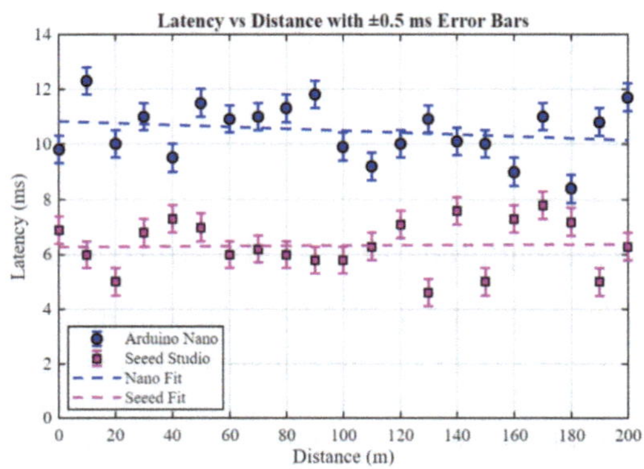

Fig. 5 Results of Average Latency vs. Distance on Xiao Seed Studio C6

3.4 Power Consumption of Seeed Studio XIAO ESP32C6

The Seeed Studio XIAO ESP32C6 power consumption (Table 2) was evaluated in two states: idle and motion-activated. During idle operation, it consumed only 0.117 μW, indicating efficient low-power performance. When motion was detected, power usage increased to 6.767 mW as the microcontroller processed data and transmitted via ESP-NOW. This brief rise reflects the energy required for wireless communication, while the one-minute sensing delay helps maintain overall energy efficiency in the smart lighting system.

Table 2 Power Consumption Measurement of Motion Detection System

Scenario	Voltage	Current	Power
Idle	1.3 mV	0.09 mA	0.117 μ W
Motion activated	2.583 V	2.62 mA	6.767 mW

For the smart switch controller, power consumption (Table 3) was measured in idle and active states. When no motion was detected, it consumed only 0.426 μW during background monitoring. Upon motion detection, power rose to 2.56 mW as the controller activated the lamp through a relay. These results show that the Seeed Studio XIAO ESP32C6 efficiently conserves energy during idle periods while providing sufficient power for reliable lighting control when triggered.

Table 3 Power Consumption Measurement of Smart Switch Controller

Scenario	Voltage	Current	Power
Idle	0.3 mV	1.42 mA	0.426 μW
Motion-activated	32.5 mV	78.9 mA	2.56 mW

4 Conclusion

The results confirm the proposed ESP32-based smart lighting system is reliable and energy-efficient for smart lighting applications. The SR501 provided the best motion detection, while the Seeed Studio XIAO ESP32C6 achieved stronger signal strength, lower latency, and minimal power use, ensuring responsive, stable, and efficient performance for modern smart building environments.

Acknowledgement. This research was supported by the Centre of Excellence for Engineering and Technology (CREaTE), Jabatan Kerja Raya (JKR) through the Ministry Contract Grant Scheme (Vot K542) and Universiti Tun Hussein Onn Malaysia (UTHM).

References

1. Widartha, V.P., Ra, I., Lee, S.-Y., Kim, C.-S.: Advancing smart lighting: A developmental approach to energy efficiency through brightness adjustment strategies. *J. Low Power Electron. Appl.* **14**(1), 6 (2024)
2. Mohankumar, A., Supramaniam, M.H., Ramachandran, R., Praveena, P.: A low-cost IOT-enabled smart lighting system and its energy-saving potential. J. Sci. Technol. **61**(3), 1–10 (2023)
3. Obioma, P., Agbodike, O., Chen, J., Wang, L.: ISLS: IoT-Based Smart Lighting System for Improving Energy Conservation in Office Buildings, arXiv preprint arXiv:2503.13474, 2025.
4. Ember, "Malaysia: Country Profile – Power Sector," Ember Climate. https://ember-climate.org (2025). Accessed: 24 Aug 2025.
5. Malek, D., Ismail, S., Ahmad, N.: IoT and sensors for sustainable buildings. *J. Adv. Res. Dyn. Control Syst.* **17**(2), 112–122 (2025)
6. Omar, N.A., Rasid, M.A., Rahman, K.A.: Smart city: Recent advances in intelligent street lighting systems based on IOT and wireless sensor networks. *J. Electr. Comput. Eng.* **2022**, 5249187 (2022)
7. Espressif Systems, ESP32-C6 Datasheet: Wi-Fi 6 & Bluetooth 5.3 SoC with RISC-V Processor, Ver. 1.3. https://www.espressif.com/sites/default/files/documentation/esp32-c6_datasheet_en.pdf (2023). Accessed: 24 Dec 2024.

Preparation of Paper Making with Waste Sugarcane Bagasse and Eggshell Powder for Application Biodegradable Straw

Wan Mohammad Irfan Hushoiri Khairul Anuar and Nik Alnur Auli Nik Yusuf(✉)

Bio-product and Bioprocessing Technology Research Group, Faculty of Bioengineering and Technology (FBKT), Universiti Malaysia Kelantan (UMK), Jeli Campus, 17600 Jeli, Kelantan, Malaysia

alnurauli@umk.edu.my

Abstract. This research investigates the creation of biodegradable paper utilizing sugarcane bagasse (SCB) as a matrix, eggshell powder (ESP) as a filler, and aloe vera gel (AVG) as a binder. The objective was to examine the physical, mechanical and water absorption characteristics of paper produced with SCB and varying concentrations of ESP (0%, 5%, 10%, 15%, and 20%). SCB fibres were produced via kraft pulping and bleaching methods, whereas ESP was milled into a fine powder. The materials were blended with AVG and moulded into thin sheets through a hand-sheet forming process. The results indicated that SCB fibres and ESP fillers enhanced mechanical properties, whilst AVG improved binding, providing smoother and more durable paper. Fourier-transform infrared spectroscopy (FTIR) study demonstrated an interaction between SCB fibres, ESP and AVG, suggesting the existence of functional groups that improve bonding. Scanning electron microscopy (SEM) pictures revealed that moderate ESP concentrations increased fibre bonding and filler dispersion, whereas higher ESP concentrations caused agglomeration. Mechanical testing, such as tensile strength and Young's modulus, revealed that tensile characteristics improved as ESP content increased, with the 15% ESP sample having the highest tensile strength. However, when the ESP concentration increased elongation at break decreased, indicating a loss of flexibility. Water absorption tests revealed that as ESP content increased water uptake increased, with the 20% ESP sample absorbing the most water due to the filler's hydrophilicity. The paper with 0% and 5% ESP is the best composition to making a water-resistant straw due to the lowest water absorption than other. In summary, incorporating ESP into SCB-based paper improves mechanical strength while sacrificing flexibility and water resistance, making it a sustainable option for applications such as biodegradable straws.

Keywords: Biodegradable paper · Sugarcane bagasse · Eggshell powder · Kraft pulping · Mechanical properties

M. A. b. Jamlos et al. (Eds.): ICSET 2025, LNNS 1897, pp. 339–346, 2026.
https://doi.org/10.1007/978-3-032-20722-7_43

1 Introduction

1.1 Agro Waste of Sugarcane Baggase

Agricultural waste (AW) is produced in millions of tonnes every year, and incorrect management and disposal have significant environmental consequences, including ecosystem damage. When appropriately controlled and utilised, AW has the potential to significantly contribute to environmental sustainability and energy security. In general, AW in the form of bagasse, bran, husks, peels, leaves, seeds, stems, stalks, etc. Agro waste as lignocellulosic biomass (LCB) is mostly composed of cellulose backbone, with hemicellulose and lignin serving as additional essential carbonaceous components (Sharma et al., 2022).

Food waste is divided into several categories, including starch-based waste, dairy waste, meat and fish waste, vegetable trimmings, fruit peels, wasted grains, and pulp garbage. Traditionally, food waste from the food processing industry is discarded, burnt, landfilled, composted, or anaerobically digested to replenish the soil. However, these food wastes are incredibly nutritious, making them ideal substrates for microbial growth and the manufacture of a variety of enzymes (Ravindran et al., 2018). Fig. 1 shows the various classifications of agro-industrial food waste used as soil amendment, fertilizer, animal feed, and in various other applications.

Fig. 1 Classification of different types of agro-industrial food waste (Ravindran et al., 2018)

Sugarcane bagasse (SCB) distinguishes out from other agricultural waste products as a highly ideal raw material for paper manufacture due to its unique set of qualities. The high cellulose content of 40 to 50% makes it ideal for producing robust and flexible paper (Gañán et al., 2016). SCB's low lignin concentration (20 to 25%) makes it easier to pulp and bleach than other agro wastes like wheat straw or rice husk, which require more chemical processing (Rodríguez et al., 2017). SCB fibre length, which ranges from 0.8 to 1.6 mm, leads to improved paper strength and quality, making it a viable alternative to wood-based fibres (Dutt et al., 2018). Furthermore, SCB is readily accessible as a byproduct of the sugar industry, making it a viable and cost-effective solution. SCB, unlike rice husk, has a low silica concentration (<3%), which prevents equipment scaling and improves pulping efficiency (Moraes et al., 2016). Its naturally pale colour also lowers the need for excessive bleaching, which reduces environmental effect (Silverio et al., 2017). Because of these combined advantages of high cellulose, low lignin, long fibres, low silica, and widespread availability with the SCB is highly accepted in the paper and pulp industry as an environmentally benign and cost-effective alternative to wood pulp. Bagasse is recognized as a fibrous residue (refer to Fig. 2) that remains following the extraction of juice from sugar cane stalks. It is composed of water, fibres, and minor quantities of soluble solids. The percentage contribution of each component fluctuates based on factors such as the variety of sugar cane, its maturity, the harvesting method, and the efficiency of the crushing plant.

Fig. 2 Extraction of juice from Sugarcane Bagasse (Wilson, 2018).

2 Methodology

Methodology of Kraft Pulping Process of Sugarcane Baggase. The air-dry bagasse (10 kg) was screened, moistened, depicted and air dried. The prepared raw materials were put into a bag and delivered to the Material Technology Lab in UMKJ for processing. The sugarcane bagasse was screened to remove sand and dust, soaked in hot water (60 °C) for 30 min and air dried to a moisture content of around 10%–15% before pulping.

To remove any lignin and hemicellulose present in the bagasse fibre, a kraft pulping process was conducted using a solution of sodium hydroxide (NaOH) with a concentration of 15% and 20% sodium sulfide (Na_2S). The SCB chips were cooked in this solution with these concentrations that are typically used in the kraft pulping process. This concentration has been selected refer from the previous study by Mamaye et al. (2020) that kraft pulping condition of SCB chip based on variation of active alkali & sulfidity that has reported. Sodium sulfide was suitable for the kraft pulping process of sugarcane bagasse, preparing the fibre for effective bleaching. Although it was not a direct bleaching agent, it worked in combination with traditional bleaching agents like sodium hypochlorite and hydrogen peroxide to improve the overall efficiency and quality of the bleaching stage. In kraft pulping, the combination of NaOH and Na_2S effectively to break down lignin and hemicellulose while preserving the cellulose in the bagasse (Doherty et al., 2010). This process helped produce strong, high-quality pulp that was easier to bleach in subsequent stages.

2.1 Physical Testing of Optical Microscope & Water Absorption Test

Microscopic testing assesses the morphology and structural qualities of paper and paper products in the paper industry. Optical microscopy was conducted at magnifications ranging from 10× to 20× to examine the fiber dispersion, surface uniformity, and filler distribution of SCB/ESP paper before further mechanical testing. This analysis was essential for ensuring even dispersion of the materials and detecting any surface inconsistencies. The samples were oven-dried at 60 °C for 24 h before being cut into 10 mm × 10 mm sections and carefully mounted on glass slides to minimize structural disturbance.

The water absorption test evaluates the paper straw ability to absorb water, which is important for applications such as agricultural mulch and erosion management. It aids in determining moisture retention capacity and probable deterioration rates in moist conditions. This test verifies that straw is suitable for numerous uses and helps forecast its performance under varied situations. Water absorption of the paper sheets was measured at room temperature (ASTM D570–98). Paper pieces measuring 2.5 × 2 cm were weighed before (wi) and after (wf) the water absorption procedure. A water absorption test was conducted to determine the amount of water absorbed by the paper straws (Ngo Thi Thuong, 2020).

3 Results & Discussion

3.1 Physical Testing of Films by Appearance, Optical Microscope & Water Absorption Test

The results of the experimental work analyzed and summarized. This chapter will discuss the results of paper that made from various composition of eggshell powder (ESP) and how different percent of ESP in sugarcane bagasse paper could give different effect towards the SCB pulp. The physical and mechanical properties was investigated, and several tests were conducted to study and describe the materials of paper characterization. In this research, the data obtained were characteristics of paper appearance, optical microscope and water absorption test were analyzed and discussed.

Paper Appearance. The visual characteristics of paper are essential factors that determine its quality, appropriateness for various uses, and market value. According to Hubbe et al. (2008), paper's visual properties, such as opacity, brightness, colour, gloss, and uniformity, are critical to its printability, aesthetic appeal, and tactile feel. Features like surface smoothness, absence of defects, and consistent colour are especially important for uses such as high-quality printing and packaging, where a flawless and uniform appearance is essential for producing sharp, clear results. The structure, material composition, and additives employed in papermaking have a direct impact on these visual qualities. For example, calcium carbonate fillers in eggshell powder improve opacity and brightness by enhancing light scattering, while the fibre network impacts surface smoothness and uniformity.

Water Absorption Test. Fig. 3 show the water absorption behavior of sugarcane bagasse (SCB) paper with various percentages of ESP content (0%, 5%, 10%, 15%, and 20%) over time. The water absorption of the paper increased with the addition of eggshell filler. Notably, the samples absorbed a significant quantity of water within the first minutes, after which the absorption rate stabilized. The results indicate that water absorption rises with increased ESP concentration, since ESP improves the paper's hydrophilic properties and porosity

Among all sample, the paper with 0% ESP sample exhibited the lowest water absorption, reaching a maximum of 98.06% after 5 min, whereas samples with 5%, 10%, 15%, and 20% ESP showed higher values of 109.09%, 115.43%, 117.28%, and 123.86%, respectively. The paper sample with 20% ESP had the greatest water absorption, reaching a maximum of 126.81% after 1 min. The water absorption remained relatively stable, oscillating between 122.84% and 123.86% for the following 4 min. Similarly, the sample with 15% ESP had a maximum water absorption of 123.7% after 1 min, but then gradually decreased to 115.09% after 4 min. In the 10% ESP sample, the peak water absorption was noted at 117.38% after 1 min, decreasing to 106.28% after 3 min and then stabilizing between 106.28% to 115.43%. Conversely, the paper sample containing 5% ESP shown a more moderate rise in water absorption, reaching a high of 109.09% after 5 min, and the sample with 0% ESP showed the lowest water absorption, peaking at 98.06% after the 5 min duration making it a suitable composition for making paper straw.

Table 1 Appearance of paper with different eggshell powder.

No	Paper Sample (g) SCB (Sugarcane bagasse), AVG (Aloe vera gel), ESP (Eggshell powder)	Appearance
1.	 20SCB3AVG0ESP	• Natural beige with obvious fibre textures.
2.	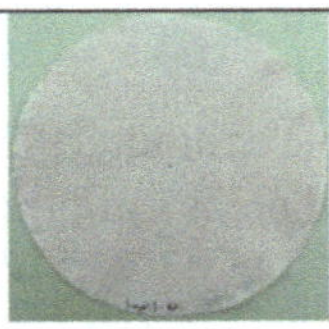 20SCB3AVG5ESP	• The colour is a little lighter and it seems more homogeneous.
3.	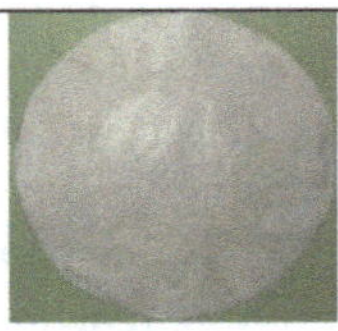 20SCB3AVG10ESP	• Noticeable increase in brightness and colour consistency.
4.	 20SCB3AVG15ESP	• Minimal imperfections, smooth and clean appearance.
5.	 20SCB3AVG20ESP	• With 20% eggshell powder, the paper turned chalky white and had the smoothest surface.

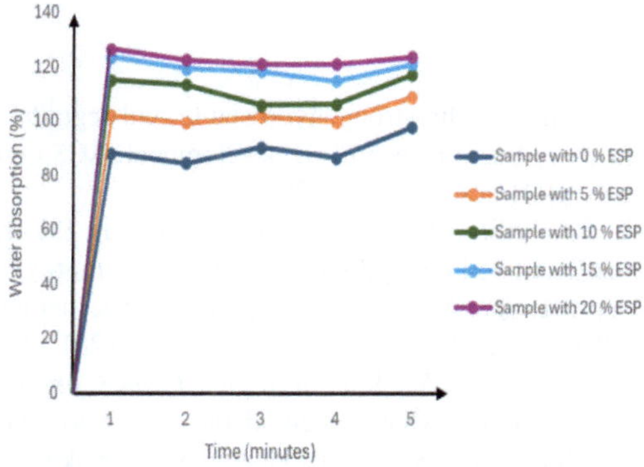

Fig. 3 Water absorption of the paper sample that made from various percent ESP.

4 Conclusion

The characteristics and analysis of physical properties of eggshell powder percentage in paper made from SCB for the application biodegradable straw are reported. Using sugarcane bagasse (SCB) and eggshell powder (ESP) as the primary materials, this study effectively illustrated the potential to produce biodegradable paper straws. By incorporating ESP as a filler and aloe vera gel (AVG) as a binder, the paper's physical and mechanical properties were substantially improved. The calcium carbonate-rich composition of ESP improved the smoothness, brightness, and opacity of the paper, whereas AVG increased fibre bonding, resulting in greater strength and uniformity that have the opaque structure by effectively filled fibre gaps.

Among the evaluated compositions, the sample containing 15% ESP demonstrated superior performance in both physical and mechanical assessments. This formulation offered improved smoothness, brightness, and opacity, as the calcium carbonate in ESP efficiently filled fibre voids, leading to a more consistent and aesthetically pleasing paper structure. Furthermore, the water absorption tests indicated that the 15% ESP composition achieved a good equilibrium between mechanical durability and water resistance, mitigating the prevalent shortcomings of paper-based straws. This ratio content providing an optimal balance between durability and flexibility.

This study revealed that agricultural waste (SCB) and food waste (eggshells) may be efficiently repurposed into profitable, eco-friendly items. The findings are consistent with the objective to create a sustainable replacement to plastic straws, indicating a possible solution to the worldwide issue of plastic pollution. By using these waste materials, the study highlights the potential to lessen environmental impact while also promoting the creation of biodegradable materials. Furthermore, the paper straws' very quick decomposition rate when compared to standard plastic straws demonstrated their biodegradability, making them a viable option for decreasing plastic waste. Overall, this study demonstrated the feasibility of using agricultural and food waste to create eco-friendly goods, therefore contributing to worldwide efforts to promote sustainability and reduce environmental pollution.

References

Agumba, D.O., Pham, D.H., Kim, J.: Ultrastrong, hydrostable, and degradable straws derived from microplastic-free thermoset films for sustainable development. ACS Omega. **8**(8), 7968–7977 (2023). https://doi.org/10.1021/acsomega.2c07797

Ahmadi, F., Zamiri, M.J., Khorvash, M., Ziaee, E., Polikarpov, I.: Pre-treatment of sugarcane bagasse with a combination of sodium hydroxide and lime for improving the ruminal degradability: optimization of process parameters using response surface methodology. J. Appl. Anim. Res. **44**(1), 287–296 (2015). https://doi.org/10.1080/09712119.2015.1031783

Alan. *What biodegradable straws do You need to Know?-Sugarcane straws!* Homestraw. https://www.homestraw.com/what-biodegradable-straws-do-you-need-to-know-sugarcane-straws/#Benefits_of_Sugarcane_Straws_Biodegradable_Drinking_Straws (2023)

Al-Sulaimani, K., Dwivedi, P.: Production of handmade papers from sugar cane bagasse and banana fibers in Oman. Int. J. Stud. Res. Technol. Manag. **5**(3), 16–20 (2017). https://doi.org/10.18510/ijsrtm.2017.534

Amin, M., Attia, M., Agwa, I., Elsakhawy, Y., El-Hassan, K.A., Abdelsalam, B.A.: Effects of sugarcane bagasse ash and nano eggshell powder on high-strength concrete properties. Case Stud. Constr. Mater. **17**, e01528 (2022)

Aragaw, T.A.: Sustainable management of drinking plastic straws is required to reduce plastic pollution: are we using them more during COVID-19? J. Hazard. Mater. Adv. **12**, 100328 (2023). https://doi.org/10.1016/j.hazadv.2023.100328

Asem, M., Jimat, D.N., Jafri, N.H.S., Nawawi, W.M.F.W., Azmin, N.F.M., Wahab, M.F.A.: Entangled cellulose nanofibers produced from sugarcane bagasse via alkaline treatment, mild acid hydrolysis assisted with ultrasonication. J. King Saud Univ. - Eng. Sci. **35**(1), 24–31 (2021). https://doi.org/10.1016/j.jksues.2021.03.003

PhyQA: An Ontology-Based Question Answering System for Formula Identification in Physics Subject

Catherine Seng Jing Hii(✉) and Chih How Bong

Faculty of Computer Science and Information Technology, University of Malaysia Sarawak, Sarawak, Malaysia
catherine.hiisj@outlook.com, chbong@unimas.my

Abstract. In today's fast-paced world, it's crucial to access accurate and relevant information, especially in physics, where complex calculations are standard. Finding answers to physics questions can be difficult and time-consuming, particularly for those without a deep understanding of the subject. This paper proposes a novel solution called PhyQA - a question-answering system that utilizes ontology in the physics domain. By leveraging the power of ontologies to provide structured knowledge about physics concepts and their relationships, PhyQA can efficiently and accurately identify calculational questions. The development of this system was motivated by the need for a tool that could help bridge the gap between the complex calculations of physics and first-time learners' understanding. PhyQA provides clear and concise answers to physics questions, making it available to a broad audience, including students, researchers, and anyone interested in the subject. Using ontologies in this system provides a new technique for organizing and structuring knowledge, allowing for a more sophisticated representation of complex concepts. Ultimately, this paper emphasizes the importance of creating a question-answering system with ontology in the physics domain. It demonstrates its potential to revolutionize how we access and understand knowledge in this subject. By providing accurate and accessible answers to calculational questions, PhyQA has the potential to impact education and research in this field significantly.

Keywords: Question-answering system · Physics · Ontology

1 Introduction

Question-answering systems (QAS) are applications of natural language processing (NLP) that are becoming increasingly popular. QAS aims to interpret and respond to inquiries from users in their native tongue [1].

The system consists of three modules: query processing, document processing, and answer processing [2]. The query processing module is a module that pre-processes the natural language question, which includes identification, classification, and reformulation. The answer-processing module extracts the final element from the previous module. QAS is an acronym for Question Answering System, which provides solutions

M. A. b. Jamlos et al. (Eds.): ICSET 2025, LNNS 1897, pp. 347–353, 2026.
https://doi.org/10.1007/978-3-032-20722-7_44

to theory-based questions like definitions, factoids, and descriptions. QAS analyses the input questions and extracts the subject, relationship, and object triples. However, the relationships between the entities in the questions are usually not further processed and need clarification. Ontology can enhance QAS by comprehending the words and sentence structure by providing additional information to an answer [3]. In the current study, the researchers use a closed domain ontology-based focusing on Physics, PhyQA.

The structure of this paper presents in the following manner: Sect. 2 presents related works for ontology-based QAS. Sect. 3 explains the PhyQA architecture, while Sect. 4 discusses the results. The last section provides a concluding statement.

2 Research Methodology

The primary objective of this study is the development of PhyQA, an ontology-driven question-answering system, designed to enhance the accurate identification and application of appropriate physics formulas for calculation-based questions. Concurrently, this project aims to develop PhyQA, a formalized physics ontology designed to capture the semantics of Malaysia's Sijil Pelajaran Malaysia physics curriculum [4].

In order to achieve the specified goals, this study is guided by a set of research inquiries:

1. How can the development of PhyQA effectively enhance its ability to identify appropriate physics formulas for calculation-based questions among secondary students in Malaysia?
2. How does integrating PhyOnto enhance the precision and accuracy of responses generated by PhyQA, especially when addressing calculation-oriented physics questions?
3. PhyQA and Fong and Bong's QAS have notable differences in their approach, encompassing distinct features and characteristics. Hence, these differences affect their effectiveness in meeting the specific needs of secondary students studying Physics in Malaysia.

The research questions presented here serve as the foundation for our investigation into enhancing physics education in Malaysia. The objective is to thoroughly examine the contributions of PhyQA and PhyOnto to enhanced learning outcomes and evaluate their effectiveness in comparison to existing models, such as Fong and Bong's QAS (2017) [5]. In essence, this study aims to develop PhyQA and PhyOnto tools to enhance the value of physics teaching by providing accurate solutions for calculation-based questions.

3 PhyQA Architecture

The implementation of PhyQA involves the utilization of an ontology to facilitate automatic question-answering. Hence, it consists of four main modules: (1) Ontology Construction, (2) Question Pre-processing, (3) Candidate Answers Retrieval, and (4) Answer Formulation. The ontology serves as a knowledge base for PhyQA to recognize and understand physics concepts and their relationships. Figure 1 illustrates the architecture of PhyQA and its modules.

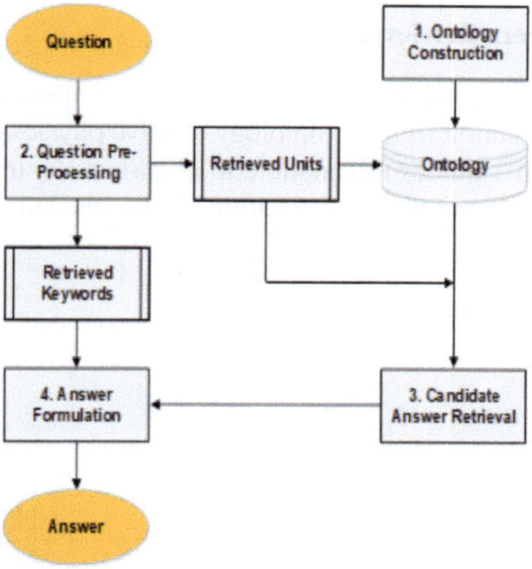

Fig. 1 The PhyQA architecture

Figure 1 depicts the process for solving calculation physics questions. Each module is outlined below.

3.1 Ontology Construction

To establish a solid foundation, various methodologies for ontology development were analyzed [6], leading to the decision to implement the "101 Method" [7]. This approach guides the creation of PhyOnto, an ontology designed to support PhyQA by capturing information about physics units and measurement instruments.

The ontology development process relies on these textbooks as the primary data source [8, 9]. The aim is to incorporate the fundamental concepts, theories, and principles emphasized in the official curriculum into the ontology, named PhyOnto. In summary, the focus is placed on Malaysia's SPM examination standard, with primary reliance on textbooks, as this examination plays a central role in the Malaysian education system.

Additionally, PhyOnto uses OWL for illustration. PhyOnto consists of 9 classes, four object properties, 153 individuals, and 704 axioms. PhyOnto can be found on GitHub. The information in PhyOnto can be retrieved using SPARQL [10], the standard language for querying RDFS data.

3.2 Question Pre-Processing

The question pre-processing module was developed with the specific goal of effectively handling various inquiries related to the field of physics. Its primary aim is to accurately respond to calculation-based questions that commonly emerge within this discipline.

The NLP techniques employed in the question pre-processing module of PhyQA include tokenization, stopword removal, lemmatization, stemming and entity recognition [11, 12].

3.3 Candidate Answers Retrieval

The candidate answer retrieval module is a necessary component of the PhyQA as it extracts pertinent information from the ontology to solve physics questions that involve calculations. Algorithm 1 guides this module to identify units in user questions while connecting them with relevant physical quantities within the ontology [13].

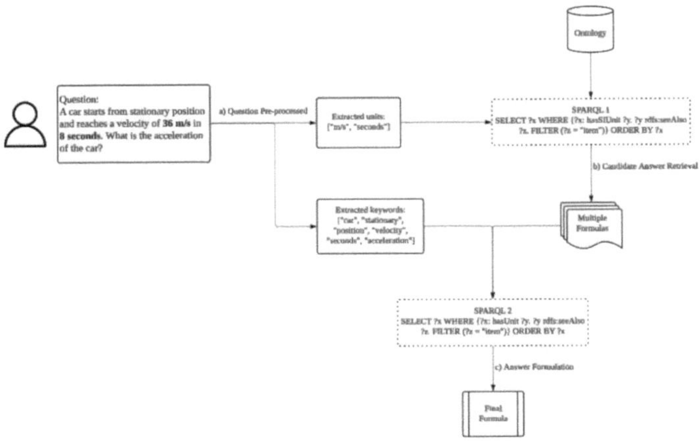

Fig. 2 The process of PhyQA on retrieving the answers from ontology using SPARQL queries.

The utilization of SPARQL queries in PhyQA, as illustrated in Fig. 2, demonstrates how the ontology, PhyOnto, is navigated to retrieve potential answers.

3.4 Answer Formulation

This section provides a detailed explanation of the process through which PhyQA offers accurate responses to inquiries about physics.

From a set of physical quantities and their corresponding formulas obtained from the SPARQL 2, an intersection operation assists in identifying the optimal answer. However, in certain circumstances, if only one physical quantity matches the retrieved units, the intersection process is unnecessary, and the corresponding formula is the answer.

Single Answer. If only one physical quantity matched the retrieved units, the corresponding formula can be directly considered as the answer. In this case, PhyQA can immediately provide the final formula to the user for solving the questions without further intersection or comparison with other formulas.

Multiple Possible Answers. If multiple physical quantities match the retrieved units, using the intersection process to identify the most relevant solution, the intersection between the results for different units is defined as the set of physical quantities with both units. For example, an intersection between the results for "m/s" and "seconds" can be defined as the set of physical quantities that have the unit "m/s" and also the unit "seconds." It appears in the query as m/s $\cap$seconds = {physics:physicalQuantity (hasUnit)∈ m/s and physicalQuantity (hasUnit)∈seconds}. There are two possible results after the intersection: one final answer or many final answers.

PhyQA's approach of leveraging an ontology-based knowledge base, utilizing an intersection process, and providing final problem-solving formulas showcases its potential as a promising question-answering system for physics education. The following section will further discuss the results of PhyQA.

4 Results and Discussion

4.1 Gold Standard

The proposed QAS is benchmarked and evaluated against a gold standard comprising 50 calculation questions and their corresponding answers. The questions were manually selected from textbooks, which served as the primary source. Consequently, only questions requiring calculations were included.

5 Evaluation Metrics

Comparing PhyQA with Fong and Bong's QAS is relevant because it is benchmark against existing Physics QAS. Fong and Bong's QAS is a QAS that was developed for physics education and aimed at answering five types of questions. Table 1 displays the results.

Table 1 Accuracy, Precision, Recall, and FI score of PhyQA.

Evaluation Metric	PhyQA	Fong and Bong
Accuracy	0.450	0.012
Precision	0.514	0.014
Recall	0.782	0.083
F1 Score	0.620	0.024

5.1 Advantages

Table 1 shows that PhyQA achieved an accuracy of 0.45, precision of 0.51, recall of 0.78, and an F1 score of 0.62. When compared to Fong and Bong's results, PhyQA outperforms them in accuracy, precision, and recall. However, it's worth noting that PhyQA's F1 score of 0.62 is higher than Fong and Bong's QAS F1 score of 0.02, indicating better overall performance regarding a balance between precision and recall.

PhyQA can provide the appropriate formula for acceleration based on the units given in the question which are "m/s" and "seconds." This capability to identify and utilize relevant formulas enhances its ability to solve Physics calculation questions. In contrast,

Fong and Bong's QAS returns a list of theories related to car crashes, indicating a limitation in its ability to provide accurate and relevant answers for physics calculations.

One advantage of PhyQA is that it provides a structured and standardized approach to answering physics calculation questions, which can be practical for educational purposes. The high recall value of PhyQA suggests that it effectively identifies relevant answers. The relatively low precision value may also include some irrelevant answers. This performance can be further improved through additional advancements to the system. Nevertheless, PhyQA's performance metrics demonstrate its potential as a reliable and effective QAS for physics education, surpassing the performance of Fong and Bong, as shown in the comparison of the evaluation metrics.

5.2 Limitation

However, a limitation of PhyQA is that it relies on the availability and accuracy of the underlying knowledge base and may need to perform better on questions that require more complex reasoning or interpretation. The system has the potential to yield false positive answers when dealing with situations where two or more physical quantities in a formulation share identical units.

Besides, some calculation questions involved diagrams, graphs, or charts. PhyQA cannot solve this type of calculation question. It has no image processing capacity to extract the parameters.

In addition, the system cannot solve questions requiring HOT skills. These questions often require multiple intermediate steps, implicit details, or inferring missing values through additional physics principles. Consequently, the necessary formulas cannot be directly applied, as key variables are not explicitly stated and must be derived via deeper conceptual reasoning. Thus, despite the presence of relevant formulas in the ontology, the system fails to generate the final answer.

6 Conclusion and Future Works

PhyQA represents a pioneering and innovative approach to addressing physics as it utilizes the capability of an ontology-based knowledge base, PhyOnto, specifically designed in this research to fulfil the requirements of solving calculation questions. The PhyQA system comprises four modules. The candidate answer retrieval module works with PhyOnto to retrieve the most relevant answers to the user's question.

This research identifies, describes, and analyzes the limitations of the PhyQA system despite its promising results. The researchers predict that PhyQA will enhance with advanced image processing techniques and merge with other QASs to expand its capabilities. This step will contribute to advancing PhyQA towards fulfilling its potential as a comprehensive physics question-answering tool, offering users a more streamlined and effective means of accessing the necessary knowledge.

Acknowledgments. The authors would like to thank University of Malaysia Sarawak for providing general support during the course of this research.

References

1. Pundge, A.M., Khillare, S.A., Mahender, C.N.: Question answering system, approaches and techniques: A review. Int. J. Comput. Appl. **141**(3), 34–39 (2016)
2. Allam, A.M.N., Haggag, M.H.: The question answering systems: A survey. Int. J. Res. Rev. Inf. Sci. (IJRRIS). **2**(3), 211–221 (2012)
3. Bouziane, A., Bouchiha, D., Doumi, N., Malki, M.: Question answering systems: survey and trends. Procedia Comput. Sci. **73**, 366–375 (2015)
4. Jing, C.H.S., How, B.C.: PhyOnto: An ontology based framework for educational physics units and measurements. Int. J. Comput. Appl. **187**(29), 31–40 (2025). https://doi.org/10.5120/ijca2025925499
5. Fong, K.S., Bong, C.H.: A hybrid question answering system based on ontology and topic modeling. J. Telecommun. Electron. Comput. Eng. **9**(2–10), 151–158 (2017)
6. Iqbal, R., Murad, M.A.A., Mustapha, A., Sharef, N.M.: An analysis of ontology engineering methodologies: A literature review. Res. J. Appl. Sci. Eng. Technol. **6**(16), 2993–3000 (2013)
7. Noy, N. F., & McGuinness, D. L. (2001). Ontology Development 101: A Guide to Creating Your First Ontology.
8. Choy, C.S., Chuan, K.K., Beng, O.H., Md Mustafa, M.K.A., Ragavan, R.: Physics Form 4. Sasbadi Sdn Bhd (2019)
9. Chuan, K.K., Choy, C.S., Bongkek, N.R., Kasron, J., Md Mustafa, M.K.A., Chakrabarty, P.K.: Physics Form 5. Penerbit Bestari Sdn Bhd (2020)
10. Prud'hommeaux, E., & Seaborne, A. SPARQL query language for RDF. World Wide Web Consortium. Retrieved from https://www.w3.org/TR/rdf-sparql-query/ (2008)
11. Elnozahy, W.A., El Khayat, G.A., Cheniti-Belcadhi, L., Said, B.: Question answering system to support University students' orientation, recruitment and retention. Procedia Comput. Sci. **164**, 56–63 (2019)
12. Jurafsky, D., Martin, J.H.: Speech and Language Processing: An Introduction to Natural Language Processing, Computational Linguistics, and Speech Recognition. Prentice Hall, Saddle River, NJ (2000)
13. Zamani, H., Dumais, S., Craswell, N., Bennett, P. and Lueck, G., Generating clarifying questions for information retrieval. In Proceedings of the Web Conference, pp. 418-428 (2020)

Performance Enhancement of Wearable Antenna Using AMC Structure

I. Adam[1(✉)], L. Zahid[1], M. F. Zambak[2], M. N. M. Yasin[3], H. A. Rahim[3], and N. H. Ramli[3]

[1] Centre of Excellence for Advanced Computing, Faculty of Electronic Engineering & Technology, Universiti Malaysia Perlis, 02600 Arau, Perlis, Malaysia
ismahayati@unimap.edu.my

[2] Teknik Elektro, Universitas Muhammadiyah Sumatera Utara, Kota Medan, Sumatera Utara 20238, Indonesia

[3] Centre of Excellence for Advanced Communication Engineering, Faculty of Electronic Engineering & Technology, Universiti Malaysia Perlis, 02600 Arau, Perlis Malaysia,, Malaysia

Abstract. This paper presents the design and performance enhancement of an elliptical monopole antenna using an Artificial Magnetic Conductor (AMC) surface. The proposed antenna, fabricated on a flexible felt substrate with *Shieldit* Super conductive material, covers a wide frequency range from 1.8 GHz to 12 GHz, making it suitable for ultra-wideband (UWB) applications. To improve its radiation characteristics, a hexagonal AMC surface is integrated beneath the antenna, resulting in significant gain and directivity enhancement due to in-phase reflection and surface wave suppression. The AMC demonstrates multiple zero-phase crossings within the 2.86 GHz to 8.6 GHz range, aligning well with key communication bands. Experimental results show notable improvements in antenna performance with AMC integration, including a gain increase from 2.35 dBi to 7.32 dBi at 2.45 GHz and from 2.96 dBi to 7.11 dBi at 5.8 GHz. These results highlight the effectiveness of AMC structures in developing compact, high-efficiency antennas for wearable and wireless applications.

Keywords: AMC · Elliptical monopole antenna · In-phase reflection · Surface wave suppression · Ultra-wideband (UWB) · Antenna gain · Wearable antennas · Wireless communication

1 Introduction

The ongoing expansion of modern wireless systems particularly in ultra wideband (UWB), wearable, and IoT applications has underscored a critical need for compact, broadband antennas with enhanced gain, efficiency, and low-profile form factors. However, conventional planar antennas face inherent challenges in achieving high radiation performance across a wide frequency range.

In recent years, Artificial Magnetic Conductors (AMCs), also known as high-impedance surfaces, have emerged as a powerful metasurface technology to address

M. A. b. Jamlos et al. (Eds.): ICSET 2025, LNNS 1897, pp. 354–360, 2026.
https://doi.org/10.1007/978-3-032-20722-7_45

these limitations. By providing in-phase reflection and effectively suppressing surface-wave modes, AMC surfaces improve the radiation pattern and gain of nearby antennas [1–3]

Pioneering works have shown that AMC-backed monopole designs can yield remarkable results. For instance, in wearable body area network (WBAN) applications, a dual-layer AMC-backed monopole achieved a maximum gain of 7.4 dB, an 82% gain increase compared to the standalone monopole, while also significantly reducing SAR [4]. Similarly, in mmWave systems, a compact rectangular-ring monopole over a modified square AMC attained a gain of 8.59 dBi nearly 77% higher than without AMC and a front-to-back ratio (FBR) exceeding 15 dB [5].

Hexagonal AMC geometries have also proven that placing a hexagon-shaped AMC beneath a microstrip patch antenna raised gain from approximately 7.45 dBi to about 7.94 dBi and directivity from 11.22 dBi to 17.29 dBi [6]

Despite these successes, AMC-backed designs often suffer from limited bandwidth since the zero-phase reflection condition exists over a narrow frequency range. Strategies such as multi-band unit cells and optimized spacing between antenna and AMC have been proposed to broaden this useful band and maximize performance improvement over wide frequency ranges [7–10].

In this work, a monopole elliptical patch antenna integrated with an AMC structure is proposed to enhance the antenna gain. The elliptical geometry is selected because its inherent asymmetry helps reduce the quality factor (Q), thereby increasing the impedance bandwidth [11]. Similar elliptical configurations have also been reported in the literature [12, 13] for achieving wideband performance.

2 Antenna Design

In this section, the proposed ellipse antenna and AMC structure are described in detail. The proposed design is inspired by a circular patch configuration incorporating with AMC structure to enhance the gain performance. Both antenna and AMC are designed in Computer Simulation Technology (CST) Microwave Studio where the substrate is from Felt with ε_r of 1.44 mm and *Shieldit Super* as conducting layer. The total size is 60×60 mm^2.

2.1 The Proposed Antenna

The proposed antenna is an ellipse monopole with dimensions of $42 \times 47 \times 1.34$ mm^3 as illustrated in Fig. 1. The half ground plane helps in achieving wide bandwidth. The S_{11} result shows in Fig. 2 (a), where the antenna resonates from 2.39 GHz up to 17 GHz, which cover UWB frequency range. The performance table presents the gain, directivity, and efficiency of the proposed AMC-backed antenna across a wide frequency range from 1.8 GHz to 12 GHz. At the lower band (1.8 GHz), the antenna exhibits a gain of 0.99 dBi, directivity of 1.98 dBi, and an efficiency of 79.7%. As the frequency increases, both gain and directivity improve significantly, reaching peak values of 4.23 dBi and 4.88 dBi respectively at 12 GHz. The antenna maintains high radiation efficiency throughout the entire band, consistently above 84%, with a maximum efficiency of 92.5% recorded at 2.45 GHz.

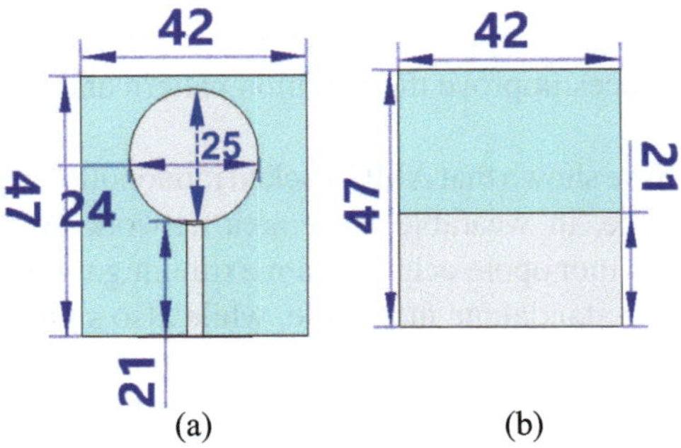

Fig. 1 The proposed antenna, (a) front and (b) back view.

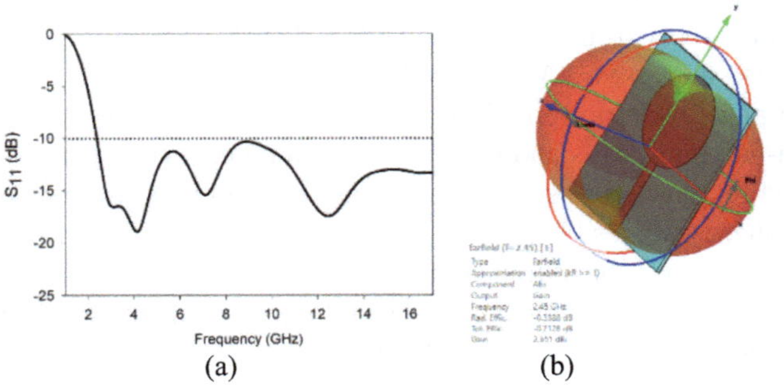

Fig. 2 The antenna performance, (a) S11 and (b) radiation pattern at 2.45 GHz

Table 1 Antenna performance at different frequencies.

Frequency (GHz)	1.8	2.45	3.5	5.2	5.8	7.5	10	12
Gain (dBi)	0.99	2.35	3.21	2.27	2.83	2.96	4.22	4.23
Directivity (dBi)	1.98	2.69	3.86	2.66	3.21	3.46	4.84	4.88
Efficiency (%)	79.7	92.5	86.2	91.5	91.6	89.1	86.7	84.3

2.2 AMC Structure

Figure 3(a) illustrates a 3 × 3 array configuration of a hexagonal AMC unit cell, arranged within a 60 mm × 60 mm surface. Each unit cell features a central hexagonal patch, contributing to the frequency-selective and reflection-phase-controlling characteristics of the surface. Each unit cell consists of a central hexagonal patch with inner and outer side lengths of approximately 6 mm and 8 mm, respectively. The reflection phase response of the AMC surface, as illustrated in Fig. 3(b), demonstrates that the structure behaves as a perfect magnetic conductor within a broad frequency range from approximately 2.86 GHz to 8.6 GHz, where the reflection phase remains within ±90°. Notably, the plot exhibits four zero-phase crossing points at approximately 3.5 GHz, 5.2 GHz, 5.8 GHz, and 7.5 GHz. These frequencies represent the operational bands where the AMC surface provides in-phase reflection, effectively suppressing surface waves and enhancing antenna performance.

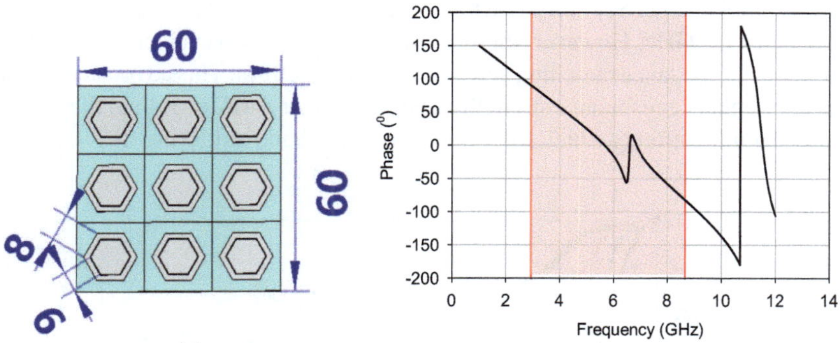

(a)

Fig. 3 (a) Hexagon AMC structure and (b) the reflection phase of the AMC.

3 Results and Discussion

In this section, the AMC structure is located underneath the antenna as illustrated in Fig. 4. This will enhance the antenna performance by providing an in-phase reflection, which minimizes destructive interference and improves radiation efficiency.

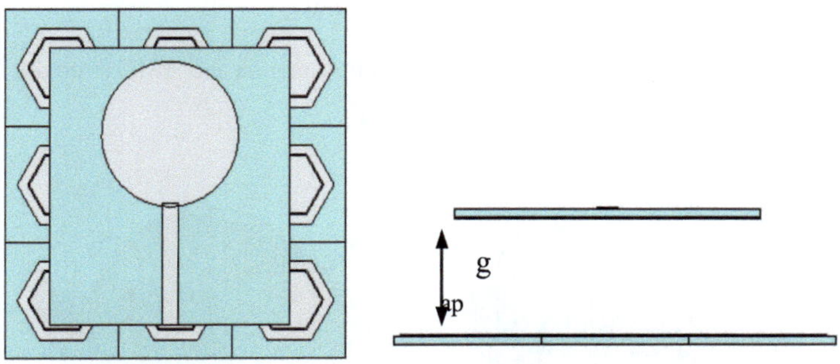

(a) (b)

Fig. 4 (a) Antenna with AMC structure and (b) side view of the antenna with AMC

The AMC surface reflects incident waves with near-zero phase shift, allowing constructive interference with the antenna's radiated waves. This leads to improved gain, reduced back radiation, and better front-to-back ratio. Additionally, the AMC suppresses surface wave propagation, which helps to confine energy in the desired direction. Figure 5 illustrates the simulated S_{11} parameter for the proposed antenna backed by the AMC surface, with varying separation distances (g) of 15 mm, 5 mm, and 3 mm between the antenna and the AMC. The S_{11} is plotted over a frequency range of 1–16 GHz, demonstrates how varying the separation gap between the antenna and the AMC surface influences impedance matching and resonant behavior. A gap of 15 mm results in a broader impedance bandwidth, with S_{11} values remaining below −10 dB from approximately 2.6 GHz to 16 GHz, except for a slight mismatch between 4.7 GHz and 5.6 GHz, where the reflection coefficient rises above −10 dB. In contrast, the smaller gaps of 5 mm and 3 mm exhibit deeper S_{11} nulls, reaching values as low as −40 dB, particularly around 8.5 GHz and 11.5 GHz. However, these configurations show narrower bandwidths and less consistent impedance matching across the entire spectrum. This advocates that while smaller gaps enhance resonance strength through stronger AMC coupling, they may also introduce detuning effects which limit the operational bandwidth (Fig. 5).

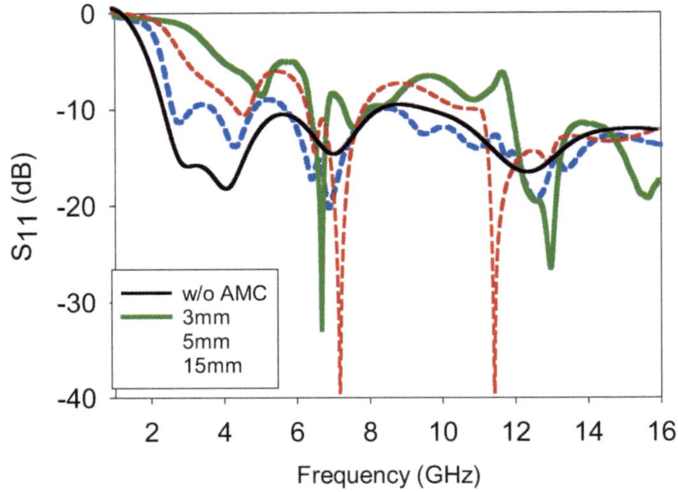

Fig. 5 S_{11} response with gap variations of the antenna and AMC structure.

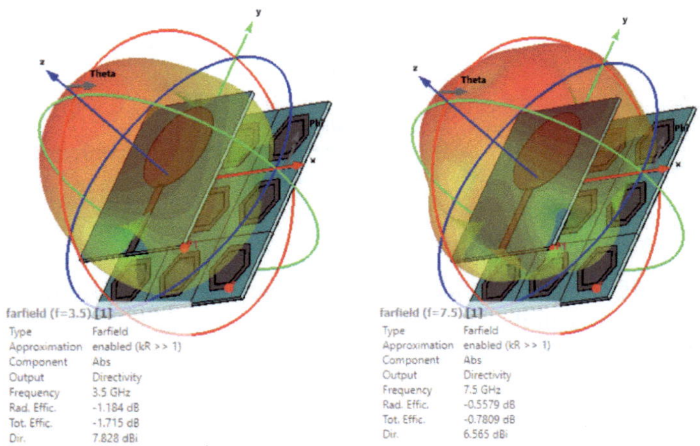

(a) 3.5 GHz (b) 7.5 GHz

Fig. 6 The radiation pattern at (a) 3.5 GHz and (b) 7.5 GHz

The integration of the AMC structure beneath the antenna significantly enhances its radiative performance particularly between 2.45 GHz to 7.5 GHz. The measured gain increases substantially compared to the antenna without AMC, with notable improvements observed at key frequencies. For instance, at 2.45 GHz, the gain rises from 2.35 dBi to 7.32 dBi, while at 5.8 GHz, it increases from 2.96 dBi to 7.11 dBi, demonstrating the AMC's effectiveness in reinforcing forward radiation through in-phase reflection. Similarly, the directivity improves across all bands, reaching 7.92 dBi at 10 GHz, indicating a more focused radiation pattern. Efficiency also remains high at critical operating bands, with values such as 90% at 2.45 GHz and 89.8% at 5.8 GHz, showing minimal loss and strong impedance matching. The results are presented in Table 2.

Table 2 The antenna with AMC structure performance.

Frequency (GHz)	1.8	2.45	3.5	5.2	5.8	7.5	10	12
Gain (dBi)	4.01	7.32	6.64	6.21	7.11	6.00	7.15	5.42
Directivity (dBi)	7.00	7.77	7.83	6.68	7.54	6.52	7.92	6.81
Efficiency (%)	50	90	76	81.8	89.8	87.9	83.7	72.7

4 Conclusion

This study has demonstrated the effectiveness of integrating a hexagonal AMC structure to enhance the performance of a broadband antenna. The reflection phase analysis confirmed that the AMC provides in-phase reflection across a wide frequency range (2.86–8.6 GHz), with multiple zero-phase crossings that enable multi-band operation. In addition to the AMC integration, this work also incorporates a detailed gap analysis to achieve optimal antenna performance. Unlike previous studies that predominantly focus on narrowband operation at specific frequencies, this design supports wideband performance suitable for UWB applications. Comparative performance evaluation further showed notable improvements in antenna gain and directivity with the AMC. At 2.45 GHz and 5.8 GHz, the antenna gain increased from 2.35 dBi and 2.96 dBi (without AMC) to 7.32 dBi and 7.11 dBi (with AMC), respectively. These enhancements are attributed to the AMC's ability to suppress surface waves while providing in-phase reflection, resulting in superior radiation characteristics and improved overall performance. Future research will focus on fabricating and experimentally validating the proposed AMC-backed antenna, including measurements in free space and on body phantoms to assess SAR compliance and real-world performance. Overall, the proposed AMC-backed antenna design proves to be a promising solution for compact, high-performance antennas in multi-band wireless and wearable applications.

Acknowledgments. The authors would like to acknowledge the support from the International Research Fund (INTERES) 9001-00746 and 9008-00033.

References

1. Xiao, Y., et al.: Multiband and low-specific-absorption-rate wearable antenna with low profile based on highly conductive graphene assembled film. IEEE Antennas Wirel. Propag. Lett. **22**(9), 2195–2199 (2023)
2. Ashyap, A., Raad, R., Tubbal, F., Khan, W.A., Abulgasem, S.: Highly bendable AMC-based antenna for wearable applications. IEEE Access. **12**, 154195–154211 (2024)
3. de Dieu Ntawangaheza, J., Sun, L., Xie, Z., Pang, Y., Zheng, Z., Rushingabigwi, G.: A single-layer low-profile broadband metasurface antenna array for sub-6 GHz 5G communication systems. IEEE Trans. Antennas Propag. **69**(4), 2061–2071 (2021)
4. Hazarika, B., Basu, B., Nandi, A.: An artificial magnetic conductor-backed monopole antenna to obtain high gain, conformability, and lower specific absorption rate for WBAN applications. Int J. RF Microw. Comput. Aided Eng. **30**(12), 1–9 (2020)
5. Kumar, A., Kumar, A.: Gain enhancement of a wideband rectangular ring monopole millimeter-wave antenna using artificial magnetic conductor structure. Int. J. RF Microw. Comput.-Aided Eng. **32**(10), e23319 (2022)
6. Abdiaziz, M., Erdem, E.: Design of hexagonal-shaped artificial magnetic conductor for gain and directivity enhancement. Int. J. Adv. Nat. Sci. Eng. Res. **8**(11), 213–220 (2024)
7. Malekpoor, H., Hamidkhani, M.: Performance enhancement of low-profile wideband multi-element MIMO arrays backed by AMC surface for vehicular wireless communications. IEEE Access. **9**, 166206–166222 (2021)
8. Yadav, M., Muquaddar, A., Yadav, R.P.: Gain enhanced dual band antenna backed by dual band AMC surface for wireless body area network applications, IEEE Indian Conference on Antennas and Propagation (InCAP), 2021.
9. Harutyunyan, A., Sen, P.: On-chip mm-Wave artificial magnetic conductor backed dipole antenna on low-ohmic substrate. In: 18th European Conference on Antennas and Propagation (EuCAP). IEEE (2024)
10. Upadhyay, N., Kaur, A., Pattanayak, A., Singh, A.: An artificial magnetic conductor backed cylindrical dielectric resonator antenna for portable wireless communications. In: IEEE Microwaves, Antennas, and Propagation Conference (MAPCON) (2024)
11. Balanis, C.A.: Antenna Theory: Analysis and Design. John wiley & sons (2016)
12. Jassim, A.K., Thaher, R.H.: Enhancement gain of broadband elliptical microstrip patch array antenna with mutual coupling for wireless communication. Indones. J. Electr. Eng. Comput. Sci. **13**(1), 401–408 (2019)
13. Zulkifli, F.Y., Wahdiyat, A.I., Maulana, T., Rambe, A.H., Nurhayati, N.: Super-wideband antenna with modified elliptical-shaped for broad spectrum capability in software-defined radio applications. e-Prime-Adv. Electr. Eng. Electron. Energy. **11**, 100933 (2025)

Balancing the Vision: A Comparative Study of ViT-B16 and ViT-B32 for Emotion Recognition on RAF-DB

Mohanad Qubtan[1(✉)], Hasimah Ali[1,2], Mohamed Elshaikh[3], and Anna Nurnazilah Chamim[4]

[1] Faculty of Electrical Engineering & Technology, Universiti Malaysia Perlis, Arau, Perlis, Malaysia
mohanadbaqubtan@studentmail.unimap.edu.my, hasimahali@unimap.edu.my, hasimah.ali@umy.ac.id
[2] Research Fellow, Universitas Muhammadiyah Yogyakarta, Yogyakarta, Indonesia
[3] Faculty of Electronic Engineering & Technology, Universiti Malaysia Perlis, Arau, Perlis, Malaysia
elshaikh@unimap.edu.my
[4] Faculty of Electrical Engineering, Universitas Muhammadiyah Yogyakarta, Yogyakarta, Indonesia
anna_nnc@umy.edu.my

Abstract. Facial Expression Recognition (FER) plays a crucial role in affective computing but remains challenging due to the presence of noisy, imbalanced, and real-world data. The Real-world Affective Faces Database (RAF-DB) is a widely used benchmark that reflects these challenges, as it contains class imbalance and subtle variations between expressions. To address this, we explore the use of Vision Transformers (ViTs), specifically ViT-B/16 and ViT-B/32, which differ in patch size and therefore in their ability to capture fine-grained facial features. Our study investigates two training regimes: using the original imbalanced dataset and applying augmentation-based balancing to improve minority-class representation. Experiments were conducted on the RAF-DB with pretrained ImageNet-21 k ViT models, where only the classification head was reinitialized. Results show that ViT-B/16 with imbalanced training achieves the highest accuracy of 84.84%, while balancing improves performance on minority classes such as Fear and Disgust, reflected in higher recall and F1-scores. These findings highlight that smaller patch sizes capture subtle expression cues more effectively and that class balancing enhances equitable recognition across categories, even if overall accuracy is slightly reduced.

Keywords: Vision Transformers (ViT) · Transfer learning · Facial expression recognition (FER)

M. A. b. Jamlos et al. (Eds.): ICSET 2025, LNNS 1897, pp. 361–367, 2026.
https://doi.org/10.1007/978-3-032-20722-7_46

1 Introduction

Facial Expression Recognition (FER) is essential in affective computing for applications such as healthcare and driver safety [22], but it remains difficult due to limited data and strong class imbalance, where emotions like happy and neutral dominate over fear and disgust. These issues reduce model generalization and real-world performance [1, 2, 21]. Vision Transformers (ViTs) have recently emerged as effective alternatives to CNNs by using self-attention to capture global context [3]. FER studies on RAF-DB [1], an imbalanced and noisy dataset, include models like MVT [4] and hybrid ViTs [5, 6], which improve robustness to imbalance and label noise. Research on ViTs for FER is still limited [3, 5], especially regarding patch size. ViT-B/16 uses 16×16 patches that capture fine details but increase computation, while ViT-B/32 uses 32×32 patches that are cheaper but may lose subtle cues [3]. Although patch size strongly affects accuracy in general vision tasks [7], its role in FER is understudied. Since FER datasets are highly imbalanced [1, 2], balanced training methods improve minority-class recognition and macro metrics [8]. To the best of our knowledge few works analyze how ViT design choices interact with balancing, motivating our comparison of ViT-B/16 and ViT-B/32 under imbalanced and balanced RAF-DB settings. This paper investigates the effects of patch size and dataset balancing on ViT models for FER. Sect. 2 reviews related work, Sect. 3 outlines the dataset and methods, Sect. 4 presents results, and Sect. 5 concludes with key findings and future directions.

2 Related Work

Early FER methods used handcrafted features like LBP [9] and HOG [10], but they struggled with pose and lighting variations. Deep learning shifted the field to CNNs, which improved robustness by learning hierarchical features [23]. RAF-DB introduced a locality-preserving CNN with 86.9% accuracy [1, 11] reported 85.7%, while AffectNet [2] showed the difficulty of large-scale FER with a 58% baseline. More advanced CNNs such as MMNet achieved 89.8% on RAF-DB using augmentation and re-weighting [12], underscoring the importance of addressing class imbalance. ViTs have recently been applied to FER, offering stronger global dependency modeling than CNNs [3]. ViT-B/16 reached 87.22% accuracy on RAF-DB with limited augmentation [5], and augmentation-based ViT transfer learning achieved 83.44% [13]. Multi-task transformer models further reached 88.6% [14], showing ViTs are competitive but still sensitive to class imbalance. Overall, CNNs and ViTs perform well on RAF-DB, but most ViT studies examine only single variants and use simple balancing. ViTs also face overfitting on small datasets, often underperforming CNNs due to limited inductive bias [15], and mitigation methods show mixed success [16]. Their quadratic attention cost increases computation [17], motivating deeper analysis of how patch size and balancing strategies can improve FER robustness.

3 Methodology

This study examines how patch size and class balancing affect Vision Transformer performance on facial expression recognition using RAF-DB.

3.1 Dataset & Preprocessing

The RAF DB dataset version used in this paper contains 13,251 images labeled across seven emotions. After splitting the data into training, validation, and test sets, we observed a strong imbalance, with Happy and Neutral appearing far more often than emotions like Fear and Disgust. To address this, we created two data settings: the original imbalanced version and a balanced version generated by augmenting minority classes using flips, rotations, zooming, and small shifts. These added samples improved diversity and strengthened representation of minority emotions. Figure 1 show the class distributions before and after balancing.

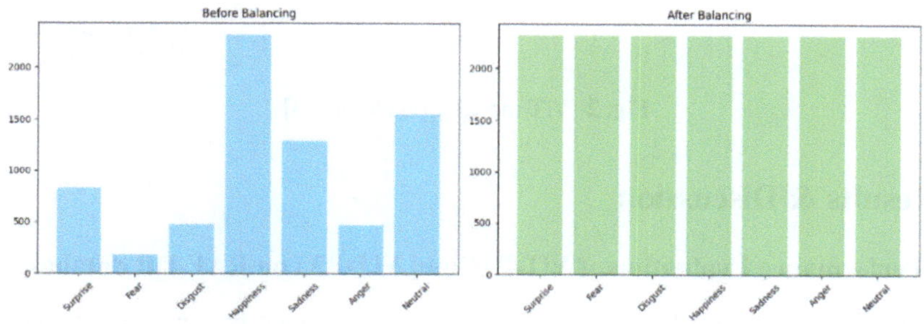

Fig. 1 Class distribution before and after balancing

3.2 Vision Transformer Model Architecture

The Vision Transformer (ViT) [3] applies the Transformer architecture to images by treating them as sequences of patches. We used two standard variants: ViT-B/16 with 16×16 patches and ViT-B/32 with 32×32 patches. All experiments used pretrained models from the HuggingFace library [18] initialized with ImageNet-21 k weights [19]. Only the classification head was reinitialized for RAF-DB's seven classes, while the backbone remained pretrained to support faster convergence and stronger generalization. Figure 2 shows the Vision Transformer model architecture.

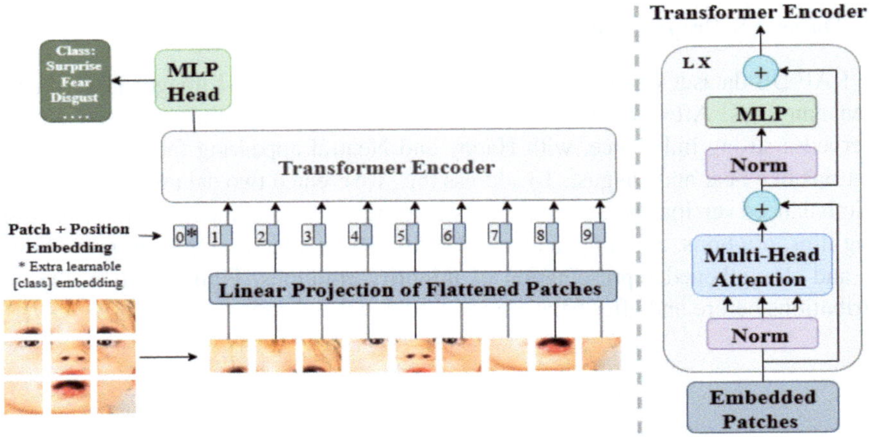

Fig. 2 ViT model architecture [3]

4 Results & Discussion

4.1 Performance Evaluation of ViT-B16 and ViT-B32 on RAF-DB database

Classification Results. To capture class-level performance beyond overall accuracy, we report precision, recall, and F1-scores for all seven emotions. Tables 1, 2, 3 and 4 present the results for ViT-B/16 and ViT-B/32 under imbalanced and balanced training conditions, highlighting how patch size and class distribution affect recognition of both majority and minority emotions.

Table 1 presents the per-class performance of the ViT-B/16 trained without balancing

Class	Precision	Recall	F1-Score	Support
Surprise	0.85	0.86	0.86	354
Fear	0.63	0.39	**0.48**	79
Disgust	0.62	0.49	**0.54**	206
Happy	0.94	0.95	**0.95**	990
Sadness	0.79	0.86	0.83	556
Anger	0.81	0.75	0.78	203
Neutral	0.80	0.84	0.82	667
Accuracy			**84.84%**	3055
Macro Avg	0.78	0.73	0.75	3055
Weighted Avg	0.84	0.84	0.84	3055

ViT-B/16 showed the best overall performance, helped by its smaller patch size. Imbalanced training gave higher accuracy but favored majority classes like happy, while

Table 2 ViT-B/16 when augmentation was used to balance classes

Class	Precision	Recall	F1-Score	Support
Surprise	0.82	0.86	0.84	329
Fear	0.67	0.50	**0.57**	74
Disgust	0.61	0.55	**0.58**	160
Happy	0.97	0.87	**0.91**	1185
Sadness	0.78	0.83	0.81	478
Anger	0.77	0.78	0.78	162
Neutral	0.75	0.87	0.80	680
Accuracy			**83.12%**	3068
Macro Avg	0.77	0.75	0.76	3068
Weighted Avg	0.84	0.83	0.83	3068

Table 3 The results of ViT-B/32 trained without balancing

Class	Precision	Recall	F1-Score	Support
Surprise	0.78	0.84	0.81	354
Fear	0.54	0.39	**0.46**	79
Disgust	0.70	0.46	**0.55**	206
Happy	0.96	0.91	**0.94**	990
Sadness	0.68	0.85	0.76	556
Anger	0.68	0.74	0.71	203
Neutral	0.80	0.75	0.77	667
Accuracy			**80%**	3055
Macro Avg	0.74	0.71	0.71	3055
Weighted Avg	0.81	0.80	0.80	3055

balancing improved minority emotions recognition like fear despite a slight accuracy drop. Overall, ViT-B/16 remained the most reliable, and balancing produced fairer results across emotions.

Table 4 The performance of ViT-B/32 when balancing was applied

Class	Precision	Recall	F1-Score	Support
Surprise	0.78	0.83	0.80	329
Fear	0.67	0.50	**0.57**	74
Disgust	0.52	0.47	**0.49**	160
Happy	0.95	0.84	**0.89**	1185
Sadness	0.77	0.80	0.78	478
Anger	0.73	0.77	0.75	162
Neutral	0.72	0.85	0.78	680
Accuracy			**80.28%**	3068
Macro Avg	0.73	0.72	0.72	3068
Weighted Avg	0.81	0.80	0.80	3068

5 Conclusion

This study evaluated Vision Transformers for facial expression recognition using the RAF DB dataset. Two variants, ViT-B/16 and ViT-B/32, were tested on both imbalanced and balanced data. ViT-B/16 showed the strongest results in all settings, indicating that smaller patch sizes capture facial details more effectively. ViT-B/16 reached 84.84% accuracy on the imbalanced dataset and 83.12% on the balanced version, where minority class performance improved despite a small accuracy drop. ViT B32 performed lower overall, reaching about 80% in both conditions, with the balanced setup offering slightly better recognition of minority emotions. Overall, balancing improves fairness across emotion classes, while ViT-B/16 remains the most effective model. The results show that patch size and dataset balancing are key factors for reliable Vision Transformer based FER systems.

Acknowledgments. This research article was funded by the Universitas Muhammadiyah Yogyakarta (UMY), Indonesia under the UMY Fellowship Program 2025.

References

1. Li, S., Deng, W., Du, J.: Reliable crowdsourcing and deep locality-preserving learning for expression recognition in the wild. In: 2017 IEEE Conference on Computer Vision and Pattern Recognition (CVPR), pp. 2584–2593. IEEE, Piscataway, NJ, USA (2017). http://whdeng.cn/RAF/model1.html
2. Mollahosseini, A., Hasani, B., Mahoor, M.H.: AffectNet: A database for facial expression, valence, and arousal computing in the wild. IEEE Trans. Affect. Comput. **10**(1), 18–31 (2017). https://doi.org/10.1109/TAFFC.2017.2740923
3. Dosovitskiy, A., *et al.*: An image is worth 16x16 words: Transformers for image recognition at scale. arXiv preprint arXiv:2010.11929. http://arxiv.org/abs/2010.11929 (2020)
4. Li, H., Sui, M., Zhao, F., Zha, Z., Wu, F.: MVT: Mask vision transformer for facial expression recognition in the wild. arXiv preprint arXiv:2106.04520. http://arxiv.org/abs/2106.04520 (2021)

5. Aouayeb, M., Hamidouche, W., Soladie, C., Kpalma, K., Seguier, R.: Learning vision transformer with squeeze and excitation for facial expression recognition. arXiv preprint arXiv:2107.03107. http://arxiv.org/abs/2107.03107 (2021)
6. Li, N., Huang, Y., Wang, Z., Fan, Z., Li, X., Xiao, Z.: Enhanced hybrid vision transformer with multi-scale feature integration and patch dropping for facial expression recognition. Sensors. **24**(13) (2024). https://doi.org/10.3390/s24134153
7. Beyer, L., Kolesnikov, A.L.D., Georgiev, D., Frosst, S., Hinton, N.: FlexiViT: One model for all patch sizes. arXiv preprint arXiv:2212.08013. http://arxiv.org/abs/2212.08013 (2023)
8. Bobojanov, S., Kim, B.M., Arabboev, M., Begmatov, S.: Comparative analysis of vision transformer models for facial emotion recognition using augmented balanced datasets. Appl. Sci. (Switz.). **13**(22) (2023). https://doi.org/10.3390/app132212271
9. Ojala, T., Pietikainen, M., Harwood, D.: A comparative study of texture measures with classification based on feature distributions. Pattern Recognit. **29**(1), 51–59 (1996)
10. Dalal, N., Triggs, B.: Histograms of oriented gradients for human detection. In: International Conference on Computer Vision & Pattern Recognition, vol. 2, pp. 886–893. IEEE, Piscataway, NJ, USA (2005) http://lear.inrialpes.fr
11. Zhang, Z., Luo, P., Loy, C.C., Tang, X.: From facial expression recognition to interpersonal relation prediction. arXiv. http://arxiv.org/abs/1609.06426 (2017)
12. Ding, H., Zhou, P., Chellappa, R.: Occlusion-adaptive deep network for robust facial expression recognition. arXiv. http://arxiv.org/abs/2005.06040 (2020)
13. Bousaid, R., El Hajji, M., Es-Saady, Y.: Facial emotions recognition using ViT and transfer learning. In: Proceedings—2022 5th International Conference on Advanced Communication Technologies and Networking, CommNet, pp. 1–6. Institute of Electrical and Electronics Engineers Inc., Marrakech, Morocco (2022) https://doi.org/10.1109/CommNet56067.2022.9993933
14. Qin, L., Zhang, Z., Zhang, J., Fu, Y., Yang, J.: SwinFace: A multi-task transformer for face recognition, expression recognition, age estimation and attribute estimation. IEEE Trans. Circuits Syst. Video Technol. (2023). https://doi.org/10.1109/TCSVT.2023.3304724
15. Zhu, H., Chen, B., Yang, C.: Understanding why ViT trains badly on small datasets: An intuitive perspective. arXiv. http://arxiv.org/abs/2302.03751 (2023)
16. Gani, H., Naseer, M., Yaqub, M.: How to train ViT on small-scale datasets? how to train vision transformer on small-scale datasets. GitHub. https://github.com/hananshafi/vits-for-small-scale-datasets (2022)
17. Lu, Z., Xie, H., Liu, C., Zhang, Y.: Bridging the gap between vision transformers and convolutional neural networks on small datasets. GitHub. https://github.com/ArieSeirack/DHVT (n.d.)
18. Wolf, T. et al.: Transformers: State-of-the-art natural language processing. GitHub. https://github.com/huggingface/ (n.d.)
19. Computer Vision and Pattern Recognition, 2009, CVPR 2009, IEEE Conference on Computer Vision and Pattern Recognition (CVPR), 20–25 June 2009, IEEE, Piscataway, NJ, USA (2009)
20. Kingma, D.P., Ba, J.: Adam: A method for stochastic optimization. arXiv. http://arxiv.org/abs/1412.6980 (2017)
21. Riyadi, S., Andriyani, A.D., Sulaiman, S.N.: Improving hate speech detection using double-layers hybrid CNN-RNN model on imbalanced dataset. IEEE Access. **12**, 159660–159668 (2024)
22. Riyadi, S., Rozan, N., Damarjati, C.: Mobile computer-assisted application for stress detection based on facial expression using modified convolutional neural network. Int. J. Comput. Digit. Syst. **17**(1), 1–11 (2025)
23. M. Y. Mustar, K. T. Putra, Y. Ardiyanto, R. O. Wiyagi, I. Arseno, H-C. Chen. A low-computational cost face recognition system for high accuracy privileged access management to support the development of smart laboratories. In: AIP Conference Proceedings, vol. 2865, (1): 020004. American Institute of Physics, Melville, NY, USA, 28 Dec (2023)

Real-Time Performance Prediction for Track Cycling Using LSTM: A Comparative Study with MLP and ANFIS

Nur Syahidatul Natrah Ahmad Suhaimi[1], Sukhairi Sudin[1,2](✉), Fathinul Syahir Ahmad Saad[1,2], Muhamad Khairul Ali Hassan[1,2], Faranadia Abdul Haris[3], and Muhammad Zulazri Hanis Mohd Nawi[1]

[1] Faculty of Electrical Engineering & Technology, University Malaysia Perlis, 02600 Arau, Perlis, Malaysia
sukhairi@unimap.edu.my

[2] Centre for Advanced Sensor Technology (CEASTech), Universiti Malaysia Perlis, 02600 Arau, Perlis, Malaysia

[3] Faculty of Electrical Engineering, Universiti Teknologi MARA Permatang Pauh, Seberang Perai, Pulau Pinang, Malaysia

Abstract. Accurate performance prediction in track cycling is a challenging time-series problem involving interacting physiological and biomechanical variables. This study compares three artificial intelligence approaches—Long Short-Term Memory (LSTM) networks, Multilayer Perceptron (MLP), and Adaptive Neuro-Fuzzy Inference System (ANFIS)—for real-time prediction of cycling performance. Data were collected from competitive cyclists using a custom wireless sensor network that captured heart rate, body temperature, pedaling power, and cadence during indoor and outdoor velodrome sessions. An LSTM model designed to handle sequential dependencies is benchmarked against MLP and ANFIS predictors. Experimental results show that LSTM consistently achieves the lowest errors for multi-step-ahead speed and finishing time predictions, with improved generalization relative to MLP and ANFIS. These findings highlight the advantages of deep learning for modeling the dynamic, nonlinear nature of cycling performance and demonstrate the potential of LSTM-based predictors for real-time athlete monitoring, training optimization, and strategic decision-making in track cycling.

Keywords: Track cycling · Performance prediction · LSTM · MLP · ANFIS · Time series · Artificial intelligence

1 Introduction

Track cycling is a high-stakes sport where race outcomes are often decided by fractions of a second, so accurate performance prediction is important for optimizing training, pacing, and in-race decisions [1, 2]. Traditional tests such as the Astrand–Ryhming test [3] and Physical Working Capacity (PWC) [4] estimate aerobic capacity from submaximal heart

M. A. b. Jamlos et al. (Eds.): ICSET 2025, LNNS 1897, pp. 368–374, 2026.
https://doi.org/10.1007/978-3-032-20722-7_47

rate responses, but they focus on isolated metrics under controlled conditions and do not capture the complex, time-varying interaction of physiological and biomechanical factors in real competition [5].

Recent advances in wireless sensor networks (WSN) and wearable technologies allow continuous monitoring of heart rate, body temperature, pedaling power, and cadence during actual cycling sessions [6, 7]. These multimodal time-series data enable AI-based performance prediction using models that can handle nonlinear and temporal relationships [8]. Previous studies have applied models such as MLP and ANFIS in various sports [9–11], but these approaches are limited in modeling long-range temporal dependencies in sequential data.

Deep learning models, particularly Long Short-Term Memory (LSTM) networks, address this limitation by preserving relevant historical information and have shown strong performance in many time-series applications [12], although their use in real-time track cycling prediction is still limited [2]. This paper compares LSTM, MLP, and ANFIS predictors for real-time track cycling performance using data collected from a custom WSN. We evaluate their ability to predict next-lap speed and finishing time and aim to inform the design of intelligent, adaptive systems for athlete monitoring and performance optimization in track cycling.

2 Related Work

2.1 Performance Prediction in Sports

AI-based performance prediction has been explored in many sports, including football, volleyball, cricket, athletics, and cycling [2, 10]. Neural networks have been used to predict match outcomes, player performance, team rankings, athlete selection, and metrics such as VO2max or fitness level [1, 4, 7, 9]. ANFIS models, which combine neural networks and fuzzy logic, can handle uncertainty and nonlinear physiological data and have been applied to classify work rate from heart rate and to model complex performance relationships [5, 11].

In cycling, most studies rely on laboratory ergometer tests or post-event analysis using static datasets and traditional statistics [3]. Real-time performance prediction using wearable sensors and AI, especially in track cycling, is still limited. Sudin et al. [6] highlighted the potential of combining WSNs with AI for real-time monitoring, but systematic comparisons between classical machine learning and deep learning models on real-world cycling data remain scarce.

2.2 Time Series Prediction with LSTM

LSTMs, a variant of recurrent neural networks (RNNs), are designed to capture long-term dependencies in sequential data using memory cells and gating mechanisms [8, 12]. In biomedical applications, LSTMs have outperformed traditional models in modeling ECG and EEG signals by better capturing temporal patterns [13]. In sports science, LSTMs have been used to predict running performance, injury risk, and workload trends, but their adoption in cycling—particularly for live, real-time scenarios—is still emerging [2, 6]. Given their capacity to integrate multiple sensor streams and model nonlinear temporal dynamics, LSTMs are promising for track cycling performance prediction.

2.3 Comparative Studies

Comparative studies across domains indicate that LSTMs often outperform MLP and ANFIS in tasks with strong temporal dependencies [10, 13]. In energy demand forecasting and biomedical signal analysis, LSTMs typically yield lower prediction errors and better generalization [8]. MLPs, however, are simpler and computationally efficient, and ANFIS offers interpretable rule-based models, which is attractive where transparency is important [11]. In sports performance prediction there remains a shortage of benchmarks that compare LSTM, MLP, and ANFIS using real-world, multimodal sensor data [5, 7]. This work contributes such a benchmark in the context of track cycling.

3 Methodology

3.1 Data Acquisition System

A custom WSN was developed to continuously monitor key physiological and biomechanical parameters during track cycling while minimizing interference with athletes' movements. Data were collected from 12 competitive cyclists (aged 24–29 years) during multiple indoor and outdoor velodrome sessions to capture different pacing strategies, fatigue profiles, and environmental conditions. Heart rate was measured using a photoplethysmography (PPG) glove sensor at 10 Hz for HRV analysis, core body temperature was monitored with a helmet-mounted thermistor (± 0.1 °C), pedaling power was measured using a Garmin Vector pedal-based power meter ($\pm 2\%$), and cadence and speed were obtained from Hall-effect sensors measuring crank revolutions per minute and wheel speed. All sensor data were synchronized and transmitted in real time via XBee Series 2 modules using the ZigBee protocol, and custom Python software provided real-time visualization, data logging, and basic integrity checks. Each session was timestamped and stored in a centralized database for later preprocessing and analysis.

3.2 Preprocessing and Feature Engineering

Raw sensor data were processed through a multi-stage preprocessing pipeline. A moving average filter (window size five) was applied to each stream to reduce high-frequency noise and artifacts. Outliers were detected using the interquartile range (IQR) method, and missing or corrupted values were filled by linear interpolation. All sensor streams were then aligned to a unified timeline to correct for transmission delays and clock drift, ensuring synchronization across channels. From the aligned data, features relevant to cycling performance were extracted, including HRV indices from the PPG signal, core body temperature, lap-averaged pedaling power, cadence, speed, and derived variables such as power-to-weight ratio and approximate rate-of-perceived-exertion proxies. These features were segmented into lap-based windows for one-step and multi-step-ahead prediction tasks and standardized to zero mean and unit variance before model training.

3.3 Predictive Models

Multilayer Perceptron (MLP) The MLP model receives engineered features (heart rate, temperature, power, cadence, speed, and derived variables) at each prediction step. The architecture consists of an input layer followed by two fully connected hidden layers with 64 and 32 neurons, respectively, each using ReLU activation. The output layer predicts either next-lap speed or finishing time. The model is trained with backpropagation to minimize mean squared error (MSE). Dropout and early stopping based on validation loss are used to reduce overfitting. While the MLP effectively maps nonlinear relationships between inputs and outputs, it does not explicitly model temporal dependencies across laps.

Adaptive Neuro-Fuzzy Inference System (ANFIS) ANFIS combines fuzzy inference with neural network learning. Input features (e.g., heart rate, power) are passed through fuzzification layers using Gaussian membership functions, which convert crisp inputs into degrees of membership. A rule layer computes firing strengths of fuzzy if—then rules, which are then normalized so their sum equals one. Defuzzification converts the normalized firing strengths into crisp outputs. Parameter learning employs a hybrid scheme: least-squares estimation for consequent parameters and gradient descent for membership function parameters. The number and type of membership functions and learning rates are selected via grid search. ANFIS is suited to cases where model interpretability and robustness to uncertain physiological signals are important.

Long Short-Term Memory (LSTM) LSTM networks extend conventional RNNs with a memory cell and three gates—forget, input, and output—that regulate information flow across time steps. At each time step t_i with x_t and hidden state h_{t-1}, the gates are:

$$f_t = \sigma\left(W_f \bullet \left[h_{t-1}, x_t\right] + b_f\right) \tag{1}$$

$$i_t = \sigma\left(W_i \bullet \left[h_{t-1}, x_t\right] + b_i\right) \tag{2}$$

$$o_t = \sigma\left(W_o \bullet \left[h_{t-1}, x_t\right] + b_o\right) \tag{3}$$

These gates determine what information to forget, what new information to store, and what to output, enabling the model to preserve relevant historical context over many time steps.

The LSTM architecture used here consists of two stacked LSTM layers with 64 and 32 units, separated by dropout layers. A final dense output layer predicts next-lap speed or finishing time. Training uses the Adam optimizer with MSE loss, a batch size of 32, and a learning rate of 0.001. This configuration allows the model to capture both short- and long-term patterns in the cycling data.

4 Experimental Setup

4.1 Test Protocol

Experiments were conducted under controlled yet realistic cycling conditions. Each of the 12 cyclists completed multiple five-lap trials on an indoor virtual reality trainer and an outdoor standard velodrome. Prior to each session, cyclists performed a standardized warm-up to reduce physiological variability. For outdoor sessions, environmental conditions (temperature, humidity, wind speed) were recorded.

During each lap, the WSN continuously captured heart rate, body temperature, pedaling power, cadence, and speed. The primary prediction tasks were one-step-ahead prediction, which forecast speed and finishing time for the next lap using data from the previous four laps; two-step-ahead prediction, which forecast two laps ahead using data from the preceding three laps; and three-step-ahead prediction, which forecast three laps ahead using data from the preceding two laps. This multi-horizon setup allowed assessment of how each model's accuracy degrades as the forecast horizon increases.

4.2 Training and Validation

The dataset was partitioned into training (70%), validation (15%), and test (15%) sets. Data from the same session were not split across training and test sets to prevent leakage. The training set was used to fit the models, the validation set to tune hyperparameters and control early stopping, and the test set for final evaluation.

To further improve robustness, five-fold cross-validation was performed. In each fold, a different subset of data served as test data, and average performance metrics across folds were reported. This procedure produced more reliable estimates of generalization performance and reduced reliance on a single random split.

4.3 Hyperparameter Tuning

Hyperparameters were tuned using grid search for each model. For the MLP, we varied the number of hidden layers (1–3), neurons per layer (16–128), learning rate (0.001–0.01), batch size (16–64), and dropout rate (0–0.5). For the LSTM, we explored the number of layers (1–3), units per layer (16–128), dropout rate (0–0.5), learning rate (0.0005–0.01), and sequence length, with early stopping applied if validation loss did not improve for 20 epochs. For ANFIS, we tuned the number and type of membership functions per input (2–5, Gaussian or triangular), learning rates (0.001–0.05), and training epochs/tolerance. The configurations with the lowest validation RMSE were selected for final.

5 Result and Discussion

5.1 Error Analysis

Figure 1 presents a comparative analysis of the prediction errors (measured as RMSE in m/s) for the LSTM, MLP, and ANFIS models under three distinct data conditions: normal data, data with outliers, and data with missing values. The grouped bar chart clearly illustrates the robustness and reliability of each model when exposed to real-world data imperfections.

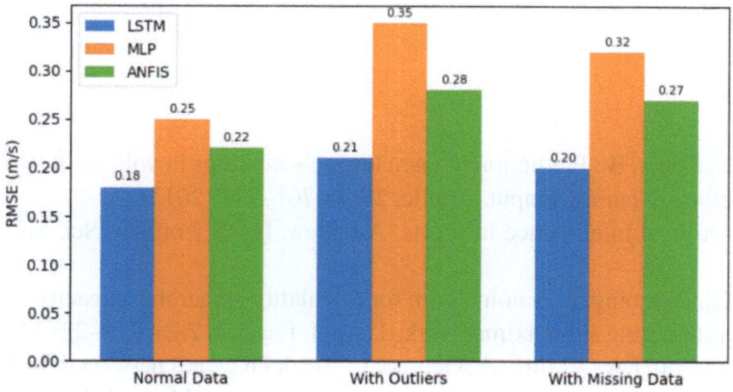

Fig. 1 Error analysis under different data conditions

Under normal data conditions, the LSTM model achieves the lowest RMSE (0.18 m/s), outperforming both MLP (0.25 m/s) and ANFIS (0.22 m/s). When outliers are introduced into the dataset, the RMSE for all models increases, but the degradation is most pronounced for the MLP model (0.35 m/s), followed by ANFIS (0.28 m/s). The LSTM model demonstrates greater resilience, with only a modest increase in error (0.21 m/s).

Likewise, when data are missing, the LSTM attains a comparatively low RMSE (0.20 m/s), while MLP (0.32 m/s) and ANFIS (0.27 m/s) exhibit notably higher errors. This indicates that the LSTM's memory and gating architecture is better suited to handling incomplete or noisy time-series inputs.

6 Conclusion

This study demonstrates that Long Short-Term Memory (LSTM) networks are highly effective for real-time track cycling prediction, consistently outperforming Multilayer Perceptron (MLP) and Adaptive Neuro-Fuzzy Inference System (ANFIS) models on next-lap speed and finishing time. Leveraging a custom wireless sensor network, we collected rich physiological and biomechanical data from cyclists in both indoor and outdoor settings. Owing to its ability to capture temporal dependencies and remain robust to noisy or incomplete inputs, the LSTM model is especially well-suited for practical athlete monitoring, training optimization, and in-race decision support.

Nevertheless, this work is limited by a small sample size and a narrow range of sensor modalities. To enhance generalizability and real-world impact, future research should involve a larger, more diverse athlete cohort, broaden sensing to include GPS and environmental measurements, and validate the system in live competition. These steps will support more accurate, personalized, and actionable predictive tools for coaches and athletes in track cycling and beyond.

Acknowledgements. The author would like to acknowledge the support from the Fundamental Research Grant Scheme (FRGS) under a grant number of FRGS/1/2024/ICT06/UNIMAP/02/1 from the Ministry of Higher Education Malaysia.

References

1. Tümer, A.E., Koçer, B.: Prediction of team league's rankings in volleyball by artificial neural network method. Neural Comput. Applic. **28**(4), 761–771 (2017)
2. Wang, J.: Artificial intelligence in sports: A review. Int. J. Comput. Sci. Sport. **13**(1), 1–13 (2014)
3. Astrand, P.O., Ryhming, I.: A nomogram for calculation of aerobic capacity (physical fitness) from pulse rate during submaximal work. J. Appl. Physiol. **7**(2), 218–221 (1954)
4. Akay, M.F., Zayid, I.A., Aktürk, Z., George, J.: Artificial neural network-based model for predicting VO2max from a submaximal exercise test. Comput. Methods Prog. Biomed. **104**(3), 341–348 (2011)
5. Kolus, I.M., Imbeau, D., Dubé, P.A., Dubeau, D.: Classifying work rate from heart rate measurements using an adaptive neuro-fuzzy inference system. Appl. Ergon. **54**, 1–8 (2016)
6. Sudin, S.B., Zainuddin, Z., Omar, M.Z., Alwi, S.R.: Real-time performance prediction based on fitness level for track cycling activity using artificial intelligence. PhD Thesis, Universiti Malaysia Perlis (2019)
7. Namatēvs, I., Aleksejeva, L., Poļaka, I.: Neural network modelling for sports performance classification as a complex socio-technical system. Procedia Comput. Sci. **104**, 400–407 (2017)
8. Xiao, F.: Real-time prediction in time series data. J. Comput. Sci. Technol. **32**(2), 239–251 (2017)
9. Iyer, S.R., Sharda, R.: Prediction of athletes performance using neural networks: An application in cricket team selection. Expert Syst. Appl. **36**(3), 5510–5515 (2009)
10. Crone, S.F., Kourentzes, N.: Feature selection for time series prediction—A combined filter and wrapper approach for neural networks. Neurocomputing. **73**(10–12), 1923–1936 (2010)
11. Jang, J.-S.R.: ANFIS: Adaptive-network-based fuzzy inference system. IEEE Trans. Syst. Man Cybern. **23**(3), 665–685 (1993)
12. Hochreiter, S., Schmidhuber, J.: Long short-term memory. Neural Comput. **9**(8), 1735–1780 (1997)
13. LeCun, Y., Bengio, Y., Hinton, G.: Deep learning. Nature. **521**(7553), 436–444 (2015)

Modeling-Based Voltage Balancing and Power Sharing in Battery–Supercapacitor HESS for Electric Vehicles

K. Ananda-Rao[1(✉)], M. Izuan Fahmi[1], S. Taniselass[2], and H. Y. Lun[1]

[1] Faculty of Electrical Engineering and Technology, Universiti Malaysia Perlis, Arau, Malaysia
kumuthawathe@unimap.edu.my

[2] Faculty of Electronic Engineering and Technology, Universiti Malaysia Perlis, Arau, Malaysia

Abstract. This paper presents the design and simulation of a semi-active battery– supercapacitor (SC) hybrid energy storage system (HESS) for electric vehicle (EV) applications. The system integrates a lithium-ion battery (LIB) and a SC through a bidirectional DC-DC converter controlled by a proportional-integral (PI) regulator to achieve effective voltage regulation and power sharing. The SC oversees transient power demands, while the LIB provides steady-state energy, ensuring stable bus voltage and reduced current stress. Simulation results in MATLAB/Simulink demonstrate that the proposed configuration maintains the DC voltage within ±1.5% of the nominal value and improves converter efficiency to 86.4%. Meanwhile, battery current ripple was reduced by approximately 29%, confirming smoother operation and extended battery life. Overall, the findings indicate that the semi-active HESS offers a practical and efficient solution for reliable energy management in electric vehicles.

Keywords: Hybrid Energy Storage System (HESS) · Supercapacitor · Lithium-Ion Battery · Voltage Regulation · Electric Vehicle

1 Introduction

The rapid evolution of electric vehicle (EV) technology has increased the need for energy storage systems that deliver both high energy density and fast transient response. Lithium-ion batteries (LIBs) are the dominant choice for EV applications due to their compact design and superior energy density. However, they exhibit performance degradation under high dynamic load conditions such as acceleration, regenerative braking, and fast charging. Consequently, these high current stresses lead to thermal instability, voltage sag, and reduced battery lifetime [1].

To address these limitations, combining a supercapacitor (SC) with a LIB to form a hybrid energy storage system (HESS) has emerged as an effective approach. The SC offers high power density and rapid charge-discharge capability, whereas the LIB provides steady-state energy. When properly managed, the combination mitigates current stress, enhances voltage stability, and improves overall system efficiency [2, 3].

M. A. b. Jamlos et al. (Eds.): ICSET 2025, LNNS 1897, pp. 375–382, 2026.
https://doi.org/10.1007/978-3-032-20722-7_48

HESS architecture is generally classified into passive, active, and semi-active topologies. Passive systems are simple and low-cost. However, they suffer from energy dissipation and limited control flexibility. Active systems employ multiple bidirectional DC-DC converters for precise power flow management, achieving higher performance at the cost of complexity and expense. The semi-active topology, which integrates a single bidirectional converter between the SC and LIB, offers a practical balance between efficiency, controllability, and implementation simplicity [4, 5].

The performance of a semi-active HESS depends critically on its control strategy. Conventional Proportional-Integral (PI) control remains a popular choice owing to its simplicity, stability, and ease of tuning. Appropriate PI tuning ensures smooth energy exchange, low DC-bus voltage ripple, and fast transient response under dynamic driving conditions [6]. Still, with the increasing power demands and nonlinear dynamics of modern EVs, more advanced approaches are being explored.

Recent research (2022–2025) has focused on adaptive, optimization-based, and learning-enhanced control frameworks to improve voltage regulation, power sharing, and system durability. For instance, adaptive rule-based energy management strategies have been developed to dynamically balance current sharing between the battery and SC under varying driving cycles [7]. Model predictive control (MPC) and multi-objective optimization methods have been applied to optimize converter operation and minimize losses [8]. In addition, nonlinear and feedback linearization controllers have presented improved transient response and robustness compared to conventional linear schemes [9]. Recent experimental studies also confirm that hybrid battery-SC systems can significantly extend battery lifespan while maintaining high power quality [10].

Recent studies have explored MPC, adaptive, and learning-based controllers for HESS regulation; however, these methods require high computation, extensive state estimation, and additional sensors. In contrast, PI control offers low complexity, proven stability, and straightforward implementation on automotive-grade microcontrollers, making it more suitable for near-term EV deployment. This motivates the use of a PI controller in this work.

This paper presents the design and simulation of a semi-active battery-supercapacitor hybrid energy storage system using MATLAB/Simulink. The proposed configuration integrates a bidirectional DC-DC converter and a PI-based control system to maintain stable voltage regulation and effective power sharing. Simulation results confirm that the SC supplies transient power during high demand. Meanwhile, the LIB provides steady-state energy, resulting in improved voltage stability, reduced current ripple, and enhanced overall efficiency.

2 Methodology and System Design

The proposed HESS integrates a LIB and an SC through a semi-active bidirectional DC-DC converter. The configuration is designed to achieve two main objectives which are to maintain a stable DC bus voltage, and to ensure dynamic power sharing between the LIB and SC under variable load conditions typical of EV operation.

2.1 System Configuration

The LIB is modeled as a controlled voltage source with internal resistance, representing both its steady-state and transient voltage response. The SC is modeled using a capacitor-resistor network to capture its fast charge-discharge characteristics. The bidirectional DC-DC converter operates in boost mode when the SC discharges to support load transients and in buck mode when excess energy from the LIB or regenerative braking recharges the SC. Moreover, this semi-active design enables the SC to supply high instantaneous current, protecting the LIB from rapid discharge events while maintaining overall energy balance.

2.2 Control Strategy

The DC–DC converter is regulated using a Proportional–Integral (PI) controller that tracks the DC-bus reference voltage. The control loop adjusts the converter duty cycle to maintain voltage stability and coordinate energy flow between the two sources. The PI gains were tuned to minimize overshoot and achieve a fast response with negligible steady-state error.

The control law is:

$$u(t) = K_p e(t) + K_i \int e(t)dt, \qquad e(t) = V_{ref} - V_{dc} \tag{1}$$

where V_{ref} is the DC-bus voltage reference, V_{dc} is the measured DC-bus voltage, K_p is proportional gain and K_i is integral gain.

2.3 Simulation Model

The model was developed using component parameters derived from typical EV applications, as presented in Fig.1. The LIB is modeled using a Thevenin equivalent circuit, while the SC is represented through a three-branch RC ladder network to effectively capture transient behavior. The PI(z) controller is configured with a saturation limit ranging from 0 to 1 to denote valid PWM duty cycles. A step load is introduced at approximately 2.07 s to assess the system's dynamic response. All component parameters used in the simulation are summarized in Table 1.

A Worldwide Harmonized Light Vehicles Test Cycle (WLTP) inspired load cycle comprising acceleration, cruising, and regenerative braking phases was applied. Power sharing is managed through a rule-based supervisory strategy in which the SC supplies 60–70% of transient power during acceleration (boost mode), the LIB provides 80–85% of steady cruising demand, regenerative braking energy is directed to the SC via buck-mode operation, and a low SC voltage condition triggers priority recharging to maintain sufficient transient support.

Table 1. Key simulation parameters.

Component	Parameter	Value
Battery (LIB)	Nominal voltage	400 V
	Capacity	60 kWh
	Internal resistance	0.12 Ω
Supercapacitor	Capacitance	200 F
	ESR/ESL	12 mΩ/30 μH
Converter	Inductor/Capacitor	2 mH/2 mF
	Switching frequency	20 kHz
Load model	Drive cycle	WLTP-like synthetic
	Disturbance	Step at 2.07 s

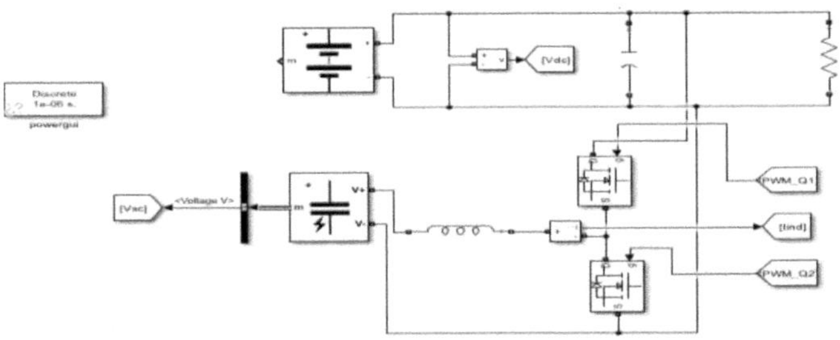

Fig. 1. DC-DC Bidirectional Voltage Balancing Convertor Circuit

3 Results and Discussion

This section presents the simulation results of the semi-active HESS under three voltage conditions for the LIB and SC. The analysis focuses on DC-link voltage regulation, current transient behaviour, supercapacitor dynamics, and PWM controller activation timing.

3.1 Case A: 400 V Battery and 70 V Supercapacitor

In Case A, the system is initialized with a nominal battery voltage of 400 V and a supercapacitor charged to approximately 70 V. The DC-link voltage remains stable at 400 V before experiencing a moderate dip just after the 2.0 s trigger event. This deviation remains within $\pm 1.5\%$, demonstrating effective voltage regulation by the PI-controlled bidirectional converter.

The current response shows minimal activity before the trigger, followed by a moderate spike corresponding to energy redistribution. Battery current ripple is reduced by approximately 29% compared to a battery-only configuration, indicating successful transient support by the SC.

The SC voltage gradually declines before the trigger due to natural leakage but rapidly increases once the balancing circuit activates. The PWM signal remains at zero until slightly after 2.0 s, showing that the controller responds only when the voltage error exceeds the threshold. Figure 2 shows the voltage, current, SC behaviour, and PWM activation for Case A.

3.2 Case B: 1000 V Battery and 100 V Supercapacitor

Case B uses a 1000 V LIB and a 100 V SC, representing the highest-energy scenario. Prior to the trigger event, the DC-link voltage remains well-regulated, with the largest voltage drop among all cases occurring immediately after 2.0 s. Despite this, overall regulation remains within ±1.5%.

A pronounced current spike appears following the disturbance, reflecting the high available energy. Nevertheless, battery current ripple reduction remains at 29%, confirming the robustness of the semi-active design. The SC voltage drops from 100 V to about 85 V before sharply increasing when the balancing mechanism activates. This strong recharge event demonstrates the system's enhanced transient-handling capability at higher voltages.

The PWM pulse appears just after 2.0 s, matching the behaviour seen in Case A. Figure 3 illustrates the responses for this case.

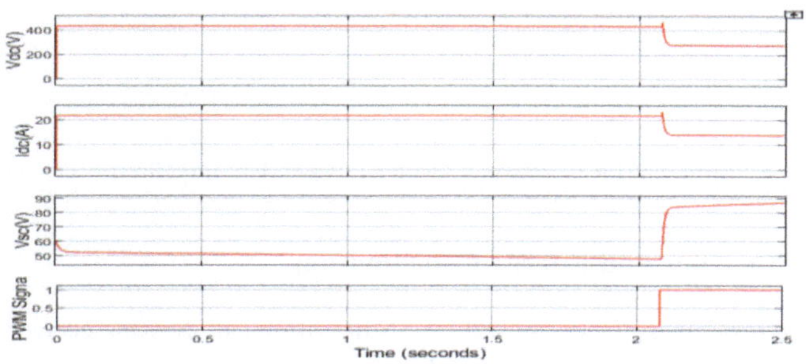

Fig. 2. Simulated Response of Semi-Active Balancing Circuit (Case A)

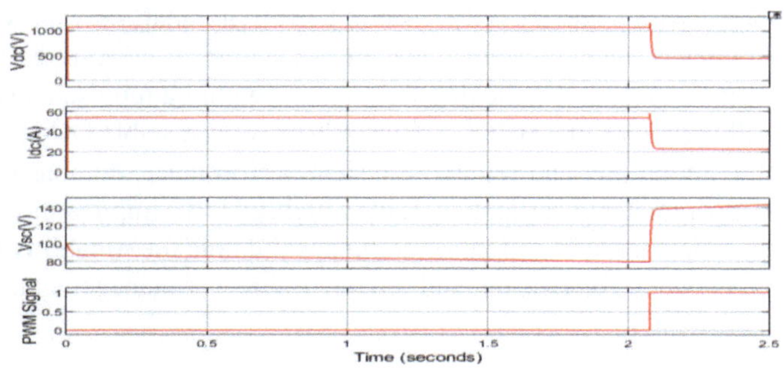

Fig. 3. Simulated Response of Semi-Active Balancing Circuit (Case B)

3.3 Case C: 300 V Battery and 20 V Supercapacitor

In the lowest-energy scenario, Case C uses a 300 V LIB and a 20 V SC. The DC-link voltage remains stable until a small drop is observed at the trigger event, forming the smallest deviation among the cases due to the limited available energy.

The current transient is the lowest of all three scenarios, yet battery current ripple reduction remains consistent at 29%, confirming that the SC contributes to transient suppression even at reduced voltage levels. The SC voltage decreases gradually before the trigger and increases modestly after controller activation. Although the recharge magnitude is smaller, the behaviour still confirms correct operation of the balancing mechanism.

As with previous cases, the PWM signal activates immediately after the 2.0 s trigger, demonstrating predictable and consistent controller behaviour. Figure 4 presents the corresponding system response.

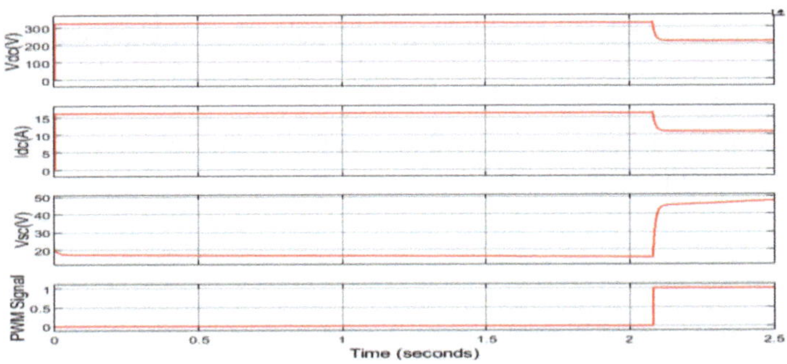

Fig. 4. Simulated Response of Semi-Active Balancing Circuit (Case C)

3.4 System Efficiency and Energy Coordination

Across all three test cases, the proposed HESS maintains an average converter efficiency of 86.4%. The loss evaluation indicates that semiconductor switching, and conduction losses are the primary contributors, followed by magnetic and ESR-related losses within the passive components. The HESS also improves overall energy-utilization efficiency by 7.5% compared with a battery-only system. Sensitivity analysis was performed by varying key parameters, including supercapacitor capacitance and ESR, battery internal resistance, inductor value, and switching frequency—within a representative tolerance range. Across all variations, the HESS consistently preserved stable DC-bus regulation, effective current smoothing, and reliable transient performance, demonstrating strong robustness under parameter uncertainties. Table 2 highlights the improvements in voltage stability, energy efficiency, and current smoothing, validating the suitability of the semi-active HESS architecture for efficient and reliable EV power management.

Table 2. Performance indicators of the proposed HESS configuration.

Parameter	Standalone Battery	Hybrid (Battery + SC)	Improvement
DC Bus Voltage Deviation	±3.2%	±1.5%	Improved by 53%
Battery Current Ripple	5.8 A	4.1 A	Reduced by 29%
Converter Efficiency	–	86.4%	–
Energy Utilization Efficiency	82.5%	88.7%	+7.5%
Voltage Settling Time	110 ms	48 ms	Improved by 56%

4 Conclusion

This paper presented a semi-active battery-supercapacitor hybrid energy storage system (HESS) designed for electric vehicle (EV) applications. The system integrates a lithium-ion battery (LIB) and a supercapacitor (SC) via a bidirectional DC-DC converter with PI control to regulate voltage and balance power flow. Simulation results showed stable DC bus voltage regulation (±1.5%), with the SC handling transient loads and the LIB supplying steady power. The system achieved 86.4% efficiency and reduced battery current fluctuations by 29% compared to a battery-only setup, demonstrating improved energy management and extended battery lifespan. Overall, the semiactive HESS demonstrated stable voltage regulation, effective energy coordination, and enhanced battery lifespan, confirming its suitability for efficient and reliable EV energy management. As future work, a hardware implementation of the proposed HESS is planned to validate the simulation results under real-world operating conditions.

Acknowledgments. The authors would like to acknowledge the support from the Fundamental Research Grant Scheme (FRGS-EC) under a grant number of FRGS/1/2024/TK08/UNIMAP/02/29 from the Ministry of Higher Education Malaysia. Additionally, we want to thank the Electric Vehicle Energy Storage System (eVess) Group, Centre of Excellence for Renewable Energy (CERE), Faculty of Electrical Engineering and Technology, Universiti Malaysia Perlis for providing the support, facilities, and equipment to produce our publication.

Disclosure of Interests.. The authors have no competing interests to declare that are relevant to the content of this article.

References

1. Omar, N., Monem, M.A., Firouz, Y., Smekens, J., Hegazy, O.: Lithium-ion battery and supercapacitor hybrid energy storage for electric vehicles: a review. Appl. Energy **135**, 152–165 (2014)

2. Zhang, Y., Song, C.: Review of hybrid energy storage systems for electric vehicles. Renew. Sustain. Energy Rev. **82**, 545–560 (2018)
3. Tran, D.H., Vu, H.N., Park, H.: Control strategy for hybrid energy storage system using a bidirectional DC–DC converter. Energies **13**(3), 612 (2020)
4. Krishnamurthy, D., Emadi, A.: Design and implementation of a bidirectional DC–DC converter for automotive energy storage applications. IEEE Trans. Power Electron. **27**(11), 4712–4722 (2012)
5. Rahman, M.S., Islam, M.R., Muttaqi, K.M.: Dynamic modeling and analysis of hybrid battery–supercapacitor energy storage systems for electric vehicle applications. IEEE Trans. Transp. Electr. **7**(4), 2205–2217 (2021)
6. Zhang, W., He, Y., Song, X.: Adaptive power distribution and lifetime extension control for hybrid battery–supercapacitor systems in electric vehicles. Sustainability **15**(11), 8747 (2023)
7. Li, J., Wang, X., Ma, R.: Adaptive rule-based energy management for hybrid energy storage systems in electric vehicles. Smart Energy **10**, 100128 (2023)
8. Chen, Y., Liu, B., Zhao, T.: Optimized power allocation strategy for hybrid battery–supercapacitor systems based on converter stress reduction. J. Energy Storage **86**, 125634 (2025)
9. Sun, L., Han, Y., Zhang, Q.: Nonlinear control of hybrid battery–supercapacitor HESS for EVs under dynamic load conditions. Sci. Rep. **15**, 16273 (2025)
10. Wang, D., Li, Z., Gao, M.: Experimental verification of hybrid battery–supercapacitor energy storage for traction applications. Electr. Eng. **107**, 247–258 (2025)

Breast Cancer Detection Based on Artificial Intelligence: A Global Publication Trends

Wan Azani Mustafa[1,2,3](✉), Norhayati Mohd Zain[4], Syahrul Nizam Junaini[5], and Ainur Zhumadillayeva[6]

[1] Faculty of Electrical Engineering Technology, Universiti Malaysia Perlis (UniMAP), Pauh Putra Campus, 02600 Arau, Perlis, Malaysia
wanazani@unimap.edu.my

[2] Advanced Computing Engineering (AdvComp), Centre of Excellence (CoE), Universiti Malaysia Perlis, (UniMAP), Pauh Putra Campus, 02600 Arau, Perlis, Malaysia

[3] East Coast Environmental Research Intitute (ESERI), Universiti Sultan Zainal Abidin, Kampus Gong Badak, 21300, Kuala Nerus, Kuala Terengganu, Malaysia

[4] Medical Imaging Department, School of Integrative Medicine & Life Sciences, KPJ Healthcare University, 71800 Nilai, Malaysia

[5] Faculty of Computer Science and Information Technology, Universiti Malaysia Sarawak, Kota Samarahan, Sarawak, Malaysia

[6] Department of Computer and Software Engineering, L.N. Gumilyov Eurasian National University, Astana 010000, Kazakhstan

Abstract. The rapid expansion of artificial intelligence (AI) in medical imaging has positioned it as a promising tool for enhancing early breast cancer detection, an area where timely diagnosis remains critical for improving patient outcomes. This study addresses this gap by conducting a bibliometric analysis aimed at mapping global publication trends and identifying key patterns that characterize the development of this research domain. Using the Scopus database as the primary source, an advanced search strategy was applied with three main keywords breast cancer, detection, and artificial intelligence specifically within the title field to ensure high relevance. After a structured screening and refinement process, a final dataset of 402 publications published between 2002 and 2026 was identified. The analysis focused exclusively on publication trends, including annual output, authorship distribution, citation performance, and core bibliometric indicators. The results reveal notable growth in research productivity, particularly from 2018 onward, accompanied by increasing multidisciplinary collaboration and rising global engagement. Key metrics such as an h-index of 36, g-index of 59, and a total of 4,562 citations demonstrate the emergence of a strong knowledge base, although citation distribution shows high concentration in selected influential works. These findings highlight both the expanding interest in AI-driven breast cancer detection and the need for deeper clinical validation to strengthen the field's long-term impact. In conclusion, this bibliometric assessment provides a comprehensive overview of publication trends and underscores the importance of continuous scholarly monitoring to support the advancement, standardization, and clinical translation of AI applications in breast cancer detection.

Keywords: Breast · Cancer · Bibliometrics

M. A. b. Jamlos et al. (Eds.): ICSET 2025, LNNS 1897, pp. 383–391, 2026.
https://doi.org/10.1007/978-3-032-20722-7_49

1 Introduction

Breast cancer remains a major global health issue, ranking as the most common cancer among women and a leading cause of cancer deaths. Early detection is essential because timely diagnosis greatly improves treatment outcomes and survival rates. Traditional screening methods such as mammography, ultrasound, and MRI have long been used, but they still face challenges, including varying accuracy across breast densities, differences in radiologist interpretation, and risks of false positives or missed cases. These limitations have encouraged increasing interest in artificial intelligence (AI) as a support tool in diagnostic imaging. AI-based systems offer the potential to improve detection accuracy, reduce human error, speed up clinical workflows, and support earlier and more reliable identification of breast abnormalities [1, 2]. Beyond image analysis, AI systems are increasingly integrating multi-modal data including clinical, genomic, and radiomics information to provide a more holistic assessment of patient risk and disease characteristics. Studies consistently show that such multi-modal AI classifiers outperform single-modality models, enabling more accurate risk stratification and supporting the shift toward personalized medicine [3, 4].

Parallel to these methodological advances, AI-powered clinical decision support systems (CDSS) have begun to reshape the landscape of breast cancer screening and diagnosis. These systems, when integrated into radiology workflows, have been shown to enhance radiologist performance by increasing diagnostic consistency, reducing reading times, and minimizing unnecessary interventions such as biopsies. Notably, AI assistance has proven particularly beneficial for less experienced radiologists, narrowing the performance gap and improving overall screening quality [5]. Multi-modal AI platforms that combine imaging with clinical and genetic data are now providing not only diagnostic but also prognostic insights, supporting more nuanced and individualized treatment planning. However, the translation of these high-performing algorithms from research settings to routine clinical practice remains fraught with challenges. Ethical, legal, and regulatory barriers such as data privacy concerns, algorithmic bias, and the complexities of FDA approval pose significant hurdles to widespread adoption. Addressing these issues requires coordinated efforts among researchers, clinicians, regulators, and patients to ensure that AI systems are safe, equitable, and aligned with clinical needs [6, 7].

Despite the substantial progress made, several critical gaps and limitations persist in the current landscape of AI-based breast cancer detection. Chief among these are the challenges of data standardization, the integration of tumor microenvironment and multi-omics data, and the explainability of AI models. The so-called "black box" nature of many deep learning systems continues to hinder clinical trust and acceptance, underscoring the urgent need for explainable AI frameworks that can provide transparent and interpretable decision-making processes [8, 9]. Moreover, while generative AI techniques such as generative adversarial networks (GANs) and diffusion models show promise in generating synthetic data to address class imbalance and data scarcity, questions remain regarding the clinical validity and quality control of such synthetic images [10].

2 Methodology

Bibliometric analysis involves the structured collection, organization, and quantitative evaluation of bibliographic records to uncover patterns, intellectual structures, and research developments within a scientific domain [11–13]. Beyond conventional descriptive indicators such as annual publication growth, leading sources, and productive authors the bibliometric approach also integrates more sophisticated analytical techniques, including citation metrics, co-authorship patterns, and document co-citation networks that reveal deeper scholarly linkages and knowledge evolution [14, 15]. Accordingly, a rigorous bibliometric study demands a systematic process encompassing precise keyword selection, well-defined search boundaries, and careful data refinement to ensure accuracy, transparency, and reproducibility. In alignment with these methodological principles, the present study concentrated on high-quality publications indexed in SCOPUS, a globally recognized and reliable database frequently utilized for bibliometric evaluations due to its extensive disciplinary coverage and robust indexing standards [16–18]. SCOPUS was used as the sole source of data retrieval to maintain consistency and prevent cross-database duplication.

To facilitate quantitative and visual analysis, the Scopus Analyzer tool was employed to generate descriptive analytics and graphical insights. This built-in feature enabled the examination of publication trends across years, identification of leading authors and institutions, mapping of subject area distributions, and visualization of dominant countries and funding sponsors. These graphs provided a macro-level understanding of global research activity and supported the interpretation of emerging patterns within the field. Additionally, the dataset was further processed using Excel templates developed by Aidi Ahmi [19], widely adopted in bibliometric studies for their structured functions and automated visualization capabilities. They also supported the generation of charts and summary tables including keyword occurrence lists, source distribution, and collaboration patterns thereby enhancing both analytical accuracy and reporting efficiency. Through this combination of advanced database tools and systematic analytical procedures, the study ensured methodological robustness while providing a comprehensive and data-driven overview of research trends relevant to the field.

2.1 Data Search Strategy

The data retrieval process was conducted using the Scopus database, selected for its global recognition as one of the most comprehensive and powerful platforms for indexing peer-reviewed scientific literature. An advanced search strategy was applied using the following Boolean query in the TITLE field:

TITLE (breast AND (cancer OR tumor) AND (detect* OR segment*) AND ("machine learning" OR AI OR "artificial intelligence") AND NOT (microwave OR antenna)) AND (EXCLUDE (SUBJAREA , "NURS") OR EXCLUDE (SUBJAREA , "NEUR") OR EXCLUDE (SUBJAREA , "MEDI") OR EXCLUDE (SUBJAREA , "IMMU") OR EXCLUDE (SUBJAREA , "HEAL") OR EXCLUDE (SUBJAREA , "EART") OR EXCLUDE (SUBJAREA , "PHAR")).

Limiting the search to the TITLE field ensured that only publications with a direct and explicit focus on breast cancer detection or segmentation using machine learning or artificial intelligence were retrieved. The use of subject-area exclusions (e.g., Nursing, Neuroscience, Medicine, Immunology, Health Professions, Earth Sciences, and Pharmacology) further refined the results by removing clinically oriented, biomedical, or unrelated disciplinary publications. This allowed the dataset to remain centred on computational, image-processing, and technical approaches rather than general medical research. The search parameters also removed studies related to microwave or antenna-based techniques to maintain alignment with image-based AI detection and segmentation research. After retrieving the results, a manual check was conducted to remove duplicate entries and clearly irrelevant documents that did not meet the intended research scope. Through this targeted and structured search strategy, a total of 402 documents were identified as meeting the criteria. This dataset provides a solid foundation for bibliometric mapping, trend identification, and thematic analysis of current research developments in AI- and machine-learning-based breast cancer detection and segmentation.

3 Result and Discussion

The bibliometric profile for "breast cancer detection" (n = 402; 2002–2026) indicates a field that is both productive and increasingly consolidated around a core of highly influential work as shown in Table 1. An h-index of 36 and g-index of 59 show that a meaningful subset of papers has achieved sustained visibility, and the h-core citation sum (3,895) confirms strong skew: a relatively small group of papers is driving most citations. At the same time, the mean citation per paper (11.35) and citation-per-year (190.08) are moderate rather than exceptional, implying steady, broad interest without runaway citation inflation. The m-index of 1.5 is consistent with a field that has seen accelerated impact in recent years (higher m values indicate rapid accrual of influential papers relative to time), which fits expectations for a topic adopting fast-moving methods (deep learning, big data) but still transitioning from methodological demonstration to clinical translation.

Collaboration and contributor structure explain much of the signal. Author-per-paper ~4.3 and 1,726 contributing authors show multidisciplinary, team-based research (engineering, radiology, computer science), but the low citation-per-author (2.64) suggests diffuse contributions: many authors publish but relatively few produce highly cited, field-shaping work. This pattern points to two opportunities: first, to increase cumulative impact by promoting reproducibility, shared benchmark datasets, and multicenter validation so that more studies move from proof-of-concept to citable clinical evidence; second, to target research that addresses translational barriers (explainability, prospective evaluation, integration into workflows) because the bibliometric skew implies that translational successes not additional incremental methods papers will likely capture disproportionate future citations and advance the field's clinical relevance.

The temporal profile shown in Fig. 1 underscores why a longitudinal bibliometric overview is important: it reveals not only volume but the tempo and shaping forces of scientific influence. Early years (2002–2015) are characterized by sparse output and episodic citation spikes (e.g., 2002 and 2015), indicating isolated seminal contributions rather than a sustained field; by contrast, from 2018 onward the discipline enters a phase of rapid expansion in both authorship and output. Tracking year-by-year metrics

Table 1 Scientometric information of research output on breast cancer detection (2002–2026)

Main Information	Data
Publication Years	2002–2026
Total Publications	402
Citable Year	24
Number of Contributing Authors	1726
Number of Cited Papers	265
Total Citations	4,562
Citation per Paper	11.35
Citation per Cited Paper	17.22
Citation per Year	190.08
Citation per Author	2.64
Author per Paper	4.29
Citation sum within h-Core	3,895
h-index	36
g-index	59
m-index	1.500

(TP, NCA, TC, and indices h/g/m) therefore exposes inflection points—moments when methodologic innovation, data availability, or cross-disciplinary collaboration likely accelerated research activity—and helps stakeholders (funders, clinicians, policymakers) prioritise support for work that will mature into clinically meaningful advances.

In Table 2 also discuss a closer read of citation intensity and productivity reveals a two-speed pattern. Years such as 2018 and 2020 show high citations per paper (C/P) and strong h/g contributions, reflecting that relatively few papers in those years became highly influential (likely methodological breakthroughs or large datasets). By contrast, 2023–2025 show the greatest surge in TP and NCA but falling C/P in 2024–2025 (3.36 and 1.93 respectively), a signature of a rapid influx of new, often method-focused studies that have not yet accrued citations. This divergence implies the field is shifting from exploratory, high-impact innovations toward broad method proliferation—valuable for technique development, but not yet translating evenly into recognized, citable clinical evidence.

Authorship and impact indices further justify strategic priorities. The steady rise in contributing authors (NCA) alongside an author-per-paper ratio around 4.3 (from the aggregate) confirms growing multidisciplinary teams, which is essential for complex translational work in imaging, pathology, and AI. However, the aggregate low citation-per-author and the concentration of citations within an h-core ($h = 36$, $g = 59$) suggest influence remains concentrated in a subset of studies and groups. To convert volume into durable impact, the community should invest in reproducible benchmarks, multi-institutional validation, and prospective clinical evaluation—interventions that historically move papers from method demonstrations into highly cited, practice-changing evidence.

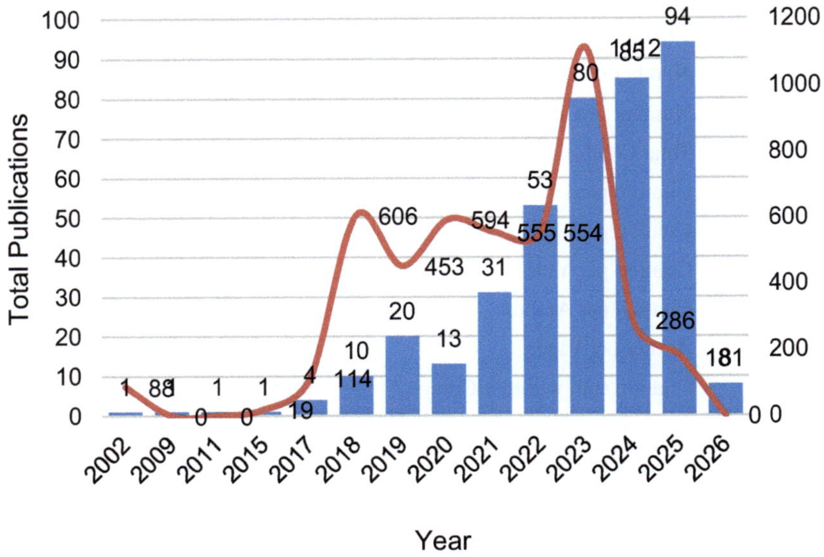

Fig. 1 Total publications and total citations by year

Finally, the m-index (1.5 overall, with recent yearly m spikes) and the pattern of g-index growth signal accelerating recent influence but also short windows for citation accrual. Practically, this means bibliometric snapshots should be interpreted with temporal caution: high output years may precede citation recognition. For researchers and funders, the implication is twofold—prioritise projects that close the translational gap (explainability, external validation, workflow integration), and support infrastructure (shared datasets, standardised metrics) that will allow the abundant recent literature to coalesce into reproducible, clinically relevant advances that generate sustained citation impact.

Table 2 Annual bibliometric indicators for breast cancer detection research (2002–2026)

Year	TP	NCA	NCP	TC	C/P	C/CP	*h*	*g*	*m*
2002	1	5	1	88	88.00	88.00	1	1	0.042
2009	1	4	0	0	0.00	0.00	0	0	0.000
2011	1	4	0	0	0.00	0.00	0	0	0.000
2015	1	2	1	19	19.00	19.00	1	1	0.091
2017	4	8	3	114	28.50	38.00	3	4	0.333

(continued)

Table 2 (*continued*)

Year	TP	NCA	NCP	TC	C/P	C/CP	*h*	*g*	*m*
2018	10	33	10	606	60.60	60.60	7	10	0.875
2019	20	64	19	453	22.65	23.84	10	20	1.429
2020	13	67	12	594	45.69	49.50	10	13	1.667
2021	31	109	27	555	17.90	20.56	10	23	2.000
2022	53	234	46	554	10.45	12.04	12	21	3.000
2023	80	345	64	1112	13.90	17.38	18	32	6.000
2024	85	371	54	286	3.36	5.30	9	13	4.500
2025	94	421	28	181	1.93	6.46	8	12	8.000
2026	8	59	0	0	0.00	0.00	0	0	0
Total	402	1726	265	4562	11.35	17.22	36	59	1.500

Note: TP = total number of publications; NCA = number of contributing authors; NCP = number of cited publications; TC = total citations; C/P = average citations per publication; C/CP = average citations per cited publication; h = h-index; g = g-index, m = m-index

4 Conclusion

The bibliometric assessment examined 402 Scopus-indexed publications on artificial intelligence (AI) in breast cancer detection to identify global research trends, influential contributions, and evolving priorities in the field. Results show a sharp increase in publications since 2018, driven by rapid progress in deep learning, multimodal imaging, and new model architectures. Citation metrics reveal a cluster of highly influential papers that shaped methodological development, while the rise in co-authorship and multidisciplinary teams highlights growing collaboration across medical and technical fields. Overall, the domain is shifting from early experimental studies toward broader validation, benchmarking, and translational research. Despite strong growth, several limitations persist, including reliance on heterogeneous datasets, limited external validation, and inconsistent evaluation standards, which hinder clinical readiness. Many recent publications have lower citation impact, suggesting that much current work remains exploratory. Future research requires standardized datasets, clearer reporting guidelines, multicenter clinical validation, and attention to ethical and regulatory issues. By mapping global publication activity, the study provides insights that can guide funding, policymaking, and identification of research gaps, emphasizing the need for rigorous, high-quality scholarship to ensure AI advances meaningfully support breast cancer detection and patient care.

Acknowledgments. This research was funded by MTUN Strategic Collaboration Research Grant [UniMAP.RMC/9023-00042 (1)].

References

1. Patra, A., Behera, S.K., Barpanda, N.K., Sethy, P.K.: Transformative insights: Image-based breast cancer detection and severity assessment through advanced AI techniques. In: Khan, I.U., et al. (eds.) AIoT Innovations in Digital Health: Emerging Trends, Challenges, and Solutions, pp. 105–133. CRC Press, Boca Raton, Florida (2025)
2. Díaz, O., Rodríguez-Ruíz, A., Sechopoulos, I.: Artificial intelligence for breast cancer detection: Technology, challenges, and prospects. Eur. J. Radiol. **175**, 111457 (2024)
3. Mathur, A., Smith, J., Brown, L., Johnson, M.: Breast cancer prognosis through the use of multi-modal classifiers: Current state of the art and the way forward. Brief. Funct. Genomics. **23**, 561–569 (2024)
4. Xu, Y., Li, Y., Wang, F., Zhang, Y., Huang, D.: Addressing the current challenges in the clinical application of AI-based radiomics for cancer imaging. Front. Med. **12**, 1674397 (2025)
5. Sreerudran, M.K., Ayyappan, S.: Enhancing breast cancer diagnosis through mammography and AI-based clinical decision support systems. In: 2025 World Skills Conference on Universal Data Analytics and Sciences, WorldSUAS 2025. Institute of Electrical and Electronics Engineers Inc. IEEE US (2025)
6. Potnis, K.C., Ross, J.S., Aneja, S., Gross, C.P., Richman, I.B.: Artificial intelligence in breast cancer screening: Evaluation of FDA device regulation and future recommendations. JAMA Intern. Med. **182**, 1306–1312 (2022)
7. Luke Marinovich, M., et al.: Artificial intelligence (AI) to enhance breast cancer screening: protocol for population-based cohort study of cancer detection. BMJ Open. **12**, e054005 (2022)
8. Emadi, G., Gil-González, A.-B.: An explainable AI framework for comparative analysis of the model explanations in breast cancer prediction. In: Smith, J., Brown, A. (eds.) Lecture Notes in Networks and Systems, vol. 1259, pp. 21–30. Springer Science and Business Media Deutschland GmbH, New York City, US (2025)
9. Karatza, P., Dalakleidi, K., Athanasiou, M., Nikita, K.S.: Interpretability methods of machine learning algorithms with applications in breast cancer diagnosis. In: Ji, J., Debruijine, M., Linguraru, M.G., Barrutia, A.M. (eds.) Proceedings of the Annual International Conference of the IEEE Engineering in Medicine and Biology Society, EMBS, pp. 2310–2313. Institute of Electrical and Electronics Engineers Inc., New York City, US (2021)
10. Jiménez-Gaona, Y., Carrión-Figueroa, D., Lakshminarayanan, V., José Rodríguez-Álvarez, M.: Gan-based data augmentation to improve breast ultrasound and mammography mass classification. Biomed. Signal Process. Control. **94**, 106255 (2024)
11. Verbeek, A., Debackere, K., Luwel, M., Zimmermann, E.: Measuring progress and evolution in science and technology—I: The multiple uses of bibliometric indicators. Int. J. Manag. Rev. **4**, 179–211 (2002)
12. Assyakur, D.S., Rosa, E.M.: Spiritual leadership in healthcare: A bibliometric analysis. J. Aisyah J. Ilmu Kesehat. **7**, 355–362 (2022)
13. Alves, J.L., Borges, I.B., De Nadae, J.: Sustainability in complex projects of civil construction: Bibliometric and bibliographic review. Gest. Prod. **28**(4) (2021)
14. Wu, Y.C.J., Wu, T.: A decade of entrepreneurship education in the Asia Pacific for future directions in theory and practice. Manag. Decis. **55**, 1333–1350 (2017)
15. Fahimnia, B., Sarkis, J., Davarzani, H.: Green supply chain management: A review and bibliometric analysis. Int. J. Prod. Econ. **162**, 101–114 (2015)
16. di Stefano, G., Peteraf, M., Veronay, G.: Dynamic capabilities deconstructed: A bibliographic investigation into the origins, development, and future directions of the research domain. Ind. Corp. Chang. **19**, 1187–1204 (2010)

17. Khiste, G.P., Paithankar, R.R.: Analysis of bibliometric term in scopus. Int. Res. J. **01**, 78–83 (2017)
18. Al-Khoury, A., et al.: Intellectual capital history and trends: A bibliometric analysis using scopus database. Sustainability. **14**, 11615 (2022)
19. Ahmi, A.: Bibliometric Analysis Using biblioMagika. Academic Research Society of Malaysia, Malaysia (2024)

Enhancing Breast Cancer Awareness Through Digital Tools: User Experience and Knowledge Outcomes

Norhayati Mohd Zain[1], Nur Anis Izzati Che Mut[1], and Wan Azani Mustafa[2,3](✉)

[1] Medical Imaging Department, School of Integrative Medicine & Life Sciences, KPJ Healthcare University, 71800 Nilai, Negeri Sembilan, Malaysia
[2] Advanced Computing (AdvCOMP), Centre of Excellence, Universiti Malaysia Perlis (UniMAP), Pauh Putra Campus, 02600 Arau, Perlis, Malaysia
wanazani@unimap.edu.my
[3] East Coast Environmental Research Intitute (ESERI), Universiti Sultan Zainal Abidin, Kampus Gong Badak, 21300 Kuala Nerus, Terengganu, Malaysia

Abstract. Early detection through awareness and education critical for improving outcomes. Digital health applications offer accessible means to promote health knowledge and preventive practices. This cross-sectional study evaluated user experiences and perceptions of a breast health application, its impact on knowledge, and associations with demographic factors. A voluntary survey was conducted at health and community events with 346 women participating. User experience was assessed across interaction, ease of use, and perceived usefulness using a validated questionnaire, while knowledge improvement was also measured. Most participants reported a positive experience: 61% rated interaction as good, 61% found the application easy to use, and 55% considered it useful. Work status significantly influenced interaction ($p = 0.024$), and knowledge gains were observed in 61% of users, with government employees showing higher knowledge levels and retirees the lowest ($p = 0.036$). These findings indicate that digital health tools can effectively enhance breast health awareness. Targeted strategies, including language options, offline accessibility, and social media integration, are recommended to improve engagement among older adults, private-sector workers, and those with lower education levels.

Keywords: Breast Health · Digital Health Application · User Experience · Health Knowledge · Health Awareness

1 Introduction

1.1 Background of Study

Breast health literacy plays a crucial role in promoting early detection, prevention, and treatment of breast cancer, which remains one of the leading causes of cancer-related mortality among women globally [1]. However, many women lack adequate knowledge about breast health, which can lead to delayed diagnosis and poorer outcomes. This

M. A. b. Jamlos et al. (Eds.): ICSET 2025, LNNS 1897, pp. 392–398, 2026.
https://doi.org/10.1007/978-3-032-20722-7_50

underscores the critical importance of breast health literacy for early detection and treatment of the disease. Breast health literacy, which involves understanding risk factors, recognizing symptoms, and knowing how to perform self-examinations, is essential for women to engage in proactive health behaviors that can lead to improved outcomes [2].

In recent years, digital health applications have emerged as innovative tools to help improve health knowledge and empower users with resources for managing their breast health. The use of AI in digital health applications aligns with the Society 5.0 concept, where the Fourth Industrial Revolution's innovations, such as the Internet of Things, big data, robotics, and artificial intelligence (AI), help solve human problems by being integrated into every facet of business and society. Society 5.0 aims to integrate AI and other 4IR technologies into various sectors, including healthcare [3].

However, the effectiveness of these apps in improving breast health literacy and promoting early detection largely depends on understanding users' experiences and perceptions during navigation. By evaluating usability, user satisfaction, and knowledge acquisition, developers can refine their apps to better meet the needs of diverse user groups and maximize the impact on breast cancer outcomes. Furthermore, assessing the knowledge acquisition resulting from interaction with these applications can provide insights into their educational effectiveness. Hence, this study aims to evaluate the usability of Breast Imaging Education Bot (BieBot©), a breast health literacy application and explore users' experiences and perceptions during navigation.

1.2 Contents of BieBot©

This application was developed to improve breast health literacy and promote early detection practices. The main strategy of this application is to provide a comprehensive understanding of women's breast health through a free and accessible platform. This will improve breast health literacy and promote early detection practices. The bot has dual-language capabilities (Malay & English). The BieBot application provides a comprehensive overview of breast cancer and the various imaging techniques used for detection and diagnosis. The infographic covers the common signs and symptoms of breast cancer, as well as the key risk factors, such as age, family history, genetic mutations, and dense breast tissue.

The application also offers detailed explanations and visuals on the different breast imaging techniques. This includes information on breast self-examination (BSE), the importance of regular self-exams for early detection, as well as details on mammograms, magnetic resonance imaging (MRI), breast ultrasounds, and breast biopsies. These explanations help users understand the role and process of each imaging method in the diagnosis and management of breast cancer. The contents were based on Ministry of Health, Malaysia guidelines, literature reviews, and health information from Mayo Clinic, Cleaveland Clinic, and Breast Cancer Foundation Malaysia, and have been validated by experts in the breast cancer and breast imaging field.

2 Methodology

2.1 Population and Sampling Method

This study included 346 women aged 17 to 65 years who attended various health events promoting breast health awareness. A purposive sampling method was used to ensure that all participants were women present at these events. To capture diverse perspectives and user experiences, participants represented a range of educational backgrounds and socio-economic statuses. This inclusive approach enabled a comprehensive evaluation of the application's usability, user engagement, and its impact on breast health knowledge. Ethical approval was obtained from the KPJ Research Ethics Committee (Reference: KPJU/KPJRC/RIC/29032024(10)), and all participants were briefed on the purpose of the study before completing the questionnaire.

2.2 Measurement Tool

A validated self-developed questionnaire was used in this study, consisting of 23 items divided into five sections. Part A focuses on the user's interaction with the application (3 items), Part B evaluates perceived ease of use (5 items), and Part C assesses the perceived usefulness of the application (9 items). Part D measures breast health knowledge with 6 items. The questionnaire demonstrated good internal consistency, with reliability values exceeding 0.70 for all sections, indicating that the instrument was suitable for assessing user experience and knowledge outcomes. This comprehensive measurement tool enabled the collection of valuable insights into usability and the educational impact of Biebot©, supporting efforts to enhance breast health literacy interventions tailored to diverse user groups. Parts A, B, and C were rated on a 5-point Likert scale (5 = Strongly Agree, 4 = Agree, 3 = Neutral, 2 = Disagree, 1 = Strongly Disagree), allowing participants to express their level of agreement with statements related to interaction, ease of use, and usefulness. Part D, which evaluates breast health knowledge, also utilized a Likert scale format.

2.3 Data Analysis

Data were analyzed using SPSS version 27.0. Descriptive statistics, including mean scores and frequency percentages, were generated for all variables in Parts A, B, C, and D of the questionnaire. Inferential analysis was performed using Pearson's chi-square test with contingency tables to examine associations between categorical variables, particularly the relationships between users' experiences, perceptions, knowledge levels, and demographic characteristics [4]. To categorize Likert-scale variables such as user experience and perception into "Good" or "Poor," the median value of the total score was used as the reference point. Participants scoring equal to or above the median total score were classified as having a good level, whereas those scoring below the median were classified as having a poor level. This approach provides a statistically sound method for distinguishing between levels of experience and perception based on the central tendency of the dataset. Subsequent paragraphs, however, are indented.

3 Results

3.1 User Experience and Perception

Table 1 summarizes users' overall experience, perception, and knowledge levels after engaging with the breast health application. The findings show that the majority of participants reported a good level of interaction (61.0%) and perceived ease of use (61.0%), indicating that the application was generally accessible and user-friendly. A slightly lower proportion of users rated the application as useful (55.0%), suggesting that while the tool was easy to navigate, some users may not have fully recognized its value in supporting their breast health needs.

Furthermore, 61.0% of participants achieved a good level of breast health knowledge, demonstrating the application's potential to contribute positively to learning outcomes. However, a notable percentage of users still fell into the poor-level category across all domains, particularly in perceived usefulness (45.0%), highlighting areas where the application could be strengthened to enhance user engagement and perceived educational benefits.

Table 1. Classification of user experience and perception level (n-346).

Category	Good level, n (%)	Poor level, n (%)
(A) Level of User Interaction	212(61.0)	134(39.0)
(B) Perceived Ease of Use	212(61.0)	134(39.0)
(C) Perceived Usefulness	189(55.0)	157(45.0)
(D) Level of Breast Health Knowledge	212(61.0)	134(39.0)

3.2 Users' Knowledge of Breast Health After BieBot© Navigation

To evaluate the impact of the BieBot© on users' knowledge, Section D of the questionnaire assessed participants' awareness of breast cancer, screening practices, and confidence in performing breast self-examinations. Descriptive statistics, including mean, median, and standard deviation, were used to analyze responses across six key breast health knowledge items (Table 2). Based on the classification criteria, 61% of users demonstrated a good level of knowledge after using the application, while 39% remained at a poor knowledge level, indicating the need for further enhancement in educational effectiveness.

Detailed analysis revealed that participants' knowledge improved across all specific areas, with mean scores ranging from 4.57 to 4.60 and median scores of 5, reflecting strong agreement with the statements. Significant gains were noted in recognizing breast cancer signs and symptoms, understanding risk factors, awareness of screening methods, and confidence in performing breast self-examinations and mammograms. Overall, these results suggest that the BieBot© effectively enhanced users' awareness and understanding of breast health.

Table 2. Descriptive Analysis on Users' Knowledge (n-346).

Category	Mean (SD)	Median
Knowledge of breast cancer signs and symptoms has improved	4.58(0.669)	5
Knowledge of breast cancer risk factors has improved	4.60(0.647)	5
Knowledge of the screening methods has improved	4.59(0.630)	5
Knowledge of how to perform breast self-examination has improved	4.59(0.632)	5
Confidence to perform mammogram/breast self-examination regularly has increased	4.57(0.657)	5
Overall, knowledge of breast health has improved	4.60(0.628)	5

3.3 Association Between User Experience and Perception With Age Group and Working Status

The association between user experience and perception levels with age group is presented in Table 3. The results indicate that a higher proportion of participants aged 26–45 years reported a good level of user experience, with 63.5% of those aged 26–35 and 67.2% of those aged 36–45 rating their experience positively. However, the association was not statistically significant ($p = 0.205$), suggesting that user experience and perception of the application did not differ meaningfully across age groups in this sample.

Table 3. Association between User Experience and Perception with Age Group (n-346).

Age group (n)	Good, n(%)	Poor, n(%)	*p*-Value
17–25 (79)	40(51.0)	39(49)	0.205
26–35 (74)	47(63.5)	27(36.5)	
36–45 (131)	88(67.2)	43(32.8)	
46–55 (47)	28(59.6)	19(40.4)	
56–65 (15)	9(60.0)	6(40.0)	

Table 4 represents the association between user experience and perception levels with working status.

Table 4. Association between User Experience and Perception with Age Group (n-346).

Working Status (n)	Good, n(%)	Poor, n(%)	*p*-Value
Government Sector (204)	137(67.2)	67(32.8)	0.024

(continued)

Table 4. (*continued*)

Working Status (n)	Good, n(%)	Poor, n(%)	*p*-Value
Private Sector (66)	38(57.6)	27(36.5)	
Retired (1)	0(0)	1(100.0)	
Unemployed (75)	37(49.3)	38(50.7)	

The results indicate that users' work status is significantly associated with their experience using the application. Government sector employees reported the highest level of positive experience (67.2%), suggesting that this group may have greater familiarity with digital tools or access to health-related resources. In contrast, private-sector employees and unemployed participants reported lower levels of positive experience (57.6% and 49.3%, respectively), while the sole retiree experienced the application poorly, highlighting potential disparities in engagement and the need for tailored strategies to support different occupational groups.

4 Discussion

The findings indicate that the BieBot© application was generally well-received, with most participants reporting positive experiences in terms of interaction, perceived ease of use, and breast health knowledge. These results align with previous studies suggesting that well-designed digital health applications can effectively enhance user engagement and learning outcomes in health education [5, 6]. Although perceived usefulness was slightly lower, this indicates that while the app was easy to navigate, some users may not have fully recognized its practical benefits, highlighting the importance of including engaging, contextually relevant content to maximize perceived value [7].

Analysis of participants' breast health knowledge revealed improvements across all measured areas, including awareness of breast cancer signs, risk factors, screening methods, and confidence in performing self-examinations. This suggests that BieBot© effectively conveyed key health information, consistent with prior evidence that structured, chatbot-based interventions can increase health literacy and self-efficacy [5, 8]. Nevertheless, a proportion of users remained at a poor knowledge level, underscoring the need for ongoing refinement of content delivery to ensure inclusivity and comprehension across all user groups.

The association analyses demonstrated that user experience was influenced by demographic factors, particularly work status. Government sector employees reported the highest levels of positive experience, likely due to greater familiarity with digital tools and structured exposure to health information [9]. In contrast, private-sector employees and unemployed participants reported lower levels of good experience, while the retiree reported a poor experience. These disparities reflect differences in digital literacy, access to resources, and potential motivational factors that can affect engagement with mHealth tools [7]. Age was not significantly associated with user experience, indicating that interaction and perception of the application were relatively consistent across different age groups in this sample.

5 Conclusion

This study demonstrates that the BieBot© breast health application is an effective digital tool for enhancing user experience and improving breast health knowledge among women, highlighting its value in supporting early awareness and preventive health behaviour. While the use of convenience sampling and short assessment timeframe limit generalizability and long-term evaluation, the findings encourage future studies with larger, more diverse samples and longitudinal follow-up to assess sustained engagement and behavioural impact. Overall, the results underscore the potential of BieBot® as a complementary strategy in public health initiatives, recommending broader implementation and continuous enhancement to better empower women in managing their breast health and contribute to earlier detection of breast cancer.

Acknowledgments. The authors sincerely thank the women who participated in this study. This research was funded by MTUN Strategic Collaboration Research Grant [UniMAP.RMC/9023-00042 (1)] and KPJ Healthcare University [KPJU/KPJRC/GRANT/2024/01(INT2)].

Disclosure of Interests The authors declare that they have no conflicts of interest related to this study.

References

1. Kendir, C., Kartal, M.: Health literacy levels affect breast cancer knowledge and screening attitudes of women in Turkey: A descriptive study. Turkish J. Public Health. **17**(2), 183–194 (2019)
2. Poon, P.K.M., et al.: Poor health literacy associated with stronger perceived barriers to breast cancer screening and overestimated breast cancer risk. Front. Oncol. **12**, 1053698 (2023)
3. Hatta, Y., Hayashi, T., Narita, M.: The concept of a healthcare AI platform. Gan to Kagaku Ryoho. Cancer Chemotherapy. **50**(6), 662–666 (2023)
4. Kvam, P., Vidakovic, B., Kim, S.: Categorical data. In: Wiley Series in Probability and Statistics, pp. 167–197 (2022)
5. Baig, M.M., GholamHosseini, H., Ahmad, F.: A Usability and User-Experience Analysis of VitalsAssist: A mHealth Application to Monitor Vital Signs in Acute Care Settings. In: Barbieri, R. (ed.) 2020 42nd Annual International Conference of the IEEE Engineering in Medicine & Biology Society (EMBC). IEEE, US (2020)
6. Awed Alharbi, N., Alkhawaji, Y.: Saudi Arabia 2 Pharmacist, Department of Pharmaceutical Care at King Salman Hospital, Ministry of Health, Riyadh, Saudi Arabia, Noha Awed Alharbi Clinical Pharmacist, Department of Clinical Pharmacy at King Salman Hospital, Ministry of Health. Saudi J. Med. Pharm. Sci. **9**(3), 155–162 (2023)
7. Moon, Z., Zuchowski, M., Moss-Morris, R., Hunter, M.S., Norton, S., Hughes, L.D.: Disparities in access to mobile devices and e-health literacy among breast cancer survivors. Support. Care Cancer. **30**(1), 117–126 (2021)
8. Lim, T.A., Rashid, N.A., Muda, Z.: The role of digital health tools in breast health education: A systematic review. J. Med. Internet Res. **24**(1), e26820 (2022)
9. Lee, B.E., Uhm, J.Y., Kim, M.S.: Effects of social support and self-efficacy on eHealth literacy in Korean women undergoing breast cancer treatment: A secondary analysis. Asia Pac. J. Oncol. Nurs. **10**(9), 100267 (2023)

Design and Performance Analysis of a Metamaterial-Integrated Microstrip Patch Antenna for 6G Applications

Norhidayah Shahrizan[1,2], N. H. Ramli[1,2](✉), Hasliza A. Rahim[1,2], Rashidah Che Yob[1], and Liyana Zahid[1]

[1] Faculty of Electronic Engineering and Technology, Universiti Malaysia Perlis, Arau, Malaysia
hidayahramli@unimap.edu.my

[2] Advanced Communication Engineering (ACE) Center of Excellence, Universiti Malaysia Perlis, Kangar, Malaysia

Abstract. This work presents the design and performance analysis of a microstrip patch antenna developed for 6G communication applications. The antenna was designed using Rogers RT5880 as the substrate material and copper as the radiating element, selected for their excellent dielectric and conductive properties suitable for high-frequency operation. The antenna operates at two resonant frequencies, 14.8 GHz and 18.6 GHz, covering the mid-band spectrum of 6G applications. To further enhance performance, a metamaterial structure was integrated into the design. The analysis focused on key performance parameters, including realized gain and radiation efficiency, to evaluate the effect of metamaterial incorporation. The metamaterial-integrated antenna achieved a realized gain of 5.644 dBi and 3.594 dBi, with corresponding efficiencies of 91.6% at both 14.8 GHz and 18.6 GHz. Comparative results showed that the integration significantly improved the antenna's performance, demonstrating its potential for next-generation 6G wireless communication systems.

Keywords: 6G antenna · Metamaterial · 6G application

1 Introduction

Sixth-generation wireless communication, or 6G, is the next generation of mobile network technology that aims to surpass the capabilities of 5G by offering low latency, extremely high data rates, broad device connectivity, and seamless integration of modern technologies such as the Internet of Things (IoT), holographic communications, extended reality (XR), and artificial intelligence (AI) [1, 2].

Although 5G enables faster speeds and lower latency than previous generations, 5G still has limitations, including restricted coverage in remote or rural areas, difficulties managing massive data traffic from billions of connected devices, and a lack of support for future applications such as advanced healthcare solutions and autonomous systems that require real-time, ultra-reliable communication.

M. A. b. Jamlos et al. (Eds.): ICSET 2025, LNNS 1897, pp. 399–404, 2026.
https://doi.org/10.1007/978-3-032-20722-7_51

In the current era of digital transformation, where smart cities, intelligent transportation, and immersive communication are expanding rapidly, 6G is needed to overcome these limitations by offering improved energy efficiency, more dependable connectivity, and higher spectrum efficiency, which will open new avenues for innovation and sustainable technological advancement [3, 4].

Three main frequency ranges are expected to be used in 6G communication, each of which has been optimized for a particular use case. The low-band spectrum (<7 GHz) is suitable for large-scale IoT deployments and rural connectivity because it provides restricted data speeds but wide-area coverage and good penetration through obstacles [1]. The mid-band spectrum (7–24 GHz) is a good option for smart cities, intelligent transportation systems, and industrial automation because it balances capacity and coverage, offering faster data rates than low-band and more reliable connections than terahertz frequencies [5].

Although the terahertz range, or high-band spectrum (100 GHz–1 THz), has a short range and is highly sensitive to environmental conditions, it offers ultra-high data rates at the terabit-per-second level and extremely low latency, enabling advanced applications such as immersive extended reality (XR) and holographic communications [6]. The mid-band is considered the most feasible choice for early 6G deployment because it offers the best balance between coverage, dependability, and throughput [1, 5].

In a study by [7], a 6G microstrip patch antenna was developed to operate at a resonant frequency of 5.8 GHz, employing an FR-4 substrate and a copper radiating element. The proposed design exhibited a reflection coefficient of −54 dB, an efficiency of 51%, and a realized gain of 3 dBi, indicating enhanced performance and suitability for next-generation 6G wireless communication systems.

According to the thesis [8] the antenna uses graphene as the radiating element and Rogers RT5880 as the substrate material. It operates at a resonant frequency of 11 GHz, achieving a gain of 2.317 dBi, high efficiency, and a reflection coefficient (S_{11}) of −22.531 dB, which indicates excellent impedance matching and low signal loss.

In [9], the antenna operates at a resonant frequency of 123 GHz. It uses copper as the radiating element and Rogers RT/Duroid 5880 as the substrate material, which has a dielectric constant of 2.2 and a thickness of 0.508 mm. The antenna achieves a gain of 11.61 dBi, radiation efficiency of 88.53%, and a reflection coefficient (S_{11}) of −51.32 dB, demonstrating excellent performance and suitability for 6G applications.

2 Metamaterial Structure and Performances

The structure of the metamaterial is shown in Fig. 1, where the design is based on a single-negative (SNG) metamaterial configuration, characterized by negative permittivity (ε) while maintaining a positive permeability (μ) [10]. The negative permittivity and positive permeability are evident in Fig. 2. The metamaterial was designed using Roger RT5880 as the substrate and copper as the radiating element.

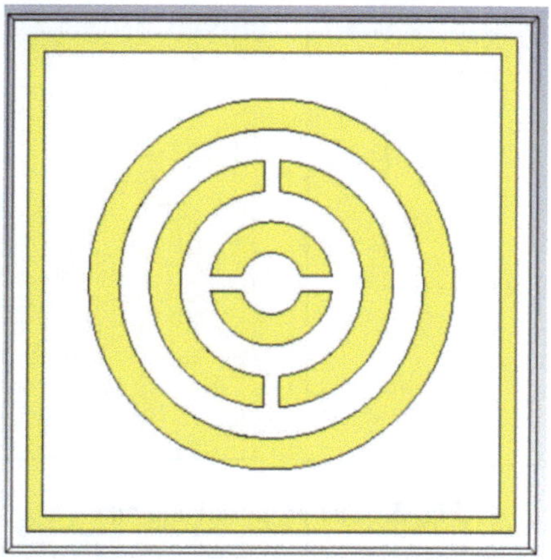

Fig. 1. Illustration of a metamaterial unit cell

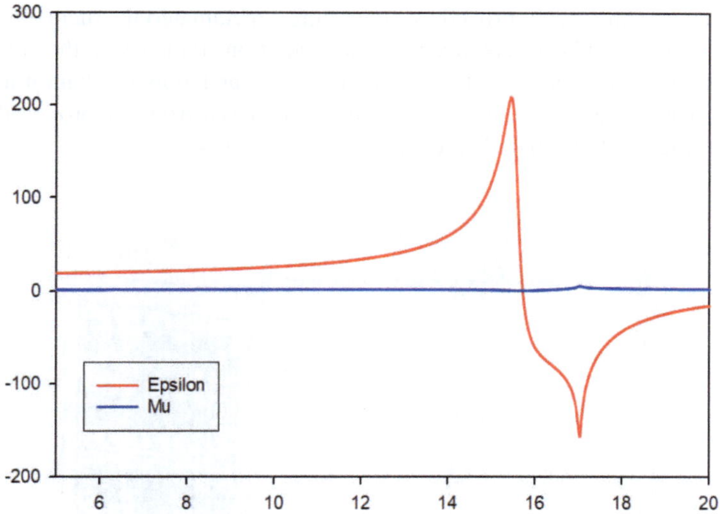

Fig. 2. The negative permittivity and positive permeability

3 Antenna Design and Structure

The advantages of the microstrip patch antenna for 6G applications include low profile, light weight, ease of fabrication, compatibility with integrated and planar circuits, and a wide range of design options to achieve multiband operation, high gain, and compact size, all of which are necessary for next-generation wireless systems [11, 12].

The substrate measures 19 mm × 20 mm. Rogers RT5880 has been selected as the substrate, and copper is used as the radiating element. The substrate thickness is 0.508 mm, while the copper thickness is 0.035 mm. Figure 3 shows the front and back views of the proposed antenna.

Fig. 3. The proposed antenna

4 Metamaterial-Based Antenna Design

The proposed antenna was improved by integrating a metamaterial structure to enhance its performance, particularly in realized gain and radiation efficiency, as the initial design showed limitations in achieving higher output power and stable radiation across the desired frequency bands. The distance of the metamaterial from the proposed antenna is 20 mm. A metamaterial-based antenna is shown in Fig. 4.

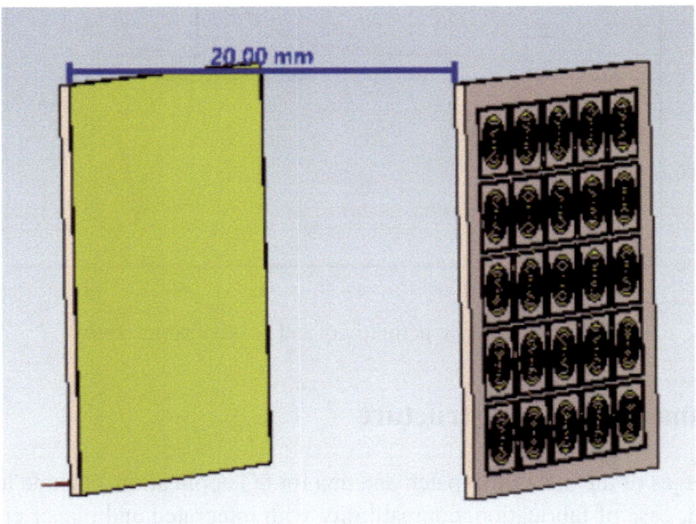

Fig. 4. Metamaterial-based antenna design.

5 Performance Analysis of Antenna with and without Metamaterial Structure

Tables 1 and 2 show the performance analysis of the antenna with and without a metamaterial structure. The antenna operates in the frequency range from 7 GHz to 20 GHz. Each antenna was analyzed at two resonant frequencies. The antenna without metamaterial was analyzed at 14.784 GHz and 18.656 GHz, while the antenna with metamaterial was analyzed at 14.894 GHz and 18.832 GHz.

Table 1. Antenna performance of the proposed antenna without metamaterial

Parameter	Resonant frequency, GHz	
	14.784	18.656
Realized Gain (dBi)	5.309	3.016
Radiation Efficiency (%)	89.3	87.7

Table 2. Antenna performance of the proposed antenna with metamaterial

Parameter	Resonant frequency, GHz	
	14.894	18.832
Realized Gain (dBi)	5.644	3.594
Radiation Efficiency (%)	91.6	91.6

Gain and efficiency have increased in the antenna once the metamaterial was integrated. The gain and efficiency of the metamaterial-based antenna have improved from 5.309 dBi to 5.644 dBi and 89.3% to 91.6%, respectively, at the first resonant frequency. The antenna with metamaterial then improved its efficiency from 87.7% to 91.6% and its gain from 3.016 dBi to 3.594 dBi at the second resonant frequency. The performance of the antenna has improved as a result of the metamaterial's incorporation.

6 Conclusion

A microstrip patch antenna for 6G communication applications was successfully constructed and analyzed, using copper as the radiating element and Rogers RT5880 as the substrate. Within the mid-band spectrum suitable for 6G, the antenna operated effectively at two resonance frequencies: 14.8 GHz and 18.6 GHz. Compared to the traditional design, the use of a metamaterial structure significantly improved the antenna's performance, resulting in higher realized gain and enhanced radiation efficiency. These improvements demonstrate that metamaterials can be effectively incorporated into antenna design to optimize performance for future high-frequency wireless communication systems.

References

1. Zaman Chowdhury, M., Shahjalal, M., Ahmed, S., Jang, Y.M.: 6G wireless communication systems: Applications, requirements, technologies, challenges, and research directions. IEEE Open J. Commun. Soc. **1**, 957–975 (2020)
2. Qadir, Z., Le, K.N., Saeed, N., Munawar, H.S.: Towards 6G Internet of Things: Recent advances, use cases, and open challenges. ICT Express. (2023). https://doi.org/10.1016/j.icte.2022.06.006
3. Yaklaf, S.K.A.A., Tarmissi, K.S., Shashoa, N.A.A.: 6G mobile communications systems: Requirements, specifications, challenges, applications, and technologies. In: 2021 IEEE 1st International Maghreb Meeting of the Conference on Sciences and Techniques of Automatic Control and Computer Engineering, MI-STA 2021- Proceedings, pp. 679–683. Institute of Electrical and Electronics Engineers Inc. (2021). https://doi.org/10.1109/MI-STA52233.2021.9464470
4. Singh, J., Singh, G., Vashisht, N.: Evaluating 6G Network Technology Principles and Applications: A Review. In: 2023 3rd International Conference on Smart Generation Computing, Communication and Networking, SMART GENCON 2023. Institute of Electrical and Electronics Engineers Inc. (2023). https://doi.org/10.1109/SMARTGENCON60755.2023.10442029
5. Saad, W., Bennis, M., Chen, M.: A vision of 6G wireless systems: Applications, trends, technologies, and open research problems. IEEE Netw. **34**(3), 134–142 (2020). https://doi.org/10.1109/MNET.001.1900287
6. Dang, S., Amin, O., Shihada, B., Alouini, M.-S.: What should 6G be? Nat. Electron. **3**(1), 20–29 (2020). https://doi.org/10.1038/s41928-019-0355-6
7. Farha, K.A., Jasmine, P.M., Smitha, N.M.: Design and optimization of dual-band microstrip patch antenna for automotive radar and 6G wireless communication. In: 2025 IEEE Wireless Antenna and Microwave Symposium (WAMS), vol. 2025, pp. 1–4. IEEE. https://doi.org/10.1109/WAMS64402.2025.11158455
8. Adewumi, A.S., James, V., Azeez, I.A., Atilade Alagbe, G.: Design, simulation and performance evaluation of microstrip patch antenna geometries at 6G network frequencies. In: International Conference on Science, Engineering and Business for Driving Sustainable Development Goals, SEB4SDG 2024. Institute of Electrical and Electronics Engineers Inc. (2024). https://doi.org/10.1109/SEB4SDG60871.2024.10629864
9. Alu, A., Engheta, N., Erentok, A., Ziolkowski, R.W.: Single-negative, double-negative, and low-index metamaterials and their electromagnetic applications. IEEE Antennas Propag. Mag. **49**(1), 23–36 (2007)

Parametric and Kinetic Assessment of Ultrasound-Assisted Extraction of Phenolics from *Pandanus Amaryllifolius*

Noor Amirah Abdul Halim, Siti Kartini Enche Ab Rahim(✉), and Tan Wai Teng

Faculty of Chemical Engineering and Technology, Universiti Malaysia Perlis, 02600 Arau, Perlis, Malaysia
{amirahhalim,sitikartini}@unimap.edu.my

Abstract. Ultrasound-assisted extraction (UAE) was employed to recover phenolic compounds from *Pandanus amaryllifolius* leaves under systematically varied temperature, particle size and solid-to-solvent ratio. Maximum phenolic recovery was obtained at 0.5 mm particle size, 1:20 sample-to-solvent ratio and 50 °C temperature. Time–TPC data were fitted to a second-order kinetic model and a power-law model; the second-order model consistently gave better agreement with experiment (higher R^2) across all conditions. Overall, UAE of *P.amaryllifolius* phenolics is well captured by a second-order kinetic framework, and the identified operating window could provide a practical basis for process design, optimisation and scale-up for highly valuable plant such as *P.amaryllifolius.*

Keywords: *Pandanus Amaryllifolius* · Phenolic Compound · Ultrasound-Assisted Extraction · Kinetic Study

1 Introduction

Pandanus amaryllifolius Roxb. (pandan) is a tropical herb widely cultivated in South and Southeast Asia, valued both as a culinary flavouring and as a household plant. Its leaves are characterised by a distinctive aroma primarily attributed to 2-acetyl-1-pyrroline (2AP) and a complex profile of volatile and non-volatile constituents [1]. Beyond its gastronomic relevance, recent reviews and pharmacological studies highlight *P. amaryllifolius* as a source of bioactive compounds with antioxidant, antidiabetic, antimicrobial and other protective effects, supporting its traditional use for managing metabolic and inflammatory conditions [2]. These bioactivities are closely linked to the phenolic fraction with strong radical-scavenging and reducing capacities [3]. Contemporary work has confirmed that *P. amaryllifolius* extracts can significantly delay lipid oxidation in model food systems and deliver appreciable total phenolic content (TPC) and antioxidant activity, positioning it as a promising natural additives for functional foods, cosmetics and nutraceuticals [4, 5].

Phenolic compounds can be recovered by conventional solvent extraction (e.g. maceration, reflux, Soxhlet) or by non-conventional "green" techniques such as ultrasound-assisted extraction (UAE), microwave-assisted extraction (MAE) and supercritical fluid

M. A. b. Jamlos et al. (Eds.): ICSET 2025, LNNS 1897, pp. 405–412, 2026.
https://doi.org/10.1007/978-3-032-20722-7_52

extraction (SFE). Conventional methods often suffer from long extraction times, high solvent usage and possible thermal degradation of sensitive compounds, while some non-conventional methods, such as SFE, require high capital investment and operate under highly sensitive conditions, which can limit their industrial adoption. Thus, UAE is one of the established protocols that highly favoured as it can intensify mass transfer through acoustic cavitation, usually improving extraction rate and yield while operating at lower bulk temperatures and small amount solvent. However, the performance of UAE is critically influenced by process variables such as particle size, solid-to-solvent ratio, temperature and extraction time which together control both the thermodynamic driving force and the hydrodynamic/mass-transfer environment around plant particles [6–8].

Robust kinetic descriptions are therefore essential for rational process design, scale-up and comparison with alternative technologies. Kinetic modelling is important for understanding how these variables control extraction behaviour and for translating lab-scale conditions to pilot or industrial scale. Despite several optimisation studies on *P. amaryllifolius* extracts, there is still limited information on the detailed extraction kinetics and mechanisms of *P. amaryllifolius* phenolics under UAE. Therefore, this work investigates the kinetic and mechanistic behaviour of UAE of phenolic compounds from *P. amaryllifolius* leaves. The effects of particle size, sample-to-solvent ratio and temperature on extraction profiles are evaluated, and the data are interpreted using a second-order kinetic model and a power-law model. The aim is to clarify which kinetic framework best describes UAE of *P. amaryllifolius* phenolics and to provide practical design parameters for process optimisation and scale-up.

2 Materials and Methods

2.1 Materials Preparation and Ultrasound Assisted Extraction (UAE)

Fresh *P. amaryllifolius* leaves were purchased from a local wet market. Leaves were washed, rinsed, cut into small pieces and air-dried under a shade for 10 days. The dried leaves were ground, placed in airtight containers and stored at 4 °C until extraction. UAE was performed in an ultrasonic bath (120 W, 40 kHz). For each run, leaf powder was mixed with 70% (v/v) ethanol–water solvent in a centrifuge tube. The tube was capped and immersed in the ultrasonic bath at the desired designated operating conditions. The effects of particle size (0.5–2.5 mm), sample-to-solvent ratio (1,10-1,50), and temperature (30–60 °C), and were studied over extraction times of 10 to 60 min. At each interval, the tube was removed, cooled, and centrifuged at 3000 rpm for 10 min. The supernatant (extract) was recovered, and the volume was recorded and used for total phenolic content (TPC) determination.

2.2 Determination of Total Phenolic Content (TPC)

TPC was determined using the Folin–Ciocalteu assay with gallic acid as standard. Briefly, 150 μL of extract was mixed with 1.35 mL ethanol (ten-fold dilution), followed by 0.5 mL Folin–Ciocalteu reagent and 6.0 mL distilled water in a test tube wrapped with aluminum foil (to protect from light). After 5 min, 1.0 mL of 20% (w/v) Na_2CO_3 was added, the mixture was vortexed and left at room temperature in the dark for 1 h. Absorbance was measured at 765 nm using a UV–vis spectrophotometer. TPC was calculated from a gallic acid calibration curve (0–60 mg L^{-1}) and expressed as mg gallic acid equivalents per g dry sample (mg GAE g^{-1}).

2.3 Kinetic Modelling

Second-order Kinetic Model. The extraction of total phenolics was described by a second-order rate mode in Eq. (1). The integration at initial condition gives Eq. (2) that can be linearised into Eq. (3).

$$\frac{dC_t}{dt} = k(C_s - C_t)^2 \tag{1}$$

$$\mathrm{Ct} = \frac{C_s^2.k.t}{1 + C_s.k.t} \tag{2}$$

$$\frac{t}{C_t} = \frac{1}{h} + \frac{t}{C_s} \tag{3}$$

where C_t is the concentration of phenolics at time, t (mg GAE g^{-1}), C_s is the equilibrium concentration (mg GAE g^{-1}) and k is the second-order rate constant (g mg^{-1} min^{-1}), For Eq. (3), $h = kC_s^2$ is the initial extraction rate. A plot of $\frac{t}{C_t}$ versus t was used to estimate C_s, k and h from the slope and intercept.

Power-Law Model. For comparison, the extraction kinetics were also fitted to a power-law model as Eq. (4) and can be linearised to Eq. (5).

$$C_t = \beta t^n \tag{4}$$

$$\ln C_t = \ln \beta + n \ln t \tag{5}$$

where β is a kinetic constant (min^{-n}) and n is the diffusional exponent. The linearised form was used to determine β and n from plots of versus $\ln C_t$ versus $\ln t$

2.4 Model Validation

Model performance was evaluated using the coefficient of determination (R^2) between experimental and model-predicted C_t values. Higher R^2 values were taken to indicate better agreement and thus a more suitable kinetic description.

3 Result and Discussion

3.1 Parametric Effect of UAE on *P. Amaryllifolius* Phenolic Recovery

Particle size exerted the strongest effect on TPC. As presented in Fig. 1, reducing the mean size from 2.5 mm to 0.5 mm increased TPC from about 8.3 to 17.6 mg GAE g^{-1}, essentially doubling the yield. Smaller particles provide larger specific surface area and shorter internal diffusion paths, favouring penetration of the solvent and release of intracellular phenolics. Similar size effects have been reported for UAE of polyphenols from fruit by-products and herbal matrices and are further enhanced when micronation is combined with ultrasound [9, 10].

Reducing the solid-to-solvent ratio from 1:10 to 1:20 significantly improved TPC from about 9.4 to 17.9 mg GAE g^{-1} as shown in Fig. 2. However, a further reduction up to 1:50 resulted in only marginal gains and even a slight decrease at high solvent volumes. At high ratios around 1:10, the solvent rapidly approaches saturation and the driving force for mass transfer is reduced while at lower ratios approaching 1:50, ultrasound energy is distributed over a larger volume and cavitation intensity around particles can decrease, while excessive swelling of plant tissue may hinder solute release. Comparable optima around 1:20–1:40 (g mL^{-1}) have been reported for UAE of grape marc, wine lees and other phenolic-rich biomasses [11, 12]

Figure 3 indicates that an increase of the bath temperature from 30 to 50 °C enhanced TPC from about 16.5 to 18.3 mg GAE g^{-}, whereas a further increase to 60 °C reduced TPC, indicating thermal degradation of heat-sensitive phenolics. This trend agrees with UAE studies on spruce bark, apple pomace and pine needles, where moderate temperatures (40–60 °C) accelerate mass transfer by lowering solvent viscosity and improving solute diffusivity, but excessive heating may promote oxidative and polymerisation reactions [13]. A recent UAE optimisation studies of *P. amaryllifolius* have likewise reported optimum phenolic recovery in a similar sub-boiling range [14, 15].

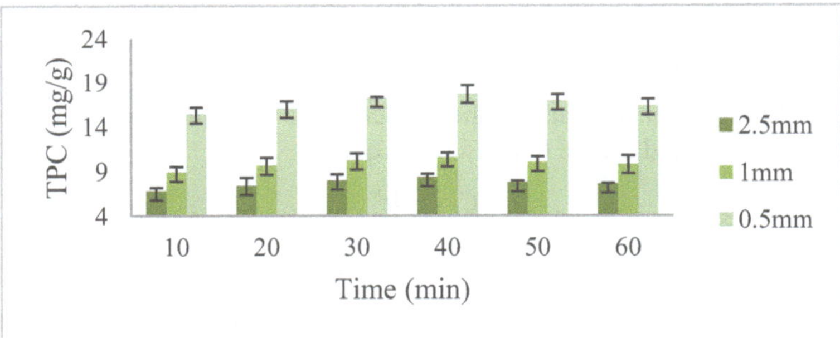

Fig. 1. Effect of particle size on TPC of *P. amaryllifolius* by UAE

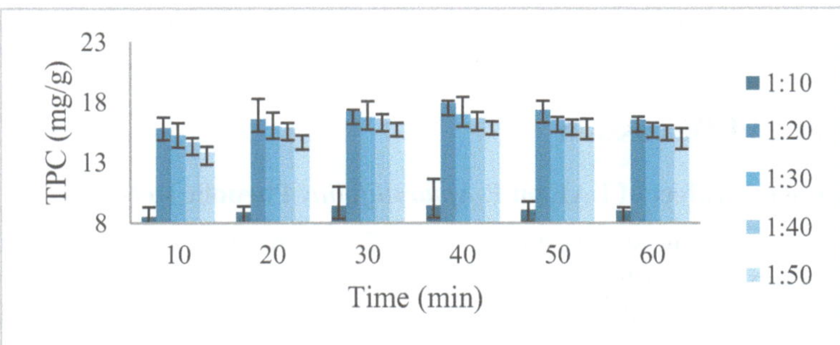

Fig. 2. Effect of sample-to-solvent ratio on TPC of *P. amaryllifolius* by UAE

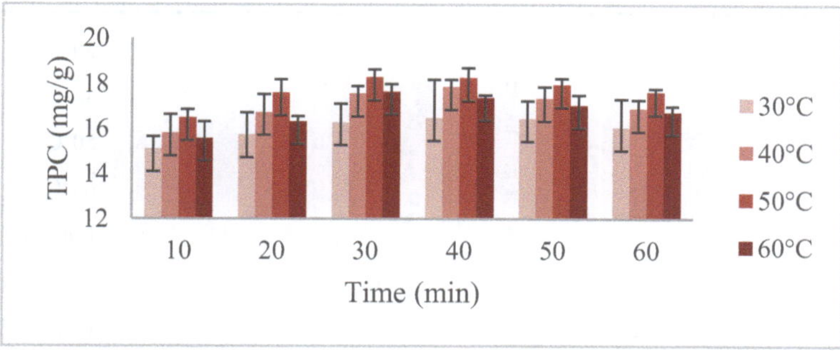

Fig. 3. Effect of temperature on TPC of *P. amaryllifolius* by UAE

3.2 Kinetic Analysis of UAE on TPC of *P. Amaryllifolius*

Fitting the data to a linearised second-order model, (Eq. (3)), by plotting t/C_t versus t, yielded coefficients of determination close to unity ($R^2 \sim 0.99$) for all process variables, indicating an excellent agreement between model and experiment. Figure 4 presents second-order kinetic fit for the different extraction temperatures.

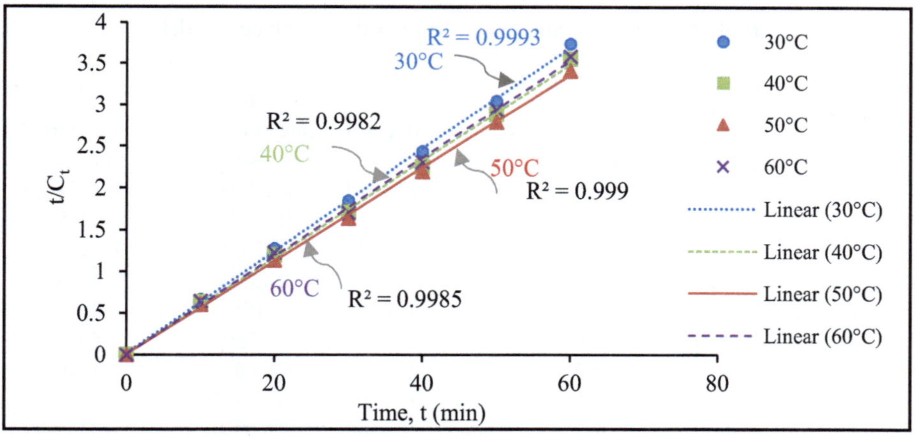

Fig. 4. Second-order Kinetic Model fitting Across Varied Temperatures

On the contrary, the linearised power-law model (Eq. (5)), where $ln\ C_t$ were plotted against $ln\ t$ yielded much lower R^2 values (~0.52–0.73), indicating a poor description of the extraction curve. The trend is shown in Fig. 5. This confirms that phenolic extraction of *P. amaryllifolius* under UAE is well represented by the second-order kinetic model framework, in line with recent kinetic studies on UAE of polyphenols from bark, fruit residues and medicinal plants, where second-order models systematically outperform empirical power-law or Peleg-type fits [9, 13, 16].

Given that the experimental data were best fitted to the second-order kinetic model than by the power-law model, the second-order model was adopted to estimate the kinetic parameters such as concentration of polyphenols at saturation (C_s), rate constant (k) and

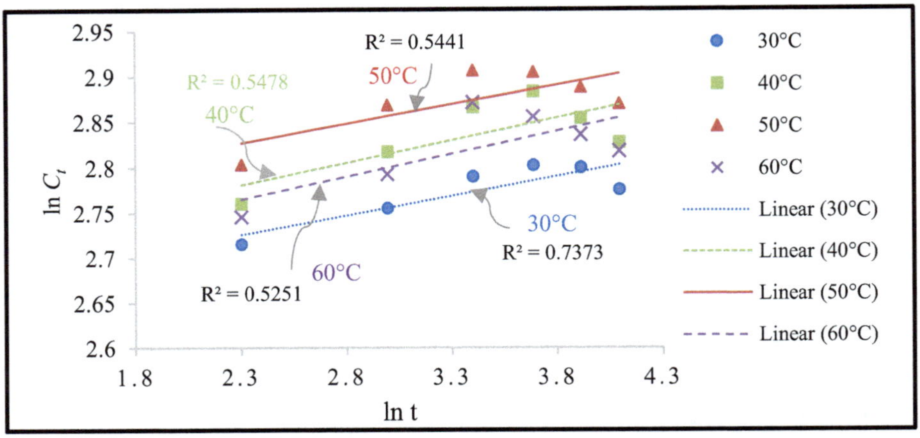

Fig. 5. Power-Law Kinetic Model fitting Across Varied Temperatures

initial extraction rate (h) at different processing parameters. All the kinetic parameters were calculated as shown in Table 1.

Table 1. Kinetic parameters of the second order kinetic model

Parameter	Range	Gradient	y-intercept	C_s (mg g^{-1})	h (mg min^{-1} g^{-1})	k (g mg^{-1} min^{-1})
Temperature (°C)	30	0.0612	0.0194	16.3399	51.5464	0.1931
	40	0.0578	0.0089	17.3010	112.3596	0.3754
	50	0.0558	0.0064	17.9211	156.2500	0.4865
	60	0.0586	0.0109	17.0648	91.7431	0.3150
Particle size (mm)	2.54	0.1279	0.0366	7.8186	27.3224	0.4470
	1.00	0.0994	0.0250	10.0604	40.0000	0.3952
	0.50	0.0596	0.0064	16.7785	156.2500	0.5550
Solid to liquid ratio	1:10	0.1093	0.01	9.1491	100.0000	1.1946
	1:20	0.0588	0.0014	17.0068	714.2857	2.4696
	1:30	0.0611	0.004	16.3666	250.0000	0.9333
	1:40	0.0619	0.0067	16.1551	149.2537	0.5719
	1:50	0.064	0.0085	15.625	117.6471	0.4819

4 Conclusion

In summary, the kinetic parameters confirm that UAE of *P. amaryllifolius* phenolics is favoured by moderate temperature (50 °C), fine particle size (0.5 mm) and an intermediate solid-to-liquid ratio (1:20), which together maximise both the extraction rate and the equilibrium phenolic concentration.

References

1. Wang, W., et al.: Botany, phytochemistry, pharmacology, and applications of Pandanus amaryllifolius Roxb.: A review. Fitoterapia. **177**, 106144 (2024). https://doi.org/10.1016/j.fitote.2024.106144
2. Harun Nurashid, H.N.H., Samsu Baharudin, A., Mohd Amin, N.A., Mad Nordin, M.F., Hafid, H.S.: Progress on the studies of Pandanus amaryllifolius, the cultivation in Malaysia and its application in various fields. Adv. Agric. Food Res. J. **6** (2025). https://doi.org/10.36877/aafrj.a0000590
3. Buddhakala, N., Yongkhamcha, B.: Phytochemical constituents, and antioxidant, antidiabetic and anti-inflammatory activities of the extracts from Pandanus amaryllifolius Roxb. J. Herb. Med. **54**, 101062 (2025). https://doi.org/10.1016/j.hermed.2025.101062
4. Ali, M.A., Mohamad Sabri, R., Khu, S.L., Nik Mahmood, N.A.: Impact of pandan leaf extract addition on the oxidative degradation of sunflower oil during microwave heating. JSM. **44**, 1159–1166 (2015). https://doi.org/10.17576/jsm-2015-4408-12
5. Balakrishnan, V.B., Husin, F., Zainol, N., Norhisham, F., Hasham, R., Yaakob, H.: Optimisation of microwave-assisted extraction of antioxidant activity from pandanus amaryllifolius leaf and evaluation of blending properties in refined bleached deodorised palm olein through response surface methodology. Adv. Agric. Food Res. J. **5** (2024). https://doi.org/10.36877/aafrj.a0000498
6. Cassol, L., Rodrigues, E., Zapata Noreña, C.P.: Extracting phenolic compounds from Hibiscus sabdariffa L. calyx using microwave assisted extraction. Ind. Crops Prod. **133**, 168–177 (2019). https://doi.org/10.1016/j.indcrop.2019.03.023
7. Sanou, A., et al.: Modelling and optimisation of ultrasound-assisted extraction of roselle phenolic compounds using the surface response method. Sci. Rep. **13**, 358 (2023). https://doi.org/10.1038/s41598-023-27434-5
8. Mehta, N., et al.: Ultrasound-assisted extraction and the encapsulation of bioactive components for food applications. Foods. **11**, 2973 (2022). https://doi.org/10.3390/foods11192973
9. Hobbi, P., et al.: Kinetic modelling of the solid–liquid extraction process of polyphenolic compounds from apple pomace: influence of solvent composition and temperature. Bioresour. Bioprocess. **8**, 114 (2021). https://doi.org/10.1186/s40643-021-00465-4
10. Lee, Y.-C., Chiang, Y.-C., Chen, M.-H., Chiang, P.-Y.: Micronization combined ultrasound-assisted extraction enhances the sustainability of polyphenols from pineapple and lemon peels utilizing acidified ethanol. Foods. **14**, 2872 (2025). https://doi.org/10.3390/foods14162872
11. Dalmau, E., Rosselló, C., Eim, V., Ratti, C., Simal, S.: Ultrasound-assisted aqueous extraction of biocompounds from orange byproduct: Experimental kinetics and modeling. Antioxidants. **9**, 352 (2020). https://doi.org/10.3390/antiox9040352
12. Jaroenkunpanit, P., Wungsintaweekul, J., Chetpattananondh, P.: Optimization, kinetics, and thermodynamics of ultrasound-assisted extraction of phytochemicals from hemp. ACS Omega. **10**, 45729–45747 (2025). https://doi.org/10.1021/acsomega.5c06069
13. Lazar, L., Talmaciu, A.I., Volf, I., Popa, V.I.: Kinetic modeling of the ultrasound-assisted extraction of polyphenols from Picea abies bark. Ultrason. Sonochem. **32**, 191–197 (2016). https://doi.org/10.1016/j.ultsonch.2016.03.009
14. Ghazali, A.M., Umar, N.B., Hussain, N.: Application of different aqueous ultrasound-assisted extraction process parameters on the quality of pandan (Pandanus amaryllifolius Roxb) leaves extract. J. Biochem. Microbiol. Biotechnol. **12**, 9–11 (2024). https://doi.org/10.54987/jobimb.v12iSP1.924

15. Oo, N., Jafari, S., Jansamutr, S., Assatarakul, K.: Sustainable production of bioactive pandan leaf extract microcapsules *via* optimized ultrasound-assisted extraction and spray drying. Sustain. Food Technol. **3**, 2144–2157 (2025). https://doi.org/10.1039/D5FB00341E
16. Reche, C., Rosselló, C., Umaña, M.M., Eim, V., Simal, S.: Mathematical modelling of ultrasound-assisted extraction kinetics of bioactive compounds from artichoke by-products. Foods. **10**, 931 (2021). https://doi.org/10.3390/foods10050931

Data-Driven Modeling of Mass Transfer for α-tocopherol Extraction using Hot Compressed Water

Amirah Nasuha Mohd Razib[1,2], Mohd Sharizan Md Sarip[1,2](✉), Nik Muhammad Azhar Nik Daud[1], Amirul Ridzuan Abu Bakar[1], Mohd Asraf Zainudin[1], Siti Kartini Enche Ab Rahim[1], Mohd Aminudin Jamlos[3], and Mohd Ahmad Hazim Abdul Aziz[4]

[1] Faculty of Chemical Engineering Technology, Universiti Malaysia Perlis (UniMAP), 02600 Arau, Perlis, Malaysia
sharizan@unimap.edu.my

[2] Centre of Excellence for Frontier Materials Research (FrontMate), Universiti Malaysia Perlis, Arau 02600, Perlis, Malaysia

[3] Faculty of Electronic Engineering Technology, Universiti Malaysia Perlis (UniMAP), 02600 Arau, Perlis, Malaysia

[4] Faculty of Food Science and Nutrition, Universiti Malaysia Sabah, Sabah Kota Kinabalu, Malaysia

Abstract. Understanding the kinetic behavior of extraction systems is essential for optimizing process design and evaluating extraction efficiency. This study investigated the mass transfer kinetics of α-tocopherol extraction using hot compressed water extraction (HCWE) process using data-drive kinetic model. Two kinetic models, first-order and second-order mass transfer models, were applied to experimental data to identify the most suitable model describing α-tocopherol extraction behavior. The analysis revealed that the second-order model provided superior fitting performance, with AARD values ranging from 2.59% to 12.73% and correlation coefficients (r^2) between 0.887 and 0.999 across all operating conditions. The equilibrium concentration (Ce) of α-tocopherol varied between 909.09 and 1666.67 g-solute/g-dried mesocarp. The extraction rate was highest within the first 10–20 min, achieving nearly 80–82% recovery, and reached equilibrium at 30 min. The effects of temperature (120–180 °C) and pressure (30–50 bar) were found to be statistically insignificant; however, the interaction between both parameters influenced the extraction trend. The findings confirm that α-tocopherol extraction in the HCWE process follows second-order kinetics and that prolonged extraction does not enhance recovery. The study provides valuable insight into optimizing HCWE conditions for thermally stable bioactive compounds such as α-tocopherol.

Keywords: Data driven · Mass transfer modeling · Hot compressed water

M. A. b. Jamlos et al. (Eds.): ICSET 2025, LNNS 1897, pp. 413–418, 2026.
https://doi.org/10.1007/978-3-032-20722-7_53

1 Introduction

In chemical processes, understanding the kinetic characteristics of a system provides essential information for process design, optimization, and economic evaluation [1]. In extraction systems, these kinetic characteristics are primarily governed by diffusion mechanisms, which often serve as the rate-limiting step that determines the overall efficiency of the extraction process. The diffusion mechanism describes how solute molecules migrate from the solid phase into the solvent, thereby influencing both the rate and extent of solute recovery.

In a typical extraction process involving plant materials, seeds, or rhizomes, mass transfer occurs according to the principles of mass transport phenomena, which explain the transfer of solute molecules from the solid matrix into the liquid solvent. This process continues until equilibrium is reached, where the solute concentration becomes uniform between the solid and liquid phases. The driving force for this movement is the concentration gradient, which can be mathematically represented by the diffusion coefficient—a parameter that indicates the solute's mobility within the medium and is influenced by temperature, solvent properties, and particle size.

Furthermore, mass transfer modeling can be performed using theoretical equations fitted with experimental data to provide a comprehensive understanding of the extraction behavior under various conditions. Such models are crucial for identifying key process parameters, such as rate constants and mass transfer coefficients, which directly affect extraction performance. These predictive models not only enhance the accuracy of process control and optimization but also support scaling-up operations and economic feasibility assessments in industrial applications.

2 Mass Transfer Modeling

In order to provide an appropriate representative model in this study, two mathematical expressions in mass transfer modeling were employed for α-tocopherol concentration. The mass transfer model are the best method to obtain the compound concentration versus time correlation and dissolution rate meanwhile the used of the diffusion model would introduced the unnecessary parameter [2]. In this study, models derived from the mass transfer model namely first order [3] and second order mass transfer model were compared to obtain the accurate and fitted model with the experimental data. In order to construct the model, few assumptions have been made: (1) the effective surface areas for solid particles are constant, (2) the model would represent α-tocopherol extraction although these component was extracted simultaneously together with other component in palm oil mesocarp during the process and (3) the constant agitation ratio.

The second order mass transfer model or known as second order rate law is being proposed in this study. The model was proven to provide an accurate representation in solid liquid extraction using water or aqueous solution which is similar with HCWE process [4]. The second order kinetic model is expressed as equation Eq. 1.

$$\frac{dc_t}{d_t} = k_2\left(c_{eq} - c_t\right)^2 \tag{1}$$

where k_2 is an absolute second order extraction rate coefficient (g-dried mesocarp/g-solute.min). This model can be integrated from $c_t = 0$ to c_t at times $t = 0$ to t respectively.

$$c_t = \frac{C_{eq}^2 k_2 t}{1 + c_{eq}^2 k_2 t} \tag{2}$$

$$\frac{t}{c_t} = \frac{1}{c_{eq}^2 k_2} + \frac{t}{c_{eq}} \tag{3}$$

The initial rate of extraction h (g-solute/g-dried mesocarp.min) can be expressed as Eq. 6 when the t and c_1 is approach 0.

$$\mathrm{h} = k_2 c_1 (ini)^2 \tag{4}$$

Eqs. 5 and 6 can be rearranged as Eq. 7.

$$\frac{\mathrm{t}}{\mathrm{c}_1} = \frac{1}{\mathrm{h}} + \frac{\mathrm{t}}{\mathrm{c}_{1(\mathrm{ini})}} \tag{5}$$

The h value, c_t, c_{eq} and k_2 is determined from the y intercept and slope of tct against t plot. The differential equation of the model was solved using Matlab R2012b (MathWorks, Inc., USA). The predictive capability for both models was evaluated based on the average absolute relative deviation, AARD (%) which is calculated using eq. 8.

$$AARD\ (\%) = \sum \frac{1}{\mathrm{n}} \left| \frac{\mathrm{y_e} - \mathrm{y_p}}{\mathrm{y_e}} \right| \times 100 \tag{6}$$

where n is the no of experimental data, ye is the experimental data, and y_p is the predicted data. The goodness of the model fit to the experimental data was evaluated using r_2. The lower value of AARD (%) indicated the better predictive capability of the model meanwhile the higher r^2 value near to unity indicated the lower deviation between predicted and experimental data.

3 Experimental Data

The experimental was conducted using hot compressed water extractor. The detail experimental procedure already published in previous report [5]. Concentration of α-tocopherol in the oil was analyzed using HPLC 1100 (Agilent, USA) with UV fluorescence detector (Agilent, USA) modified from Puah et al., 2007 [6]. A Zorbax RX-SIL normal phase silica column (5 μm, 250 mm × 4.6 mm i.d.) (Agilent, USA) with mobile phase of n-hexane (Friendemann Schmidt, Germany):THF(Friendemann Schmidt, Germany): 2-propanol (Friendemann Schmidt, Germany) (1000:60:4, by volume) and flow rate of 1.0 $\mathrm{mLmin^{-1}}$ were used. The compounds were detected at an emission wavelength of 326 nm and excitation wavelength of 292 nm.

4 Result and Discussion

Based on the mass transfer modeling analysis show that the process can be represent by second order mass transfer modeling and high deviation is observed for the first order mass transfer model. This phenomenon is also similar for the modeling of other minor component of crude palm oil namely β-carotene [5]. The analysis on second order mass transfer was tabulated in Table 1 with the kinetic models coefficient, AARD and r^2. As shown in Table 1, second order kinetic model exhibits the good fitting performance with the AARD value of 2.59 to 12.73% and r^2 value ranges from 0.887 to 0.999 for all operating conditions. The C_e for α-tocopherol is varied betweem 909.09 to 1666.67 g-solute/g-dried mesocarp.

Table 1. Second order mass transfer coefficient, AARD and r2 value for α-tocopherol concentration.

Temperature (°C)	Pressure (bar)	k_2 (g-dried mesocarp/g-solute.min) x 10^4	C_e (g-solute/g-dried mesocarp)	h (g-solute/g-dried mesocarp.min)	AARD (%)	r^2
120	30	0.61	1428.57	125.00	7.13	0.973
140	30	0.85	1428.57	172.41	4.96	0.986
160	30	0.53	909.09	434.78	2.59	0.999
180	30	1.17	1428.57	238.10	5.77	0.992
120	40	0.67	1666.67	185.19	6.61	0.986
140	40	0.33	1428.57	67.11	12.73	0.887
160	40	0.78	1250.00	121.95	8.64	0.968
180	40	1.69	1428.57	344.83	4.07	0.997
120	50	1.84	1111.11	227.27	3.46	0.996
140	50	0.51	1666.67	142.86	8.01	0.977
160	50	0.92	1666.67	256.41	6.43	0.993
180	50	1.69	1111.11	208.33	3.90	0.994

The highest AARD is 12.73% at the temperature of 140 °C and pressure of 40 bar. However, the error is still acceptable which reach below 15%. Based on this analysis, the α-tocopherol extraction using HCW process follows the second order mass transfer model. Meanwhile, the first order mass transfer model gave a high deviation and therefore is not suitable to be used to represent this process. In fact, the first order mass transfer model suggests that extraction rate is constant over the duration of extraction which is too simplistic an assumption and is not representative in most application of solid-liquid extraction process. The cumulative α-tocopherol extracted using first and second order kinetic models with experimental data at different temperature and pressure is depicted in Fig. 1. In alls the condition, the higher α-tocopherol extraction is obtained in the first 10 or 20 mins of extraction for most of the operating conditions. For example, at 160 °C and 30 bar, the high extraction rate was obtained in the first 10 mins of extraction where almost 80% is extracted. Meanwhile, at 120 °C and 30 bar, the high extraction rate in observed in the first 20 mins where almost 82% compounds is extracted. Futhermore, in all condition, the maximum recoverable for this compounds are achieved in the 30 mins of the extraction.

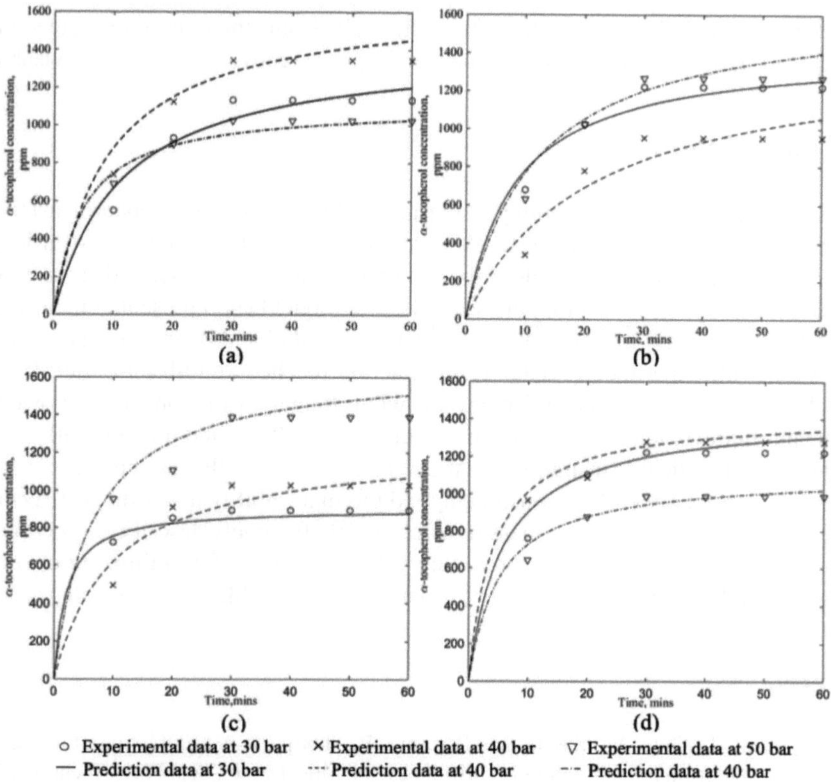

Fig. 1. Cummulative α-tocopherol in HCWE process per unit time at a) 120 °C b) 140 °C c)160 °C d)180 °C

It was also noted from Fig. 1, apart from temperature, pressure also affects the α-tocopherol concentration in the CPO. However, the effect of pressure corresponds to the extraction temperature, reflecting the unique characteristics of the HCWE process.

At 120 °C, the concentration of α-tocopherol increased from 1135.57 to 1346.20 ± 0.08 ppm as the pressure rose from 30 to 40 bar. After that, its concentration slightly decreased to 1025.06 ± 0.08 ppm when the pressure increased from 40 to 50 bar. At 140 °C, the concentration of α-tocopherol slightly decreased from 1222.89 to 1024.90 ± 0.04 ppm as the pressure increased from 30 to 40 bar, then increased again to 1267.05 ± 0.04 ppm at 50 bar. A similar trend was observed at 160 °C, where the maximum concentration of 1388.19 ± 0.09 ppm was achieved at 50 bar. At 180 °C, the concentration of this compound decreased from 1222.07 to 987.15 ± 0.01 ppm as the pressure increased from 30 to 50 bar. The effects of pressure and temperature were found to be statistically insignificant, as indicated by $F(2,16) = 2.96$, $p < 0.08$ and $F(3,16) = 2.05$, $p < 0.147$.

In conclusion, the findings of this work indicate that the optimal α-tocopherol yield under the respective experimental conditions was achieved after 30 min of extraction.

Therefore, although this component is more thermally stable than carotenoids, extending the extraction time beyond this point does not improve recovery and is not advisable.

5 Conclusion

This study examined the modeling of extraction kinetics for α-tocopherol using the hot compressed water extraction (HCWE) process. Two kinetic models—first-order and second order were evaluated to describe the extraction behavior derived from experimental data. Results showed that the second-order model best represented the process, with AARD values of 2.59–12.73% and r^2 between 0.887 and 0.999, indicating strong correlation with experimental data. The extraction reached equilibrium after 30 min, with about 80–82% α-tocopherol recovery, while longer extraction times showed no significant improvement. Although temperature (120–180 °C) and pressure (30–50 bar) effects were statistically insignificant, their interaction slightly influenced extraction performance. α-tocopherol demonstrated good thermal stability under the tested conditions. Overall, the second-order mass transfer model effectively described α-tocopherol extraction in the HCWE process, providing valuable insights for optimizing operating parameters and improving sustainable extraction of heat-stable bioactive compounds.

References

1. Anekpankul, T., Goto, M., Sasaki, M., Pavasant, P., Shotipruk, A.: Extraction of anti-cancer damnacanthal from roots of morinda citrifolia by subcritical water. Sep. Purif. Technol. **55**(3), 343–349 (2007)
2. Cussler, E.: Diffusion: Mass Transfer in Fluid Systems. Cambridge University Press, New York (1984)
3. Spiro, M., Jago, D.S.: Kinetics and equilibria of tea infusion. Part 3.-Rotating-disc experiments interpreted by a steady-state model. J. Chemical Soc., Faraday Trans. 1: Phys. Chem. Condensed Phases. **78**(1), 295–305 (1982)
4. Xi, J., Luo, S.: Pressure-enhanced solid liquid extraction of rutin from Chinese scholar-tree flower: Kinetic modeling of influential factors. Sep. Purif. Technol. **156**, 809–816 (2015)
5. Md Sarip, M.S., Morad, N.A., Mohd Nawi, M.A.H., Abdul Aziz, M.K., Syed Jaapar, S.Z.: β-carotene enrichment in crude palm oil using subcritical water extraction and its relationship with the solubility. Food Chem. Adv. **2**, 100245 (2023)
6. Puah, C.W., Choo, Y.M., Ma, A.N., Chuah, C.H.: The effect of physical refining on palm Vitamin E (Tocopherol, Tocotrienol and Tocomonoenol). American J. Appl. Sci. **4**(6), 374–377 (2007)

Techniques for Enhancing Antenna Performance: A Comprehensive Review

Shakhirul Bin Mat Salleh(✉), Liyana Binti Zahid, Norfatihah Binti Bahari, Ahmad Ashraf Bin Abdul Halim, and Mohd Hafizuddin Bin Mat

Faculty of Electronic Engineering and Technology, Universiti Malaysia Perlis, Pauh Putra Campus, 02600 Arau, Perlis, Malaysia
shakhirulsalleh@unimap.edu.my

Abstract. Microstrip patch antennas are widely used in modern wireless communication systems due to their compact size, low cost, and ease of fabrication. However, inherent limitations such as narrow bandwidth, low gain, and reduced efficiency necessitate the development of performance enhancement techniques. This review paper presents a comprehensive overview of recent advancements aimed at improving key antenna parameters, including bandwidth, gain, directivity, and impedance matching. Various structural modifications and design methods are discussed, including slotted and slit patches for current path manipulation, antenna arrays for directional gain improvement, parasitic element integration for impedance matching, and reflective layers for radiation pattern control. Additional approaches such as air gaps, defected ground structures (DGS), and metasurface reflectors are also examined for their ability to suppress surface waves and enhance overall efficiency. Each technique is evaluated based on its design principle, operational mechanism, and reported performance improvements from literature.

Keywords: Microstrip patch antenna · antenna performance enhancement · antenna techniques

1 Introduction

Microstrip antennas have gained widespread use owing to their small size, cost-effectiveness, flexibility, and compatibility with Radio Frequency (RF) devices. They present an array of feeding options, different polarization possibilities, the ability to operate across multiple bands, and the capability to steer the beam [1]. Although microstrip antennas offer certain advantages, such as their compact size and ease of integration, they are often limited in their application due to their narrow bandwidth and low gain. Additionally, the performance of microstrip antennas is negatively affected by three types of losses: dielectric loss, conductor loss, and surface wave loss [2].

The quality of substrate and conductor materials used in the manufacturing process determines the extent of dielectric and conductor losses. The use of high-quality materials can help reduce these losses, resulting in increased antenna gain. Additionally, suppressing surface waves can further enhance antenna performance by preventing

M. A. b. Jamlos et al. (Eds.): ICSET 2025, LNNS 1897, pp. 419–427, 2026.
https://doi.org/10.1007/978-3-032-20722-7_54

power from being trapped in the substrate layer and decreasing efficiency. The propagation of surface waves is influenced by the permittivity and thickness of the substrate, which are therefore crucial factors to consider [3]. Apart from that, this paper aims to review various techniques used by previous researchers to enhance antenna performance.

2 Ground Plane Slot Antennas

Printed slot antennas, specifically the printed wide-slot monopole antennas, are becoming increasingly popular for their suitability in wideband and Ultra-Wideband (UWB) applications. The antennas are composed of a monopole and a slot that are integrated into a metal frame functioning as a ground plane. They are attractive due to their ability to offer extensive impedance bandwidths and low near-field coupling, which aids in mitigating Electromagnetic Compatibility (EMC) and Electromagnetic Interference (EMI) problems with neighbouring objects and devices. Nevertheless, the requirement for antennas that are compact, low-profile, and have wideband performance usually leads to a compromise between antenna size and bandwidth. In order to tackle this challenge, various methods have been proposed in literature to reduce the size of slot antennas while preserving their wideband performance. These approaches involve the utilization of different slot geometries, including square, rectangular, triangular, circular, and elliptical, as well as various excitation techniques such as cross, fork, triangle, square, and arc monopoles.

One notable example is a microstrip-fed antenna with a rectangular tuning stub and a tapered slot, proposed in [4], as shown in Fig. 1. The structure has a dimension of approximately 22 mm × 24 mm × 1.6 mm (0.52 λo × 0.57 λo × 0.04 λo, with λo being the free-space wavelength at the central frequency), and it is fabricated on a FR4 Rogers substrate with a dielectric constant of 4.6 and a loss tangent of 0.02. By selecting appropriate geometries for the monopole and slot, the operational bandwidth of the antenna was optimized to range from 3 to 11.2 GHz (FBW = 115.5%), making it suitable for UWB applications. In Fig. 1, it can be observed that the rectangular monopole exhibited superior coupling as compared to the rounded monopoles. Furthermore, the tapered slot showed a wider impedance bandwidth compared to the circular, elliptical, and square slots due to longer current paths along the tapered slot edges. To improve the coupling between the radiating slot and microstrip feeding line, the gap between the monopole and metal frame was precisely adjusted. The antenna has a maximum realized gain ranging of 5.4 dBi at 9.8 GHz and its radiation patterns display low cross-polarization levels (−20 dB) at lower frequencies. However, as the frequency increases, the polarization purity decreases.

(a) Front view

(b) Back view

Fig. 1. Example of a printed wide-slot monopole antenna [4].

3 Slotted Patch

A popular technique for improving bandwidth involves cutting slots into a half-wavelength patch at the desired resonance frequency [2]. The inclusion of slots in microstrip antennas can enhance impedance matching, especially at higher frequencies. These slots alter the current distribution on the patch, resulting in modifications to both the input impedance and the length of the current path [5]. The addition of slots to a microstrip antenna generates new resonance frequencies. To achieve a wider bandwidth, it is essential to optimize the position and dimensions of the slots. This disturbance causes the first two broadside radiation modes of the radiating patch to converge, resulting in the convergence of the resonance frequencies [6].

The use of cutting slots on a radiating patch is a useful technique to reduce the size of a microstrip antenna while simultaneously creating multiple frequency bands within the desired range. By optimizing the slot configuration, the desired resonant mode can be achieved. Various slot shapes, such as U-slot, H-slot, T-slot, E-slot, and others, can be employed. For instance, an E-shaped slot etched into a rectangular patch has been found to produce a bandwidth of 30.3% [7]. A U-shaped (refer Fig. 2) slot has been shown to produce an impedance bandwidth of 26% [8].

The bandwidth of a microstrip antenna is dependent on its quality factor. Therefore, when slots are introduced into the patch, it leads to a reduction in the amount of energy stored beneath the patch, resulting in a degradation of the quality factor, and a consequent increase in radiation [1]. Ahsan et al. [9] designed a circular patch with mirrored horizontal slots, as depicted in Fig. 3. This configuration has the ability to achieve lower resonance frequencies due to the extended current distribution with the fundamental resonant mode. However, designing a slot patch antenna can be a challenging task, requiring a significant amount of time to determine the appropriate size and position of the slots to achieve the desired operating frequency.

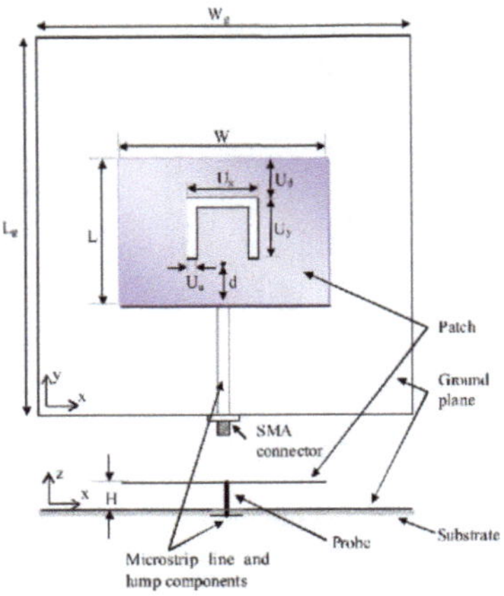

Fig. 2. U-shape slot antenna [8].

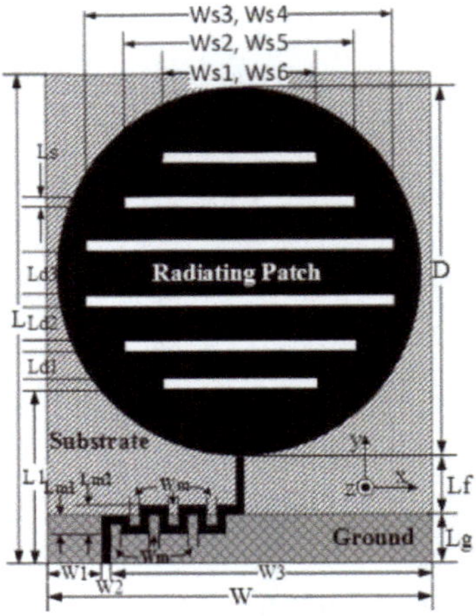

Fig. 3. Circular patch antenna with horizontal slots [9].

4 Slit Antenna

The slit antenna is designed by removing a portion of the original patch at its edges, unlike the slot antenna, which is created within the patch area (see Fig. 4). The previous researcher are successful improved the impedance matching, bandwidth and gain by implemented the slit antennas technique [10–14]. The implementation of slits on the patch radiator serves to enhance impedance matching, particularly at higher frequencies. The introduction of these slits modifies the current distribution on the radiator, which in turn alters the current path length and impedance at the input point. The addition of slits to the patch antenna also generates a new resonance frequency [15]. With the addition of additional slits, multiple resonance regions can be created, which, following optimization, result in an improved bandwidth.

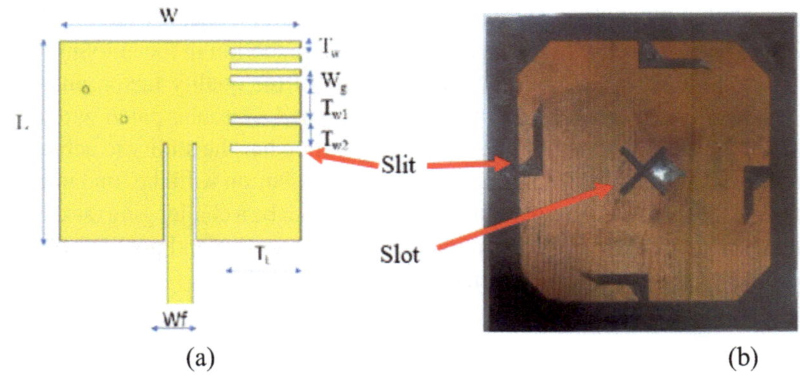

Fig. 4. Examples of (a) Slit antenna [13] and (b) Slits and slot antenna [12].

5 Array Antenna

The utilization of multiple elements in the form of an array is the most widely used method to attain high antenna gain. Individual microstrip antenna elements typically have a wide radiation pattern and low gain, leading to a preference for antennas with directive radiation in many applications. This is achieved by arranging the radiating elements in various geometrical structures. The complete emitted field of the array can be calculated by adding up the vectors, assuming that each element has an independent current and neglecting any coupling that may occur depending on the distance between the elements. In order to produce specific directional patterns, the fields must be combined constructively in the intended directions and destructively cancelled in other directions.

The performance of an antenna array is influenced by several important factors, including the arrangement of the array, the pattern of the radiating elements, the amplitude and phase of excitation, and the spacing between elements [16–21]. An example of a 16 microstrip array is illustrated in Fig. 5 [22]. The design of array antennas presents a formidable challenge in the form of the feeding network, which is essential for exciting the array. The complexity of the design and the substantial space requirements of the feeding network can make the design process particularly challenging.

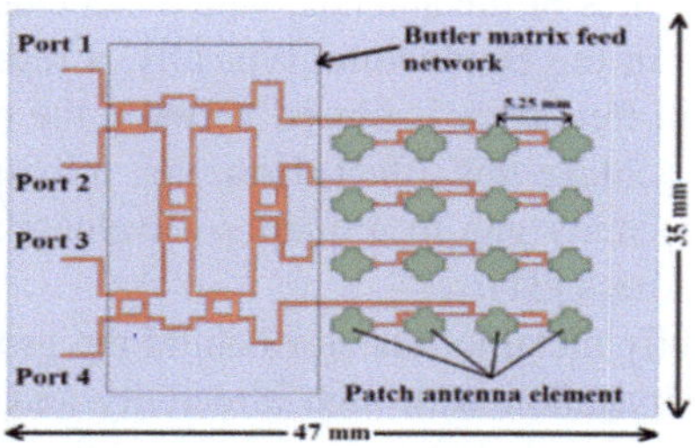

Fig. 5. 16 - elements antenna array [22].

6 Parasitic Antenna

The parasitic patch technique is employed to enhance the bandwidth and gain microstrip patch antennas. Two configurations of this technique are available, the coplanar technique [23–27] and the stacked technique [28–30] as shown in Fig. 6. The coplanar approach involves arranging several patches on one plane above a dielectric substrate, designating one as the main patch that receives stimulation. In contrast, the stacked method places one patch on top of another, with a dielectric layer separating them and enabling multiple patches to share a common aperture area. Parasitic stacks with a lower dielectric constant are frequently included above the radiating patch in a stacked antenna setup to decrease the overall effective permittivity of the multi-layered antenna, which enhances gain [2]. By utilizing multi-layered microstrip antennas, the impact of surface waves is reduced while the size (length and width) is maintained but antenna thickness will increase significantly.

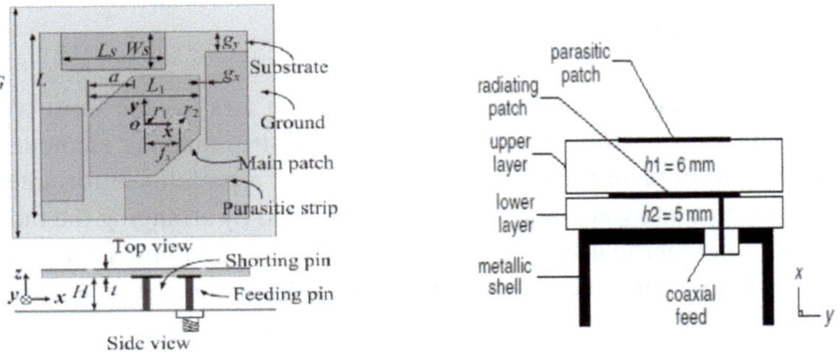

(a) (b)

Fig. 6. Examples of parasitic technique design (a) coplanar technique [27] and (b) stacked technique [28].

7 Reflecting Layer

Employing reflective surfaces, such as reflector planes or partial reflectors, is a popular technique to boost the directivity and gain of microstrip antennas, but it comes at the cost of reduced operating bandwidth. This is due to the reflector altering the radiation pattern and resonant properties of the antenna, which can cause a shift in the resonant frequency and a decrease in the bandwidth. Typically, a resonant cavity with a half-wavelength at the resonant frequency is placed between the reflective surface and the metallic ground plane to achieve this effect. However, utilizing an artificial magnetic conductor can reduce the cavity gap to a quarter-wavelength, but it is impossible to entirely eliminate the thickness [31].

The placement of a fully metallic reflector behind the ground plane is intended to achieve unidirectional propagation with high directivity and a superior front-to-back ratio [32]. Furthermore, for a wide-band slot antenna, numerous metal back reflectors have been created. The diagram of the microstrip antenna with a reflector is illustrated in Fig. 7, where the reflector is composed of a substrate covered with a conductive layer, like a perfect electric conductor (PEC).

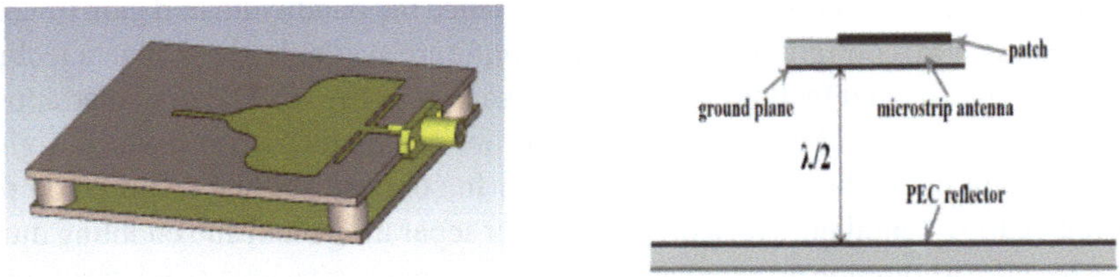

(a) (b)

Fig. 7. Examples of reflector technique (a) reflector metal placed at the back [32] and (b) the PEC reflector configuration [2].

To boost performance, a meta-surface reflector (MSR) structure is utilized to reflect undesirable back radiation by artificially engineering the metallic reflector surface. This design relies on a mesh grid array configuration, which was first introduced by Enoch et al. [33]. The meta-surface is a two-dimensional metamaterial that provides a structure with less loss and a smaller physical footprint, enabling artificial control of microwaves. The fundamental structure of the MSR comprises a periodic planar metallic surface that interacts with the incoming electromagnetic beam.

The metallic elements serve as an inductive grid, while the spaces between them serve as a capacitive grid. The copper elements of the MSR produce currents in phase with those of the microstrip antenna, leading to more directed emission [34, 35]. As a result of this effect, the MSR can decrease back radiation and enhance directivity. The interaction between the microstrip antenna and the MSR structure must be carefully assessed to account for mutual coupling, and the separation distance must be examined to optimize impedance matching. The MSR can be placed in between patch and ground or at the top of the patch (see Fig. 8) [36].

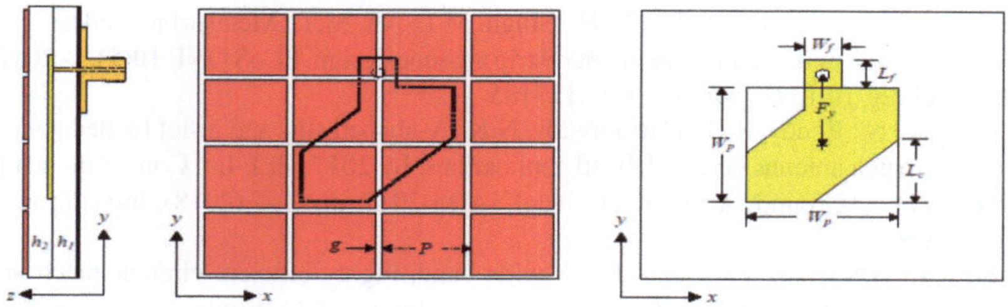

(a) (b) (c)

Fig. 8. Configuration of MSR (a) side view, (b) top view, and (c) driven patch [36].

8 Conclusion

This review has summarized various techniques used to enhance antenna performance, including slotted designs, parasitic structures, array configurations, and the use of reflectors or defected ground structures. The paper contributes by providing a clear comparison of these methods and highlighting their effectiveness in improving bandwidth, gain, and radiation efficiency. For future work, combining two or more enhancement techniques such as integrating parasitic elements with metasurfaces or slotted patches with defected ground planes can further boost antenna performance while maintaining compactness. This hybrid approach, supported by AI-based optimization and advanced materials, offers great potential for developing high-performance antennas suited for next-generation wireless communication systems.

References

1. Balanis, C.A.: Antenna Theory: Analysis and Design (2010)
2. Mahendran, K., Dhivya, K., Prasanniya, V.: Microstrip patch antenna enhancement techniques: a survey. Int. J. Eng. Appl. Sci. Technol. **04**(11), 245–249 (2020). https://doi.org/10.33564/ijeast.2020.v04i11.042

3. Aktar, M.N., Shahin Uddin, M., Amin, M.R., Ali, M.M.: Enhanced gain and bandwidth of patch antenna using EBG substrates. Int. J. Wirel. Mob. Networks **3**(1), 62–69 (2011). https://doi.org/10.5121/ijwmn.2011.3106
4. Azim, R., Tariqul, M.: Printed wide slot ultra-wideband antenna. Adv. Microstrip Antennas Recent Appl. (2013). https://doi.org/10.5772/51961
5. Chen, D., Cheng, C.H.: A novel compact ultra-wideband (UWB) wide slot antenna with via holes. Prog. Electromagn. Res. **94**, 343–349 (2009). https://doi.org/10.2528/PIER09062306
6. Han, L., Guo, L., Ma, R., Zhang, W.: Frequency reconfigurable U-slot microstrip patch antenna. WIT Trans. Inf. Commun. Technol. **49**, 423–430 (2014). https://doi.org/10.2495/ICIE130491
7. Yang, F., Zhang, X.X., Ye, X., Rahmat-Samii, Y.: Wide-band E-shaped patch antennas for wireless communications. IEEE Trans. Antennas Propag. **49**(7), 1094–1100 (2001). https://doi.org/10.1109/8.933489
8. Wi, S.H., Lee, Y.S., Yook, J.G.: Wideband microstrip patch antenna with U-shaped parasitic elements. IEEE Trans. Antennas Propag. **55**(4), 1196–1199 (2007). https://doi.org/10.1109/TAP.2007.893427
9. Ahsan, M.R., Islam, M.T., Ullah, M.H., Singh, M.J., Ali, M.T.: Metasurface reflector (MSR) loading for high performance small microstrip antenna design. PLoS ONE **10**(5), 1–20 (2015). https://doi.org/10.1371/journal.pone.0127185
10. Rajasekhar, N., Reddy, R.R., Darimireddy, N.K.: V-shaped slits and a slot loaded pentagonal boundary patch antennas for wideband applications. In: 2017 IEEE Int. Conf. Antenna Innov. Mod. Technol. Ground, Aircr. Satell. Appl. iAIM 2017, pp. 1–5 (2018). https://doi.org/10.1109/IAIM.2017.8402600
11. Garg, R.K., Dwivedi, S., Tomar, R.: A wide band, highly-efficient circular patch antenna with split triangular-shape slit. In: Proc. - 2018 8th Int. Conf. Commun. Syst. Netw. Technol. CSNT 2018, pp. 1–5 (2018). https://doi.org/10.1109/CSNT.2018.8820273
12. Rao, K.S., Jahagirdar, D.R., Ramakrishna, D.: Compact broadband asymmetric slit Circularly Polarized Microstrip Patch Antenna for GPS and GLONASS applications. In: 2017 IEEE Int. Conf. Antenna Innov. Mod. Technol. Ground, Aircr. Satell. Appl. iAIM 2017, pp. 1–3 (2018). https://doi.org/10.1109/IAIM.2017.8402558
13. Arulmurugan, S., Sureshkumar, T.R., Alex, Z.C.: Compact wearable microstrip patch antenna for 2.4 GHz using loaded slits and shorting pins. In: 2021 IEEE Int. Conf. Emerg. Trends Ind. 4.0, ETI 4.0 2021, pp. 1–5 (2021). https://doi.org/10.1109/ETI4.051663.2021.9619435
14. Elakaş, A., Imrak, G.A., Şencan, M., Imeci, S.T., Durak, T.: Microstrip patch antenna with triangular slits. In: 2018 Int. Appl. Comput. Electromagn. Soc. Symp. Denver, ACES-Denver 2018, pp. 12–13 (2018). https://doi.org/10.23919/ROPACES.2018.8364311
15. Kumar, A., Gupta, N., P.C.: Gain and bandwidth enhancement techniques in microstrip patch antennas - a review. Int. J. Comput. Appl. **148**(7), 9–14 (2016). https://doi.org/10.5120/ijca2016911207
16. Station, W.B., Wang, Z., Zhang, G., Yin, Y., Wu, J.: Design of a dual-band high-gain antenna array for **13**, 1721–1724 (2015)
17. Jihao, L.: Antenna on board package for 5G millimeter wave phased array antenna. In: 2020 21st Int. Conf. Electron. Packag. Technol. ICEPT 2020 (2020). https://doi.org/10.1109/ICEPT50128.2020.9202652
18. Huang, G.C., Iskander, M.F., Hoque, M., Goodall, S.R., Bocskor, T., Chen, S.: Antenna array design and system for directional networking. In: 2014 IEEE Conf. Antenna Meas. Appl. CAMA 2014 **14**, pp. 1141–1144 (2014). https://doi.org/10.1109/CAMA.2014.7003446
19. Elsheakh, D.N., Iskander, M.F., Abdallah, E.A., Elsadek, H.A., Elhenawy, H.: Microstrip array antenna with new 2D-electromagnetic band gap structure shapes to reduce harmonics and mutual coupling. Prog. Electromagn. Res. C **12**, 203–213 (2010). https://doi.org/10.2528/PIERC09112008

20. Li, W.Y., Chung, W., Chou, J.H.: Highly-integrated wideband 28 GHz and 39 GHz array antennas for 5G mobile phone applications. In: 2020 IEEE Int. Symp. Antennas Propag. North Am. Radio Sci. Meet. IEEECONF 2020 - Proc., pp. 1581–1582 (2020). https://doi.org/10.1109/IEEECONF35879.2020.9330360
21. Lian, R., Wang, Z., Yin, Y., Wu, J., Song, X.: Design of a low-profile dual-polarized stepped slot antenna array for base station. IEEE Antennas Wirel. Propag. Lett. **15**, 362–365 (2016). https://doi.org/10.1109/LAWP.2015.2446193
22. Klionovski, K., Shamim, A., Sharawi, M.S.: 5G antenna array with wide-angle beam steering and dual linear polarizations. In: 2017 IEEE Antennas Propag. Soc. Int. Symp. Proc., pp. 1469–1470 (2017). https://doi.org/10.1109/APUSNCURSINRSM.2017.8072777
23. Choi, E.C., Lee, J.W., Lee, T.K., Lee, W.K.: Circularly polarized S-band satellite antenna with parasitic elements and its arrays. IEEE Antennas Wirel. Propag. Lett. **13**, 1689–1692 (2014). https://doi.org/10.1109/LAWP.2014.2347998
24. Kim, J.W., Ham, H.J., Woo, J.M., Lee, D.K.: A dual-band microstrip antenna using a T-shaped parasitic element. In: 2014 20th Int. Conf. Microwaves, Radar Wirel. Commun. MIKON 2014, vol. 2010, pp. 7–9 (2014). https://doi.org/10.1109/MIKON.2014.6899899
25. Alias, H., Ali, M.T., Subahir, S., Ya'Acob, N., Sulaiman, M.A.: Aperture coupled microstrip antenna array integrated with DGS and parasitic elements. In: IEEE Symp. Wirel. Technol. Appl. ISWTA, pp. 259–263 (2013). https://doi.org/10.1109/ISWTA.2013.6688783
26. Zhang, H.Y., Zhang, F.S., Zhang, F., Li, T., Li, C.: Bandwidth enhancement of a horizontally polarized omnidirectional antenna by adding parasitic strips. IEEE Antennas Wirel. Propag. Lett. **16**, 880–883 (2017). https://doi.org/10.1109/LAWP.2016.2613875
27. Hu, W., Tang, Z.Y., Fei, P., Yin, Y.Z.: Broadband circularly polarized Z-shaped dipole antenna with parasitic strips. Int. J. RF Microw. Comput. Eng. **27**(1), 559–562 (2017). https://doi.org/10.1002/mmce.21052
28. Jiang, H., Xue, Z., Li, W., Ren, W.: Broad beamwidth stacked patch antenna with wide circularly polarised bandwidth. Electron. Lett. **51**(1), 10–12 (2015). https://doi.org/10.1049/el.2014.3783
29. Guo, L., Tang, M.C.: A low-profile dual-polarized patch antenna with bandwidth enhanced by stacked parasitic elements. In: 2018 Int. Conf. Microw. Millim. Wave Technol. ICMMT 2018 - Proc., pp. 1–3 (2018). https://doi.org/10.1109/ICMMT.2018.8563348
30. Dong, J., Wang, A., Wang, P., Hou, Y.: A novel stacked wideband microstrip patch antenna with U-Shaped parasitic elements. In: ISAPE 2008 - 8th Int. Symp. Antennas, Propag. EM Theory Proc., pp. 185–188 (2008). https://doi.org/10.1109/ISAPE.2008.4735172
31. Mekki, A.S., Hamidon, M.N., Ismail, A., Alhawari, A.R.H.: Gain enhancement of a microstrip patch antenna using a reflecting layer. Int. J. Antennas Propag. **2015** (2015). https://doi.org/10.1155/2015/975263
32. Ge, Y., Esselle, K.P., Bird, T.S.: The use of simple thin partially reflective surfaces with positive reflection phase gradients to design wideband, low-profile EBG resonator antennas. IEEE Trans. Antennas Propag. **60**(2 PART 2), 743–750 (2012). https://doi.org/10.1109/TAP.2011.2173113
33. Enoch, S., Tayeb, G., Sabouroux, P., Guérin, N., Vincent, P.: A metamaterial for directive emission. Phys. Rev. Lett. **89**(21) (2002). https://doi.org/10.1103/PhysRevLett.89.213902
34. Metamaterials, W.U., Amir, S., Saghanezhad, H., Atlasbaf, Z.: Miniaturized dual-band CPW-fed antennas loaded. IEEE Antennas Wirel. Propag. Lett. **14**, 658–661 (2015)
35. Fernandez, O., Gomez, A., Basterrechea, J., Vegas, A.: Asymmetric Chiral Metamaterial superstrate for patch antenna polarization transformation. In: 2016 10th Eur. Conf. Antennas Propagation, EuCAP 2016, pp. 5–8 (2016). https://doi.org/10.1109/EuCAP.2016.7481902
36. Ta, S.X., Park, I.: A circularly polarized patch antenna loaded with metamaterial. In: 2015 9th Int. Congr. Adv. Electromagn. Mater. Microwaves Opt. METAMATERIALS 2015, pp. 466–468 (2015). https://doi.org/10.1109/MetaMaterials.2015.7342489

A Computational Modelling and Simulation Approach for Slope Volume Estimation Using UAV-Photogrammetry-Derived Raster and TIN Data

Shahrul Izwan Sukri[1(✉)], Ahmad Razali Yusoff[2], Mohd Farid Mohd Ariff[3], Nursyafiqah Jessica Georgeson[2], Radzuan Sa'ari[4], and Mohd Radhie Mohd Salleh[4]

[1] Geospatial Imaging and Information Research Group (Gi2RG), Faculty of Built Environment and Surveying, Universiti Teknologi Malaysia, 81310 Skudai, Johor, Malaysia
s.izwan1999@graduate.utm.my

[2] Department of Geotechnical and Transportation, Faculty of Civil Engineering, Universiti Teknologi Malaysia, 81310 Skudai, Johor, Malaysia
a.razali@utm.my

[3] UTM-PDRM Geospatial Forensics Satellite Laboratory, Faculty of Built Environment and Surveying, Universiti Teknologi Malaysia, 81310 Skudai, Johor, Malaysia
mfaridma@utm.my

[4] Department of Water and Environmental Engineering, Faculty of Civil Engineering, Universiti Teknologi Malaysia, 81310 Skudai, Johor, Malaysia
{radzuans,mohdradhie}@utm.my

Abstract. Accurate slope volume estimation is essential in modern engineering systems for earthwork analysis, slope management, and geotechnical applications. This study presents a computational modeling approach for slope volume estimation using UAV-derived photogrammetric data, where raster- and triangulated irregular network (TIN)-based representations are treated as alternative surface models. The study was conducted at Ibnu Sina, Universiti Teknologi Malaysia, using UAV imagery acquired with a DJI Phantom 4 and processed via Structure-from-Motion to generate digital elevation models. Volume computation was performed in ArcGIS using the Surface Volume tool for both raster and TIN inputs. The raster-based model yielded a volume of 51,399.90 m^3, while the TIN-based model produced 51,767.60 m^3, resulting in an absolute difference of 367.70 m^3 (0.7%). This small variation indicates that both surface representations provide reliable slope volume estimates. Raster models offer computational simplicity, whereas TIN models better preserve surface detail in complex terrain. The results confirm that surface representation influences computed slope volumes and should be selected based on analysis scale and terrain complexity.

Keywords: UAV-Photogrammetry · Volume Estimation · Slope Monitoring and Management

M. A. b. Jamlos et al. (Eds.): ICSET 2025, LNNS 1897, pp. 428–433, 2026.
https://doi.org/10.1007/978-3-032-20722-7_55

1 Introduction

Accurate characterization of slope topography is fundamental to geotechnical, environmental, and engineering applications such as landslide hazard assessment, slope stability design, earthwork estimation, and infrastructure development. Inadequate slope characterization has been linked to construction delays, cost overruns, and safety risks in large infrastructure projects, according to Tsachouridis et al. [1]. Consequently, reliable slope geometry and volume estimation are critical requirements for geotechnical assessment, hydrological modelling, and construction management. Conventional slope monitoring techniques, including total station surveys, terrestrial laser scanning (TLS), and airborne LiDAR, have long been used to generate topographic models, presented by Regmi et al. [2]. However, these methods are often constrained by high operational costs, limited accessibility, low revisit frequency, and safety concerns, particularly in hazardous environments such as unstable slopes and open-pit mine highwalls, explained by Kinde et al. [3].

In recent years, unmanned aerial vehicle (UAV)–based photogrammetry has emerged as a cost-effective and flexible alternative for high-resolution terrain mapping. By processing overlapping aerial imagery using Structure-from-Motion (SfM) algorithms, UAV photogrammetry can generate dense point clouds and digital elevation models (DEMs) capable of representing slope geometry with high accuracy, according to Lee et al. [4]. UAV-derived surface models are increasingly applied to slope-related volume estimation tasks, including cut-and-fill analysis and earthwork reconciliation, with studies demonstrating that volumetric results obtained using DEM-of-Difference (DoD) or surface-to-surface comparison workflows are comparable to those derived from conventional surveying techniques.

Despite the rapid adoption of UAV photogrammetry, the accuracy of volume estimation remains strongly influenced by the choice of computational surface representation. Raster-based grid models and triangulated irregular network (TIN) models are the most used surface representations in practice; however, they differ fundamentally in surface discretization, boundary handling, and preservation of slope discontinuities. Comparative evaluations between raster- and TIN-based volume computation methods remain limited, particularly for steep and bench-dominated engineered slopes where accurate representation of breaklines is critical, explained by Kregar et al. [5]. Within the context of modern engineering systems, slope volume estimation can therefore be framed as a computational surface modeling problem. This study addresses the identified gap by comparatively evaluating raster- and TIN-based volume estimation methods using UAV-photogrammetry-derived surface models over an engineered embankment slope, supporting informed selection of surface modeling strategies for reliable slope volume estimation, according to Guan et al. [6].

2 Methodology

This study estimates slope volume using UAV-derived photogrammetric data, formulated as a computational surface modeling problem. The overall workflow, illustrated in Fig. 1, comprises data acquisition, photogrammetric processing, and volumetric computation using alternative surface representations.

The study area is an engineered slope at Ibnu Sina, Universiti Teknologi Malaysia, covering approximately 26,600 m^2, a typical engineered embankment slope commonly encountered in infrastructure corridors. UAV data were acquired using a DJI Phantom 4 flown at 60 m altitude with 85% forward and 80% side overlap, capturing 180 overlapping images. Five ground control points (GCPs), surveyed using a Topcon HiPer HR GNSS receiver, were used to improve georeferencing accuracy.

UAV imagery was processed in Agisoft Metashape Professional using a standard Structure-from-Motion workflow to generate a digital terrain model (DTM). The DTM was exported in GeoTIFF format and analyzed in ArcGIS Desktop. Volume estimation was performed using the Surface Volume tool with two computational surface representations: raster-based and triangulated irregular network (TIN)-based inputs. Both methods used the same reference plane defined by the minimum terrain elevation (64.731 m), enabling a consistent comparison of volumetric results.

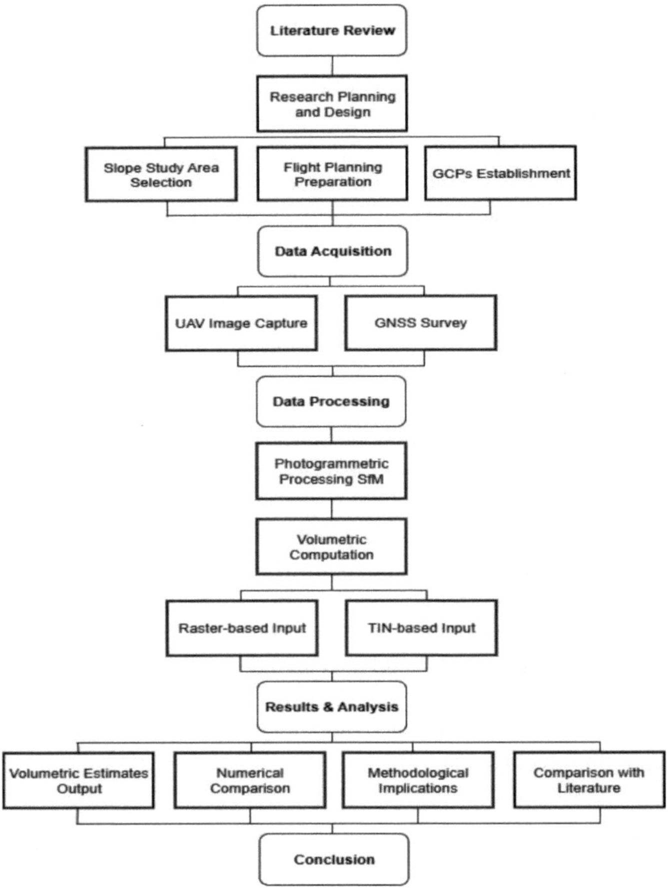

Fig. 1. Flowchart overall of the methodology co

3 Results

To simulate the behavior of different surface modeling strategies, raster and TIN models were generated from identical UAV datasets and evaluated under consistent processing conditions. This study evaluated and compared both volumetric estimations from raster-based input and TIN-based input. Methodological insights are provided to assess the overall output. Comparison to literature is also made to give validations of the produced outcome.

3.1 Volumetric Outputs

UAV photogrammetry processing in Agisoft Metashape produced a digital terrain model (DTM) of the study slope. Two surface representations were evaluated in ArcGIS Desktop for volume estimation: (i) raster input using the DTM exported as GeoTIFF, and (ii) TIN input derived from the raster using 3D Analyst tools. Both surfaces were analyzed using the Surface Volume tool relative to a constant reference plane defined by the minimum terrain elevation (64.731 m).

The volumetric results are summarized in Table 1. The TIN-based estimation yielded a slightly higher volume (51,767.60 m^3) compared to the raster-based result (51,399.90 m^3), a trend commonly observed when triangulated surfaces preserve local slope breaks and geometric variability, mentioned by Alsayed et al. [7].

Table 1. Slope volume estimation results.

Input Type	Volume (m^3)	2D Area (m^2)	3D Area (m^2)
Raster	51,399.90	6372.06	6881.33
TIN	51,767.60	6372.06	6723.31

3.2 Numerical Comparison

From the estimated volume of both raster and TIN approach, an absolute difference, ΔV and relative difference, RD can be concurred. With volumes of raster and TIN denoted as V_{ras} and V_{tin} respectively, the absolute difference can be achieved through the subtraction of both volumes. The absolute difference yields 367.70 m^3.

Relative difference can be concurred with having the ratio between absolute difference and the estimated volume of raster (acts as a baseline). The relative difference yields at 0.7%. Given the total volume of over 51,000 m^3, this discrepancy is negligible, indicating strong consistency between the two approaches. This negligence is addressed in Subsection 3.4.

The results yield is presented in the following Table 2.

Table 2. Numerical comparison.

Input	Value
Absolute difference, ΔV	367.70 m^3
Relative difference, RD	0.7%

3.3 Methodological Implications and Literature Comparison

The observed variation reflects differences in surface representation rather than processing parameters, as both methods use an identical reference plane. Raster-based models represent terrain as a uniform grid, which may smooth localized variations and under-represent sharp slope features, presented by Bailey et al. [8]. In contrast, TIN-based models preserve breaklines and slope discontinuities through irregular triangulation, often resulting in slightly higher volume estimates, according to Alsayed et al. [7].

Previous studies report similar discrepancies between raster- and TIN-based volume estimation methods. Pepe et al. [9] observed differences of approximately 2% for bench-dominated terrain and less than 1% for linear slopes, while Sankey et al. [10] reported average volume estimation errors of 1.7–2.3% across multiple surface representations. The 0.7% variation observed in this study falls well within these documented ranges, supporting the reliability of both raster and TIN surface models for slope volume estimation when consistent data and processing parameters are applied.

4 Conclusions

In conclusion, this study presented a computational modeling and simulation framework for slope volume estimation using UAV-derived photogrammetric data, where raster- and triangulated irregular network (TIN)-based representations were treated as alternative surface models. The comparative analysis demonstrated that the choice of computational surface model has a measurable influence on slope volume estimation results, reflecting differences in surface discretization and geometric representation. The simulation findings indicate that raster-based models exhibit sensitivity to grid resolution, while TIN-based models provide greater geometric adaptability for complex terrain surfaces

From an engineering systems perspective, the results emphasize the importance of selecting appropriate surface modeling strategies for reliable slope volume estimation in earthwork planning and geotechnical applications. The proposed framework supports informed decision-making in UAV-based terrain analysis and can be extended to integrate intelligent monitoring and modeling systems for slope assessment and management. It is hopeful for in future study, with the advancement in AI, criteria of input for volumetric computation can benefit from this study.

Acknowledgments. This research was funded by a grant from Ministry of Higher Education of Malaysia (FRGS Grant R.J130000.7822.5F752) and Universiti Teknologi Malaysia (UTMER Q.J130000.3822.42 J03).

Disclosure of Interests The authors have no competing interests to declare that are relevant to the content of this article.

References

1. Tsachouridis, S., Pavloudakis, F., Sachpazis, C., Tsioukas, V.: Monitoring slope stability: A comprehensive review of UAV applications in open-pit mining. Land. **14**(6), 1193 (2025). https://doi.org/10.3390/land14061193
2. Regmi, S., Dahal, R.: Consequences of slope instability and existing practices of mitigation in hydropower projects of Nepal. Geoenviron. Disasters. **11**, 26 (2024). https://doi.org/10.1186/s40677-024-00289-2
3. Kinde, M., Getahun, E., Jothimani, M.: Geotechnical and slope stability analysis in the landslide-prone area: A case study in Sawla–Laska Road Sector, Southern Ethiopia. Sci. Afr. **23**, e02071 (2024). https://doi.org/10.1016/j.sciaf.2024.e02071
4. Lee, K., Lee, W.H.: Earthwork volume calculation, 3D model generation, and comparative evaluation using vertical and high-oblique images acquired by unmanned aerial vehicles. Aerospace. **9**(10), 606 (2022). https://doi.org/10.3390/aerospace9100606
5. Kregar, K., Trajkovski, K.K.: Combining UAV photogrammetry and TLS for change detection on slovenian coastal cliffs. Drones. **9**(4), 228 (2025). https://doi.org/10.3390/drones9040228
6. Guan, S., Zhu, Z., Wang, G.: A review on UAV-based remote sensing technologies for construction and civil applications. Drones. **6**(5), 117 (2022). https://doi.org/10.3390/drones6050117
7. Alsayed, A., Nabawy, M.R.A.: Stockpile volume estimation in open and confined environments: A review. Drones. **7**(8), 537 (2023). https://doi.org/10.3390/drones7080537
8. Bailey, G., Li, Y., McKinney, N., Yoder, D., Wright, W., Washington-Allen, R.: Las2DoD: Change detection based on digital elevation models derived from dense point clouds with spatially varied uncertainty. Remote Sens. **14**(7), 1537 (2022). https://doi.org/10.3390/rs14071537
9. Pepe, M., Domenica, C., Nicola, Z.: 4D geomatics monitoring of a quarry for the calculation of extracted volumes by TIN and grid model: Contribute of UAV photogrammetry. Geographia Technica. **16**(Special Issue), 1–14 (2021). https://doi.org/10.21163/GT_2021.163.01
10. Sankey, J. B., Caster, J., Bransky, N., Fuest, S., Sesnie, S., and Bedford, A. 2024. Lidar estimation of storage capacity for managed water resources used by desert bighorn sheep (*Ovis canadensis Mexicana*) at Cabeza Prieta national wildlife refuge, Arizona. No. 2024–1046. US Geol. Surv. doi:https://doi.org/10.3133/ofr20241046

Modelling and Simulation of Hybrid PSO–ACO Based PID Controller for Precision Robotic Arm Position Control

Wan Faizura Wan Tarmizi[1](✉), Fahmi Samsuri[1], and Muhammad Luqman Muhd Zain[2]

[1] Fakulti Teknologi Kejuruteraan Elektrik dan Elektronik, Universiti Malaysia Pahang, Al - Sultan Abdullah, 26600 Pekan Al-Sultan Abdullah (UMPSA), Lebuh Persiaran Tun Khalil Yaakob, 26300 Kuantan, Pahang, Malaysia
wanfaizura@hotmail.com, fahmi@umpsa.edu.my

[2] Fakulti Teknologi Kejuruteraan Elektrik & Automasi, University College TATI, Teluk Kalong, 24000 Kemaman, Terengganu, Malaysia

Abstract. This paper presents the modelling and simulation of a hybrid Particle Swarm Optimization–Ant Colony Optimization (PSO–ACO) algorithm-based PID controller for precision robotic arm position control. The research begins with the mathematical modelling of a DC servomotor, followed by the implementation of a conventional PID controller as the baseline. To address the limitations of traditional tuning methods, PSO and ACO are each applied separately to optimize the PID gains and then integrated into a hybrid framework to combine their respective advantages. Simulation results demonstrate that while individual PSO and ACO tuning improve the system's dynamic response by reducing rise time, settling time, overshoot, and steady-state error, the proposed hybrid PSO–ACO method outperforms both, delivering faster, more stable, and more accurate position control. The findings confirm that the hybrid strategy offers a robust and efficient solution for high-precision robotic arm applications, providing clear benefits for advanced industrial automation and robotics research.

Keywords: Robotic Arm · Optimization · PID · PSO · ACO

1 Introduction

Robotic arms are increasingly valuable in areas such as manufacturing, health, and service sectors, where they perform repeated actions demanding high precision. Robotic arms are essential in manufacturing, healthcare, and automation, where precision and stability are crucial. PID controllers remain the industry standard due to their simplicity but tuning them for nonlinear multi-DOF robotic arms remains challenging. To overcome these challenges, metaheuristic algorithms such as PSO and ACO have been explored in prior work, usually independently. This research introduces a novel cooperative hybrid PSO–ACO approach for PID tuning, leveraging PSO's exploration to generate candidate solutions and ACO's exploitation to refine them. This hybridization is shown to improve

M. A. b. Jamlos et al. (Eds.): ICSET 2025, LNNS 1897, pp. 434–452, 2026.
https://doi.org/10.1007/978-3-032-20722-7_56

trajectory tracking accuracy, reduce overshoot, and accelerate convergence, marking the first such application to the Braccio robotic arm. The functioning is determined by the performance of the control system, and one of the most commonly used control systems in industry is the PID (Proportional-Integral-Derivative) system because of its simple design and trustworthy performance [1]. But for the best results, especially in systems with complex nonlinear dynamics, it's important to tune the PID parameters. To address this problem, various optimization techniques have been explored by prior studies that are hoping. Robotic arm control methods are diverse, including PID control, model-based control, and hybrid schemes. The PID controller, the basic feedback controller, has been investigated extensively with the development of accurate and stable systems. Combining proportional, integral, and derivative control actions, the PID controller reduces the difference between the actual position and the set position on the fly. The proportional term is a function of the current error, the integral term eliminates any long-term error, and the derivative term provides a rate of future errors to increase stability. This control mode has shown value in industrial automation, robot control, and automotive applications.

The hybrid control algorithms based on fuzzy logic, genetic algorithms, PSO, and ACO were suggested in previous research [2–5]. Jialin et al., [6] designed a fuzzy PID controller for a multi-joint manipulator, dynamically adjusting parameters through fuzzy logic to achieve optimal performance for both manufacturing and medical purposes. Hardware-in-the-loop (HiL) control via MATLAB Simulink was accomplished to give practicality to the proposed process for real-life experimentation. Othman and Hussein [7] also developed a PID-fuzzy control system for a flexible robotic arm, which resulted in a 9% error ratio and 8% overshoot, thus enabling more accuracy in the robot motion. Ghani et al., [8] further investigated fuzzy logic controllers for manipulators, finding that they reduced steady-state error by 44% compared to conventional PID controllers. The research emphasized the adaptability of fuzzy logic in handling nonlinearities and disturbances. Meanwhile, Razali et al., [9] proposed a PID controller using fuzzy theory, improving mobile robot path tracking. This approach showcased the benefits of adaptive tuning, enabling real-time adjustments for varying operating conditions, thus reducing manual tuning efforts. Traditional PID controllers often struggle with nonlinearities, uncertainties, and changing environmental conditions, limiting their effectiveness in advanced robotic applications.

Traditional PID controllers often struggle with nonlinearities, uncertainties, and changing environmental conditions, limiting their effectiveness in advanced robotic applications. Addressing this challenge, Vuong et al., [10] proposed a PID controller for DC servomotors using the Ziegler-Nichols method, a traditional tuning approach. Studies show that nearly 90% of industries rely on PID controllers, but conventional tuning methods often struggle with complex dynamics. Heuristic techniques like Genetic Algorithms (GA) and Particle Swarm Optimization (PSO) have emerged as better alternatives. Research shows that GA and PSO improve key performance metrics, including rise time, settling time, overshoot, and steady-state error. Simulations confirm their superiority over Ziegler-Nichols' tuning. These methods enhance control loop stability and provide more precise parameter tuning. Experimental results validate their effectiveness, proving that GA and PSO outperform traditional methods. Future research will focus on multi-objective optimization, further improving PID controller performance.

To address this challenge, Saifuddin et al., [11] implemented a PID controller for a 5°-of-freedom (DOF) robotic arm, utilizing MATLAB simulations to assess control effectiveness. Their findings highlighted significant error reduction and improved task precision. Abdul Karim and Thamrin [12] extended this work by integrating PID control into a 6-DOF robotic arm, effectively reducing angular errors from 47° to 0° and enhancing positioning accuracy. Innovative PID enhancements continue to emerge, as demonstrated by Zahira et al., [13], who introduced Euler's Gamma function-based PID controller for a two-DOF manipulator. This approach addressed the limitations of traditional PID control, yielding higher precision and smoother motion control. Ji et al., [14] proposed an optimization-based iterative learning algorithm for 6-DOF robotic arm path planning, integrating fast scaling-based random tree and bidirectional scaling balanced connected bi-tree methods for efficient trajectory generation. The algorithm improves convergence speed and enhances trajectory programming performance. Results show it quickly finds optimal solutions while reducing computation time. The study highlights the significance of kinematic analysis and trajectory optimization in robotic applications.

The previous study highlights that while PID controllers have been a basis in robotic arm control due to their simplicity and reliability, they face significant limitations when dealing with nonlinearities, time-variant dynamics, and environmental uncertainties common in real-world applications. Traditional PID tuning methods, such as the Ziegler-Nichols approach, often fail to provide optimal performance in complex systems, leading to issues such as overshoot, prolonged settling times, and steady-state errors. In response, various hybrid control strategies [15–18] have been proposed that integrate fuzzy logic, genetic algorithms, and heuristic optimization methods like PSO and ACO to improve adaptability and control precision. Despite these advancements, gaps remain in systematically comparing the performance of nature-inspired optimization techniques for PID tuning in robotic arm position control, particularly in dynamically complex and multi-degree-of-freedom systems.

To address this, metaheuristic optimization techniques have been applied, but most studies focus on single optimizers (e.g., PSO or ACO) independently. In this work, we introduce a novel hybrid PSO–ACO PID tuning strategy, which combines PSO's global search capability with ACO's local exploitation strength, resulting in faster convergence and improved controller robustness. This is the first application of hybrid PSO–ACO to PID tuning for the Braccio 5-DOF robotic arm, making this study a unique contribution to precision control research. The objective of this study is to address these gaps by optimizing the PID controller parameters for robotic arm position control using two prominent bio-inspired algorithms: Particle Swarm Optimization (PSO) and Ant Colony Optimization (ACO). This research aims to evaluate and compare the effectiveness of these methods in enhancing system response metrics such as rise time, overshoot, settling time, and steady-state error. Through simulation-based analysis, the study seeks to demonstrate how intelligent tuning algorithms can outperform traditional methods in managing the complexities of robotic arm dynamics. Ultimately, the goal is to provide a robust, adaptable, and precise PID control strategy that can be implemented in real-world robotic applications across various industries.

2 Kinematics of the Braccio Robotic Arm

A Braccio robotic arm that consists of five main joints: base rotation, shoulder, elbow, wrist flexion, and wrist rotation, along with the end effector (gripper), as shown in Fig. 1. Each joint of the robotic arm is connected by a link of a specific length, as indicated in the Table 1: base to shoulder (0.0715 m), shoulder to elbow (0.125 m), elbow to wrist flex (0.125 m), wrist flex to wrist rotation (0.06 m), and wrist rotation to gripper (0.132 m) as shown in Table 1.

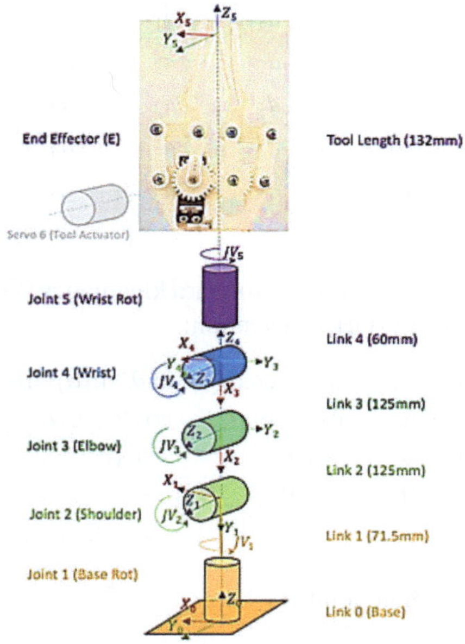

Fig. 1. A Braccio Robotic Arm [19]

Table 1. Braccio Robotic Arm Parameter.

Joint (i)	Description	Approximate Length (α_i) (m)
1	Base to Shoulder	0.0715
2	Shoulder to Elbow	0.125
3	Elbow to Wrist Flex	0.125
4	Wrist Flex to Wrist Rotation	0.06
5	Wrist Rotation to Gripper	0.132

2.1 DH Transform for Robotic Arm

The kinematic model is derived using Denavit-Hartenberg (DH) parameters, with joint link lengths and twists defined in Table 2. The forward kinematics is obtained by multiplying homogeneous transformation matrices.

Table 2. DH Table for Robotic Arm.

Joint (i)	θ_i(degree)	d_i(mm)	a_i(mm)	α_i(degree)
1	θ_1	d_1	a_1	α_1
2	θ_2	d_2	a_2	α_2
3	θ_3	d_3	a_3	α_3
4	θ_4	d_4	a_4	α_4
5	θ_5	d_5	a_5	α_5

For a 5-DOF Braccio robotic arm, the forward kinematics (Eq. 1) can be represented using the Denavit–Hartenberg (DH) convention:

$$T_i^{i-1} = \begin{bmatrix} \cos\theta_i & -\sin\theta_i\cos\alpha_i & \sin\theta_i\sin\alpha_i & a_i\cos\theta_i \\ \sin\theta_i & \cos\theta_i\cos\alpha_i & -\cos\theta_i\sin\alpha_i & a_i\sin\theta_i \\ 0 & \sin\alpha_i & \cos\alpha_i & d_i \\ 0 & 0 & 0 & 1 \end{bmatrix} \tag{1}$$

Where,

- $\theta_i =$ Joint angle (for revolute joint)
- $d_i =$ Link offset
- $a_i =$ Link length
- $\alpha_i =$ Link twist

Hence, the forward kinematics transformation matrix (Eq. 2) is produced:

$$T_i^{i-1} = T_1^0 \bullet T_2^1 \bullet T_3^2 \bullet T_4^3 \bullet T_5^4 \tag{2}$$

The joint angle $\boldsymbol{\theta_i(t)}$ or its Laplace transform $\boldsymbol{\theta_i}\,(s)$ is produced by the DC servomotor that actuates joint $\boldsymbol{i}$.

3 Methodology

3.1 DC Servomotor Model

The DC servomotor is modelled using its electrical and mechanical equations to accurately represent its dynamic behavior for position control. In this study, the DC servomotor is used to control the robotic arm's joints and produce smooth and controlled motions. Fig. 2 shows the electrical model of a DC servomotor. The motor accurately controls the angular position, speed, and torque for applications requiring accurate and responsive movement. The parameters of the DC servomotor are shown in Table 3.

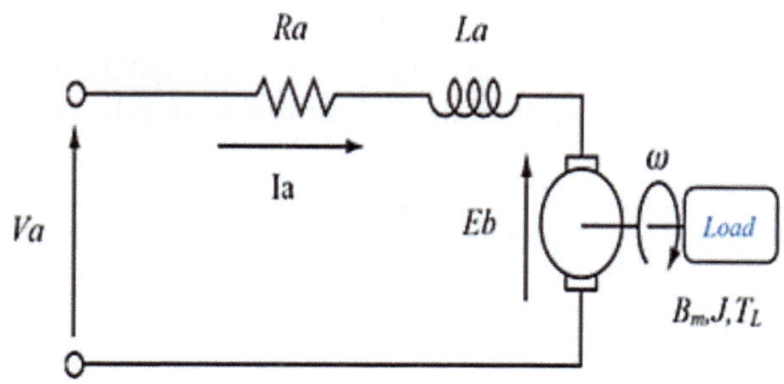

Fig. 2. Electrical Model of DC Servomotor [20]

Table 3. DC Servomotor Parameters.

Parameter	Value
Moment of inertia	$J_m = 0.0148\ kgm^2$
Friction coefficient	$b = 0.01\ N.ms$
Back EMF constant	$k_e = 0.048 \frac{V}{ms^{-1}}$
Torque constant	$k_t = 0.048 \frac{Nm}{A}$
Electric resistance	$R = 1\ ohm$
Electric inductance	$L = 0.5\ H$

3.1.1 Electrical Equations

The equation of the DC servomotor is referred to [20]. Equation (3) relates the applied voltage (V) to the armature current (i), considering the resistance (Ri), inductance (L), and back electromotive force (e). It helps study electrical dynamics and engine control.

$$V(t) = L\frac{di(t)}{dt} + Ri(t) + e(t) \tag{3}$$

Equation (4) relates electromotive force (e) to the product of motor constant (Ke), angular velocity (ω). The equation is fundamental in the control and design of motors.

$$e(t) = k_e\omega(t) \tag{4}$$

3.1.2 Transfer Function of DC Servomotor

In Eq. (5), the speed of the DC motor model can be described by its transfer function. For a DC motor, the typical transfer function is:

$$\text{Gs} = \frac{K}{(Js + b)(Ls + R) + K^2} \tag{5}$$

In SI units, the motor torque and back EMF constants are equal; that is,$kt = ke$. Therefore, we will use it to represent both the motor torque and the back EMF constant. Equation (6) is the transfer function of the DC servomotor for the robot arm.

$$G(s) = \frac{0.048}{0.0074s^2 + 0.0198s + 0.2404} \tag{6}$$

3.2 Research Methodology

Figure 3 shows the research methodology flowchart. The study starts by creating a mathematical model of the DC servomotor to describe its behavior accurately. Then, a PID controller is used to control the motor's position. To improve its performance, the PID gains are tuned using Particle Swarm Optimization (PSO) and Ant Colony Optimization (ACO). After testing each method separately, they are combined into a hybrid PSO–ACO approach to get better results. Finally, the standard, individual, and hybrid controllers are tested and compared in simulations to check improvements in rise time, settling time, overshoot, and steady-state error.

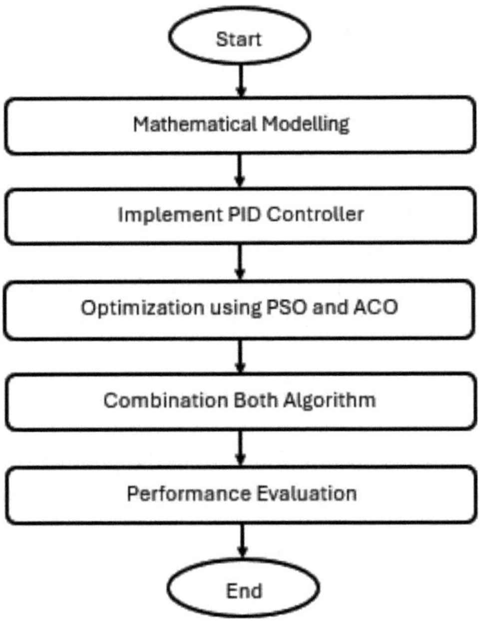

Fig. 3. Research Methodology

3.3 Control Methods

In this study, there are two control methods used for the optimization of the PID, which are Particle Swarm Optimization (PSO) and Ant Colony Optimization (ACO) algorithms.

This method is crucial for maintaining stability and precision in robotic movements. The effectiveness of position control depends on proper PID tuning, which balances response speed, overshoot, and steady-state error. The position control of a DC servomotor in a robotic arm ensures that each joint reaches a desired angle accurately.

3.3.1 PID Controller

A PID controller stands for proportional-integral-derivative controller, is widely used to control DC servomotors because it combines three actions: the proportional part reduces the difference between the target and actual positions, the integral part removes steady-state error by accounting for accumulated past errors, and the derivative part anticipates future errors by observing how fast the error is changing, which helps limit overshoot and improve system stability. The control law is given by:

$$u(t) = K_P e(t) + K_i \int e(t)dt + K_d \frac{de(t)}{dt} \tag{7}$$

Where,

- $u(t)$ is a control signal (voltage to the motor)
- $e(t)$ is the error signal
- K_P, K_i and K_d are the gains of PID

Figure 4 shows the block diagram of a robotic arm. The process begins with the desired position as the input. Subsequently, the operation proceeds through the controller and DC motor. Finally, the output is generated, and the data from the desired and actual positions are compared. Table 4 shows different values of PID parameters using the Ziegler-Nichols method for controlling the speed of a DC motor. Where *Ku* is the ultimate gain and *Tu* is the ultimate period.

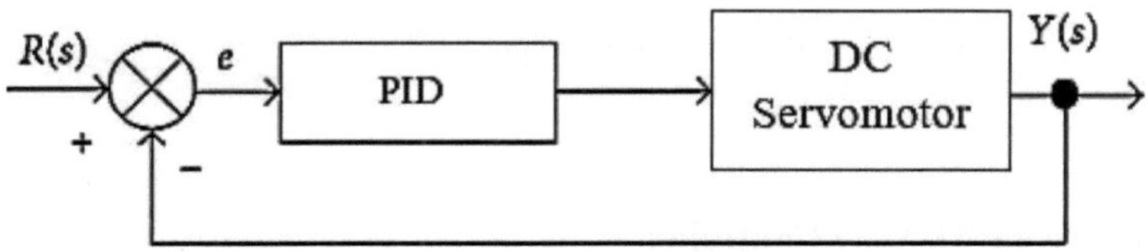

Fig. 4. PID Block Diagram

Table 4. Ziegler-Nichols (ZN) Parameter.

Type of Controller	***Kp***	***Ki***	***Kd***
P controller	0.5*Ku*	–	–
PD Controller	0.8*Ku*	–	*Tu*/8
PI Controller	0.45*Ku*	1.2*Kp*/*Tu*	–
PID Controller	0.6*Ku*	2*Kp*/*Tu*	*KpTu*/8

3.3.2 Adaptive PID-PSO Controller

Figure 5 shows the PID-PSO block diagram of a robotic arm. The process begins with the desired position as the input. Subsequently, the operation proceeds through the controller and DC motor. Finally, the output is generated, and the data from the desired and actual positions are compared. The parameter of the PSO algorithm used in the study is shown in Table 5.

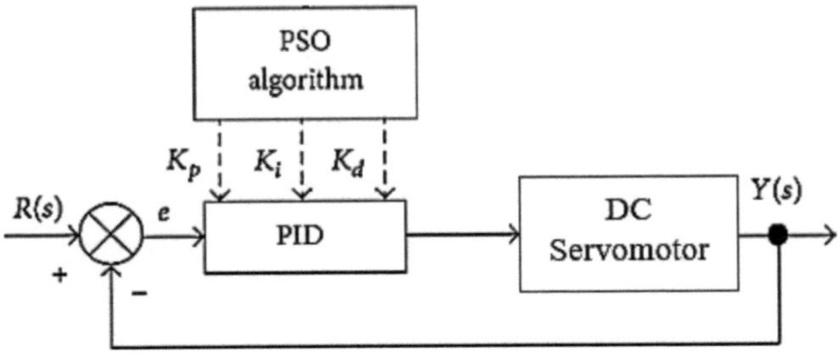

Fig. 5. PID-PSO Block Diagram

Table 5. PSO Algorithm Parameter

Variable	Value	Description
nParticles	30	Swarm Size (Number of particles)
maxIter	100	Maximum number of iterations
c1	2	Cognitive Coefficient
c2	2	Social Coefficient
w	0.9	Inertia Weight

3.3.3 Adaptive PID-ACO Controller

Figure 6 shows the PID-ACO block diagram of a robotic arm. The operation PID-ACO is similar to PID-PSO but has different parameter settings. The parameter of the PSO algorithm used is shown in Table 6.

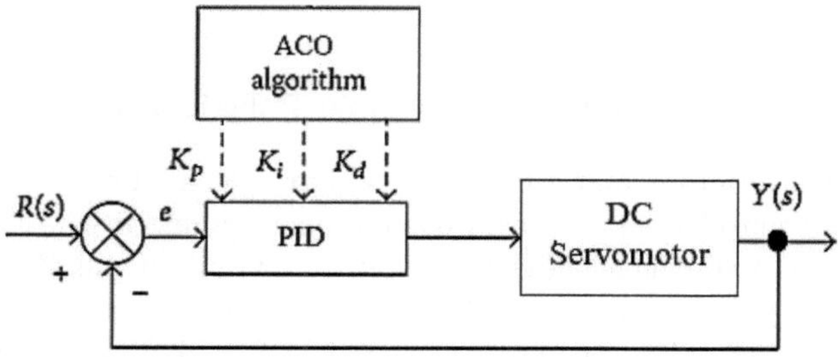

Fig. 6. PID-ACO Block Diagram

Table 6. ACO Algorithm Parameter

Variable	Value	Description
NumAnt	10	Number of Ants
NumIter	50	Number of Iterations
α	1	Alfa
β	2	Beta
ρ	0.5	Pheromone evaporation rate
Q	1	Pheromone deposit factor

3.3.4 Hybrid PID-PSO-ACO Controller

Figure 7 shows the PID combining PSO and ACO block diagram of a robotic arm. The operation PID-ACO is similar to PID-PSO but has different parameter settings.

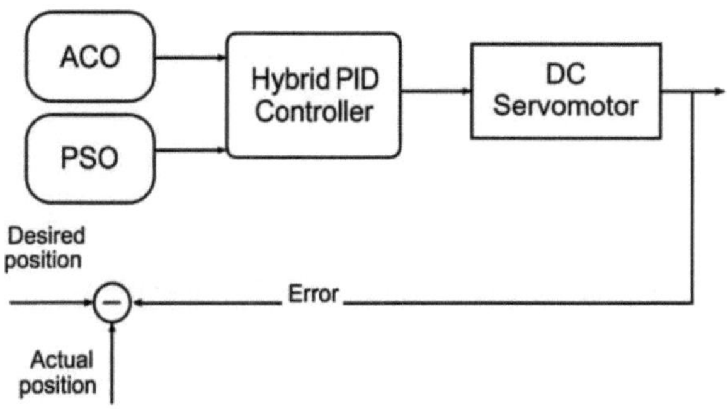

Fig. 7. Hybrid PID-PSO-ACO Block Diagram

4 Results

4.1 Experimentation on Position Control of the Robotic Arm

In this study, the position control of a robotic arm is tested using four approaches: a standard PID controller, an adaptive PID controller optimized with Particle Swarm Optimization (PID-PSO), an adaptive PID controller optimized with Ant Colony Optimization (PID-ACO), and a hybrid PID controller that combines both PSO and ACO techniques (PID-PSO-ACO). The performance of each method is evaluated and compared to determine how effectively they enhance the accuracy and control of the motor's position.

4.1.1 Results of the PID (references) using Ziegler-Nichols

Figure 8 shows the step response of a system using five different PID controller settings. From PID 1 to PID 5, the PID gains (*Kp*, *Ki*, *Kd*) increase, improving the system's performance with faster rise and settling times. No overshoot or steady-state error is seen. PID 5, which has the highest gains ($Kp = 210$, $Ki = 140$, $Kd = 78.75$), achieves the fastest response with a rise time of $0.005s$ and a settling time of $0.009s$, as shown in Table 7. For comparison, PID gains tuned by the Ziegler-Nichols method often cause more overshoot and longer settling times, showing its limits for complex systems. In this study, all PID sets maintain zero overshoot and steady-state error, proving that higher gains from advanced tuning methods help reach the target faster without losing stability or accuracy.

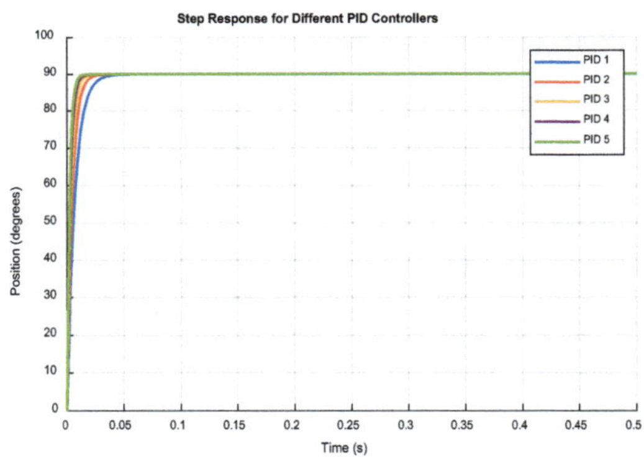

Fig. 8. Results of the PID using Ziegler-Nichols

Table 7. The results of the PID gains using Ziegler-Nichols

PID SET (Sample 1)	*Kp*	*Ki*	*Kd*	Rise Time (*s*)	Settling Time (*s*)	Overshoot (%)	Steady-State Error
PID 1	90	60	33.75	0.002	0.015	0.027	0
PID 2	120	80	45	0.014	0.010	0.018	0
PID 3	150	100	56.25	0.006	0.008	0.013	0
PID 4	180	120	67.5	0.002	0.006	0.011	0
PID 5	210	140	78.75	0.013	0.005	0.009	0

4.1.2 Results of the PID-PSO

Figures 9, 10 and 11 show the system response of a robotic arm using five different PID-PSO configurations (PID 1 to PID 5) from sample 1 to sample 3, each based on optimized parameters. The graph compares how quickly and accurately the arm reaches the target angle by showing rise time, settling time, overshoot, and steady-state error. The PID–PSO results show that the PSO algorithm effectively tunes the PID gains for the 5-DOF robotic arm to achieve very fast rise and settling times, with most configurations maintaining zero steady-state error and minimal or no overshoot. In all three samples, the rise times reach as low as 0.002*s*, and the settling times are mostly within 0.003*s*, demonstrating quick convergence to the desired position, is shown in Table 8. While a few PID sets produce moderate overshoot (around 10–16%), the majority maintain smooth responses, indicating that PSO consistently finds gain combinations that balance speed and stability. Overall, the results confirm that PSO-based tuning reliably improves the system's trajectory tracking performance compared to traditional tuning, providing high precision and robust dynamic behavior.

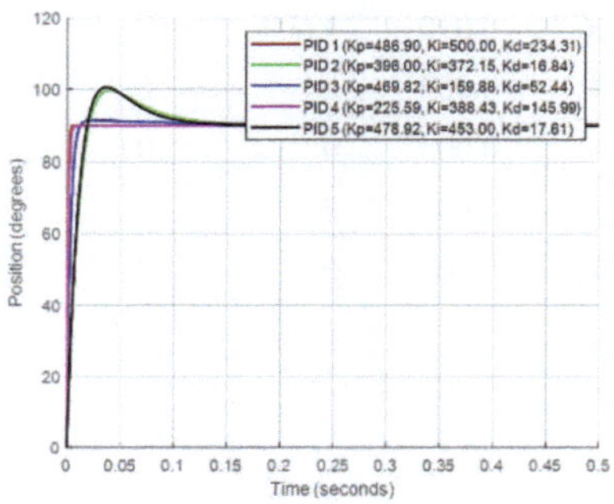

Fig. 9. Results of the PID-PSO (Sample 1)

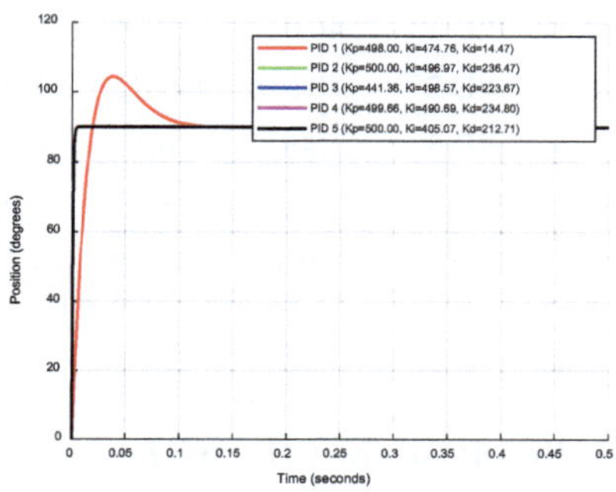

Fig. 10. Results of the PID-PSO (Sample 2)

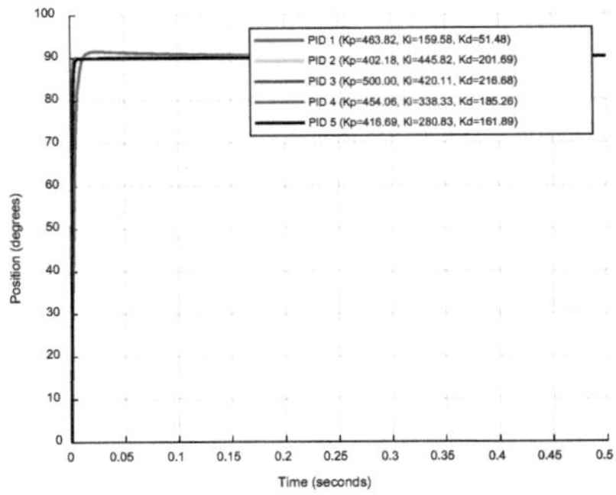

Fig. 11. Results of the PID-PSO (Sample 3)

Table 8. The results of the PID-PSO

Samples	PID SET	*Kp*	*Ki*	*Kd*	Rise Time (s)	Settling Time (s)	Overshoot (%)	Steady-State Error
Sample 1	PID 1	486.9	500	234.31	0.002	0.003	0	0
	PID 2	396	372.15	16.84	0.014	0.110	10.747	0
	PID 3	469.82	159.88	52.44	0.006	0.010	1.571	0
	PID 4	225.59	388.43	145.99	0.002	0.004	0	0
	PID 5	478.92	453	17.61	0.013	0.099	11.704	0
Sample 2	PID 1	498	474.76	14.47	0.014	0.092	15.768	0
	PID 2	500	496.97	236.47	0.002	0.003	0	0
	PID 3	441.36	498.57	223.67	0.002	0.003	0	0
	PID 4	499.66	490.69	234.8	0.002	0.003	0	0
	PID 5	500	405.07	212.71	0.002	0.003	0	0
Sample 3	PID 1	463.82	159.58	51.48	0.006	0.010	1.609	0
	PID 2	402.18	445.82	201.69	0.002	0.003	0	0
	PID 3	500	420.11	216.68	0.002	0.003	0	0
	PID 4	454.06	338.33	185.26	0.002	0.003	0	0
	PID 5	416.69	280.83	161.89	0.002	0.004	0	0

4.1.3 Results of the PID-ACO

Figures 12, 13 and 14 show the system responses of PID controllers tuned using Ant Colony Optimization (ACO), with the corresponding tuning parameters and performance metrics presented. The PID–ACO results demonstrate that the ACO algorithm successfully tunes the PID gains to achieve consistently fast and stable position control. Across all three samples, the rise times remain very short (mostly $0.003 - 0.006s$) with settling times also quick ($0.006 - 0.012s$), while all configurations maintain zero overshoot and zero steady-state error, as shown in Table 9. The PID gains optimized by ACO vary moderately but always produce smooth, well-damped responses, showing that ACO can find gain combinations that avoid oscillations without sacrificing response speed. This consistent absence of overshoot across all sets highlights the ACO's ability to generate more conservative but reliable tuning, ensuring precise tracking and robust stability for the robotic arm's position control tasks.

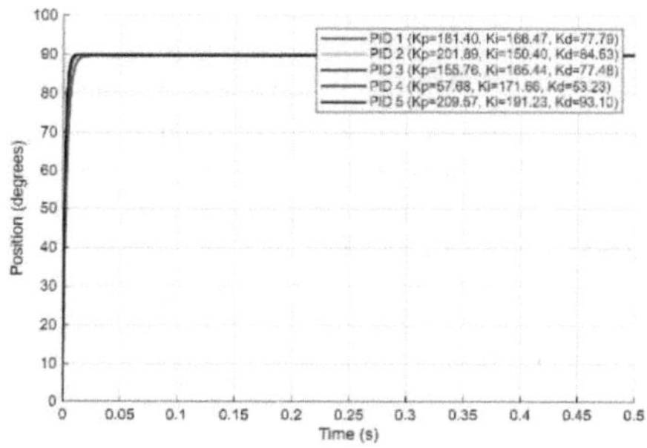

Fig. 12. Results of the PID-ACO (Sample 1)

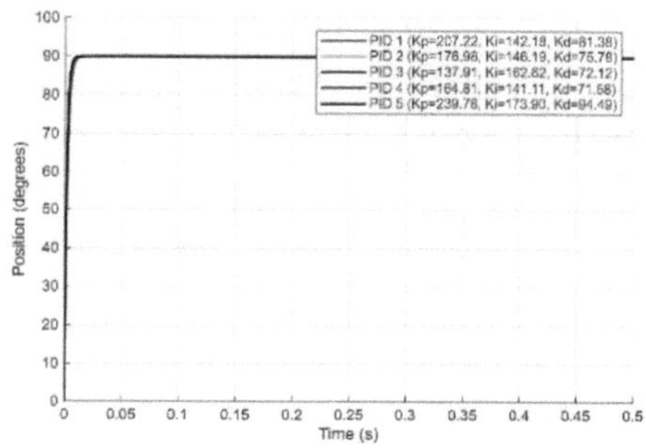

Fig. 13. Results of the PID-ACO (Sample 2)

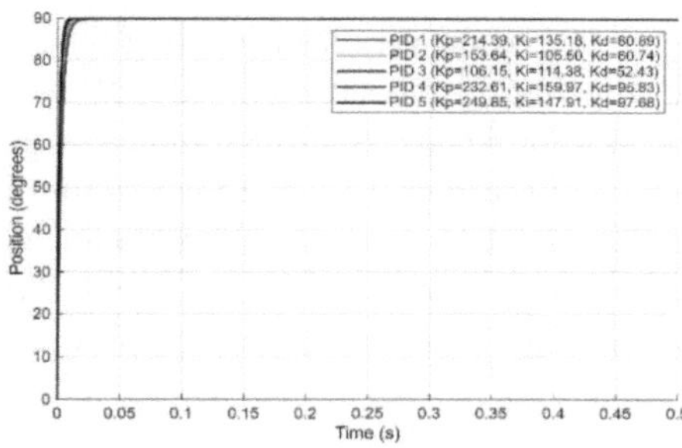

Fig. 14. Results of the PID-ACO (Sample 3)

Table 9. The results of the PID-ACO

Samples	PID SET	*Kp*	*Ki*	*Kd*	Rise Time (*s*)	Settling Time (*s*)	Overshoot (%)	Steady-State Error
Sample 1	PID 1	161.4	166.47	77.788	0.004	0.008	0	0
	PID 2	201.89	150.4	84.527	0.004	0.007	0	0
	PID 3	155.76	165.44	77.481	0.004	0.008	0	0
	PID 4	57.679	171.66	53.226	0.006	0.012	0	0
	PID 5	209.57	191.23	93.095	0.003	0.006	0	0

(*continued*)

Table 9. (*continued*)

Samples	PID SET	*Kp*	*Ki*	*Kd*	Rise Time (s)	Settling Time (s)	Overshoot (%)	Steady-State Error
Sample 2	PID 1	207.22	142.18	81.385	0.004	0.007	0	0
	PID 2	176.98	146.19	75.782	0.004	0.008	0	0
	PID 3	137.91	162.62	72.124	0.004	0.008	0	0
	PID 4	164.81	141.11	71.577	0.004	0.008	0	0
	PID 5	239.78	173.9	94.491	0.003	0.006	0	0
Sample 3	PID 1	214.39	135.18	80.891	0.004	0.008	0	0
	PID 2	153.64	105.5	60.737	0.005	0.010	0	0
	PID 3	106.15	114.38	52.433	0.006	0.011	0	0
	PID 4	232.61	159.97	95.829	0.003	0.006	0	0
	PID 5	249.85	147.91	97.676	0.003	0.006	0	0

4.1.4 Results of the hybrid PID-PSO-ACO

The hybrid PID–PSO–ACO results show excellent position control performance, combining the strengths of both optimization methods to fine-tune the PID gains as shown in Figs. 15, 16 and 17. Across all samples, the controllers achieve extremely fast rise times (as low as $0.001 - 0.002s$) and very short settling times ($0.002 - 0.004s$) with zero overshoot and no steady-state error is shown in Table 10. The optimized gain values are relatively high, especially for *Kp* and *Ki*, which indicates that the hybrid approach pushes the controller to be more responsive while maintaining full stability and precision. Compared to using PSO or ACO alone, the hybrid tuning consistently delivers the fastest and smoothest responses, proving that combining both algorithms provides better exploration and exploitation of the search space for complex position control tasks.

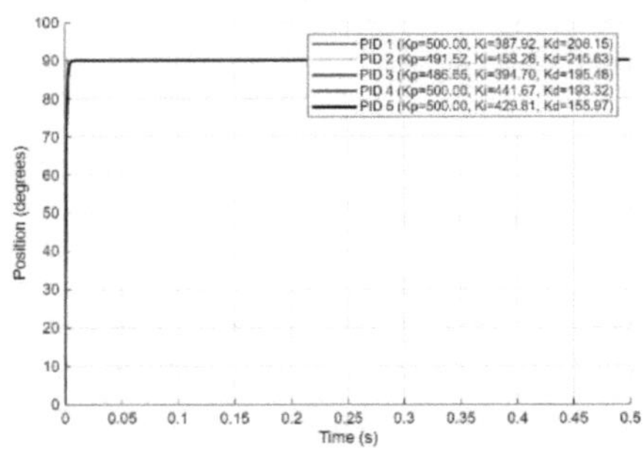

Fig. 15. Results - PID-PSO-ACO (Sample 1)

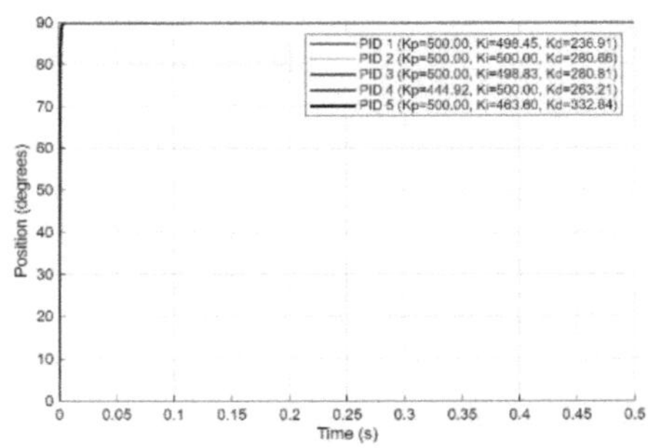

Fig. 16. Results - PID-PSO-ACO (Sample 2)

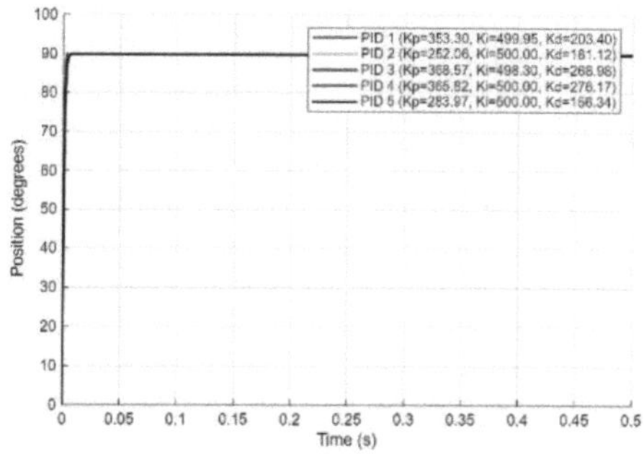

Fig. 17. Results - PID-PSO-ACO (Sample 3)

Table 10. The results of the hybrid PID-PSO-ACO

Samples	PID SET	*Kp*	*Ki*	*Kd*	Rise Time (s)	Settling Time (s)	Overshoot (%)	Steady-State Error
Sample 1	PID 1	500	387.92	208.15	0.002	0.003	0	0
	PID 2	491.52	458.26	245.63	0.002	0.003	0	0
	PID 3	486.65	394.7	195.48	0.002	0.003	0	0
	PID 4	500	441.67	193.32	0.002	0.003	0	0
	PID 5	500	429.81	155.97	0.002	0.003	0	0
Sample 2	PID 1	500	498.45	236.91	0.002	0.003	0	0
	PID 2	500	500	280.66	0.001	0.002	0	0
	PID 3	500	498.83	280.81	0.001	0.002	0	0
	PID 4	444.92	500	263.21	0.001	0.002	0	0
	PID 5	500	463.6	332.84	0.001	0.002	0	0
Sample 3	PID 1	353.3	499.95	203.4	0.002	0.003	0	0
	PID 2	252.06	500	181.12	0.002	0.004	0	0
	PID 3	368.57	498.3	268.98	0.001	0.002	0	0
	PID 4	365.82	500	276.17	0.001	0.002	0	0
	PID 5	283.97	500	156.34	0.002	0.004	0	0

4.2 Comparison of the Average Results of Control Methods

Overall, the average results show that the hybrid PID controller combining both ACO and PSO (PPA) achieves the best overall performance, as shown in Fig. 18, with the fastest rise time ($0.001s$) and the shortest settling time ($0.003s$), while maintaining zero overshoot and steady-state error. The standalone PID–ACO also performs well, eliminating overshoot and improving response times compared to the reference PID, but it does not reach the extreme speed of the hybrid method.

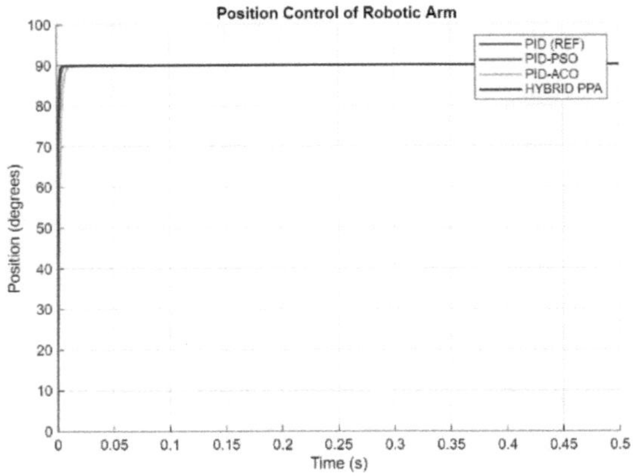

Fig. 18. Average Results Comparison of Control Methods

In contrast, the PID–PSO controller achieves a faster response than the reference but still allows an average overshoot of about 2.8% and a longer settling time (0.024s). The conventional PID tuned by standard methods (REF) has the slowest rise time (0.013s) and shows a slight overshoot (0.009%) as shown in Table 11. Overall, the results clearly demonstrate that the hybrid PSO–ACO tuning achieves the best balance of fast, stable, and precise position control by combining the advantages of both optimization techniques.

Table 11. Average Results Comparison of Control Methods

Control Methods	*Kp*	*Ki*	*Kd*	Rise Time (s)	Settling Time (s)	Overshoot (%)	Steady-State Error
PID (REF)	210	140	78.75	0.013	0.005	0.009	0
PID-PSO	448.87	392.28	147.09	0.005	0.024	2.759	0
PID-ACO	177.98	151.60	77.94	0.004	0.008	0	0
HYBRID PID-PSO-ACO	436.45	471.43	231.93	0.001	0.003	0	0

5 Conclusions

The research on optimizing PID controllers for robotic arm position control demonstrates that combining Particle Swarm Optimization (PSO) and Ant Colony Optimization (ACO) into a hybrid approach provides significantly better performance than using either algorithm alone. While PSO and ACO individually improve the traditional PID controller by reducing rise time, settling time, overshoot, and steady-state error, the hybrid method effectively combines their strengths. PSO enables fast global exploration of the search space, while ACO strengthens convergence toward optimal parameter regions through adaptive learning. The comparative results confirm that the hybrid PID–PSO–ACO (PPA) controller achieves more consistent, precise, and efficient position control, making it a highly promising solution for advanced robotic arm applications. In practical robotic applications, overshoot and settling time have direct implications on safety, precision, and efficiency. High overshoot can lead to excessive joint excursions, causing additional mechanical stress, increasing wear on gears and bearings, and potentially leading to collisions with nearby objects. Minimizing overshoot ensures that the arm stops precisely at the target point, critical in assembly or pick-and-place operations. Likewise, shorter settling time allows the robot to stabilize faster after a movement, reducing cycle time and improving throughput in high-speed production lines. Thus, optimizing both parameters directly improves operational reliability and productivity.

Acknowledgement. This research was funded by a grant from Universiti Malaysia Pahang Al-Sultan Abdullah (UMPSA Distinguished Research Grant: RDU233022). We would like to express our gratitude to University College TATI for providing exceptional research facilities and additional funding that have greatly contributed to the success of this study. Their support and resources have been invaluable in advancing our research efforts.

References

1. Naing, W.W., Thanlyin, M., Aung, K.Z., Thike, A.: Position control of 3-DOF articulated robot arm using PID controller. Int. J. Sci. Eng. Appl. **7**(9), 254–260 (2018). https://doi.org/10.7753/IJSEA0709.1001
2. Liu, Y., et al.: Self-tuning control of manipulator positioning based on fuzzy PID and PSO algorithm. Front. Bioeng. Biotechnol. **9**, 817723 (2022). https://doi.org/10.3389/fbioe.2021.817723
3. Zhang, Z., Bai, D.: Optimization of improved PID control strategy based on genetic algorithm. J. Phys.: Conf. Ser. **2417**(1), 012025 (2022). https://doi.org/10.1088/1742-6596/2417/1/012025
4. Bi, M.: Control of robot arm motion using trapezoid fuzzy two-degree-of-freedom PID algorithm. Symmetry. **12**(4), 665 (2020). https://doi.org/10.3390/sym12040665
5. Baghli, F.Z., Lakhal, Y., El Kadi, Y.A.: The efficiency of an optimized PID controller based on ant colony algorithm (ACO-PID) for the position control of a multi-articulated system. J. Robot. Control. **4**(3), 289–298 (2023). https://doi.org/10.18196/jrc.v4i3.17709
6. Lin, J., Liu, X., Ren, Z.: Fuzzy PID control for multi-joint robotic arm. In: 2022 IEEE 20th International Conference on Industrial Informatics (INDIN), pp. 723–728. IEEE, Piscataway, NJ (2022). https://doi.org/10.1109/INDIN51773.2022.9976122
7. Othman, T.T., Hussein, F.S.: Modeling and fuzzy control of flexible manipulator link with moving end effector. J. Phys.: Conf. Ser. **2467**(1), 012026 (2023). https://doi.org/10.1088/1742-6596/2467/1/012026
8. Ghani, N.M., Abdul, A.O., Hashim, A.A.A., Nasir, A.N.K.: Comparative analysis of PID and fuzzy logic controllers for position control in double-link robotic manipulators. J. Intell. Syst. Control. **2**(4), 183–196 (2023). https://doi.org/10.56578/jisc020401
9. Razali, M.R., Faudzi, A.A.M., Shamsudin, A.U.: Position and Angular Control Using Fuzzy Tuned PID Controller for Mobile Robot Path Tracking. In: 2022 IEEE 5th International Symposium in Robotics and Manufacturing Automation (ROMA), pp. 1–6. IEEE, Piscataway, NJ (2022). https://doi.org/10.1109/ROMA55875.2022.9915684

10. Quoc, D.V., Minh, N.D.T., Thi, T.T., Tien, L.N., Quang, M.N., Thanh, B.N.T.: Tuning PID controller bases on random search algorithms. Int. J. Eng. Res. Technol. **10**(1), 73–77 (2021). https://doi.org/10.17577/IJERTV10IS010038
11. Suhaimin, S.C., Azmi, N.L., Rahman, M.M., Yusof, H.M.: Analysis of point-to-point robotic arm control using PID controller. In: 2019 7th International Conference on Mechatronics Engineering (ICOM), pp. 1–6. IEEE, Piscataway, NJ (2019). https://doi.org/10.1109/ICOM47790.2019.8952036
12. Karim, M.Z.B.A., Thamrin, N.M.: Servo motor controller using PID and graphical user interface on Raspberry Pi for robotic arm. J. Phys.: Conf. Ser. **2319**(1), 012015 (2022). https://doi.org/10.1088/1742-6596/2319/1/012015
13. Zahira, O., Hamza, A., Mamou, M.A., Mourad, B.: Improvement of the precision of a two-degree-of-freedom manipulator arm PID controller. In: 2023 IEEE International Conference on Artificial Intelligence & Green Energy (ICAIGE), pp. 1–4. IEEE, Piscataway, NJ (2023). https://doi.org/10.1109/ICAIGE58321.2023.10346405
14. Ji, H., Li, L.: The six-degree-of-freedom robotic arm optimization iterative learning algorithm improves trajectory planning. In: 2022 2nd International Conference on Computer Science, Electronic Information Engineering and Intelligent Control Technology (CEI), pp. 627–630. IEEE, Piscataway, NJ (2022). https://doi.org/10.1109/CEI57409.2022.9950149
15. Al-Mola, M.H., Abdelmaksoud, S.I.: Performance improvement of a space robotic manipulator using fuzzy-based PID controller. In: 2023 International Telecommunications Conference (ITC-Egypt), pp. 524–529. IEEE, Piscataway, NJ (2023). https://doi.org/10.1109/ITC-Egypt58155.2023.10206133
16. Lee, S.-F., Lu, H.-C., Hung, T.-H.: Optimal design GA-based fuzzy PID controllers. In: IEEE International Conference on Systems, Man and Cybernetics, pp. 886–888. IEEE, Piscataway, NJ (2002). https://doi.org/10.1109/ICSMC.2002.1173412
17. Wu, M.-C., Jhong, B.-G., Chen, M.-Y.: Apply adaptive neural network PID controllers for a 6DOF robotic arm. In: 2022 International Conference on System Science and Engineering (ICSSE), pp. 79–83. IEEE, Piscataway, NJ (2022). https://doi.org/10.1109/ICSSE55923.2022.9948251
18. Luo, S., Chu, D., Li, Q., He, Y.: Inverse kinematics solution of 6-DOF manipulator based on multi-objective full-parameter optimization PSO algorithm. Front. Neurorobot. **16**, 791796 (2022). https://doi.org/10.3389/fnbot.2022.791796
19. Rowan, A. Arduino (Tinkerkit) Braccio robot arm + kinematics. Hackster.io, 29 August. [Online]. https://www.hackster.io/rpatterson/arduino-tinkerkit-braccio-robot-arm-kinematics-1a8303 (2021). Accessed 27 May 2025
20. Morales, G.R., Canto, S.E., Hernandez, V.G.: Speed control of DC servo motor under comparison with PID tuner control and neural network control using Simulink and ESP32. In: 2023 IEEE International Conference on Engineering Veracruz (ICEV), pp. 1–8. IEEE, Piscataway, NJ (2023). https://doi.org/10.1109/ICEV59168.2023.10329699

Portable AI-IoT Air Quality Monitoring System Using Hybrid Rule-Based and Isolation Forest Analytics

Muhammad Fairuz bin Imran, Ahmad Najmuddin Muizz bin Mohd Razali, Muhammad Akram bin Malik Radzuan, Muhammad Irfan bin Nazarudin, Mohd Hafizuddin Mat(✉), Nur Hidayah Ramli, Shakhirul Mat Salleh, and Nurul Husna Abd Wahab

Faculty of Electronic Engineering & Technology, Universiti Malaysia Perlis (UniMAP), Arau, Malaysia
hafizuddinmat@unimap.edu.my

Abstract. Air pollution remains a significant environmental and health concern, while conventional monitoring stations are often expensive, stationary, and lack real-time alerting capabilities. This work presents a low-cost, portable AI-IoT air quality monitoring system based on a Raspberry Pi 4B. The system integrates a DHT22 module for temperature and humidity sensing, an MQ135 gas sensor for air quality assessment, and an OLED module for local visualization. Sensor data are transmitted to a Firebase Realtime Database and analyzed using a hybrid anomaly-detection strategy combining rule-based thresholds with an Isolation Forest model. A web dashboard enables real-time monitoring, and automated Telegram notifications provide immediate alerts. Experimental evaluation using 424 data samples demonstrated reliable operation with sub-2-s latency, effective anomaly detection (42 anomalies and 3 fire-related events), and stable cloud synchronization. The proposed solution delivers a unified, scalable, and low-cost platform suitable for residential, industrial, and smart-city applications. Future work includes GPS integration, predictive analytics, and energy-efficient mobile hardware.

Keywords: Edge Computing · Anomaly Detection · Environmental Monitoring · Hybrid Analytics

1 Introduction

Recent advancements in AI-IoT air quality monitoring have transitioned from expensive, static stations to affordable, portable systems utilizing platforms like Raspberry Pi and ESP8266 [1, 3, 6, 10]. These unified solutions integrate low-cost multi-sensor arrays (e.g., MQ-series, PM2.5) to capture real-time environmental data [2, 11, 12], which is processed locally or via cloud platforms like ThingSpeak and Firebase for visualization [1, 4, 5, 7]. Furthermore, the embedding of machine learning algorithms such as LSTM, XGBoost, and Kalman Filters enables precise anomaly detection and predictive capability [8, 9, 13, 14], facilitating automated alerting systems that enhance public health management [1, 2, 5, 12].

M. A. b. Jamlos et al. (Eds.): ICSET 2025, LNNS 1897, pp. 453–458, 2026.
https://doi.org/10.1007/978-3-032-20722-7_57

2 System Architecture

As shown in Fig. 1, the system is built around a Raspberry Pi 4 Model B, interfaced with a DHT22 sensor, MQ135 gas module, and an OLED display. Sensor data are uploaded to Firebase and visualized on a web dashboard. The system supports local display, cloud monitoring, ML anomaly detection, and Telegram alerting.

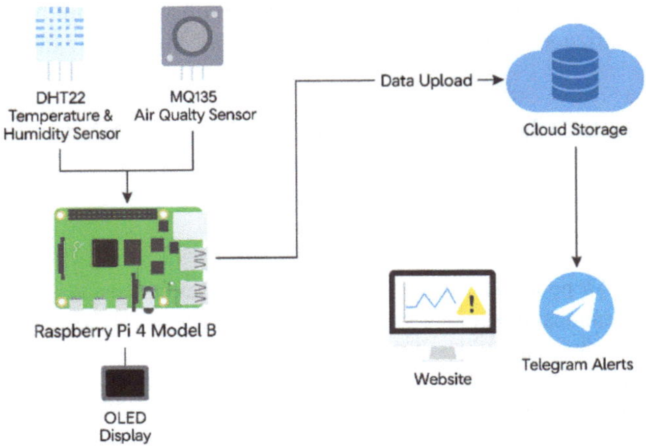

Fig. 1. Overall system architecture of the proposed IoT environmental monitoring and alerting platform.

3 Hardware Setup

The system hardware includes a Raspberry Pi 4B, DHT22, MQ135, and OLED display connected through GPIO and I2C. Components were selected for compatibility, reliability, and ease of integration as in Table 1.

Table 1. Project Components.

Component	Description	GPIO Pin
Raspberry Pi	Single-board-computer	–
DHT 22	Temperature & Humidity Sensor	GPIO 4
MQ135	Air Quality Sensor (Digital Out)	GPIO 17
OLED Display	I2C Communication	SCL: GPIO3
		SDA: GPIO2

The raspberry Pi was powered by external plug-point which could produce 5 V supply. The components were selected based on compatibility and reliability with Raspberry Pi's platform.

4 Software Framework

The software workflow includes Raspberry Pi OS configuration, Python development via Geany, and installation of required libraries (Fig. 2). The main program handles sensor acquisition, cloud upload, OLED display updates and Telegram alerting.

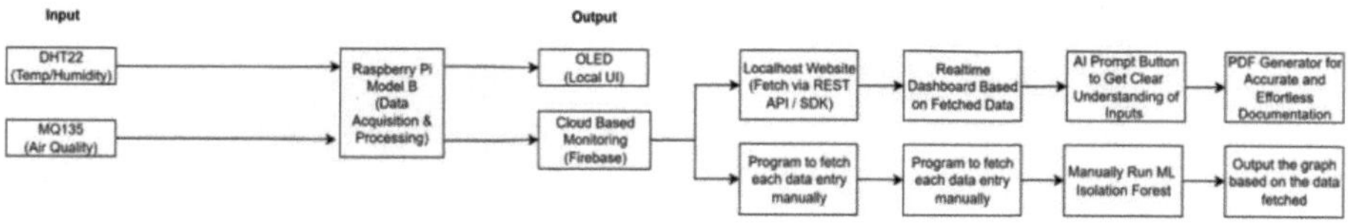

Fig. 2. Block Diagram.

5 Data Pipeline and Machine Learning

A total of 424 data entries were collected and analyzed. Rule-based logic detects extreme temperature and humidity conditions, while an Isolation Forest model identifies complex anomalies. Outputs are stored with classification labels. The workflow initiates with the extraction and cleaning of Firebase sensor logs, followed by the normalization of categorical air quality data into a binary numeric format. The system employs a hybrid detection logic where a rule-based check first filters for potential fire conditions, specifically temperatures exceeding 40 °C before subjecting the remaining unflagged data to the Isolation Forest model trained on temperature, humidity, and air quality features. Finally, the processed results are exported into an Excel file containing both the original readings and the generated classifications, complete with embedded visualization charts to aid in the interpretation of the model's performance.

6 AI Recommendations

The system integrates Gemini 1.5 Flash API as the AI engine to process real-time sensor data from temperature, humidity, and air quality inputs. Through carefully designed prompt engineering and API interactions, the AI generates context-aware recommendations such as 'stay hydrated' or 'close the windows'. Unlike static rule-based alerts, this approach leverages AI reasoning to adapt suggestions dynamically, providing a more personalized and intelligent decision-support layer. This integration demonstrates not only data monitoring but also how AI can transform sensor readings into actionable environmental and health insights.

7 Web Dashboard and Reporting

A real-time dashboard built using Firebase SDK visualizes temperature, humidity, and air quality. It features color-coded indicators, time-series data, and responsive design. Users can generate PDF reports summarizing conditions and system activity through a dedicated reporting tool.

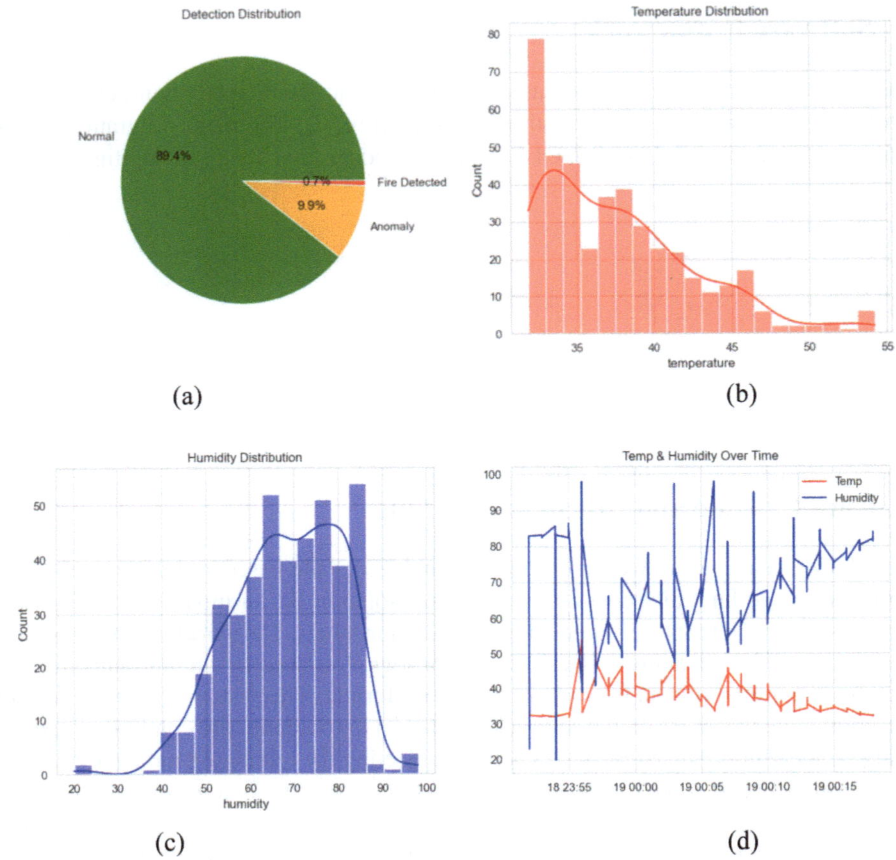

Fig. 3. Analysis of Environmental Sensor Data and System States. (a) The distribution of detection classes (Normal, Anomaly, Fire Detected). (b) The frequency distribution of temperature values. (c) The frequency distribution of humidity values. (d) The temporal variation of temperature and humidity readings over the recorded timeline.

8 Results and Discussion

Out of 424 samples, 379 were normal, 42 anomalies were detected by Isolation Forest, and 3 were classified as fire related. Cloud update latency remained below 2 s, and alerts were dispatched reliably. Figure 3 shows Analysis of Environmental Sensor Data and System States. The experimental results demonstrate the system's classification distribution, where 89.4% of instances were identified as 'Normal', while 'Anomaly' and 'Fire Detected' states accounted for 9.9% and 0.7%, respectively. The environmental data analysis reveals a temperature concentration between 33–40 units and humidity levels predominantly ranging from 60–90 units. Furthermore, the time-series evaluation highlights dynamic sensor responses, characterized by significant humidity spikes and corresponding temperature fluctuations, verifying the system's sensitivity in monitoring diverse environmental conditions.

9 Conclusion

This study successfully demonstrates a cost-effective and portable AI-IoT framework for real-time air quality monitoring using Raspberry Pi by integrating DHT22 and MQ135 sensors, Firebase cloud storage, and a hybrid anomaly detection scheme consisting of rule-based thresholds and Isolation Forest. Experimental evaluation validated the system's capability to rapidly detect environmental anomalies, including fire-related events, with sub-2-s latency, while the inclusion of a real-time dashboard, automated Telegram alerts, and Prophet-based time-series forecasting significantly enhances its usability and scalability compared to traditional stationary solutions. Uniquely combining affordability, mobility, and AI-driven intelligence, this platform provides a robust foundation for future enhancements such as GPS integration for location-aware deployments, deep learning predictive analytics, and energy-efficient hardware implementations. Overall, the proposed framework contributes a practical, extensible approach to smart environmental monitoring that offers immediate value to end-users and supports large-scale smart-city applications.

References

1. R, P.: Real-time air quality monitoring using low-cost sensors – IoT system. Int. J. Sci. Res. Eng. Manag. **12**(2), 9–15 (2025)
2. Usha, S., Teja, S., Dilip, V., Reddy, S.: Machine learning-driven system for real-time air quality monitoring and prediction. Int. J. Sci. Res. Comput. Sci. Eng. Inf. Technol. **10**(2), 45–52 (2025)
3. Jabbar, W., Subramaniam, T., Ong, A., Shu'Ib, M., Wu, W., De Oliveira, M.: LoRaWAN-based IoT system implementation for long-range outdoor air quality monitoring. Internet Things. **19**, 100540 (2022)
4. Mani, G., Viswanadhapalli, J., Sriramalakshmi, P.: AI powered IoT based real-time air pollution monitoring and forecasting. J. Phys. Conf. Ser. **2115**, 012002 (2021)
5. Sunitha, M., Adilakshmi, T., Prasanna, M.L.: Vayu – An IoT powered indoor air quality monitoring system. J. Inf. Syst. Eng. Manag. **10**(34s), 5910 (2025)
6. Zhang, H., Srinivasan, R., Ganesan, V.: Low cost, multi-pollutant sensing system using raspberry pi for indoor air quality monitoring. Sustainability. **13**(1), 370 (2021)
7. Karnati, H.: IoT-based air quality monitoring system with machine learning for accurate and real-time data analysis. *arXiv preprint.* , abs/2307.00580 abs/2307.00580 (2023)
8. Lai, X., Yang, T., Wang, Z., Chen, P.: IoT implementation of kalman filter to improve accuracy of air quality monitoring and prediction. Appl. Sci. **9**(9), 1831 (2019)
9. Pazhanivel, D., Velu, A., Palaniappan, B.: Design and enhancement of a fog-enabled air quality monitoring and prediction system: An optimized lightweight deep learning model for a smart fog environmental gateway. Sensors. **24**(3), 1125 (2024)
10. Aashiq, M., et al.: An IoT-based handheld environmental and air quality monitoring station. Acta IMEKO. **12**(3), 1–8 (2023)
11. Kairuz-Cabrera, D., Hernandez-Rodriguez, V., Schalm, O., Martinez, A., Laso, P., Alejo-Sánchez, D.: Development of a unified IoT platform for assessing meteorological and air quality data in a tropical environment. Sensors. **24**(5), 1765 (2024)
12. García, L., García-Sánchez, A., Asorey-Cacheda, R., García-Haro, J., Zúñiga-Cañón, C.: Smart air quality monitoring IoT-based infrastructure for industrial environments. Sensors. **22**(9), 3460 (2022)

13. Reis, M., Serôdio, C.: Edge AI for real-time anomaly detection in smart homes. Future Internet. **17**(5), 179 (2025)
14. Rajesh, M., Babu, R., Moorthy, U., Easwaramoorthy, S.: Machine learning-driven framework for real-time air quality assessment and predictive environmental health risk mapping. Sci. Rep. **15**(1), 1–16 (2025)

Tunable Dielectric Resonator Matching Model for Stability Enhancement in 5 GHz Microwave Amplifiers

R. C. Yob[1(✉)], S. Z. Ibrahim[2], L. Zahid[1], N. H. Ramli[1], M. A. Jamlos[1], and N. S. M. Suhaimi[3]

[1] Dept. of Electronic, Faculty of Electronic Engineering and Technology (FKTEN), Arau, Perlis, Malaysia
rashidahcheyob@unimap.edu.my

[2] Dept. of Computer, Faculty of Intelligent Computing (FKC), Universiti Malaysia Perlis (UniMAP), 02600 Arau, Perlis, Malaysia

[3] Dept. of Software Engineering and Information System, Faculty of Computer Science and Information Technology, Universiti Putra Malaysia (UPM), 43400 UPM, Serdang, Selangor, Malaysia

Abstract. Stability and impedance matching are decisive factors in achieving reliable performance in microwave amplifiers, particularly because transistor stability characteristics vary with frequency. An amplifier that is stable at its design frequency may exhibit oscillatory behavior outside the intended operating band, which degrades performance and can lead to circuit failure. To address this challenge, a tunable dielectric resonator matching model (DRMM) is proposed for enhancing the stability of microwave amplifiers operating at 5 GHz. The proposed approach employs an inhomogeneous multi-permittivity cylindrical dielectric resonator (MPCDR) coupled to parallel microstrip lines, providing post-fabrication frequency adjustability and improved impedance matching. A comprehensive parametric study is carried out using CST Microwave Studio (CST-MW) to investigate the effects of coupling spacing, curved microstrip configuration, and angular position of the DR. The optimal configuration is achieved with a coupling of 19 mm and a curved microstrip angle of 155°, yielding excellent return loss and a stable resonance at 5 GHz. The results demonstrated that the proposed DRMM effectively suppresses unwanted oscillations while maintaining good matching characteristics, thereby improving amplifier robustness and reliability. This tunable and compact solution is well suited for next generation 5G and emerging 6G wireless transceiver applications.

Keywords: Dielectric Resonator Matching · Microwave Amplifier Stability · Impedance Matching · Tunable Resonator · 5 GHz

1 Introduction

Signal amplification is considered a most important function of the microwave circuits [1]. Low-noise amplifiers (LNAs) and power amplifiers (PAs) are widely investigated for emerging applications, including the Internet of Things (IOTs) [2–5] and high-speed

M. A. b. Jamlos et al. (Eds.): ICSET 2025, LNNS 1897, pp. 459–466, 2026.
https://doi.org/10.1007/978-3-032-20722-7_58

communication systems [1, 6]. The 5 GHz frequency band is often selected due to its advantages in wider bandwidth (BW) and lower interference. Maintaining power efficiency, a low loss, and a high-quality factor (Q-factor) is ensured through the stability element and matching component implementation. Proper implementation of stability elements and matching components is vital to maintain high efficiency and a high Q-factor [6–9]. Instability is a key consideration because input and output matching networks are generally frequency dependent [1]. The amplifier may be stable at the design frequency but potentially unstable at other frequencies.

The necessity of implementing a proper stability factor is crucial for achieving optimum performance [1, 6–10]. Conventional techniques are known to be difficult or impossible to adjust after fabrication [11, 12]. Oscillation occurs if the transistor operates in an unstable region, causing circuit failure. Consequently, a frequency variable mechanism is required to control the operating frequency and ensure the circuit remains within the stable region. The operating frequency can shift by varying the placement of DR with fixed physical parameters [13]. Thus, the DR is implemented to fulfill this necessity, which offers inherent features like a high Q-factor, compact size, and good temperature stability [14–16]. The DR is used as the DRMM to overcome the issues of fixed parameters and self-oscillation. The configuration and characteristics of the DR were evaluated. This model operates as a variable frequency element, as illustrated in Fig. 1.

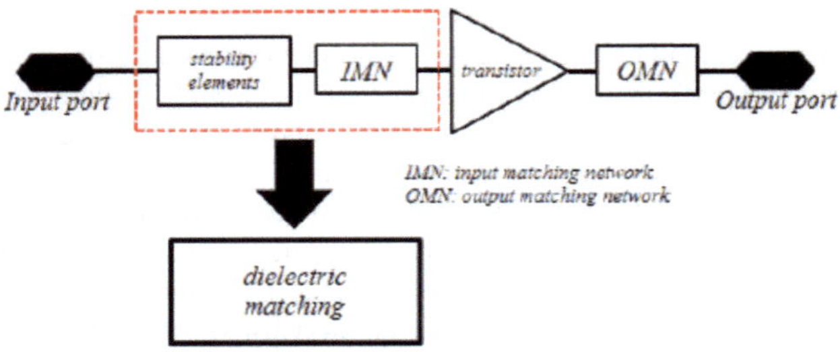

Fig. 1. Block diagram of a microwave amplifier configuration.

By utilizing different DR materials [15–17], the DRMM can control the operating frequency to sustain stability. The DRMM utilizes an inhomogeneous multi-permittivity cylindrical DR (MPCDR) [15, 16] combined with a microstrip-based coupling structure to achieve controllable resonance at 5 GHz. This study introduces a tunable DRMM based on MPCDR integrated with curved parallel microstrip lines, enabling post-fabrication frequency control and enhanced stability at 5 GHz. A detailed parametric investigation is performed to identify the optimal configuration that provides stable operation and superior impedance matching.

2 Parametric Study of Dielectric Resonator Matching Model

2.1 Dielectric Resonator Materials and Configuration

A DR is an unmetallized ceramic component with a high dielectric constant, ε_r and low loss. The frequency of an isolated CDR operating in the $TE_{01\delta}$ mode is dictated by its dimensions of the diameter, *d*, height, h_r and dielectric constant, ε_r as in (1) [15].

$$f_{GHz} = \frac{34}{d_{mm}\sqrt{\varepsilon_r}}\left(\frac{d}{h_r} + 3.45\right) \tag{1}$$

Initial dimensions calculated for a 5 GHz resonant frequency were a diameter, *d* of 9 mm and a height, h_r of 18 mm. The RO4003C board was selected as the substrate due to its low dielectric tolerance and low loss, suitable for C-band operation. The investigation utilized an inhomogeneous or MPCDR, which was composed of 90° pie-shaped sectors of two different dielectric materials [15, 16], which are Magnesium Titanate Oxide ($MgTiO_3$) with ε_r of 15.72 and Cobalt Titanate Oxide ($CoTiO_3$) with ε_r of 9.85. High permittivity materials contribute to strong coupling and high Q-factors, while low permittivity materials tend toward wider bandwidth operation. In addition, the parallel microstrip line was chosen as the feeding mechanism, as it provides an optimum power level with more stability than the series configuration.

2.2 Parametric Study of the Dielectric Resonator

The parametric study of the proposed CDR was primarily undertaken to identify the influence of physical changes and subsequently obtain the optimal CDR configuration suitable for dielectric matching. This investigation is starting by examining seven spacing distances, ranging from 10 mm to 25 mm, between the inhomogeneous CDR and the parallel microstrip line to characterize the coupling efficiency. The references of the coupling distance are based on the half-width of the overall dimension of the DR. Then, the distances between the coupling coefficients are analyzed by adding and subtracting the spacing between the parallel microstrip lines. Following this, the influence of the curve configuration of the parallel microstrip lines at λ/4 wavelength was studied. Four configurations of 65°, 90°, 155° and 165° were evaluated. These parametric studies are illustrated in Fig. 2.

Then, the next investigation of the parametric study of the modelling DR is the angular position, ranging from 0° to 55° as shown in Fig. 3. Initially, the angular position is referred to the longitudinal axis of the parallel microstrip line. Then, the different values of the angular position are changed or rotated in an anticlockwise direction. All the parameters of this parametric study are based on typical and empirical investigations.

The simulation tool of the CST-MW software is used to analyze the parametric study. The best configuration will be used to be incorporated as the stability element and matching component on a microwave amplifier.

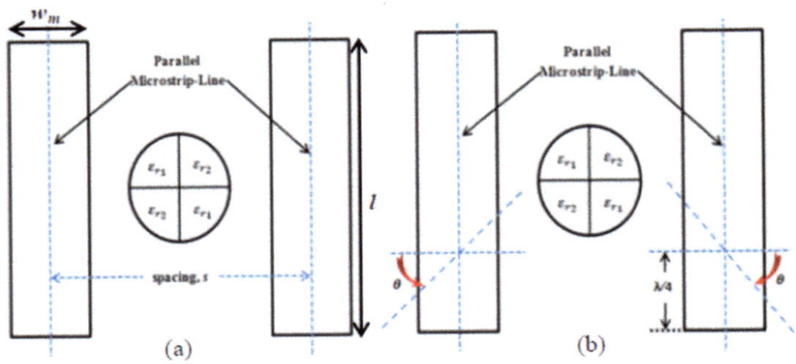

Fig. 2. The parametric study of the parallel microstrip lines: (a) spacing distances and (b) curve configurations.

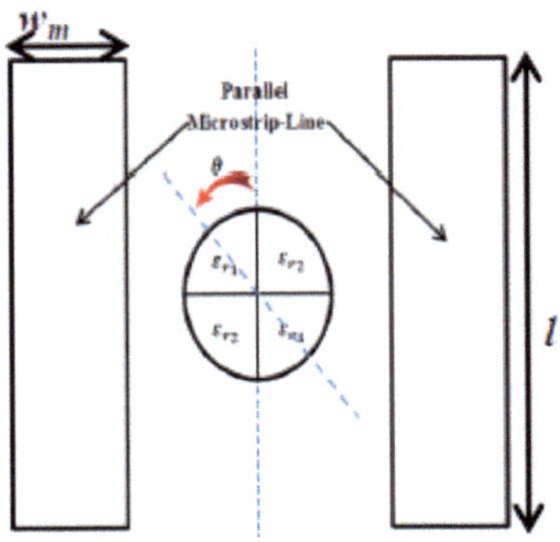

Fig. 3. Parametric study of various curved microstrip line configurations at λ/4 wavelength.

3 Results and Discussions

The simulated S-parameters responses of the seven evaluated spacing distances or coupling coefficients between the parallel microstrip lines and the inhomogeneous CDR and for the $S_{11}S_{22}$ waveguide port, as illustrated in Fig. 4.

The best performance of the return loss, S_{11} is obtained at 17 mm and 19 mm of the coupling spacing, around −18 dB. Thus, it is shown the good agreement between the expected theoretical, which it should be lower than −10 dB. However, the operating frequency is not accurately at 5 GHz. The obtained results indicate that improper coupling spacing between the inhomogeneous CDR and the parallel microstrip lines leads to poor performance. Therefore, an optimal distance must be determined between the CDR and the microstrip line. Excessive or insufficient spacing also increases the insertion loss, S_{21}. A higher insertion loss reflects a weaker coupling coefficient between the inhomogeneous CDR and the parallel microstrip lines.

The investigation of the curved configuration of the parallel microstrip lines is considering the best different coupling spacing of 17 mm and 19 mm. The obtained results of this investigation are shown in Fig. 5.

In general, the optimal results are represented by the 90° and 155° curved configurations for coupling spacing of 17 mm and 19 mm, which are around −18 dB to −20 dB.

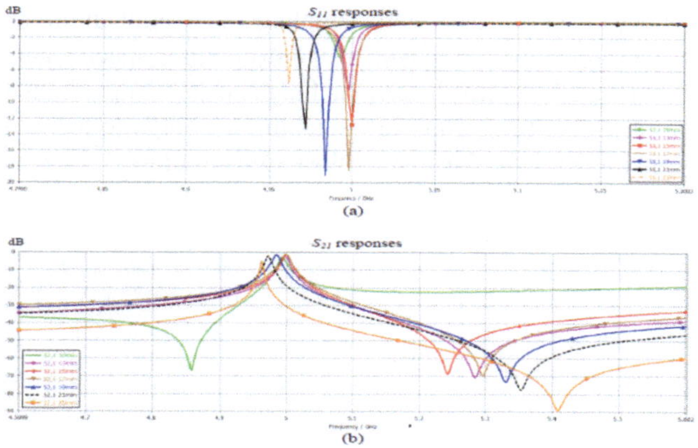

Fig. 4. S-parameters for various coupling spacings: (a) S_{11} and (b) S_{21} responses of $S_{11}S_{22}$ waveguide port.

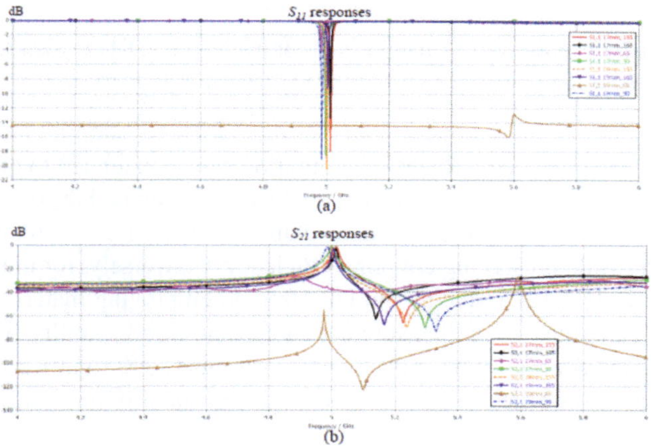

Fig. 5. S-parameters for various curved configurations at λ/4 wavelength: (a) S_{11} and (b) S_{21} responses of $S_{11}S_{22}$ waveguide port.

However, only the 155° curve configuration with 19 mm of coupling spacing is presented as an accurate operation at 5 GHz of resonant frequency. Therefore, this configuration of the proposed inhomogeneous CDR is identified as the optimal design that provides superior performance. It is consequently selected for the subsequent evaluation of the angular positioning of the inhomogeneous CDR.

The maximum electromagnetic coupling from the feeding structure is ensured by the optimal angular positioning of the inhomogeneous CDR. The obtained results of this investigation are presented in Fig. 6. The minimal impact on its resonant frequency and S-parameter performance is obtained by the angular position of the inhomogeneous CDR. Optimal results are achieved without angular position adjustment, yielding an S_{11} of −20.538 dB and S_{21} of −2.180 dB. The comprehensive analysis confirms that the coupling spacing, curved configuration, and angular positioning of the inhomogeneous significantly influence the electromagnetic coupling and overall amplifier performance. The optimal configuration is achieved with a 19 mm coupling spacing and a 155° curved structure, providing excellent impedance matching and stable operation at 5 GHz. This capability allows it to fine-tune amplifier matching networks to achieve stable, high-performance operation without redesigning the entire structure.

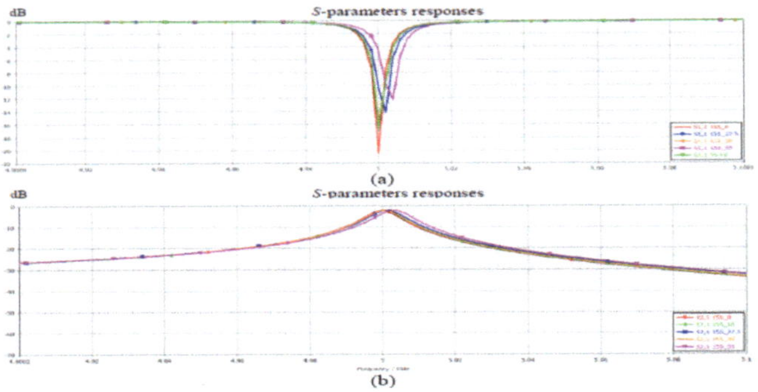

Fig. 6. S-parameters for various angular positions of the proposed CDR at λ/4 wavelength with 19 mm spacing and a 155° curve: (a) S_{11} and (b) S_{21} responses of the $S_{11}S_{22}$ waveguide port.

In addition, the proposed DRMM configuration can be extended to other frequency bands by adjusting the DR dimensions and dielectric constants. For lower frequencies, such as 3.5 GHz, increasing the resonator diameter and height proportionally maintains resonance while preserving coupling efficiency. Conversely, for higher frequency operation around 10 GHz, reducing the DR size and employing materials with higher permittivity, greater than 20, will ensure the compactness and stability. This scalability allows the DRMM concept to be applied across various C-, X-, and Ku-band amplifier designs, offering a versatile platform for tunable, high-stability microwave systems.

4 Conclusion

This paper has presented a tunable DRMM for enhancing the stability of 5 GHz microwave amplifiers. The proposed approach employs an inhomogeneous MPCDR coupled to curved parallel microstrip lines, providing both frequency tunability and improved impedance matching. Through extensive parametric analysis, an optimal configuration consisting of a 19 mm coupling spacing and a 155° curved microstrip structure is identified. This configuration achieves superior matching performance and stable resonance at 5 GHz. The proposed DRMM offers significant advantages, including post-fabrication adjustability, reduced risk of unwanted oscillations, and straightforward integration into compact microwave circuits. These features are well suited for integration into next-generation 5G and emerging 6G microwave amplifier systems.

Acknowledgement. The work was supported and funded by the Fundamental Research Grant-Earlier Career (FRGS-EC) Scheme (FRGS/1/2024/TK07/UNIMAP/02/30). The authors would like to acknowledge the FKTEN and FKC at UniMAP for their facilities, resources, and continuous support. Extend appreciation to colleagues and technical staff for their assistance throughout the project. ChatGPT was utilized for language refinement and grammar.

References

1. Pozar, D.M.: Microwave Engineering, 3rd edn. Wiley, New York (2005)
2. Dayarnab, B., Athul, K., Mitradip, B.: Low Noise Amplifier for the Wearable IoT Sensing System. In: 2022 IEEE Global Conf. on Comp., Power, and Comm. Tech. (GlobConPT), India Habitat Centre, Lodhi Road, , New Delhi, India, pp. 1–5. IEEE, Piscataway, NJ (2022)
3. Rouchen, L., Tomas, M., Michael, B., Anming, G., Songbin, G.: Piezoelectric RF resonant voltage amplifier for IoT applications. In: 2016 IEEE MTT-S International Microwave Symposium (IMS), San Francisco, CA, USA, 1–4 May 2016, pp. 1–4. IEEE, Piscataway, NJ (2016)
4. Godfrey, A.A., Bruno, J.S., Gerhard, P.H., Abu-Mahfouz, A.M.: A survey on 5G network for the internet of things: Communication technologies and challenges. IEEE Access. **6**, 3619–3647 (2017)
5. Rajeshwari, V.V.: Design of a power amplifier for NB-IoT. Master Thesis. Department of Electrical and Information Technology, Faculty of Engineering, LTH, Lund University, Sweden. (2018)
6. Inder, J.B.: Fundamentals of RF and Microwave Transistor. John Wiley & Sons, Inc, Hoboken, NJ (2009)
7. Hasan, S.H., Bukhari, S.A.A., Saboor, H.: Enhancing linear output and stability in Class C UHF power amplifiers using coplanar waveguide impedance matching. In: 2025 International Conference on Communication Technologies (ComTech), Rawalpindi, Pakistan, 1–4 May 2025, pp. 1–4. IEEE, Piscataway, NJ (2025)
8. Zaid, M. et al.: Optimizing low noise amplifiers: A two-stage approach for improved noise figure and stability. IEEE Access. **12**, 53475–53484 (2024)
9. Le, B.H., Tran, T.A., Zimmer, G.: Optimum input two-port for stabilization of transistors. In: 2013 International Conference on Advanced Technologies for Communications (ATC'13), Ho Chi Minh City, Vietnam, pp.531–534. IEEE, Piscataway, NJ (2013)
10. Shreyasi, S., Kushal, S., Chnadar, J.: Analysis of different matching techniques for microwave amplifier. Int. J. of. Eng. Trends and Tech. (IJETT). **9**(2), 625–632 (2014)
11. Dellier, S., Gourseyrol, R., Soubercaze-Pun, G., Collantes, J.-M., Anakabe, A., Narendra, K.: Stability analysis of microwave ciruits. In: 2012 IEEE 13th Annual Wireless and Microwave Technology Conference (WAMICON), Cocoa Beach, FL, USA, 15–17 April 2012, pp. 1–5. IEEE, Piscataway, NJ (2012)
12. Su, G., Sun, L., Wen, H., Yu, Z., Nan, Z.: An estimable stability method applied to power amplifier design. In: IEEE Electrical Design of Adv. Packaging and Sys. Symposium (EDAPS), Hanzhou, China, 1–4. IEEE, Piscataway, NJ (2011)
13. Jain, A., Hannurkar, P.R., Pathak, S.K., Biswas, A., Srivastva, M.: Improved performance of two-way power divider using dielectric resonator. Microw. Opt. Tech. Letters. **56**(4), 858–861 (2014)
14. Manisha, B., Sreelakshmi, K., Dora, R.K.: Dielectric Resonator Oscillator. Int. J. Adv. Res. Ideas Innov. Technol. **7**(4), 71–73 (2021)

15. Ullah, U., et al.: A novel multi-permittivity cylindrical dielectric resonator antenna for wideband applications. Radioengineering. **23**, 1071–1076 (2014)
16. Ullah, U., Ali, W.F.F.W., Ain, M.F., Mahyuddin, N.M., Ahmad, Z.A.: Design a novel dielectric resonator antenna using $MgTiO_3$-$CaTiO_3$ for wideband applications. Mater. Des. **85**, 396–403 (2015)
17. Baolin, G., Xiaokang, L., Yiwen, W., Qing, Y.: A hydantion epoxy resin gel casting of $MgTiO_3$–$CaTiO_3$ ceramics and its application for 5G/6G waveguide filter. Ceram. Int. **51**, 48075–48083 (2025)

Attention-Enhanced Fully Convolutional Network for Lane Detection in Intelligent Transportation Systems

Noor Jannah Zakaria[1(✉)], Wan Noor Hamiza Wan Ali[1], Adam Mohd Khairuddin[1], Normaisharah Mamat[1], and Rabatul Aduni Sulaiman[2]

[1] Department of Intelligence Informatics, Faculty of Artificial Intelligence, Universiti Teknologi Malaysia, 54100 Kuala Lumpur, Malaysia
noor.jannah@utm.my

[2] Software Engineering Department, Faculty of Computer Science and Information Technology, Universiti Tun Hussein Onn Malaysia, Parit Raja, Johor 86400, Malaysia

Abstract. Traffic accidents continue to rise worldwide, increasing the demand for safer and more intelligent vehicle systems. Lane detection is a critical component of Advanced Driver Assistance Systems (ADAS), yet the use of attention mechanisms in convolutional lane detection models remains limited. This study introduces a Fully Convolutional Network (FCN) enhanced with a channel attention module to improve lane feature representation and detection accuracy. The model performs dense prediction without fully connected layers, while the attention module selectively emphasizes informative channel features. Experiments on a custom onboard-camera dataset and the KITTI benchmark show that the proposed method achieves over 96% accuracy and ranks among the top three models on the KITTI leaderboard. These findings highlight the effectiveness of integrating attention mechanisms in FCNs for robust lane detection in intelligent transportation systems.

Keywords: Lane Detection · Advanced Driver Assistance System (ADAS) · Deep Learning · Fully Convolutional Network · Attention Mechanism

1 Introduction

Road traffic accidents remain a major global safety issue, causing 1.19 million deaths and 20–50 million injuries annually, with over 90% of fatalities in low- and middle-income countries [1]. Human error—fatigue, distraction, and cognitive overload is a primary contributor [2]. To address this, modern vehicles increasingly use active safety technologies like Advanced Driver Assistance Systems (ADAS), including Adaptive Cruise Control, Lane Departure Warning, Lane Keeping Assist, Automatic Emergency Braking, and Blind Spot Monitoring [3]. Lane detection is central to these systems, providing continuous information on lane geometry and vehicle alignment [4]. Recent automaker innovations, such as Mercedes-Benz's Automatic Lane Change and Honda Sensing 360+, rely on multi-modal sensors but face challenges in poor visibility or harsh

M. A. b. Jamlos et al. (Eds.): ICSET 2025, LNNS 1897, pp. 467–475, 2026.
https://doi.org/10.1007/978-3-032-20722-7_59

conditions [5]. Vision-based deep learning offers a cost-effective alternative, enhancing lane detection accuracy and scene understanding [6]. The EU now mandates ADAS features including Lane Keeping, AEB, and drowsiness detection in new vehicles [7]. Despite progress, lane detection is still difficult under occlusion, poor lighting, or faded markings. Traditional methods and standard CNNs often fail in such conditions [8]. This study proposes a Fully Convolutional Network with a Channel Attention Mechanism, enabling pixel-level predictions while emphasizing salient features, improving robustness and accuracy in real-world scenarios.

2 Literature Review

Deep learning (DL) has played a major role in advancing lane detection, with Convolutional Neural Networks (CNNs) demonstrating strong capability in learning hierarchical spatial features from raw images [9], supported by scalable architectures that became prominent following AlexNet [10] and deeper models such as ResNet with skip connections [69]. Beyond CNNs, several architectures including RNNs, LSTMs for temporal analysis [11], FCNs for per-pixel segmentation [13] and encoder–decoder models for image-to-image prediction [13] have been widely adopted. These enable both semantic and instance-level segmentation of lane markings [14], while autoencoders provide compressed feature representations. Recent studies further show that integrating DL with geometric modeling, classical ML, or attention mechanisms can improve robustness by emphasizing salient lane features and suppressing irrelevant background noise. A variety of stand-alone DL models have been introduced, including ConvLSTM-based temporal–spatial processing [15], encoder–decoder segmentation networks [16], dual-decoder architectures such as LaneNet for binary and instance segmentation [17], and weakly supervised strategies for pseudo-label generation [18]. Enhanced encoder–decoder methods using VGG-16 backbones [19], atrous convolution with SPP [20], or integrated frameworks such as NVIDIA DriveWorks for combined perception [21] have also been demonstrated. Additional contributions include structured lane decoding [22], YOLO-based marker detection [23], and compact CNNs for lane geometry extraction [24]. Hybrid DL–geometric approaches combine CNNs with procedures such as edge detection, Hough Transform, perspective mapping, clustering, curve fitting, and FPGA acceleration [25], while DL–ML pipelines integrate CNN feature extraction with SVM-based classification or post-processing refinement [26]. Two-stage DL architectures further enhance reliability through sequential feature extraction, temporal modeling, ensemble fusion, and refined probability mapping [27]. Despite these advancements, lane detection remains challenging under faded markings, shadows, low illumination, and adverse weather, where early CNN and purely geometric approaches often fail. Existing attention-enhanced models, though effective, tend to be computationally heavy and still insufficient for safety-critical ADAS performance. Addressing these limitations, this research introduces a Channel Attention Enhanced Fully Convolutional Network that incorporates a lightweight channel-attention mechanism to recalibrate feature responses, strengthen discriminative cues, and suppress noise. Through comprehensive evaluation on benchmark datasets, this study contributes an efficient, robust, and scalable lane detection framework that advances DL-based ADAS solutions.

3 Methodology

3.1 Public Online Dataset

The proposed lane detection algorithm is trained and evaluated on the KITTI Vision Benchmark Suite, a widely used dataset in autonomous driving research. KITTI provides diverse real-world driving scenarios, including urban streets and highways, under varied conditions such as occlusions, shadows, and lane markings. Its standardized evaluation framework enables fair comparisons, encourages methodological innovation, and supports reproducibility. The lane detection component includes three scene types: Urban Unmarked (UU), Urban Marked (UM), and Urban Multiple Marked (UMM), comprising 579 images (289 training with ground truth, 290 test). Images are captured using dual Point Grey Flea 2 cameras with varifocal lenses, accompanied by calibration files. Performance is evaluated using KITTI's standard metrics: Precision, Recall, and Maximum F1-Measure. Table 1 summarizes image distribution.

Table 1. The specifications of the KITTI dataset

Scene category	Training	Testing
UU	98	100
UM	95	96
UMM	96	94
URBAN (ALL)	289	290

3.2 Data Collection and Pre-Processing

In addition to public datasets, a custom dataset was collected in Malaysia to improve the generalization of the proposed lane detection model. Data were recorded along highways (DUKE–LATAR) and urban streets (Sultan Yahya Petra to Cheras), capturing structured, predictable lanes as well as complex urban conditions. Highways were prioritized for clear lane markings, minimal occlusions, and safer data collection, while urban areas ensured robustness against diverse road structures and traffic. Drivers maintained legal speeds (50–110 km/h) during recording.

The data collection procedure involves capturing image sequences or video footage of driving environments using a vehicle-mounted camera, which are later used to train the deep learning model for lane detection tasks. A high-quality camera capable of capturing detailed visual features is essential for this purpose; therefore, the e-CAM21 CUMI1290 MOD camera was selected for its imaging capabilities. To minimize vibrations and maintain recording stability, the camera was securely mounted under the rearview mirror. The camera was angled at approximately 150°, providing a forward-looking field of view covering a distance of approximately 1.5 to 2 m from the front of the vehicle. This configuration ensures the visibility of standard lane widths, which range between 2.75 to 3.5 m. During data acquisition, the vehicle was driven along the designated routes while recording was initiated and maintained at a consistent frame rate and resolution, ensuring uniformity across the dataset. The dataset consisted of video recordings at 25 fps and 1280×720 resolution, decomposed into individual RGB frames. To reduce redundancy, one frame was randomly selected from every ten. Erroneous or uninformative frames such as those with occluded lanes, high traffic, construction zones, toll areas, or degraded lane markings were removed to improve data quality and prevent model mislearning. Data collection was scheduled outside peak traffic hours (06:00–10:00, 16:00–20:00) to minimize occlusions. The curated dataset, comprising RGB images and pixel-wise lane masks, was split into training (80%) and validation (20%) sets. Images were downscaled to 160×80 pixels, preserving critical lane features while reducing computational load. The dataset includes highway-day and urban-day scenes, with Table 2 summarizing image distribution.

Table 2. The specifications of the DUKE-LATAR-SYPSCS dataset

Category	Training	Testing
Highway-day	500	150
Urban-day	500	150
TOTAL	1000	300

3.3 Fully Convolutional Network with Integrated Attention Mechanism

The proposed Fully Convolutional Network (FCN) is designed to enhance spatial feature representation for accurate lane detection. To improve performance, a channel attention mechanism is integrated, enabling the model to emphasize critical lane features while suppressing irrelevant information. The overall architecture, illustrated in Fig. 1, consists of a normalization layer, seven convolutional layers, three max-pooling layers, ten dropout layers, seven deconvolutional layers, three unpooling layers, and six attention layers. Both channel attention and efficient channel attention mechanisms are incorporated to enhance feature representation and improve detection accuracy. The model is evaluated on both experimental and public datasets to ensure robustness and generalization.

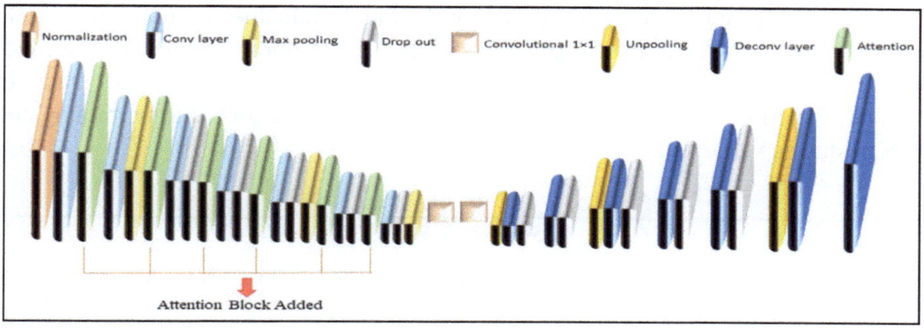

Fig. 1. FCN lane detection network model.

The input to the network is a color image of size 80 × 160 pixels (RGB format). Image patches were generated from training images and annotations and randomly shuffled to increase sample diversity. The FCN encoder begins with a 3 × 3 convolutional layers with 64 filters and a stride of 7, followed by normalization. Max-pooling (2 × 2) is applied after the second, fifth, and seventh convolutional layers to reduce feature map resolution and memory usage. Dropout (rate = 0.2) prevents overfitting, while sequential attention mechanisms—including channel attention and efficient channel attention enhance feature learning without dimensionality reduction. The decoder restores spatial dimensions via un-pooling and transposed convolutions. A 1 × 1 convolution reduces feature channels and deepens the network. The final deconvolution generates a grayscale lane image from the green channel, which is merged with zeroed red and blue channels to produce the RGB output with lanes highlighted in green. The network propagates through convolutional, pooling, dropout, and attention layers during encoding and through corresponding deconvolution, un-pooling, and attention layers during decoding. The FCN supports varying input sizes and classifies regions into lane and non-lane categories. Input images are initially downsampled, with intermediate feature map sizes adjusted at each convolution and pooling stage. During decoding, upsampling and deconvolution progressively restore the original image size of 80 × 160 pixels. Model training uses the Adam optimizer for adaptive learning, with performance evaluated via accuracy, loss (MSE), precision, recall, and F1-Measure. Ground truth lane images contain two lane markings and serve as target outputs, while testing is conducted on a separate dataset. Figure 1 illustrates the image flow through the FCN architecture.

4 Results and Discussion

4.1 Model Performance on the Training and Validation Sets Using the Public Online Dataset

The performance of the proposed Fully Convolutional Network (FCN) was evaluated across three experimental configurations: Experiment 1 (FCN without attention), Experiment 2 (FCN with channel attention), and Experiment 3 (FCN with efficient channel attention). Based on the experimental results, Experiment 3 demonstrated superior performance, achieving the highest accuracy and lowest loss values. As illustrated in Fig. 2,

Experiment 3 yielded a training accuracy of 96.13% and a validation accuracy of 96.20%, outperforming both Experiment 2 and Experiment 1. Experiment 2 reported the second-best performance, with 96.00% training accuracy and 96.15% validation accuracy. In contrast, Experiment 1 recorded the lowest accuracy, with 92.50% during training and 92.62% during validation.

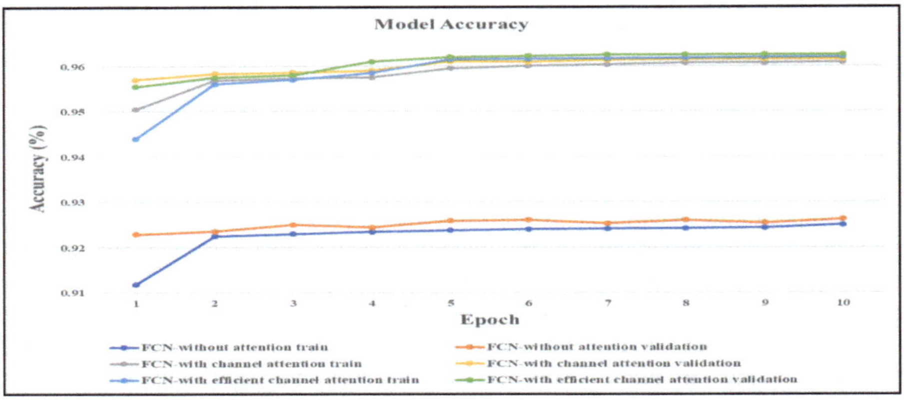

Fig. 2. Model accuracy of the training and validation KITTI dataset.

Furthermore, Fig. 2 presents the corresponding loss values for each configuration. Experiment 3 again achieved the most favorable outcome, with the lowest training and validation losses of 0.0015 and 0.0014, respectively. Experiment 2 exhibited slightly higher losses, with 0.0045 (training) and 0.0035 (validation). Meanwhile, Experiment 1 had the highest training and validation losses, recorded at 0.0055 and 0.0048, respectively. By integrating attention modules, the FCN architecture was able to more effectively highlight salient spatial features relevant to lane detection while suppressing background noise. This improved the accuracy of feature mapping and supported the learning of more discriminative features in the convolutional layers. As a result, the attention-enhanced FCN outperformed the baseline FCN in both accuracy and loss metrics, thereby validating the attention mechanism as a critical component in enhancing deep learning-based lane detection systems.

4.2 Model Performance on the Training and Validation Sets Using the Collected Dataset

To evaluate the robustness of the proposed approach on custom-collected data, three additional experiments were conducted: Experiment 4 (FCN without attention), Experiment 5 (FCN with channel attention), and Experiment 6 (FCN with efficient channel attention). The training and validation performance curves are illustrated in Fig. 3, respectively. The results indicate that Experiment 6 achieved the best overall performance, followed closely by Experiment 5, while Experiment 4 exhibited the lowest accuracy and highest loss.

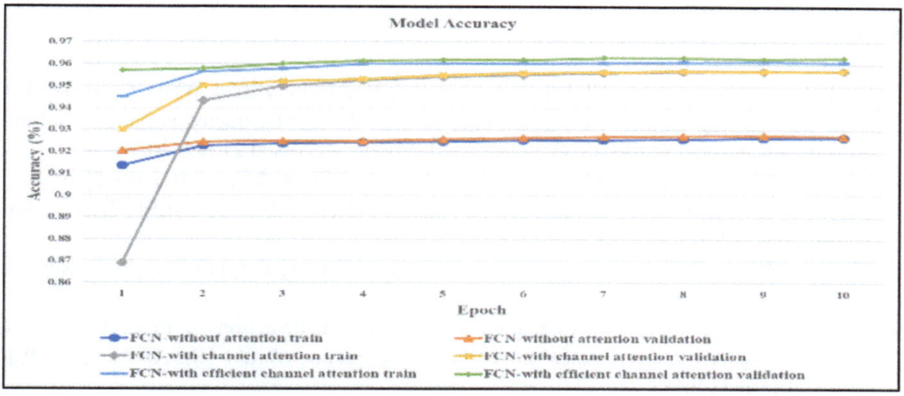

Fig. 3. Model accuracy of the training and validation DUKE-LATAR SYPSCS dataset.

By incorporating channel-wise and efficient channel attention blocks at the convolutional layers, the model learns to emphasize the most critical features for lane detection. This not only results in higher accuracy but also ensures reduced training loss and better generalization to unseen data. Furthermore, the attention mechanism enriches the feature representation by introducing global context into the network's local receptive fields, allowing the model to focus more effectively on lane line structures while suppressing irrelevant background information. In conclusion, the experimental results on the collected dataset clearly demonstrate that the integration of attention mechanisms significantly improves the learning dynamics and detection performance of the FCN model, making it more suitable for real-world deployment in intelligent transportation systems.

5 Conclusion

This study proposed a novel Fully Convolutional Network (FCN) with a channel-type attention mechanism to improve lane detection in ADAS, achieving over 96% accuracy and meeting all research objectives: optimizing deep learning parameters, incorporating channel-based attention to enhance local feature extraction, and validating the model on the KITTI benchmark, ranking among the top three in performance. The contributions include (1) integrating attention with FCN to overcome traditional lane detection limitations, (2) improving feature representation for sparse and thin lane markings, and (3) demonstrating robust performance on KITTI and DUKE-LATAR-SYPSCS datasets. Limitations include reduced performance under occlusion, hyperparameter sensitivity, slight lane boundary shifts, and higher computational cost. Future work will focus on improving robustness under complex conditions, integrating multiple attention mechanisms, and enhancing real-time performance and interpretability for practical deployment.

Acknowledgments. This research was not funded by any grant.

Disclosure of Interests The authors have no competing interests to declare that are relevant to the content of this article.

References

1. World Health Organization: Global Status Report on Road Safety 2024. WHO, Geneva (2024)
2. National Highway Traffic Safety Administration (NHTSA), "Distracted Driving," [Online]. https://www.nhtsa.gov/risky-driving/distracted-driving. Accessed 21 Feb 2026
3. European Commission, "Intelligent Transport Systems," 2024. Accessed 21 Feb 2026
4. Li, H., et al.: Lane detection for driver assistance systems: A survey. IEEE Trans. Intell. Transp. Syst. **22**(5), 2342–2357 (2021)
5. Mercedes-Benz, "Mercedes-benz expands automated driving capabilities." Press Release. 2024. Accessed 21 Feb. 2026
6. Pan, X., et al.: Spatial as deep: Spatial CNN for traffic lane detection. In: Proceedings of the Thirty-Second AAAI Conference on Artificial Intelligence (AAAI-18). AAAI Press, Palo Alto, California (2018)
7. European Union General Vehicle Safety Regulation. Regulation (EU) 2019/2144. Official Journal of the European Union. 2019. Accessed 21 Feb 2026
8. Aly, R., Bouaziz, S.: Limitations of traditional lane detection in challenging scenarios. Int. J. Automot. Technol. **24**(5), 789–798 (2023)
9. Li, H., Chen, Q., Wang, X., Li, L.: Lane detection for driver assistance systems: A survey. IEEE Trans. Intell. Transp. Syst. **22**(5), 2342–2357 (2021)
10. Zhang, J., Qiu, M., Wang, L.: Real-time lane detection using deep neural networks. IEEE Access. **8**, 68294–68303 (2020)
11. K. Simonyan and A. Zisserman, "Very Deep Convolutional Networks for Large-Scale Image Recognition," arXiv:1409.1556, 2015
12. Krizhevsky, A., Sutskever, I., Hinton, G.: ImageNet classification with deep convolutional neural networks. Commun. ACM. **60**(6), 84–90 (2017)
13. He, K., Zhang, X., Ren, S., Sun, J.: Deep residual learning for image recognition. In: Proceedings of the IEEE Conference on Computer Vision and Pattern Recognition (CVPR 2016), pp. 770–778. IEEE, Las Vegas, NV (2016)
14. Shi, X., et al.: Convolutional LSTM network: A machine learning approach for precipitation nowcasting. In: Proceedings of the 28th International Conference on Neural Information Processing Systems (NIPS 2015). MIT Press, Montreal, Canada (2015)
15. Long, J., Shelhamer, E., Darrell, T.: Fully convolutional networks for semantic segmentation. In: Proceedings of the IEEE Conference on Computer Vision and Pattern Recognition (CVPR 2015). IEEE, Boston, MA (2015)
16. Badrinarayanan, V., Kendall, A., Cipolla, R.: SegNet: A deep convolutional encoder-decoder architecture for semantic pixel-wise segmentation. IEEE Trans. Pattern Anal. Mach. Intell. **39**(12), 2481–2495 (2017)
17. Zhao, J., et al.: Deep learning based lane detection for autonomous driving. Sensors. **18**(12), 4271 (2018)
18. Wu, Z., et al.: Lane detection based on ConvLSTM networks. In: Proceedings of the IEEE Intelligent Vehicles Symposium (IV 2018). IEEE, Changshu, China (2018)
19. Noh, H., Hong, S., Han, B.: Learning deconvolution network for semantic segmentation. In: Proceedings of the IEEE International Conference on Computer Vision (ICCV 2015). IEEE, Santiago, Chile (2015)
20. Neven, A., De Brabandere, B., Georgoulis, S., Proesmans, M., Van Gool, L.: Towards end-to-end lane detection: an instance segmentation approach. In: Proceedings of the IEEE Intelligent Vehicles Symposium (IV 2018). IEEE, Changshu, China (2018)
21. Lee, M., Park, H., Kim, D.: Weakly-supervised road marking detection using multimodal data. Sensors. **21**(5), 1685 (2021)

22. Y. Huval et al., "An empirical evaluation of deep learning on highway driving," arXiv:1504.01716, 2015
23. Sun, Y., et al.: LaneNet: Real-time lane detection networks. In: Proceedings of the IEEE Intelligent Vehicles Symposium (IV 2020). IEEE, Las Vegas, NV (2020)
24. Kemsaram, N., Das, A., Dubbelman, G.: Robust lane detection and free space estimation using deep learning and DriveWorks. In: Proceedings of the IEEE International Conference on Intelligent Transportation Systems (ITSC 2021). IEEE, Indianapolis, IN (2021)
25. J. Philion, "FastDraw: Addressing the long tail of lane detection by adapting a sequential prediction network," arXiv:1905.03780, 2019
26. Dawam, E.S., Feng, X.: YOLOv3-based road surface marker detection for autonomous vehicles. In: Proceedings of the IEEE International Conference on Vehicular Electronics and Safety (ICVES 2022). IEEE, Istanbul, Turkey (2022)
27. Yang, L., et al.: Key-point-based lane detection using multi-stage CNN. In: Proceedings of the IEEE Intelligent Vehicles Symposium (IV 2020). IEEE, Las Vegas, NV (2020)

AI Enabled Wireless Power Transfer: A Research Landscape and Trend Analysis

Mohd Hafizuddin Mat[1,2](✉), Nurul Husna Abd Wahab[1], Siti Nur Aqilah Kamarudin[1], and Mohd Nazri A. Karim[1,2]

[1] Faculty of Electronic Engineering & Technology, Universiti Malaysia Perlis (UniMAP), Pauh Putra Campus, 02600 Arau, Perlis, Malaysia
hafizuddinmat@unimap.edu.my

[2] Advanced Communication Engineering (ACE), Centre of Excellence (CoE), Dera Ghazi Khan, Punjab, Pakistan

Abstract. This paper presents a bibliometric and science mapping assessment of research at the intersection of wireless power transfer (WPT) and artificial intelligence (AI). Records retrieved from the Scopus database using a structured title field query combining major WPT modalities and AI or computational intelligence methods, with subject area exclusions to focus on engineering and technical literature. The final corpus comprised 261 documents published between 2011 and 2026 across 109 sources. Descriptive indicators show sustained growth (7.6% annual growth rate) and strong recency (average document age 2.85 years), alongside a highly collaborative authorship structure (954 authors; 4.47 co-authors per document; 22.99% international co authorship). A logistic life cycle model provided a moderate fit to cumulative output ($R^2 = 0.691$), estimating peak growth around 2023.6 and an asymptotic saturation level of approximately 275 publications within the defined scope. The thematic map indicates that inductive power transmission and energy transfer function as central motor themes, while deep learning and reinforcement learning appear as specialized niche themes, and IPT, inductance, and particle swarm optimization represent foundational basic themes. Overall, the findings suggest a field transitioning from rapid expansion toward consolidation, with increasing emphasis on AI enabled optimization, adaptive control, and robust WPT system design.

Keywords: WPT · Artificial intelligence · Bibliometric

1 Introduction

Wireless power transfer (WPT) has moved from laboratory demonstrations to practical charging infrastructure, particularly for electric mobility, where contactless charging improves convenience, environmental sailing, and operational safety. Current high-power systems increasingly emphasize power density, modularity, and scalability, but they remain constrained by electromagnetic coupler design, compensation tuning, misalignment sensitivity, and the need to maintain safe field exposure while preserving

M. A. b. Jamlos et al. (Eds.): ICSET 2025, LNNS 1897, pp. 476–482, 2026.
https://doi.org/10.1007/978-3-032-20722-7_60

efficiency across realistic operating conditions [1]. At the same time, WPT spans both near field and far field regimes. Near field inductive and resonant solutions dominate EV and industrial charging, whereas far field radio frequency and microwave approaches depend critically on rectifier and energy harvesting behavior, waveform design, and compliance with safety regulations [2]. These properties create a coupled design space where geometry, materials, control, sensing, and rectification nonlinearities interact and are difficult to optimize purely with hand tuned models.

Recent work increasingly applies artificial intelligence (AI) and computational intelligence to reduce this complexity by enabling data driven modeling, accelerated optimization, and adaptive control. Examples include reinforcement learning for magnetic core structural optimization [9] and machine learning driven ferrite core design for EV charging pads [10]. In multi transmitter and multi receiver WPT architectures, metaheuristics such as particle swarm optimization have used to derive coil designs that improve anti offset performance while supporting coordinated control objectives [3]. For dynamic scenarios, deep learning is being explored as a control enabler in time varying WPT conditions [5], while sensor less or reduced sensor estimation of coupling and load characteristics can be achieved using transmitter side measurements and supervised learning models [6]. Complementarily, hybrid frameworks that combine global search (for example genetic algorithms), surrogate models (digital twins), and reinforcement learning for real time tuning have proposed to maintain high power transfer efficiency under misalignment and load variation in EV charging contexts [7]. This paper provides a research landscape and trend analysis of AI enabled WPT using a bibliometric and thematic mapping workflow based on records retrieved from major indexing databases (Scopus and Web of Science) within the 2020 to 2026 window. The study aims to (i) quantify the growth and topical structure of the field, (ii) identify dominant and emerging AI method families (for example deep learning, reinforcement learning, Bayesian and evolutionary optimization), and (iii) highlight under explored opportunities for robust, safe, and deployable WPT systems.

2 Methodology

This study employed a bibliometric research design using records retrieved from the Scopus database. A total of 261 bibliographic records exported forming the analytical corpus. The retrieved dataset spans the publication years 2011 to 2026, as indicated by the publication year metadata attached to each record. Notably, the presence of records dated beyond the current calendar year can occur in Scopus exports when items indexed as early access or articles in press and assigned a future publication year.

2.1 Metadata Fields and Document Typology

The exported Scopus records included standard bibliographic metadata required for performance and science mapping analyses, including publication year, source title, document type, author names, author affiliations, citation counts, reference lists, and keyword fields. Document type classification used to differentiate publication channels and dissemination modes. In this corpus, the dominant document categories were journal articles (140 records) and conference papers (115 records), with a small remainder distributed across less frequent types (for example book chapters, reviews, short surveys, letters, and errata). These classifications used as grouping variables in subsequent descriptive analyses.

2.2 Source and Outlet Analysis

Publication outlets analyzed by counting the number of distinct sources represented in the corpus. The dataset includes 109 unique sources (journals, conference proceedings, and other outlets), and source level productivity assessed through frequency distributions of documents per outlet. This procedure supports identification of dispersion versus concentration of publications across venues and provides a basis for later identification of core sources if required (for example via ranked lists or Bradford type concentration checks).

Scholarly impact assessed using citation counts available in the Scopus metadata at the time of export. Average citations per document computed (11.41 citations per document) to provide a corpus level estimate of citation uptake. In parallel, the intellectual base of the field approximated by aggregating the cited references across the dataset (2014 references in total), enabling a descriptive account of how extensively the corpus draws upon prior scholarship. Because citation accumulation is time dependent, citation indicators interpreted alongside the document age metric to ensure that the recency of publications explicitly accounted for in impact assessment.

To support subsequent thematic mapping and content analysis, keyword fields extracted and summarized. The dataset contains 1827 unique index keyword entries (reported as "Keywords Plus (ID)" in the bibliometric output) and 696 unique author supplied keywords (DE). These fields retained for downstream analyses such as keyword co-occurrence networks, thematic clustering, trend detection across time, and topic evolution mapping. The separation between index keywords and author keywords preserved to allow comparison between database assigned descriptors and author framed terminology.

2.3 Analytical Workflow Summary

In summary, the methodological workflow consisted of: (1) exporting a Scopus bibliographic dataset (n = 261) covering 2011 to 2026, (2) cleaning and standardizing key metadata fields, (3) computing descriptive performance indicators (growth rate and document age), (4) quantifying source dispersion and document typology, (5) computing collaboration indicators (team size, single authorship, and international co authorship), (6) estimating citation impact and reference base size, and (7) preparing keyword fields for thematic analysis.

2.4 Scopus Search Strategy and Final Query

A structured bibliographic search conducted in the Scopus database using a title restricted Boolean query to maximize precision and ensure that retrieved records explicitly addressed both wireless power transfer and artificial intelligence related methods. The search executed on 10 December 2025. The final query used in Scopus was (TITLE (("wireless power transfer" OR "wireless charging" OR WPT OR "inductive power transfer" OR IPT OR "magnetic resonance coupling" OR "capacitive power transfer" OR CPT OR "microwave power transfer" OR rectenna*) AND ("artificial intelligence" OR "machine learning" OR "deep learning" OR "neural network*" OR "reinforcement learning" OR "genetic algorithm*" OR GA OR "particle swarm" OR PSO OR "differential evolution" OR "swarm intelligence" OR "ant colony" OR "Bayesian optim*")) AND NOT TITLE (CPT)) AND (EXCLUDE (SUBJAREA, "ARTS") OR EXCLUDE (SUBJAREA, "IMMU") OR EXCLUDE (SUBJAREA, "VETE") OR EXCLUDE (SUBJAREA, "NURS") OR EXCLUDE (SUBJAREA, "PSYC") OR EXCLUDE (SUBJAREA, "PHAR") OR EXCLUDE (SUBJAREA, "BUSI") OR EXCLUDE (SUBJAREA, "BIOC") OR EXCLUDE (SUBJAREA, "AGRI") OR EXCLUDE (SUBJAREA, "CENG") OR EXCLUDE (SUBJAREA, "CHEM") OR EXCLUDE (SUBJAREA, "NEUR") OR EXCLUDE (SUBJAREA, "ECON") OR EXCLUDE (SUBJAREA, "SOCI")).

3 Results and Discussion

Figure 1 shows a strong upward publication trend from 2011 to 2025. Output is minimal in 2011 to 2014 (0 to 1 article per year), increases modestly from 2015 to 2018 (3 to 9 articles), then accelerates from 2019 onward, rising from 21 (2019) and 23 (2020) to 32 (2022) and 35 (2023).

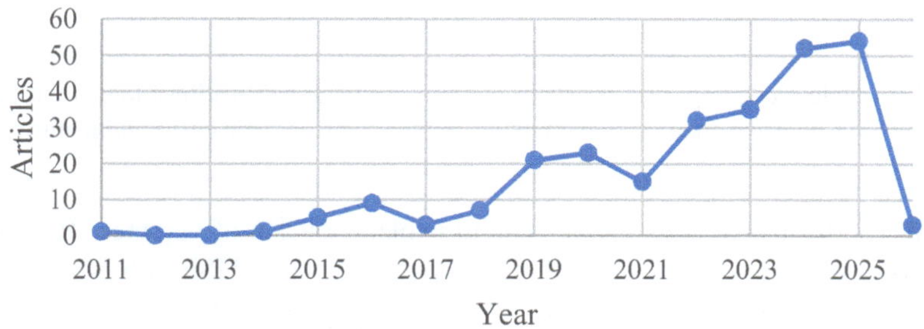

Fig. 1. Annual Scientific Production.

The peak occurs in 2024 and 2025 with 52 and 54 articles, indicating rapid recent growth in AI related wireless power transfer research. The sharp drop in 2026 (3 articles) is due to incomplete indexing or partial year coverage rather than a real decline.

3.1 Logistic Growth Model

The life cycle analysis indicates (Fig. 2) an S shaped cumulative growth pattern and a bell-shaped annual output pattern, consistent with a field that moved from early emergence to rapid expansion and is now approaching maturity. The logistic model estimates the inflection point at tm = 2023.56, implying that the fastest cumulative growth and the modelled peak annual output occurred around mid 2023. The estimated carrying capacity is K = 274.64 publications, suggesting the total cumulative output is likely to stabilize approximately 275 publications within the defined query scope. The expansion speed is relatively high, with delta_t = 6.71 years to move from 10% to 90% of K, indicating a short and intense growth window.

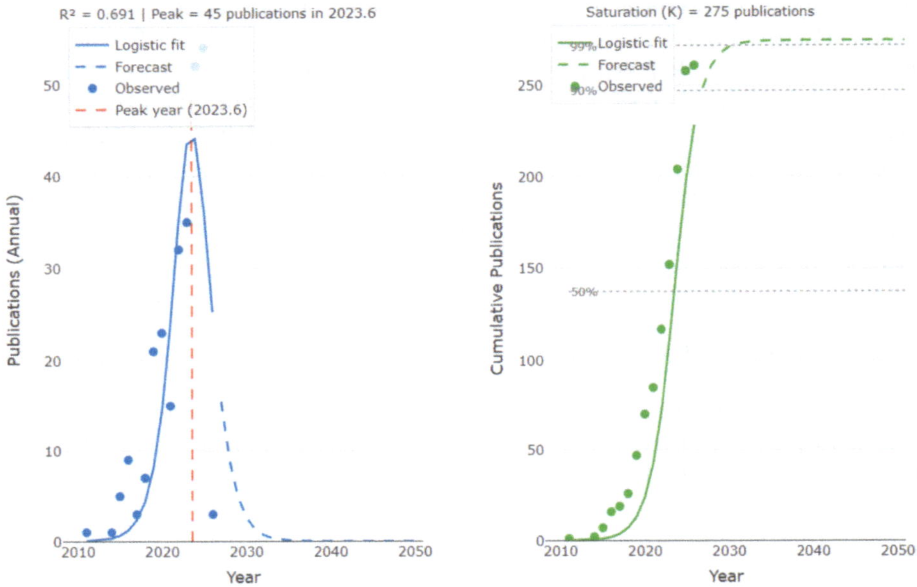

(a) (b)

Fig. 2. Life cycle modelling. (a) Life cycle analysis of annual publications: logistic fit and peak timing and (b) Life cycle analysis of cumulative publications: logistic curve and maturity phase.

Model performance is moderate. The fit explains 69.1% of variance ($R^2 = 0.691$) and the typical annual deviation is about 9.8 publications (RMSE = 9.82). Deviations are most visible near the peak, where observed annual counts exceed the smooth fitted peak (observed peak around the mid 2020s versus a model peak near 45), implying a sharper and more concentrated surge than the logistic trend captures. The low value in 2026 also departs strongly from the fitted decline and is more plausibly attributable to incomplete indexing or year coverage effects rather than a genuine structural collapse. Overall, the results support the interpretation that this topic has passed its strongest expansion phase and is entering a consolidation stage, where growth slows and output approaches a plateau. For research positioning, this typically implies fewer broad foundational contributions and more opportunities in synthesis, optimizations of established methods, and specialized sub problems within the existing knowledge base.

3.2 Clustering Map

Figure 3 Strategic thematic map showing the conceptual structure of the retrieved Scopus corpus based on keyword co-occurrence clustering. The horizontal axis represents theme relevance (centrality) and the vertical axis represents theme development (density). Motor themes (high centrality, high density) are dominated by core engineering topics such as wireless power transfer, energy transfer, and inductive power transmission.

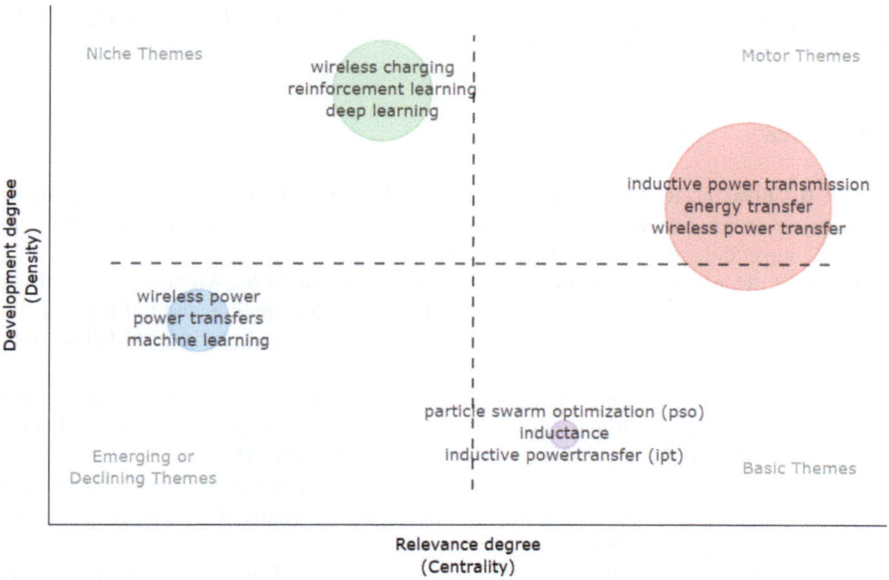

Fig. 3. Thematic map of research themes in AI enabled wireless power transfer literature.

Niche themes (low centrality, high density) include specialized AI approaches such as wireless charging with reinforcement learning and deep learning. Basic themes (high centrality, low density) reflect foundational and widely connected topics, including inductive power transfer (IPT), inductance, and particle swarm optimization (PSO). Emerging or declining themes (low centrality, low density) include more weakly consolidated clusters associated with wireless power and machine learning. Bubble size indicates the relative weight of each theme cluster in the dataset.

4 Conclusion

This study mapped research at the intersection of wireless power transfer (WPT) and artificial intelligence (AI) using a Scopus based bibliometric workflow combining productivity indicators, trend analysis, logistic life cycle modelling, and thematic mapping. Overall, the field shows rapid recent expansion, strong collaboration, and early signs of consolidation. The dataset contained 261 documents from 109 sources (2011 to 2026), with 7.6% annual growth and a low average age of 2.85 years. Outputs were journal articles (140) and conference papers (115). Collaboration was high (954 authors, 4.47 co-authors per paper, 6 single authored papers) with 22.99% international co authorship. Average impact was 11.41 citations per document. The logistic model showed a moderate fit ($R^2 = 0.691$; RMSE = 9.82), estimating peak growth at tm = 2023.6 and a plateau near $K \approx 275$ cumulative publications within the query scope. The thematic map identified WPT, energy transfer, and inductive power transmission as central themes, with deep learning and reinforcement learning as specialized themes; IPT, inductance, and PSO appeared as broadly relevant foundational themes. Future work should extend coverage beyond Scopus by integrating IEEE Xplore and Dimensions, and by including patents and standards to capture translational activity.

References

1. Feng, H., Tavakoli, R., Onar, O.C., Pantic, Z.: Advances in high-power wireless charging systems: overview and design considerations. IEEE Trans. Transp. Electrific. **6**(3), 886–919 (2020). https://doi.org/10.1109/TTE.2020.3012543
2. Clerckx, B., Huang, K., Varshney, L.R., Ulukus, S., Alouini, M.S.: Wireless power transfer for future networks: signal processing, machine learning, computing, and sensing. IEEE J. Sel. Topics Signal Process. **15**(5), 1060–1094 (2021). https://doi.org/10.1109/JSTSP.2021.3098478
3. Liu, J., Min, Y., Gao, J., Yang, A., Zhou, J.: Design and optimization of a modular wireless power system based on multiple transmitters and multiple receivers architecture. Front. Energy Res. **10**, 896575 (2022). https://doi.org/10.3389/fenrg.2022.896575
4. Xing, L., Liu, M., Du, L., et al.: Optimal path planning for wireless power transfer robot using area division deep reinforcement learning. Wireless Power Transf. **9**, 21885 (2022). https://doi.org/10.1155/2022/9921885
5. Bertoluzzo, M., Di Barba, P., Forzan, M., Mognaschi, M.E., Sieni, E.: A deep learning approach to improve the control of dynamic wireless power transfer systems. Energies. **16**(23), 7865 (2023). https://doi.org/10.3390/en16237865
6. Kim, M., Niada, W.Y.E.F., Park, S.: Predicting receiver characteristics without sensors in an LC–LC tuned wireless power transfer system using machine learning. Sensors. **24**(2), 501 (2024). https://doi.org/10.3390/s24020501
7. Ben Fadhel, Y., Bouattour, G., Bouchaala, D., Derbel, N., Kanoun, O.: Intelligent optimization and real-time control of wireless power transfer for electric vehicles. Electronics. **14**(22), 4478 (2025)
8. Wang, L., Wang, Y., Hu, S., et al.: Machine learning assisted long-range wireless power transfer. Prog. Electromagn. Res. **183**, 59–66 (2025). https://doi.org/10.2528/PIER25020801
9. Choi, B.-G., Lee, E.S., Kim, Y.-S.: Optimal structure design of ferromagnetic cores in wireless power transfer by reinforcement learning. IEEE Access. **8**, 179295–179306 (2020). https://doi.org/10.1109/ACCESS.2020.3027765
10. Choi, B.-G., Kim, Y.-S.: New structure design of ferrite cores for wireless electric vehicle charging by machine learning. IEEE Trans. Ind. Electron. **68**(12), 12162–12172 (2021). https://doi.org/10.1109/TIE.2020.3047041

Wearable Frequency Reconfigurable Antenna for 5G Applications

Shahrul Shoeb[1], Norhidayah Shahrizan[1], N. H. Ramli[1,2(✉)], and Mohd Hafizuddin Mat[1,2]

[1] Faculty of Electronic Engineering & Technology, Universiti Malaysia Perlis, Arau, Perlis, Malaysia
hidayahramli@unimap.edu.my

[2] Advanced Communication Engineering (ACE) Center of Excellence, Universiti Malaysia Perlis, Arau, Perlis, Malaysia

Abstract. The article discusses the development of a wearable reconfigurable antenna for 5G applications. By adjusting the switching state, the suggested antenna, which incorporates PIN diodes and varactors using cotton as a substrate, can alter its working frequency. The non switch antenna operates at 5.8 GHz and uses copper as PIN diodes to achieve frequency reconfigurability. The reconfigurable antenna can operate at 2.642 GHz, 4.496 GHz, and 7 GHz, enabling flexibility for varied 5G applications. With a tiny footprint of 55 mm × 50 mm × 1.07 mm, the antenna design is efficient and compact, making it perfect for wearable technology and portable devices in the developing 5G environment.

Keywords: Wearable antenna · Reconfigurable antenna · PIN diode · 5G application

1 Introduction

5G is the fifth-generation technology standard for cellular networks, which cellphone operators began building everywhere around the world in 2019, and is designed to eventually replace the 4G networks that currently connect cellphones to the internet. Smart cities, self-driving cars, virtual reality, and improved health care are just some of the new applications it enables. 5G, with big data transmission speeds, will allow real-time communication and "remote surgery and self-driving cars." The modem is also compatible with ultra-large MIMO, the multi-antenna technology designed to improve capacity, and millimeter wave communication. Yet, roadblocks in the form of infrastructure and security would persist even with the latest developments in machine learning, blockchain, as well as high-level communications. In general, 5G is expected to reshape the industry and our lives with its more efficient and reliable connections [1, 2].

In applications such as 5G systems, which require dynamic frequency choice, the frequency reconfigurable antennas (FRAs) have played an important role in these types of modern communications. These antennas can change their emission pattern, operating frequency, and, in some cases, polarization. Numerous switching mechanisms can

M. A. b. Jamlos et al. (Eds.): ICSET 2025, LNNS 1897, pp. 483–488, 2026.
https://doi.org/10.1007/978-3-032-20722-7_61

be used to achieve reconfigurability, such as PIN diodes, which change the resonant frequency by adjusting the electrical characteristics of the respective antenna. One well-known means of providing several operation frequency bands is to employ PIN diodes in the radiating patch of the antenna, so that it can be switched between the off state and the on state [3, 4].

In this work [5], a small-size frequency-tunable monopole antenna for sub-6 GHz 5G applications is proposed. The antenna can work in four frequency bands: 2 GHz, 2.45 GHz, 3.1 GHz, and 3.4 GHz, with the necessary frequency selectivity by a lumped element switch. The design was etched on an FR-4-based modified hexagonal-shaped patch, which has a small dimension of 37 × 35 × 1.6 mm. Single-band operation of the antenna for WLAN at 2.45 GHz or 3.1 GHz, and dual-band mode operations are obtained at 2 GHz and 3.4 GHz, making it a promising candidate for potential usage in 5G communication systems. Numerical studies based on CST Microwave Studio indicate a high efficiency (more than 85%), a good gain (higher than 1.5 dB), and a VSWR of less than 2 for all the operating bands. Due to the compactness of the design and wide bandwidth, it is a good candidate for portable and wearable devices used in new technologies.

In this article, a frequency reconfigurable antenna for 5G applications such as N78 and N79 bands (3.4–3.6 GHz and 4.8–5.0 GHz) is introduced. The frequency reconfigurability in the antenna is achieved by varying the length of an active dipole, wherein diodes are employed to switch it. It works at 3.5 GHz in its Yagi-Uda antenna mode and at 4.9 GHz in the short backfire antenna mode. This configuration can be adopted to achieve compact size, low cost of the antenna, and high gain and directivity for indoor multi-system application scenarios, like in a shopping mall or aboard a ship. The presented antenna offers a compact solution for the coverage of the important 5G bands [6].

2 Antenna Design and Specification

This project proposes a wearable frequency-reconfigurable metamaterial that operates at 5.8 GHz and 7 GHz frequencies. The antenna used is a microstrip antenna, and to achieve the desired frequency characteristics, a copper tape is employed to adjust the frequencies. The antenna specification can be seen in Table 1.

The antenna is designed using a 0.035 mm copper layer as the radiating element, supported by a 1 mm-thick cotton substrate that provides structural stability. The substrate's overall dimensions of 55 mm by 50 mm guarantee a small and effective design for the antenna's operation. The material and size combination aims to produce a lightweight, flexible antenna that is ideal for wearable technologies. The antenna design is shown in Fig. 1.

Table 1. Antenna Specification

Antenna Performance	Frequency Reconfigurable Antenna
Type of antenna	Microstrip Antenna
Operating Frequency reconfigurable	5.8 GHZ and 7 GHz
Reflection Coefficient, S11	< -10 dB
Efficiency	> 50%
Antenna size	55 mm × 50 mm × 1.07 mm
Reconfiguration Mechanism	Copper Tape

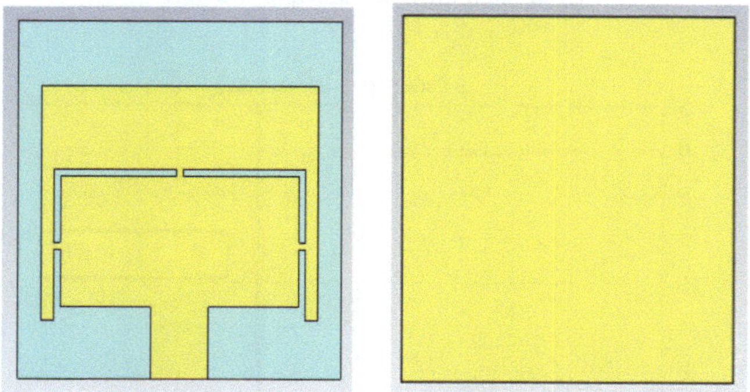
(a) (b)

Fig. 1. Reconfigurable antenna design (a) Front view (b) Back view

3 Antenna Performance

The antenna performance is examined by simulating both the design without the switching element and the proposed configuration in CST Microwave Studio, followed by an analysis of the resulting characteristics. The introduction of triple-conductor tape acting as a switch permits four alternative modes. Excitation is achieved by using the inset feeding technique. The features of the various modes are investigated in detail after turning on the switch.

3.1 Non-switched Antenna Performance

Figure 2 shows the non- switched antenna structure while Fig. 3 presents the simulated S11 parameter of the antenna across the 2–8 GHz frequency span. The antenna exhibits a distinct resonance at 5.78 GHz, corresponding to the minimum reflection coefficient observed in the response. At this resonant frequency, the S11 magnitude reaches −27.30 dB, indicating excellent impedance matching between the feed line and the radiating structure.

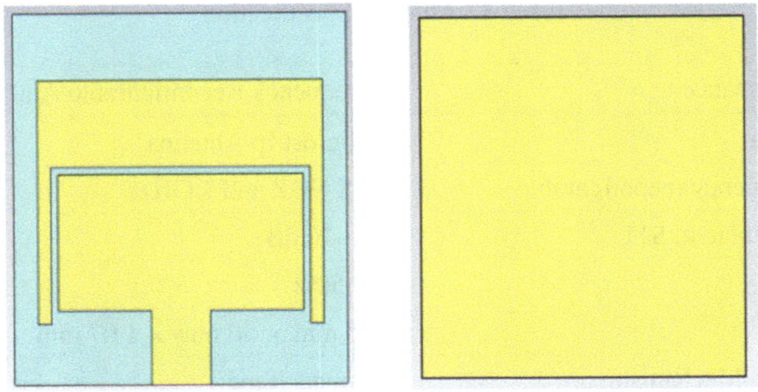
(a) (b)

Fig. 2. Non-switched antenna structure

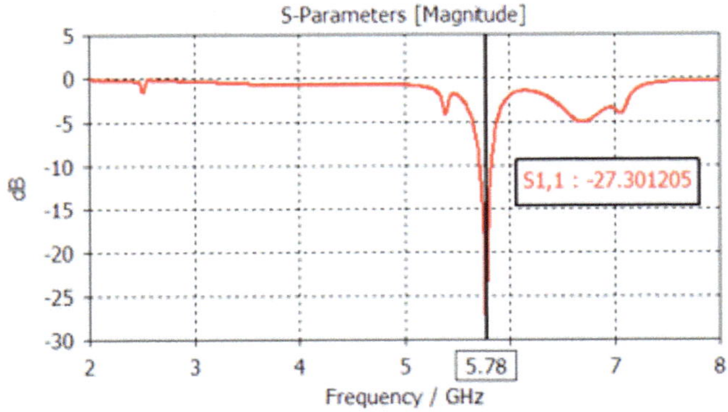

Fig. 3. Reflection coefficient of non-switched antenna

3.2 Antenna Performance with Four Different Switches

The antenna was evaluated under four switching configurations, and the results demonstrate distinct reconfigurable behavior in terms of reflection coefficient, operating frequency, gain, and radiation characteristics as shown in Table 2. In the baseline state (no switch activated), the antenna resonates at 5.78 GHz with an S11 of −27.30 dB and achieves the highest gain of 7.90 dBi, indicating excellent impedance matching. Activating Switch 1 shifts the resonance to 2.64 GHz with an S_{11} of approximately −23 dB and a gain of 5.19 dBi, while Switch 2 produces a mid-band resonance at 4.50 GHz with an S11 of −18.6 dB and a 6.54 dBi gain. When all switches are activated, the antenna also resonates near 4.50 GHz but with a degraded S11 of about −8.4 dB, resulting in reduced matching efficiency despite a similar broadside radiation pattern. Overall, the results confirm that the switching mechanism effectively reconfigures the operating band while maintaining stable radiation characteristics across all states.

Table 2. Antenna performance for four switching conditions

<table>
<tr><th>NO</th><th>Switch Condition</th><th>Antenna Structure</th><th>S-parameter</th></tr>
<tr><td>1</td><td>No Switch</td><td>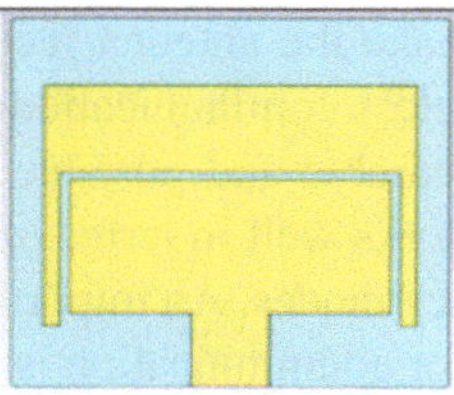</td><td>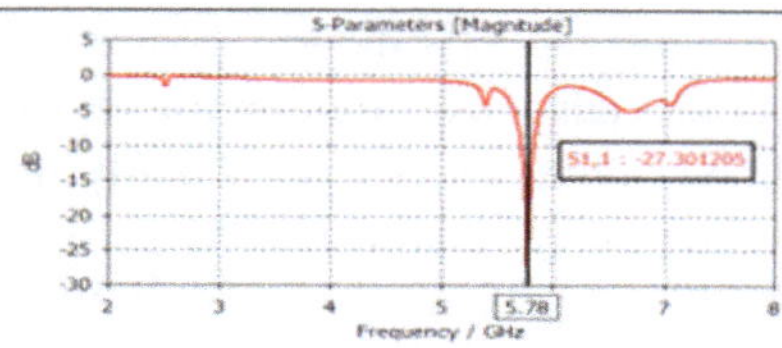
</td></tr>
<tr><td>2</td><td>Switch 1 ON</td><td>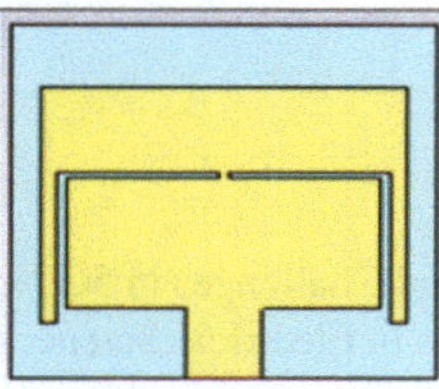</td><td>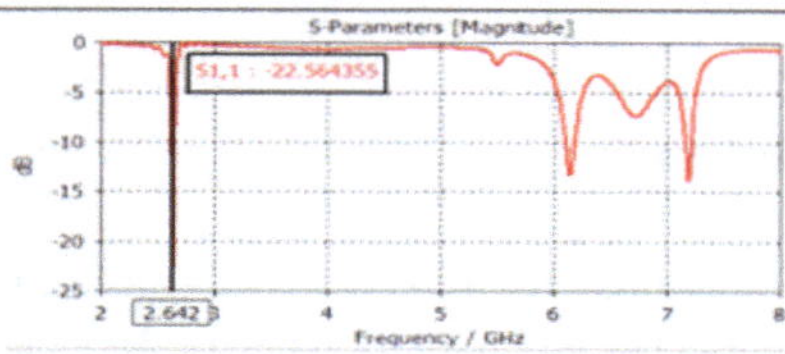
</td></tr>
<tr><td>3</td><td>Switch 2 ON</td><td>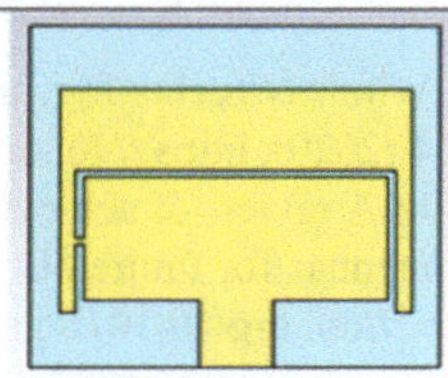</td><td>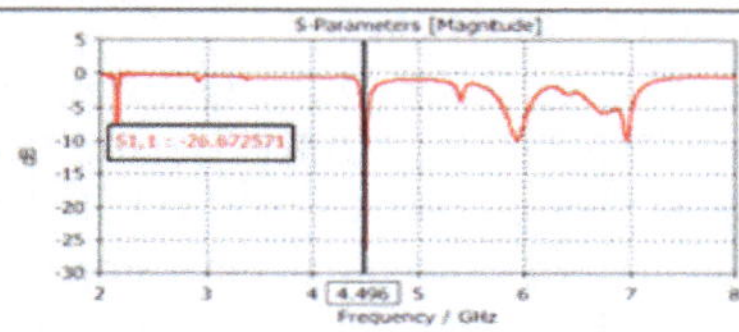
</td></tr>
<tr><td>4</td><td>Switch 3 ON</td><td>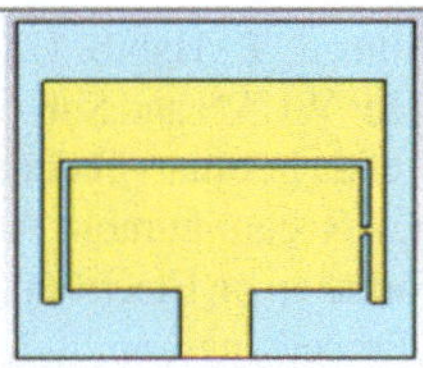</td><td>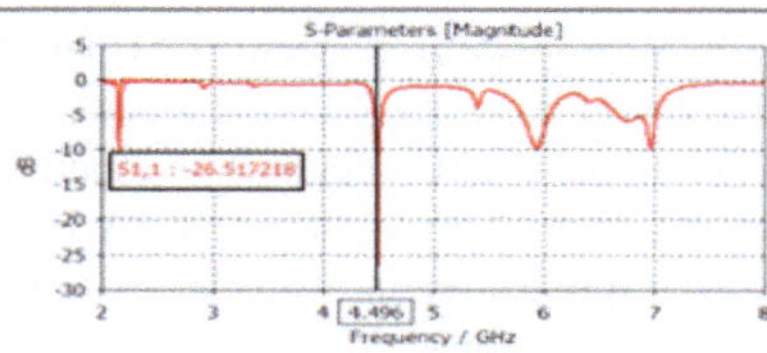
</td></tr>
<tr><td>5</td><td>All Switch ON</td><td>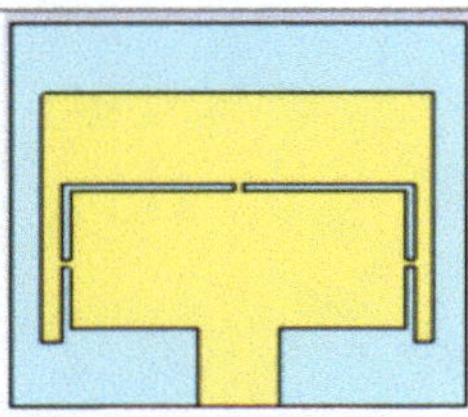</td><td>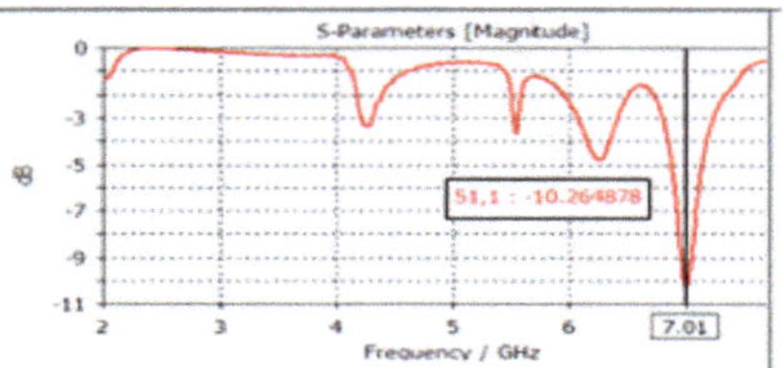
</td></tr>
</table>

4 Conclusion

A wearable, frequency-reconfigurable antenna for 5G applications is shown in this paper. It can function at several frequencies, such as 2.642 GHz, 4.496 GHz, and 7.01 GHz. Copper tape and PIN diodes are used to make the antenna reconfigurable, providing a small and adaptable solution for next wireless communication systems. The antenna is a good option for 5G, particularly in wearable and portable devices, according to the simulation results, which show that it works well in terms of reflection coefficient, bandwidth, and efficiency across all frequency modes. An important benefit for meeting the varied demands of modern technology for communication is the antenna's capacity to adjust to different frequency bands without compromising performance.

References

1. Sindhushree, K., Naik, D.C.: Advancements and Challenges in 5G Networks. In: International Conference on Smart Systems for Applications in Electrical Sciences, ICSSES 2023, pp. 1–6. Institute of Electrical and Electronics Engineers Inc. (2023). https://doi.org/10.1109/ICSSES58299.2023.10201156
2. Patra, G.R.R., Sharma, K.K.: Overview of 5G Wireless Technology and its Revolution. Int. J. Electr. Eng. Technol. (IJEET). **11**(10), 158–164 (2020). https://doi.org/10.34218/IJEET.11
3. Kamran Shereen, M., Khattak, M.I., Nebhen, J.: A review of achieving frequency reconfiguration through switching in microstrip patch antennas for future 5G applications. Alex. Eng. J. **61**, 29–40 (Jan. 01, 2022, Elsevier B.V). https://doi.org/10.1016/j.aej.2021.04.105
4. Abdulhussein, A.J., Farhan, M.J., Ali, G.M.: Design and implementation of a frequency reconfigurable antenna using PIN switch for sub-6 GHz applications. Open Eng. **13**(1), 20220453 (Jan. 2023). https://doi.org/10.1515/eng-2022-0453
5. Ullah, S., Ahmad, I., Raheem, Y., Ullah, S., Ahmad, T., Habib, U.: Hexagonal shaped CPW Feed based Frequency Reconfigurable Antenna for WLAN and Sub-6 GHz 5G applications. In: 2020 International Conference on Emerging Trends in Smart Technologies (ICETST) (2020)
6. Sun, K., Ji, Y., Sun, X., Guo, N.: A Frequency Reconfigurable Antenna Used for 5G-N78 and N79 application. In: IEEE International Workshop on Electromagnetics: Applications and Student Innovation Competition, iWEM 2021- Proceedings, pp. 1–3. Institute of Electrical and Electronics Engineers Inc., Guangzhou, China (2021). https://doi.org/10.1109/iWEM53379.2021.9790418

Optimization of Polyacrylamide in a Sodium Alginate Hydrogel for Wound Healing Applications

Nur Afiqah binti Shamsur and Siti Nur Aishah binti Mat Yusuf(✉)

Faculty of Chemical Engineering & Technology, Universiti Malaysia Perlis, Kompleks Pusat Pengajian Jejawi, 02600 Arau, Perlis, Malaysia
nuraishahyusuf@unimap.edu.my

Abstract. Polyacrylamide is a synthetic polymer composed of acrylamide subunits, while sodium alginate is a natural polymer derived from brown seaweed. Hydrogels, as three-dimensional water-insoluble polymer networks, possess excellent water absorption and retention capabilities, making them suitable for wound-healing applications. This study aimed to optimize the formulation of a cross-linked hydrogel by combining polyacrylamide and sodium alginate for enhanced wound-healing performance. The optimization process focused on evaluating the effects of varying polyacrylamide concentrations and peppermint oil infusion for antimicrobial activity. The hydrogel containing 15% (w/v) polyacrylamide exhibited the optimal swelling ratio (130%), indicating superior water absorption capacity, along with excellent adhesion (0.054%) and springiness (0.871%). In contrast, the 20% (w/v) formulation achieved the highest cohesiveness (0.771%), reflecting a denser internal network. Based on overall performance, the 15% (w/v) hydrogel was selected as the optimized matrix for peppermint oil infusion. The amount of oil added affected the swelling ratio, where higher oil content enhanced swelling due to reduced crosslinking density and increased hydrophobicity. Antimicrobial testing revealed clear zones of inhibition against both Gram-positive and Gram-negative bacteria, confirming the optimized hydrogel's antimicrobial efficacy. Overall, this study successfully optimized the polymer ratio and peppermint oil content to develop a hydrogel with improved physical integrity, swelling behavior, and antimicrobial properties for potential wound-healing applications.

Keywords: Sodium alginate · polyacrylamide · hydrogel · peppermint oil · wound healing

1 Introduction

Existing research on alginate, polyacrylamide (PAM) hydrogels, and peppermint oil in wound healing demonstrates both their individual and combined potential for novel wound care treatments. Alginate hydrogels, generated from brown seaweed, have been widely explored because to their high biocompatibility, moisture retention, and capacity to form gels in the presence of calcium ions, making them suitable for wound dressing [1].

M. A. b. Jamlos et al. (Eds.): ICSET 2025, LNNS 1897, pp. 489–494, 2026.
https://doi.org/10.1007/978-3-032-20722-7_62

PAM, a synthetic polymer, has been studied for mechanical strength and stability in hydrogel compositions. Polyacrylamide hydrogels may be adjusted for specific purposes by altering their cross-linking density, producing materials with desirable features such as high-water content and adjustable mechanical strength. These properties make polyacrylamide an appealing component for composite hydrogels intended for wound healing [2].

Studies on essential oils, such as peppermint, have also shown that they possessed antibacterial activity and promote tissue regeneration and provide pain relief, making them excellent additives to wound healing compositions. Recent research has focused on the production of composite hydrogels that take use of the capabilities of each constituent [3]. The aims of this research were to prepare antibacterial hydrogel based on sodium alginate and polyacrylamide with loaded peppermint oil, analyses the mechanical properties and investigate its antimicrobial activity.

2 Methodology

2.1 Alginate-Polyacrylamide Hydrogel Synthesis

Firstly, 2.5 g of sodium alginate was added in 50 ml of distilled water. The solution was stirred until homogenous. Next, 5 g of polyacrylamide was added in 50 ml of distilled water. The solution was stirred until homogenous. Both solutions were combined and stirred until homogenous. Calcium chloride was used to crosslink the hydrogel and left for 30 min. The $CaCl_2$ was then removed after 30 min. The steps were repeated with different percentage of polyacrylamide which were 7.5 g and 10 g of polyacrylamide.

2.2 Infusion of Peppermint Oil

10 μl of peppermint oil was infused in 10 ml of synthesis hydrogel. The hydrogel was then stirred until homogenous. The hydrogel was then immersed in calcium chloride solution for 30 min. After 30 min the calcium chloride solution was collected to use for UV-vis test. The steps were repeated using 50 μl and 100 μl of peppermint oil [4].

2.3 Characterisation of the Synthesised Hydrogel

2.3.1 Swelling Ratio

The initial weight of the hydrogel was recorded before starting the test. Hydrogels were placed into the beaker containing 100 ml distilled water. The hydrogels were allowed to swell for 30 min. The swollen hydrogels were removed carefully and weighed again. The swelling ratio of the hydrogel samples were calculated using Eq. 1 [5].

$$Swelling\ ratio = \frac{Ws - Wd}{Wd}\ x\ 100\% \quad (1)$$

In which, W_s is the weight of swelling hydrogel W_d is the weight of dry hydrogel.

2.3.2 Mechanical Analysis

Texture analyzer was used to obtain the elasticity, cohesiveness and adhesiveness of the hydrogel. The texture analyzer was equipped with with ½ spherical probe with 8 mm of diameter. The probe was squeezed at a speed of 1 mm/s for arrange of 5 mm into the hydrogel and removed. The pretest and post-test speeds were both 0.1 mm/s, and the force for the probe to start was 0.02 N. The time interval of 20 was set between the two compression cycles [6].

2.4 Antimicrobial Study

The antimicrobial activities of the hydrogels were evaluated using *Bacillus subtilis* (gram positive bacteria) and *Escherichia coli* (gram negative bacteria). The tested bacteria were diluted to1.5 × 10^8 CFU/mL and swabbed onto Mueller Hinton Agar. The samples used for antimicrobial were synthesised hydrogel, infused hydrogel, antibiotic, peppermint oil, and distilled water. The plates were incubated at 37 °C for 48 h [7].

3 Results & Discussion

3.1 Swelling Ratio

The swelling ratio reflects a hydrogel's ability to absorb and hold water or other solvents. This swelling capacity is a distinguishing feature of hydrogels and is critical to many of their uses [8]. Figure 1 shows the swelling ratio of synthesized hydrogel at different percentage of PAM (%).

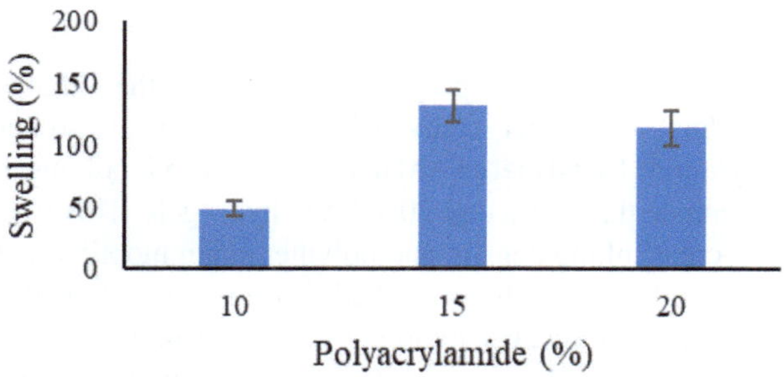

Fig. 1. Swelling percentage of synthesised hydrogel at different polyacrylamide percentage.

From Fig. 1, the highest swelling ratio is hydrogel with 15% w/v of polyacrylamide. The hydrogel has the largest swelling ratio due to its unique structural and chemical features, which make it extremely effective at absorbing and holding water or other solvents [9]. The hydrogel's crosslinked polymer network contains several binding sites for water molecules, resulting in widespread hydration and swelling. Furthermore, the hydrogel's porosity and large surface area help to explain its amazing swelling capability by allowing solvent molecules to diffuse quickly into its structure [10]. Furthermore, polymer composition, degree of crosslinking, and ambient variables can all influence the hydrogel's swelling behaviour [11]. Thus, 15% w/v of polyacrylamide was selected for infusion of peppermint oil.

3.2 Mechanical Analysis

Figure 2 shows the adhesion, springiness and cohesiveness of hydrogel at different percentage of PAM.

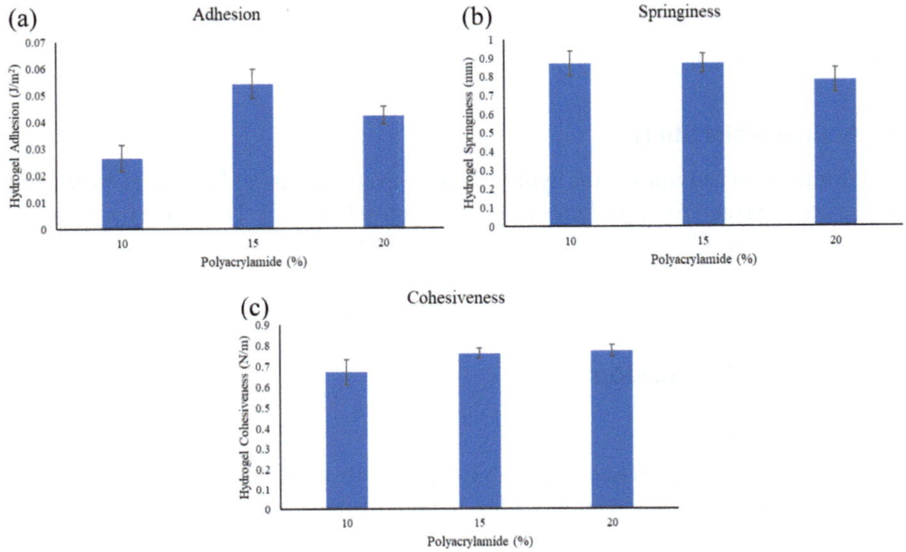

Fig. 2. The mechanical analysis of the synthesised hydrogel, (a) adhesion, (b) springiness, (c) cohesiveness.

Hydrogel with 15% w/v of polyacrylamide has the highest average of adhesion (0.054 J/m^2). Previous study in the mechanical properties of poly-dopamine/polyacrylamide found that a PAM hydrogel with 15% polymer content demonstrated greater adherence than 10% and 20% PAM hydrogels. This was attributable to an ideal balance of crosslinking density and polymer chain mobility at a concentration of 15%. In contrast, the 10% PAM hydrogel had a lower crosslinking density and fewer sticky functional groups, but the 20% hydrogel was too crosslinked, restricting fluidity and contact with the hydrogel. Hence, the 15% PAM hydrogel formulation achieved the best balance of swelling, crosslinking, and adhesive functioning, yielding the maximum adhesive strength among the concentrations examined [12].

Springiness is the ability of a material to return to its original shape and size after being stretch, compressed, or deformed. The study of springiness properties of polyacrylamide hydrogels discovered that PAM hydrogels with lower polymer concentrations, such as 10% and 15%, showed more significant changes in swelling ratio and local elastic modulus over time than higher concentrations, such as 20%. This shows that the 15% PAM hydrogel may have more dynamic and sensitive mechanical properties. This shows that the hydrogel can deforms under stress and the hydrogel is able to respond to small changes in applied stress [13].

The cohesiveness of hydrogel is the highest in 20% w/v of PAM. The hydrogel has the maximum cohesiveness because of its strong internal network structure and optimised crosslinking density. This closely woven network, produced by strong covalent and hydrogen connections, keeps the polymer chains well-integrated and resistant to stress, and allows the hydrogel to regain its shape following deformation, showing strong internal connection [14]. These features work together to guarantee that the hydrogel preserves its form and consistency, demonstrating high cohesiveness [15].

3.3 Antimicrobial Study

Table 1 shows the summary of bacterial inhibition zone. It shows that *B. subtilis* has larger inhibition zones for compared to *E. coli.*

Table 1. Microbial inhition zone.

	Inhibition zone (mm)	
	B. subtilis	*E. coli*
Peppermint oil	6.67 ± 0.5	6.3 ± 0.577
Synthesised hydrogel	0	0
Infused Hydrogel	12.67 ± 1.15	8.7 ± 0.577
Positive control	13.33 ± 0.56	12.67 ± 1.15
Negative control	0	0

[a] Data are represented as mean ± SD, n = 3

The inhibition was contributed to the content of catechin and neomenthol from peppermint [16]. Previous study also showed inhibition from peppermint oil as it contains menthol with alcohol functional group and menthone with ketone functional group [17].

4 Conclusion

In conclusion, the study successfully synthesised an alginate-polyacrylamide hydrogel infused with peppermint oil. The 15% w/v polyacrylamide hydrogel has ideal characteristics for wound healing applications. However, the loss in porosity produced by peppermint oil infusion requires further consideration based on the individual use situation.

References

1. Solberg, A.: Alginate blocks and block polysaccharides: A Review. Macromol. Symp. **408**(1), 2200072 (2023)
2. Tangri, A.: Polyacrylamide based Hydrogels: Synthesis, characterization and applications. Int. J. Pharm. Chem. Biol. Sci. **2014**(4), 951–959 (2014)
3. Shelepova, O.V.: The history of the introduction of peppermint (Mentha × piperita L.) in Imperial Russia. BIO Web Conf. **38**, 00115 (2021)

4. Zhang, K.: Protocol efficiently measuring the swelling rate of hydrogels. MethodsX. **7**, 100779 (2020)
5. Richbourg, N.R.: The swollen polymer network hypothesis: Quantitative models of hydrogel swelling, stiffness, and solute transport. Prog. Polym. Sci. **105**, 101243 (2020)
6. Szulc-Musioł, B.: Formulation and evaluation of hydrogels based on sodium alginate and cellulose derivatives with quercetin for topical application. Appl. Sci. **13**(13), 7826 (2023)
7. Farasati Far, B.: Enhanced antibacterial activity of porous chitosan-based hydrogels crosslinked with gelatin and metal ions. Sci. Rep. **14**, 7505 (2024)
8. Park, H.: Effect of Swelling Ratio of Injectable Hydrogel Composites on Chondrogenic Differentiation of Encapsulated Rabbit Marrow Mesenchymal Stem Cells In Vitro. Biomacromolecules. **10**(3), 541–546 (2009)
9. Duan, J.: Preparation and Drug Release Properties of a Thermo Sensitive GA Hydrogel. Polymers. **13**(1), 119 (2020)
10. Parhi, R.: Cross-Linked Hydrogel for Pharmaceutical Applications: A Review. Adv. Pharm. Bull. **7**(4), 515–530 (2017)
11. Thang, N.H.: Polymer-based hydrogels applied in drug delivery: An overview. Gels. **9**(7), 523 (2023)
12. Peng, L.: Dramatic improvement in the mechanical properties of polydopamine/polyacrylamide hydrogel mediated human amniotic membrane. RSC Adv. **13**(6), 3635–3642 (2023)
13. Subramani, R.: The influence of swelling on elastic properties of polyacrylamide hydrogels. Front. Mater. **7**, 8 (2020)
14. Song, R.: Enhanced strength for double network hydrogel adhesive through cohesion-adhesion balance. Adv. Funct. Mater. **34**(23), 2313322 (2024)
15. Xue, B.: Hydrogel tapes for fault-tolerant strong wet adhesion. Nat. Commun. **12**(1), 7156 (2021)
16. Osawa, K.: The antibacterial activities of peppermint oil and green tea polyphenols, alone and in combination, against enterohemorrhagic Escherichia coli. Biocontrol Sci. **4**(1), 1–7 (1999)
17. Yunilawati, R.: Peppermint oil loaded on recycled paper as an antibacterial label for shrimp freshness. IOP Conf. Ser. Earth Environ. Sci. **739**, 012065 (2021)

Computational Text-Analysis of the Safety Culture of Building Safety Accidents: A Case Study of the Wanda Plaza (Changsu) Underground Parking Pipe Collapse

Lin Longxing[1,2], Suriani Ismail[1(✉)], and Nur Aira Abd Rahim[1]

[1] Department of Professional Development and Continuing Education, Faculty of Educational Studies, UPM, Serdang, Malaysia
gs68469@student.upm.edu.my, suriani_ismail@upm.edu.my
[2] Zhejiang Kejia Engineering Consulting Co., Ltd., Hangzhou, Zhejiang Province, China

Abstract. This study explores the relationship between safety culture and construction safety accidents through a case study of the "5·6" pipe collapse in Changshu Wanda Plaza's underground parking lot. By reviewing safety culture literature and analyzing the official accident investigation report, it identifies six context-specific safety culture elements in Chinese construction projects: regulatory stringency, risk perception and management, safety training and education, responsibility attribution and accountability, safety culture atmosphere, and technical support and innovation. The findings highlight the critical role of these elements in accident prevention and provide a framework for improving construction safety culture.

Keywords: Accident Investigation Report · Safety Culture · Project Safety Culture Elements · Case Analysis

1 Introduction

Safety culture, defined as shared beliefs, values, and norms shaping safety prioritization and implementation, is pivotal in high-risk industries like construction (Pidgeon, 1991; Reason, 1997). Despite its recognized importance, few studies conduct in-depth single-case analyses to derive safety culture elements. This study uses the Changshu Wanda Plaza pipe collapse (three fatalities) to examine how safety culture deficits contribute to accidents, aiming to refine safety culture evaluation and improvement strategies for construction projects.

M. A. b. Jamlos et al. (Eds.): ICSET 2025, LNNS 1897, pp. 495–500, 2026.
https://doi.org/10.1007/978-3-032-20722-7_63

2 Literature Review

2.1 Safety Culture

Safety Scholars have conceptualized safety culture as a multidimensional construct: Cooper (2000) proposed psychological, behavioral, and situational dimensions; Zohar (1980) linked safety climate to employee behavior; and Schein (1992) emphasized deep-seated cultural assumptions. For construction, Choudhry et al. (2007) highlighted its dynamic, context-specific nature, while Neal and Griffin (2002) underscored its impact on safety performance. Based on these characteristics and roles, safety culture can be defined as: a collection of shared beliefs, values, and behavioral norms about safety within an organization, influencing members' attitudes and behaviors through psychological, behavioral, and situational dimensions, and enhancing overall safety performance. Safety culture exhibits unique characteristics and roles at different developmental stages and in various cultural contexts.

2.2 Models of Safety Culture Element

Key models include Cooper's (2000) three-dimensional framework, Parker et al.'s (2018) policy-practice-belief model, and Zhang et al.'s (2017) cross-cultural model. These emphasize leadership, employee participation, and systemic integration but lack context-specific adaptation for Chinese construction.

2.3 Construction Project Safety Culture

Accident reports provide critical qualitative data on safety culture implementation. Studies (e.g., Chen & Wang, 2022; Zhang & Niu, 2022) use reports to identify causes like inadequate supervision and poor training, but gaps remain in linking report findings to safety culture elements.

3 Methodology

3.1 Case Background and Record

The "5·6" accident (May 6) involved the collapse of an underground parking lot's air conditioning cooling water pipeline in Changshu Wanda Plaza, killing three. Direct causes: improper selection/installation of expansion bolts (insufficient strength). Indirect causes: failures by construction, supervision, design, and government entities.

3.2 Data Collection and Processing

TData were extracted from the official investigation report, transcribed, categorized (direct/indirect causes, responsible parties, recommendations), and coded (Table 1).

Table 1. Collection of Structured Data from the Official Report

Code	Category	Information
Immediate cause	Technical	Improper selection and installation of expansion bolts led to inadequate strength of No.3 support bracket, resulting in collapse of cooling water pipeline.
	Human	Failure to prepare construction scheme for HVAC project as required, and lack of supervision on construction quality.
	Organizational	Construction unit directly designated subcontractor, and general contractor failed to implement unified management on construction units.
Indirect cause	Organizational culture	Supervisory unit failed to detect and stop irregularities of subcontractor.
	Leadership	Architectural design unit did not specify calculation requirements for construction unit.
	Training	Inadequate technical training for construction unit.
Responsible party	Management	Architectural design unit, construction unit, and supervisory unit failed to fulfill supervisory responsibilities.
	Execution	HVAC construction unit did not carry out construction as required.
Recommendations	Safety policy	Improve construction technical standards and strengthen quality supervision.
	Training plan	Enhance technical training for construction units.

3.3 Reliability and Validity

Triangulation (cross-verifying with news and official statements), expert member checking, and thick case description ensured rigor. Ethical compliance included anonymizing sensitive data and maintaining analytical objectivity.

4 Findings and Discussion

Hofstede's cultural diversity theory provides a method for analyzing organizational behavior and cultural differences. Based on this theory, we can redefine several new dimensions to explain the occurrence of the "5.6" large collapse of the pipeline in the underground parking lot of Wanda Plaza in Changshu City. These new dimensions include the strictness of regulations and systems, risk perception and management, safety training and education, responsibility attribution and accountability, safety culture atmosphere, and technical support and innovation.

4.1 Regulatory Stringency

The strictness of the organization in safety regulations and systems, including the detail of safety regulations, the intensity of implementation, and the consequences of violating regulations. In this accident, the calculation book for the expansion bolts of the support and hanger provided by Jiangsu Qi'an Construction Engineering Co., Ltd. only considered the static load and water flow load, but did not consider the vibration and other dynamic loads generated by the car during the use of the underground parking lot. In addition, the calculation book was not reviewed by the supervision unit, nor was it submitted to the general contractor or design unit for approval. This reflects that the implementation of regulations and systems is not strict, resulting in safety hazards not being discovered and resolved in a timely manner.

4.2 Risk Perception and Management

The degree of risk perception of organizations and individuals, and the ability to take corresponding measures to manage and reduce risks. The accident investigation showed that the construction and management parties had insufficient risk perception and management. For example, the problem of insufficient diameter of the expansion bolts of the support brackets and the removal of the bolt heads was not discovered, which led to the collapse of the cooling water pipe due to vibration and dynamic load during the operation stage. In addition, the workers and team leaders relied on experience and blind self-confidence during the construction process, and failed to effectively conduct technical briefings and feedback, further increasing the risk.

4.3 Safety Training and Education

The organization's investment in safety training and education, including the frequency of employee training, the breadth and depth of training content. In this accident, the on-site construction personnel and supervisors lacked sufficient safety training and professional knowledge, resulting in construction without in-depth design drawings and construction plans, and failed to identify and correct errors in the construction process. In particular, the feedback link of the team and workers on safety management was missing, and the PDCA cycle was not completed, making it difficult for the safety management culture to form a closed loop.

4.4 Responsibility Attribution and Accountability

After the accident, the organization's attribution of responsibility and the degree of perfection of the accountability mechanism. The accident investigation report clearly identified the responsibilities of multiple parties, including the project manager, construction worker, supervisor, etc., but the attribution of the team and subcontractor, workers and team leaders was not pursued. This led to the general contractor's quality management, purchasers and supervisors ignoring the importance of on-site workers and team leaders, failing to timely feedback and solve on-site problems, resulting in a situation of decentralized responsibility and imperfect accountability mechanism.

4.5 Safety Culture Climate

The degree of attention paid to safety within the organization and the overall safety atmosphere formed. The report shows that the supervision of regulatory agencies such as Changshu Housing and Urban-Rural Development Bureau and Quality Supervision Station is not comprehensive, timely and effective, the safety culture atmosphere within the organization is weak, and a good safety management system and awareness have not been formed. This led to the failure to effectively monitor and manage the dynamic load of the underground parking lot during the operation phase, which ultimately led to the accident.

4.6 Technical Support and Innovation

The organization's support for safety technology and the degree of innovation in safety management technology. Insufficient technical support and a lack of innovation led to serious issues in the technical calculation and installation of support and hangers, and construction technical problems were not identified and addressed promptly. Especially during the operational stage, there was a lack of technical support and innovative management measures to address the dynamic load in underground parking lots, resulting in a failure to eliminate safety hazards in a timely manner.

Through the analysis of the above dimensions, we can more comprehensively understand the causes of the "May 6" major collapse of the underground parking lot pipeline in Wanda Plaza, Changshu" and the diversity of safety culture behind it. These dimensions provide a framework for evaluating and improving an organization's safety culture to prevent similar accidents.

Acknowledgments. My deepest gratitude goes first to my supervisor, SURIANI ISMAIL, and co-supervisor, NUR AIRA ABD RAHIM, for their unwavering guidance throughout my academic journey. Their rigorous academic spirit, which demands precision in every analysis and clarity in every argument, has shaped my research approach.

I am also deeply grateful to the organizing committee of the second International Conference on Science, Engineering and Technology Application (ICOSETA 2025) for providing a prestigious platform to present this research. Their efforts to curate academic exchange opportunities and foster dialogue among professionals in the sciences, engineering, and technology have significantly contributed to the refinement of this work.

As the chief engineer and supervision engineer at Zhejiang Kejia Engineering Consulting Co., Ltd.. To my beloved son—thank you for being my anchor. Raising you as a single father has taught me the true meaning of responsibility, patience, and resilience. Your laughter and healthy growth remind me daily of the importance of safety and care in every aspect of life, a lesson that has deeply informed this study on safety culture. I also wish to express my heartfelt gratitude to Daisy, whose presence in my life for over two years remains an unforgettable chapter. The love we shared became a profound source of motivation, driving me to persevere through the challenges of doctoral studies and professional responsibilities. Though our paths diverged, the warmth and strength from that period continue to resonate, and for that, I am truly grateful. This work is a testament to all who have walked alongside me in academia, work, and life.

References

Choudhry, R.M., Fang, D., Mohamed, S.: The nature of safety culture: A survey of the state-of-the-art. Saf. Sci. **45**(10), 993–1012 (2007)

Chen, Z., Wang, T.: Accident cause importance assessment method based on text mining and complex network: taking large-scale housing and municipal accidents as an example. China Occup. Saf. Sci. Technol., 224–230 (2022)

Cooper, M.D.: Towards a model of safety culture. Saf. Sci. **36**(2), 111–136 (2000)

Neal, A., Griffin, M.A.: Safety climate and safety behavior. Aust. J. Manag. **27**(1_suppl), 67–75 (2002)

Parker, D., Lawrie, M., Hudson, P.: Safety culture and climate: Integrating psychological and systems perspectives. Springer International Publishing, Cham (2018)

Pidgeon, N.: Safety culture and risk management in organizations. J. Cross-Cult. Psychol. **22**(1), 129–140 (1991)

Reason, J.: Managing the risks of organizational accidents. Ashgate, Aldershot (1997)

Schein, E.H.: Organizational culture and leadership (2nd ed.). Jossey-Bass, San Francisco (1992)

Zhang, J., Yang, D., Sheng, Y.: Cross-cultural comparison on safety culture and safety climate. Saf. Sci. **99**, 221–229 (2017)

Zhang, H., Niu, Y.: How does accident investigation promote risk prevention?—An empirical analysis based on 167 accident investigation reports. Adm. Forum **2**, 62–73 (2022)

Zohar, D.: Safety climate in industrial organizations. J. Appl. Psychol. **65**(1), 96–102 (1980)

Assessment of Skid Resistance on Road Markings Towards Pavement Surface Conditions and Temperatures

Nora Sheda Mohd Zulkiffli[1](✉), Azzuhana Roslan[2], Rizati Hamidun[3], Syed Tajul Malik Syed Tajul Arif[1], Mohd Shafie Nemmang[1], and Sharifah Allyana Syed Mohamed Rahim[1]

[1] Traffic and Infrastructure Unit (TRAFIK), Malaysian Institute of Road Safety Research (MIROS), 43000 Kajang, Selangor, Malaysia
norasheda@miros.gov.my

[2] Sustainable Mobility and Transport Unit (MAMPAN), Malaysian Institute of Road Safety Research (MIROS), 43000 Kajang, Selangor, Malaysia

[3] Road Safety Engineering and Environment Centre (REER), Malaysian Institute of Road Safety Research (MIROS), 43000 Kajang, Selangor, Malaysia

Abstract. Road markings play a substantial role in conveying essential information to road users without diverting their attention from the road, and are commonly used for traffic guidance, lane demarcation, and road safety enhancement. Effective road markings must offer good visibility and adequate skid resistance, even under wet conditions. This study assesses the skid resistance of nine road marking types at various locations in Klang Valley under dry and wet surfaces, and varying surface temperatures using a British Pendulum Apparatus. Markings were grouped into longitudinal, transverse, other, and no markings. Results show a significant drop in SRV under wet conditions, ranging from 19% to 32%. Although some values have decreased, they still meet the minimum safety threshold requirements. A weak correlation between surface temperature and SRV was observed. These findings suggest incorporating routine SRV testing into regular maintenance planning, particularly under wet conditions, and further research is recommended to explore other factors that may influence skid resistance.

Keywords: Skid Resistance · Road Markings · Skid Resistance Value

1 Introduction

Road traffic safety is a continuous global concern, particularly for vulnerable road users such as motorcyclists, who are reported to be 27–28 times more likely to die in a traffic crash compared to car occupants [1, 2]. Poor road surface conditions, mostly during bad weather conditions, are a major cause of loss-of-control crashes [3]. Skid resistance, which is the frictional force that prevents tires from slipping, plays a vital role in preventing such crashes. However, road marking materials and conditions of road markings vary significantly, which affects their durability, visibility, and friction [4], and reduced

M. A. b. Jamlos et al. (Eds.): ICSET 2025, LNNS 1897, pp. 501–507, 2026.
https://doi.org/10.1007/978-3-032-20722-7_64

skid resistance heightened risks during accelerating, cornering, or braking on slippery or wet road surfaces [4–6]. Furthermore, worn-out markings can further hinder motorcyclists' ability to maintain control, emphasizing the importance of maintaining good visibility and maintain adequate skid resistance as recommended by Jabatan Kerja Raya [8]. Skid Resistance Value (SRV) is a widely accepted metric for evaluating surface grip, and decreased friction under wet conditions significantly increases skidding risk [9].

The Transport and Road Research Laboratory (TRRL) (1969), as cited by Suhaimi et al. [10], outlining the minimum requirements of skid resistance for urban roads SRV of 55 and stated that skid resistance consists of two major components, which are adhesion and hysteresis, which depend on pavement's microtexture (fine surface roughness) and macrotexture (larger surface irregularities).

Surface temperature is another important factor influencing SRV. Previous studies by Bao et.al, [11] and others [7, 12, 13] indicate that higher pavement temperature reduce SRV due to binder softening and lower surface friction. Similarly, VicRoads [14] reports that as surface temperature increases reduce hysteresis in viscoelastic tyre and bitumen materials. In Malaysia's hot, humid, and monsoon climate both surface temperature and condition strongly influence skid resistance. Enhancing surface texture through aggregate selection, surface treatments, or maintenance can improve SRV and reduce crash risk [10, 15, 16], and regular maintenance, such as surface cleaning, gravel sealing, micro surfacing, and ultra-thin overlays, is able to restore and increase pavement skid resistance, which is important for road safety, especially under wet conditions [17–19].

A preliminary study was conducted to assess how road marking conditions and road surface temperature influence SRV in a selected urban area in Selangor, Malaysia, emphasizing several underexplored aspects. The study aims to determine minimum SRV threshold conditions necessary for safe road operation and to determine the relationship between road marking surface temperature and SRV, anticipating that higher temperatures reduce skid resistance. Findings are expected to inform safety recommendations for the road markings maintenance, ensuring markings remain visible and functional over time, despite wear and tear due to traffic, weather conditions, and other degrading factors. Way forward, the results could guide road safety policies and improve infrastructure standards for road markings.

2 Methodology

SRV data was collected across nine (9) road markings in the Klang Valley, grouped into longitudinal, transverse, other, and no markings. Each marking was tested under both dry and wet surface conditions. The road surface temperature was recorded during each test. A skid resistance apparatus was used to measure friction on a road markings surface by using the British Pendulum Apparatus and surface temperature was recorded with an OMRON MC-720.

2.1 Categories of Road Markings

By referring to the guidelines by Jabatan Kerja Raya [8], the road markings were classified into four (4) main categories, consisting of longitudinal markings, transversemarkings, other markings, and no marking, as in Table 1:

Table 1. Road markings classifications.

No.	Road Marking	Marking Details	Item
1.	Longitudinal	Broken, unbroken or combined lines guiding traffic and restricting overtaking in hazardous areas	Continuity lines, Discontinuity lines and Edge lines
2.	Transverse	Run across traffic direction to prompt actions like stopping or yielding	Pedestrian Crossing and Transverse Yellow Bars
3.	Other	Other markings applied on the pavement	Arrow, Hatching, Wording and Yellow Boxes
4.	No Marking	Pavement without painted lines or symbols	Unpainted roadway

2.2 Data Analysis

Descriptive statistics, including mean, median, minimum, maximum, and standard deviation, were used to compute SRV for dry versus wet conditions. Spearman's rank correlation test was used to assess the relationship between surface temperature and SRV. The correlation coefficient measures the strength of the relationship between surface temperature and SRV. The correlation coefficient derived from the sample, $r = \pm 1$ for a perfect relationship to 0 for no relationship, with negative values indicating a negative linear relationship and positive values indicating a positive linear relationship [20, 21].

3 Results

The skid resistance was tested based on dry and wet conditions, which resulted summarized in Table 2. The findings showed the average SRV for dry conditions is substantially higher as compared to wet conditions, confirming that there is a reduction in skid resistance when moisture is present. Although standard deviations are similar for both conditions, the lower SRV value on wet surfaces highlights a higher risk of skidding and also reduces the braking efficiency.

Table 2. Performance of SRV by road conditions.

Surface Condition	Average SRV	Min SRV	Max SRV	Std. Dev
Dry	86.05	67.10	104.20	7.96
Wet	65.51	48.43	86.77	7.79

Figure 1 presents a bar chart of SRV for each road marking under dry and wet conditions, showing a clear SRV decline in wet conditions across all road marking types, particularly on unmarked roadway. According to Suhaimi et al., [10] indicated that the threshold for a minimum skid resistance is 55 for urban areas. Despite the reduction in wet conditions, all measured SRV values remain above this threshold.

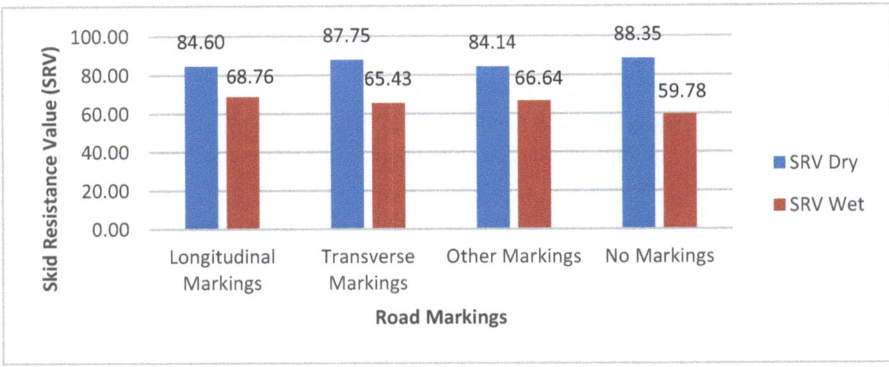

Fig. 1. Skid resistance value (SRV) by road markings and road surface conditions.

Other than that, correlation analysis was conducted to explore the relationship between temperature and SRV. Figure 2 demonstrates a scatter plot illustrating the weak correlation between surface temperature and SRV under different road surface conditions.

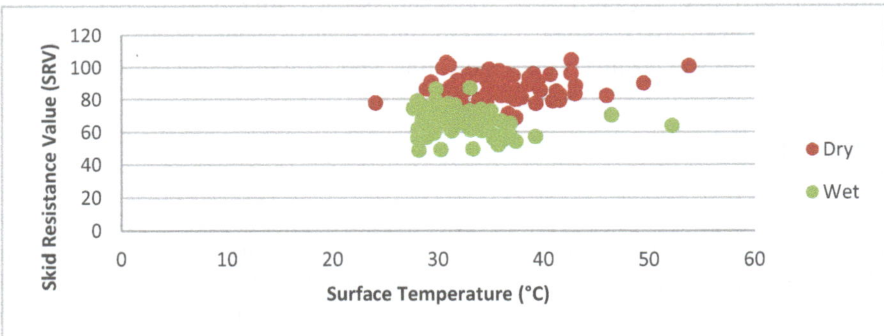

Fig. 2. Skid resistance value (SRV) versus surface temperature.

The scatter plot shows surface temperature values between 25 °C and 40 °C and SRV values between 45 and 105. The SRV is consistently higher on dry surfaces (red points), ranging from 75 to 105, reflecting greater friction. In contrast, wet surfaces (green points) mostly range from 50 to 75 SRV, indicating reduced skid resistance in wet conditions. During on-site measurements, most surface temperatures (for both dry and wet) are range between 28 °C and 40 °C. However, no clear trend of SRV and temperature was observed, as values are scattered horizontally, indicating that temperature may not be the dominant factor affecting SRV in this preliminary dataset. The Spearman correlation analysis showed a weak positive relationship with Dry SRV ($r = 0.0866$, $p = 0.445$)

and a very weak negative relationship with Wet SRV ($r = -0.1413$, $p = 0.211$). Both p-values exceed 0.05, indicating no statistically significant relationship. Therefore, no strong evidence of a consistent effect of temperature on SRV under either condition.

In summary, dry conditions yield higher SRV, indicating better grip, while wet conditions significantly reduce SRV and increase skid risk; however, the threshold skid resistance value still meets the requirements. Conversely, temperature does not show a strong direct effect on SRV within the tested range, whereas wet surface conditions clearly have a major impact on skid resistance.

4 Discussions

Findings confirm that SRV significantly reduces in wet conditions, with an average reduction ranging from 19% to 32%, consistent with previous studies [12–14] revealing that surface water reduces tyre–pavement friction due to the lubricating effect of water films and the risk of partial aquaplaning. Despite the drop, measured values remained above the minimum threshold levels, showing that the studied road sections still provide adequate skid resistance. But the magnitude of SRV reduction is also noteworthy. A drop of up to 32% (average SRV 59.78) represents a substantial decrease in available friction, which may elevate the risk of skidding during emergency braking or sudden manoeuvres. Suhaimi et al., [10] also mentioned that crash risk significantly increased when SRV falls below the average threshold of 52.31, specifically under wet conditions. This draws attention to the importance of continuous monitoring, as values close to the threshold may quickly deteriorate beyond safe limits under further wear and tear, contamination from sand, dirt, oil, or traffic polishing effects.

Although temperature showed a weak positive correlation with Dry SRV ($r = 0.0866$, $p = 0.445$) and Wet SRV ($r = -0.1413$, $p = 0.211$), these results were not statistically significant, indicating no strong influence of temperature on SRV under normal conditions. These finding aligns with previous studies [10, 22], which suggest temperature effects are more pronounced at extreme conditions or when combined with other variables such as aggregate type, binder properties, or weather variation. Urban pavement materials in Klang Valley are relatively resilient to normal temperature fluctuations, and the influence of water film and texture depth overshadows the effect of surface temperature. From a real-world perspective, while the temperature showed only a weak effect in this study, prolonged exposure to high temperatures could gradually alter binder stiffness and aggregate polishing resistance, as highlighted in studies by [12–14, 22]. The SRV values remaining above safety thresholds provide reassurance for current maintenance practices but also emphasize the need for proactive intervention before values approach critical limits.

Overall, this finding supports the importance of maintaining adequate skid resistance through routine monitoring of road markings and surface texture, particularly under wet conditions where friction loss is most severe. Future research could expand on these findings by including a broader range of factors influencing friction and skidding such as vehicle speed, texture, water depth and tyre characteristics, weather effects, road geometry, surface contamination, surfacing aggregates and surfacing type or age should not be overlooked as highlighted by VicRoads, [14] to better understand the combined and long-term effects on skid resistance performance.

5 Conclusions

This paper highlights the reduction of skid resistance under wet conditions with SRV dropping from 19% to 32%, and it is recommended that the sites have regular SRV monitoring to ensure safety thresholds are maintained. Therefore, road authorities should prioritize repainting the road markings and improving road surfaces after SRV drops to less than 60. The weak correlation observed between pavement surface temperature and SRV indicates that, within the prevailing temperature ranges of the study area, temperature alone may not be a dominant factor in influencing skid resistance. Nevertheless, other factors that influence friction and skidding should not be overlooked. Overall, wet-weather performance remains the most critical determinant of skid resistance, and proactive monitoring and targeted maintenance strategies can help ensure that skid resistance remains above safety thresholds, thereby reducing the likelihood of wet-weather skidding crashes and supporting safer urban mobility. Further research is recommended to explore other factors that may influence skid resistance.

Acknowledgements. This research was funded by a grant from the Malaysian Institute of Road Safety Research (RE103233). The authors sincerely thank all team members for providing the necessary data within the constrained time frame.

Disclosure of Interests The authors have no competing interests to declare that are relevant to the content of this article.

References

1. Islam, M.: The effect of motorcyclists' age on injury severities in single-motorcycle crashes with unobserved heterogeneity. J. Saf. Res. **77**, 125–138 (2021)
2. Tiwari, G.: Difficult safety problems. Int. J. Inj. Control Saf. Promot. **23**(2), 113–114 (2016)
3. Malin, F., Norros, I., Innamaa, S.: Accident risk of road and weather conditions on different road types. Accid. Anal. Prev. **122**, 181–188 (2019)
4. Mayora, J.M.P., Piña, R.J.: An assessment of the skid resistance effect on traffic safety under wet-pavement conditions. Accid. Anal. Prev. **41**(4), 881–886 (2009)
5. Rasol, M., et al.: Progress and monitoring opportunities of skid resistance in road transport: A critical review and road sensors. Remote Sens. **13**(18), 3729 (2021)
6. Fwa, T.F.: Skid resistance determination for pavement management and wet-weather road safety. Int. J. Transp. Sci. Technol. **6**(3), 217–227 (2017)
7. Wu, D., Hours, M., Martin, J.L.: Risk factors for motorcycle loss-of-control crashes. Traffic Inj. Prev. **19**(4), 433–439 (2018)
8. Jabatan Kerja Raya Malaysia: Arahan Teknik (Jalan) 2D/85 (Revision 2019): Manual on Traffic Control Devices, Road Marking and Delineation. Cawangan Jalan, Ibu Pejabat Jabatan Kerja Raya Malaysia (2019)
9. Merritt, D.K., Lyon, C., Persaud, B.: Evaluation of pavement safety performance (Report No. FHWA-HRT-14-065). Federal Highway Administration, U.S. Department of Transportation (2015). https://rosap.ntl.bts.gov/view/dot/35858
10. Suhaimi, A., & Baharuddin, A. R.: Threshold value of skid resistance and texture depth for Malaysia road. In IOP Conference Series: Materials Science and Engineering (Vol. 512, No. 1, p. 012035). IOP Publishing (2019)

11. Bao, J., Zhao, H., Jiang, Y., Li, S.: International perspectives on skid resistance requirements for pavement markings: A comprehensive synthesis and analysis. Lubricants. **13**(1), 29 (2025)
12. Dan, H.C., He, L.H., Xu, B.: Experimental investigation on skid resistance of asphalt pavement under various slippery conditions. Int. J. Pavement Eng. **18**(6), 485–499 (2017)
13. Wei, J., Zhang, Z., He, Y., Tan, Q., Yang, X., Wang, D.: Study on the skid resistance deterioration behavior of the SMA pavement. Sustainability. **14**(5), 2864 (2022)
14. VicRoads.Technical Note TN110: Measurement and Interpretation of Skid Resistance Using a SCRIM® Machine. VicRoads (2018). https://www.scribd.com/document/385384294/Technical-Note-TN-110-Measurement-and-Interpretation-of-Skid-Resistance-Using-a-SCRIM-Machine. Accessed 07 Apr 2026
15. Kumar, A., Gupta, A.: Review of factors controlling skid resistance at tire-pavement interface. Adv. Civ. Eng. **1**, 2733054 (2021)
16. Pranjić, I., Deluka-Tibljaš, A., Cuculić, M., Šurdonja, S.: Influence of pavement surface macrotexture on pavement skid resistance. Transp. Res. Procedia. **45**, 747–754 (2020)
17. Zhan, Y. et al.: Pavement preventive maintenance decision-making for high antiwear and optimized skid resistance performance. Constr. Build. Mater. **400**(4), 132757 (2023)
18. Mahmoud, A.A., Ali, A.A.E.L.M., Mostafa, A., Ibrahim, M.F.: Studying the effect of contaminated asphalt pavement surfaces on skid resistance. Innov. Infrastruct. Solut. **8**(2), 82 (2023)
19. Yan, C., Zhang, Z., Zhang, Y.: Research on skid resistance performance assessment and rehabilitation decision-making technology of asphalt pavement. In: E3S Web of Conferences, vol. 490, p. 03017. EDP Sciences (2024)
20. Shamshuritawati, S.: Statistics for Non-Statisticians: Basic Guide to SPSS. 1st edn. Muna Management Services & Enterprise, Kedah, Malaysia (2017)
21. Ali Abd Al-Hameed, K.: Spearman's correlation coefficient in statistical analysis. Int. J. Nonlinear Anal. Appl. **13**(1), 3249–3255 (2022)
22. Bullas, J. C.: Tyres, road surfaces and reducing accidents—a review. Report prepared for The AA Foundation for Road Safety Research and the County Surveyors' Society (2004)

Enhancing Riboflavin Gastroretentive Drug Delivery Using an Effervescent Matrix and Computational Mixture Design Approach

F. N. A. Puad, M. S. Sapiyan, N. S. A. N. Aris, and M. A. Azhar(✉)

Faculty of Chemical and Process Engineering Technology, Universiti Malaysia Pahang Al-Sultan Abdullah, Lebuh Persiaran Tun Khalil Yaakob, 26300 Kuantan, Pahang, Malaysia
akmalazhar@umpsa.edu.my

Abstract. Riboflavin, an essential vitamin, exhibits limited gastrointestinal absorption and a short elimination half-life, posing significant challenges for effective drug delivery. This study aimed to formulate gastroretentive floating matrix tablets of riboflavin to improve its bioavailability by prolonging its residence time in the stomach. A mixture design approach was employed to investigate the effect of different concentrations of polymers (HPMC K75M) and gas-forming agents (sodium bicarbonate and citric acid) on floating properties. Tablets were produced using the direct compression technique and assessed for their physical properties and in vitro buoyancy. The optimized formulation containing 36% HPMC K75M and 24% sodium bicarbonate demonstrated superior floating properties, with a floating lag time of 0.93 min and a total floating time of 352 min. Statistical analysis confirmed the model's accuracy, with p-values of <0.005 for key variables and adjusted R^2 values of 0.9618 and 0.8315 for floating lag time and floating duration, respectively. The developed gastroretentive floating tablet formulation improves the bioavailability and therapeutic efficacy of riboflavin, offering a promising approach for the oral delivery of compounds with similar absorption challenges.

Keywords: Riboflavin · Effervescent matrices · Gastroretentive drug delivery · Floating tablets · HPMC K75M

1 Introduction

GRDDS had been developed to enhance the drug bioavailability, to deliver drug at predictable release profiles, to reduce unwanted effects by site specific delivery of the drug, whenever desired. GRDDS provide significant benefits when compared to traditional dosage forms by extending gastric residence time for drug absorption, especially in cases where drugs have narrow absorption windows in the upper gastrointestinal tract [1].

Drug delivery using tablet floating systems are one of the most suitable types of GRDDS. These systems are kept floating in the stomach by decreasing the density of the formulation compared to gastric fluids. Floating system are generally classified into two types, namely, non-effervescent (swelling) systems that depend on gel-forming

M. A. b. Jamlos et al. (Eds.): ICSET 2025, LNNS 1897, pp. 508–520, 2026.
https://doi.org/10.1007/978-3-032-20722-7_65

water-soluble or insoluble polymers or effervescent systems that produce gas by chemical reaction with gastric fluids. These approaches are designed to attain buoyancy and prolonged gastric retention thereby improving the therapeutic efficacy [2].

Riboflavin is a vital vitamin that has been applied for the treatment of riboflavin deficiency and riboflavin-related syndromes, and it faces challenges in oral delivery due to its low gastrointestinal absorption and a short elimination half-life. Such issues may need novel formulation strategies to improve its bioavailability. The floating tablets provide a potential solution to overcome this problem, as they increase the gastric retention time, leading to gradually release of riboflavin over 12 h [3].

This study employs a novel combination of effervescent and swelling agents to optimize floating properties and achieve superior gastric retention. Using a mixture design approach, we investigated the influence of functional excipients, including HPMC K75M as a swelling polymer and sodium bicarbonate as a gas-generating agent, on buoyancy and drug release characteristics. The findings aim to establish a robust formulation that enhances riboflavin bioavailability while offering potential applicability to other drugs with similar pharmacokinetic limitations [4].

2 Methodology

2.1 Materials

The pharmaceuticals, including riboflavin and talc, were purchased from Take It Global Sdn. Bhd. Magnesium stearate was obtained from Sigma-Aldrich, HPMC K75M from DChemie Chemical Supplies, and lactose from HMBG Chemical.

2.2 Pre-evaluation of Powder Blend

The precompression test consists of several assessments: the angle of repose, which measures the cohesive behavior of the powder; tap and bulk density, which measure compressibility and flowability; and Hausner's ratio, which calculates powder flow characteristics.

Angle of Repose. This test is commonly measured through the employment of the funnel method. In this method, a precisely weighed powder is loaded into a funnel. The funnel's position is adjusted to ensure that its apex is in direct contact with the highest point of the powder blend. Subsequently, the powdered mixture is permitted to descend unobstructed via the funnel onto the underlying surface [5]. The diameter of the resulting powder cone is recorded, and the angle of repose is calculated using a specific formula, as shown in Eq. (1).

$$\tan \theta = \frac{H}{r} \tag{1}$$

Where,
= angle of repose
H = vertical height of cone in cm
r = base radius of powder cone in cm

Bulk Density. The apparent bulk density is calculated by dividing the overall mass of the powder by its bulk volume. In the experiment, the granules were accurately weighed and transferred into a 100 ml graduated cylinder [6]. The powder was properly levelled with minimal force, and its volume was quantified. The apparent bulk density was determined by applying Eq. (2), resulting in a value expressed in grams per millilitre.

$$Bulk\ Density = \frac{Weight\ of\ samples\ (g)}{Volume\ occupied\ by\ the\ sample\ (mL)} \tag{2}$$

Tapped Density. The powder tap density is defined as the proportion between the total weight of the powder and its compacted volume after tapping. The granules were carefully dispensed into a glass funnel and gradually transferred into a 100 ml graduated cylinder [6]. The cylinder was repeatedly struck until a uniform volume was attained. The sample's volume after 100 taps were measured and the tapped density was computed using Eq. (3)

$$Tapped\ Density = \frac{Weight\ of\ samples\ (g)}{Volume\ occupied\ by\ the\ sample\ (mL)} \tag{3}$$

Compressibility Index. Compressibility index was used to assess the powder blend's compaction properties [7]. The formula for the index is provided below in Eq. (4)

$$Carr\ index\ (\%) = \frac{Tapped\ density - bulk\ density}{Tapped\ Density} \times 100 \tag{4}$$

Hausner Ratio. The ratio of tapped density to bulk density was employed to evaluate the interparticle friction and predict the flow characteristics of the powder [7]. Hausner's ratio was calculated using Eq. (5) provided below

$$Hausner's\ Ratio = \frac{Tapped\ Density}{Bulk\ density} \tag{5}$$

2.3 Tablet Preparation

This research employed the direct compression technique to prepare floating matrix tablets of Riboflavin. All ingredients were accurately weighed according to the quantities specified in the table to make up 11 formulations. The blend was transferred into a plastic bag and subsequently agitated for 5 min to get a uniform consistency [8]. All the weighed formulations undergo pre-compression examinations including tapped density, bulk density, and angle of repose. As a result, Carr's Index and Hausner's ratio were determined. The blend was geometrically ground using a mortar and pestle for 15 min to achieve a homogeneous powder mixture. On a manual punch tabletting machine size of 17 mm, tablets were compressed directly to produce the desired floating tablets of 1000 mg tablet mass.

2.4 Experimental Design Using DOE

The components of effervescent as well as their respective ratios were calculated using the Design of Expert methodology (DOE). Mixture design using simple lattice was utilised to assess the impact of various functional polymers as excipients on floating time and floating lag time. Three levels (−1, 0, 1) were used to analyse each element in the arrangement. The estimated ridge analysis was performed to determine the best combinations to create a product that is acceptable and to estimate the product that is most acceptable and has the longest floating time and shortest floating lag time [9]. The ratios of the components were written as fractions of the combination with a total of one (X1 + X2 + X3 = 60%). Table 1 shows the composition of 11 formulations of floating tablets, each with a total tablet weight of 1000 mg.

Table 1. Composition of different riboflavin floating tablet formulations prepared using an effervescent matrix approach with HPMC K75M, sodium bicarbonate, and citric acid

Ingredient (%)	Fm1	Fm2	Fm3	Fm4	Fm5	Fm6	Fm7	Fm8	Fm9	Fm10	Fm11
Riboflavin	20	20	20	20	20	20	20	20	20	20	20
HPMC K75M (X1)	0	60	60	30	10	40	0	0	10	30	20
Sodium bicarbonate (X2)	0	0	0	30	40	10	60	30	10	0	20
Citric acid (X3)	60	0	0	0	10	10	0	30	40	30	20
Magnesium stearate	3	3	3	3	3	3	3	3	3	3	3
Lactose	14	14	14	14	14	14	14	14	14	14	14
Talc	3	3	3	3	3	3	3	3	3	3	3

Three numerical components making up 60% were HPMC K75M (1), sodium bicarbonate (2)and citric acid (3). Each tablet contained a fixed concentration of tabletting excipients, contributing to tablet compressing ability, along with 20% of riboflavin. The dependent variables investigated were the floating lag time (1) and floating time (2), both calculated concerning a 1000 mg tablet mass. The effect of various polymers and gas-generating agents as swelling and floating agents on the floating time and lag floating time was studied.

2.5 Buoyancy Evaluation Studies

This test measures both the floating lag time and the overall floating duration. 993 ml of deionized water was prepared in a glass bottle. 2 g of sodium chloride and 7 ml of hydrochloric acid were dissolved until the final stock solution became 1 L. The medium was maintained at 37 ± 0.5 °C, and the tablet was placed in a 250 ml glass vessel containing 150 ml of 0.1 N HCl solution (pH 1.2). The medium was maintaned at 37 ± 0.5 °C, and the tablet was immersed in a 1 L container with 0.1 N HCl solution and the pH range between 1.2 to 1.5. The floating lag time is the timeframe it takes for the tablet to ascend to the surface of the dissolving liquid, while the overall floating time is the period in which the tablet remains buoyant until it completely dissolves in the medium [10].

2.6 Post-compression of Floating Tablets

The following physicochemical tests were conducted to evaluate the tablets.

Weight Variation. A random sample of twenty pills was selected and their individual weights were assessed. The weights of the tablets were then compared to the calculated average weight. According to this approach, the weight of no more than two pills should exceed the pharmacopoeia restrictions by more than ±5% [10].

Friability Test. The tablets' friability was assessed using a friabilator [10]. The tablets were exposed to both abrasion and shock by placing them in a rotating plastic container at a speed of 25 rpm for a duration of 4 min. Additionally, a tablet was dropped from a height of 6 in. during each revolution. The pills underwent a process of reweighing. The tablets were cleaned of dust using a delicate muslin cloth and then weighed again. The desired level of friability was determined to be less than 1%.

$$\%of\ Friability = \frac{Initial\ weight - final\ weight}{Initial\ weight} \times 100 \tag{6}$$

Thickness. A Sotax hardness tester was used to determine the thickness of randomly 10 selected tablets. Mean and standard deviation were evaluated. The acceptable range for deviation in the tablet formulation is set at ±5% of their determined values, ensuring that the tablets do not deviate beyond this specified limit [11].

Hardness. The force necessary to fracture a tablet during compression is referred to as the tablet's hardness or crushing strength. For this investigation, a total of 10 tablets were chosen at random and placed individually in a hardness tester. The resulting hardness of the tablets was measured and reported in Newtons (N). According to USP criteria, a good tablet physically has a hardness between 4 and 10 kg [12].

3 Results and Discussions

The riboflavin-containing tablets were formulated using the direct compression method, and various evaluation parameters were assessed. The Design-Expert software proposed 11 different formulations for optimization. The composition of excipients was set at 60%, including three key functional polymers. Other fixed components included riboflavin at 20% as the active ingredient, 7.5% citric acid, 14% lactose serving as a filler, and 3% each of magnesium stearate and talc acting as lubricants.

3.1 Analysis of Mixture Design and Variance

A simplex lattice approach within the Design-Expert software was utilized to optimize tablet formulations, focusing on three functional polymers as excipients for improved gastrointestinal retention (X1: HPMC K75M, X2: sodium bicarbonate, and X3: citric acid). The study included one replicate and ten model points, with the results presented in Table 2. The response variables analyzed were Y1 (floating lag time) and Y2 (floating duration).

Table 2. The simple lattice design and results of the experiment of effervescent matrices with HPMC K75M

Formulation	X1	X2	X3	Y1	Y2
1	0	0	0.60	156.63	156.63
2	0.60	0	0	28.30	28.30
3	0.60	0	0	28.39	28.39
4	0.30	0.30	0	4.20	401.07
5	0.10	0.40	0.10	64.20	64.20
6	0.40	0.10	0.10	0.49	26.13
7	0	0.60	0	267.57	267.57
8	0	0.30	0.30	0.09	81.50
9	0.10	0.10	0.40	3.66	8.21
10	0.30	0	0.30	36.63	36.63
11	0.2	0.20	0.20	0.18	25.72

[a] X1 = HPMC K75M, X2 = Sodium Bicarbonate, X3 = Citric Acid

Formulations 2 and 3 in Table 2 are regarded as replication models in this experiment. A replication model is crucial for assessing the accuracy of estimates and providing supplementary insights on variations in background processes. Replication models are useful for estimating pure error and providing supplementary insights into variations in background processes. A lack-of-fit test can be used to assess the suitability and efficiency of the provided models. The software indicates a highly significant p-value (<0.005) for Y1 in quadratic models.

As shown in Table 3, the suggested model by the software is quadratic for Y1 and a special cubic for Y2. This is due to the lowest sequential p-value for quadratic in Y1, which is 0.0003 and 0.0249 for Y2, whereas the lack of fit p-value is 0.0026 and 0.0008 for Y1 and Y2, respectively. Although the linear model has the lowest p-value for lack of fit in Y2, it has an insignificant value for Y1. Adjusted R2 for Y1 and Y2 are 0.9618 and 0.8315, respectively. This shows the model can predict the outcome by as much as 96% and 83% for the quadratic model in Y1 and the special cubic model in Y2, respectively. In some cases, a higher adjusted R-squared value suggests a stronger alignment between the regression model and the data (Table 3).

The resulting quadratic (Y1) and special cubic (Y2) regression equations are as follows:

$$Y1 = 30A + 269.69B + 149.37C - 515.7AB - 182.85AC - 811.42BC \quad (7)$$

Table 3. Comparison of sequential models and assessment of lack of fit

Type of model	Sequential p-value		Lack of fit p-value		Adjusted R2		Predicted R2	
	Y1	Y2	Y1	Y2	Y1	Y2	Y1	Y2
Linear	0.271	0.2477	0.0006	0.0004	0.0981	0.1181	−1.3474	−0.3789
Quadratic	0.0003	0.1086	0.0026	0.0005	0.9618	0.5338	0.6938	−1.5283
Special cubic	0.0008	0.0249	0.0029	0.0008	0.9718	0.8315	0.6443	−0.5976
Cubic	0.0043	0.1507	0.0046	0.0005	0.9946	0.8	0.3695	−22.166

$$\begin{aligned} Y2 =& 25.99A + 247.62B + 163.63C + 958.4AB \\ &- 221.59AC - 546.3BC - 5171.43ABC \end{aligned} \tag{8}$$

Based on the equation, component A is HPMC K75M, B is sodium bicarbonate and component C is citric acid. The analysis from Table 4 highlighted the significance of the linear mixture interaction BC (pBC = 0.0001). There is an interaction between sodium bicarbonate and citric acid which has the most significant impact on Y1 (floating lag time). Additionally, interactions AB (pAB = 0.0007) and AC (pAC = 0.0469) also significantly impacted to Y1. For Y2 (floating duration), interaction AB was significant (pAB = 0.0158). However, combinations AC and BC had no significant effect, as their p-values were greater than 0.05. The ternary interaction ABC, however, was notable for Y2 with a p-value of 0.035 ($p < 0.05$), indicating its influence on floating duration.

Table 4. Analysis of variance (ANOVA) for the quadratic (Y1) and special cubic (Y2) model

Source	Y1		Y2	
	F-value	p-value	F-value	p-value
Model	51.4	0.0003	9.22	0.0249
Linear Mixture	36.48	0.0010	8.74	0.0347
AB	54.74	0.0007	16.17	0.0158
AC	6.88	0.0469	0.8646	0.4051
BC	122.19	0.0001	4.76	0.0946
ABC	–	–	9.83	0.0350
Residual				
Lack of fit	84290.73	0.0026	8.720E+05	0.0008

3.2 Optimization and Effects of Mixture Components on Formulation

The contour plots and 3D response surfaces effectively visualized the interplay among the mixture components for the floating tablet formulations.

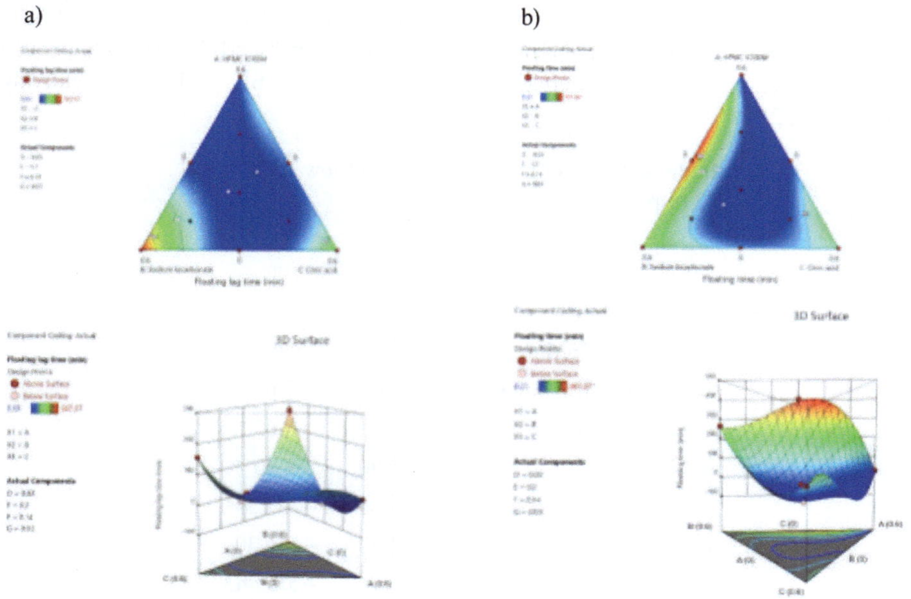

Fig. 1. Surface and contour plot of mixture analysis for a) floating lag time and b) floating time

Figure 1a demonstrates that vertex B exhibits a high value of Y1, indicating that a greater proportion of sodium bicarbonate at this point is necessary to achieve the greatest floating lag time of the riboflavin tablet. When it comes into contact with gastric acid in the stomach, the liberated carbon dioxide gas from sodium bicarbonate gets trapped within the tablet, creating a gas-filled chamber. The presence of the generated carbon dioxide gas creates a buoyant force within the tablet. This buoyant force counteracts the gravitational force, allowing the tablet to remain afloat on top of the gastric contents in the stomach [13].

For Fig. 1b, a higher proportion of HPMC K75M (component A), moderate sodium bicarbonate (component B), and minimal citric acid (component C) yielded extended floating times (Y2). This is due to the formation of a robust gel layer by HPMC K75M. The gel matrix helps trap the carbon dioxide gas produced by sodium bicarbonate, contributing to the maintenance of buoyancy. We found that increasing HPMC K75M concentration from 10% to 36% significantly improved floating time. This is consistent with findings by Senjoti et al. [14] who reported a similar trend using HPMC K100M in floating tablets of metformin, though our formulation achieved longer floating times at comparable polymer concentrations. However, we found that incorporating citric acid did not further improve floating behavior for riboflavin tablets. Our results also show that 24% sodium bicarbonate provided optimal floating properties, which is in agreement with the results of Nigusse et al. [15] who found that gastroretentive tablets of ranitidine exhibited optimal floating at 20–25% sodium bicarbonate.

Based on Table 5, the coefficients for Y1 revealed the highest contribution from sodium bicarbonate (B), followed by citric acid (C) and HPMC K75M (A). The binary mixture of sodium bicarbonate and citric acid (BC) minimized floating lag time. For Y2, sodium bicarbonate was the most influential factor, followed by citric acid and HPMC K75M. The binary combination of HPMC K75M and sodium bicarbonate (AB) achieved the longest floating duration, whereas the combinations AC, BC, and ABC negatively influenced Y2.

Table 5. Effect of mixture component toward Y1 and Y2 responses

Coefficient	Response function	
Component	Y1	Y2
A-HPMC K75M	29.20	25.99
B-Sodium Bicarbonate	268.14	247.62
C-Citric Acid	147.83	164.63
AB	−515.70	958.40
AC	−182.85	−221.59
BC	−811.42	−546.30
ABC	–	−5171.43

3.3 Model Validation

The desirability function was used to identify the three most optimal formulations for floating tablets, as shown in Table 6.

Table 6. The optimal excipient combination and prediction results

Formulation	A: HPMC K75M	B: Sodium Bicarbonate	C: Citric Acid	Y1: Floating lag time (mins)	Y2: Floating time	Desirability
1	0.360	0.240	0.000	1.000	344.635	0.473
2	0.334	0.251	0.015	0.999	307.097	0.420
3	0.310	0.260	0.030	1.000	259.151	0.365

The optimized formula consisted of 36% HPMC K75M, 24% sodium bicarbonate, and 0% citric acid, accounting for 60% of the excipient blend. The fixed components included 20% riboflavin, 14% lactose, 3% talc, and 3% magnesium stearate. The predicted responses for this formulation were a floating lag time of 0.93 min and a floating duration of 352 min. Experimental validation using a t-test confirmed no significant differences between predicted and observed values for floating lag time ($p = 0.11$) and floating duration ($p = 0.27$), as shown in Table 7. These results demonstrate that the simplex lattice design reliably predicts optimal formulations for riboflavin floating tablets.

Table 7. T-test of optimized floating tablets

Formulation		Y1			Y2	
	F1	F2	F3	F1	F2	F3
Obtained	0.93	0.81	0.37	352.00	336.00	287.00
Predicted	1.00	0.99	1.00	344.64	307.10	259.15
p-value		0.11			0.27	

3.4 Flow Behavior

The resulting powder mixture's flow characteristics were assessed using factors such as the angle of repose, bulk density, tapped density, compressibility index, and Hausner's ratio. Table 8 presents the pre-compression parameter findings. All of the formulations were determined to have compressibility indices and angles of repose that were commensurate with favourable flow characteristics outlined in the official USP pharmacopoeia, making them ideal for compression into floating tablets [11].

Table 8. Pre-compression characterization of the powder blend

Formulation	Effervescent matrices with HPMC K75M		
	F1	F2	F3
Compressibility Index	0.366	0.366	0.366
Hausner's Ratio	1.578	1.578	1.578
Bulk Density	0.527	0.527	0.527
Tapped Density	0.832	0.832	0.832

3.5 Swelling Characteristics and Buoyancy Behavior

HPMC K75M in the context of floating tablets, were analysed to establish there is a relationship between the drug's release characteristics and the processes by which it is released. The density of the formulated floating tablets was lower than that of gastric juice in the stomach (1.004 g/cm^3). To assess the swelling characteristics, a 0.1 N HCl solution was employed. During the buoyancy test, the formulated tablets were submerged in gastric fluids, surfaced within the fluid medium, and expanded as a result of hydration facilitated by gel-forming polymers. The prepared formulations exhibit a swelling property and effervescent effects. Among the tested formulations, F1 demonstrated the longest floating time (352 min) and an optimal floating lag time (0.93 min). This superior performance can be attributed to the higher concentration of

HPMC K75M (36%), which formed a strong gel matrix upon hydration, effectively entrapping CO_2 gas generated from sodium bicarbonate. This delayed water penetration and prolonged buoyancy. Our optimized formulation achieved a floating lag time of 0.93 min, which is significantly shorter than the 2.5 min reported by Masareddy et al. for riboflavin floating tablets using HPMC K4M [16]. Additionally, our total floating time of 352 min exceeds the 240 min reported in their study, demonstrating improved gastroretention potential.

The swelling of HPMC K75M led to the development of a dense gel barrier that trapped carbon dioxide produced by sodium bicarbonate, causing a slowdown and an increase in the diffusion path length for drug molecules. The subsequent development of a viscous gel layer further prolonged the floating time as shown in Fig. 2. This gel layer acts as a barrier, slowing drug diffusion and maintaining tablet buoyancy.

Initial time After 9 seconds After 1 hour After 5 hours

Fig. 2. Pictorial representation of buoyancy of floating tablet

3.6 Post Compression Properties

The physical attributes of tablets, including diameter, thickness, hardness, weight variation, and friability, were assessed using the prescribed methods as stated in USP pharmacopoeia. When compressing tablets, the die and punches are used to determine their diameter and form. All recorded data fell within the specified range of official values and are tabulated in Table 9. The tablet was consistently of a consistent weight and thickness, according to the data. There was no breakage or degradation of tablet integrity during transit or handling, according to the data.

Table 9. Physicochemical evaluation parameters of Riboflavin tablets

Formulation	Effervescent matrices with HPMC K75M		
	F1	F2	F3
Diameter (mm)	16.63	16.58	16.39

(continued)

Table 9. *(continued)*

Formulation	Effervescent matrices with HPMC K75M		
	F1	F2	F3
Hardness (N)	46	59	47
Thickness (mm)	3.79	3.69	3.74
Friability (%)	0.06	0.04	0.04

4 Conclusion

In conclusion, this study successfully developed and optimized a gastroretentive floating tablet formulation for sustained release of riboflavin. The optimized formulation, comprising 36% HPMC K75M and 24% sodium bicarbonate, demonstrated excellent floating properties with a minimal lag time of 0.93 min and an extended floating duration of 352 min. These characteristics, combined with the sustained drug release profile observed over 12 h, suggest significant potential for improving riboflavin bioavailability by prolonging gastric retention time. The mixture design effectively identified the optimal polymer and gas-generating agent combination for gastroretentive formulation. The incorporation of HPMC K75Mas a gel-forming agent, along with sodium bicarbonate as an effervescence-inducing component, resulted in a robust gel layer that effectively entrapped carbon dioxide, maintaining tablet buoyancy and controlling drug release.

While this study provides promising results, further in vivo investigations are warranted to confirm the enhanced bioavailability and therapeutic efficacy of the developed formulation. Additionally, future research could explore the applicability of this gastroretentive system to other drugs with similar absorption challenges or limited absorption windows in the upper gastrointestinal tract.

In summary, this study advances the understanding of gastroretentive drug delivery systems and offers a potential strategy for improving the oral delivery of riboflavin and similar compounds with limited gastrointestinal absorption.

Acknowledgments. The authors would like to thank the Ministry of Higher Education for providing financial support under Fundamental Research Grant Scheme – Early Career Researcher (FRGS-EC) No. FRGS/1/2024/TK09/UMPSA/02/2 (University reference: RDU243721) dan Tabung Persidangan Dalam Negara.

Disclosure of Interests The authors have no competing interests to declare that are relevant to the content of this article.

References

1. Das, S., Kaur, S., Rai, V.K.: Gastro-retentive drug delivery systems: A recent update on clinical pertinence and drug delivery. Drug Deliv. Transl. Res. **11**(4), 1435–1453 (2021). https://doi.org/10.1007/s13346-020-00875-5

2. Suradkar, P., Mishra, R., Nandgude, T.: Overview on trends in development of gastroretentive drug delivery system. Res. J. Pharm. Technol. **12**(11), 5565–5570 (2019). https://doi.org/10.5958/0974-360X.2019.00975.2
3. Kumar, A., Sharma, A.K., Dutt, R.: A review of gastro-retentive drug delivery systems for antidiabetics and its present status. Res. J. Pharm. Technol. **14**(1), 538–546 (2021). https://doi.org/10.5958/0974-360X.2021.00098.6
4. Pund, A.U., Shendge, R.S., Pote, A.K.: Current approaches on gastroretentive drug delivery systems. J. Drug Deliv. Ther. **10**(1), 139–146 (2020). https://doi.org/10.22270/jddt.v10i1.3803
5. Kedar, S.M., Chaudhari, P.M., Patil, R.N.: Formulation and evaluation of sustain release tablet of salbutamol sulphate by wet granulation method using release retarding agent. Research J. Pharm. and Tech. **5**(2), 222–226 (2012)
6. Raghuwanshi, C., Agarwal, S.: Formulation and Evaluation of Floating Matrix Tablet of Riboflavin. Int. J. Sci. Res. **8**(6), 1887–1891 (2019). https://doi.org/10.21275/ART20198747
7. Aslani, A., Jahangiri, H.: Formulation, characterization and physicochemical evaluation of ranitidine effervescent tablets. Adv. Pharm. Bull. **3**(2), 315–322 (2013). https://doi.org/10.5681/apb.2013.051
8. Bomma, R., Swamy Naidu, R., Yamsani, M., Veerabrahma, K.: Development and evaluation of gastroretentive norfloxacin floating tablets. Acta Pharma. **59**(2), 211–221 (2009). https://doi.org/10.2478/v10007-009-0019-6
9. Diós, P., et al.: Preformulation studies and optimization of sodium alginate based floating drug delivery system for eradication of Helicobacter pylori. Eur. J. Pharm. Biopharm. **96**, 196–206 (2015). https://doi.org/10.1016/j.ejpb.2015.07.020
10. Albadry, A.A., Ali, W.K., Al-saady, F.A.A.: Formulation and evaluation of prochlorperazine maleate sustained release floating tablet. Int. J. Pharm. Pharm. Sci. (IJPPS). **9**(2), 89–98 (2017)
11. Rao, M.Y.: Formulation and evaluation of fenoverine floating tablets. Asian J. Pharm. Clin. Res. (AJPCR). **14**, 175–180 (2021). https://doi.org/10.22159/ajpcr.2021v14i4.40999
12. Kurniawan, D.W., Triyanto, D.A., Utami, V.V.F.R.: Sustained release of captopril tablet with floating system using a cross-linked matrix of alginate. Kartika: J. Ilm. Farm. **6**(1), 11 (2019). https://doi.org/10.26874/kjif.v6i1.123
13. Rahim, S.A., Carter, P.A., Elkordy, A.A.: Design and evaluation of effervescent floating tablets based on hydroxyethyl cellulose and sodium alginate using pentoxifylline as a model drug. Drug Des. Devel. Ther. **9**, 1843–1857 (2015). https://doi.org/10.2147/DDDT.S78717
14. Senjoti, F., Mahmood, S., Md Jaffri, J., Kumar Mandal, U.: Design and in-vitro evaluation of sustained release floating tablets of metformin HCL based on effervescence and swelling. Iran. J. Pharm. Res. **15**(1), 53–70 (2015) [Online]. Available: https://www.researchgate.net/publication/283300223
15. Nigusse, B., Gebre-Mariam, T., Belete, A.: Design, development and optimization of sustained release floating, bioadhesive and swellable matrix tablet of ranitidine hydrochloride. PLoS One. **16**(6), 1–25 (2021). https://doi.org/10.1371/journal.pone.0253391
16. Masareddy, R., Kumar Yellanki, S., Patil, B.R., Manvi, F.V.: Development and evaluation of floating matrix tablets of riboflavin. Int. J. Pharm Tech Res. CODEN. **2**(2), 1439–1445 (2010)

Finite Element Method Simulation of Structural Behavior in Reinforced Concrete Bridge Girders

Arman Setiawan[1,2](✉), Rudy Djamaluddin[1], and Rusdi Usman Latief[1]

[1] Hasanuddin University, South Sulawesi Gowa 92171, Indonesia
arman.rafif.setiawan@gmail.com
[2] Bosowa University, South Sulawesi Makassar 90231, Indonesia

Abstract. Reinforced concrete is widely used for short–medium span bridges, but linear-elastic theoretical methods remain conservative and cannot fully capture nonlinear concrete–steel interaction, stress redistribution, or crack formation. This study analyzes the flexural behavior of reinforced concrete beams through theoretical calculations and FEM modeling. Three beams (150 × 250 × 3300 mm) with 30 MPa concrete and 420 MPa steel, each with different tensile reinforcement, were simulated using LUSAS. Parameters evaluated included ultimate load, load–deflection, strain, crack patterns, and failure modes. FEM consistently predicts higher capacity and strain, while increased reinforcement improves strength, stiffness, and ductility but produces faster collapse. Mechanical performance is classified into six load levels. Overall, FEM provides a more realistic beam response and supports bridge condition assessment.

Keywords: Reinforced concrete beams · Finite Element Method (FEM) · Flexural Performance · Crack patterns · Structural assessement

1 Introduction

Reinforced concrete is widely used because it is strong and economical, but large construction activities have an impact on the environment [1]. Bridges require excellent design and construction to avoid failure due to technical errors, overloading, and extreme events. Although reinforced concrete is commonly used for short to medium spans [2], codes such as AASHTO LRFD, Eurocode 2, and SNI still rely on linear elastic assumptions that cannot model strain redistribution and crack development [3–5]. The Finite Element Method (FEM) has become an essential tool for predicting structural behavior, validating laboratory and field tests, and analyzing nonlinear responses, crack patterns, and flexural capacity more accurately [6–16]. FEM also supports capacity, service life, and rehabilitation evaluations through integration with manual analysis and NDT [17].

However, previous studies are still limited due to the use of conservative theoretical approaches, the lack of evaluation of crack patterns and stiffness changes at each load stage, and the lack of correlation with the SNI 2847:2019 standard [18]. This study offers a more comprehensive evaluation through nonlinear modeling of concrete-steel interaction, classification of six levels of structural performance, and validation against SNI,

M. A. b. Jamlos et al. (Eds.): ICSET 2025, LNNS 1897, pp. 521–528, 2026.
https://doi.org/10.1007/978-3-032-20722-7_66

making the results relevant for monitoring, maintenance, and improving infrastructure safety. The objectives of the study are to compare the flexural parameters of reinforced concrete beams from theoretical calculations and FEM simulations, and to assess the influence of crack patterns and collapse mechanisms on performance parameters based on load levels.

2 Theoretical Approach and Research Design

2.1 Theoretical Approach

The flexible behavior of reinforced concrete beams changes as the load increases, starting from initial cracking to plastic deformation of the steel. There are four main stages: initial cracking when the concrete loses its tensile strength, stress redistribution when the steel begins to dominate, cracking when the steel reaches its yield point and the concrete compression approaches its capacity, and the peak stage before the strength decreases and the structure collapses. Changes in stress-strain distribution indicate concrete-steel interaction, with peak moment, plastic moment, and cracking moment (Mcr) as important parameters [19]. Theoretical calculations are used to obtain cracking, yield, and ultimate moments, as well as other parameters such as deflection and strain (Fig. 1).

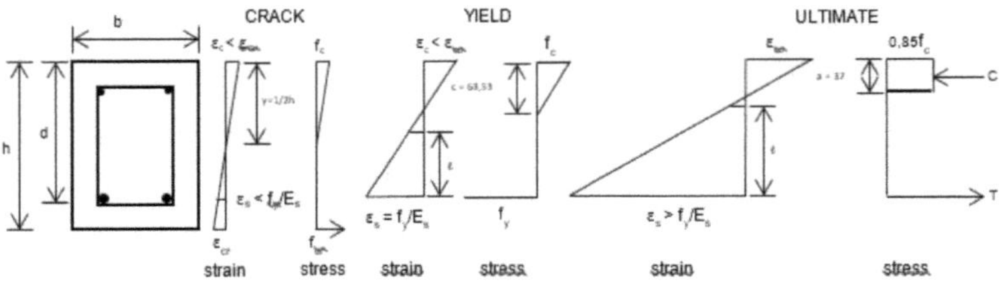

Fig. 1. Strain and Stress Distribution in the Crack, Pre-Yield, Yield, and Ultimate Stages

The Finite Element Method (FEM) is used to solve engineering problems with complex geometries and loads that are difficult to handle analytically. Various studies by M.Z. Yusof (2015), Yeong et al. (2022), and Ismail et al. (2021) show that FEM is capable of accurately predicting stress distribution and failure patterns, so it can be used as a reference for further numerical research.

2.2 Research Design

To determine the dimensions and reinforcement of the beams, the ultimate strength design method was used. In this study, there were three types of beams with dimensions of 150 × 250 × 3300 mm. The planned concrete quality of the beams was f'c 30 MPa and fy 420 MPa. The beam variations and details can be seen in Table 1 and Figs. 2 and 3.

Table 1. Research beam variations

No.	Symbol	Type Beam Variation
1	BLFR	Flexural Minimum reinforcement
2	BLFP	Flexural Required reinforcement
3	BLFT	Flexural Maximum reinforcement

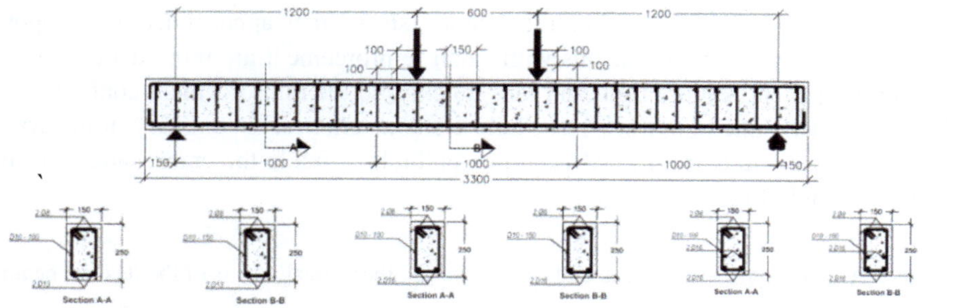

(a) Beam details BLFR (b) Beam details BLFP (c) Beam details BLFT

Fig. 2. Details of research beams

3 Results dan Discussion

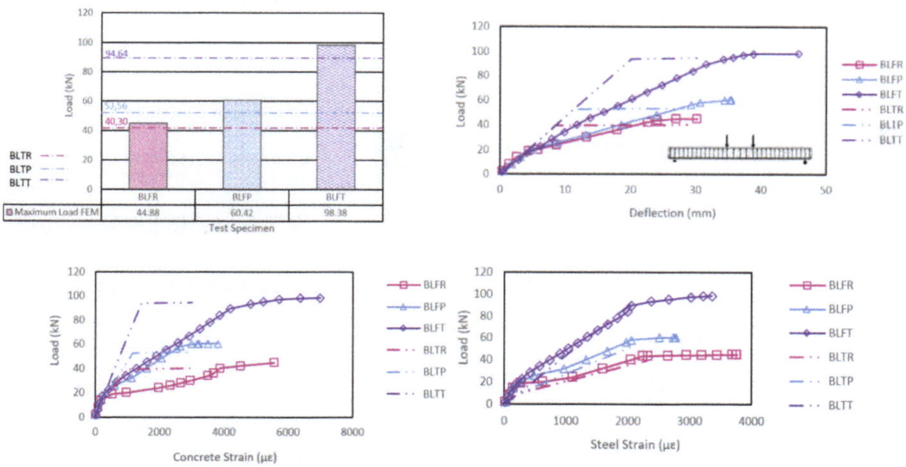

Fig. 3. Experimental test results chart

The maximum load analysis shows that the number and diameter of tensile reinforcement greatly affect the capacity of the beam, where FEM always produces a greater capacity than theory. Beams with small reinforcement have the lowest capacity, and increasing the reinforcement increases the capacity, because theory is conservative while FEM models the concrete-reinforcement interaction in more detail. The load-deflection curve shows a pattern of increasing load until it decreases, with FEM capacity always higher; FEM critical deflection is smaller so that the beam appears stiffer in the early stages. The concrete load-strain relationship shows that larger reinforcement increases capacity and ultimate strain, with FEM values being more realistic than linear-elastic theory. Steel strain is also higher in FEM, due to more accurate nonlinear modeling and material interaction, while theory provides conservative safety limits.

The crack pattern begins with a flexural crack in the lower center of the beam due to tension, then develops upward until a diagonal shear crack appears near the support, especially in sparse stirrups. Beams with small reinforcement are more ductile, while large reinforcement is more rigid and fails quickly, so the reinforcement configuration affects the crack pattern, stiffness, and failure characteristics. A summary of the crack length and area results from FEM is shown in Table 2, while the crack pattern of the beam is shown in Fig. 4.

Table 2. Crack length in the FEM analysis results and theoretical results of the flexural beam

FEM result				Theoretical results				l_{cr} FEM/Theo (%)		
Specimen	Crack Length (mm)			Specimen	Crack Length (mm)					
	Cr_{cr}	Cr_{yd}	Cr_{ult}		Cr_{cr}	Cr_{yd}	Cr_{ult}	Cr_{cr}	Cr_{yd}	Cr_{ult}
BLFR	27	190	205	**BLTR**	0	141	173	–	1.34	1.18
BLFP	29	193	210	**BLTP**	0	128	157	–	1.51	1.34
BLFT	32	202	220	**BLTT**	0	117	163	–	1.73	1.35

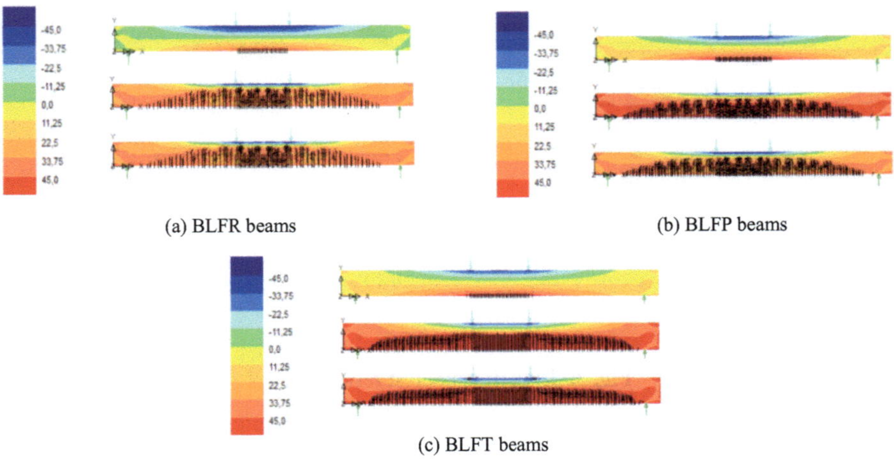

(a) BLFR beams

(b) BLFP beams

(c) BLFT beams

Fig. 4. Crack pattern of beams at crack, yield, and ultimate

The load-deflection curve illustrates the response of the structure through six levels of damage, from initial to ultimate conditions, including initial strain, cracking, increased steel-concrete deformation, and decreased stiffness. By integrating field data and FEM, the behavior of the bridge can be predicted to determine the plastic and ultimate limits. These stages also help assess the actual condition of the bridge based on deformation and cracking symptoms (Fig. 5 and Table 3).

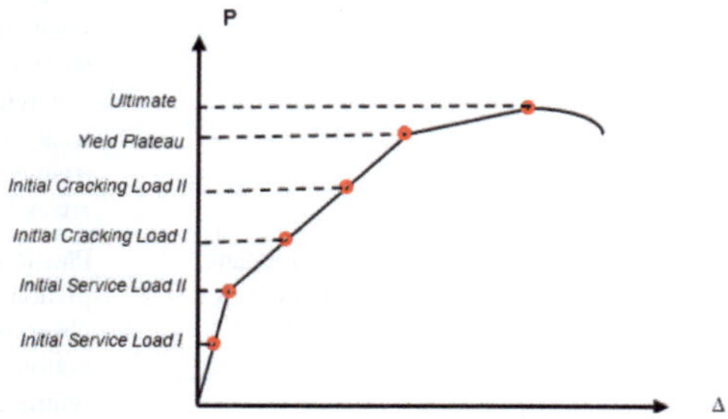

Fig. 5. Load-deflection relationship curve and damage level of reinforced concrete beams

Table 3. Mechanical Performance Parameters

LOADING LEVEL	APPLIED LOAD (%)	CRACK LENGTH (%)	CRACK ZONE (%)		STRUCTURAL PERFORMANCE PARAMETERS	MATERIAL RESPONSE (STEEL & CONCRETE)	COMBINED STRUCTURAL EVALUATION
			Vertical	Diagonal			
Initial Service Load I (Prelinear Service)	0–9	0–3	0–2	–	Structure is still in uncracked and in elastic behavior	Strain begins to appear in tensile reinforcement; concrete remains within elastic limits	Flexural and shear members still functioning; no visual damage
Initial Service Load II (Crack Initiation)	10–19	4–22	3–4	–	Initial cracks appear	Initial cracks appear in the bottom tensile zone; concrete in shear zone remains active	Early flexural damage signs; shear beams do not yet show structural symptoms
Initial Cracking Load I (Progressive Cracking)	20–40	23–30	5–12	–	Measured crack development	Tensile strain in the bottom reinforcement increases; the shear concrete begins to exhibit slight deformation	Cracks propagate upward; deformation becomes visible in the web

(*continued*)

Table 3. *(continued)*

LOADING LEVEL	APPLIED LOAD (%)	CRACK LENGTH (%)	CRACK ZONE (%)		STRUCTURAL PERFORMANCE PARAMETERS	MATERIAL RESPONSE (STEEL & CONCRETE)	COMBINED STRUCTURAL EVALUATION
			Vertical	Diagonal			
Initial Cracking Load II (Pre-Yield Transition)	41–96	31–50	13–35	–	Reduction in structural stiffness	Tensile reinforcement fully engaged; shear strain increases; concrete weakens in tensile and web zones	Structure exhibits nonlinear behavior; weak regions begin to spread
Yield Plateau	91–99	73–93	42–54	0–28	Significant deformation is visible	Plastic strain is predominantly observed in the bottom tensile reinforcement; shear strain is mainly concentrated in the stirrups (shear reinforcement); the concrete and web experience damage due to shear forces	Vertical and shear cracks develop simultaneously; deformation in the web and support becomes integrated
Ultimate	100	100	>60	>28	Structural collapse	Maximum strain occurs in the bottom tensile reinforcement and shear reinforcement; failure due to the combined action of shear and flexure is inevitable	The structure reaches the threshold of local collapse; deformation becomes irreversible; the shear zone is crushed

4 Conclusions

FEM has been proven to provide a more realistic description of the behavior of reinforced concrete beams than theoretical calculations, with higher capacity, strain, and crack patterns. Variations in the number and diameter of reinforcements have a significant effect on flexural capacity, stiffness, ductility, and failure patterns. The six load levels generated form the basis for evaluating structural performance and are useful for monitoring bridge conditions, so that the integration of theory, FEM, and design standards improves the accuracy of assessments and supports infrastructure maintenance strategies.

References

1. Djamaluddin, R., Irmawaty, R., Fakhruddin, Yamaguchi, K.: Flexural behavior of repaired reinforced concrete beams due to corrosion of steel reinforcement using grouting and FRP sheet strengthening. Civ. Eng. J. **10**(1), 222–233 (2024). https://doi.org/10.28991/CEJ-2024-010-01-014
2. Oliveira Júnior, A.L.R., Ribeiro, F.L.A., Silva, C.: Structural efficiency of reinforced concrete beams in transportation infrastructure. Mens Agit. **16**(1), 42–51 (2021)
3. Rahman, M.A., Haque, M.M.: Experimental investigation and finite element analysis of reinforced concrete rectangular beams. Int. J. Civ. Eng. Technol. **9**(10), 176–185 (2018)
4. Revanna, A.M., Kumar, S., Ramesh, B.: Ultimate load capacity of reinforced concrete beams. J. Appl. Math. Phys. **8**(11), 2429–2440 (2020)
5. Suryanto, A.B., Prakoso, A., Setiawan, A.: Progressive cracking behavior in RC beams: Experimental and analytical study. Case Stud. Constr. Mater. **11**, e00264 (2019)
6. Zienkiewicz, O.C., Taylor, R.L.: The Finite Element Method: Volume 1: The Basis, 6th edn. Butterworth-Heinemann, Oxford (2005)
7. Jaspart, J.P., Weynand, K., Picaut, P.: Finite element analysis for structural engineering and laboratory tests. J. Struct. Eng. **129**(3), 397–404 (2003)
8. Oñate, E.: Structural Analysis with the Finite Element Method: Linear Statics. Springer Series in Computational Mechanics. Springer, Barcelona (2009)
9. Meghashree, K., G., R.C.: Non-destructive testing in RC bridge deck evaluation: A comparative approach. Int. Adv. Res. J. Sci. Eng. Technol. **9**(9), 123–128 (2022). https://doi.org/10.17148/IARJSET.2022.9922
10. Salgado, R., Guner, S.: Nonlinear finite element analysis of reinforced concrete bridges. J. Bridge Eng. **25**(2), 04020003 (2020)
11. Mahmoud, A.M.M.H., Naser, M.: Experimental and finite element investigation of cracking in RC beams. J. Adv. Concr. Technol. **22**(3), 145–160 (2024). https://doi.org/10.1186/s40069-024-00709-5
12. Aslan, S., Erkan, İ.H.: Flexural behavior of reinforced concrete beams with openings: Experimental and analytical study. Rev. la Construcci6n. **24**(1), 5–20 (2025). https://doi.org/10.7764/RDLC.24.1.5
13. Filippoupolitis, A.A., Giannopoulos, G., Vrochidou, E.: Numerical modeling of RC beams using FEM. In: Papadrakakis, M., Papadopoulos, V., Stefanou, G., Plevris, V. (eds.) Proceedings of the VII European Congress on Computational Methods in Applied Sciences and Engineering (ECCOMAS 2016), vol. 1, pp. 11638–11649. National Technical University of Athens, Athens (2016)
14. Fu, A.Z., Zhang, Y., Li, J.: FEM analysis of impact load on reinforced concrete beams: Comparison with manual calculations. Appl. Sci. **15**(2), 1234 (2025). https://doi.org/10.12913/22998624/199869
15. Filippoupolitis, A.A., Giannopoulos, G., Vrochidou, E.: Numerical modeling of RC beams using FEM. In: Papadrakakis, M., Papadopoulos, V., Stefanou, G., Plevris, V. (eds.) ECCOMAS Congress 2016: Proceedings of the VII European Congress on Computational Methods in Applied Sciences and Engineering, pp. 1–12. National Technical University of Athens, Athens (2016). https://doi.org/10.48550/arXiv.1402.4180
16. Ammari, A.H., Garnier, J., Jing, W., Kang, H., Wahab, A.: Mathematical and statistical methods for electrical impedance tomography. J. Math. Imaging Vis. **48**(2), 1–21 (2014). https://doi.org/10.48550/arXiv.1405.4582
17. Gamal, M.: Integrated framework for bridge assessment using manual, experimental, and FEM approaches. Infrastructures. **7**(3), 36 (2022). https://doi.org/10.3390/infrastructures3030036

18. Badan, K., Nasional, S.: SNI_2847_2019_Persyaratan_Beton_Struktur, Badan Standarisasi Nas. no. 8, (2019)
19. Charif, A.A., Mourad, S.M., Khan, M.I.: Flexural behavior of beams reinforced with steel bars exceeding the nominal yield strength. Lat. Am. J. Solids Struct. **13**(6), 945–958 (2016). https://doi.org/10.1590/1679-78251683

G-BWM-Based Evaluation of Key Factors Driving the Success of Project Management Solutions

Rosa Micale[1](✉), Giada La Scalia[2], Concetta Manuela La Fata[2], and Antonio Giallanza[1]

[1] Department of Engineering, University of Messina, 98166 Messina, Sicily, Italy
rosa.micale@unime.it

[2] Department of Engineering, University of Palermo, 90128 Palermo, Sicily, Italy

Abstract. In today's competitive markets, Project Managers rely on project management tools to guide teams through complex tasks, tight deadlines, and collaborations. In this context, Project Management Information Systems (PMIS) have emerged as crucial tools for planning, organizing, and controlling projects of varying complexities across different sectors, offering tools such as time optimization, resources allocation, and risk management. This paper identifies the most relevant PMIS features and their relative importance, through a literature review, commercial software evaluations, and expert interviews. Their relative importance was quantified using the Group Best Worst Method (G-BWM). Planning plays a central role, where Scheduling, Budget Tracking, and Risk Management are vital elements. This is followed by Control and Risk/Issue Management, while Document Management, Communication, and Utility result less important. The results support researchers and professionals in PMIS selection and development, enabling informed decisions aligned with real user needs.

Keywords: Best Worst Method (BWM) · Project Management Information Systems (PMIS) · multi-experts · decision making

1 Introduction

In today's rapidly evolving and competitive business landscape, project management has become increasingly complex [1]. Efficient project execution gains strategic importance, prompting the adoption of advanced tools to support Project Managers (PMs) throughout all project phases and align with business strategies and methodologies to ensure successful project outcomes. Nenni *et al.*, [2] linked the low success rate of many projects (~ 35%) to the underutilization of project management technologies. Integrating Project Management Information Systems (PMIS) into project management practices could enhance project success. PMISs include a set of tools and techniques crucial for managing a project throughout its lifecycle, providing automated solutions that enhance key project management processes, including scheduling, monitoring, communication,

M. A. b. Jamlos et al. (Eds.): ICSET 2025, LNNS 1897, pp. 529–536, 2026.
https://doi.org/10.1007/978-3-032-20722-7_67

and data management [3, 4]. Numerous studies have examined the application and evaluation of PMISs. Alkudary and Gardiner [5] showed that integrating blockchain into PMISs increases data traceability, reliability, and security. Besouw and Bond-Barnard [6] highlighted the lack of a universal PMIS and proposed a Smart PMIS model, while Micale et al. [7] identified planning and control as the most critical features. Given the wide variety of PMIS software, selecting the most appropriate solution can be challenging and requires a clear identification and evaluation of the key features such software should possess [7]. Based on this consideration, this study aims to identify, define, and evaluate the critical success factors that should be incorporated into project management software. To achieve this objective, a comprehensive literature review and an analysis of commercial software were conducted. This selection process was refined through experts' consultation, which involved consolidating similar functionalities, eliminating redundancies, and removing less important features. As a result, six features and twenty-three sub-features were identified, and their relative importance determined using the Group Best Worst Method (G-BWM) [8]. Data were gathered through a questionnaire distributed to experts, including academics, consultants, PMs, and members of the Project Management Institute Southern Italy Chapter (PMI SIC). This study provides valuable insights for both practitioners and researchers: it can support decision-makers in selecting the most suitable PMIS to manage complex projects and offer guidance for software developers seeking to design more effective solutions.

2 Key Factors: Identification and Definition

This study aims to identify, define, and evaluate the critical success factors to be integrated into project management software. An extensive review of literature and commercial tools lead to the identification of about 90 features, later refined through input from experts (4 academics, 4 consultants, 4 PMs, and 4 PMI SIC members). Redundant or less relevant functionalities were removed, and similar ones consolidated, resulting in the identification of six key factors and twenty-three sub-factors, systematically organized into the hierarchical structure, as in Fig. 1.

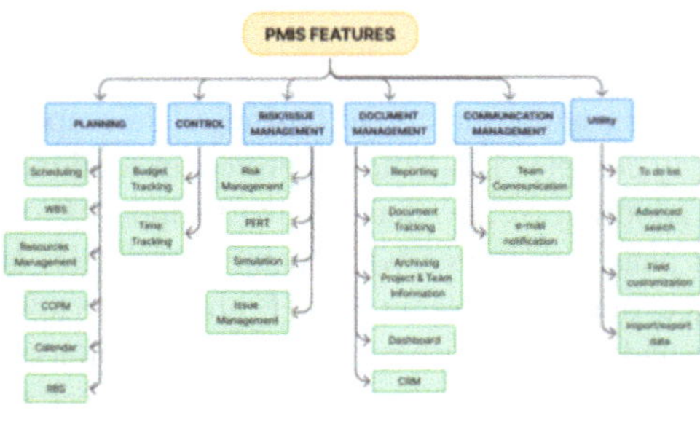

Fig. 1. Hierarchical structure of key factors

Below is a brief description of the considered features and sub-features:

i. Planning (K_1): it includes tools for project scheduling, activities visualization, and resource management. The identified sub-features are:
   a. Scheduling ($K_{1.1}$): it considers tools for planning and scheduling, including CPM, milestone representation, and the Gantt chart for activity tracking and timeline management [3, 9].
   b. Work Breakdown Structure (WBS) ($K_{1.2}$): it consists of a hierarchical decomposition of the project into smaller, manageable components [9].
   c. Resources management ($K_{1.3}$): it ensures efficient allocation and utilization of resources, elimination of overloads, costs planning, and workload visualization [3, 9, 10].
   d. Critical Chain Project Management (CCPM) ($K_{1.4}$): it considers resource availability alongside task dependencies [7].
   e. Calendar ($K_{1.5}$): it allows the creation of project, activities and resources calendars [10, 12].
   f. Resources Breakdown Structure (RBS) ($K_{1.6}$): it consists of a hierarchical organization of project resources by type or function [7].

ii. Control (K_2): it includes tools to manage and control budget, work progress, and monitor performance against planned objectives. The identified sub-features are:
   a. Budget tracking ($K_{2.1}$): it allows monitoring and controlling activity or project costs by comparing actual and planned values and generating indices and/or indicators (e.g., Cost Variance – CV and/or Cost Performance index – CPI) [3, 9, 11].
   b. Time tracking ($K_{2.2}$): it allows monitoring and controlling the work spent on activities or project by comparing actual and planned schedule, and generating indices and/or indicators (e.g., Schedule Variance - SV and/or Schedule Performance Index - SPI) [3, 9].

iii. Risk/Issue management (K_3): it includes tools for identifying, assessing, and managing potential risks and issues throughout the project lifecycle. The identified sub-features are:
   a. Risk management ($K_{3.1}$): it includes the identification, assessment, and prioritization of project risks, including tools such as SWOT analysis and risk register [3, 9, 10].
   b. Program Evaluation and Review Technique (PERT) ($K_{3.2}$): it allows to estimate project completion time considering uncertainty in activity durations [9].
   c. Simulation ($K_{3.3}$): it enables the simulation of random variables to estimate outcomes and assess risks in complex scenarios (e.g., Monte Carlo simulation) [7].
   d. Issue management ($K_{3.4}$): it includes the project issues management, including tools such as issue register and issue log [9].

iv. Document management (K_4): it includes tools for handling project documentation and reporting. The identified sub-features are:
   a. Reporting ($K_{4.1}$): it includes tools for generating and organizing project information [9, 10].
   b. Document tracking ($K_{4.2}$): it enables the monitoring of project documents and revisions [12].

c. Archiving project and team information ($K_{4.3}$): it allows the storage and organization of project and team data [12].
d. Dashboard ($K_{4.4}$): it allows the visualization of project performance and customized [10].
e. Customer Relationship Management (CRM) ($K_{4.5}$): it enables managing customer interactions, storing their information and categorizing them [9].

v. Communication management (K_5): it includes tools that enabling collaboration and information exchange among team members. The identified sub-features are:
a. Team communication ($K_{5.1}$): it facilitates synchronous (real-time interaction, e.g., chat, video/audio calls) and asynchronous (delayed interaction, e.g., forums, and messages) communication among team members [3, 9, 10, 12].
b. e-mail notification ($K_{5.2}$): it automatically sends updates and alerts via email to keep stakeholders informed about progress, assignments, deadlines, and other key activities [3].

vi. Utility (K_6): it includes tools for managing data. The identified sub-features are:
a. to do list ($K_{6.1}$): it allows users to have a clear overview of pending activities [7].
b. advances search ($K_{6.2}$): it allows to perform refined searches, utilizing various filters and criteria to quickly locate specific information or items based on custom parameters [10, 12].
c. field customization ($K_{6.3}$): it enables the creation and customization of fields to adapt data to specific requirements [12].
d. Import/export data ($K_{6.4}$): it allows data import/export in various formats (e.g., excel) [7].

3 Group Best Worst Method (G-BWM)

The following outlines the Group Best Worst Method [8]:

i. select a panel of l experts {E_1, E_2, ..., E_l} and establish q criteria for evaluation {K_1, K_2, ..., K_q};
ii. each expert E_i determines the most important (B) and the least important (W) criteria;
iii. each expert E_i determines the following vectors (1):

$$G_{BOE_i} = \left[g_{B1,E_i}, \ldots, g_{Bq,E_i}\right] \text{ and } G_{OWE_i} = \left[g_{1W,E_i}, \ldots, g_{qW,E_i}\right] \quad (1)$$

iv. where $[g_{Bj,E_i}]$ and $[g_{jW,E_i}]$ are the judgments of expert E_i given using a scale from 1 (equal significance) to 9 (highest priority);
v. cluster the experts into group C_s (s = 1, ..., Z) if they select the same best (B) and worst (W) criteria. If l_s denotes the number of experts in the group C_s, then $l = l_1 + \ldots + l_Z$;
vi. use the geometric mean to calculate g_{Bj,C_s} and g_{jW,C_s} as reported in Eq. (2):

$$g_{Bj,C_s} = \sqrt[l_s]{\prod_{i \in C_s} g_{Bj,E_i}} \quad \forall j, \forall C_s \quad \text{and} \quad g_{jW,C_s} = \sqrt[l_s]{\prod_{i \in C_s} g_{jW,E_i}} \quad \forall j, \forall C_s \quad (2)$$

vii. achieve the criteria weight vector $(p_{1,C_s}, \ldots, p_{q,C_s})$ for each C_s. The optimal criteria weights must satisfy: $\frac{p_{B,C_s}}{p_{j,C_s}} = g_{Bj,C_s}$ and $\frac{p_{j,C_s}}{p_{W,C_s}} = g_{jW,C_s}$. When full consistency is not met and, if there are more than three criteria, multiple optimal solutions may arise [13]. To obtain a single optimal solution, the linearization model (3) has to be considered:

Min ξ

$$\text{s.t.}\begin{cases} \left|p_{B,C_s} - p_{j,C_s} \bullet g_{Bj,C_s}\right| \leq \xi \\ \left|p_{j,C_s} - p_{W,C_s} \bullet g_{jW,C_s}\right| \leq \xi \\ \sum_{j=1}^{q} p_{j,C_s} = 1 \qquad \forall C_s \\ p_{j,C_s} \geq 0 \qquad \forall j, \forall C_s \end{cases} \tag{3}$$

viii. Compute for each criterion the final weigh using Eq. (4):

$$p_j = \frac{\sum_{s=1}^{Z} \left(p_{j,C_s} \bullet l_s\right)}{l} \qquad \forall j \tag{4}$$

Moreover, to evaluate the consistency level of experts, the Input-based Consistency Ratio (CRI) [14] is employed. Acceptable inconsistency thresholds are defined by Liang *et al.*, [14].

4 Case Study

The G-BWM method was applied seven times to determine the weights of features (criteria) and sub-features (sub-criteria): once for criteria and six times for sub-criteria. A panel of 16 experts (4 academics, 4 consultants, 4 PMs, 4 PMI-SIC members) was interviewed. Each expert selected the most and least important criteria and conducted pairwise comparisons using the nine-point scale to derive G_{BOE_i} and G_{OWE_i}. The same procedure was applied to each group of sub-criteria. Consistency of evaluations was assessed, with only one Risk/Issue Management sub-criteria assessment exceeding the threshold and requiring revision. Experts were clustered based on their identical best and worst criteria, then g_{Bj,C_s} and g_{jW,C_s} were computed using Eq. (2). Data were input into the BWM solver to evaluate the relative importance of each C_s and group of criteria/sub-criteria. For Control and Communication Management sub-criteria, involving only two criteria, was sufficient to consider that $g_{BW,C_s} = \frac{p_{B,C_s}}{p_{W,C_s}}$ and $p_{B,C_s} + p_{W,C_s} = 1$. Group weights were obtained using Eq. (4), and the resulting criteria and sub-criteria weights based on the hierarchy in Fig. 1 are shown in Fig. 2:

Experts identified Planning as the most important criterion (0.405), followed by Control (0.188) and Risk/Issue Management (0.155), essential for effective monitoring and risk management. Document Management (0.09), Communication Management (0.087), and Utility (0.075) received lower weights. Among sub-criteria, Scheduling ranks first (0.151), followed by Budget Tracking (0.125) and Risk Management (0.08). These elements play a crucial role in ensuring effective project planning, costs control, and risk assessment. WBS (0.075) and Resource Management (0.066) support task clarity and resources optimization. In Communication Management sub-criteria, automatic

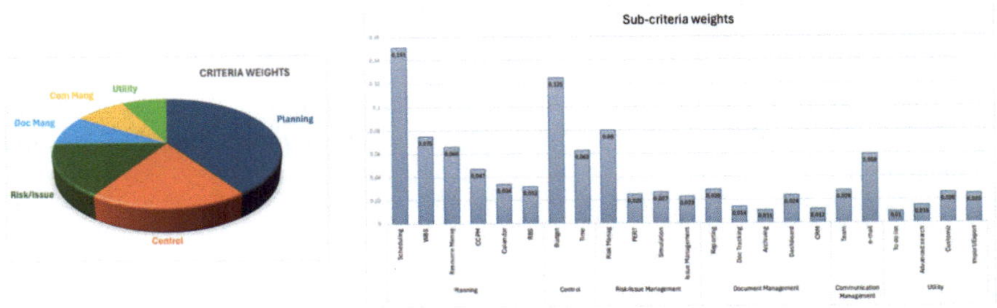

Fig. 2. Criteria and sub-criteria weights

email updates are prioritized over chat and video/audio calls, emphasizing the need for clear, documented communication. In Utility category, the most important sub-criteria were customization (0.026), enabling users to tailor the software to their specific needs, followed by the ability to import and export data in various formats (0.025), enabling easy integration with other systems. Reporting features (0.029) allow companies to analyse data effectively, supporting informed decision-making. Conversely, sub-features such as to-do list, CRM, and basic information storage are not considered essential for project success.

5 Conclusions

Today, companies operate in an increasingly dynamic market that demands rapid delivery, cost efficiency, and customization, making projects more complex to manage. This presents a challenge that businesses must inevitably face. In this context, the use of software tools that assist PMs in project management is crucial to ensuring project success, making their proper selection essential. The numerous software available on the market offer a variety of features, making the selection process even more challenging. Based on this, this study, through a literature review, commercial software analysis and interviews with experts, identified six key features, and twenty-three sub-features that should be included in project management software. The G-BWM was applied to assess the relative importance of these factors, with data collected from a panel of sixteen experts (academics, project management consultants, PMs, and members of the PMI-SIC). The results indicate the centrality of Planning in project management, as experts unanimously identified it as the most critical feature, followed by Control and Risk Management. Cost control and risk analysis play a crucial role in identifying potential risks, enabling timely intervention with appropriate corrective actions, preventing delays, and enhancing the likelihood of project success. Additionally, features such as Customization and Data Import/Export, and reporting were also recognized as vital for tailoring software to specific company needs, facilitating the integration with other systems and supporting decision-making by providing valuable insights. Conversely, sub-features such as To do lists and CRM were regarded as less significant, suggesting that the primary focus should be on core project management functionalities rather than supplementary tools. Future research could explore the possibility of integrating project management software technologies with Artificial Intelligence (AI) and machine learning. Additionally, a comparative analysis of different software solutions in real-world projects could provide valuable insights into their effectiveness. Lastly, expanding the expert panel to include professionals from diverse industries may also help refine feature prioritization.

Acknowledgement. This research was not funded by any grant.

References

1. McKevitt, D., Carbery, R., Lyons, A.: A profession but not a career? Work identity and career satisfaction in project management. Int. J. Proj. Manag. **35**(8), 1673–1682 (2017)
2. Nenni, M.E., De Felice, F., De Luca, C., Forcina, A.: How artificial intelligence will transform project management in the age of digitization: A systematic literature review. Manag. Rev. Quartely. **75**, 1669–1716 (2025). https://doi.org/10.1007/s11301-024-00418-z
3. Braglia, M., Frosolini, M.: An integrated approach to implement project management information systems within the extended enterprise. Int. J. Proj. Manag. **32**(1), 18–29 (2014). https://doi.org/10.1016/j.ijproman.2012.12.003
4. Weinert, A., Banas, R.: Analysis of the functions of project management information systems – traditional approach. Mod. Manag. Rev. **29**(1), 59–67 (2024). https://doi.org/10.7862/rz.2024.mmr.05
5. Alkudary, R., Gardiner, P.: Utilizing blockchain to enhance project management information systems: Insights into project portfolio success, knowledge management and learning capabilities. Int. J. Manag. Proj. Bus. **17**(4–5), 731–754 (2024). https://doi.org/10.1108/IJMPB-01-2024-0021
6. Besouw, J., Bond-Barnard, T.: Smart project management information systems (SPMIS) for engineering projects – project performance monitoring & reporting. Int. J. Inf. Syst. Proj. Manag. **9**(1), 78–97 (2021). https://doi.org/10.12821/ijispm090104
7. Micale, R., La Fata, C.M., Lombardo, A., La Scalia, G.: Project management infor-mation systems (PMISs): A statistical-based analysis for the evaluation of software package features. Appl. Sci. **11**(23), 11233 (2021). https://doi.org/10.3390/app112311233
8. Haseli, G., Sheikh, R., Wang, J., Tomaskova, H., Tirkolaee, E.B.: A novel approach for group decision making based on the best–worst method (G-BWM): Application to supply chain management. Mathematics. **9**(16), 1881 (2021). https://doi.org/10.3390/math9161881
9. Taniguchi, A., Onosato, M.: Use of project management information system to initiate the quality gate process for ERP implementation. Int. J. Inf. Technol. Comput. Sci. **12**, 1–10 (2017). https://doi.org/10.5815/ijitcs.2017.12.01
10. Zambare, P., Dhawale, A.: Project management information system in construction industry. Int. J. Eng. Sci. Res. Technol. **6, no. (7)**, 54–60 (2017). https://doi.org/10.5281/zenodo.822964
11. Certa, A., Enea, M., Giallanza, A.: Chapter 5 - A synthetic measure for the assessment of the project performance. In: Taticchi, P. (ed.) Business Performance Measurement and Management: New Contexts, Themes and Challenges, pp. 167–180. Springer, Berlin, Germany (2020). https://doi.org/10.1007/978-3-642-04800-5_11
12. Nitithamyong, P., Skibniewsky, M.J.: Web-based construction project management systems: How to make them successful? Autom. Constr. **13**(4), 491–506 (2004). https://doi.org/10.1016/j.autcon.2004.02.003

13. Scalia, L., Giada, M., Rosa, L.F., Manuela, C., Sciutba, L., Settanni, L.: Evaluating criteria for assessing sustainable bread preparated with olive oil mill wastewater: A multi-criteria-based approach. Future Foods. **12**(2025), 100817 (2025). https://doi.org/10.1016/j.fufo.2025.100817
14. Liang, F., Brunelli, M., Rezaei, J.: Consistency issues in the best worst method: Measurements and thresholds. Omega. **96**, 102175 (2020). https://doi.org/10.1016/j.omega.2019.102175

Hydrophobic Chain Functionalization Enhances Lipid Membrane Localization of a Luminescent Ruthenium(Ii) Complex

Lin QianQian[1](✉) and Wang JiaJIing[2]

[1] Shibaura Institute of Technology, 3-7-5 Toyosu, Koto-ku, Tokyo 135-8548, Japan
675476916a@gmail.com

[2] The University of Kitakyushu (Hibikino Campus), 1-1 Hibikino, Wakamatsu-ku, Kita-kyushu, Fukuoka 808-0135, Japan

Abstract. Fluorescent probes are essential for visualizing molecular dynamics. Ruthenium(II) polypyridyl complexes offer strong luminescence and high stability, making them useful for bioimaging. Here, we synthesized a $Ru(bpy)_3$ derivative modified with a long alkyl chain to improve its membrane incorporation. The compound was obtained through oxidation, amidation, and coordination steps, and confirmed by ^1H NMR and TLC. Liposome experiments showed that the alkylated complex produced much stronger red luminescence within lipid bilayers than unmodified $Ru(bpy)_3$, demonstrating enhanced membrane affinity. This highlights hydrophobic chain modification as an effective strategy for improving membrane-targeted fluorescent probes.

Keywords: Luminescent ruthenium complex · Liposome · Hydrophobic chain · Membrane targeting · Bioimaging

1 Introduction

Fluorescent probes are essential for non-invasive, real-time visualization of biomolecules in living cells, and since Tsien's calcium probe Quin-2 in 1980 [1], many probes have been developed for ions, ROS, and other biological targets [2]. Ruthenium(II) polypyridyl complexes are attractive due to their long-lived MLCT emission, high photostability, and oxygen sensitivity, making them suitable for time-resolved and oxygen-sensing imaging [3, 4]. However, traditional $Ru(bpy)_3$ complexes have poor lipid-membrane affinity, limiting their use in membrane-targeted imaging, and effective strategies for improving membrane localization remain lacking. This study therefore designs a long-chain alkylated $Ru(bpy)_3$ derivative with enhanced hydrophobicity and examines its incorporation into lipid bilayers using NMR and fluorescence microscopy.

2 Methodology

2.1 Instruments and Apparatus

Nuclear magnetic resonance (NMR) spectra were recorded using a JEOL JNM-LA300 spectrometer operating at 300 MHz with DMSO-d_6 as the solvent. Thin-layer chromatography (TLC) was performed on silica gel plates (Merck) using a chloroform/methanol

M. A. b. Jamlos et al. (Eds.): ICSET 2025, LNNS 1897, pp. 537–542, 2026.
https://doi.org/10.1007/978-3-032-20722-7_68

(9:1) solvent system. Fluorescence imaging was conducted using an Olympus BX51 fluorescence microscope equipped with a 40 × objective lens and a CCD camera. All fluorescence images were obtained under identical exposure conditions using excitation at 470–495 nm and emission detection above 510 nm.

2.2 Experimental Reagents

Lecithin, a major phospholipid component of biological membranes, was used to prepare model liposomes for evaluating the membrane incorporation of the ruthenium complexes.

4, 4′-Dimethyl-2, 2′-bipyridine (Wako Pure Chemical Industries, Ltd., $C_{12}H_{12}N_2$), potassium chromate (K_2CrO_4), thionyl chloride ($SOCl_2$), dodecylamine ($CH_2(CH_2)_{11}NH_2$), triethylamine ($C_6H_{15}N$), tetrahydrofuran (THF, C_4H_8O), and tris(2, 2′-bipyridyl)ruthenium(II) dichloride hexahydrate ($C_{30}H_{36}Cl_2N_6O_6Ru$) were all purchased from Wako Pure Chemical Industries and used without further purification.

2.3 Synthesis of the Alkylated Bipyridine Ligand

4, 4′-Dimethyl-2, 2′-bipyridine was oxidized with potassium chromate in sulfuric acid to yield 4, 4′-dicarboxy-2, 2′-bipyridine. The resulting dicarboxylic acid was converted into its acid chloride by reaction with thionyl chloride under a nitrogen atmosphere. Subsequently, the acid chloride was reacted with dodecylamine in dehydrated tetrahydrofuran (THF) in the presence of triethylamine to form the target alkylated bipyridine ligand.

Figure 1 — Synthetic Route

Figure 1 shows the synthetic route for the preparation of the alkylated bipyridine ligand and its ruthenium(II) complex.

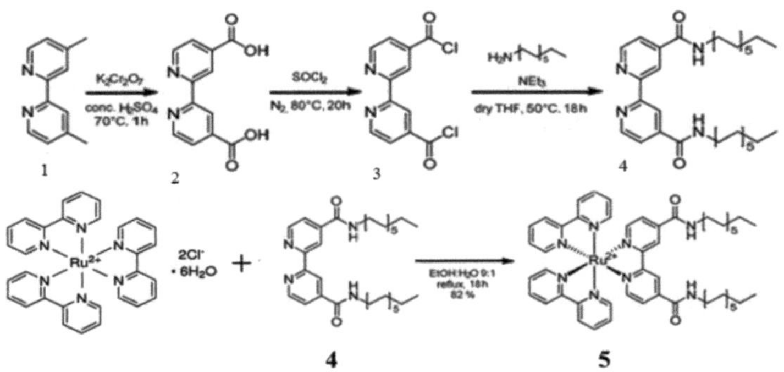

Fig. 1. Synthetic route of the alkylated bipyridine ligand and subsequent coordination to $Ru(bpy)_3{}^{2+}$ to yield the hydrophobic complex.

2.4 Synthesis of the Ruthenium Complex

The alkylated bipyridine ligand was coordinated to tris(2, 2′-bipyridyl)ruthenium(II) dichloride hexahydrate in an ethanol/water (1:9) solution. The mixture was refluxed under nitrogen at 70 °C for 12 h. After solvent removal, the dark red solid obtained was dried under reduced pressure.

2.5 Characterization

^{1}H NMR analysis of the synthesized [2, 2′-bipyridine]- 4, 4′-dicarboxamide ligand was performed using a JEOL JNM-LA300 spectrometer (300 MHz) with DMSO-d_6 as the solvent. As shown in (Fig. 2), the aromatic proton signals appear between δ 7.7–8.8 ppm, while the aliphatic proton peaks are observed around δ 0.8–1.5 ppm. The presence of solvent peaks corresponding to DMSO and residual H_2O is also evident,confirming that the synthesized compound possesses the expected structural characteristics.

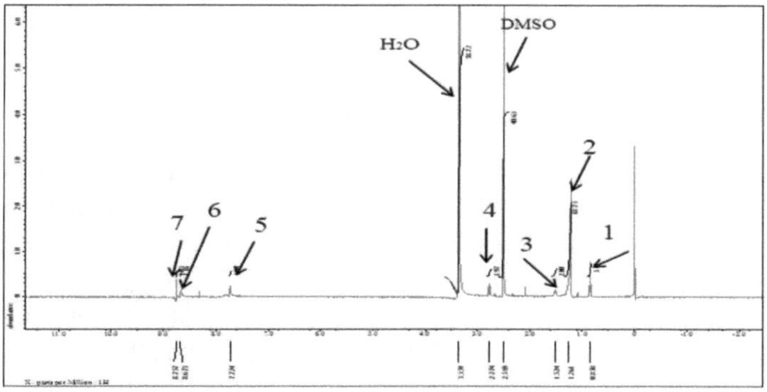

Fig. 2. ^{1}H NMR spectrum of [2, 2′-bipyridine]- 4, 4′-dicarboxamide in DMSO-d_6, showing characteristic proton signals and solvent peaks (DMSO and H_2O).

Table 1. ^{1}H NMR chemical shifts and proton assignments of the alkylated bipyridine ligand.

chemical shift (δ) (ppm)	coupling constant (J)	integration	proton(s)
0.850	d 3H	3.17	1
1.261	s $(2H)_9$	18.13	2
1.524	quin 2H	2.00	3
2.510	m	40.63	DMSO
2.774	d 2H	1.97	4
3.359	s	58.12	H_2O
7.724	d 1H	1.34	5
8.673	d 1H	1.16	6
8.757	s 1H	1.03	7

2.6 Liposome Preparation

Liposomes were prepared using the Bangham method [5]. Lecithin and the ruthenium complex (or unmodified $Ru(bpy)_3$ for comparison) were dissolved in chloroform/ethanol. The solvent was evaporated under reduced pressure to form a thin lipid film. The film was hydrated with 50 mM HEPES buffer (pH 7.0) above the phase transition temperature and vortexed to form multilamellar vesicles.

2.7 Fluorescence Microscopy

Fluorescence imaging of liposomes was carried out using an Olympus BX51 fluorescence microscope equipped with a 40 × objective lens and a CCD camera. Excitation light was provided through a 470–495 nm filter, and fluorescence emission above 510 nm was collected using a long-pass filter.

All fluorescence images were obtained under identical exposure and gain conditions to ensure consistent comparison between unmodified and alkylated ruthenium complexes.

3 Results

3.1 Characterization of the Ligand and Complex

The alkylated bipyridine ligand was successfully synthesized and confirmed by thin-layer chromatography (TLC) and ^{1}H nuclear magnetic resonance (NMR) spectroscopy. TLC analysis showed a retention factor (R_f) of 0.87 in chloroform/methanol (9:1), distinct from that of the starting material 4, 4′-dimethyl-2, 2′-bipyridine.

^{1}H NMR spectra (Fig. 3) exhibited characteristic aromatic proton peaks at δ 7.7–8.8 ppm and aliphatic signals between δ 0.8–1.5 ppm, confirming the successful introduction of a long-chain alkyl group. The corresponding chemical shifts, coupling constants (J), and proton integrations are summarized in Table 1.

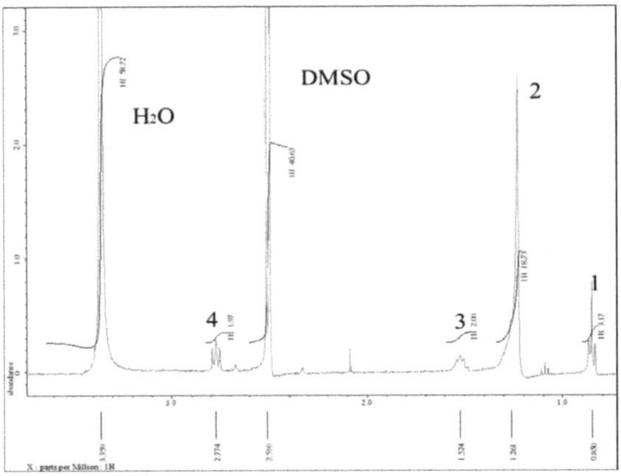

Fig. 3. An enlarged view of the NMR spectrum (Fig. 2) showing the chemical shift range from 0 to 4 ppm.

3.2 Fluorescence Microscopy of Liposomes

To investigate membrane incorporation, liposomes were prepared by the Bangham method and loaded with either the unmodified $Ru(bpy)_3$ complex or the newly synthesized alkylated complex. Bright-field and fluorescence microscopy revealed that liposomes containing unmodified $Ru(bpy)_3$ exhibited only weak red luminescence, mostly dispersed in the surrounding aqueous medium (Fig. 4). In contrast, liposomes loaded with the alkylated ruthenium complex showed markedly stronger and more localized red fluorescence at the lipid bilayer (Fig. 5), demonstrating improved membrane affinity.

Fig. 4. Bright-field image (400 × magnification) of liposomes containing tris(2, 2′-bipyridyl)dichlororuthenium(II) hexahydrate.

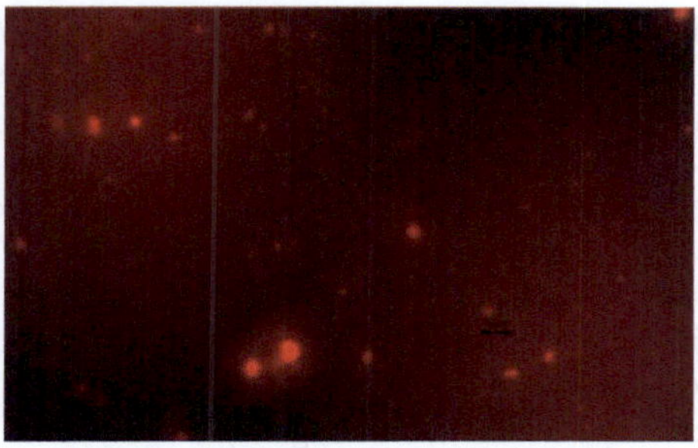

Fig. 5. Red fluorescence image (400 × magnification) of liposomes containing tris(2, 2′-bipyridyl)dichlororuthenium(II) hexahydrate.

3.3 Discussion of Membrane Incorporation

The enhanced fluorescence at the lipid bilayer observed for the alkylated complex is attributed to the hydrophobic effect of the dodecyl chain, which increases the affinity of the ruthenium complex for lipid membranes. Hydrophobic interactions likely reduce leakage of the complex into the aqueous phase and stabilize its localization within the bilayer, resulting in higher membrane-associated luminescence. This structural modification strategy could be applied to other metal complexes to improve their suitability as membrane-targeted fluorescent probes.

4 Conclusions

In this study, a long-chain alkylated ruthenium(II) polypyridyl complex was successfully synthesized to improve the lipid membrane targeting capability of $Ru(bpy)_3$ derivatives. The synthetic route involved oxidation of 4, 4′-dimethyl-2, 2′-bipyridine, amidation with dodecylamine, and subsequent coordination to ruthenium(II). Structural analysis by thin-layer chromatography (TLC) and ^1H nuclear magnetic resonance (NMR) confirmed the successful introduction of the hydrophobic alkyl chain and complex formation.

Fluorescence microscopy of liposomes revealed that the alkylated ruthenium complex exhibited significantly stronger and more localized red luminescence at the lipid bilayer compared with the unmodified $Ru(bpy)_3$ complex. This enhancement is attributed to the increased hydrophobicity of the modified complex, which promotes stable incorporation into lipid membranes.

These findings demonstrate that hydrophobic chain modification is an effective design strategy for improving the membrane affinity and bioimaging performance of metal complex-based fluorescent probes. The approach presented here may be further extended to other transition metal complexes for the development of next-generation, membrane-targeted luminescent imaging agents.

References

1. Tsien, R.Y.: New calcium indicators and buffers with high selectivity against magnesium and protons: design, synthesis, and properties of prototype structures. Biochemistry **19**(11), 2396–2404 (1980). https://doi.org/10.1021/bi00552a018
2. Sun, W., Guo, S., Chong, H., Fan, J., Peng, X.: Recent development of chemosensors based on cyanine platforms. Chem. Rev. **116**(14), 7768–7817 (2016). https://doi.org/10.1021/acs.chemrev.6b00001
3. Lo, K.K.W., Choi, A.W.T., Law, W.H.T.: Applications of luminescent inorganic and organometallic transition metal complexes as biomolecular and cellular probes. Dalton Transactions **41**(20), 6021–6047 (2012). https://doi.org/10.1039/C2DT11892K
4. Baggaley, E., Weinstein, J.A., Williams, J.G.: Lighting the way to see inside the live cell with luminescent transition metal complexes. Coordination Chem. Rev. **256**(15–16), 1762–1785 (2012). https://doi.org/10.1016/j.ccr.2012.03.018
5. Bangham, A.D., Standish, M.M., Watkins, J.C.: Diffusion of univalent ions across the lamellae of swollen phospholipids. J. Mol. Biol. **13**(1), 238–252 (1965). https://doi.org/10.1016/S0022-2836(65)80093-6

Intelligent Analysis of Chitosan Coagulant Efficiency in Turbid Sea Water Treatment

Wan Nur Syazwani Wan Azhal and Siti Amira Othman(✉)

Faculty of Applied Sciences and Technology, Universiti Tun Hussein Onn Malaysia, 86400 Pagoh, Johor, Malaysia
sitiamira@uthm.edu.my

Abstract. Chitosan, a natural polysaccharide derived from the deacetylation of chitin, has demonstrated potential as an eco-friendly coagulant for the treatment of turbid seawater. Turbidity, caused by suspended organic and inorganic particles, can adversely affect seawater quality and disrupt aquatic ecosystems by limiting light penetration and oxygen absorption. This study investigates the efficiency of chitosan in reducing seawater turbidity using an intelligent analysis approach. Various solvents were evaluated to determine the optimal medium for chitosan dilution, while seawater samples with differing pH levels were treated to assess coagulation performance. Changes in water color and chitosan mass before and after treatment were recorded. UV-Visible spectroscopy provided quantitative analysis of turbidity reduction, and field-emission scanning electron microscopy (FESEM) coupled with energy-dispersive X-ray spectroscopy (EDX) characterized the morphology and elemental composition of chitosan and coagulated samples. Results revealed the optimal solvent for chitosan application and highlighted its significant role in improving water clarity. The study demonstrates the integration of computational intelligence in evaluating and optimizing chitosan-based coagulation for sustainable seawater treatment.

Keywords: Chitosan · Coagulant · Treatment · Turbidity · Seawater

1 Introduction

Chitosan is a natural polysaccharide widely used as a pharmaceutical excipient. Chitosan is derived from a natural product, chitin, which is relatively abundant in shrimp shells. Chitin is a non-toxic, biodegradable polymer of high molecular weight. An enormous number of living organisms synthesise it. After cellulose, it is the most common polysaccharide found in nature. Like cellulose, chitin is a fibre, and in addition, it presents exceptional chemical and biological qualities that can be used in many industrial and medical applications [1].

Although chitin and chitosan have the same chemical structure, chitin has more applications and becomes more useful when transformed into chitosan. Chitosan can be obtained by the partial deacetylation of chitin, a natural polymer composed of randomly linked β-(1- 4)-D-glucosamine. This is achieved by removing sufficient acetyl groups

M. A. b. Jamlos et al. (Eds.): ICSET 2025, LNNS 1897, pp. 543–548, 2026.
https://doi.org/10.1007/978-3-032-20722-7_70

($COCH_3$) to make the molecule soluble in most diluted acids. This deacetylation process releases amine groups (NH_2) and confers a cationic characteristic to the chitosan. This is especially interesting in an acidic environment where the majority of polysaccharides are usually neutral or negatively charged. Lately, chitosan has been utilised in various fields, including biomedical studies, biotechnology, the textile industry, and many others [2]. Increasing research on chitosan has been conducted to maximise its benefits and applications in various fields. Chitosan is frequently used in chemical and medical studies due to its ability to aid in weight loss, lower cholesterol levels, and its promise as a biopolymer in drug delivery systems [3].

2 Material

Chitosan powder, sea water, Hydrochloric acid (0.2M, 0.4M, 0.6M, 0.8M, 1.0M), Sodium hydroxide (0.2M, 0.4M, 0.6M, 0.8M and 1.0M), Sea water obtained from Sungai Lurus, Senggarang, acetone (0.1M), distilled water, Chitosan (pnnm) weight 5g hydropilized stired at 4 °C. Chitosan was diluted in hydrochloric acid with 0.2 M, 0.4 M, 0.6 M, 0.8 M, and 1.0 M. The characterisation of the samples was performed using two different types of instruments, including Ultraviolet-Visible Spectroscopy (UV-Vis) and Field Emission Scanning Electron Microscope (FESEM).

3 Results and Discussion

3.1 UV-Visible Spectroscopy Analysis

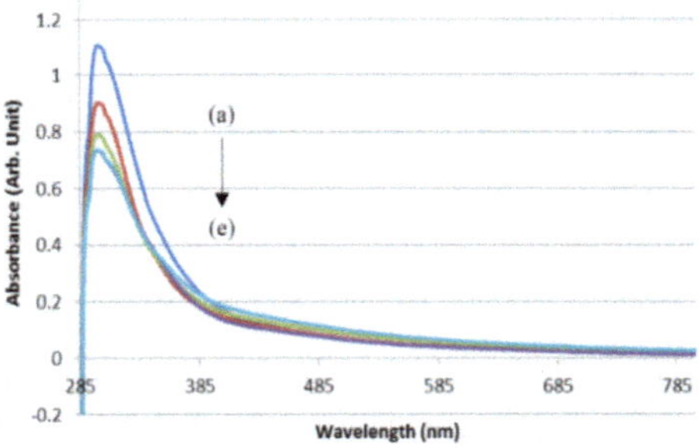

Fig. 1. Graph of absorbance against wavelength for chitosan diluted in different molarities of hydrochloric acid (a) 0.2M (b) 0.4M (c) 0.6 M (d) 0.8 M (e) 1.0 M

Figure 1 shows the graph of absorbance against wavelength for chitosan dilutions in different molarities of hydrochloric acid. The pattern of the graph indicates that as the molarity of hydrochloric acid increased, the light absorption decreased [4]. It might be due to the number of molecules present in each sample of different molarity. There is a sharp peak between wavelengths of 285nm and 385nm, where the maximum absorbance was obtained at approximately 295nm. The maximum of light absorption for each sample appears in the same wavelength range. However, the values of absorbance depended on the amount of chitosan that was completely soluble in the hydrochloric acid. The decrease in light absorbance value from 0.2 M to 1.0 M showed that the overall trend of the absorbance against wavelength graph decreases as the molarity of hydrochloric acid, the solvent for chitosan, increases. Based on the graph, 0.2 M is placed at the highest line, followed by 0.4 M, 0.6 M, 0.8 M, and 1.0 M. The greater light absorbance value for the sample of 0.2 M hydrochloric acid indicated that there were more molecules in this sample compared to the other samples, since the amount of light absorbance represents the number of molecules in a sample [5].

Coagulation and Treatment of Turbid Sea Water. Three chitosan solutions were prepared for three beakers of turbid seawater with different pH values: pH 3.0, pH 6.7, and pH 9.0. The original pH of seawater was 6.7. A certain amount of hydrochloric acid and sodium hydroxide was added to the seawater to alter the pH value [6]. The initial turbidity of the seawater was observed, and an image was captured using a phone camera for comparison with the final turbidity.

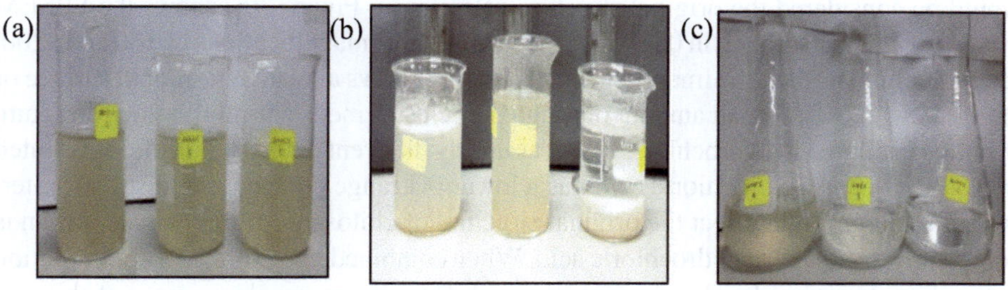

Fig. 2. (a) Initial turbidity of sea water samples (b) Coagulation and flocculation (c) Final turbidity of sea water

Figure 2(a) shows the initial turbidity of each different pH of seawater. The seawater samples with pH levels of 3.0, 6.7, and 9.0 were labelled as Sample A, Sample B, and Sample C, respectively. After that, coagulant was added to each beaker of seawater, and it was stirred at a constant speed for 30 min until coagulation formed. The beakers were then left to sit at room temperature for a couple of hours to ensure flocculation settled to the bottom before being filtered [7]. Figure 2(b) shows the formation of coagulation and flocculation of chitosan and the particles in the water. The solutions were filtered using filter paper to separate the coagulation formed with the seawater. The final turbidity of seawater can be observed after the filtration process is done. Thus, the changes in turbidity of each can were observed, and the most clear treated water was identified [8]. Based on Fig. 2(b), coagulation formed the most in sample C, which was the sample of seawater of pH 9, while the coagulation showed the least in sample A, which was the sample of water of pH 3.

Figure 2(c) shows the final turbidity of the seawater after treatment with chitosan. This water was obtained after the chitosan coagulation had been removed through a filtration process [8]. Based on the figure, sample C, which was turbid seawater with a pH of 9.0, showed the most transparent colour. Compared to the other two samples, it indicated the most turbidity removal. Sample A, which was turbid seawater with a pH of 3.0, had the lowest turbidity removal since it had the lowest clarity. The pH of the seawater after treatment with chitosan would be measured again. This study aimed to determine whether the use of chitosan as a coagulant can alter the pH of treated seawater. The final mass of chitosan, or the mass of chitosan after it has been used as a coagulant, would be the weight [9].

Field Emission Scanning Electron Microscope (FESEM) Analysis. Figure 3(a) shows the FESEM images of chitosan diluted in distilled water at magnification of x50 and x200, respectively. Based on the images, the morphology reveals a rough and uneven surface of chitosan, yet each particle retains an almost identical shape. As mentioned earlier, chitosan is insoluble in water and retains its original shape even after being diluted in distilled water. Since there were no other elements except chitosan and water, this sample would be considered the original structure of chitosan. Figure 3(b) shows the FESEM images of chitosan diluted in 0.2 M hydrochloric acid at magnifications of x50 and x200, respectively. Based on the images, the morphology shows a relatively smooth surface of chitosan, with only a small amount of rough particles formed. The morphology structure of chitosan diluted in hydrochloric acid was utterly different from that of chitosan diluted in distilled water. Hydrochloric acid has a lower pH range compared to distilled water; thus, its acidity might affect the original structure of chitosan. Additionally, chitosan is completely dissolved in hydrochloric acid. When combined, they formed a thick solution as the chitosan molecules absorbed into the acid. Thus, the morphology image of chitosan diluted in hydrochloric acid was smoother than that of chitosan diluted in distilled water. Since the chitosan was diluted entirely, the particles also became smaller and softer [10].

Figure 3(c) shows the FESEM images of chitosan diluted in 1.0 M sodium hydroxide at magnifications of x50 and x200, respectively. Based on the images, the morphology of the very rough surface of chitosan is evident. The particles were positioned close to each other, but they had a very rough shape [11]. The morphology structure of chitosan diluted in sodium hydroxide was utterly different from that of chitosan diluted in hydrochloric acid.

(a)
(b)
(c)
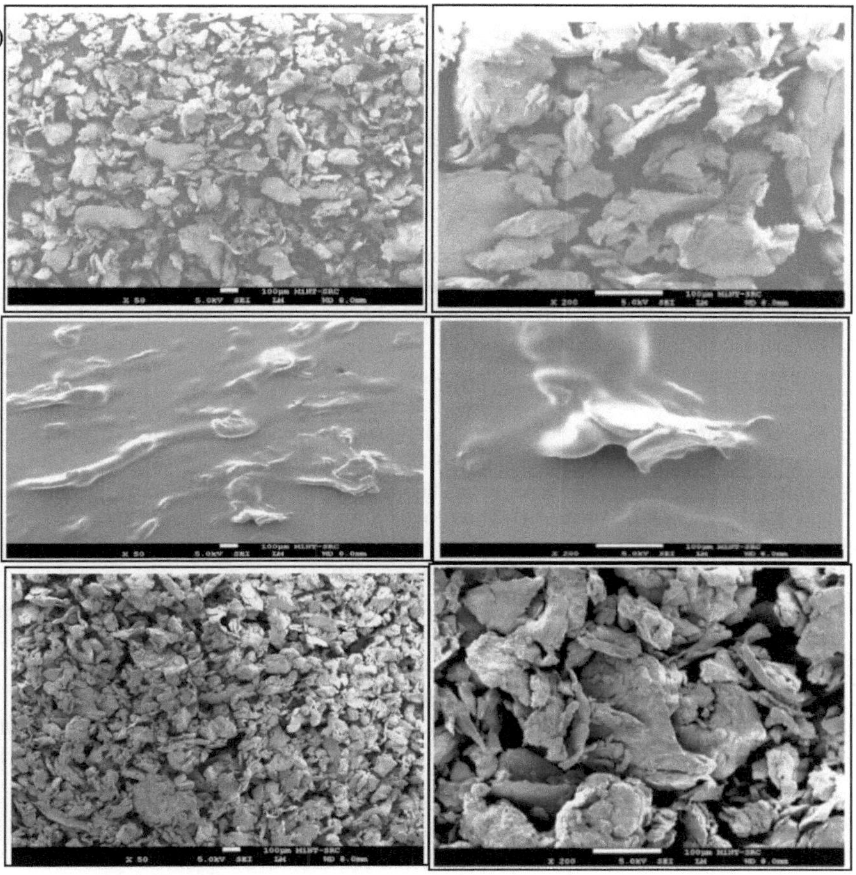

Fig. 3. FESEM image of chitosan in (a) distilled water (b) 0.2 M HCl (c) 1.0 M NaOH

4 Conclusion

Based on the results and data obtained from this research, it can be concluded that the solubility of chitosan in a solvent depends on the pH and molarity of the solvent. In this research, chitosan proved its ability and role as a coagulant in treating turbid seawater. The role and capability of chitosan to coagulate water turbidity were studied and observed through this research by preparing samples of seawater with different pH values. Chitosan began to coagulate after being mixed with seawater and stirred at a constant speed for 30 min. From the coagulation process, flocculation begins and settles to the bottom of the beaker, forming an emulsion that can then be separated from the treated water. The chitosan's ability to coagulate water turbidity was actually dependent on the pH level of the turbid seawater.

Acknowledgement. The authors would like to thank the Faculty of Applied Sciences and Technology for the facilities provided, which made the research possible.

References

1. Rinaudo: Chitin and chitosan: Properties and applications. Progr. Polym. Sci. **31**, 603–632 (2006)
2. Chen, C.-Y., Wu, C.-Y., Chung, Y.-C.: The coagulation characteristics of humic acid by using acid-soluble chitosan, water-soluble chitosan, and chitosan coagulant mixtures. Environ. Technol. **36**(9–12), 1141–1146 (2015)
3. Bina, B., Mehdinejad, M.H., Nikaeen, M., Attar, H.M.: Effectiveness of chitosan as natural coagulant aid in treating turbid waters. Iran. J. Environ. Health Sci. Eng. **6**(4), 247–252 (2009)
4. Annadurai, G., Sung, S.S., Lee, D.J.: Simultaneous removal of turbidity and humic acid from high turbidity stormwater. Adv. Environ. Res. **12**(3–6), 513–665 (2004)
5. Bergamasco, R., Bouchard, C., da Silva, F.V., Reis, M.H.M., Fagundes-Klen, M.R.: An application of chitosan as a coagulant/flocculant in a microfiltration process of natural water. Desalination **245**(1–3), 205–213 (2009)
6. Keeley, J., Jarvis, P., Smith, A.D., Judd, S.J.: Coagulant recovery and reuse for drinking water treatment. Water Res. **88**, 502–509 (2016)
7. Sahu, O.P., Chaudhari, P.K.: Review on chemical treatment of industrial wastewater. J. Appl. Sci. Environ. Manage. **17**(2), 241–257 (2013)
8. Ramli, S.F., Aziz, H.A.: Use of ferric chloride and chitosan as coagulant to remove turbidity and color from landfill leachate. In: Proceedings of [Conference Name], pp. 1163–1167 (2015)
9. LeChevallier, M.W., Evans, T.M., Seidler, R.J.: Effect of turbidity on chlorination efficiency and bacterial persistence in drinking water. Appl. Environ. Microbiol. **42**(1), 159–167 (1981)
10. Zemmouri, H., Drouiche, M., Sayeh, A., Lounici, H., Mameri, N.: Chitosan application for treatment of Beni-Amrane's water dam. Energy Procedia **36**, 558–564 (2013)
11. Zemmouri, H., Drouiche, M., Sayeh, A., Lounici, H., Mameri, N.: Coagulation–flocculation test of Keddara's water dam using chitosan and sulfate aluminium. Procedia Eng. **33**, 254–260 (2012)

Palm Oil Fuel Ash as a Self-Healing Agent in Cement Systems: Phase 1 Material Characterisation and Healing Potential Index Modelling

Salinah Dullah[1(✉)], Hidayati Asrah[1(✉)], Habib Musa Mohamad[1], Habibah Ghazali[2], and Aslina Baharum[3]

[1] Civil Engineering, Faculty of Engineering, Universiti Malaysia Sabah, Jalan UMS, 88400 Kota Kinabalu, Sabah, Malaysia
{salinahd,hidayati}@ums.edu.my

[2] College of Engineering and Science, Victoria University, Footscray Park Campus, Ballarat Road, Footscray, P.O. Box 14428, Melbourne, VIC 8001, Australia

[3] Department of The Design School, Faculty of Innovation and Technology, Taylor's University, No. 1, Jalan Taylor's, Subang Jaya 47500, Selangor, Malaysia

Abstract. This study investigates the effect of controlled grinding duration on the physical, chemical, and microstructural activation of palm oil fuel ash (POFA) for self-healing cementitious applications. Raw POFA was processed at three grinding durations (2.00, 2.25 and 2.50 h) and characterised using PSA, specific gravity, BET surface area, XRD, XRF, LOI and SEM-EDX analyses. Results showed a progressive refinement in fineness, with G-2.25 achieving the smallest median particle size ($D_{50} \approx 9.9$ µm) and the most favourable combination of surface area (39.79 m^2/g), mesoporosity, and amorphous silica content (~61%). Chemical composition met ASTM C618 Class N requirements, and SEM-EDX confirmed uniform particle morphology at this condition. A Healing Potential Index (HPI) was developed to integrate multiple activation parameters, with G-2.25 achieving the highest value (HPI = 42.86), indicating an optimum balance of reactivity and material stability. These findings identify 2.25 h as the optimal activation duration and establish a quantitative basis for selecting POFA variants for Phase 2 self-healing performance evaluation.

Keywords: Palm Oil Fuel Ash (POFA) · Self-healing cementitious systems · Supplementary cementitious materials · Pozzolanic reactivity · Characterisation · Sustainable concrete

1 Introduction

Improving the durability of cement-based materials remains a focal challenge for the construction industry, particularly in reducing the long-term impact of cracking. It remains a promising direction to gain deeper understanding of self-healing mechanisms within concrete so that small cracks are repaired before they propagate.

M. A. b. Jamlos et al. (Eds.): ICSET 2025, LNNS 1897, pp. 549–557, 2026.
https://doi.org/10.1007/978-3-032-20722-7_71

Palm Oil Fuel Ash (POFA) is produced as a residue in palm-oil-mill boilers. In recent decades, researchers have started to acknowledge its potential as a supplementary cementitious material (SCM) because of the high content of reactive silica [1–5]. It was discovered when the ash is finely ground, it behaves as a micro-filler that tightens the particle packing and refines the pore structure, which in turn supports the development of additional C-S-H gel [6]. Several studies also note that longer grinding times increase surface area and accelerate the pozzolanic reaction, leading to higher compressive strength [4, 7]. Interestingly, the ultrafine and nanosized ashes show the greatest Strength Activity Index (SAI) values and noticeably better durability [7, 8]. Despite this, relatively few studies have examined how controlled preparation variables such as particle size distribution, surface area, mineralogy, and chemical composition including ASTM C618 criteria such as $SiO_2 + Al_2O_3 + Fe_2O_3$ ($\geq$70%), SO_3 ($\leq$4%), and Loss on Ignition ($\leq$10%) influence POFA's potential for self-healing. BET surface area analysis can quantify the textural properties that determine reaction rates [9, 10], while X-ray diffraction (XRD) can reveal both crystalline phases and amorphous content that are critical indicators of pozzolanic reactivity [11]. Yet, these techniques have rarely been applied in a coordinated manner to POFA within a self-healing framework. Most prior investigations have focused on cement replacement levels rather than processing parameters [5, 12, 13]. For example, Shahid et al. [12] finding that replacing 10% of cement with POFA under water curing achieved the highest healing rate, through measurement ultrasonic pulse velocity recovery. This study introduces a simple parameter-based screening model (HPI) to support early material selection, the Healing Potential Index (HPI) shown in Eq. (1), designed to integrate critical parameters, amorphous SiO_2 content, BET surface area, particle fineness (D_{50}), and LOI into a single indicator.

This Phase 1 research addresses that gap by examining unground and ultrafine POFA produced via controlled grinding times. The materials are evaluated for particle size distribution, specific gravity, BET surface area, XRD-derived mineralogy, and XRF-measured oxide composition (including SO_3 content) against ASTM C618 [14] chemical requirements. The goal is to pinpoint the combination of properties that maximise pozzolanic reactivity and self-healing potential, providing the scientific basis for Phase 2. The next phase will validate reactivity through SAI testing in accordance with ASTM C311 [22] and incorporate advanced microstructural analysis to better understand the mechanisms of autogenous crack sealing. Accordingly, this study combines experimental characterisation with a performance-oriented parameter modelling approach to guide material selection for self-healing cementitious systems.

2 Methodology

2.1 Materials

Palm Oil Fuel Ash (POFA) was sourced from a palm oil mill in Lumadan, Sabah, Malaysia.

2.2 Sample Preparation and Grinding

Raw POFA was oven-dried at 105 Â± 5Â°C for 24 h [15, 16] and sieved through a 150 μm mesh to produce unground POFA (Ug). Grinding was conducted using a planetary ball mill at 300 rpm with 600 g batches. Grinding durations of 2.00, 2.25 and 2.50 h produced G-2.00, G-2.25 and G-2.50, respectively (Fig. 1)

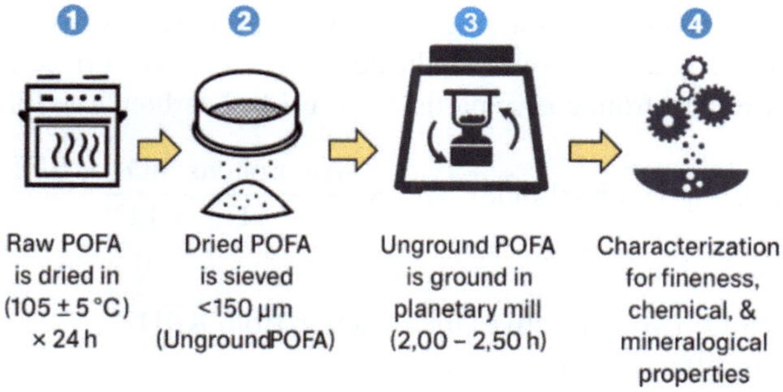

Fig. 1. Preparation process of ground and ultrafine palm oil fuel ash (POFA) (1) Oven drying; (2) Sieving <150 μm; (3) Grinding 2.00–2.50 h; (4) Characterisation

2.3 Particle Size Analysis (PSA)

Particle size was measured using Laser Particle Size Analysis (LPSA) and Dynamic Light Scattering (DLS) to capture both bulk and ultrafine fractions [19].

2.4 Physical and Chemical Characterisation

Specific gravity (ASTM C188), BET surface area, and oxide composition (XRF; ASTM E1621) were measured to evaluate particle densification and pozzolanic suitability. Loss on Ignition (ASTM C311) was used to quantify residual carbon. SEM–EDX was conducted to observe morphological refinement and elemental distribution after grinding [4, 9, 10, 14].

2.5 Scanning Electron Microscopy-Energy Dispersive X-Ray (SEM-EDX)

SEM-EDX was used to observe particle morphology and elemental composition before and after grinding [6, 16, 18, 19].

2.6 Performance-Oriented Parameter Modelling and Selection

A performance-oriented parameter modelling approach was adopted to integrate experimental outputs and identify the optimum POFA grinding condition for self-healing applications. Parameters were evaluated against ASTM C618 [14] chemical requirements and consolidated using the Healing Potential Index (HPI). The optimum condition for Phase 2 was selected based on the following criteria:

i. Lowest D_{50} value without evidence of re-agglomeration [16, 17]
ii. Highest effective BET surface area before over-grinding
iii. Highest amorphous SiO_2 content and compliance with ASTM C618 limits [14]
iv. Acceptable LOI and favourable SEM-EDX morphology

2.7 Healing Potential Index (HPI)

The Healing Potential Index (HPI) was formulated to consolidate four key reactivity-related parameters, amorphous silica content, BET surface area, particle fineness (D_{50}), and loss on ignition (LOI) into a single deterministic descriptor of theoretical self-healing efficiency. Higher HPI values indicate greater reactive surface availability and lower inhibitory effects from coarse particles or residual carbon. The formula is:

$$\text{Healing Potential Index, HPI} = \frac{\text{Amorphous SiO}_2 \times \text{BET}}{D_{50} \times \text{LOI}} \tag{1}$$

Where:
Amorphous SiO_2 (%) = reactive silica fraction from XRD
BET (m^2/g) = effective surface area
D_{50} (μm) = median particle size
LOI (%) = carbon-related mass loss affecting reactivity

3 Results

3.1 Particle Size Distribution and Specific Gravity (SG)

Particle size decreased with increasing grinding duration, consistent with mechanical activation behaviour reported in SCM studies [16]. G-2.25 achieved the smallest fineness ($D_{50} \approx 9.9$ μm; DLS < 1 μm), indicating the presence of reactive ultrafine particles [18–20]. G-2.50 showed slight re-agglomeration. Specific gravity increased with fineness, from 1.958 (Unground) to 2.410 at 2.25 h, peaking at 2.541 at 2.50 h. G-2.25 represents the optimum densification condition before re-agglomeration [21].

3.2 BET Surface Area and Pore Structure

BET results (Table 1) showed that G-2.25 achieved the highest effective surface area (39.79 m^2/g), an optimum range associated with improved pozzolanic activation [9]. Surface area decreased at 2.50 h due to particle re-agglomeration.

Table 1. Surface and Pore Structure Characteristics of POFA at Different Grinding Durations

Sample	BET Surface Area (m^2/g)	Total Pore Volume (cm^3/g)	BJH Pore Width (Å)	Observation
OPC	0.20	0.006	965	Dense reference cement exhibiting negligible surface area; the apparent large pore width reflects sparse capillary voids rather than true mesoporosity, serving as a non-pozzolanic baseline
Ug	56.37	0.069	116	High apparent surface area arising from combustion-derived porous carbon residues and internal ash voids, producing a mesoporous texture in a coarse, pre-activation state where reactive silica remains poorly dispersed and only partially effective for pozzolanic activation
G-2.00	32.53	0.049	144	Initial grinding reduces unstable carbon derived porosity and widens pore channels, indicating partial activation accompanied by the onset of particle densification.

(*continued*)

Table 1. (*continued*)

Sample	BET Surface Area (m^2/g)	Total Pore Volume (cm^3/g)	BJH Pore Width (Å)	Observation
G-2.25	39.79	0.052	132	Peak effective surface area with balanced mesoporosity, reflecting optimal dispersion of reactive amorphous silica and maximised accessibility for pozzolanic reactions.
G-2.50	24.41	0.042	163	Pronounced surface-area loss and pore coarsening due to re-agglomeration of ultrafine particles during over-grinding, limiting effective reactive surface availability.

3.3 XRD Analysis

All samples exhibited quartz and cristobalite peaks with a broad amorphous hump (20°–35° 2θ). G-2.25 showed the highest amorphous silica content (~ 61%), which is the primary reactive phase contributing to pozzolanic behaviour [11].

3.4 Chemical Composition (XRF and LOI)

All POFA samples satisfied ASTM C618 and BS EN 450-1 requirements [14, 23]. G-2.25 recorded the highest combined oxides (70.19%) and acceptable LOI (6.16%), supporting its suitability as a reactive Class N pozzolan [14].

3.5 SEM-EDX Morphological and Elemental Characterisation

SEM-EDX (Fig. 2) showed progressive refinement with grinding. G-2.25 displayed the most uniform particles with clearer Si–Al enrichment, consistent with its higher reactivity.

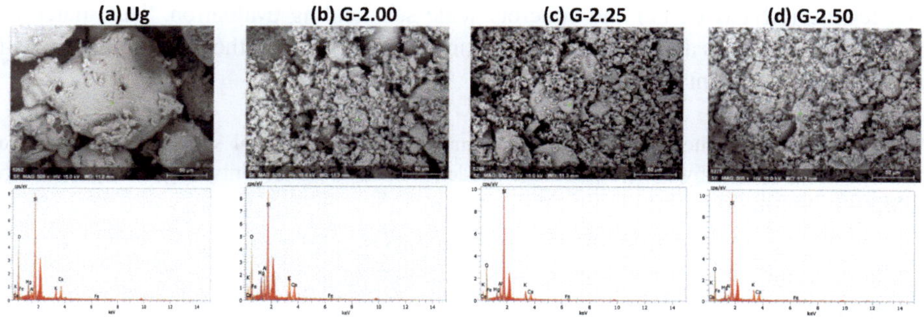

Fig. 2. Composite SEM-EDX analysis of Palm Oil Fuel Ash (POFA) at different grinding durations (Ug, G-2.00, G-2.25, G-2.50). SEM images were captured at 500× magnification with an accelerating voltage of 15.0 kV and working distance of ~11 mm. Progressive grinding refined the particle morphology, producing denser and more uniform surfaces, while EDX spectra showed stronger Si and Al peaks with reduced Ca intensity, confirming enhanced surface reactivity

3.6 Optimum Selection and Healing Potential Index (HPI)

HPI values (Table 2) showed that G-2.25 achieved the highest healing potential (HPI = 42.86). This aligns with studies highlighting the importance of integrating fineness, surface area, and reactive silica in pozzolanic performance models [19].

Table 2. HPI values calculated for each POFA variant

Sample	Grinding Time (h)	Amorphous SiO_2 (%)	BET (m^2/g)	D_{50} (μm)	LOI (%)	HPI
Ug	0.00	61.72	56.37	30.68	5.64	20.11
G-2.00	2.00	65.05	32.53	10.94	4.55	42.51
G-2.25	2.25	65.43	39.79	9.86	6.16	42.86
G-2.50	2.50	65.26	24.41	9.86	4.49	35.98

4 Conclusions

Grinding POFA for 2.25 h (G-2.25) produced the optimum activation condition, yielding a median particle size of ~9.9 μm and the highest Healing Potential Index (HPI = 42.86). The chemical composition ($SiO_2 + Al_2O_3 + Fe_2O_3$ = 70.19%, LOI = 6.16%, SO_3 = 0.23%) met ASTM C618 Class N criteria. XRD confirmed a substantial amorphous silica fraction, while BET surface area (39.79 m^2/g) and balanced mesoporosity reflected effective activation. SEM–EDX showed uniformly refined particles at this duration. The proposed HPI provides a concise framework for integrating key reactivity indicators and identifying optimal activation conditions. Phase 2 will validate this framework through

SAI testing (ASTM C311) and concrete-scale self-healing evaluation. This material-based index serves as a preliminary predictor, with validation at the structural scale still required in subsequent phases.

Acknowledgement. The author gratefully acknowledges the financial support for conference participation provided by the Postgraduate Student Research Fund, Universiti Malaysia Sabah (UMS).

References

1. Edvardsen, C.: Water permeability and autogenous healing of cracks in concrete. ACI Mater. J. **96**(4), 448–454 (1999). https://doi.org/10.14359/653
2. Van Tittelboom, K., De Belie, N.: Self-healing in cementitious materials—A review. Materials. **6**(6), 2182–2217 (2013). https://doi.org/10.3390/ma6062182
3. Snoeck, D., De Belie, N.: Autogenous healing in strain-hardening cementitious materials with and without superabsorbent polymers: An 8-year study. Front. Mater. **6**, 48 (2019). https://doi.org/10.3389/fmats.2019.00048
4. Chindaprasirt, P., Rukzon, S., Sirivivatnanon, V.: Resistance to chloride penetration of blended portland cement mortar containing palm oil fuel ash, rice husk ash and fly ash. Constr. Build. Mater. **22**(5), 932–938 (2008). https://doi.org/10.1016/j.conbuildmat.2006.12.001
5. Awal, A.S.M.A., Hussin, M.W.: The effectiveness of palm oil fuel ash in preventing expansion due to alkali–silica reaction. Cem. Concr. Compos. **19**(4), 367–372 (1997). https://doi.org/10.1016/S0958-9465(97)00034-6
6. Tangchirapat, W., Jaturapitakkul, C., Chindaprasirt, P.: Use of palm oil fuel ash as a supplementary cementitious material for producing high-strength concrete. Constr. Build. Mater. **23**(7), 2641–2646 (2009). https://doi.org/10.1016/j.conbuildmat.2009.01.008
7. Mohammed, A.N., Megat Johari, M.A., Zeyad, A.M., Tayeh, B.A., Yusuf, M.O.: Improving the engineering and fluid transport properties of ultra-high-strength concrete utilizing ultrafine palm oil fuel ash. J. Adv. Concr. Technol. **12**, 127–137 (2014). https://doi.org/10.3151/jact.12.127
8. Rukzon, S., Chindaprasirt, P.: Strength and chloride penetration of blended portland cement mortar containing palm oil fuel ash and fly ash. Int. J. Miner. Metall. Mater. **16**(6), 714–720 (2009). https://doi.org/10.1016/S1674-4799(09)60083-2
9. Brunauer, S., Emmett, P.H., Teller, E.: Adsorption of gases in multimolecular layers. J. Am. Chem. Soc. **60**(2), 309–319 (1938). https://doi.org/10.1021/ja01269a023
10. Sing, K.S.W.: Reporting physisorption data for gas/solid systems. Pure Appl. Chem. **57**(4), 603–619 (1985). https://doi.org/10.1351/pac198557040603
11. Taylor, H.F.W.: Cement Chemistry, 2nd edn. Thomas Telford, London (1997). https://doi.org/10.1680/cc.25929
12. Shahid, K.A., Jaafar, M.F.M., Yahaya, F.M.: Self-healing behaviour of pre-cracked POFA-concretes in different curing conditions. J. Mech. Eng. Sci. **7**, 1127–1235 (2014) https://doi.org/10.15282/jmes.7.2014.22.0120
13. Jackson, M.D., et al.: Material and elastic properties of al-tobermorite in ancient roman seawater concrete. Am. Ceram. Soc. **102**(7), 1435–1450 (2013). https://doi.org/10.1111/jace.12407
14. ASTM International: Annual Book of ASTM Standards. ASTM International, West Conshohocken, PA (2017–2022) Standards referenced: C188-17, C311-18, C618-22, E1621-13
15. Neville, A.M.: Properties of Concrete, 5th edn. Pearson Education Limited, Harlow (2011)

16. Cordeiro, G.C., Toledo Filho, R.D., Fairbairn, E.M.R.: Ultrafine grinding of sugar cane bagasse ash for application as pozzolanic admixture in concrete. Cem. Concr. Res. **39**(2), 110–115 (2009). https://doi.org/10.1016/j.cemconres.2008.11.005
17. Cordeiro, G.C., Toledo Filho, R.D., Fairbairn, E.M.R., Tavares, L.M.M., Oliveira, C.H.: Influence of mechanical grinding on the pozzolanic activity of residual sugarcane bagasse ash. In: Vázquez, E., Hendriks, C.F., Janssen, G.M.T. (eds.) International RILEM Conference on the Use of Recycled Materials in Buildings and Structures (PRO 40), pp. 731–740. RILEM Publications SARL, Barcelona, Spain (2004, November) Retrieved from https://www.researchgate.net/publication/237842672
18. Mehta, P.K., Monteiro, P.J.M.: Concrete: Microstructure, Properties, and Materials, 4th edn. McGraw-Hill, New York (2014) SBN: 0071462899, 9780071462891
19. Abu Aisheh, Y.I.: Palm oil fuel ash as a sustainable supplementary cementitious material for concrete: A state-of-the-art review. Case Stud. Constr. Mater. **18**, e01770 (2023). https://doi.org/10.1016/j.cscm.2023.e01770
20. Payá, J., Monzó, J., Borrachero, M.V., Velázquez, S.: Evaluation of the pozzolanic activity of fluid catalytic cracking catalyst residue (FC3R): Thermogravimetric analysis studies on FC3R–Portland cement pastes. Cem. Concr. Res. **33**(4), 603–609 (2003). https://doi.org/10.1016/S0008-8846(02)01026-8
21. Sata, V., Jaturapitakkul, C., Kiattikomol, K.: Utilization of palm oil fuel ash in high-strength concrete. J. Mater. Civ. Eng. **16**(6), 623–628 (2004). https://doi.org/10.1061/(ASCE)0899-1561(2004)16:6(623)
22. ASTM C311/C311M-13: Standard Test Methods for Sampling and Testing Fly Ash or Natural Pozzolans for Use in Portland-Cement Concrete. ASTM International, West Conshohocken, PA (2013)
23. BS EN 450-1:2012: Fly Ash for Concrete - Part 1: Definition, Specifications and Conformity Criteria. British Standards Institution, London, UK (2012)

Pozzolanic Reactivity of Palm Oil Fuel Ash in Mortar: Insights from Strength Activity Index and Pozzolanic Effect

Satriani, Muhammad Wihardi Tjaronge, and Muhammad Akbar Caronge(✉)

Department of Civil Engineering, Faculty of Engineering, Hasanuddin University, Jl. Poros Malino Km. 6, Gowa, South Sulawesi 92171, Indonesia
ma-caronge@unhas.ac.id

Abstract. This paper examines the pozzolanic reactivity and mechanical performance of mortar incorporating Palm Oil Fuel Ash (POFA). Mortar mixes with 0%, 10%, and 20% POFA were evaluated for compressive strength at 28 and 90 days, and pozzolanic activity was assessed using Strength Activity Index (SAI) and pozzolanic contribution indices (*R*, *K*, *P*), representing the pozzolanic effect. Results show decreased early-age strength but notable long-term enhancement, where POFA-10 and POFA-20 surpassed the control at 90 days with SAI values of 107.30% and 104.44%, respectively. Pozzolanic contribution increased significantly with curing age, confirming the sustained reactivity of POFA. These findings demonstrate that POFA satisfies ASTM C618 requirements and offers a sustainable, energy-efficient alternative for mortar production.

Keywords: Palm Oil Fuel Ash · Pozzolanic Reactivity · Strength Activity Index

1 Introduction

The construction sector is a major global consumer of natural resources and energy, with cement production contributing approximately 7–8% of anthropogenic CO_2 emissions [1, 2, 3, 4]. Rapid urbanization, particularly in developing regions, continues to increase demand for cement-based materials, intensifying the need for sustainable alternatives that reconcile environmental responsibility with structural performance. Consequently, international policy frameworks increasingly emphasize low-carbon binders and supplementary cementitious materials (SCMs) as key decarbonization strategies in construction [1]. Palm Oil Fuel Ash (POFA), a silica-rich by-product of palm oil mill power plants, has gained attention as a potential SCM due to its abundance in Southeast Asia and its ability to enhance mechanical and durability properties of mortar and concrete [5]

However, the practical use of POFA is constrained by its low early reactivity in raw form, commonly associated with coarse particle size, unburned carbon, and heterogeneous chemical phases. While grinding or calcination can improve POFA performance, these treatments require additional energy and compromise sustainability benefit [6]. highlighting the need for low-energy activation approaches.

M. A. b. Jamlos et al. (Eds.): ICSET 2025, LNNS 1897, pp. 558–564, 2026.
https://doi.org/10.1007/978-3-032-20722-7_72

Recent studies indicate that physical sieving to obtain ultrafine POFA fractions passing the ASTM No. 325 sieve (<45 μm) can enhance packing density, pozzolanic reactivity, and long-term strength without energy-intensive processing [5, 7]. Nevertheless, comprehensive evaluations integrating mechanical performance, Strength Activity Index (SAI), and pozzolanic contribution indices (*R*, *K*, *P*) remain limited.

Moreover, although ASTM C618 focuses on SAI at 28 days, recent research underscores the importance of extended curing periods (≥90 days) to capture the delayed pozzolanic behavior of SCMs such as POFA [8, 9]. To address this gap, the present study investigates sieved ultrafine POFA (100% passing No. 325 sieve) as a low-energy SCM in mortar by assessing compressive strength development, SAI compliance, and quantitative pozzolanic contribution at 28 and 90 days, thereby proposing a sustainable pathway for agro-industrial waste valorization in green construction.

2 Methodology

2.1 Materials and POFA Processing

Ordinary Portland Cement Type I, sand, water, and Palm Oil Fuel Ash (POFA) were used. POFA was collected from a palm oil biomass power plant, air-dried, and sieved so that 100% passed the ASTM No. 325 sieve (≤45 μm), ensuring a low-energy ultrafine fraction suitable for pozzolanic evaluation [5]. The raw materials are illustrated in Fig. 1.

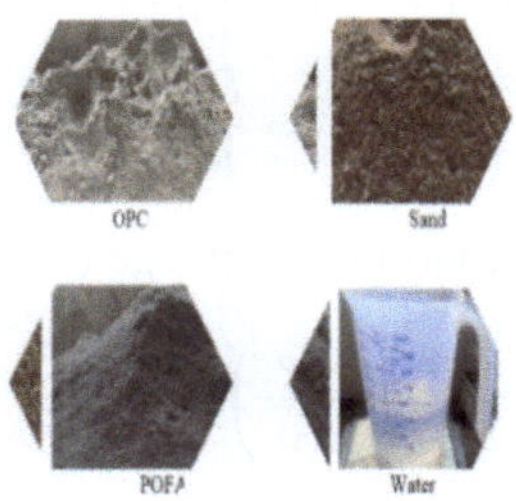

Fig. 1. Raw materials used in mortar production: OPC, Sand, POFA, and Water.

2.2 Mix Design and Specimen Preparation

Three mortar mixtures were prepared: a control (0% POFA) and mixtures with 10% and 20% cement replacement by POFA. All mixes followed ASTM C109 proportions with a constant water-to-binder ratio of 0,484. Cube specimens (50 × 50 × 50 mm) were cast and water-cured [10].

2.3 Testing and Analysis

Compressive strength was measured at 28 and 90 days and calculated using Eq. (1).

$$f = \frac{P_{\max}}{A} \quad (1)$$

where: f is the compressive strength (MPa), P_{max} is the maximum applied load (N) and A is the cross-sectional area of the specimen (mm^2).

Pozzolanic performance was assessed using the Strength Activity Index (SAI) accordance with ASTM C618, as expressed in Eq. (2).

$$SAI\ (\%) = \frac{Compressive\ Strenght\ of\ POFA\ Mix}{Compressive\ Strenght\ of\ Control\ Mix} \times 100\% \quad (2)$$

The pozzolanic contribution to compressive strength was quantified following the analytical framework proposed by Pu (1999) and refined by Yu et al. (2003), which distinguishes strength contributions from cement hydration and pozzolanic reactions. Three indices were employed: the specific strength ratio (R), the specific strength index (K), and the pozzolanic contribution percentage (P). [11]

The specific strength ratio (R) is defined as:

$$R = \frac{f}{q} \quad (3)$$

Where q represents the fraction of cementitious material. Two cases are considered: R_C is the contribution of unit cement and R_M is contribution of unit mineral admixture. The pure pozzolanic contribution (R_P) is calculated as:

$$R_P = R_M \text{--} R_C \quad (4)$$

The index of specific strength (K) is expressed as:

$$K = \frac{R_M}{R_C} \quad (5)$$

Finally, the pozzolanic contribution percentage (P) is defined as:

$$P = \left(\frac{R_P}{R_M}\right) \times 100\% \quad (6)$$

3 Results and Discussion

3.1 Compressive Strength at 28 and 90 Days

The compressive strength results of mortar incorporating sieved ultrafine Palm Oil Fuel Ash (POFA) at curing ages of 28 and 90 days are presented in Tables 1 and 2. At 28 days, the control mixture achieved a compressive strength of 31,80 MPa, while POFA-10 and POFA-20 reached 30,30 MPa and 29,20 MPa, respectively. Although slightly lower than the control, both POFA mixtures exhibited comparable strength levels, indicating the onset of pozzolanic activity and a reduced dilution effect at this age. At 90 days, a clear performance enhancement was observed: POFA-10 and POFA-20 surpassed the control strength of 33,60 MPa by reaching 36,05 MPa and 35,09 MPa, respectively. This improvement confirms the significant long-term contribution of sieved ultrafine POFA to strength development (Table 2).

Table 1. Indices of Pozzolanic Effect and Mixed Content of POFA (28 days)

Mixture	f (N/mm^2)	SAI (%)	POFA (%)	OPC (%)	R_M (N/mm^2)	R_P (N/mm^2)	K	P (%)
Control	31,80	–	–	100,00	0,318	0,00	1,00	–
POFA-10	30,30	95,38%	10,00	90,00	0,337	0,02	1,06	5,64
POFA-20	29,20	91,79%	20,00	80,00	0,365	0,05	1,15	12,84

Table 2. Indices of Pozzolanic Effect and Mixed Content of POFA (90 days)

Mixture	f (N/mm^2)	SAI (%)	POFA (%)	OPC (%)	R_M (N/mm^2)	R_P (N/mm^2)	K	P (%)
Control	33,60	–	–	100,00	0,336	–	1,00	–
POFA-10	36,05	107,30%	10,00	90,00	0,401	0,06	1,19	16,12
POFA-20	35,09	104,44%	20,00	80,00	0,439	0,10	1,31	23,40

3.2 Strength Activity Index (SAI)

The Strength Activity Index values further substantiate the compressive strength trends. At 28 days, SAI values exceeded 90% for POFA-10 (95,38%) and POFA-20 (91,79%), satisfying the ASTM C618 acceptance criterion for supplementary cementitious materials. At 90 days, SAI values increased markedly to 107,30% for POFA-10 and 104,44% for POFA-20, demonstrating that mortars containing sieved POFA can outperform the control mixture at extended curing ages. The relationship between compressive strength and SAI is illustrated in Fig. 2, highlighting the progressive enhancement of pozzolanic reactivity over time.

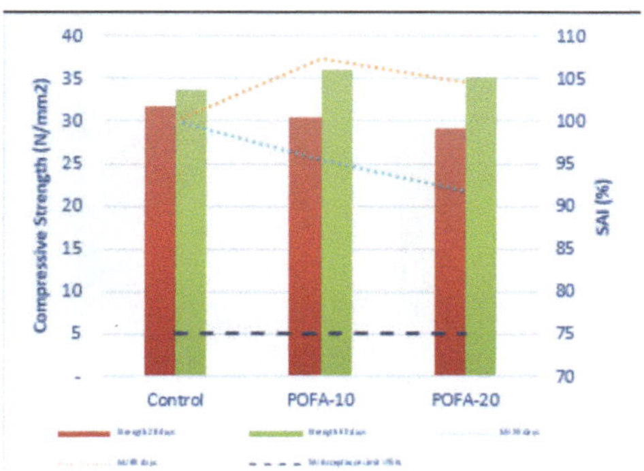

Fig. 2. Compressive Strength Vs SAI at 28 and 90 days

3.3 Pozzolanic Contribution and Discussion

The pozzolanic contribution indices (*R*, *K*, and *P*) provide mechanistic insight into the observed strength gains. At 28 days, *P* values increased to 5,64% for POFA-10 and 12.84% for POFA-20, indicating a measurable pozzolanic contribution. At 90 days, P rose substantially to 16,12% and 23,40%, respectively, while *K* values exceeded unity, confirming that POFA contributed more efficiently to strength per unit mass than cement (Tables 1–2). Figures 3 and 4 demonstrate the strong correlation between compressive strength, SAI, and *P*, reflecting the synergistic effects of micro-filler action and secondary calcium–silicate–hydrate formation. Although POFA-20 exhibited the highest pozzolanic contribution, POFA-10 provided the most balanced performance, supporting the identification of moderate replacement levels as optimal for sustainable mortar design.

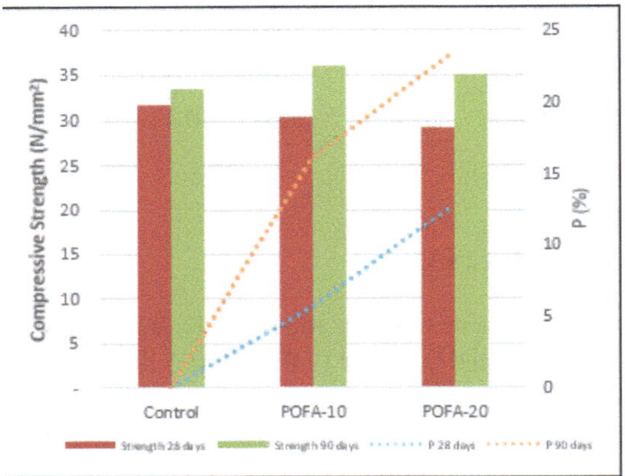

Fig. 3. Compressive Strength Vs Pozzolanic Effect Index at 28 and 90 days

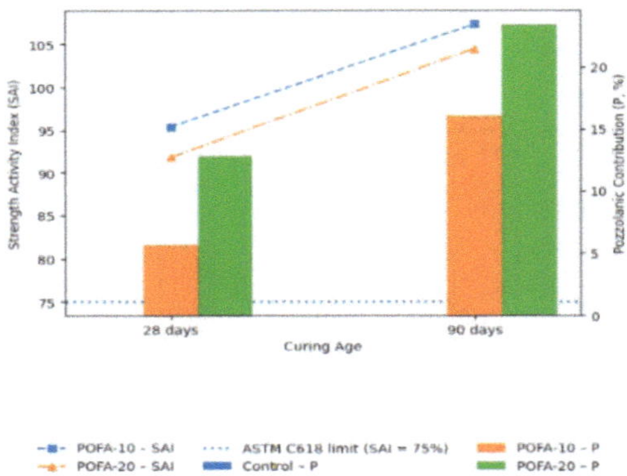

Fig. 4. SAI Vs Pozzolanic Effect Index at 28 and 90 days

4 Conclusion

This study confirms that sieved ultrafine Palm Oil Fuel Ash (POFA) is a technically viable and environmentally efficient supplementary cementitious material for mortar. While early-age strength was reduced due to dilution and delayed pozzolanic kinetics, significant long-term enhancement was achieved. At 90 days, mortars with 10% and 20% POFA exceeded the control in compressive strength, with Strength Activity Index values above 100% and pozzolanic contribution reaching up to 23,40%. The results indicate that long-term strength development is governed by synergistic micro-filler effects and sustained pozzolanic reactions, with 10% POFA identified as the optimal replacement level. This study provides quantitative evidence that low-energy sieving, without grinding or calcination, effectively unlocks the pozzolanic potential of POFA, supporting sustainable binder development, while future research should address durability and microstructural performance under aggressive conditions.

Acknowledgments. The authors would like to express their sincere gratitude to *The Centre for Higher Education Funding and Assessment and The Indonesia Endowment Funds for Education (LPDP)* Ministry of Higher Education, Science, and Technology of Republic Indonesia, for the financial support provided through the doctoral scholarship program. Appreciation is also extended to *Center for Higher Education Funding and Assessment* for the research collaboration and technical facilities that greatly contributed to the success of this study.

Disclosure of Interests The authors declare that they have no competing interests relevant to the content of this article.

References

1. T. Breakthrough Agenda Report, The Breakthrough Agenda Report 2024 2024.
2. R. M. Andrew, "Global CO_2 emissions from cement production, 1928-2018," Ther. Nova 20, 2019, Copernicus GmbH. doi:https://doi.org/10.5194/essd-11-1675-2019.
3. Wu, S., et al.: Global CO_2 uptake by cement materials accounts 1930–2023. Sci. Data. **11**(1), 1409 (2024). https://doi.org/10.1038/s41597-024-04234-8
4. International Energy Agency (IEA), "Cement," https://www.iea.org/energy-system/industry/cement.
5. Liew, Y.X., Saad, S.A., Anand, N., Tee, K.F., Chin, S.C.: Evaluating the impact of reducing POFA's particle fineness on its pozzolanic reactivity and mortar strength. J. Mater. Sci.: Mater. Electron. **19**(1) (2024). https://doi.org/10.1186/s40712-024-00169-5
6. Mohd Sufian, A.N.B., Rahman, M.R., Said, K.A.B.M., Bakri, M.K.B.: A critical review of various types of palm oil fuel ash (POFA) utilization in enhancing concrete and mortar properties. J. Build. Rehabil. **10**, 124 (2025). https://doi.org/10.1007/s41024-025-00622-8
7. Amran, M., et al.: Palm oil fuel ash-based eco-friendly concrete composite: A critical review of the long-term properties. MDPI. **14**(22), 7074 (2021). https://doi.org/10.3390/ma14227074
8. Ahmed, A.: Assessing the effects of supplementary cementitious materials on concrete properties: A review. Discover Civ. Eng. **1**(1), 145 (2024). https://doi.org/10.1007/s44290-024-00154-z

9. ASTM International: Specification for Coal Fly Ash and Raw or Calcined Natural Pozzolan for Use in Concrete. ASTM International, West Conshohocken, PA (2015). https://doi.org/10.1520/C0618-15
10. L.-H. Yu, H. Ou, and L.-L. Lee Standard Test Method for Compressive Strength of Hydraulic Cement Mortars (Using 2-in. or [50-mm] Cube Specimens) 1, Investigation on pozzolanic effect of perlite powder in concrete.

Preliminary Study of UH-Pod Stability as an Innovative Armor Unit for Rubble Mound Breakwaters

Rusmawati(✉), Muhammad Arsyad Thaha, and Riswal Karamma

Hasanuddin University, 92171 Gowa, South Sulawesi, Indonesia
rusmawati0990@gmail.com

Abstract. Indonesian coastal areas face high pressure due to human activities, climate change, and sea wave intensity, which triggers coastal abrasion and erosion. Coastal protection structures such as breakwaters are the main solution for dissipating wave energy before it reaches the coastline. However, the use of conventional armor units like Tetrapod, Dolos, Accropode, and Xbloc still has limitations in stability efficiency, large material volume requirements, and installation difficulties. Therefore, an innovative design for a protective block with high stability, geometric efficiency, and ease of production is needed. This study is a literature review aimed at identifying the concept of armor layer stability in rubble mound breakwaters and analyzing the potential development of a new shape called UH-Pod (Universitas Hasanuddin Pod) as an alternative armor unit. Based on the review, armor layer stability is influenced by several main parameters, including wave height, block specific weight, slope gradient, and unit dimension ratio. Furthermore, the geometric shape and degree of interlocking between units are crucial factors in enhancing structural efficiency. The UH-Pod, with its symmetrical H-shaped design, offers potential advantages in the form of modularity, ease of installation, and inter-unit interlocking capability. The results of the review indicate that the UH-Pod concept can be an innovative alternative for breakwater armor layer design that is efficient, stable, and oriented towards environmental sustainability. This study provides a conceptual basis for subsequent experimental research on the formulation of the UH-Pod stability formula in the form of dimensionless parameters.

Keyword: Breakwater · Rubble Mound · Armor layer · UH-Pod · Stability

1 Introduction

1.1 Background

Coastal areas have strategic value in economic, social, and ecological development. However, these areas are also vulnerable to threats such as ocean waves, coastal abrasion, and infrastructure damage caused by extreme oceanographic phenomena. One of the main problems of coastal damage is caused by waves [1]. Waves can generate energy that shapes the coastline, creates currents and sediment transport in both cross-shore and longshore directions, and exerts forces on coastal structures [2]. Therefore, an effective coastal defense structure with minimal environmental impact is needed [3].

M. A. b. Jamlos et al. (Eds.): ICSET 2025, LNNS 1897, pp. 565–573, 2026.
https://doi.org/10.1007/978-3-032-20722-7_73

The most dominant criterion of coastal damage is erosion/abrasion, which causes significant shoreline changes [1]. Abrasion occurs due to wave energy striking the land and currents carrying sediment material, leading to shoreline retreat [4]. To prevent abrasion, several types of coastal structures can be utilized, including breakwaters, groins, and revetments [5]. Coastal damage caused by waves can be mitigated by constructing breakwaters to reduce wave energy before it reaches the shore [6].

Breakwaters are structures designed to protect harbor waters from wave disturbances [7]. They are not only recognized as protective structures against erosion and abrasion but are also built to break waves by dissipating their energy [8]. Furthermore, breakwaters reduce wave height by allowing waves to break through the porous armor layer [9].

Breakwaters can be classified into two types: emerged breakwaters and submerged breakwaters [9]. Installing submerged breakwaters can reduce incoming wave height by up to 30% [10]. Breakwaters are also available in several types, designed based on site-specific needs. One of the most common types is the rubble mound breakwater, whose primary function is to dissipate incoming wave energy and which can be constructed using various kinds and sizes of armor units [11].

The armor layer is installed on the breakwater structure to ensure its stability. Several types of armor units can be used on sloped breakwaters, including Tetrapod and Dolos [12]. The use of Tetrapods is more efficient than natural stones because they are lighter, easier to produce and install in the field, and provide higher stability in wave dissipation due to their three interlocking legs [13]. A study by [14] showed that the BPPT-lock unit had lower damage percentages and better stability than Tetrapods. Another study [15] comparing A-Jack, Tetrapod, BPPT-Lock, and Antifer armor units found that A-Jack had higher stability compared to the other types.

Research by [16], which tested three types of armor layers-stone, concrete cube, and Tetrapod-found that Tetrapod armor had the lowest damage rate and highest stability compared to stone and concrete cubes. However, the use of large natural stones for erosion and sedimentation control in coastal areas faces challenges due to irregular shapes, which can shift and create gaps between the stones [17].

However, these shapes have limitations in terms of large material volume requirements, mold complexity, and installation difficulties in the field. In response to these problems, a new innovation in the form of a concrete protective unit named UH-Pod (Universitas Hasanuddin Pod) was developed. The UH-Pod design has a symmetrical geometric shape resembling the letter H, which is designed to increase inter-unit interlocking and material efficiency.

This study aims to systematically review the literature related to the stability of concrete armor layers on rubble mound breakwaters, compare the stability coefficient (KD) values of various units that have been developed, and assess the potential of UH-Pod as an innovative new shape that is more efficient, stable, and environmentally friendly. This study also serves as a conceptual basis for subsequent experimental research in the hydraulic laboratory.

The research gap lies in the limitations of conventional armor units in terms of shape efficiency, material requirements, and stability against extreme waves. Hence, a new modular and stable innovation is needed. The UH-Pod, with its symmetrical and modular H-shaped design, is expected to be more stable and efficient than conventional units.

The urgency of this study is driven by the severe threats faced by Indonesia's coastal areas, which are vulnerable to abrasion, erosion, and infrastructure damage caused by ocean wave energy. Thus, effective coastal defenses are essential. Moreover, improvements in design efficiency are necessary. The UH-Pod is proposed to enhance stability, interlocking, and efficiency in both production and installation compared to conventional units.

1.2 Literature Review

1.2.1 Waves

Ocean waves are generated by the interaction between wind and the sea surface, resulting in the propagation of energy in the form of surface waves. The main wave parameters used in coastal structure design are wave height (H), period (T), and wavelength (L).

Waves play an important role in coastal dynamics and coastal engineering, as the energy they carry directly influences processes such as erosion, sedimentation, and the stability of coastal protection structures. Waves significantly affect the stability of armor units [18].

Wind waves are formed when wind blows over the sea surface, transferring energy to the water through friction. The magnitude of waves formed is influenced by three primary factors: wind speed, wind duration, and the fetch length (distance over which the wind blows). Wind waves, often referred to as waves and tides, are among the most important types of waves in coastal engineering [2].

1.2.2 Coastal Structurec

Coastal structures are engineering constructions built on or near the shoreline to protect coastal zones from natural threats such as abrasion, tidal flooding, and high waves. To prevent abrasion, options such as breakwaters, groins, and revetments can be employed [5].

These structures play a vital role in maintaining shoreline stability while protecting assets and infrastructure in coastal areas. They are essential in mitigating the impacts of ocean waves and rapid shoreline changes. By reducing wave energy reaching the shore, coastal structures help preserve land stability and prevent infrastructure damage and land loss [19]. Additionally, they contribute to sediment management, preventing excessive erosion or siltation.

1.2.3 Breakwater

Breakwaters are one of the most widely used coastal structures for shoreline protection [20]. They are designed to dissipate wave energy before it reaches the shoreline or harbor area. Breakwaters act as barriers that absorb part of the wave energy [8], separating offshore waters from harbor waters to ensure calmer conditions within ports [2].

The primary functions of breakwaters are to reduce the destructive impact of ocean waves, prevent coastal erosion, reduce siltation in harbor basins, and protect coastal structures. Furthermore, breakwaters create calm waters behind them, enabling safer berthing and loading-unloading activities [14].

Breakwaters function as maritime structures that protect harbors, coastal industries, or shoreline settlements from the impacts of large waves. There are two main types of breakwaters:

- Rubble mound breakwater: constructed using rock or precast concrete units.
- Vertical breakwater: constructed using massive concrete structures.

Breakwaters reduce wave heights through three mechanisms:

- Energy absorption: waves lose energy as they interact with rough or porous surfaces.
- Wave reflection – part of the wave energy is reflected back to the sea by the vertical or sloping face of the breakwater.
- Dissipated wave propagation – waves passing beyond the breakwater carry less energy, reducing erosion and abrasion behind the structure.

1.2.4 Armor Layer

The armor layer of a breakwater is one of the most critical components, serving as the outer protective layer that resists wave pressure and reduces structural damage caused by hydrodynamic loads. This layer typically consists of several material types, ranging from natural stones to systematically arranged concrete blocks, to achieve optimal stability and protection.

As the outermost layer, the armor dissipates wave energy and protects the underlying layers. Its main function is to prevent the displacement of the core material due to wave forces. Some commonly used armor units for sloped breakwaters include Tetrapods and Dolos [12].

Typical armor materials include:

- Natural rock (riprap): Widely used due to availability and ability to absorb wave energy. However, irregular shapes can lead to movement and erosion if not properly placed.
- Concrete units: More durable, and can be shaped into special modules such as Tetrapods and Dolosse, designed for interlocking and efficient load distribution.

The layout and thickness of the armor layer are crucial for ensuring stability. Typically, it consists of several layers, from the rough inner core to the outer protective layer of special concrete blocks [19]. The thickness of the armor layer must withstand maximum design wave forces over the structure's service life. Design standards depend on wave height, period, and environmental conditions such as tides and currents.

1.2.5 Breakwater Armor Unit

Artificial armor blocks are precast concrete elements with special shapes, designed to form the protective layer of a breakwater. These blocks dissipate and break wave energy, reducing erosion and damage to the coast or harbor. Common types include:

Tetrapod

The Tetrapod is a precast concrete unit shaped with four legs (pods) that spread symmetrically from a single center. This shape is specifically designed to dissipate sea wave energy and reduce the direct impact force on coastal or protective structures. First developed by Laboratoire Dauphinois d'Hydraulique in France in the early 1950s, the tetrapod became a pioneer in the use of non-conventional concrete blocks as armor layer of rubble mound breakwaters.

Dolos

The Dolos was developed in South Africa and has a shape resembling an anchor with two curved arms. The Dolos is a concrete unit created with a unique shape, like the letter "T" and two long intersecting arms forming a pattern resembling a non-symmetrical "H" or "X", with openings between the units. The unique shape results in a high interlocking effect. This unit provides good interlocking, but its large weight and mold complexity lead to high production costs and difficulties in field placement.

Tribar

The Tribar is a precast concrete protective unit designed for use in the armor layer of breakwater structures and other coastal protection structures. Its shape resembles three cylindrical rods that are mutually perpendicular, forming a three-dimensional star-shaped structure.

Accropode dan Core-Loc

Both units were developed to overcome the limitations of previous shapes. The Accropode uses a monolithic design with a wide base to increase stability, while the Core-Loc introduced a single-layer armor placement system. Although efficient, licensing and patenting processes become obstacles to their use in some countries.

Xbloc

Developed in the Netherlands, the Xbloc stands out for its asymmetrical shape and strong inter-unit interlocking capability. Its advantage is high stability with less material. However, its installation requires high precision and precise orientation calculation.

A-Jack

An armor unit with six interlocking legs. It can be placed uniformly or randomly. A-Jack units have been widely used in breakwaters and other shoreline protection structures [21].

1.2.6 Armor Stability

The stability of the breakwater armor layer is a crucial aspect of structural design, as the armor must withstand hydrodynamic forces from waves without displacement or damage. Stability calculations form the basis for determining the size, weight, and shape of armor units.

The stability coefficient (K_D) is an empirical parameter used in the design of armor units for rubble mound breakwater structures. The K_D value indicates the efficiency of a unit's shape in resisting wave forces without displacement or failure. The higher the

K_D value, the greater the unit's ability to withstand wave energy with a smaller weight or volume.

Stability of the armor layer is expressed by Hudson's Formula [2]:

$$K_D = \frac{\gamma_r H^3}{W(S_r - 1)^3 cot\theta} \quad (1)$$

$$S_r = \frac{\gamma_r}{\gamma_a} \quad (2)$$

where:

W: block weight (kg or kN)

γ_r : unit weight of armor material (kN/m^3)

γ_a : unit weight of seawater (kN/m^3), typically 9.81 kN/m^3

H: design wave height (m)

K_D: stability coefficient, depending on the shape of the armor unit (natural or artificial stone), surface roughness, angularity, interlocking, and wave breaking condition

S_r: relative specific gravity of block

θ: breakwater slope angle

2 Methodology

This study presents a systematic literature review that aims to identify, evaluate, and synthesize previous research on the stability of armor layers in rubble mound breakwaters, while also examining the positioning and innovative potential of UH-Pod (Universitas Hasanuddin Pod) as a new protective unit.

3 Results

3.1 Comparative Analysis of Stability Between Armor Units

Based on the results of the literature analysis, the development of armor unit shapes shows a significant increase in efficiency and stability over time. The development of armor unit design shows a significant increase in terms of efficiency, stability, and placement effectiveness. Conventional rock and concrete units such as Tetrapod and Dolos have relatively low stability coefficient (K_D) values, ranging from 6 to 10, thus requiring a larger unit weight to achieve the same stability compared to modern units.

With advances in hydraulic technology, units such as Accropode, Core-Loc, and Xbloc were developed to increase K_D through a geometric approach and a single layer armor placement system. Research results by [22] show that stability efficiency can increase by up to 50–70% with a single layer placement system, provided that inter-unit interlocking is strong and installation orientation is controlled.

From the comparison results, it can be concluded that the K_D value increases with the degree of shape symmetry, interlocking capability, and efficient arrangement orientation. Design innovations that emphasize three-dimensional geometry and uniform load distribution have been shown to enhance unit stability without significantly increasing the concrete mass.

Table 1. Comparison of Stability Coefficient (KD) values

Unit Type	Arrangement Method	Interlocking Mechanism	K_D Value
Natural Stone	Two layers	Friction and self-weight	2–4
Tetrapod	Two layers	Moderate interlocking	6–8
Dolos	Two layers	Strong interlockingt	8–10
Accropode	Single layers	Self-interlocking	12–15
Core-Loc	Single layers	Strong self-interlocking	14–16
Xbloc	Single layers	3D self-interlocking	15–18
A-Jack	Two layers/random	Random interlocking	9–11
UH-Pod	Single layers	Three-way self-interlocking	-

3.2 Relationship Between Geometric Shape and Stability Coefficient (K_D)

Geometric shape is a key factor determining the performance of armor layer stability. Units with large contact areas and the ability to interlock with neighboring units tend to exhibit higher K_D values [23, 24]. Units such as Xbloc and Core-Loc achieve K_D values above 15 due to their optimized self-interlocking and vertical-horizontal load distribution.

In general, the relationship between geometric shape and stability can be explained through three aspects:

- Contact Surface Area: The larger the contact area between units, the greater the frictional force resisting shear induced by waves.
- Three-Dimensional Interlocking (3D Interlocking Mechanism): Interlocking in three directions (x, y, and z) enhances resistance to both rotational and translational displacement.
- Balance of Weight and Moment of Inertia: Units with a low center of mass and evenly distributed weight are more stable against overturning moments caused by wave pressure.

In this context, the UH-Pod design, which resembles the letter "H," shows high stability potential as it fulfills all three aspects. The two vertical bars and one horizontal bar of the UH-Pod enable three-way interlocking, similar to the A-Jack but with greater symmetry. This shape also creates spaces between units that allow wave energy dissipation while increasing the porosity of the armor layer.

4 Conclusion

Based on a review of various scientific references, several key conclusions can be drawn:

Evolution of Armor Unit Design: The development of armor unit designs shows an evolution from conventional shapes such as Tetrapod and Dolos, which employ double-layer systems with simple interlocking, to modern shapes such as Accropode, Core-Loc, and Xbloc, which utilize single-layer systems with three-dimensional self-interlocking. This shift directly increases the stability coefficient K_D from an average of 6–8 in older units to 14–18 in modern units.

Influence of Geometry and Arrangement: K_D values are highly influenced by the geometric shape and arrangement method. The better a unit can interlock with neighboring units and distribute loads evenly, the higher the resulting K_D value. Other parameters, such as slope angle, wave height, and material density, also affect the stability of the armor layer.

Potential of UH-Pod: The UH-Pod is expected to offer better stability, more efficient installation, and ecological development potential. It is designed to enhance interlocking, simplify installation, and reduce the required material volume.

References

1. Kahiking, F., Jansen, T., Jasin, M.I.: Study on coastal damage in Buloh beach, Mandolang district, Minahasa. TEKNO. **20**, 1123–1133 (2022)
2. Triatmodjo, B.: Coastal Engineering. Beta Offset, Yogyakarta (1999)
3. Saengsupavanich, C., Arif, H., Shin, L., Anthony, D.: Heliyon Environmental impact of submerged and emerged breakwaters. Heliyon. **8**, e12626 (2022). https://doi.org/10.1016/j.heliyon.2022.e12626
4. Sappetau, I.A., Tanan, B., Bungin, E.R.: Design of breakwater at Auki beach, Biak Numfor regency Papua. Paulus Civ. Eng. J. **2**(3), 161–170 (2020)
5. Panto, P.T.: Detailed Design of Coastal Protection Structures. **1**(1), 28–48 (2024)
6. Rachmanda, F., Tugiono, S., Zakaria, A., Herison, A.: Stability analysis of submerged rubble mound breakwater. Jurnal Rekayasa Sipil dan Desain. **10**(2), 385–394 (2022)
7. Mayanda, I.N., Chayati, C., Rahmanto, A.D.: Evaluation of breakwater planning against waves at Pasongsongan fishing port, East Java. Case study: Pasongsongan fishing port, East Java. Jurnal Teknik Gradien. **15**(02), 10–19 (2023)
8. Gomes, A., Pinho, J.L.S., Valente, T., Antunes, S.: Performance assessment of a semi-circular breakwater through CFD modelling. J. Mar. Sci. Eng. **8**(3), 1–16 (2020)
9. Hariati, F., Muhammad, F., Taqwa, L.: Conceptual study of breakwaters. Jurnal Komposit: Jurnal Ilmu-Ilmu Teknik Sipil. **5**(1), 31–37 (2020)
10. Husain, F., Paroka, D., Rahman, S., Lamputang, P.: Use of submerged breakwaters to reduce abrasion on Lamputang island. Jurnal Pengabdian Masyarakat Teknik. **3**(2), 65–70 (2021). https://doi.org/10.24853/jpmt.3.2.65-70
11. Clavero, M., Díaz-Carrasco, P., Losada, M.A.: Bulk wave dissipation in the armor layer of slope rock and cube armored breakwaters. J. Mar. Sci. Eng. **8**(3), 152 (2020). https://doi.org/10.3390/jmse8030152
12. Khasanah, M., Sugeng, H., Rikha, W.: Physical model simulation of breakwater effectiveness using Tetrapod and Dolos. Jurnal Teknik Hidraulik. **13**(1), 17–26 (2022)
13. Zalukhu, S., Sipil, J.T., Immanuel, U.K., Yogyakarta, D.I.: Comparison of Tetrapod and Rubble mound usage as Jetty Armor layer at Progo river Estuary, Bantul regency. In: Proc. Prosiding Seminar Nasional Teknik Sipil UMS, pp. 374–382 (2025)
14. Setyandito, O., et al.: Physical model test on the stability of breakwater Armour units at Sanur port. J. Coast. Eng. Manag. **4**(2), 139–149 (2021)

15. Sujantoko, S., Pangestu, P.A., Silva, F.G.K., Ahmad, R.P., Firman, M.R.: Test of a 2D physical model of seawall stability at various slope angles with A-Jack Armour units. E3S Web Conf. **576**, 1–11 (2024). https://doi.org/10.1051/e3sconf/202457602005
16. Azhar, R.M., Larasari, A., Husrin, S., Bachtiar, H.: Hydraulic responses and stability of Armor layer. J. Tek. Hidraul. **10**(1), 29–42 (2019). https://doi.org/10.32679/jth.v10i1.603
17. Gunawan, G., et al.: Analysis of wave breaking on L-type Jetty structure (Case study: Purus Beach, Padang City). Sci. J. Civ. Eng. **17**(2), 144–152 (2020)
18. Maula, A.I.: Experimental Study on the Stability of Hexaloc-2 Armor Unit in Breakwater Structure with Variations in Slope, Wave Height, and Wave Period. Sepuluh Nopember Institute of Technology (2023)
19. Komar, P.D.: The Modeling of processes and morphology in the coastal zone—Reflections on the maturity of our science. Shore & Beach. **66**, 10–22 (1998)
20. Ginting, J.W., Putra, I.A.I.D.R., Simanjuntak, E.M.: Physical model of wave energy dissipation using pile breakwater. J. Tek. Hidraul. **10**(1), 55–70 (2019). https://doi.org/10.32679/jth.v10i1.606
21. E. K. Syahlan, Numerical Modeling of Hydrodynamics for A-Jack Armor on Seawall, 2022, [Online]. Available: https://repository.its.ac.id/115905/1/04311840000011-Undergraduate_Thesis.pdf
22. S. Van der Meer J. W., & Sigurdarson, Manual on the Use of Concrete Armour Units, 2017.
23. Hudson, R.Y.: Laboratory investigation of rubble-mound breakwaters. J. Waterw. Harb. Div., 93–121 (1959)
24. Van der Meer, J.W.: Rock Slopes and Gravel Beaches under Wave Attack. Delft Univ. Technol. (1988)

Squeezed Light Dynamics in PT-Symmetric Coupler via Positive-P Representation

Ameer Syahiran Bin Azhan and Abdel-Baset M. A. Ibrahim(✉)

Faculty of Applied Sciences, Universiti Teknologi MARA (UiTM), 40450 Shah Alam, Selangor, Malaysia
abdelbaset@uitm.edu.my

Abstract. Squeezed states of light are key resources in modern quantum technologies, enabling measurements beyond classical limits and supporting applications in precision sensing and secure communication. Generating strong and stable squeezing, however, remains challenging in open quantum systems where loss and gain compete. Here, we investigate a PT-symmetric Kerr coupler with balanced amplification and attenuation as a platform for enhanced squeezing. The system is modeled by transforming the Lindblad master equation into a Fokker–Planck equation using the Positive-P representation and then into stochastic differential equations for accurate numerical simulation. Our results show that squeezing is strongly influenced by system parameters: increased coupling raises the oscillation frequency, while higher Kerr nonlinearity enhances squeezing but reduces stability. The exceptional point emerges as a critical operating regime where squeezing reaches its maximum, while stability degrades in the broken-symmetry phase. These findings demonstrate that optimal performance occurs when parameters are tuned near the exceptional point. The study also highlights the capability of the Positive-P method to capture full quantum fluctuations and offers practical guidance for designing non-Hermitian photonic structures for quantum applications.

Keywords: Quantum optics · Squeezed states of light · PT-symmetric coupler

1 Introduction

Squeezed states of light are key resources in modern quantum technologies such as quantum communication [1] and sensing [2], as they enable sensitivities beyond the standard quantum limit. These states arise when fluctuations in one quadrature are reduced below shot noise while the conjugate quadrature expands, as required by Heisenberg's uncertainty principle. Their ability to enhance measurement precision [3] and support quantum information processing continues to motivate efforts to realize strong and controllable squeezing.

Nonlinear couplers and waveguide systems have emerged as versatile platforms for generating squeezed light. Raman-induced squeezing has been explored using phase-space methods [4], while single-mode squeezing and its detection have also been

M. A. b. Jamlos et al. (Eds.): ICSET 2025, LNNS 1897, pp. 574–579, 2026.
https://doi.org/10.1007/978-3-032-20722-7_74

demonstrated [5]. Broadband and multichannel squeezing has been reported in Kerr and coupled-channel systems [6], including the influence of competing nonlinearities. Additional work has examined harmonic-generation–induced squeezing [7], and higher-order couplers have revealed rich squeezing and other quantum features [8]. These studies highlight the flexibility of nonlinear couplers and motivate the investigation of squeezing in PT-symmetric Kerr structures.

Non-Hermitian optical systems exhibiting parity–time (PT) symmetry have recently attracted significant attention as platforms for manipulating nonclassical light [9]. PT symmetry can be implemented in coupled waveguides with balanced gain and loss [10], as illustrated in Fig. 1. Below the exceptional point (EP), eigenvalues are real and dynamics remain stable, while crossing the EP introduces symmetry breaking and enhanced sensitivity [11]. These properties make PT-symmetric couplers promising candidates for controlling squeezing and related quantum effects. Although PT-symmetric systems have shown entanglement, antibunching, and quantum Zeno effects [12], systematic studies of how exceptional points influence squeezing in nonlinear Kerr couplers remain limited. Many analytical approaches rely on approximations that break down near the EP, where nonlinearities and quantum noise become significant [13].

In this work, we investigate squeezing dynamics in a PT-symmetric Kerr coupler using the Positive-P representation. This method transforms the Lindblad master equation into stochastic differential equations that capture full quantum fluctuations without truncation. We study how coupling strength, input amplitude, Kerr nonlinearity, and gain-loss balance affect the degree and stability of squeezing, with particular emphasis on the exceptional point. Our results show that squeezing is maximized near the EP but becomes less stable in the broken-symmetry phase. These findings provide insight into the interplay between PT symmetry and nonlinear interactions and offer guidance for optimizing squeezing in non-Hermitian photonic systems.

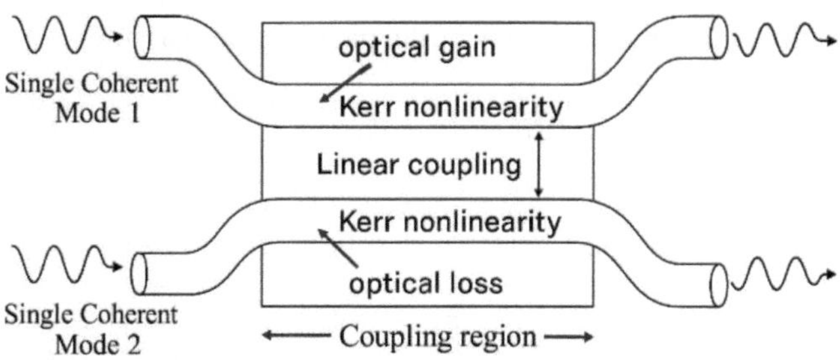

Fig. 1. Schematic illustration of PT- Symmetric coupler with Kerr nonlinearity

2 Mathematical Model and Methods

We consider a PT-symmetric optical coupler consisting of two single-mode Kerr waveguides driven by coherent inputs at frequency ω. The system Hamiltonian [14], including free evolution, linear coupling, Kerr nonlinearity, and balanced gain-loss, is

$$\frac{\hat{H}}{\hbar} = \sum_{j=1}^{2} \omega \hat{a}_j^{\dagger} \hat{a}_j + k\left(\hat{a}_1^{\dagger} \hat{a}_2 + \hat{a}_2^{\dagger} \hat{a}_1\right) + g \sum_{j=1}^{2} \hat{a}_j^{\dagger 2} \hat{a}_j^2 + i\gamma \hat{a}_1^{\dagger} \hat{a}_1 - i\gamma \hat{a}_2^{\dagger} \hat{a}_2 \tag{1}$$

Here, $\hat{a}_j$ and $\hat{a}_j^{\dagger}$ $(j = 1{,}2)$ denote the bosonic operators of mode j. The linear term describes photon exchange between the two waveguides at rate k, the nonlinear term accounts for Kerr self-phase modulation with coefficient g, and the balanced gain $(+i\gamma)$ and loss $(-i\gamma)$ implement PT symmetry [15]. Since the system is open, its evolution obeys the Lindblad master equation

$$\frac{d\hat{\rho}}{dt} = -\frac{i}{\hbar}[\hat{H}, \hat{\rho}] + \sum_j \mathrm{L}_j[\hat{\rho}], \quad \mathrm{L}_j[\hat{\rho}] = \sigma\left(\hat{a}_j \hat{\rho} \hat{a}_j^{\dagger} - \frac{1}{2}\left(\hat{a}_j^{\dagger} \hat{a}_j \hat{\rho} + \hat{\rho} \hat{a}_j^{\dagger} \hat{a}_j\right)\right) \tag{2}$$

where σ is the damping rate. This operator equation forms the basis for the Positive-P representation adopted below. To map the quantum evolution into c-number stochastic equations while retaining exact operator ordering, we employ the Positive-P representation [16], a phase-space method well suited for nonlinear and non-Hermitian systems [13]. The correspondence rules between operators and c-number variables $_j$, $_j$ are $\hat{a}_j^{\dagger} \hat{\rho} = \left(\beta_j - \frac{\partial}{\partial \alpha_j}\right) P$, $\hat{a}_j \hat{\rho} = \alpha_j P$, $\hat{\rho} \hat{a}_j = \left(\alpha_j - \frac{\partial}{\partial \beta_j}\right) P$, $\hat{\rho} \hat{a}_j^{\dagger} = \beta_j P$ where $(_j, _j)$ is the Positive-P distribution. Substituting the Hamiltonian from Eq. (1) into Eq. (2) and utilizing these expressions yields a Fokker–Planck equation whose drift and diffusion terms are directly convertible into Itô stochastic differential equations (SDEs) [17]. The resulting SDEs governing the dynamics of the phase-space variables are

$$\frac{d\alpha_1}{dt} = -i(\omega + i\gamma)\alpha_1 - 2ig\beta_1\alpha_1^2 - ik\alpha_2 - \frac{1}{2}\sigma\alpha_1 + \alpha_1\sqrt{-2ig}\,\eta_1(t) \tag{3}$$

$$\frac{d\beta_1}{dt} = i(\omega + i\gamma)\beta_1 + 2ig\alpha_1\beta_1^2 + ik\beta_2 - \frac{1}{2}\sigma\beta_1 + \beta_1\sqrt{2ig}\,\eta_2(t) \tag{4}$$

$$\frac{d\beta_2}{dt} = i(\omega - i\gamma)\beta_2 + 2ig\alpha_2\beta_2^2 + ik\beta_1 - \frac{1}{2}\sigma\beta_2 + \beta_2\sqrt{2ig}\,\eta_2(t) \tag{5}$$

$$\frac{d\alpha_2}{dt} = -i(\omega - i\gamma)\alpha_2 - 2ig\beta_2\alpha_2^2 - ik\alpha_1 - \frac{1}{2}\sigma\alpha_2 + \alpha_2\sqrt{-2ig}\,\eta_1(t) \tag{6}$$

where () are independent real Gaussian noises. These equations capture both the deterministic dynamics induced by linear coupling, Kerr nonlinearity, and gain-loss imbalance, as well as the quantum fluctuations introduced through multiplicative noise. All parameters may be made dimensionless by scaling with the common optical frequency ω, without altering the structure of Eqs. (3)–(6). To quantify squeezing, we evaluate field quadratures for each mode, $\hat{X}_j = \frac{1}{2}\left(\hat{a}_j + \hat{a}_j^{\dagger}\right)$, $\hat{Y}_j = \frac{1}{2i}\left(\hat{a}_j - \hat{a}_j^{\dagger}\right)$ whose variances satisfy the vacuum limits $\left\langle(\Delta\hat{X}_j)^2\right\rangle = \left\langle(\Delta\hat{Y}_j)^2\right\rangle = \frac{1}{4}$. Squeezing occurs when either variance falls below 1/4 in accordance with the uncertainty principle. In the Positive-P representation, the quadrature variances become stochastic averages of $_j$ and $_j$:

$$\begin{bmatrix}\left\langle(\Delta X_j)^2\right\rangle \\ \left\langle(\Delta Y_j)^2\right\rangle\end{bmatrix} = \frac{1}{4}\left(1 + 2\langle\beta_j\alpha_j\rangle - 2\langle\beta_j\rangle\langle\alpha_j\rangle \pm \left[\left\langle\alpha_i^2\right\rangle - \langle\alpha_j\rangle^2 + \left\langle\beta_j{}^2\right\rangle - \langle\beta_j\rangle^2\right]\right) \tag{7}$$

Finally, the degree of squeezing is expressed in decibels as

$$S_{X\,j} = 10\log_{10}\left(\frac{\left\langle(\Delta X_j)^2\right\rangle}{1/4}\right),\ S_{Y\,j} = 10\log_{10}\left(\frac{\left\langle(\Delta Y_j)^2\right\rangle}{1/4}\right) \tag{8}$$

where negative values indicate noise reduction below the shot-noise limit. These measures are used in the next section to assess how system parameters such as coupling, input conditions, Kerr nonlinearity, and gain-loss ratio influence the squeezing dynamics.

3 Results and Discussion

The squeezing dynamics of the PT-symmetric Kerr coupler were analyzed by solving the stochastic differential Eqs. (3)–(6) in the Positive-P representation. Numerical simulations were carried out under different parameter conditions to investigate how coupling strength, initial input amplitudes, Kerr nonlinearity, and the gain-loss ratio influence the quadrature variances and the degree of squeezing. Figure 2 shows the quadrature variance of mode 1 for two different linear coupling strengths: (a) strong coupling ($\widetilde{k} = 0.8$) and (b) weak coupling ($\widetilde{k} = 0.1$). In Fig. 2(a), the variance exhibits rapid oscillations, reflecting enhanced photon exchange between the two coupled modes. Stronger coupling accelerates the redistribution of quantum fluctuations, resulting in a higher oscillation frequency. By contrast, in Fig. 2(b), the oscillations are slower and less frequent, since weak coupling limits the rate of energy transfer. These results demonstrate that the coupling constant directly controls the frequency of quadrature variance oscillations, and hence the temporal structure of squeezing dynamics.

Figure 3 shows the degree of squeezing in decibels before, at, and after the exceptional point (EP). In the unbroken phase ($\widetilde{\gamma} = 0.2$, $\widetilde{k} = 0.3$), squeezing is moderate (-2 to -3 dB) with stable oscillations. At the EP ($\widetilde{\gamma} = \widetilde{k} = 0.3$), squeezing is significantly enhanced (-6 to -10 dB), demonstrating that critical gain-loss balance maximizes nonclassical correlations. Beyond the EP ($\widetilde{\gamma} = 0.4$, $\widetilde{k} = 0.3$), squeezing becomes deeper (exceeding -10 dB) but unstable, with irregular oscillations caused by broken symmetry. These results confirm that the EP is a key parameter regime: it enables maximum squeezing but also introduces dynamical fragility. Taken together, these results provide a systematic picture of squeezing in PT-symmetric Kerr couplers. Linear coupling governs the frequency of fluctuation exchange, Kerr nonlinearity controls the depth and stability of squeezing, and the exceptional point sets the ultimate balance between strong squeezing and dynamical robustness. This comprehensive analysis shows how parameter tuning can optimize squeezing performance and clarifies the trade-offs between depth of squeezing and system stability.

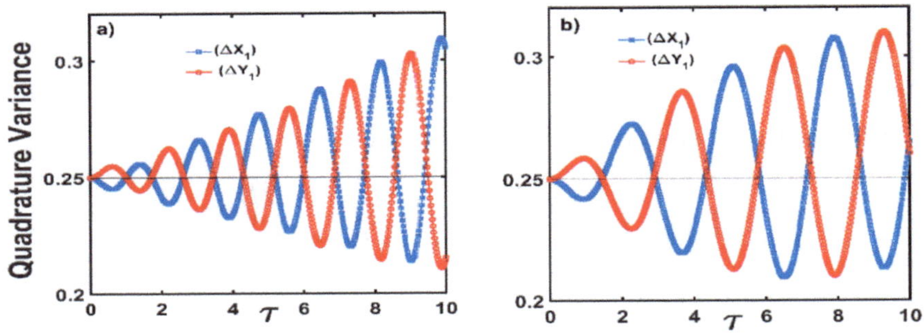

Fig. 2. Quadrature variance for single-mode squeezing for linear coupling a) $\tilde{k} = 0.8$ and b) $\tilde{k} = 0.1$. Other input parameters are fixed at $\tilde{\omega} = 1$, $\tilde{\gamma} = 0.01$, $\tilde{g} = 0.0001, \tilde{\sigma} = 5 \times 10^{-6}, \alpha_1 = 10, \alpha_2 = 10$.

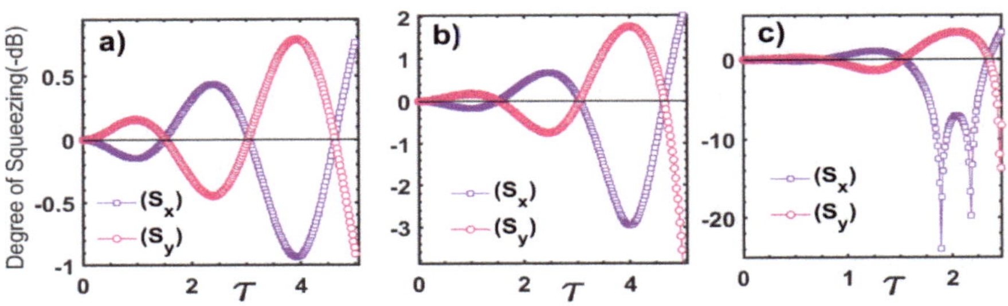

Fig. 3. Degree of squeezing for a) before EP, b) at EP, and c) after EP. Other parameters are $\tilde{\omega} = 1, \tilde{\sigma} = 5 \times 10^{-6}, \tilde{g} = 0.0001, \alpha_1 = 10, \alpha_2 = 10$

4 Conclusion

In this work, we have investigated quadrature squeezing in a PT-symmetric Kerr coupler using the Positive-P representation. By transforming the Lindblad master equation into stochastic differential equations and solving them numerically, we obtained a complete quantum description of the system's dynamics. Our results demonstrate that stronger linear coupling accelerates the oscillation of quadrature variances, while weaker coupling slows down fluctuation transfer. The exceptional point maximizes the degree of squeezing, reaching values beyond -10 dB, but at the cost of stability once PT symmetry is broken. These findings highlight the central role of exceptional points in shaping quantum fluctuations in non-Hermitian systems. The results not only deepen our understanding of squeezing in PT-symmetric couplers but also provide practical guidelines for designing nonclassical light sources for precision measurement and quantum communication.

Acknowledgments. Abdel-Baset M. A. Ibrahim acknowledges the financial support from the Ministry of Higher Education (MOHE, Malaysia) through the grant FRGS/1/2021/STG07/UITM/02/10.

Disclosure of Interests The authors have no competing interests to declare that are relevant to the content of this article.

References

1. Gisin, N., Ribordy, G., Tittel, W., Zbinden, H.: Quantum cryptography. Rev. Mod. Phys. **74**, 145–195 (2002)
2. Malitesta, M., Smerzi, A., Pezzè, L.: Distributed quantum sensing with squeezed-vacuum light in a configurable array of Mach-Zehnder interferometers. Phys. Rev. A. **108**, 032621 (2023)
3. Vahlbruch, H., Mehmet, M., Danzmann, K., Schnabel, R.: Detection of 15 dB squeezed states of light and their application for the absolute calibration of photoelectric quantum efficiency. Phys. Rev. Lett. **117**, 110801 (2016)
4. Hanapi, M.S.M., Ibrahim, A.-B.M., Julius, R., Choudhury, P.K.: Raman-induced compound-mode squeezing in a nonlinear coupler implementing Wigner representation and the analytical method. J. Opt. Soc. Am. B. **42**, 368–375 (2025)
5. Hanapi, M.S.M., Ibrahim, A.-B.M., Julius, R.: On the detection of single-mode squeezed states in Kerr Nonlinear Coupler. J. Phys. Conf. Ser. **2075**, 012002 (2021)
6. Julius, R., Ibrahim, A.-B.M., Choudhury, P.K., Alias, A.N., Abd Halim, M.S.: Quantum features of nonlinear coupler with competing nonlinearity. Sci. Rep. **12**, 8245 (2022)
7. Hanapi, M.S.M., Ibrahim, A.-B.M.A., Julius, R., Choudhury, P.K., Eleuch, H.: Nonclassical light in a three-waveguide coupler with second-order nonlinearity. EPJ Quantum Technol. **11**, 51 (2024)
8. Julius, R., Ibrahim, A.-B.M.A., Choudhury, P.K., Eleuch, H.: On the nonclassical dynamics of cavity-assisted four-channel nonlinear coupler. Chin. Phys. B. **27**, 114206 (2018)
9. El-Ganainy, R., Makris, K.G., Khajavikhan, M., Musslimani, Z.H., Rotter, S., Christodoulides, D.N.: Non-Hermitian physics and PT symmetry. Nat. Phys. **14**, 11–19 (2018)
10. Rüter, C.E., Makris, K.G., El-Ganainy, R., Christodoulides, D.N., Segev, M., Kip, D.: Observation of parity–time symmetry in optics. Nat. Phys. **6**, 192–195 (2010)
11. Miri, M.-A., Alù, A.: Exceptional points in optics and photonics. Science. **363**, eaar7709 (2019)
12. Wen, J., Jiang, X., Jiang, L., Xiao, M.: Parity-time symmetry in optical microcavity systems. J. Phys. B Atomic Mol. Phys. **51**, 222001 (2018)
13. Gilchrist, A., Gardiner, C., Drummond, P.: Positive P representation: Application and validity. Phys. Rev. A. **55**, 3014–3032 (1997)
14. Peřinová, V., Lukš, A., Křepelka, J.: Quantum description of a PT-symmetric nonlinear directional coupler. JOSA B. **36**, 855–861 (2019)
15. Peřina Jr., J., Lukš, A.: Quantum behavior of a PT-symmetric two-mode system with cross-Kerr nonlinearity. Symmetry. **11**, 1020 (2019)
16. Drummond, P.D., Gardiner, C.W.: Generalised P-representations in quantum optics. J. Phys. Math. Gen. **13**, 2353 (1980)
17. Gardiner, C.W.: Handbook of Stochastic Methods. Springer, Berlin (1985)

Strength Activity Index of Thermally Treated De-oiled Bleaching Earth: Characterisation and Regression Analysis

Elffie Yunus, Hidayati Asrah(✉), Habib Musa Mohamad, Asmawan Mohd Sarman, and Ahmad Nurfaidhi Rizalman

Civil Engineering Program, Faculty of Engineering, Universiti Malaysia Sabah, 84000 Kota Kinabalu, Sabah, Malaysia
hidayati@ums.edu.my

Abstract. The disposal of de-oiled bleaching earth (DOBE), a waste by-product of the palm oil industry, presents environmental challenges due to its high organic content and limited reuse. This study investigates the potential of DOBE as a supplementary material in cement-based applications by evaluating its pozzolanic activity through the strength activity index (SAI) at 7 and 28 days. Linear regression was also used to assess the relationship between curing time and SAI development. DOBE obtained from a palm oil mill was sieved, ground, and thermally treated at 700 °C for 120 min to reduce organic matter and enhance reactivity. The treated material was characterised using XRD, SEM, XRF, EDX, and particle size distribution analysis. Thermal treatment significantly lowered the loss on ignition and improved the physical and chemical properties of DOBE. Mortar containing treated DOBE achieved an SAI of 92% at 7 days and 95% at 28 days, showing improved performance with curing time. In contrast, untreated DOBE showed a reduction in SAI over time. Linear regression analysis further confirmed these trends, with treated DOBE displaying a positive slope of +0.143% per day, indicating increasing pozzolanic activity, while untreated DOBE exhibited a negative slope of −0.714% per day, reflecting declining performance. Overall, the findings show that properly treated DOBE has strong potential as a pozzolanic material for partial cement replacement. Its use not only enhances mortar performance but also supports sustainable waste management and greener construction practices by converting an industrial by-product into a valuable resource.

Keywords: De-Oiled Bleaching Earth · Pozzolanic reactivity · Supplementary cementitious material · Heat treatment

1 Introduction

The global edible oil industry produces about 2.5 million tonnes of spent bleaching earth (SBE) each year [1]. In Malaysia, the palm oil industry has grown rapidly and remains a key part of the national economy. In 2023, crude palm oil production increased to 18.5 million tonnes, slightly higher than in 2022 [2]. As one of the world's leading producers, Malaysia helps meet global demand for palm oil used in food, cosmetics, and various industrial products [3].

M. A. b. Jamlos et al. (Eds.): ICSET 2025, LNNS 1897, pp. 580–587, 2026.
https://doi.org/10.1007/978-3-032-20722-7_75

However, managing SBE is a major challenge. SBE can retain 30–40% oil by weight [4], and this leftover oil makes it highly prone to spontaneous combustion because it oxidises quickly on the clay's large surface area. This creates safety, handling, and environmental problems. SBE also decomposes very slowly and cannot be safely disposed of in landfills [5]. Due to these risks, SBE is classified as hazardous waste in the European Union [6]. To reduce waste and recover some of the trapped oil, SBE can be processed to produce de-oiled bleaching earth (DOBE), which contains only 1–3% residual oil. Common supplementary cementitious materials (SCMs) used in concrete include ground granulated blast-furnace slag [7], fly ash [8], silica fume [9] and eco-processed pozzolan [10, 11].

While these materials are well studied, research on using DOBE as an SCM is still very limited. Early studies show that DOBE's high silica content makes it suitable for partial cement replacement, but impurities, crystalline phases, and residual oil reduce its reactivity [12]. In particular, the relationship between curing time and SAI has not been explored, and no previous research has applied linear regression to quantify this relationship. This is an important gap because establishing a regression-based correlation allows researchers to model and predict pozzolanic performance over time, providing a clearer understanding of DOBE's reactivity trend. To address this gap, this study evaluates the SAI, material properties, and changes in SAI over curing time for both raw and treated DOBE. By comparing untreated and treated samples, this research provides new evidence on DOBE's suitability as an SCM and highlights its potential to reduce cement use, lower CO_2 emissions, and support sustainable waste management in the palm oil industry.

2 Methodology

This study investigates the pozzolanic activity of DOBE as a 20% cement replacement in mortar. Material characterisation included particle size distribution, chemical composition, mineral phases, and microstructure. Mortar specimens followed ASTM C311 for SAI determination. The following sections provide detailed descriptions of the materials used and the experimental procedures undertaken.

2.1 Materials

This study used *Cap Gunung* Portland Cement Brand that meets Malaysian standard MS 522 and BS EN 197–1 criteria. The DOBE was supplied by a palm oil refinery in Lahad Datu, Sabah, in a pellet form. DOBE was pre-treated by drying it in an oven at 110 °C for 24 h. This is to remove excess carbon and unburned organic matter to avoid potential negative effects on hydration. After drying, the DOBE was then ground using a Los Angeles grinding mill and then sieved to achieve a particle size of 150 μm. For treated DOBE, an additional heat treatment was conducted in a furnace at 700 °C for 2 h. This thermal process effectively removed the remaining unburned residues, resulting in a visible colour change from greyish to light brownish. The type of sand used in this research is river sand from Kota Belud with a specific gravity of 2.2, and the water absorption reading is 3.2%.

2.2 Methods

2.2.1 Characterisation

The particle size distribution of DOBE was determined using a laser diffraction particle size analyser. The chemical oxide composition, mineralogical phases, and microstructural features of DOBE were analysed using X-ray fluorescence (XRF), X-ray diffraction (XRD), and scanning electron microscopy (SEM), respectively.

2.2.2 Strength Activity Index and Linear Regression Analysis

For the mix preparation, 20% of the cement was replaced with DOBE to make the test mortar samples. The control mix used 1000 g of cement, 2750 g of sand, and 484 g of water. The test mix used 1600 g of cement, 400 g of DOBE, and 5500 g of sand, with enough water added to achieve a flow of 115% of the control mix. To evaluate pozzolanic activity, compressive strength tests were carried out according to ASTM C311. Cube specimens (50 × 50 × 50 mm) were cast and kept in a moist room at 23 ± 2 °C for 24 h. After demoulding, the cubes were cured in saturated lime water until testing. Compressive strength was measured at 7 and 28 days using the average of three specimens, and the strength activity index (SAI) was calculated based on ASTM C311. Linear regression was used to measure how the SAI changed with curing time using the equation $SAI = a + bt$.

3 Results and Discussions

3.1 Physical Properties

Table 1 reveals the physical properties of OPC, untreated DOBE and treated DOBE. The mean particle size d_{50} of OPC, untreated and treated DOBE were 15.32, 16.02 and 19.02 μm, respectively. The particle size, d_{90} of OPC, untreated DOBE and treated DOBE were 44.34, 62.2 and 71.58 μm, respectively. Based on the results, the specific gravity of OPC, untreated DOBE and treated DOBE were 3.13, 2.21 and 2.45, respectively. OPC has the highest specific gravity. Meanwhile, treated DOBE has a slightly higher specific gravity compared to untreated DOBE.

Table 1. Physical properties of OPC, untreated and treated DOBE.

Physical Properties	Particle size d_{10}(μm)	Mean Particle size d_{50}(μm)	Particle size d_{90}(μm)	Specific gravity
OPC	2.89	15.32	44.34	3.13
Untreated DOBE	2.41	16.02	62.2	2.21
Treated DOBE	2.79	19.02	71.58	2.45

3.2 Chemical Properties

Table 2 shows the chemical compositions of the OPC as well as the treated and untreated DOBE. The XRF result reveals that the untreated DOBE is composed of SiO_2 at 51.44% and the combination value of SiO_2, aluminium oxide (Al_2O_3), and iron oxide (Fe_2O_3) was 70.55%. Meanwhile, the combination value of SiO_2, aluminium oxide (Al_2O_3), and iron oxide (Fe_2O_3) for treated DOBE was 70.19%. The loss on ignition of untreated DOBE was 4.47% which is less than 6% as requirement specified in the ASTM C618 standard. Based on the chemical composition results, DOBE can be categorised as class F pozzolan.

Table 2. Chemical composition of OPC, untreated and treated DOBE. Place the caption above the table.

Chemical compositions (%)	OPC	Untreated DOBE	Treated DOBE
Silicon dioxide (SiO_2)	20.99	51.44	51.47
Aluminium oxide (Al_2O_3)	3.93	9.64	9.83
Ferric Oxide (Fe_2O_3)	3.99	9.47	8.89
Calcium oxide (CaO)	63.88	12.35	12.18
Magnesium Oxide (MgO)	1.60	6.03	6.57
S + A + F	-	70.55	70.19
LOI	2.19	20.89	4.47

3.3 Mineralogical Properties

X-ray diffraction (XRD) examination was carried out to identify the crystalline phases found in both untreated and treated DOBE and to assess the impact of thermal treatment on their mineralogical composition. Fig. 1 reveals that the untreated DOBE has increased crystallinity, with major phases including quartz (SiO_2), calcite ($CaCO_3$), and clay-based minerals. A strong peak at approximately 26.6° 2θ was observed, which is typically associated with quartz. In contrast, the treated DOBE displayed reduced crystalline peak intensity and a large hump between 16° and 32° 2θ, indicating the production of amorphous phases due to heat deterioration of clay structures. Notably, phases such as calcite and sulphide-related chemicals were reduced or missing in the treated sample, indicating breakdown or phase shift. These findings demonstrate that thermal treatment increases the amorphous content of DOBE, which improves its reactivity when employed as a supplementary cementitious material.

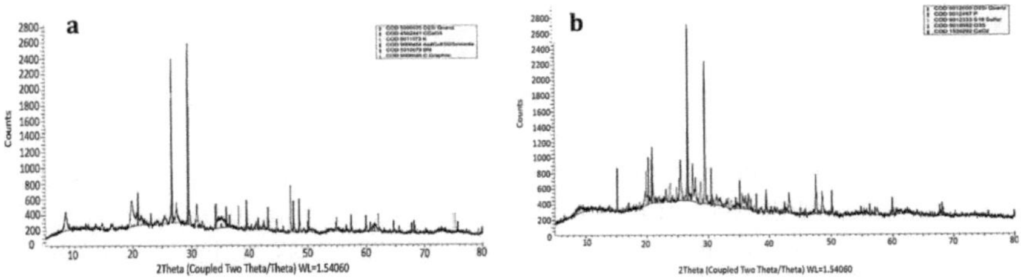

Fig. 1. XRD pattern of (a) untreated DOBE; (b) treated DOBE

3.4 Microstructural Properties

The scanning electron microscope (SEM) image of OPC and DOBE is shown in Fig. 2. The result shows that OPC has an irregular shape. Meanwhile, the untreated and treated DOBE particles exhibit distinct morphological differences, as observed under scanning electron microscopy. The untreated DOBE particles display a more irregular and rough surface texture, often characterised by agglomerated or coated appearances due to residual oil and impurities. In contrast, the treated DOBE particles exhibit a smoother and more fragmented surface, indicating improved surface cleanliness and partial structural breakdown as a result of the treatment process. As for the particle's shape, the untreated DOBE looks more like amorphous blobs, while the treated DOBE is visible in angular and flaky forms as it undergoes heat treatment.

Fig. 2. Micrograph images (a) Untreated DOBE; (b) Treated DOBE; (c) OPC

3.5 Strength Activity Index

The strength activity index (SAI) in Fig. 3 shows distinct temporal behaviours between untreated and treated DOBE. Linear regression analysis (Table 3) reveals contrasting trends: untreated DOBE exhibits a negative slope of -0.714% per day (SAI $= 105-0.714$ t), while treated DOBE demonstrates a positive slope of $+0.143\%$ per day (SAI $= 91 + 0.143$ t). For untreated DOBE, the SAI decreased from 100% at 7 days to 85% at 28 days, representing a 15% reduction over the curing period. This declining trend suggests that while untreated DOBE demonstrates strong initial reactivity, possibly due to physical filling effects, its long-term pozzolanic performance deteriorates significantly. The negative regression slope indicates that untreated DOBE possesses limited sustained pozzolanic reactivity, likely due to the presence of unburned carbon. Based on the linear model, untreated DOBE is projected to fall below the ASTM C618 minimum requirement of 75% at approximately 56 days, indicating inadequate long-term performance as a supplementary cementitious material.

In contrast, treated DOBE showed improved performance, with SAI increasing from 92% at 7 days to 95% at 28 days (3% gain). The positive regression slope confirms that thermal treatment enhances sustained pozzolanic reactivity over extended curing periods. Heating the DOBE removes organic materials, increases amorphous content, and breaks down clay minerals into more reactive phases [13], enabling continuous pozzolanic reactions with calcium hydroxide from cement hydration. The linear model suggests that treated DOBE maintains SAI values well above the ASTM C618 threshold throughout extended curing, demonstrating its suitability as a long-term pozzolanic material [14]. This sustained performance improvement validates the effectiveness of thermal treatment in converting waste de-oiled bleaching earth into a viable supplementary cementitious material.

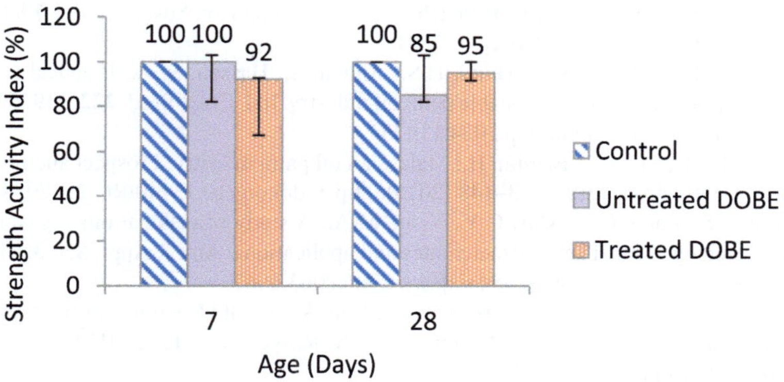

Fig. 3. Strength activity index (SAI) at 7 and 28 days

Table 3. Linear regression parameters and SAI values of untreated and treated DOBE at 7 and 28 days.

DOBE Type	Linear equation	Slope (% per day)	SAI at 7d (%)	SAI at 28d (%)
Untreated	SAI = 105–0.714 t	−0.714	100	85
Treated	SAI = 91 + 0.143 t	+0.143	92	95

4 Conclusions

The results show that treated DOBE had a much lower loss on ignition (4.47%), meaning most carbon and organic materials were removed. The treatment also improved its surface area and kept it in an amorphous form, which is important for pozzolanic activity. The strength activity index of the treated DOBE met the ASTM C168 requirements for pozzolanic materials. Linear regression analysis confirmed that treated DOBE exhibited a positive slope of +0.143% per day (SAI = 91 + 0.143 t), while untreated DOBE showed a declining trend with a negative slope of −0.714% per day (SAI = 105–0.714 t), demonstrating the effectiveness of thermal treatment in sustaining pozzolanic reactivity over time. After 28 days, the compressive strength of the mix with treated DOBE was close to that of the control mix. The results show that treated DOBE displayed pozzolanic behaviour, as the strength kept increasing over time.

Acknowledgement. This research was not funded by any grant.

References

1. Delgado-Plana, P., et al.: Effect of activating solution modulus on the synthesis of sustainable geopolymer binders using spent oil bleaching earths as precursor. Sustainability. **13**(13), 7501 (2021). https://doi.org/10.3390/su13137501
2. Zakaria, M.R., Farid, M.A.A., Hafid, H.S., Andou, Y., Hassan, M.A.: Practical role of oil palm fronds in Malaysia's sustainable palm oil industry. Ind. Crop. Prod. **222**, 119753 (2024). https://doi.org/10.1016/j.indcrop.2024.119753
3. Alam, A.S.A.F., Er, A.C., Begum, H.: Malaysian oil palm industry: Prospect and problem. J. Food Agric. Environ. **13**(2), 143–148 (2015). https://doi.org/10.13140/RG.2.1.2958.0326
4. Kheang, L.S., Foon, C.S., May, C.Y., Ngan, M.A.: A study of residual oils recovered from spent bleaching earth: Their characteristics and applications. Am. J. Appl. Sci. **3**(10), 2063–2067 (2006). https://doi.org/10.3844/ajassp.2006.2063.2067
5. Abdelbasir, S.M., Shehab, A.I., Abdel, M.A., Khalek.: Spent bleaching earth; recycling and utilization techniques: A review. Resour. Conserv. Recycl. Adv. **17**, 200124 (2023). https://doi.org/10.1016/j.rcradv.2022.200124
6. Beshara, A., Cheeseman, C.R.: Reuse of spent bleaching earth by polymerisation of residual organics. Waste Manag. **34**(10), 1770–1774 (2014). https://doi.org/10.1016/j.wasman.2014.04.021
7. Zhang, S., et al.: Effect of blast furnace slag substitution for cement in carbon-reduced and low-cost solidified/stabilized cementitious materials. Green Smart Min. Eng. **1**(1), 104–117 (2024). https://doi.org/10.1016/j.gsme.2024.01.001
8. Sunarno, Y.: The influence of fly ash characteristics as a supplementary cementitious material (SCM) on concrete's mechanical properties. Materials Research Proceedings. **48**, 587–596 (2025). https://doi.org/10.21741/9781644903414-64
9. Azizi, M., Samimi, K.: Effect of silica fume on Self-compacting Earth Concrete: Compressive strength, durability and microstructural studies. Constr. Build. Mater. **472**, 140815 (2025). https://doi.org/10.1016/j.conbuildmat.2025.140815
10. Kho, J. H. "Incorporation of eco process pozzolan (EPP) as partial cement replacement and superplasticisers in concrete." In IOP Conference Series: Earth and Environmental Science, vol. 682, no. 1, p. 012014. IOP Publishing, 2021. doi:https://doi.org/10.1088/1755-1315/682/1/012014
11. Ahmat, A.M., et al.: Assessment of sustainable eco-processed pozzolan (EPP) from palm oil industry as a fly ash replacement in geopolymer concrete. Constr. Build. Mater. **387**, 131424 (2023). https://doi.org/10.1016/j.conbuildmat.2023.131424
12. Kadhim, A., Sadique, M., Al-Mufti, R., Hashim, K.: Developing one-part alkali-activated metakaolin/natural pozzolan binders using lime waste. Adv. Cem. Res. **33**(8), 342–356 (2021). https://doi.org/10.1680/jadcr.19.00118

13. Scrivener, K., Martirena, F., Bishnoi, S., Maity, S.: Calcined clay limestone cements (LC3). Cem. Concr. Res. **114**, 49–56 (2018). https://doi.org/10.1016/j.cemconres.2017.08.017
14. Lim, C.: Enhancing the pozzolanic reactivity of spent bleaching earth ash (SBEA) in binary blended cement mortar through calcination. ASM Sci. J. **17**, 1–11 (2022). https://doi.org/10.32802/asmscj.2022.1292

Study on the Strength and Durability of Nickel Slag-Based Concrete Blocks with the Addition of HDPE (High-Density Polyethylene) Plastic

Ahmad(✉), Rita Irmawaty, and Herman Parung

Department of Civil Engineering, Faculty of Engineering, Hasanuddin University, 92171 Gowa, South Sulawesi, Indonesia
ahmad.dslng@gmail.com
https://www.unhas.ac.id

Abstract. The accelerating pace of global infrastructure development has substantially increased the consumption of cement and natural aggregates, leading to higher carbon emissions and depletion of river-sand resources. Concurrently, large quantities of nickel slag, an industrial by-product from Indonesia's rapidly expanding nickel sector remain underutilized despite their technical potential as fine-aggregate substitutes. In parallel, high-density polyethylene (HDPE) waste continues to accumulate, posing long-term environmental challenges.

This study investigates the utilization of nickel slag as a fine-aggregate replacement and evaluates, through literature analysis, the potential performance of surface-roughened (cement pre-coated) HDPE as a lightweight modifier in dry-cast concrete blocks. Phase A experimentally examined sand replacement with nickel slag (0–100%) at a constant water-to-cement ratio (W/C) of 0.52, while Phase B synthesizes previous findings regarding HDPE incorporation (2–10%) at the optimum slag level.

The experimental results demonstrated that density increased from 1785 to 2544 kg·m^{-3}, water absorption decreased from 14.50% to 5.24%, and compressive strength improved from 3.96 to 13.20 MPa, with the optimum performance achieved at 70% slag substitution. Literature consensus further indicates that up to 10% cement-coated HDPE can reduce self-weight by approximately 6–10% while maintaining compressive strength (~12 MPa) and water absorption below 10%. All mixtures complied with SNI 03–0349-1989 requirements. The combined valorization of nickel slag and HDPE presents a technically viable and environmentally sustainable approach to masonry production, aligning with circular-economy principles and Indonesia's national waste-reduction policies.

Keywords: Dry-cast blocks · Nickel slag · HDPE · Cement pre-coating

1 Introduction

Concrete blocks, conventionally produced via the dry-cast method, are widely adopted in low-rise and residential construction due to their dimensional uniformity, cost efficiency, and satisfactory thermal performance. Their mechanical integrity and durability

M. A. b. Jamlos et al. (Eds.): ICSET 2025, LNNS 1897, pp. 588–595, 2026.
https://doi.org/10.1007/978-3-032-20722-7_76

primarily depend on aggregate gradation, compaction, and water control. In Indonesia, performance standards for such blocks are governed by SNI 03–0349-1989, which specifies a minimum compressive strength of 25 kg·cm^{-2} and a maximum 25% water absorption.

The demand for fine aggregates continues to rise in tandem with urban development, resulting in excessive river-sand extraction and associated ecological degradation (Abbas & Qureshi, 2025). To mitigate these impacts, substituting natural aggregates with industrial by-products has emerged as a sustainable alternative. Nickel slag—a dense, angular, and chemically stable by-product of nickel smelting—contains oxides such as FeO, SiO_2, Al_2O_3, and MgO that contribute to improved packing density and potential pozzolanic reactivity (Wu et al., 2018; He et al., 2024). Indonesia produces over 13 million tonnes of nickel slag annually (Bethary, 2022), which has been classified as non-hazardous waste under Government Regulation No. 22/2021, thus enabling its reuse in construction materials (Fig. 1).

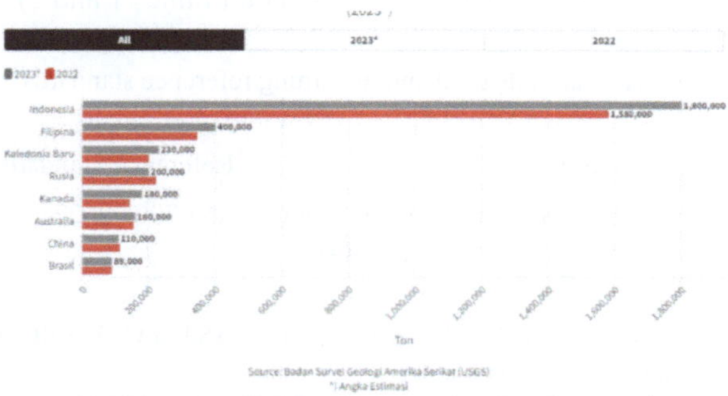

Fig. 1. Top 8 Nickel Producing Countries in the world (2023)

Simultaneously, the accumulation of HDPE waste, largely derived from packaging materials, presents both environmental and resource challenges. Although HDPE offers advantages in terms of low density and durability, its hydrophobic nature limits adhesion with cementitious matrices. Surface modification, particularly through cement pre-coating has proven effective in enhancing interfacial bonding (Rojas et al., 2025; Swinnerton et al., 2024). Several studies (Coppola et al., 2018; Darwin et al., 2019) reported that properly treated recycled plastics can partially replace fine aggregates without compromising strength.

This study aims to (i) evaluate the mechanical and durability performance of concrete blocks incorporating nickel slag as fine aggregate and (ii) review, through existing literature, the potential performance of cement-coated HDPE as a lightweight additive at the optimum slag substitution ratio.

2 Methodology

2.1 Material and Standards

According to SNI 03–0349-1989, concrete blocks are masonry units manufactured from a mixture of Portland cement, fine aggregate, and water. Two principal classifications are recognized:

a) Solid blocks, which are void-free and therefore exhibit higher density and compressive strength suitable for structural applications; and
b) Hollow blocks, which include core cavities to reduce self-weight and improve thermal performance, are typically used for non-structural walls.

Gross compressive strength is defined as the maximum load at failure divided by the gross bearing area including voids and edge recesses to reflect realistic load-carrying capacity in service. Dimensional tolerances and the maximum allowable water absorption of 25% are also prescribed in this standard (Tables 1 and 2).

Table 1. Materials used and governing reference standards

Material / Test	Description	Reference Standard
Cement	Portland Composite Cement (CaO 60–65%, SiO_2 17–25%, Al_2O_3 3–8%)	SNI 7064:2022
Fine Aggregate	Natural sand (FM 1.93) and nickel slag (FM 3.39, SG 3.10)	ASTM C33; SNI 03–2834-2000
HDPE	Recycled particles (0.88 $g \cdot cm^{-3}$) pre-coated with cement paste (1:0.4 W/C)	ASTM C136
Water	Potable, free of impurities	SNI 7656:2012
Testing	Density, absorption, compressive strength, strain	SNI 1970:2016; SNI 1973:2008; SNI 1974:2011; ASTM E83–23

2.2 Experimental Design

This study employed a controlled dry-cast production process to evaluate (1) the feasibility of replacing natural sand with nickel slag, and (2) the effect of incorporating HDPE waste as a lightweight modifier. The experimental design comprised two sequential phases (Fig. 2):

Table 2. Experimental design

Phase	Objective	Mix Variables	Parameters Evaluated
A	Assess replacement of natural sand with nickel slag	0%, 50%, 70%, 100% slag	Density, water absorption, compressive strength
B	Review potential HDPE addition at optimum slag content	2–10% HDPE by volume	Expected density reduction and strength performance

Specimens (400 × 100 × 100 mm) were cured for 28 days. HDPE particles were coated with cement paste (Fig. 3) to enhance interfacial adhesion and reduce hydrophobicity.

Fig. 2. Specimen Model

Fig. 3. Cement Pre-Coating

This simple surface-coating technique, or alternatively partial melting of HDPE, has been shown to enhance interfacial bonding with the cement matrix (Rojas et al., 2025; Swinnerton et al., 2024).

Material characterization and testing procedures were conducted in accordance with relevant ASTM standards, including ASTM C33/C33M-18 (2018) for aggregate specifications, ASTM C136/C136M-19 (2019) for sieve analysis, and ASTM E83-23 (2023) for strain measurement and extensometer verification.

3 Results and Discussion

3.1 Fine-Aggregate Characterization

Nickel slag exhibited a higher specific gravity (3.10) and lower water absorption (0.72%) compared to river sand (2.60 and 11.4%, respectively). The higher density and angular morphology of slag particles are conducive to better particle packing and lower capillary porosity, thereby improving mechanical performance and reducing water demand during mixing.

3.2 Phase A — Nickel-Slag Substitution

3.2.1 Density

As shown in Fig. 3, density increased progressively with higher slag content, from 1785 kg·m^{-3} (M1) to 2544 kg·m^{-3} (M4). This trend is attributed to the slag's higher specific gravity and enhanced packing efficiency. The 70% replacement (M3) provided the most favorable balance between density and mechanical strength (Fig. 4).

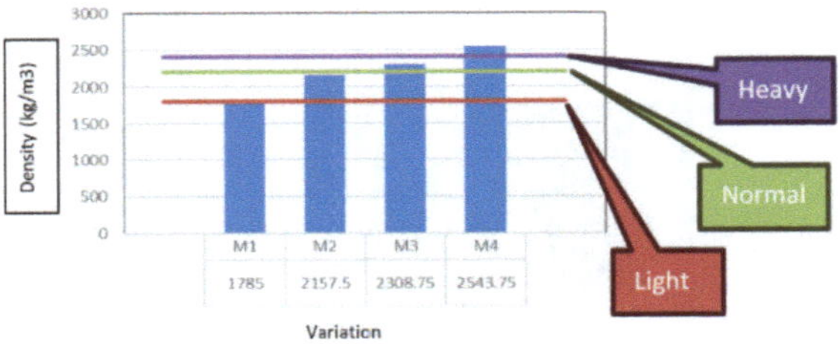

Fig. 4. Density and Specimen Variation

3.2.2 Water Absorption

Water absorption declined sharply from 14.5% (M1) to 5.24% (M4) (Fig. 5), confirming the densifying effect of slag particles. All values remained well below the SNI limit (25%), indicating enhanced impermeability and durability.

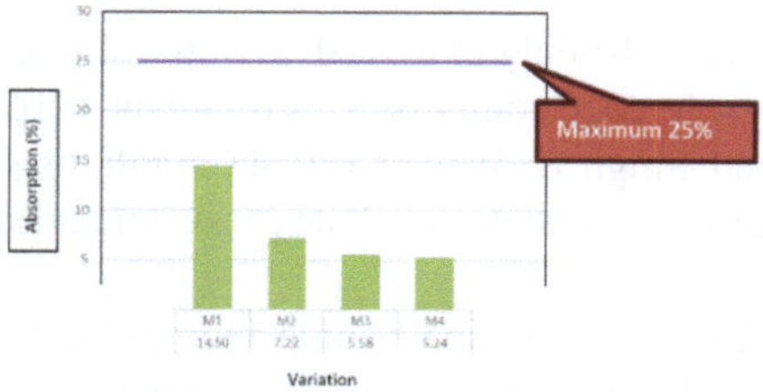

Fig. 5. Water Absorption and Specimen Variations

This interpretation aligns with the broader observation that elevated porosity degrades strength and elevates permeability, underscoring the need to control water content, aggregate grading, and fines packing in mix design (Muda et al., 2023).

3.2.3 Compressive Strength

Compressive strength increased from 3.96 MPa to 13.20 MPa with slag substitution (Fig. 6). The enhancement is primarily due to improved particle interlock and interfacial transition zone (ITZ) densification. The 70% slag mix (M3) achieved 11.95 MPa, satisfying SNI Class I requirements ($\geq$ 10 MPa).

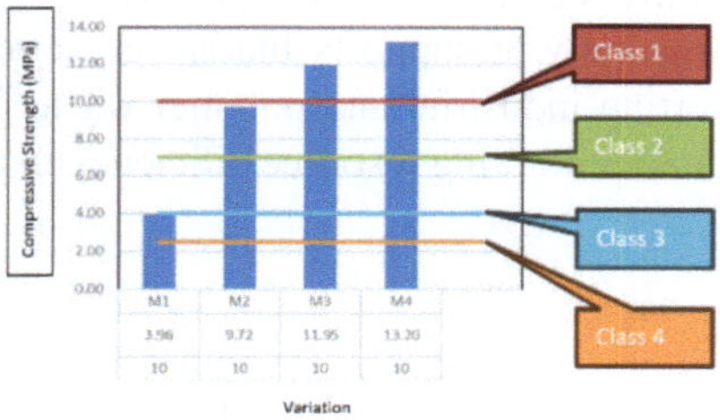

Fig. 6. Compressive Strength and Specimen Variations

These results align with previous studies demonstrating improved compactness and strength in slag-based mixtures (Wu et al., 2018; Wang et al., 2023).

3.3 Phase B — Prospective Influence of HDPE Incorporation (Literature-Based Review)

Although experimental verification of HDPE addition was not conducted within this phase, substantial literature evidence provides insight into its potential performance. Darwin et al. (2019) and Coppola et al. (2018) demonstrated that replacing 5–15% of fine aggregates with recycled plastic can lower density by up to 30% while maintaining adequate mechanical strength. Babafemi et al. (2018) and Mohammed et al. (2020) reported that the mechanical response is strongly governed by surface roughness and interfacial adhesion.

Unmodified HDPE generally weakens the ITZ due to its hydrophobic surface (Swinnerton et al., 2024). However, cement pre-coating effectively enhances mechanical interlocking, leading to superior stress transfer and improved bonding (Rojas et al., 2025;

Yooprasertchai et al., 2024). El-Seidy et al. (2023) showed that cement-coated plastics can recover up to 95% of reference compressive strength at a 10% dosage. Furthermore, Subhani et al. (2024) and Singh & Walker (2024) highlighted that surface-modified HDPE minimizes water absorption and contributes to sustainability through plastic waste upcycling.

Based on these studies, it is anticipated that incorporating ≤10% cement-coated HDPE into a 70% nickel-slag matrix could reduce density by approximately 6–10%, maintain compressive strength near 12 MPa, and limit absorption to below 10% (Alani et al., 2024; Rojas et al., 2025). Future experimental validation is recommended to substantiate these projections.

These findings are consistent with recent studies on nickel slag utilization and sustainable construction materials, which highlight the improved mechanical performance and durability of slag-based systems (Irmawaty et al., 2023; Li et al., 2024; Yang et al., 2025).

4 Sustainability Implications

The synergistic utilization of nickel slag and HDPE embodies a high-value recycling approach that addresses two critical waste streams simultaneously. From an engineering perspective, this strategy optimizes density-strength performance while reducing reliance on natural sand. Environmentally, it supports Indonesia's circular-economy roadmap under PP 22/2021 by diverting industrial and polymer waste from disposal pathways, lowering embodied CO_2, and promoting resource efficiency in masonry manufacturing.

5 Conclusion

a) Replacing natural sand with nickel slag (0–100%) increased density (1785 → 2544 kg·m^{-3}), reduced water absorption (14.5 → 5.24%), and enhanced compressive strength (3.96 → 13.2 MPa).
b) The optimum mix (70% slag) achieved 11.95 MPa compressive strength and met SNI Class I specifications, representing an ideal balance between strength and self-weight.
c) Literature suggests that ≤10% cement-coated HDPE can reduce density by 6–10% while maintaining ~12 MPa compressive strength and < 10% water absorption, offering a viable route for lightweight, durable, and sustainable blocks.
d) The integration of nickel slag and HDPE supports circular-economy goals by valorizing industrial by-products and plastic waste within the construction sector.

Acknowledgments The authors acknowledge the support of the Civil Engineering Laboratory for providing facilities for material preparation and testing, as well as the technical staff for their assistance during specimen fabrication and laboratory work. The authors also appreciate previous researchers whose studies formed the basis of the literature review presented in this paper.

Disclosure of Interests The authors declare that there are no known financial or personal relationships that could have influenced the work reported in this paper.

References

Abbas, A., Qureshi, H.: Utilization of industrial by-products as sustainable alternatives to natural aggregates in construction materials. J. Clean. Constr. Mater. **45**(2), 112–125 (2025)

Alani, M.: Enhanced interfacial bonding performance of modified plastic aggregates in cementitious composites. In: Materials Today: Proceedings (2024)

ASTM C33/C33M-18: Standard Specification for Concrete Aggregates. ASTM International (2018)

ASTM C136/C136M-19: Standard Test Method for Sieve Analysis of Fine and Coarse Aggregates. ASTM International (2019)

ASTM E83-23: Standard Practice for Verification and Classification of Extensometer Systems. ASTM International (2023)

Babafemi, A.J., Khennane, A., Shukla, S.K.: Engineering properties of concrete containing recycled plastic as partial replacement of sand. Constr. Build. Mater. **173**, 272–285 (2018)

Bethary, R.T.: Analysis of nickel slag utilization as an additive in concrete production in Indonesia. J. Constr. Mater. **10**(3), 144–152 (2022)

Coppola, L., Bellezze, T., Belli, A., Pauri, M.: Recycled plastics as fine aggregate in concrete mixtures: Influence of treatment and particle size. Mater. Today Proc. **5**(11), 23372–23379 (2018)

Darwin, P., Kuranchie, F., Nikraz, H.: Mechanical properties of concrete containing recycled plastic waste aggregate. Procedia Structural Integrity. **17**, 50–57 (2019)

Government of the Republic of Indonesia. (2021). Government Regulation No. 22/2021 concerning Environmental Protection and Management.

He, X., et al.: Preparation and properties of nickel–iron slag bricks. Case Stud. Constr. Mater. **22**, e03103 (2024)

Irmawaty, R., Caronge, M.A., Tjaronge, M.W., Abdurrahman, M.A., Ahmad, S.B.: Compressive strength and corrosion behavior of steel bars embedded in concrete produced with ferronickel slag aggregate and fly ash: An experimental study. Innov. Infrastruct. Solut. **8**, 200 (2023)

Li, B., Liu, Z., Sun, Q., Yang, L.: Preparation of alkali-activated nickel slag-based cemented tailings backfill: Workability, strength characteristics, localized deformation and hydration mechanism. Constr. Build. Mater. **411**, 134639 (2024)

Muda, M., Salleh, N., Rahman, M.M.: Pore structure and water absorption behavior of lightweight concrete using recycled materials. Mater. Today Proc. **72**(4), 1834–1842 (2023)

Rojas, M., Torres, D., Velásquez, C.: Performance of cement-coated plastic aggregates in eco-bricks: Interfacial bonding and durability assessment. Cleaner Construction Materials. **8**, 100214 (2025)

Singh, K., Walker, P.: Circular economy perspectives on polymer waste utilization in construction materials. J. Build. Eng. **87**, 108496 (2024)

Subhani, H.A., Khushnood, R.A., Shakeel, S.: Synthesis of recycled bricks containing mixed plastic waste and foundry sand: Physico-mechanical investigation. Constr. Build. Mater. **416**, 135197 (2024)

Swinnerton, A., Paul, M., George, R.: Effect of recycled polymer inclusion on dry-cast masonry performance: A review. J. Sustain. Constr. Mater. **12**(1), 55–71 (2024)

Wang, F., Zhang, X., Liu, Y.: Thermo-mechanical activation and reactivity of high-magnesium nickel slag for sustainable binders. J. Clean. Prod. **421**, 138546 (2023)

Wu, K., Ju, Y., Liu, J., Wu, P., Li, T.: Utilization of nickel slag as raw material in the production of cement and concrete. Constr. Build. Mater. **170**, 132–140 (2018)

Yang, L., et al.: Recent advances in recycling and upcycling of hazardous waste in construction materials. J. Environ. Manag. **356**, 121617 (2025)

Yooprasertchai, C., et al.: Surface modification of recycled plastics to enhance ITZ bonding in cementitious composites. Proceedings, Materials Today (2024)

Environmentally Friendly Concrete Innovation: The Use of Nickel Slag as Aggregate to Improve SCC Concrete Performance

Fitriyanti(✉), Rita Irmawaty, and Fakhruddin

Hasanuddin University, 92171 Gowa, South Sulawesi, Indonesia
fitrisubhan0778@gmail.com

Abstract. Nickel slag, a byproduct from the smelting of nickel ore, is used as a substitute for fine aggregate (sand) and coarse aggregate (gravel) in the production of SCC (Self-Compacting Concrete). The main objective is to find the optimal composition that can produce high-strength concrete while reducing industrial waste. Nickel slag was chosen because its composition and physical properties are similar to sand. The utilization of this waste aligns with the increasing global infrastructure needs, which demand construction materials that are safe, durable, and environmentally friendly. Although previous research indicated that using nickel slag as a coarse aggregate substitute in structural concrete with a design compressive strength ≥20 MPa is not recommended because it can lower the compressive strength, other studies found that combining nickel slag with additives like fly ash can improve the mechanical strength and density of concrete. This research uses a quantitative approach with a descriptive analysis method and focuses on laboratory experiments. The process includes literature review, sample preparation, and testing to achieve the research goals. The results show that utilizing nickel slag can improve the mechanical performance of concrete, including strength and corrosion resistance. The use of SCC concrete made with nickel slag also increases construction efficiency due to its ability to flow and consolidate easily without requiring mechanical compaction. Thus, nickel slag has the potential to become a key component in promoting sustainable construction practices and a circular economy, transforming waste into a valuable resource. The results of the concrete flexural strength test showed irregularities, where there was an increase in strength and the nickel slag had a fineness modulus that made it possible to be used as a substitute for sand aggregate.

Keyword: SCC Concrete · Waste · Nickel Slag · Corrosion

1 Introduction

Infrastructure holds a crucial role in supporting and driving a nation's economic growth [1]. The increasing global demand for infrastructure, along with population growth, is a major challenge faced by many countries worldwide. As the population grows, especially in urban areas, the need for adequate infrastructure, such as housing, roads,

M. A. b. Jamlos et al. (Eds.): ICSET 2025, LNNS 1897, pp. 596–604, 2026.
https://doi.org/10.1007/978-3-032-20722-7_77

bridges, transportation systems, and basic utilities like clean water and electricity, also increases. Infrastructure projects like buildings and transportation carry many risks [2]. This increase is necessary not only to support daily life but also to facilitate economic growth and reduce social inequality. With the population projected to continue growing, especially in developing countries, greater planning and investment in infrastructure are needed to ensure the sustainable and efficient availability of services for the global community.

Construction materials are the substances used to build infrastructure, with various types selected based on their function. Concrete, steel, wood, and asphalt are examples of commonly used materials. Concrete, for instance, is often chosen for its strength and load resistance, as well as its ability to be formed into various shapes. Concrete is the most frequently used construction material in the industry [3]. The materials used must meet specific quality standards so that the built infrastructure is safe, durable, and environmentally friendly. Innovation in construction materials, such as the use of environmentally friendly materials or industrial waste, is becoming increasingly popular to support sustainable development.

SCC (Self-Compacting Concrete) is an innovative type of concrete designed with the ability to flow and fill formwork automatically without needing compaction or external tools during application [4]. The main advantage of SCC is its ability to reduce working time and labor, as well as improve the reliability and quality of the final concrete product, because the compaction process is usually prone to errors during application [5]. SCC also has added benefits such as reducing noise during the construction process and reducing the need for compacting equipment. Thanks to its self-flowing ability, SCC is highly suitable for construction with complex architectural designs, such as high-rise buildings, bridges, and precast structures. SCC is an innovative finding. The main difference between conventional concrete and SCC is that the latter can compact itself without vibration, while conventional concrete is stiff and difficult to work with, despite having a low water-cement factor (WCF) when fresh.

Corrosion is a familiar word to us. Corrosion is the natural degradation or destruction process of a metal that occurs due to an electrochemical reaction between the metal and its environment, such as air and water, creating undesirable new compounds. Corrosion that occurs on rebar reduces the quality of the iron or metal [6]. Concrete is a primary material that often experiences corrosion, especially in reinforced concrete, which reduces the service life of the structure [7]. Reinforced concrete consists of two materials, rebar and concrete, used together [8]. The rebar must have the same bond stress as the concrete component. An aggressive environment causes damage to the concrete, including corrosion on the reinforcing steel, which produces rust compounds (Fe2O3nH2O) [9]. Corrosion in this context typically occurs in the concrete or the reinforcing iron, and it is hoped that the utilization of nickel slag waste for making concrete, especially SCC concrete, can reduce corrosion in the SCC concrete [10].

The Decree of the Ministry of Environment No. B-6152/Dep.IV/LH/08/2010 states that nickel slag is not categorized as Hazardous and Toxic Waste (B3), where the content of harmful metals is even lower than that of the surrounding soil [11]. Nickel slag is waste produced by steel mills during the smelting of iron and steel [12]. The addition of iron slag to Self-Compacting Concrete (SCC) improves compressive strength and flowability. The addition of iron slag, which contains CaO, SiO2, and MgO, increases the flowability and compressive strength of concrete, especially after 14 days, due to the fully reactive pozzolanic properties of the iron slag. The use of waste materials like nickel slag in SCC concrete can not only improve concrete performance but also contribute to environmental sustainability by reducing industrial waste. The processing of nickel ore through a pyrometallurgy process yields a byproduct called Nickel Slag [13] (Fig. 1).

Fig. 1. Slag Nikel PT. HNI

The problem of nickel slag utilization is very serious. This problem can be reduced by using nickel slag as a construction material mix in concrete [12]. The concrete production process is relatively easy and is considered more economical. However, the normal concrete pouring process faces challenges when the distance between the rebars is too close. The rapid development of the industry from year to year indicates very fast and meaningful progress for the Indonesian economy, but the resulting impact of industrial activities is waste [14].

Visually, nickel slag has a physical shape resembling both coarse and fine aggregate. The most common chemical composition in nickel slag is silica, with a percentage of 41.47%, which is expected to strengthen the hydration process and increase the concrete's compressive strength [15]. Although nickel slag is classified as B3 Waste (Hazardous and Toxic Waste), which originates from various human activities such as industry, households, agriculture, and health, B3 Waste can contain harmful chemical substances such as heavy metals, pesticides, pharmaceuticals, toxic chemicals, and others [16].

Based on previous research, nickel slag production reaches 13 tons per year, while tailing slurry is only 25.6 tons. Nickel slag has great potential for recycling for various needs, especially for construction materials, because it contains beneficial metals like nickel, copper, and cobalt. Furthermore, about 70% of nickel slag contains 41.47% Silica, 30.44% Oxide, and 2.58% Alumina [17] (Figs. 2 and 3).

The main challenge in using SCC (Self-Compacting Concrete) lies in optimizing its material composition to ensure optimal mechanical performance and environmental resistance. To achieve good flow without external compaction, SCC requires the right proportion of additives such as superplasticizer and fines (fine particles), which is often difficult to achieve without compromising strength or durability. Moreover, incorrect mix

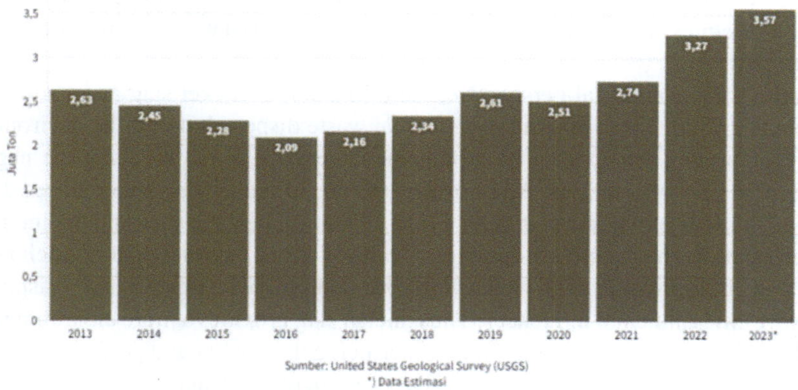

Fig. 2. Word Nickel Producing (2013–2023)

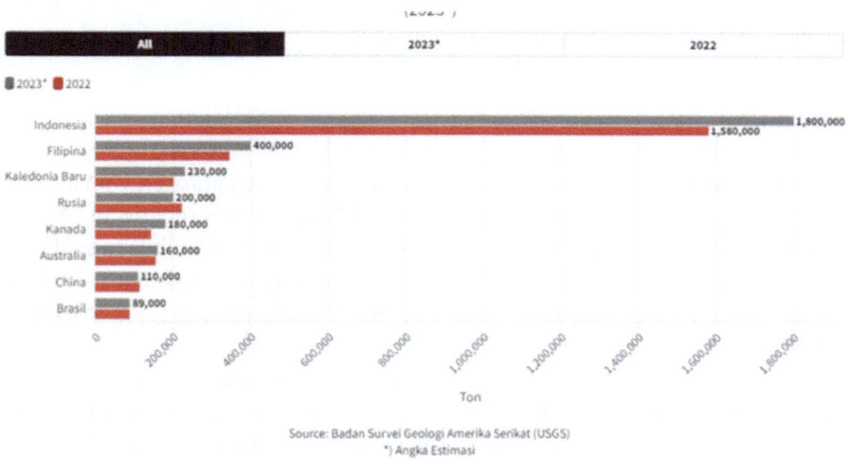

Fig. 3. Top 8 Nickel Producing Countries in the word (2023)

proportioning can cause problems like segregation or quality differences on the concrete surface. Therefore, innovation in developing binding materials, material substitution, and environmentally friendly additives is very important to ensure SCC maintains high structural strength, resistance to environmental factors like chemical attack or corrosion, and sustainable production efficiency. Further research on this material composition is needed so that SCC concrete can be widely applied with stable and environmentally friendly performance.

The objective of this research is to determine the characteristics of Nickel Slag that can be used as a substitute or additive in the production of SCC Concrete. About 75%–80% of concrete consists of aggregates (like gravel and sand) and water mixing [18]. In concrete design, concrete performance is indicated by strength and durability; these two variables are very important [19]. High-strength concrete is concrete that has a compressive strength above normal concrete. SNI (Indonesian National Standard) 03–6468-2000 defines high-strength concrete as concrete material with required compressive strength [20].

From an environmental perspective, the utilization of nickel slag as an additive in concrete helps reduce the amount of industrial waste disposed of into the environment, thereby supporting the concept of sustainable development and reducing the negative impact on the ecosystem [21]. Additionally, the use of nickel slag also reduces dependency on limited natural raw materials. From a technical performance standpoint, nickel slag can increase the concrete's durability against various external factors, such as corrosion and chemical attack, as well as improve compressive strength and resistance to aggressive environmental influences. Thus, nickel slag provides a more environmentally friendly and economical solution without sacrificing the quality and performance of the concrete. Previous research using nickel slag as a concrete mix material found that its use as a construction material will reduce dependency on natural raw materials like natural sand and gravel. The use of nickel slag as a substitute material is because it has high strength when used for concrete production. In general, concrete curing is a factor that significantly influences the concrete's strength [15].

2 Methodology

This research uses a quantitative and experimental diagnostic approach with a focus on laboratory and field work. Laboratory data was collected at the Structure and Materials Laboratory of Hasanuddin University. The hypothesis in this research is to evaluate the use of nickel slag in the composition of SCC concrete that is resistant to corrosion. Nickel slag is used as a material to determine the strength and corrosion resistance of the SCC concrete.

The nickel slag waste is combined as a fine aggregate substitute for sand and a coarse aggregate substitute for gravel. It will be proportioned with other materials to form a suitable composition that serves as an alternative substitute for natural sand and gravel. The test results are expected to provide scientific renewal and better knowledge for the future, so that industrial waste will decrease yearly, and the use of natural sand, which is also a diminishing natural resource, can be reduced as infrastructure development continues to grow.

Data Collection Stages:

1. Literature Review: As a source to obtain secondary data and additional reading material to enrich knowledge about SCC concrete and the characteristics of nickel slag used as a substitute material for SCC concrete.
2. Sample Preparation: Nickel slag is used as a substitute for fine aggregate (sand) and coarse aggregate (gravel) to find the right composition for the constituent material of high-strength SCC concrete.
3. Laboratory Testing: Involves making and testing a number of samples as primary data used in analyzing and examining the results to achieve the research objectives.

Research Tools and Materials:

1. Sieves, Shaker Machine, Digital Scale, Oven, Los Angeles Machine, Measuring Glass.
2. Research Material: Nickel Slag

Sieve Test Steps for Fine Aggregate (Nickel Slag):

1. Weigh the necessary sieves, which are those that pass through sieve No. 4 and are retained on sieve 100.
2. Add the material to be used.
3. Perform the sieving for 10–15 min.
4. Weigh again and record the results.
5. Clean the sieves

This research uses a quantitative approach with a descriptive analysis method to examine the composition of nickel slag used as a substitute material for high-strength concrete constituents. The data analysis technique in this research is processing, analyzing, and applying the correct or SNI-compliant composition for SCC concrete to achieve the intended research objectives. The research approach is quantitative analysis focusing on the strength of ordinary concrete and SCC concrete made with nickel slag. The research method aims to provide scientific renewal regarding material substitution, where nickel slag waste is used as a substitute for natural sand and gravel so that industrial waste is reduced, and the use of natural sand can also be lessened with this material substitution. The design of the test specimens in this research consists of cube-shaped concrete with dimensions of 5 cm × 5 cm × 5 cm, for the compressive strength test preparation at ages 7, 21, and 28 days, with 3 samples made for each concrete age. There are also cylinder shapes with dimensions of 10 cm × 20 cm for the Corrosion test, totaling 16 samples using dia 16 deformed iron rebar.

The following tests will be performed on the samples:

1. Concrete Compression Test
2. Modulus of Elasticity Test
3. Corrosion Test

3 Results

Nickel slag has a varied chemical composition depending on the type of ore and the smelting process, but it is generally dominated by silica (SiO2) and magnesium oxide (MgO), which can reach a significant percentage, as well as iron (Fe) in the form of oxides. Furthermore, this material also contains small amounts of oxides from various other elements such as aluminum (Al2O3), calcium (CaO), chromium (Cr), manganese (Mn), and even residual nickel (Ni) that was not fully extracted, along with other trace elements.

The high silica content makes nickel slag attractive for various applications, including as an aggregate substitute in construction or as a silica source, although the presence of certain heavy metals in varying concentrations requires strict evaluation to ensure its safety in reuse.

The results of the research on nickel slag used as a substitute material (fine aggregate) or additive in the production of high-strength concrete are as follows (Fig. 4):

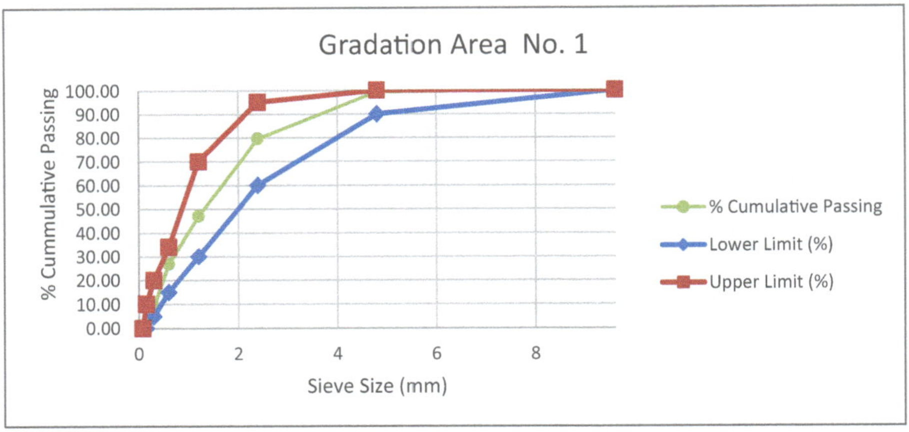

Fig. 4 . Gradation Area For Nickel Slag

The result of the nickel slag abrasion test shows that it has a fineness modulus that falls within gradation 1, which makes it possible to be used as a substitute for sand aggregate in the Self-Compacting Concrete (SCC) mix (Fig. 5).

Fig. 5. Nickel Slag After Abrasion

The use of nickel slag in SCC concrete can improve its mechanical performance. This contributes to an increase in the concrete's strength and durability, making it more resistant to pressure and harsh environmental conditions, such as corrosion. Moreover, because of the similarity of nickel slag's properties to sand, SCC concrete based on this material has the ability to flow and fill formwork easily without needing mechanical compaction. It is used as a substitute for fine aggregate (sand) and coarse aggregate (gravel) in concrete production. This material has a composition and physical properties similar to sand, making it a very feasible material for use in construction. The utilization of nickel slag is especially beneficial in the production of high-strength concrete. Nickel slag is a substitute material for fine aggregate because it has an absorption value of >40%, allowing it to be used as a fine aggregate by grinding it.

4 Conclusions

In this research, the nickel slag used is within gradation No. 1. The nickel slag used for the test specimens originates from PT. HNI Bantaeng, South Sulawesi. The utilization of this industrial waste not only helps reduce solid waste but also preserves natural resources and reduces the ecological impact of sand mining. Thus, nickel slag has the potential to be a key component in promoting sustainable construction practices and a circular economy, transforming waste into a valuable resource that can be converted into a substitute material for concrete mixing components. Nickel slag has a composition and physical properties similar to sand material. Therefore, its use as a construction material is highly feasible.

Increased Concrete Mechanical Performance: SCC (Self-Compacting Concrete) containing nickel slag can improve its mechanical performance. This contributes to an increase in the concrete's strength and durability, making SCC concrete made with nickel slag more resistant to pressure and harsh environmental conditions, such as corrosion. Higher Construction Efficiency: SCC concrete made with nickel slag has the ability to flow and fill formwork easily without requiring mechanical compaction. This reduces the time and labor needed during the casting process, thereby increasing construction efficiency. With its ease of formwork filling, this concrete is highly suitable for projects with complex architectural designs and hard-to-reach spaces, such as precast structures and finely detailed reinforced concrete. Nickel slag as a material in concrete production offers a promising and sustainable solution.

This material, which is a byproduct of nickel ore smelting, has a composition and physical properties similar to natural sand. Its use can improve the mechanical performance of concrete, including its strength and resistance to corrosion and harsh environmental conditions. Furthermore, by utilizing nickel slag, the volume of industrial solid waste can be reduced, natural resources can be preserved, and the circular economy and sustainable construction can be promoted. In the world, especially in Indonesia, the utilization of industrial waste has not been maximized due to several obstacles. Among them, the industries that produce nickel slag still face challenges in finding or solving what the waste should be made into and produced as. Nickel slag is a substitute material for fine aggregate because it has an absorption value of >40%, allowing it to be used as a fine aggregate by grinding it. Although much research on nickel slag has been conducted, the production of products from this waste is still rare. It is hoped that this research will provide a way to reduce this industrial waste, in this case, nickel slag.

References

1. Kahiking, F., Jansen, T., Jasin, M.I.: Study on coastal damage in Buloh Beach, Mandolang District, Minahasa. TEKNO. **20**, 1123–1133 (2022)
2. Triatmodjo, B.: Coastal Engineering. Beta Offset, Yogyakarta (1999)

3. Saengsupavanich, C., Arif, H., Shin, L., Anthony, D.: Environmental impact of submerged and emerged breakwaters. Heliyon. **8**(November), e12626 (2022). https://doi.org/10.1016/j.heliyon.2022.e12626
4. Sappetau, I.A., Tanan, B., Bungin, E.R.: Design of breakwater at Auki Beach, Biak Numfor Regency, Papua. Paulus Civil Engineering Journal. **2**(3), 161–170 (2020)
5. Panto, P.T.: Detailed design of coastal protection structures. **1**(1), 28–48 (2024)
6. Rachmanda, F., Tugiono, S., Zakaria, A., Herison, A.: Stability analysis of submerged rubble mound breakwater. J. Rekayasa Sipil Desain. **10**(2), 385–394 (2022)
7. Mayanda, I.N., Chayati, C., Rahmanto, A.D.: Evaluation of breakwater planning against waves at Pasongsongan Fishing port, East Java. case study: Pasongsongan fishing port, East Java. J. Teknik Gradien. **15**(02), 10–19 (2023)
8. Gomes, A., Pinho, J.L.S., Valente, T., Antunes, S.: Performance assessment of a semi-circular breakwater through CFD modelling. J. Mar. Sci. Eng., 1–16 (2020)
9. Hariati, F., Muhammad, F., Taqwa, L.: Conceptual study of breakwaters. J. Komposit: J. Ilmu-Ilmu Tek. Sipil. **5**(1), 31–37 (2020)
10. Husain, F., Paroka, D., Rahman, S., Lamputang, P.: Use of submerged breakwaters to reduce Abrasion on Lamputang island. J. Pengabdian Masyarakat Teknik. **3**, 65–70 (2021). https://doi.org/10.24853/jpmt.3.2.65-70
11. Clavero, M., Díaz-Carrasco, P., Losada, M.A.: Bulk wave dissipation in the armor layer of slope rock and cube armored breakwaters. J. Mar. Sci. Eng. **8**(3), 152 (2020). https://doi.org/10.3390/jmse8030152
12. Khasanah, M., Sugeng, H., Rikha, W.: Physical model simulation of breakwater effectiveness using tetrapod and dolos. Jurnal Teknik Hidraulik. **13**, 17–26 (2022)
13. Zalukhu, S., Sipil, J.T., Immanuel, U.K., Yogyakarta, D.I.: Comparison of tetrapod and rubble mound usage as jetty armor layer at Progo river estuary, Bantul regency. In: Proceedings of the Seminar Nasional Teknik Sipil UMS, pp. 374–382 (2025)
14. Setyandito, O., et al.: Physical model test on the stability of breakwater armour units at Sanur port. Jurnal Ilmiah Desain & Konstruksi. **20**, 139–149 (2021)
15. Sujantoko, S., Pangestu, P.A., Silva, F.G.K., Ahmad, R.P., Firman, M.R.: Test of a 2D physical model of seawall stability at various slope angles with A-Jack armour units. E3S Web Conf. **576**, 02005 (2024). https://doi.org/10.1051/e3sconf/202457602005
16. Azhar, R.M., Larasari, A., Husrin, S., Bachtiar, H.: Hydraulic responses and stability of armor layer. J. Tek. Hidraul. **10**(1), 29–42 (2019). https://doi.org/10.32679/jth.v10i1.603
17. Gunawan, G., et al.: Analysis of wave breaking on L-type jetty structure (case study: Purus Beach, Padang City). Sci. J. Civ. Eng. **17**(2), 144–152 (2020)
18. Maula, A.I.: Experimental Study on the Stability of Hexaloc-2 Armor Unit in Breakwater Structure with Variations in Slope, Wave Height, and Wave Period. Sepuluh Nopember Institute of Technology (2023)
19. Komar, P.D.: The modeling of processes and morphology in the Coastal Zone—Reflections on the maturity of our science. Shore & Beach. **66**, 10–22 (1998)
20. Ginting, J.W., Putra, I.A.I.D.R., Simanjuntak, E.M.: Physical model of wave energy dissipation using pile breakwater. J. Tek. Hidraul. **10**(1), 55–70 (2019). https://doi.org/10.32679/jth.v10i1.606
21. E. K. Syahlan, "Numerical Modeling of Hydrodynamics for A-Jack Armor on Seawall," 2022., https://repository.its.ac.id/115905/1/04311840000011-Undergraduate_Thesis.pdf
22. Van der Meer J.W., Sigurdarson S., "Manual on the Use of Concrete Armour Units," 2017.
23. Hudson, R.Y.: Laboratory investigation of rubble-mound breakwaters. J. Waterw. Harb. Div., 93–121 (1959)
24. Van der Meer, J.W.: Rock Slopes and Gravel Beaches under Wave Attack. Delft Univ. Technol (1988)

The Effect of Baffled Reactors with and without Floating Treatment Wetlands on Domestic Wastewater Remediation

Mostafa Mohamed Ahmed Mohamed(✉), Husna Takaijudin, and Teh Hee Min

Department of Civil & Environmental Engineering, Universiti Teknologi PETRONAS, 32610, Seri Iskandar, Perak Darul Ridzuan, Malaysia
mostafa.alradhi@gmail.com

Abstract. Floating treatment wetlands (FTWs) are one of the nature-based solutions (NBS) distinguished by their efficacy and installation simplicity. They comprise plants that develop in a floating system, maintaining the plant roots in continuous contact with wastewater, hence facilitating pollutant removal through multiple mechanisms. This experimental study aims to examine the FTWs efficiency using Vetiver grass in nutrient removal, including ammonia (NH_3-N), total nitrogen (TN) and total phosphorus (TP). For water quality analysis, Nessler method (8038) was employed to measure NH_3-N concentration. The effect of baffled reactors with and without FTWs is investigated to purify domestic wastewater over 10 days. The results showed that throughout all water parameters, cells with FTWs show a substantial reduction in pollutant concentrations during the 10-day period. The effluent concentrations for cells with FTWs often achieve significantly lower levels by day 10 in contrast to cells without FTWs, signifying enhanced treatment efficacy. This study contributes to the growing domain of NBS by analyzing the FTWs effectiveness in domestic wastewater management.

Keywords: Floating Treatment Wetlands · NBS · Vetiver Grass · Water Quality Analysis · Domestic Wastewater

1 Introduction

The population growth and urbanization expansion in developing nations have heightened the need for freshwater to meet increasing domestic and industrial demands. These consequently result in increased wastewater production [1, 2, 3]. In many instances, wastewater is discharged into water bodies without pre-treatment or monitoring, resulting in pollution, exacerbating freshwater scarcity, and frequently leading to the formation of sludge deposits in watercourses [4, 5]. The release of untreated effluents also results in diminished dissolved oxygen (DO), elevated nitrates, suspended solids, and other organic pollutants in these water sources [6, 7]. This subsequently affects both surface and underground freshwater supplies. Moreover, domestic wastewater is a significant contributor to surface water contamination due to its constituents, which encompass ammonia nitrogen, organic matter, bacteria, biochemical oxygen demand (BOD), and various other compounds [8, 9]. Consequently, supplementary treatment is necessary to reduce pollutant levels.

M. A. b. Jamlos et al. (Eds.): ICSET 2025, LNNS 1897, pp. 605–610, 2026.
https://doi.org/10.1007/978-3-032-20722-7_78

In recent years, a variety of technologies have been designed related to wastewater treatment and water reuse, as they are crucial topics for a more environmentally friendly approach to water management [10]. Among them are floating treatment wetlands (FTWs), a type of marshy wetland comprising emergent vegetation in a buoyant structure. The shoots and crowns of the plants emerge above the water level, while the roots penetrate into the water column, forming a vast subaqueous root system [11].

The aim of this experimental study is to investigate the FTWs efficiency using Vetiver grass in nutrient removal, including NH_3-N, TN and TP. This paper contributes to the growing domain of NBS by comprehensively analyzing the efficacy of FTWs in domestic wastewater management and investigating the interconnections among macrophytes, substrates, and microbial communities within a wastewater treatment context.

2 Methodology

2.1 Development of FTWs Systems

Basically, the selection of materials and components of FTWs is based on their availability, cost-effectiveness, and adaptability to the tropical climate of Malaysia, where the system is exposed to the surrounding uncontrolled environments.

After selecting FTWs materials, the establishing and designing steps are adopted based on the study of Arslan et al., [12]. Therefore, FTWs systems are conducted at Universiti Teknologi PETRONAS (UTP), block 21, as indicated in Fig. 1.

Fig. 1. Steps for establishing and developing FTWs models.

3 Results

3.1 The Effect of Baffled Reactor Performance with and without FTWs on Domestic Wastewater Treatment

NH_3-N Removal. On day 0, the influent NH_3 concentration measured 12.14 mg/L for the cells with FTWs, while for cells without FTWs measured 11.48 mg/L as depicted in Fig. 2. The concentrations in the cells with FTWs exhibit a progressive decline over time. By day 10, the effluent concentration decreases significantly to 1.38 mg/L, denoting efficient remediation. Whereas for the cells without FTWs, the concentration diminishes from 11.48 mg/L at day 0 to 1.12 mg/L by day 10, indicating a steady decrease in NH_3 levels, but at a slower rate than in the cells with FTWs. On the other hand, the NH_3 removal efficiency for the cells with FTWs progressively enhances with time, reaching 88.63% by day 10, while the cells without FTWs reaching 90.24%.

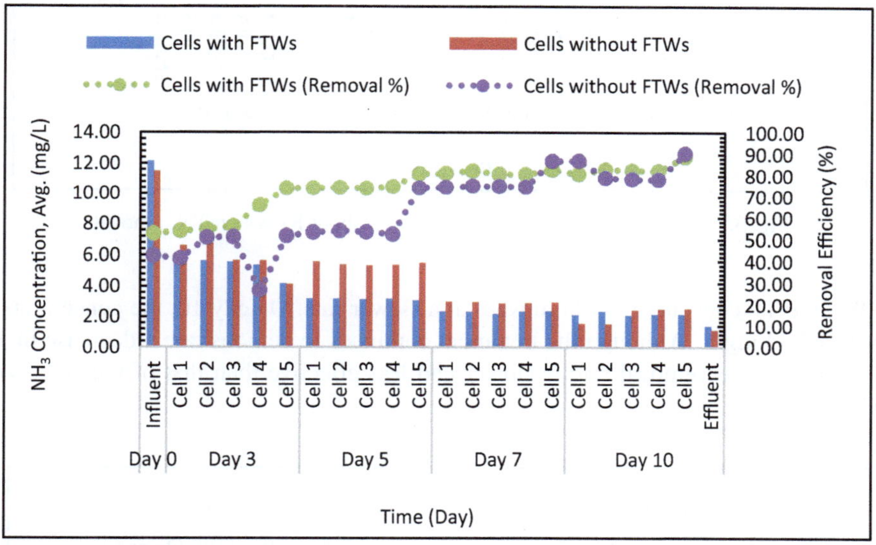

Fig. 2. Effect of baffled reactors with and without FTWs on NH_3 removal.

TN Removal. As presented in Fig. 3, in the case of cells with FTWs, TN concentration decreases markedly. By day 10, the effluent concentration decreases to 3.10 mg/L, underscoring the crucial function of FTWs in nitrogen elimination. This indicates that the FTWs are facilitating a reduction in TN via biological mechanisms, potentially through increased microbial activity or plant absorption. Whereas the cells without FTWs exhibit a reduction in TN concentration, reaching 3.60 mg/L by day 10. Nonetheless, this drop is less significant than that observed in cells with FTWs, suggesting that FTWs enhance the efficiency of the nitrogen removal process. On the other hand, the TN removal efficiency for the cells with FTWs gradually enhances with time, reaching 82.81% by day 10, while the cells without FTWs reach 82.09%.

TP Removal. On day 0, the influent TP concentration was recorded at 9.97 mg/L for the cells with FTWs, while the cells without FTWs exhibited a value of 10.43 mg/L, as illustrated in Fig. 4. Moreover, cells with FTWs demonstrate a consistent decline in TP concentration throughout the treatment period. On day 10, the effluent TP concentration is 0.98 mg/L, indicating substantial removal of TP. Conversely, in cells without

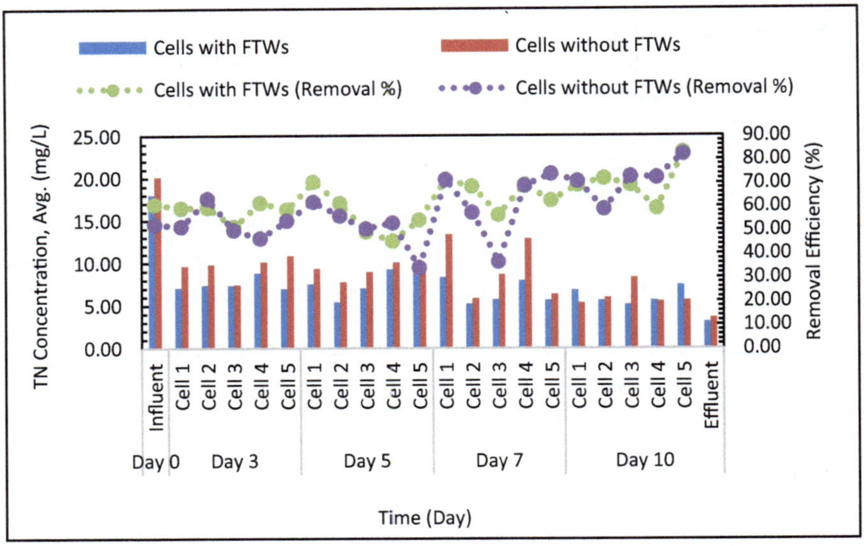

Fig. 3. Effect of baffled reactors with and without FTWs on TN removal.

FTWs, the TP concentration diminishes at a slower rate. On day 10, the concentration reaches 2.32 mg/L. Although the treatment continues to demonstrate a decrease in TP concentration, its removal rate is slower and less effective than that of the cells with FTWs.

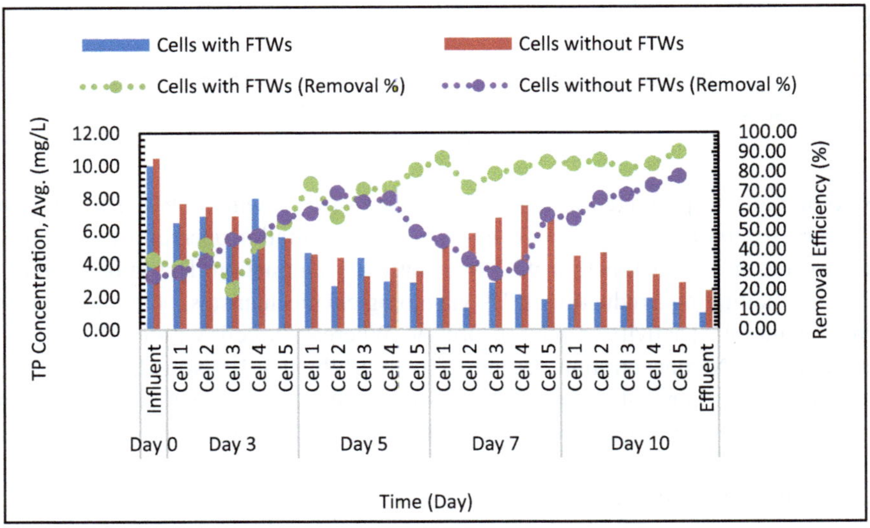

Fig. 4. Effect of baffled reactors with and without FTWs on TP removal.

4 Conclusions

FTWs are one of the nature-based solutions (NBS) distinguished by their effectiveness, maintenance, and simplicity of installation. The effect of baffled reactors with and without FTWs is investigated to purify domestic wastewater over 10 days. The results showed that throughout all water parameters, including NH_3-N, TN, and TP, cells with FTWs show a substantial reduction in pollutant concentrations during the 10-day duration. The effluent concentrations for cells with FTWs often attain significantly lower levels by day 10 in contrast to cells without FTWs, signifying enhanced treatment efficacy. Cells without FTWs demonstrate reduced rates of pollutant concentration reduction, indicating that FTWs considerably augment the treatment process, particularly for nutrient elimination. This study contributes to the growing domain of NBS by comprehensively analyzing the FTWs efficiency in domestic wastewater management and investigating the interconnections among macrophytes, substrates, and microbial communities in a wastewater treatment context.

Acknowledgments. This project is fully funded by FRGS/1/2023/TK06/UTP/03/1 under Ministry of Higher Education (MOHE). We also acknowledge Universiti Teknologi PETRONAS for their support in this project.

Disclosure of Interests The authors declare that they have no known competing financial interests or personal relationships that could have appeared to influence the work reported in this paper.

References

1. Dar, S.H., Kumawat, D.M., Singh, N., Wani, K.A.: Sewage treatment potential of water hyacinth (Eichhornia crassipes). Res. J. Environ. Sci. **5**(4), 377–385 (2011). https://doi.org/10.3923/rjes.2011.377.385
2. Zhang, D.Q., Jinadasa, K.B.S.N., Gersberg, R.M., Liu, Y., Ng, W.J., Tan, S.K.: Application of constructed wetlands for wastewater treatment in developing countries - A review of recent developments (2000–2013). J. Environ. Manag. **141**, 116–131 (2014). https://doi.org/10.1016/j.jenvman.2014.03.015
3. Shingare, R.P., Thawale, P.R., Raghunathan, K., Mishra, A., Kumar, S.: Constructed wetland for wastewater reuse: Role and efficiency in removing enteric pathogens. J. Environ. Manag. **246**, 444–461 (2019). https://doi.org/10.1016/j.jenvman.2019.05.157
4. Rahman, M.E., et al.: Design, operation and optimization of constructed wetland for removal of pollutant. Int. J. Environ. Res. Public Health. **17**(22), 1–40 (2020). https://doi.org/10.3390/ijerph17228339
5. Ajibade, F.O., Adeniran, K.A., Egbuna, C.K.: Phytoremediation efficiencies of water hyacinth in removing heavy metals in domestic sewage (A Case Study of University of Ilorin, Nigeria). Int. J. Eng. Sci. **2**(12), 16–27 (2013). https://doi.org/10.6084/m9.figshare.940965
6. Stefanakis, A.I., Bardiau, M., Trajano, D., Couceiro, F., Williams, J.B., Taylor, H.: Presence of bacteria and bacteriophages in full-scale trickling filters and an aerated constructed wetland. Sci. Total Environ. **659**, 1135–1145 (2019). https://doi.org/10.1016/j.scitotenv.2018.12.415
7. Ingrao, C., Failla, S., Arcidiacono, C.: A comprehensive review of environmental and operational issues of constructed wetland systems. Curr. Opin. Environ. Sci. Heal. **13**, 35–45 (2019). https://doi.org/10.1016/j.coesh.2019.10.007

8. Cheng, H., Zhu, Q., Xing, Z.: Adsorption of ammonia nitrogen in low temperature domestic wastewater by modification bentonite. J. Clean. Prod. **233**, 720–730 (2019). https://doi.org/10.1016/j.jclepro.2019.06.079
9. Othman, M.F., Hasan, H.A., Muhamad, M.H., Babaqi, B.S.: Biopolishing of Domestic Wastewater Using Polyvinyl Alcohol – Supported Biofilm of Bacterial Strain Bacillus velezensis Isolate JB7. J. Ecol. Eng. **24**(8), 33–42 (2023). https://doi.org/10.12911/22998993/165780
10. Afzal, M., et al.: Floating treatment wetlands as a suitable option for large-scale wastewater treatment. Nat. Sustain. **2**(9), 863–871 (2019). https://doi.org/10.1038/s41893-019-0350-y
11. Pavlineri, N., Skoulikidis, N.T., Tsihrintzis, V.A.: Constructed floating wetlands: A review of research, design, operation and management aspects, and data meta-analysis. Chem. Eng. J. **308**, 1120–1132 (2016). https://doi.org/10.1016/j.cej.2016.09.140
12. Arslan, M., Iqbal, S., Islam, E., Gamal El-Din, M., Afzal, M.: A protocol to establish low-cost floating treatment wetlands for large-scale wastewater reclamation. STAR Protoc. **4**(4), 102671 (2023). https://doi.org/10.1016/j.xpro.2023.102671

Trends and Research Directions of Digital Twin in the Naval Domain: A Bibliometric Study

Antonio Giallanza(✉), Parand Khani, and Rosa Micale

Department of Engineering, University of Messina, 98166 Messina, Sicily, Italy
antonio.giallanza@unime.it

Abstract. The adoption of Digital Twin (DT) technologies in the naval domain is accelerating, reflecting a growing interest in enhancing ship design, operation, and maintenance through digitalization. This study conducts a comprehensive bibliometric analysis to map the state of the art in order to identify emerging trends in this field. A total of 118 peer-reviewed publications were retrieved from the Scopus database and analyzed using VOSviewer for performance analysis and science mapping. The study applies techniques such as keyword co-occurrence, co-authorship networks, and bibliographic coupling to identify influential contributions. Results indicate a significant increase in scholarly output from 2021 onwards, with a concentration of research activity in China and United State of America. Prominent themes include sustainability aspect, DT integration with AI and IoT, and predictive maintenance issues. However, persistent gaps remain in standardization, interoperability, and real-time implementation. This work offers a structured overview of the knowledge landscape, providing guidance for future research and supporting strategic decisions for stakeholders in the maritime and shipbuilding industries.

Keywords: Digital Twin · Shipping 4.0 · Bibliometric Analysis

1 Introduction

The advent of Industry 4.0 technologies, such as the Internet of Things (IoT), Artificial Intelligence (AI), and Digital Twins (DT), is profoundly transforming the manufacturing sector towards greater sustainability [1]. Among the various domains impacted, the maritime sector has attracted growing attention [2–4]. Practical applications, including IoT-enabled monitoring for marine thrusters, have demonstrated significant improvements in design-test cycle efficiency [5, 6]. Within this evolving landscape, the DT has emerged as a unifying paradigm that integrates IoT, high-fidelity simulation, and AI for comprehensive lifecycle management [7–9].

Recent advances highlight the versatility of DT applications. For structural integrity, DTs successfully reconstruct global stress fields from sparse hull data [7] and monitor real-time deformations utilizing the inverse Finite Element Method [12]. In complex offshore scenarios, DTs accurately estimate the dynamic behavior of ship-connected systems, such as risers on spread-moored vessels [10]. At the operational level, DTs enable

M. A. b. Jamlos et al. (Eds.): ICSET 2025, LNNS 1897, pp. 611–618, 2026.
https://doi.org/10.1007/978-3-032-20722-7_79

real-time monitoring of marine diesel generators [11], facilitate virtual-real mapping for autonomous test vessels [9], and provide advanced frameworks for path planning in Autonomous Maritime Surface Vessels [13]. Furthermore, DTs can unify heterogeneous data to address multi-physics challenges, as demonstrated in naval ship operation models [14]. Despite these promising developments in digital design and condition-based maintenance, critical challenges regarding data governance, interoperability, and verification persist [8].

In response to this evolving landscape, the present study proposes a preliminary domain-specific bibliometric analysis of DT applications in the maritime sector. Over recent years, bibliometric and systematic literature reviews have been widely adopted across various domains to quantitatively map the evolution of scientific research, identify thematic trends, and support evidence-based decision-making [15–17]. Several studies have focused on technology-driven fields, including cobots, AI, cyber-physical systems, and other Industry 4.0 enabling technologies [18–20]. These analyses have proven instrumental in revealing research trajectories, innovation patterns, and emerging application areas within highly dynamic and interdisciplinary contexts. As the application of DT technologies in the maritime sector continues to expand, the complexity and interdisciplinarity of the field call for structured bibliometric investigation to synthesize existing research, identify emerging research areas, and support informed decision-making in both academic and industrial contexts. A bibliometric analysis offers a structured and quantitative overview of the research landscape, including publication trends, geographical and institutional distribution of contributions, and collaboration networks among authors, countries, and organizations. Moreover, the analysis of keywords and their co-occurrence patterns enable the identification of dominant research themes, emerging clusters, and a preliminary understanding of areas of growing interest. These insights provide a robust foundation for assessing the current state of the art, uncovering potential research gaps, and guiding future methodological and application-oriented developments in the field. Specifically, this paper provides a rigorously and structured mapping of DT applications in the maritime domain.

The remainder of the paper is structured as follows. Section 2 outlines the main stages of the implemented methodology, while Section 3 presents the results and the research mapping derived from the bibliometric analysis. Section 4 discusses the key findings, draws conclusions, and highlights potential directions for future research.

2 Methodology

This study employs a systematic bibliometric approach to investigate the scientific landscape of DT applications in the maritime domain. The methodology comprises four sequential phases: (i) database selection and query definition, (ii) data collection and refinement, (iii) bibliometric analysis, and (iv) interpretation of the results (Fig. 1).

Phase 1—Database Selection and Query definition: Scopus was selected as the data source due to its extensive coverage of peer-reviewed scientific publications in engineering and technology fields. The search query was carefully designed to capture relevant publications by combining keywords such as "Digital Twin", "Naval", "Maritime", "Ship", along with their variants, using Boolean operators. The search was restricted to journal articles, conference papers, and review papers written in English.

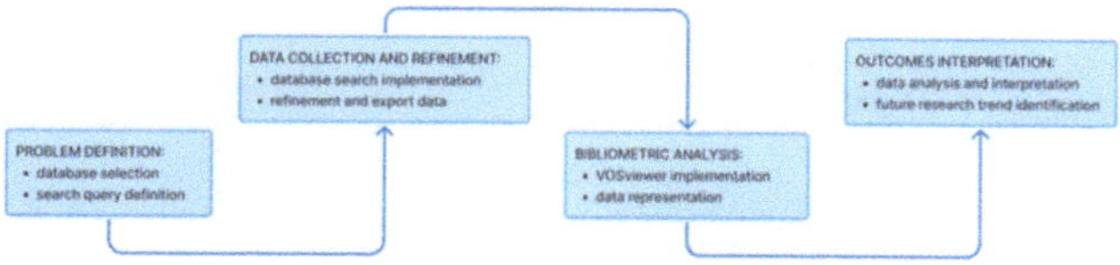

Fig. 1. Methodological framework

Phase 2—Data Collection and Refinement: The search was conducted in March 2025, yielding an initial set of 257 documents published between 2011 to 2025. A manual screening process was carried out to remove duplicate and publications outside the maritime context. The final dataset, exported in CSV format, included metadata such as authors, titles, abstracts, keywords, and citation counts.

Phase 3—Bibliometric Analysis Using VOSviewer: The refined dataset was analysed using VOSviewer (version 1.6.20) to generate quantitative insights. Three types of bibliometric mapping were performed: (1) co-authorship analysis, to identify collaboration networks; (2) keyword co-occurrence analysis, to reveal dominant thematic areas and emerging research trends; and (3) bibliographic coupling, to explore intellectual linkages among documents. In order to focus on impactful contributions, the analysis included only keywords with a minimum of five occurrences and documents with at least ten citations.

Phase 4—Interpretation and Thematic Clustering: The visual maps and network clusters produced by VOSviewer were interpreted to identify key research themes, emerging fronts, and knowledge gaps. Color-coded clusters facilitated thematic classification, while node size and link strength provided insight into the influence and connectivity of authors, institutions, and keywords.

3 Research Mapping Through Bibliometric Analysis

To map the global research landscape concerning the application of DT in the maritime sector, a structured bibliometric analysis was performed using a stepwise methodology, as illustrated in Fig. 2. This approach facilitates the identification of the most relevant studies on the topic, thereby enhancing the accuracy and overall quality of the bibliometric dataset used in the analysis. This section discusses the results obtained from the surface pressure measurement study. The effects of angle of attack, Reynolds number and leading edge bluntness are discussed in the next sub section.

Figure 3 illustrates the growing interest over time in the application of DT technology within the maritime domain. The earliest contribution date to 2019, with only two publications. From 2019 to 2023, scientific output showed steady yet moderate growth, marked by gradual annual increases. A significant rise is observed in 2024, with 90 publications, more than double the number published in 2023, marking the highest annual output in the period examined. Although data for 2025 remain incomplete, the trend indicates a publication volume exceeding all years prior to 2024, further confirming the ongoing expansion of research activity in this field.

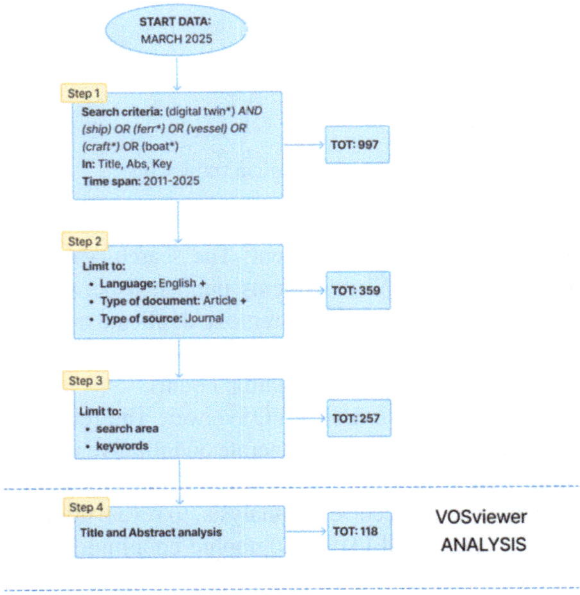

Fig. 2. Step-by-step procedure for conducting the bibliometric analysis

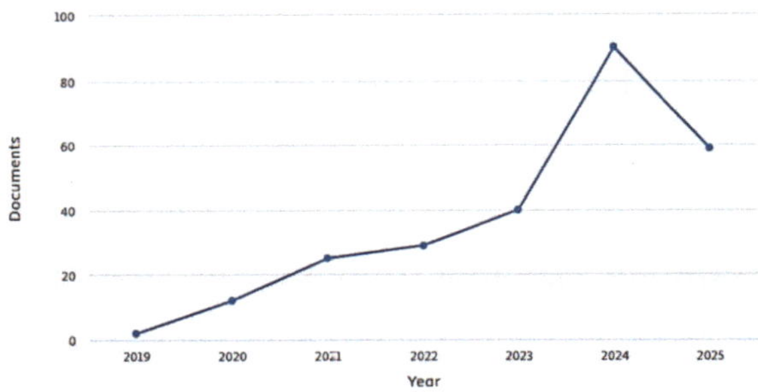

Fig. 3. Document trend by years

Figure 4 presents the geographical distribution of scientific output, focusing on the ten most productive countries in the field. The data clearly highlight China as the predominant contributor, with a publication volume nearly double that of the second-ranking country, the United States. This dominance reflects China's strategic strategic investment in maritime digitalization. Beyond the leading two countries, there is a sharp decline in publication volume, with Denmark, Greece, and Hong Kong occupying the lowest positions within the top ten, each contributing eight documents. These results indicate a highly asymmetrical global research landscape, with China and the United States collectively dominating the field.

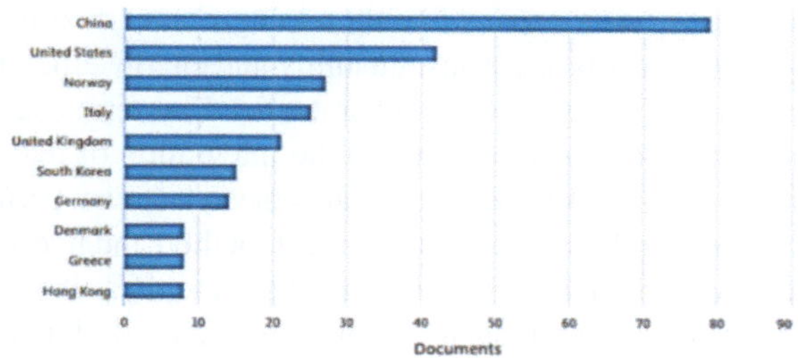

Fig. 4. Country trend by years

To ensure thematic consistency, the dataset obtained in Step 3 was further refined through a careful reading of titles and abstracts in step 4. This process led to the selection of 118 articles, which were subsequently analyzed using VOSviewer (https://www.vosviewer.com/). This software enabled the generation of advanced visualizations, providing a deeper understanding of collaboration dynamics and thematic focus areas. The key findings are summarized below.

The network visualization presented in Fig. 5 not only highlights the most active countries but also, unlike Fig. 3, reveals the nature and strength of their international partnerships. China once again emerges as the dominant leader, as evidenced by its large node size and extensive collaborative links, positioning it at the core of the network. The United States also plays a crucial role as a major hub, demonstrating substantial international engagement. Countries such as Germany and the United Kingdom act as strategic bridges, linking different regional clusters and facilitating cross-regional knowledge exchange. Within Europe, Italy emerges as the leader of the red cluster, maintaining a close collaborative relationship with the Czech Republic. Similarly, Norway leads a northern cluster that includes Germany, Poland, and Slovakia, illustrating a well-structured regional grouping. At the periphery, countries like Croatia, Slovakia, and Poland exhibit fewer and weaker international connections, indicating more localized research activity and limited integration into the broader global network (Fig. 5).

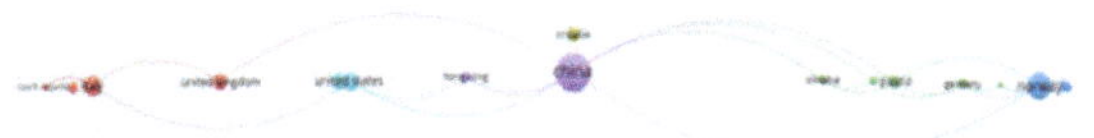

Fig. 5. Country collaboration network

Figure 6 maps the co-occurrence of research keywords, offering a thematic overview of the literature. At the core lies the term DT, which acts as the central hub connecting multiple research directions, particularly in relation to ships and real-time systems. The spatial proximity and frequency of co-occurring terms allow for the identification of several distinct yet interconnected research clusters. A first thematic area revolves around optimization and maintenance, with keywords such as reliability, life cycle, and shipping maintenance, highlighting the operational focus of DT technologies in extending asset performance and reducing downtime. A second cluster emphasizes marine propulsion and emissions, where terms like diesel engines, carbon dioxide, and combustion reflect ongoing concerns about sustainability and energy efficiency. Closely tied to this is the cluster on AI and prediction, characterized by terms such as neural networks, forecasting, and algorithms, pointing to the integration of data-driven methods for enhanced decision-making. Another significant research stream pertains to modeling and simulation, with keywords such as finite element method and hydrodynamic model, underscoring the importance of virtual testing and digital replicas in maritime engineering. Finally, a cluster around smart industry themes, Industry 4.0, shipyards, and additive manufacturing, indicates a growing intersection between maritime innovation and the broader shift toward intelligent, connected industrial systems. Overall, the keyword co-occurrence map confirms that DT acts as a central node linking engineering design, AI, and sustainability, demonstrating its pivotal role in shaping the digital transformation of the maritime domain.

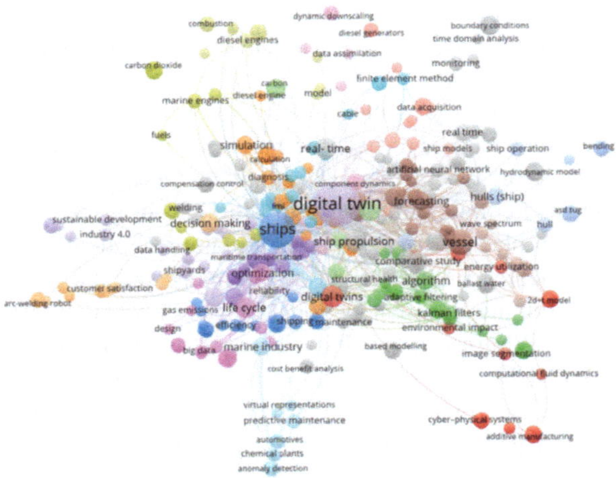

Fig. 6. Keyword co-occurrence network

4 Discussions and Conclusions

This study provided a structured mapping of the current state of the art on Digital Twins (DTs) in the maritime domain. The analysis is based on keyword co-occurrence, co-authorship, and bibliographic coupling using a final dataset of 118 publications processed with VOSviewer. Findings indicate a marked acceleration in scientific output starting in 2021, with a peak in 2024 (90 publications). Although data for 2025 are partial, the continued increasing trend suggests a highly dynamic and maturing field. In terms of geographical distribution, China leads in publication volume, followed by the United States. At the institutional level, the Norwegian University of Science and Technology (NTNU) emerges prominently. The collaboration network reveals influential authors and their interconnections, highlighting knowledge flow across institutions. DTs emerge as a central paradigm integrating multiple research directions, including optimization and maintenance throughout the vessel life cycle; propulsion and emissions with emphasis on diesel engines and CO_2; AI and predictive modelling; high-fidelity simulation (e.g., FEM and hydrodynamic models); and broader Industry 4.0 applications, encompassing shipyards and advanced manufacturing. The convergence of IoT instrumentation, high-fidelity simulation, data assimilation, and AI methods positions the DTs as a unifying construct spanning design, production, operations, and maintenance. Despite the growing body of literature, important knowledge gaps remain, underscoring the necessity for further in-depth research and analysis. Existing frameworks are still sufficient to scale DTs beyond pilot applications. Moreover, significant efforts are required to achieve high-fidelity coupling between the DT and the physical system, as well as to address open challenges related to data governance, verification, and validation. The observed geographical concentration of research output and the asymmetry in the collaboration network outline key priorities for the next phase of scientific and industrial development.

References

1. Jamwal, A., Agrawal, R., Sharma, M., Giallanza, A.: Industry 4.0 technologies for manufacturing sustainability: A systematic review and future research directions. *Appl. Sci.* (Switzerland). **11**(12), 5725 (2021). https://doi.org/10.3390/app11125725
2. Aiello, G., Giallanza, A., Vacante, S., Fasoli, S., Mascarella, G.: Propulsion monitoring system for digitized ship management: Preliminary results from a case study. Procedia Manuf. **42**, 16–23 (2020). https://doi.org/10.1016/j.promfg.2020.02.018
3. Aiello, G., Giallanza, A., Mascarella, G.: Towards shipping 4.0. A preliminary gap analysis. Procedia Manuf. **42**, 24–29 (2020). https://doi.org/10.1016/j.promfg.2020.02.019
4. Modgil, S., Singh, R.K., Banerjee, M., Gupta, S.: Mapping industry 4.0 technologies for maritime logistics: perspectives from practitioners in India. Supply Chain Forum Int. J., 1–17 (2025). https://doi.org/10.1080/16258312.2025.2557788
5. Giallanza, A., Aiello, G., Marannano, G., Nigrelli, V.: Industry 4.0: Smart test bench for shipbuilding industry. Int. J. Interact. Des. Manuf. **14**, 1525–1533 (2020). https://doi.org/10.1007/s12008-020-00739-9
6. Giallanza, A., Aiello, G., Marannano, G.: Industry 4.0: Advanced digital solutions implemented on a close power loop test bench. Procedia Comput. Sci. **180**, 93–101 (2021). https://doi.org/10.1016/j.procs.2021.01.133
7. Fujikubo, M., et al.: A digital twin for ship structures—R&D project in Japan. Data-Cent. Eng. **5**, e7 (2024). https://doi.org/10.1017/dce.2024.3
8. Lv, Z., Lv, H., Fridenfalk, M.: Digital twins in the marine industry. Electronics. **12**(9), 2025 (2023). https://doi.org/10.3390/electronics12092025
9. Xia, G., et al.: Real-time digital twin of autonomous ships based on virtualphysical mapping model. Phys. Fluids. **36**(8), 087109 (2024). https://doi.org/10.1063/5.0222332
10. Jin, C., et al.: Real-time dynamic and structural behavior estimation of a steel lazy wave riser through finite-element-based digital twin and hull-motion sensor. Appl. Ocean Res. **150**, 104137 (2024). https://doi.org/10.1016/j.apor.2024.104137
11. Dong, X., Huang, J., Luo, N., Wenshan, H., Lei, Z.: Design and implementation of digital twin diesel generator systems. Energies. **16**(18), 6422 (2023). https://doi.org/10.3390/en16186422
12. Wei, P., Li, C., Jiang, Z., Wang, D.: Real-time digital twin of ship structure deformation field based on the inverse finite element method. J. Mar. Sci. Eng. **12**(2), 257 (2024). https://doi.org/10.3390/jmse12020257

13. Raza, M., Prokopova, H., Huseynzade, S., Azimi, S., Lafond, S.: Towards integrated digital-twins: An application framework for autonomous maritime surface vessel development. J. Mar. Sci. Eng. **10**(10), 1469 (2022). https://doi.org/10.3390/jmse10101469
14. Choi, J., Moon, S., Min, S.: Digital twin simulation modeling process with system dynamics: An application to naval ship operation. Int. J. Robust Nonlinear Control. **33**(16), 10136–10150 (2023). https://doi.org/10.1002/rnc.6893
15. Han, Y., Chu, L.: A systematic review and bibliometric analysis for maritime emergency management. J. Sea Res. **205**, 102585 (2025). https://doi.org/10.1016/j.seares.2025.102585
16. Fata, L., Manuela, C., Giallanza, A., La Scalia, G., Micale, R.: Toward acceptance of human-robot collaboration in industrial settings: a bibliometric and systematic literature review. Int. J. Adv. Manuf. Technol. **139**, 2139–2160 (2025). https://doi.org/10.1007/s00170-025-16039-z
17. Zhong, B., Yang, J., Chen, Y., Fang, H.: Research hotspots and development trends of virtual power plant in the energy transition: A bibliometric analysis. Energy Rep. **14**, 371–385 (2025). https://doi.org/10.1016/j.egyr.2025.06.023
18. Giallanza, A., La Scalia, G., Micale, R., La Fata, C.M.: Occupational health and safety issues in human-robot collaboration: State of the art and open challenges. Saf. Sci. **169**, 106313 (2024). https://doi.org/10.1016/j.ssci.2023.106313
19. Park, C.: Addressing challenges for the effective adoption of artificial intelligence in the energy sector. Sustainability. **17**(13), 5764 (2025). https://doi.org/10.3390/su17135764
20. Wang, L., Hsu, H.-H.: IoT technology in maritime logistics management: exploration of data analysis methods. Discov. Internet Things. **5**, 66 (2025). https://doi.org/10.1007/s43926-025-00167-9

Flexural Strength and Failure Time Analysis of Concrete with Silica Fume as Partial Cement Replacement

Dina Heldita, Muhammad Wihardi Tjaronge, and Muhammad Akbar Caronge(✉)

Hasanuddin University, Gowa 92171, South Sulawesi, Indonesia
ma-caronge@unhas.ac.id

Abstract. Flexural strength is a critical mechanical parameter in the design of rigid pavements and structural elements subjected to dominant bending stresses. In recent years, pozzolanic materials such as silica fume (SF) have attracted significant interest not only for enhancing mechanical performance but also for reducing Portland cement consumption and related carbon emissions. While the benefits of SF on compressive strength and durability are well documented, comprehensive studies addressing its influence on flexural behavior and time-to-failure characteristics remain limited. This study investigates the effect of SF as a partial cement replacement on the flexural strength and failure time of concrete. Concrete mixtures with 0% (Control), 7%, 10%, and 12% SF were prepared using PCC and a water–cement ratio of 0.33. Flexural tests were performed at 7, 28 and 90 days following ASTM C78/C78M. Maximum load and the corresponding time to peak load were recorded using a digital data acquisition system. The results demonstrate that 7% SF provides the most notable improvement in flexural strength, attributed to the micro-filler effect and pozzolanic formation of additional C–S–H, which densifies the microstructure and enhances the ITZ. Conversely, higher dosages (10%–12%) did not yield further gains and showed slight reductions due to increased water demand, potential SF agglomeration, and micro-void formation. Failure-time analysis revealed longer pre-peak durations at 7 days for mixtures containing 7%–10% SF, whereas at 28 and 90 days, peak load was reached more quickly due to matrix stiffening and a reduced fracture process zone.

Keywords: Silica Fume · Flexural Strength · Failure Time

1 Introduction

Flexural strength is a key mechanical parameter of concrete in the design of rigid pavements and other structural components exposed to predominant flexural loading [1]. Unlike compressive strength, flexural characteristics provide an indication of the ability of concrete to withstand indirect tensile stresses induced by bending moments, thus serving as an important indicator in evaluating resistance to initial cracking and the service life of structures [2]. In highway pavements, flexural strength values are used as the basis for determining the thickness of concrete slabs to ensure their capacity to resist repeated

M. A. b. Jamlos et al. (Eds.): ICSET 2025, LNNS 1897, pp. 619–628, 2026.
https://doi.org/10.1007/978-3-032-20722-7_80

traffic loads without premature failure [3]. A thorough comprehension of the flexural strength behavior of concrete, along with its failure time under loading, is therefore an essential element in material engineering and structural design.

In the last decade, the use of pozzolanic admixtures such as silica fume (SF) has gained increasing attention as a means of enhancing mechanical performance [4]. In addition to improving mechanical performance, SF also reduces dependence on Portland cement, which is known to contribute significantly to global CO_2 emissions [5]. Silica fume, a by-product of the ferrosilicon industry, possesses ultrafine particle size (0.1–0.2 μm) and high pozzolanic reactivity, which can generate additional C–S–H, improve microstructural density, and enhance the interfacial transition zone (ITZ) between the cement paste and the aggregate [6, 7]. Several studies have demonstrated that the addition of SF at an optimum proportion can enhance flexural strength, compressive strength, and concrete durability. Han et al. and Imam et al. reported that concrete containing silica fume exhibited improved mechanical performance [8, 9]. However, the implications for flexural strength characteristics and the relationship between load and failure time have not been consistently explained.

The research gap that arises is that most of the literature focuses on compressive strength or endurance aspect of silica fume based. Concrete, while in-depth investigations of flexural response and the relationship between peak load and failure time remain relatively limited (Bhanja & Sengupta [10]). In fact, in real structural applications, cracking phenomena and time to failure have direct implications for structural reliability, particularly under repeated or long-term loading conditions. Moreover, variations in SF replacement levels often produce nonlinear effects on mechanical properties: moderate dosages (7–10%) have been reported to significantly increase strength, while excessive dosages (>12%) may reduce workability, cause segregation, and weaken the structure due to the formation of micro-voids [2, 11].

Therefore, this study aims to comprehensively analyze the flexural strength of concrete incorporating silica fume as a partial cement replacement, while also evaluating the failure time characteristics until peak load is reached. This study is significant not only because it contributes fundamental insights into the mechanisms of flexural performance improvement by SF, but also because it expands the empirical database regarding the correlation between flexural strength and failure time, which is still seldom documented. These findings are expected to support the design of high-performance concrete for long-service-life structural applications while also contributing to the decarbonization efforts of the construction industry.

2 Methodology

2.1 Research Design

This study was conducted in the form of an experimental investigation in the laboratory with the objective of evaluating the influence of silica fume (SF) as a partial substitute for cement on the flexural strength of concrete as well as its failure time characteristics. The SF replacement levels were set at 0% (control), 7%, 10%, and 12% of the total cement weight. The selection of this range was based on previous studies reporting an improvement in mechanical performance at moderate dosages (5–10%), but a tendency to decline at higher dosages due to the reduction of effective cement and the formation of micro-porosity [2, 11].

2.2 Materials

This study employed Portland Composite Cement (PCC) as the primary binder, which complies with national standards. River sand functioned as the fine aggregate, while crushed stone with a maximum size of 20 mm served the coarse aggregate. The partial replacement material, silica fume, was obtained as a by-product of the ferrosilicon industry [9]. This material is in the form of fine powder and is predominantly composed of reactive silica ($SiO_2 > 90\%$) [12]. Mixing water was sourced locally and met the requirements for structural concrete.

2.3 Mix Design

The concrete mix was prepared using the trial-and-error method using a water and cement ratio of 0.33. The mix design calculations considered the partial substitution of cement with silica fume according to the specified percentages. The mix proportions are presented in Table 1.

Table 1. Concrete Mix Design

Materials	Mixture			
	C	7%SF	10%SF	12%SF
Water	162.83	162.09	161.78	161.57
Cement	493.31	456.70	441.11	430.75
Silica Fume	0.00	34.38	49.01	58.74
Fine Aggregate	789.99	786.41	784.88	783.87
Coarse Aggregate	965.59	961.20	959.34	958.10
Superplasticizer	3.95	3.93	3.92	3.92

2.4 Specimen Preparation

Beam specimens measuring 100 mm × 100 mm × 400 mm were prepared for flexural testing in compliance with ASTM C78/C78M-22 [13]. The mixing process was carried out using a pan mixer until a homogeneous mixture was achieved, after which the fresh concrete was poured into molds and compacted with a table vibrator. Following 24 h, the specimens were removed from the molds and subsequently cured in a water tank maintained at 23 ± 2 °C until the specified testing ages of 7, 28, and 90 days.

2.5 Flexural Strength Test

The flexural strength test was conducted following ASTM C78/C78M-22 [13]. A Universal Testing Machine (UTM) with a capacity of 1000 kN was employed to apply the load incrementally. The test arrangement for flexural strength is illustrated in Fig. 1. During testing, load data (kN) and the time to reach peak load were recorded using a data acquisition system. Flexural strength was calculated using the following equation:

$$F_r = \frac{PL}{bd^2}$$

In this equation, Fr represents the flexural strength expressed in N/mm^2, while P denotes the maximum applied load in Newtons. The variable L refers to the span length in millimeters, B indicates the specimen width, and D corresponds to the specimen depth.

Fig. 1. Flexural strength test setup

2.6 Failure Time Analysis

This study also assessed failure time, defined as the time interval required for the specimen to reach maximum load from the beginning of loading. Jureje et al. Reported that one of the qualitative approaches to evaluate failure mechanisms is the analysis of loading response time behavior, as outlined in the scientific literature [14]. This parameter was automatically recorded by the testing machine data system in seconds (s). The measurement of failure time provides an overview of the rate of initial crack formation, crack propagation, and ultimate collapse of the concrete. This analysis is important to evaluate the extent to which the incorporating of silica fume (SF) influences the resistance of concrete to the rate of loading, as well as its relation to ductility behavior and crack resistance under flexural conditions.

3 Result

3.1 Flexural Strength

Figure 2 presents the outcomes of flexural strength test of concrete with various mixture compositions, showing a trend consistent with the pozzolanic mechanism and the micro-filler effect of silica fume (SF). In the control mixture without SF, the flexural strength value remained at a baseline level, reflecting the primary role of Portland cement hydration in forming the binding paste. However, the addition of 7% SF demonstrated a significant increase in flexural strength. This phenomenon confirms that concrete incorporating silica fume can function as an SCM. This can also be explained by two main mechanisms: the micro-filler effect of fine SF particles that fill capillary pores and refine the microstructure, and the pozzolanic reaction between SF and calcium hydroxide producing additional calcium silicate hydrate (C–S–H) [8]. The microstructural densification resulting from these mechanisms improves the interfacial transition zone (ITZ) between the cement paste and aggregates, which forms a more complex crack propagation path and raises the energy needed for crack initiation and growth [15]. This condition explains why flexural strength at a moderate SF dosage, specifically the 7% replacement level in this study, tends to show a marked increase compared to the control, as also reported in recent experimental studies published in reputable international journals by Zhao et al. and Hong et al. [6, 7].

Furthermore, when the SF dosage was increased to 10% and 12%, flexural strength no longer showed improvement and even tended to decrease compared to the 7% SF level. This phenomenon can be attributed to several factors, including increased specific surface area resulting in higher water demand, the risk of SF particle agglomeration due to suboptimal dispersion, and the potential formation of micro-voids that weaken the internal structure of concrete [11]. Thus, although theoretically the addition of SF continues to increase the amount of C–S–H, in practice, at higher levels some SF particles may act merely as inert fillers without contributing to chemical strengthening. This aligns with the findings of [2, 11], which state that excessive SF dosage can reduce workability, increase the risk of segregation, and produce concrete with unstable mechanical properties.

Overall, the pattern observed in the graph confirms that the optimal SF dosage for improving flexural strength is below 10%, specifically 7% SF in this study. This level of cement replacement with SF not only provides a balance between microstructural densification and workability but also aligns with various previous studies reporting that mechanical strength improvements, including flexural strength due to SF, tend to reach an optimum at moderate dosages, while excessive use is counterproductive. The practical implication of these findings is the importance of controlling the proportion and quality of SF dispersion in the concrete mix, including the use of polycarboxylate-based superplasticizers (PCE) to address workability issues. With an appropriate mix design strategy, SF at the optimum dosage not only enhances flexural strength but also improves the long-term resistance of concrete against cracking under repeated flexural loading, thereby supporting the application of sustainable concrete in modern construction.

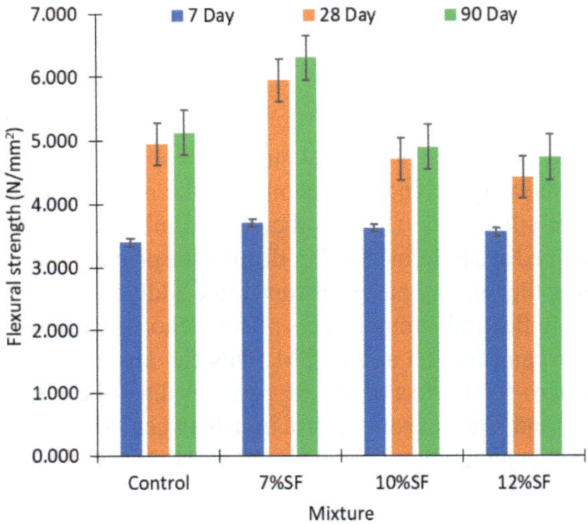

Fig. 2. Flexural Strength (7, 28, 90 days)

3.2 Time to Reach Peak Flexural

Figure 3 shows the time required for each concrete mixture to reach peak flexural strength. At the age of 7 days, mixtures with 7–10% silica fume (SF) generally required more time to reach peak flexural load compared to the control because two early mechanisms worked simultaneously: the micro-filler effect of ultrafine SF particles filled capillary pores and improved the ITZ, delaying crack coalescence, while the pozzolanic reaction, which had just begun, produced secondary C–S–H that had not yet fully maximized stiffness, this combination prolonged the pre-peak phase before failure. Recent experimental reviews indicate that SF dosage around 10% densifies the microstructure and enhances mechanical/crack resistance properties, namely by increasing tensile/splitting strength and fracture energy, which causes the load–time curve to rise for a longer duration at early ages before reaching the peak. Findings on ITZ densification and increases in modulus/fracture parameters with SF are consistent with previous studies on concrete [11, 16].

With the progression of curing age to 28 and 90 days, the pozzolanic reaction of SF became more pronounced, leading to the generation of greater amounts of secondary C–S–H, in which the matrix and ITZ became stiffer and the FPZ shrank, producing a steeper initial response and causing the peak to be reached more quickly (shorter time to peak) despite the higher peak strength. From the perspective of fracture mechanics, the increase in stiffness and the shrinkage of the FPZ raised brittleness [17]. This phenomenon has been reported for SF concrete and high-strength concrete. SF increases strength but accelerates the transition to crack localization, thereby shortening the pre-peak duration in three-point flexural testing under constant loading rates. Thus, the reduction in time to peak from 28 to 90 days, as shown in the graph, is a direct manifestation of the higher modulus and smaller FPZ (Fracture Process Zone) in matured SF concrete.

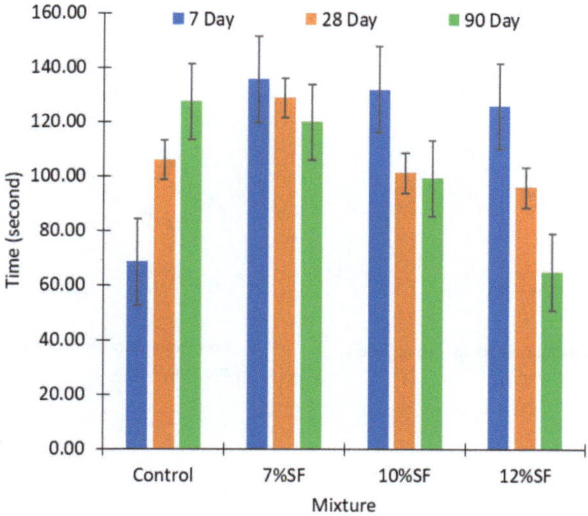

Fig. 3. Loading Time to Reach Peak Flexural Strength (7, 28, 90 days)

3.3 Relationship Between Flexural Strength and Time

Figure 4 shows the relationship between flexural strength and loading time (7, 28, and 90 days). The "flexural strength–time" graph illustrates that the addition of silica fume (SF), particularly at 7% and 10%, increases peak flexural strength and at 7 days tends to extend the peak load phase compared to the control. Mechanistically, ultrafine silica fume particles play a dual role in improving the properties of concrete. First, through the micro-filler effect, these particles fill capillary pores, resulting in a denser matrix and a more homogeneous interfacial transition zone (ITZ) between cement paste and aggregates. Second, the pozzolanic property of SF enables its reaction with calcium hydroxide ($Ca(OH)_2$), a product of cement hydration, to form secondary C–S–H. The combination of these mechanisms strengthens the internal structure of concrete and enhances its long-term durability. This combination delays crack coalescence, allowing the load curve to rise for a longer duration before failure. Many previous studies have reported that cement replacement with SF up to 10% effectively increases modulus/stiffness and flexural strength through microstructural densification, consistent with the increase in C–S–H observed at early to intermediate ages [9, 18, 19].

At 28 and 90 days, SF-containing mixtures in the graph reached peak load more quickly, although greater amounts of secondary C–S–H were present, leading to higher elastic modulus and a smaller fracture process zone (FPZ). The material became stiffer and more abrupt, accelerating the transition to peak load under the same loading protocol, a typical consequence of higher-strength concrete according to fracture mechanics/size-effect theory. At 12% SF, the shortening of time tended to be the most pronounced and variable due to the risk of agglomeration or inadequate dispersion in densified SF (HDSF), which reduces effective reactivity and introduces micro-defects that trigger early crack localization. Therefore, the control of superplasticizer becomes crucial when the SF dosage is increased [20].

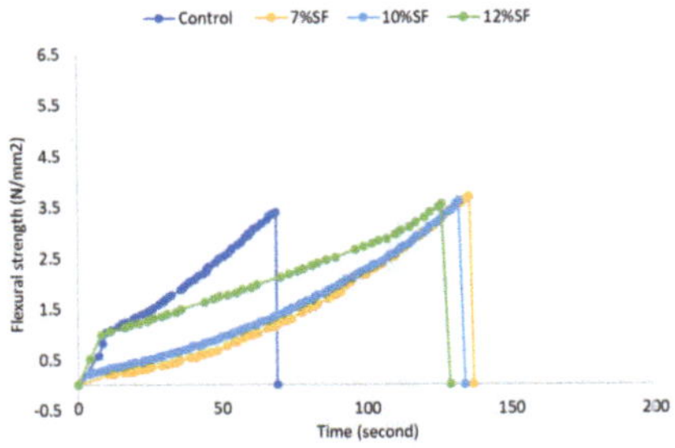

(a) Relationship of flexural strength to time at 7 days

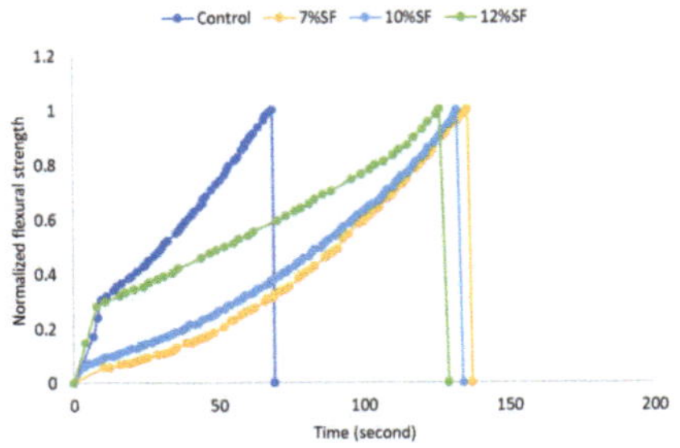

(b) Correlation between Normalized flexural strength and time 7 days

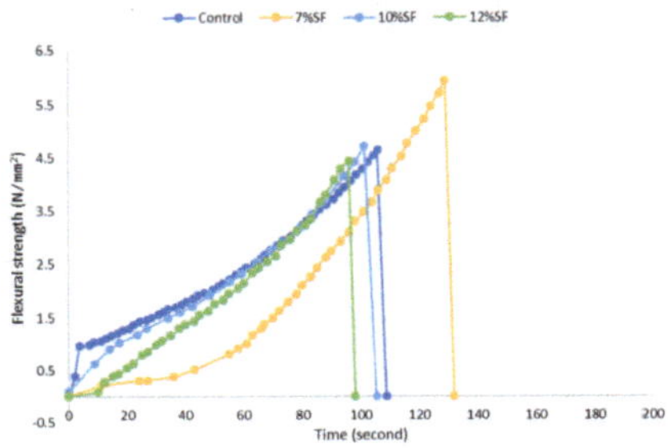

(c) Relationship of flexural strength to time at 28 days

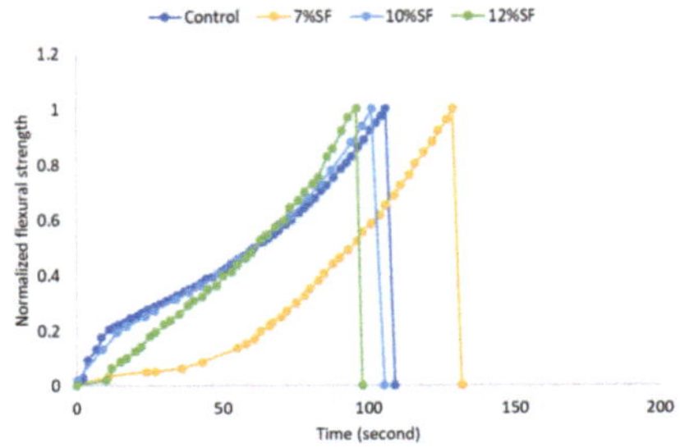

(d) Correlation between Normalized flexural strength and time 28 days

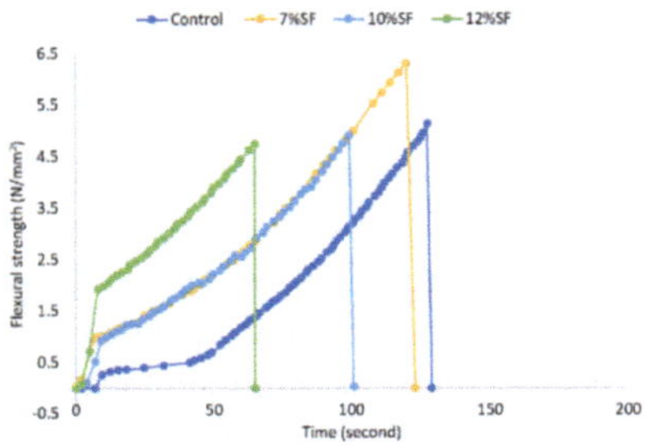

(e) Relationship of flexural strength to time at 90 days

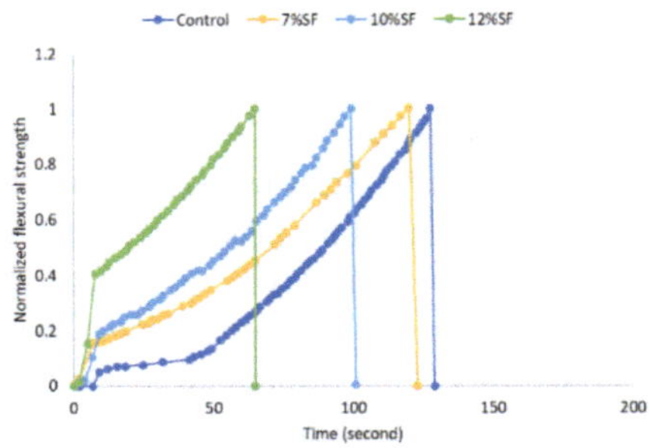

(f) Correlation between Normalized flexural strength and time 90 days

Fig. 4. Correlation between flexural strength and time (7, 28, 90 days)

4 Conclusions

The incorporation of 7% SF was found to be the most effective in enhancing the flexural strength of concrete. This strengthening mechanism originates from the combination of the micro-filler effect and the pozzolanic reaction that produces secondary C–S–H, leading to a more compact microstructure, an improved ITZ, and a more intricate crack propagation path. SF substitution above 10% did not provide significant improvement in flexural strength and even tended to reduce performance due to higher water demand, the potential for agglomeration, and the formation of micro-voids.

At 7 days, mixtures with 7–10% SF prolonged the pre-peak phase before failure, delaying crack coalescence. However, at 28–90 days, the more advanced pozzolanic reaction stiffened the matrix and reduced the FPZ, causing the peak load to be reached more quickly even though flexural strength values increased. This indicates a tendency toward greater brittleness with age.

Acknowledgments. The author gratefully extends appreciation to the laboratory staff and technical team for the essential support provided throughout the experimental process. Special acknowledgment is also given to colleagues and fellow researchers who have shared valuable insights, offered constructive suggestions, and provided continuous motivation during the course of this study. Their dedication has substantially enriched the scientific quality of this research.

(Note: "This research was not funded by any grant")

References

1. Salama, M.I., Elayat, A., Reda, M., Elsamak, G.: Influence of concrete type on rigid pavement behavior under static loads. Innov. Infrastruct. Solut. **9**(1), 1–11 (2024). https://doi.org/10.1007/s41062-023-01316-1
2. Aïtcin, P.C.: Supplementary cementitious materials and blended cements. In: Science and Technology of Concrete Admixtures, pp. 53–73. Elsevier Ltd (2016). https://doi.org/10.1016/B978-0-08-100693-1.00004-7
3. Ardana, M.D.W., Ariawan, I.M.A.: The significance of concrete slab flexural strength inference variation based on its compression strength characteristics in apron pavement analysis and design. MATEC Web of Conferences. **276**, 01038 (2019). https://doi.org/10.1051/matecconf/201927601038
4. Panjehpour, M., Abdullah, A., Ali, A., Demirboga, R.: A review for characterization of silica fume and its effects on concrete properties. Int. J. Sust. Constr. Eng. Technol. **2**(2), 1–7 (2011)
5. Scrivener, K.L., John, V.M., Gartner, E.M.: Eco-efficient cements: Potential economically viable solutions for a low-CO2 cement-based materials industry. Cem. Concr. Res. **114**, no. February, 2–26 (2018). https://doi.org/10.1016/j.cemconres.2018.03.015
6. Zhao, S., Zhang, Q.: Effect of silica fume in concrete on mechanical properties and dynamic behaviors under impact loading. Materials. **12**(19), 3263 (2019). https://doi.org/10.3390/ma12193263
7. Hong, G., Oh, S., Choi, S., Chin, W.J., Kim, Y.J., Song, C.: Correlation between the compressive strength and ultrasonic pulse velocity of cement mortars blended with silica fume: An analysis of microstructure and hydration kinetics. Materials. **14**(10), 2476 (2021). https://doi.org/10.3390/ma14102476
8. Han, C., Pang, J., Hu, S., Yang, C.: Study on static and dynamic mechanical properties and microstructure of silica fume-polypropylene fiber modified rubber concrete. Sci. Rep. **14**(1), 1–18 (2024). https://doi.org/10.1038/s41598-024-63341-z
9. Imam, A., Kumar, V., Srivastava, V.: Review study towards effect of Silica Fume on the fresh and hardened properties of concrete. Adv. Concr. Constr. **6**(2), 145–157 (2018). https://doi.org/10.12989/acc.2018.6.2.145
10. Bentz, D.P., Hansen, A.S., Guynn, J.M.: Optimization of cement and fly ash particle sizes to produce sustainable concretes. Cem. Concr. Compos. **33**(8), 824–831 (2011). https://doi.org/10.1016/j.cemconcomp.2011.04.008

11. Hamada, H.M., et al.: Effect of silica fume on the properties of sustainable cement concrete. J. Mater. Res. Technol. **24**, 8887–8908 (2023). https://doi.org/10.1016/j.jmrt.2023.05.147
12. Sounthararajan, V.M., Srinivasan, K., Sivakumar, A.: Micro filler effects of silica-fume on the setting and hardened properties of concrete. Res. J. Appl. Sci. Eng. Technol. **6**(14), 2649–2654 (2013). https://doi.org/10.19026/rjaset.6.3753
13. Standard test method for flexural strength of concrete (using simple beam with third-point loading). In: ASTM International: West, vol. 13, no. Reapproved, pp. 5–7, Conshohocken, PA, USA (2022). https://doi.org/10.1520/C0078
14. Jureje, U., Tjaronge, M.W., Caronge, M.A.: Basic engineering properties of concrete with refractory brick as coarse aggregate: Compressive stress-time relationship assessment. Int. J. Eng., Trans. B: Appl. **37**(5), 931–940 (2024). https://doi.org/10.5829/ije.2024.37.05b.11
15. Li, D., Jin, L., Du, X., Liu, J.: A theoretical method to determine the tortuous crack length and the mechanical parameters of concrete in direct tension—A particle size effect analysis. Eng. Fract. Mech. **197**, 128–150 (2018). https://doi.org/10.1016/j.engfracmech.2018.04.039
16. Smarzewski, P.: Influence of silica fume on mechanical and fracture properties of high performance concrete. Procedia Struct. Integr. **17**, 5–12 (2019). https://doi.org/10.1016/j.prostr.2019.08.002
17. Golewski, G.L.: Mechanical properties and brittleness of concrete made by combined fly ash, silica fume and nanosilica with ordinary Portland cement. AIMS Mater Sci. **10**(3), 390–404 (2023). https://doi.org/10.3934/matersci.2023021
18. Amudhavalli, N.K., Mathew, J.: Effect of silica fume on strength and durability parameters of concrete. Int. J. Eng. Sci. Emerg. Technol. **3**(1), 28–35 (2012)
19. Uzbas, B., Aydin, A.C.: Microstructural analysis of silica fume concrete with scanning electron microscopy and X-Ray diffraction. Eng. Technol. Appl. Sci. Res. **10**(3), 5845–5850 (2020). https://doi.org/10.48084/etasr.3288
20. Šoukal, F., et al.: The effects of silica fume and superplasticizer type on the properties and microstructure of reactive powder concrete. Materials. **16**(20), 6670 (2023). https://doi.org/10.3390/ma16206670

FRP Prestressed Concrete Beams Based on FEM Simulation

Asri Mulya Setiawan(✉), Rudy Djamaluddin, Fakhruddin, and Muh. Fajrul Miftah R

Department of Civil Engineering, Faculty of Engineering, Hasanuddin University, 92171 Gowa, South Sulawesi, Indonesia
klanmulyasetiawan@gmail.com

Abstract. The use of the Finite Element Method in solving structural modeling is one way to complete an analysis by taking into account the nonlinear properties of materials, the relationship between concrete and steel reinforcement, and to produce simulations of failures such as cracks. This study aims to provide insight into structural engineering using the Finite Element Method and to determine the use of CFRP strip prestressing in increasing the capacity of reinforced concrete beams. The research was conducted using FEM-based applications by comparing three models, namely BN (non-FRP reinforced concrete beams), BFRP (reinforced concrete beams with CFRP strip reinforcement), and BPFRP (reinforced concrete beams with prestressed CFRP strip reinforcement). Results. Modeling using the Finite Element method showed a significant comparison to the modeling performed, where in the BN beam modeling (non-reinforced reinforced concrete beam), it was found that the maximum load was 271.243 kN, and in the BFRP modeling (reinforced concrete beams with CFRP strip reinforcement), the maximum load was 316.961 kN, an increase of 16.85%. Then, in the BPFRP modeling (reinforced concrete beams with prestressed CFRP strip reinforcement), the maximum load was 408.782 kN. There was an increase in the beam's ability to withstand maximum loads, which was 50.70% compared to the BN model and 28.97% compared to the BFRP model. Modeling using the Finite Element Method (FEM), which was carried out by adding external reinforcement, namely prestressing on a CFRP, greatly affected the maximum load value, the concrete strain value, and also the value of the steel strain.

Keywords: Reinforced Concrete Beams · CFRP STrip · Finite Element Method · FRP Prestressed

1 Introduction and Theoretical Approach

One method that can be used to improve the strength of reinforced concrete is to use reinforcement techniques. The use of reinforcement in reinforced concrete beams has been proven effective in increasing the load capacity and service life of structures. There are many recent innovations in this field, such as concrete jacketing, steel jacketing, or fiber-reinforced polymer (FRP) reinforcement [1]. FRP is known for its high strength and light weight. FRP reinforcement can take the form of FRP strips installed on the surface of concrete beams to increase flexural and shear strength.

M. A. b. Jamlos et al. (Eds.): ICSET 2025, LNNS 1897, pp. 629–636, 2026.
https://doi.org/10.1007/978-3-032-20722-7_81

The installation of FRP layers is based on the type of load applied. Application on beams is often done on the underside, where it is intended to withstand flexural loads. Meanwhile, the use of FRP on the sides is intended to withstand shear loads, so that the installation position of the FRP is adjusted as needed [2].

The Finite Element Method (FEM) is one of the numerical methods that utilizes matrix operations to solve physical problems. Another method is the analytical method, which requires a mathematical equation that models physical behavior.

The research conducted by [3] aims to expand the understanding of the performance of reinforced concrete beams reinforced with prestressed FRP sheets through an experimental approach and numerical simulation based on the finite element method. The results show that there is an increase in parameters related to beam performance as long as the bond quality between the FRP and the beam can be maintained. Conversely, if the bond is inadequate, there is a risk of accelerating delamination.

1.1 Carbon Fiber Reinforced Polymer (CFRP) Strip

Carbon Fiber Reinforced Polymer (CFRP) strips are lightweight composite materials with very high tensile strength that are widely used to reinforce reinforced concrete structural elements.

Another study conducted by [4] shows that CFRP strips are effective in increasing the flexural strength and stiffness of reinforced concrete beams, even when subjected to repeated loading. The study also confirms that CFRP strips are able to maintain long-term structural performance, making them a reliable reinforcement solution for structures subjected to cyclic or dynamic loads (Fig. 1; Table 1).

Fig. 1. Carbon Fiber Reinforced Polymer (CFRP) Strips

Table 1. Characteristics of CFRP (*Carbon Fiber Reinforced Polymers*) Strip

Characteristics CFRP STRIP	
Characteristics Materials	**Test design**
Standard Tensile Strength (ASTM D3039)	2800 MPa
Tensile Modulus (ASTM D3039)	1,6 × 10^5 Mpa
Maximum Strain (ASTM D3039)	1,6%

(*continued*)

Table 1. (*continued*)

Characteristics CFRP STRIP	
Characteristics Materials	**Test design**
Shear Strength (ASTM D2344)	80 Mpa
Flexural Strength (ASTM D7264)	1600 Mpa
Carbon Fiber Content (%)	$\geq$ 65%
Thickness	2.0 m

1.2 FRP Prestressed

The application of external forces with the aim of strengthening the value of a reinforced concrete structure is carried out on FRP (Prestressed FRP) [5]. The reinforcement carried out on reinforced beams with the application of prestressed FRP is able to reduce tensile stress on the internal reinforcement steel, thereby increasing the ability to withstand loads and reducing cracks and deflections in concrete.

[3] also applied prestressing to carbon fiber-reinforced polymer (CFRP) sheets at 20% to prevent early debonding and found that prestressed CFRP was more effective than non-prestressed CFRP.

1.3 Research Design

The dimensions and reinforcement of the beams were analyzed using the ultimate strength design method. There are three beam variations in this study with beam dimensions of 500 $\times$ 700 $\times$ 5500 mm with planned concrete quality of f'c 40 MPa and steel quality of fy 420 MPa (Fig. 2).

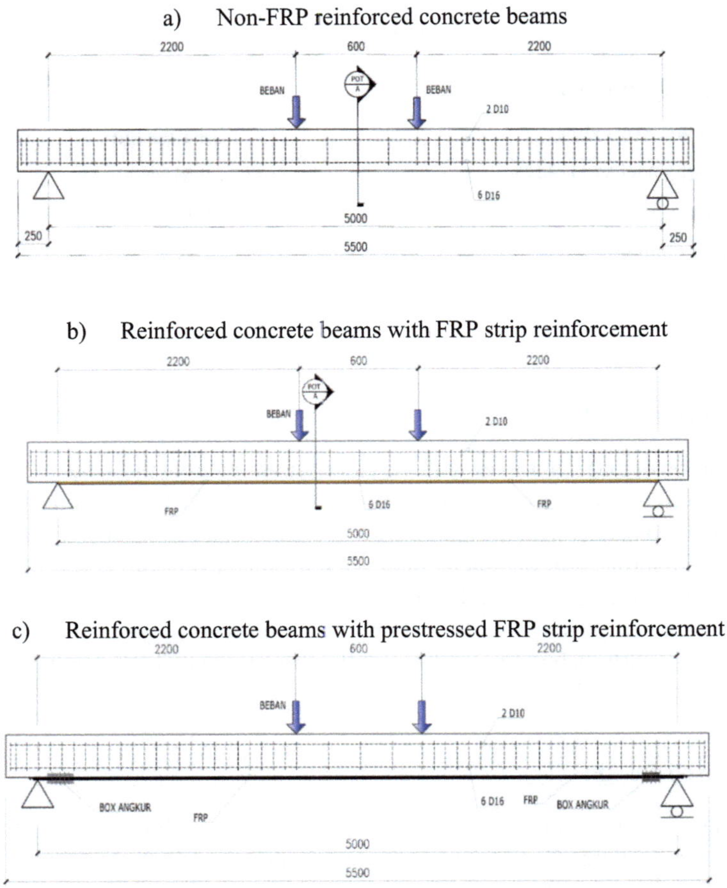

Fig. 2. Test Object Design

1.4 Result and Discussion

Load and deflection relationship (Fig. 3)

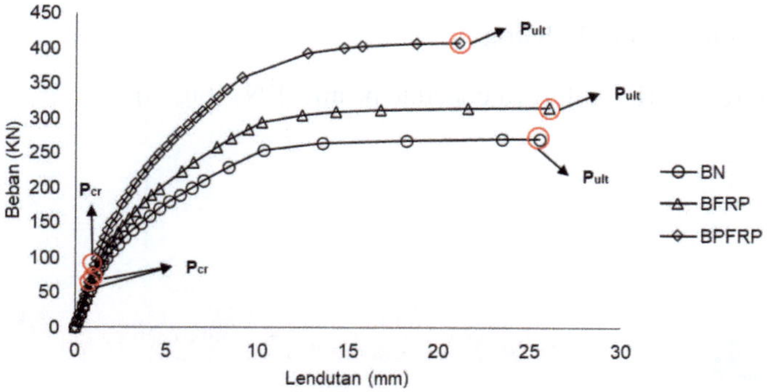

Fig. 3. Load-deflection composite graph

Load and Strain relationship (Fig. 4)

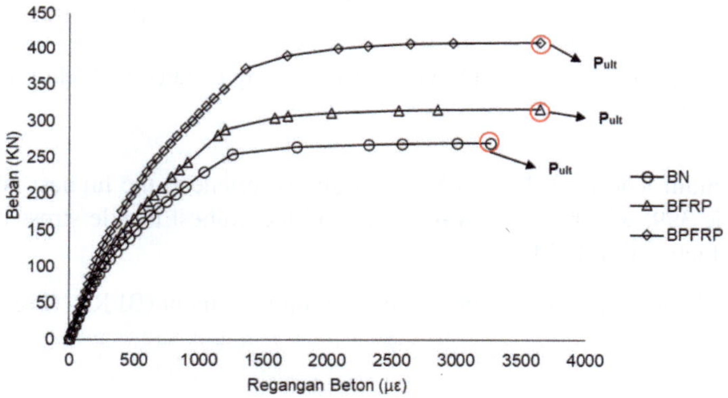

Fig. 4. Load-strain composite graph

Load and Strain Steel relationship (Fig. 5)

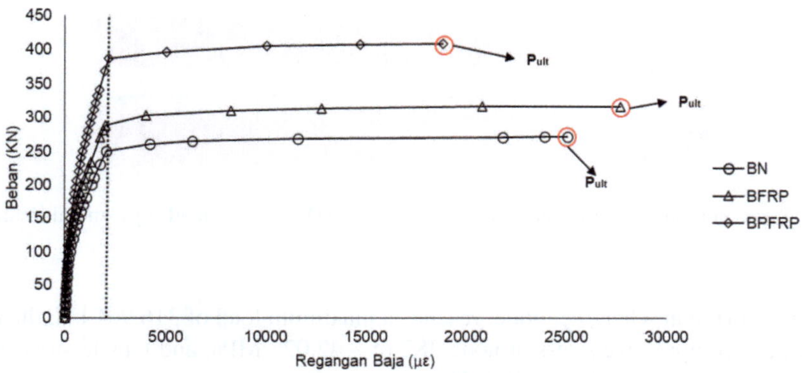

Fig. 5. Load-strain Steel composite graph

1.5 Stress Contours in Modeling

a) Non-reinforced FRP reinforced concrete beams (BN) (Fig. 6)

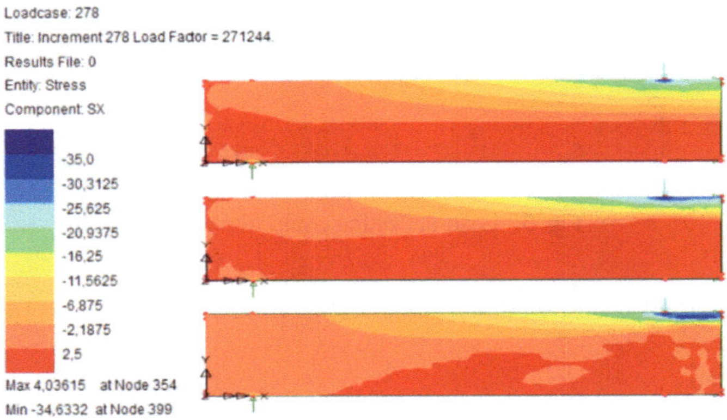

Fig. 6. Stress contours of BN Model beams under Cracking Load (Pcr), Yield Load (Py), and Ultimate Load (Pu)

At maximum load of 271.244 kN, the beam experienced the highest compressive stress at node 399, which was −34.63 MPa, and the highest tensile stress occurred at node 354, which was 4.03 MPa.

b) Reinforced concrete beams with CFRP strip reinforcement (BFRP) (Fig. 7)

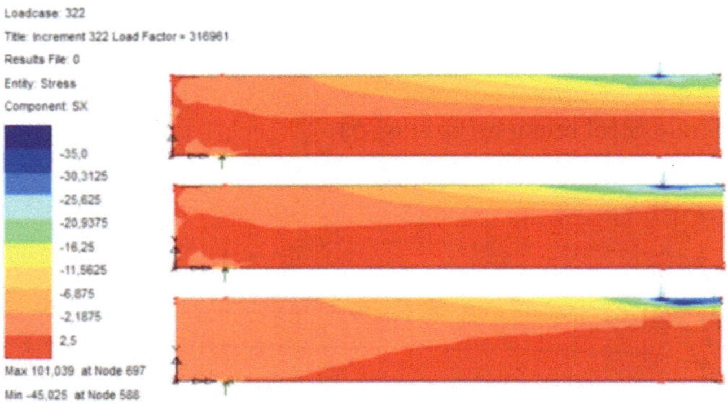

Fig. 7. Stress contours of BFRP beams under crack load (Pcr), yield load (Py), and ultimate load (Pu)

It is known that when the model reaches a maximum load of 316.961 kN, the beam experiences compressive stress at node 382 of −42.025 MPa, and tensile stress in the CFRP area at node 505 of 101.039 MPa.

c) Reinforced concrete beams with CFRP strip prestressing reinforcement (BPFRP)

The maximum tensile stress d with the modulus was 30.429 MPa at node 591, which indicates the effectiveness of prestressed CFRP in reducing the tensile stress of the concrete below (before cracking) and resisting deflection. Then, the compressive stress that occurred at node 53 was −40.18 MPa. (Fig. 8)

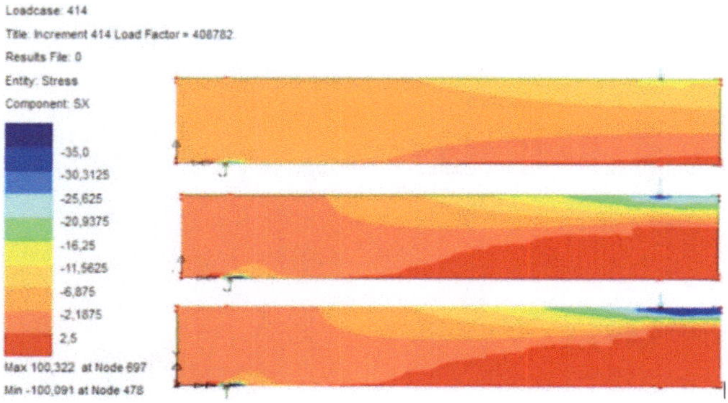

Fig. 8. Stress contours of BPFRP beams under crack load (Pcr), yield load (Py), and ultimate load (Pu)

2 Conclusions

The modeling results using the Finite Element method show a significant comparison to the modeling performed where in the BN beam modeling (non-reinforced reinforced concrete beam), it is known that the maximum load is 271.243 kN, and in the BFRP (reinforced concrete beam with CFRP strip reinforcement) modeling, the maximum load was 316.961 kN, an increase of 16.85%. Then, in the BPFRP (reinforced concrete beam with prestressed CFRP strip reinforcement) modeling, the maximum load was 408.782 kN. There was an increase in the beam's ability to withstand maximum loads, namely 50.70% compared to the BN model and 28.97% compared to the BFRP model.

References

1. Mansur, A. Z., Djamaluddin, R., Parung, H., & Irmawaty, R. (2022). Beban Ultimit Balok Beton Bertulang Hasil Perbaikan Dan Perkuatan Menggunakan FRP.
2. Harahap, S., Putri, P.Y., Putra, R.R., Andayono, T., Atika, L.: Perkuatan struktur beton dengan metode FRP pada bangunan gedung. Innov. J. Soc. Sci. Res. **4**(3), 9200–9214 (2024)
3. Nguyen, M.H., Ho, M.H., Tran, V.R., Phan, H.N., Nguyen, D.D.N., Huynh, P.N.: Flexural performance of reinforced concrete beams strengthened with prestressed FRP sheets: Experiments and numerical investigations. Case Stud. Constr. Mater. **21**, E03475 (2024). https://doi.org/10.1016/J.Cscm.2024.E03475

4. Askar, M.K., Hassan, A.F., Al-Kamaki, Y.S.S.: Flexural and shear strengthening of reinforced concrete beams using Frp composites: A state of the art. Case Stud. Constr. Mater. **17**, E01189 (2022). https://doi.org/10.1016/J.Cscm.2022.E01189
5. Moshiri, N. et al.: Flexural strengthening of RC slabs with nonprestressed and prestressed CFRP strips using EBROG method. Compos. Part B Eng. **201**, 108359 (2020). https://doi.org/10.1016/J.Compositesb.2020.108359

Practical Research on the Application of PLC Technology in Mechanical and Electrical Control Devices

Ge Kun(✉)

Department of Electrical and Electronic Engineering, Faculty of Engineering, Universiti Malaysia Sabah, 88400 Kota Kinabalu, Sabah, Malaysia
gekun@vip.qq.com

Abstract. With the continuous development of the machinery industry, the government has begun to actively focus on the advancement and application of mechatronics, introducing numerous supportive policies and appropriately increasing funding to promote sustainable and stable growth in the sector. For PLC technology, its effective application in mechanical-electrical management not only enhances the efficiency and level of mechanical automation production but also drives the innovation and development of mechanical-electrical automation technology. This, in turn, lays a solid foundation for the long-term progress of the machinery industry. Accordingly, this paper conducts an in-depth exploration of the application of PLC technology in mechanical-electrical control devices.

Keywords: Mechanical · Control device · PLC technology

1 Introduction

1.1 PLC Technology

PLC, also known as Programmable Logic Controller, is designed to automate the control of switchgear, mechanical-electrical systems, and other equipment, thereby improving operational efficiency and performance. This technology employs a digital control system as its foundational framework for achieving control functions. By storing programmed instructions in memory, it sends commands to target devices to accomplish automated control objectives. The structure of a PLC control system is illustrated in Fig. 1.

As shown in the figure, the PLC control system primarily consists of three key components: the input module, CPU module, and output module. Among these, the CPU module serves as the core of the system, closely interfacing with the programming device. Through the coordinated operation of these three modules, the system achieves comprehensive automated control.

M. A. b. Jamlos et al. (Eds.): ICSET 2025, LNNS 1897, pp. 637–642, 2026.
https://doi.org/10.1007/978-3-032-20722-7_82

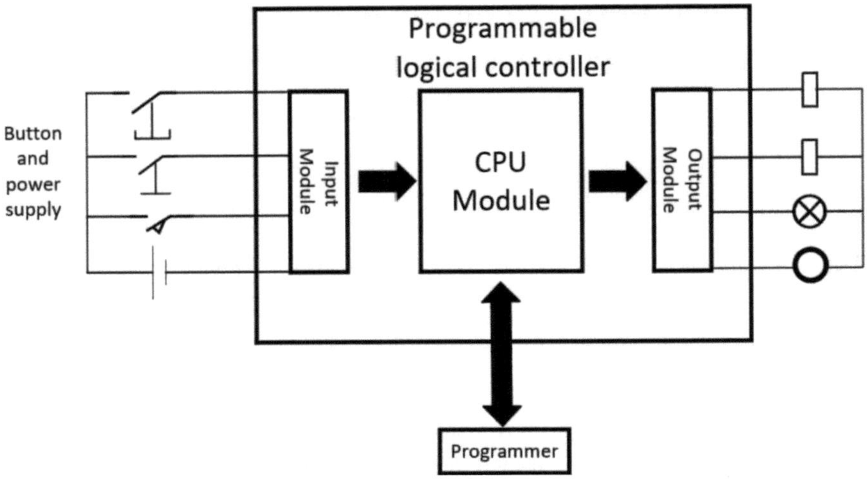

Fig. 1. Composition of PLC Control System

1.2 Advantages

Firstly, the effective application of PLC technology promotes the realization of mechatronics in mechanical manufacturing. By integrating computer technology and automation technology, it enhances quality, significantly reduces power consumption, and offers greater convenience in installation and usage. Therefore, the application of PLC technology can substantially decrease the footprint and weight of control devices while ensuring stable and reliable system performance. To a large extent, it optimizes and improves control systems, further advancing the goal of achieving mechatronics in electrical control systems. Secondly, PLC technology can be applied to various types of electrical control systems. This is primarily because PLCs are capable of automatically processing operational data, enabling detailed analysis and processing of data information to optimize and improve production. Thirdly, PLC technology possesses strong anti-interference capabilities. By fully utilizing integrated circuits, it effectively resists various disturbances, thereby enhancing the performance of electrical control systems.

2 PLC Technology in Mechanical and Electrical Equipment

2.1 DCS Technology

DCS (Distributed Control System) technology plays a crucial role in PLC applications. Its functionality relies on a well-structured DCS system, which must be designed according to practical requirements.

Also known as a distributed control system, DCS utilizes microprocessors to achieve decentralized control of mechanical and electrical operations, thereby improving the efficiency of electrical control. Additionally, this system supports data sharing, allowing users to remotely monitor equipment operation in real time via the DCS platform. This capability ensures stable and orderly equipment performance, providing a solid foundation for industrial automation.

2.2 FCS Technology

FCS (Fieldbus Control System) technology, also known as fieldbus control technology, offers significant advantages in enhancing the operational safety of mechanical and electrical systems. When applied in electrical control devices, FCS technology can effectively classify risk levels and further enable intelligent control of target equipment, making it a relatively advanced control solution. Additionally, FCS supports data sharing and remote control, which has made it a highly preferred choice in mechanical and electrical automation applications.

3 Application Analysis of PLC Technology in Mechanical and Electrical Control Devices

3.1 Air Compressor

Traditional mechanical and electrical control devices rely on microcontrollers to achieve their functions, which often introduce significant interference to mechanical systems. The control capabilities of microcontrollers are limited, leading to various shortcomings in practical applications. To address these issues, PLC technology can be effectively utilized to enhance functionality.

In air compressor systems, PLC technology can be rationally applied during the design phase to achieve comprehensive and scientific control. Moreover, as an advanced programmable control technology, PLC enables the transformation of air compressors into automated control systems—converting them into command-driven compression control systems. This significantly improves the overall performance and efficiency of the system.

3.2 System Design

The effective application of PLC technology runs through the entire device process, from design to production. Based on the specific model and working environment of the device, a rational application plan for PLC technology should be designed, with emphasis on the scientific rationality and applicability of the planning.

In terms of technical design, a thorough analysis should be conducted on existing issues to ensure the accuracy and reliability of data, which is essential for achieving effective control. When integrating design with device implementation, compatibility between the two must be comprehensively analyzed to avoid operational obstacles during equipment runtime.

This systematic approach ensures seamless integration of PLC technology, optimizing both performance and stability in mechanical and electrical control systems.

3.3 Analog Control

In production systems, analog signals are among the most unpredictable variables, with speed being one of the most critical parameters in the process. Proper control of analog signals significantly impacts the entire equipment production process.

Therefore, it is essential to fully leverage PLC technology to enhance analog signal management, facilitating their digital transformation. This approach enables effective analog control, ensuring the stability and reliability of equipment operation. By digitizing analog signals, PLC systems provide precise and responsive regulation, optimizing overall production efficiency and system dependability.

This technological application ensures that mechanical and electrical control systems maintain consistent performance even when handling complex, real-time analog variables.

3.4 Switching Logic

In mechanical and electrical control devices, electrical control relies on the coordinated operation of multiple switches to provide the necessary power support for mechanical systems. By effectively applying PLC technology, logical switch control can be achieved, ensuring the rational and efficient utilization of device functions.

The introduction of PLC technology into relay protection devices enables precise switch control, significantly enhancing the automation level and operational efficiency of relay protectors. Moreover, PLC technology can replace traditional relay protection systems with integrated chip technology, thereby expanding system functionality and flexibility.

Therefore, during device design, it is essential to strengthen fine management of switching logic to achieve versatile applications in mechanical and electrical control systems. This approach not only improves control accuracy but also optimizes system adaptability for diverse industrial scenarios.

By leveraging PLC-based logic control, mechanical and electrical systems can achieve higher reliability, smarter operation, and multifunctional integration, meeting the demands of modern industrial automation.

3.5 Distributed Control System

In most mechanical systems, distributed control systems provide indispensable functionalities. Through PLC technology, these systems can further enhance centralized control capabilities. In practical applications, PLC allows for the modular decomposition of control objects, enabling strict categorization and management of operational processes.

By adopting PLC technology, distributed systems can efficiently control various objects through programmed instructions while maintaining comprehensive oversight of different operational sequences. This significantly improves production efficiency and product quality.

When implementing distributed control systems, real-time data transmission-based monitoring is essential. The integration of PLC technology ensures reliable data communication, forming the foundation for orderly and optimized production workflows. This approach not only guarantees operational stability but also enables intelligent coordination across decentralized production units.

Through PLC-enhanced distributed control, mechanical systems achieve higher automation precision, better process synchronization, and more robust production management – driving industrial operations toward greater efficiency and quality standards.

3.6 Centralized Control System

The effective application of PLC technology in centralized control systems enables comprehensive management of mechanical and electrical control devices, achieving optimal performance under PLC-based supervision. In mechanical systems, centralized control plays a vital role by ensuring integrated functionality and coordinated operation.

To guarantee the reliability and safety of mechanical and electrical equipment, it is essential to incorporate master control switches within the system. Furthermore, implementing PLC technology throughout the system design enhances both operational efficiency of machinery and overall production productivity.

By leveraging PLC in centralized control architectures, industrial systems gain:

- ·Unified command and monitoring capabilities
- ·Enhanced process coordination across subsystems
- ·Improved response precision for critical operations

This integration represents an optimal solution for modern industrial automation where systematic control, operational stability, and production quality are paramount requirements.

4 Conclusions

In summary, PLC technology integrates computer technology, automation technology, and various advanced techniques, demonstrating unique advantages. With its strong anti-interference capability, PLC has been widely applied in electrical control systems across diverse types and environments.

The application of PLC technology has realized the automation and intellectualization of electrical control systems, significantly enhancing overall system performance and operational effectiveness. Under the new circumstances of continuous scientific and technological progress, China has prioritized mechanical manufacturing. The practical implementation of PLC technology will not only promote its maturation and iterative optimization, but also substantially improve the control precision and operational outcomes of mechanical-electrical systems.

This technological advancement represents a critical pathway for upgrading industrial control capabilities and achieving intelligent manufacturing objectives in contemporary production environments.

References

1. Xiaojing, D.: "The role and application of plc technology in mechanical and electrical control devices" [J]. Digit. Technol. Appl. **1**, 13 (2017)
2. Liang, W.: "Discussion on application of PLC technology in mechanical and electrical control devices" [J]. China Mach. **10**, 219 (2014)

3. Dianjiang, L., Ling, W.: "Research on application of PLC technology in mechanical and electrical control devices" [J]. South. Agric. Mach. **8**, 47 (2015)
4. Qiushi, W.: "Application and implementation of PLC technology in mechanical and electrical control devices" [J]. Mod. Manuf. Technol. Equip. **2**, 170–171 (2017)

Author Index

M. A. b. Jamlos et al. (Eds.): ICSET 2025, LNNS 1897, pp. 643–646, 2026.
https://doi.org/10.1007/978-3-032-20722-7

MIX
Papier aus verantwortungsvollen Quellen
Paper from responsible sources
FSC® C105338

If you have any concerns about our products,
you can contact us on
ProductSafety@springernature.com

In case Publisher is established outside the EU,
the EU authorized representative is:
Springer Nature Customer Service Center GmbH
Europaplatz 3, 69115 Heidelberg, Germany

Printed by Libri Plureos GmbH
in Hamburg, Germany